Collective Motion and Phase Transitions in Nuclear Systems

Collective Motion and Phase Transitions in Nuclear Systems

Proceedings of the
Predeal International Summer School in Nuclear Physics
Predeal, Romania 28 August – 9 September 2006

Editors

A A Raduta, V Baran,
University of Bucharest & NIPNE-HH, Romania

A C Gheorghe and I Ursu
NIPNE-HH, Romania

World Scientific

NEW JERSEY · LONDON · SINGAPORE · BEIJING · SHANGHAI · HONG KONG · TAIPEI · CHENNAI

Published by

World Scientific Publishing Co. Pte. Ltd.
5 Toh Tuck Link, Singapore 596224
USA office: 27 Warren Street, Suite 401-402, Hackensack, NJ 07601
UK office: 57 Shelton Street, Covent Garden, London WC2H 9HE

British Library Cataloguing-in-Publication Data
A catalogue record for this book is available from the British Library.

COLLECTIVE MOTION AND PHASE TRANSITIONS IN NUCLEAR SYSTEMS
Proceedings of the Predeal International Summer School in Nuclear Physics

ISBN-13 978-981-270-083-4
ISBN-10 981-270-083-8

Printed in Singapore by World Scientific Printers (S) Pte Ltd

PREFACE

This book contains the lectures given at the International Summer School octive Motion and Phase Transitions in Nuclear Systems held in Predeal, Romania, in the period of August 28- September 9, 2006. The Nuclear Physics Schools organised by Institute of Physics and Nuclear Engineering are well known to physicists throughout the world due to their long history and good reputation. The first edition took place in Bucharest in 1964 and after five-year break the series continued regularily every second year in Predeal. For some administrative reasons, in 1978, the location of the school was changed from Predeal to Poiana Brasov. In 1991, however, we moved it back to Predeal and that happened since we wanted to restore the good tradition and also to recover the excellent conditions we had before.

This edition is organized by the University of Bucharest in collaboration with the Intstitute of Physics and Nuclear Engineering. The efforts were sharred by the two important institutions mainly because some of the people from the Organizing Committee moved to University, aiming at having closer contacts with students.

The school from this year was devoted to the study of nuclear structure and dynamics of nuclear systems and their constituents. As shown in the table of contents, the chosen subjects cover a large area of modern nuclear physics. Various phases of nuclear matter at low and intermediate energiers were studied. Nuclear structure subjects were always a central issue of nuclear physics. In our school some hot subjects like critical points in nuclear shape phase transitions, octupole deformed nuclei, approaches going beyond the meanfield approximation, nuclear molecules, exotic nuclei and neutron stars, synthesis of superheavy elements, relativistic covariant descriptions, recieved the deserved attention. Various features of multifragmentation processes were interpreted both in statistical and dynamical models. Also the particle production processes in heavy ion collitions are investigated. Two extensive lectures on dark matter and various scenarious for its detection were delivered. A possible link to certain mechanisms for neutrinoless double beta decay has been discussed. A QCD self-consistent description for dilepton production has been presented. Recent results for the proton structure investigated by (e,e′) experiments at HERA was presented. The team of speakers was constituted from distinguished professors from different important corners of the world. They presented, with high competence, the most recent results in their fields and sketched appealing perspectives.

I am happy to mention that the scientific climate was very good and that the lectures stimulated an active participation of the audience.

I hope that the near future will positively evaluate the benefit provided by the present school to participants. Having in mind the hot discussions of participants during the lectures, the large volume of exchanged scientific information, the established new collaborations, the common research plans sketched for the near future, we may assert that the main scope of the present school has been accomplished. The young physicists listened outstanding professors speaking about their results as well as about the open problems in their fields and due to these facts they returned to their home institutes with an increased optimism.

In order to allow the young physicists, who were not able to attend the school, to have access to the scientific information transfered there, I tried my best to make the proceedings ready for publication, in a reasonable short time. As a matter of fact this is the only reason we missed the lecture of Dr. N.V. Zamfir on critical points of phase transitions.

Besides invited lectures, many short communications were given. These will be collected in a special issue of Romanian Journal of Physics were some extended review papers are also invited.

I hope that the present volume will be very useful to a large cathegory of nuclear physicists. Also, I am convinced that the scientific level of the lectures, the academic atmosphere and the beauty of the mountains surrounding the place are three decisive attractors for participants to the next edition.

A. A. Raduta

Organising Committee

A. A. Raduta, Director
A. C. Gheorghe, Vice-Director
V. Baran, Scientific Secretary
I. I. Ursu, Scientific Secretary
Aurora Anitoaiei, Technical Secretary
Denise Cringanu, Technical Secretary
Alexandra Olteanu, Financial Expert
Adrian Socolov, Designer

List of Participants

Francisc Aaron, *Univ. Bucharest, Fac. of Physics, Romania*
M. Avrigeanu, *Horia Hulubei National Institute of Physics and Nuclear Engineering, Bucharest, Romania*
V. Avrigeanu, *Horia Hulubei National Institute of Physics and Nuclear Engineering, Bucharest, Romania*
V. Baran, *Horia Hulubei National Institute of Physics and Nuclear Engineering, Bucharest, Romania*
Sergey Bastrukov, *JINR-Dubna, Russia*
P. G. Bizetti, *INFN-Firenze, Italy*
D. Bonatsos, *Institute of Nuclear Physics, National Centre for Scientific Research, Demokritos, Greece*
Radu Budaca, *Faculty of Physics, Bucharest Univ., Romania*
T. Buervenich, *Frankfurt/Main Univ., Germany*
Petrica Buganu, *Faculty of Physics, Bucharest Univ., Romania*
Florin Buzatu, *Horia Hulubei National Institute of Physics and Nuclear Engineering, Bucharest, Romania*
Florin Carstoiu, *Horia Hulubei National Institute of Physics and Nuclear Engineering, Bucharest, Romania*
A. Covello, *Univ. Napoli, Italy*
I. Cotaescu, *Univ. Bucharest, Fac. of Physics, Romania*

Doru Sabin Delion, *Horia Hulubei National Institute of Physics and Nuclear Engineering, Bucharest, Romania*
C. Diaconu, *IN2P3, Marseille, France*
Nguyen Dinh Dang, *Nishina Center for Accelerator-Based Science,Riken, 2-1 Hirosawa, Wako City, Saitama, Japan*
Massimo Di Toro, *LSN, INFN, Catania, Italy*
Albert Escuderos, *Univ.of Madrid, Spain;aescuder@physics.rutgers.edu*
C. Fuchs, *Inst. of Theor. Phys., Univ. of Tuebingen, Germany*
Andreea Fugaru, *Faculty of Physics, Bucharest Univ., Romania*
Radu Gherghescu, *Horia Hulubei National Institute of Physics and Nuclear Engineering, Bucharest, Romania*
Jose Maria Gomez, *Autonoma Univ. of Madrid, Spain*
Francesca Gulminelli, *LPC-Caen, France*
Dan Gurban, *Horia Hulubei National Institute of Physics and Nuclear Engineering, Bucharest, Romania*
Victor E. Iacob, *Texas A&M Univ., College Station, USA*
Amilcar Ionescu, *Horia Hulubei National Institute of Physics and Nuclear Engineering, Bucharest, Romania*
Ari Jokinen, *Dept.of Phys.,Univ. Jyvaskyla, Finland*
Jenni Kotila, *Univ. Jyvaskyla, Dept.of Physics, P.O.Box 35 (YFL), Finland*
M. Krivoruchenko, *Dept.of Physics, Univ. Genova, I-16146, Italy; Inst. for Theor.Experimental Phys., 117259, Moscow, Russia*
G. Lalazisis, *University of Thessaloniki, Greece*
Nicola Lo Iudice, *Univ.Napoli, Dep.of Phys. Sci., Nat .Inst. of Nucl. Phys., Mostra d'Oltremare, Pad. 19, 80-125, Italy*
Tatiana Mikhaylova, *JINR, DLTP, Dubna. Moscow Reg., Russia*
Calin Miron, *Fac. of Phys., Univ. Bucharest, Romania*
Serban Misicu, *Horia Hulubei National Institute of Physics and Nuclear Engineering, Bucharest, Romania*
Andrei Neacsu, *Horia Hulubei National Institute of Physics and Nuclear Engineering, Bucharest, Romania*

Ninel Nica, *Texas A&M Univ., Cyclotron Inst.College Station, USA 77843-3366*
Aura-Catalina Obreja, *Horia Hulubei National Institute of Physics and Nuclear Engineering, Bucharest, Romania*
Dimitries Petrellis, *Institute of Nuclear Physics, National Centre for Scientific Research, Demokritos, Greece*
A. Petrovici, *Horia Hulubei National Institute of Physics and Nuclear Engineering, Bucharest, Romania*
Oliver Plohl, *Univ.Tuebingen, Germany; plohl@tphys.physik.uni-tuebingen.de*
Dorin Poenaru, *Horia Hulubei National Institute of Physics and Nuclear Engineering, Bucharest, Romania*
Bogdan Popovici, *Horia Hulubei National Institute of Physics and Nuclear Engineering, Bucharest, Romania*
Oana Georgeta Radu, *Horia Hulubei National Institute of Physics and Nuclear Engineering, Bucharest, Romania*
Apolodor Aristotel Raduta, *Horia Hulubei National Institute of Physics and Nuclear Engineering, Bucharest, Romania*
Alexandru Horia Raduta, *Horia Hulubei National Institute of Physics and Nuclear Engineering, Bucharest, Romania*
Cristian-Mircea Raduta, *Horia Hulubei National Institute of Physics and Nuclear Engineering, Bucharest, Romania*
Faustin Laurentiu Roman, *Horia Hulubei National Institute of Physics and Nuclear Engineering, Bucharest, Romania*
Adriana Sandru, *Horia Hulubei National Institute of Physics and Nuclear Engineering, Bucharest, Romania*
Aurel Sandulescu, *Horia Hulubei National Institute of Physics and Nuclear Engineering, Bucharest, Romania*
Nicolaie Sandulescu, *Horia Hulubei National Institute of Physics and Nuclear Engineering, Bucharest, Romania*
Roberto Sartorio, *Univ.Federico II, Napoli, Italy*
Werner Scheid, *Inst.fur Theor.Phys.der Justus Liebig Univ. Giessen, Germany*

Andrei Silisteanu, *Horia Hulubei National Institute of Physics and Nuclear Engineering, Bucharest, Romania*
Ion Silisteanu, *Horia Hulubei National Institute of Physics and Nuclear Engineering, Bucharest, Romania*
J. Suhonen, *Dept. of Phys., Univ. Jyvaskyla, Finland*
Shneidman Timur, *JINR, BLTP, Dubna, Moscow Reg.141980, Russia*
Livius Trache, *Texas A&M Univ., College Station, USA*
Ioan Ursu, *Horia Hulubei National Institute of Physics and Nuclear Engineering, Bucharest, Romania*
John Vergados, *Joannina Univ., Greece*
H. Wolter, *Univ.Muenchen, Germany*
Sara Wuenschel, *Cyclotron Inst., Texas A&M Univ., USA*
S. Yennello, *Texas A&M Univ., College Station*
Ibrahim Yigitoglu, *University of Istambul, Turkey*
N. V. Zamfir, *Horia Hulubei National Institute of Physics and Nuclear Engineering, Bucharest, Romania*
L. Zamick, *Dept.of Phys.&Astronomy, Rutgers Univ. Piscataway, New Jersey 08854, USA*

CONTENTS

I.3 Relativistic Nuclear Structure

II. NUCLEAR MULTIFRAGMENTATION AND EQUATION OF STATE

III. ALPHA DECAY, NUCLEAR REACTIONS, COLD FISSION AND NUCLEAR FUSION

IV. DARK MATTER, DOUBLE BETA DECAY, AND POSSIBLE MECHANISMS FOR DILEPTON PRODUCTION

V. THEORETICAL AND EXPERIMENTAL RESULTS ON THE PROTON STRUCTURE

I. NUCLEAR STRUCTURE

I.1 Phenomenological Approaches

Description of nuclear octupole and quadrupole deformation close to the axial symmetry and phase transitions in the octupole mode

P.G. Bizzeti* and A.M. Bizzeti–Sona

Dipartimento di Fisica, Universitá di Firenze and INFN, Sezione di Firenze
Via G. Sansone 1, I 50019 Sesto Fiorentino (Firenze), Italy
**E-mail: bizzeti@fi.infn.it*
www.unifi.it - www.fi.infn.it

The dynamics of nuclear collective motion is investigated in the case of reflection-asymmetric deformation, with the purpose of describing the critical point of phase transitions between different shapes. The model is based on the Bohr hydrodynamical approach and employs a new parametrization of the octupole and quadrupole degrees of freedom, valid for nuclei close to the axial symmetry. Three particular cases are discussed in some detail: octupole critical point in nuclei which already possess a permanent quadrupole deformation (^{226}Th); octupole vibrations in nuclei at the X(5) critical point of quadrupole mode (^{150}Nd, ^{152}Sm); and critical point in both quadrupole and octupole modes (^{224}Ra, ^{224}Th). Results are compared with experimental data.

Keywords: Nuclear collective models; Octupole excitations; Phase transitions

1. Introduction

My talk will concern the description of quadrupole and octupole degrees of freedom, in the frame of Bohr hydrodynamical model, and in conditions close to the axial symmetry. The main motivation of this work was to provide a theoretical frame to investigate the possible phase transitions in nuclear shapes involving the octupole degrees of freedom, in a way analogous to the one used to describe the phase transitions in the quadrupole shapes. The subject of quantum phase transitions has been introduced by Prof. Zamfir at the very beginning of this School, and phase transitions in the quadrupole modes have been extensively reviewed yesterday by Prof. Bonatsos.

In my first lecture, I will discuss the basis of the model and introduce the formalism that will be used to treat, in the second lecture, a few particular cases: the phase transition from octupole vibrations to stable octupole deformation in nuclei with permanent quadrupole deformation; the octupole vibrations in nuclei at the X(5) critical point; and finally the phase transition from quadrupole–octupole vibrations to stable quadrupole–octupole deformation (*i.e.*, the reflection asymmetric rotor).

But, as a first point, I will remind you a few points concerning the basic assumptions and limitations of the original Bohr model, which will obviously concern also its extension to the octupole deformation.

The distinctive feature of our approach is the use of an intrinsic frame referred to the principal axes of the tensor of inertia of the deformed nucleus. In our opinion, properties of phase transition can be better described in this scheme than with one of the many models that have been introduced in the past to describe reflection-asymmetric deformation. Here, I will only remind some of them shortly.

In principle, algebraic models (like the spdf Interacting Boson Model [1–4], or the Extended Coherent State Model [5] discussed in this School by Prof. A.Raduta, or various forms of Cluster models [6,7]) are perfectly able to describe the entire region of transitional nuclei between the two extreme simple cases, including the critical point of the phase transition. But just because all of them are described on the same footing, it will be difficult to identify the particular features of the phase transition point. This is the reason why Iachello himself uses a geometrical–model basis to discuss the phase transitions between the IBM limiting symmetries U(5) and O(6) or U(5) and SU(3).

In the geometrical approach, the most complete treatment of the octupole degrees of freedom has been proposed by Donner and Greiner [8]. Like the original Bohr model, it defines a non–inertial intrinsic frame, which however is not referred to the principal axes of the overall tensor of inertia, but to the principal axes of the quadrupole alone [8–10]. This approach would therefore coincide with ours only when the octupole amplitudes are very small compared to quadrupole. A number of alternative models choose to work in a reduced space, usually limited to axial quadrupole and octupole deformations [9,11–17]. In this case, part of the dynamical variables (those concerning non axial degrees of freedom) are frozen to zero from the very beginning. As we shall see, such a choice has consequences also on the differential equations describing the allowed axial modes. Finally, a parametrization of the octupole mode alone in its own intrinsic frame (referred to the principal axes of its tensor of inertia) has been reported by Wexler and Dussel [18]. We shall return to this point in the Sec. 4.

2. The basis of the Bohr model

The hydrodynamical model of nuclear collective motion was introduced by A. Bohr in a famous paper of 1952 [19] concerning the coupling of collective motion to the single particle degrees of freedom. Here we are interested in the first part of the paper, where the collective motion is described as the irrotational motion of a drop of *homogeneous* liquid, induced by *small* deformations of the surface. An implicit assumption is that the typical excitation energies of the collective modes be small compared to those of single-particle excitations. Therefore, phenomena concerning a relatively large excitation energy and/or angular momentum, as band crossing / backbending, band termination, isovector excitations (Giant Dipole Resonance) cannot be described in this simple frame. Instead, the model has proved to be very useful to describe the new symmetries at the critical point of phase transitions, and we can expect that it will be so also for its extension to the octupole modes. Phenomena involving higher energies and angular momenta, as the "beat pattern" in the parity straggling of alternate parity bands remain outside our possibilities. Other lectures of this School do treat them by means of more suitable models.

Our starting point is the equation of the nuclear surface, that is given in spherical coordinates as

$$R(\theta,\phi) = R_0\Big[1+\sum_{\lambda}\sum_{\mu=-\lambda,\lambda} \alpha_\mu^{(\lambda)} Y^*_{\lambda,\mu}(\theta,\phi)\Big]\,, \qquad \alpha_\mu^{(\lambda)^*} = (-)^\mu \alpha_{-\mu}^{(\lambda)}\,.$$

With this definition, the $\alpha_\mu^{(\lambda)}$ are components of an irreducible tensor of rank λ. The amplitudes $\alpha_\mu^{(\lambda)}$ are assumed to be small. Why? Consider, *e.g.*, the nuclear volume V

$$V = \int d\Omega \int_0^{R(\theta,\phi)} r^2\, dr = R_0^3 \left[\frac{4\pi}{3} + \sqrt{4\pi}\alpha_0^{(0)} + \sum_{\lambda,\mu}\left|\alpha_\mu^{(\lambda)}\right|^2 + ...\right]\,.$$

Higher order terms are neglected, and this already means that the $\alpha_\mu^{(\lambda)}$ are assumed to be small. Up to the second order, to keep the volume constant one should put $\alpha_0^{(0)} = -(4\pi)^{-1/2}\sum_{\lambda,\mu}|\alpha_\mu^{(\lambda)}|^2$. If, and only if, all the amplitudes $\alpha_\mu^{(\lambda)}$ are small enough to neglect their modulus squared, we can assume $\alpha_0^{(0)} = 0$ with the nuclear volume *approximately* conserved.

In a similar way it is possible to show that, if even and odd values of λ are present in the sum, then to keep fixed the center-of-mass position it is necessary to assume a well definite form for the amplitudes $\alpha_\mu^{(1)}$ as bilinear combinations of amplitudes with $\lambda \geq 2$. Again, if all these amplitudes are small enough, we can assume $\alpha_\mu^{(1)} \approx 0$.

3. The Bohr model with Quadrupole and Octupole

Our next step is the description of quadrupole and octupole deformation in a proper intrinsic frame, referred to the principal axes of the overall tensor of inertia. The equation of the nuclear surface is now:

$$r(\theta,\phi) = R_0\left[1 + \sum_{\lambda=2,3} \quad \sum_{\mu=-\lambda,\lambda} \alpha_\mu^{(\lambda)} Y^*_{\lambda,\mu}(\theta,\phi)\right] .$$

In the Bohr hydrodynamical model, the classical expression of the kinetic energy is

$$T = \frac{1}{2}B_2 \sum_{\mu=-2,2} \left|\dot{\alpha}_\mu^{(2)}\right|^2 + \frac{1}{2}B_3 \sum_{\mu=-3,3} \left|\dot{\alpha}_\mu^{(3)}\right|^2$$

with constant B_2, B_3. For a classical drop, such an expression is obtained assuming irrotational flow. We now express $\alpha_\mu^{(\lambda)}$ in terms of the amplitudes $a_\nu^{(\lambda)}$ in a (non-inertial) intrinsic frame:

$$\sqrt{B_\lambda}\ \alpha_\mu^{(\lambda)} = \sum_\nu a_\nu^{(\lambda)} D_{\mu\nu}^{(\lambda)^*}(\theta_i)$$

where $D^{(\lambda)}(\theta_i)$ are the Wigner matrices and θ_i the Euler angles. In order to simplify the notation in the following, we include the inertia coefficient B_λ in the definition of the collective variables $a_\mu^{(\lambda)}$. The classical expression of the kinetic energy in terms of time derivatives of the intrinsic variables $a_\mu^{(\lambda)}$ and of the intrinsic components q_i of the angular velocity, is now the sum of a *vibrational* term T_{vib}, a *rotational* term T_{rot} and a *coupling* term T_{coup}, which is not present in the case of pure quadrupole deformation:

$$T_{\text{vib}} = \frac{1}{2}\sum_{\lambda,\mu} \left|\dot{a}_\mu^{(\lambda)}\right|^2 \qquad\qquad T_{\text{rot}} = \frac{1}{2}\sum_{k,k'} q_k q_{k'} \mathcal{J}_{kk'}$$

$$T_{\text{coup}} = i\sum_\lambda \sqrt{3(2\lambda+1)}\left[q^{(1)} \otimes \left[a^{(\lambda)} \otimes \dot{a}^{(\lambda)}\right]^{(1)}\right]_0^{(0)} .$$

The diagonal and non–diagonal components of the tensor of inertia are

$$\mathcal{J}_{\frac{1}{2}} = \sum_\lambda \left\{ C_0(\lambda)\left[a^{(\lambda)} \otimes a^{(\lambda)}\right]_0^{(0)} + \frac{1}{\sqrt{6}} C_2(\lambda)\left[a^{(\lambda)} \otimes a^{(\lambda)}\right]_0^{(2)} \right.$$

$$\left. \mp C_2(\lambda)\text{Re}\left[a^{(\lambda)} \otimes a^{(\lambda)}\right]_2^{(2)} \right\}$$

$$\mathcal{J}_3 = \sum_\lambda \left\{ C_0(\lambda)\left[a^{(\lambda)} \otimes a^{(\lambda)}\right]_0^{(0)} - \sqrt{\frac{2}{3}}\, C_2(\lambda)\left[a^{(\lambda)} \otimes a^{(\lambda)}\right]_0^{(2)} \right\}$$

$$\mathcal{J}_{13}+i\,\mathcal{J}_{23} = \sum_\lambda C_2(\lambda)\left[a^{(\lambda)}\otimes a^{(\lambda)}\right]^{(2)}_1$$

$$\mathcal{J}_{12} = \sum_\lambda C_2(\lambda)\ \mathrm{Im}\left[a^{(\lambda)}\otimes a^{(\lambda)}\right]^{(2)}_2$$

with
$$C_0(\lambda) = (-1)^\lambda\ \lambda(\lambda+1)\sqrt{2\lambda+1}\ /3$$
$$C_2(\lambda) = (-1)^{\lambda+1}\ \sqrt{\lambda(\lambda+1)(2\lambda+3)(4\lambda^2-1)/30}\ .$$

We do not give here the proof of these relations, that can be found, *e.g.* in the book by Eisenberg and Greiner [10].

4. Intrinsic amplitudes for quadrupole and octupole

In the spirit of the original Bohr paper, we decide to use as intrinsic frame the one referred to the principal axes of the overall tensor of inertia, and – preferably – define a parametrization that automatically implies vanishing of the three products of inertia $\mathcal{J}_{12}$, $\mathcal{J}_{13}$, and $\mathcal{J}_{23}$.

We have seen that in the case of a pure quadrupole deformation, this result is obtained with the Bohr parametrization

$$a_0^{(2)} = \beta\cos\gamma$$
$$a_1^{(2)} = a_{-1}^{(2)} = 0$$
$$a_2^{(2)} = a_{-2}^{(2)} = (1/\sqrt{2})\beta\sin\gamma\ .$$

In the case of a pure octupole deformation, without any contribution of quadrupole, this is also possible, with the parametrization proposed by Wexler and Dussel [18]. We can introduce a very similar one:

$$a_0^{(3)} = \beta_3\ \cos\gamma_3$$
$$a_1^{(3)} = -(5/2)\ (X+iY)\ \sin\gamma_3$$
$$a_2^{(3)} = \sqrt{1/2}\ \beta_3\ \sin\gamma_3$$
$$a_3^{(3)} = X\left[\cos\gamma_3 + (\sqrt{15}/2)\ \sin\gamma_3\right] + i\ Y\left[\cos\gamma_3 - (\sqrt{15}/2)\ \sin\gamma_3\right]\ ,$$

with $a_{-\mu}^{(3)} = (-)^\mu a_\mu^{(3)*}$. Also with this choice, the tensor of inertia turns out to be diagonal for a pure octupole deformation. To the intrinsic parameters (2 for the pure quadrupole, 4 for the pure octupole) one must add the three Euler angles to obtain the right number of parameters (5 or 7) needed to describe the nuclear deformation. We observe that in both cases the amplitudes with $\mu = \pm 2$ are real and those with $\mu = \pm 1$ are either zero (for pure quadrupole) or small of the second order, if we consider small of the first order other amplitudes with $\mu \neq 0$.

Instead, in the presence of both quadrupole and octupole deformations, also the amplitudes with $\mu = \pm 1$ and the imaginary part of those with $\mu = \pm 2$ must be considered. They are, however, not independent of one another, due to the requirement that

$$\mathcal{J}_{12} = \mathcal{J}_{13} = \mathcal{J}_{23} = 0 \tag{1}$$

The Eqs. 1 are non linear, but if we assume that non–axial amplitudes are small (of the first order) compared to the axial ones and we neglect the 2nd order terms, they reduce to the linear equations

$$\begin{aligned} \mathcal{J}_{13} + i\mathcal{J}_{23} &= \sqrt{6}\left(\beta_2 a_1^{(2)} + \sqrt{2}\,\beta_3 a_1^{(3)}\right) = 0 \\ \mathcal{J}_{12} &= -2\sqrt{6}\left(\beta_2 \mathrm{Im}\ a_2^{(2)} + \sqrt{5}\,\beta_3 \mathrm{Im}\ a_2^{(3)}\right) = 0 \end{aligned} \tag{2}$$

which are authomatically verified if one defines

$$\begin{aligned} a_1^{(2)} &= -c_1\,\sqrt{2}\,\beta_3\,(\eta_c + i\zeta_c) & a_1^{(3)} &= c_1\,\beta_2\,(\eta_c + i\zeta_c) \\ \mathrm{Im}\ a_2^{(2)} &= -c_2\,\sqrt{5}\,\beta_3\,\xi_c & \mathrm{Im}\ a_2^{(3)} &= c_2\,\beta_2\,\xi_c \end{aligned} \tag{3}$$

where c_1 and c_2 are arbitrary functions of β_2 and β_3.

With this choice, at the first order in the "small" quantities $\gamma_2,\ \gamma_3, X,\ Y,\ \eta_c,\ \zeta_c,\ \xi_c$, the intrinsic amplitudes of quadrupole and octupole deformation are

$$\begin{aligned} a_0^{(2)} &= \beta_2 \cos\gamma_2 \approx \beta_2\left[1 - (1/2)\ \gamma_2^2\right] \\ a_1^{(2)} &= -c_1\,\sqrt{2}\,\beta_3\ (\eta_c + i\zeta_c) \\ a_2^{(2)} &= \sqrt{1/2}\,\beta_2\ \sin\gamma_2\ - ic_2\,\sqrt{5}\,\beta_3\,\xi_c \approx \sqrt{1/2}\,\beta_2\,\gamma_2 - ic_2\,\sqrt{5}\,\beta_3\,\xi_c \\ a_0^{(3)} &= \beta_3\ \cos\gamma_3 \approx \beta_3\left[1 - (1/2)\ \gamma_3^2\right] \\ a_1^{(3)} &= -(5/2)\ [X + iY]\ \sin\gamma_3 + c_1\,\beta_2\ (\eta_c + i\zeta_c) \approx c_1\,\beta_2\ (\eta_c + i\zeta_c) \\ a_2^{(3)} &= \sqrt{1/2}\,\beta_3 \sin\gamma_3 + ic_2\,\beta_2\,\xi_c \approx \sqrt{1/2}\,\beta_3\,\gamma_3 + ic_2\,\beta_2\,\xi_c \\ a_3^{(3)} &= X\left[\cos\gamma_3 + \sqrt{15/4}\ \sin\gamma_3\right] + iY\left[\cos\gamma_3 - \sqrt{15/4}\ \sin\gamma_3\right] \approx X + iY \end{aligned} \tag{4}$$

With this choice, the three products of inertia $\mathcal{J}_{\kappa\kappa'}$, and also the diagonal term $\mathcal{J}_{33}$, are small of the second order in the "small" variables $\gamma_2,\ \gamma_3,\ X,\ Y,\ \xi_c,\ \eta_c,\ \zeta_c$.

In the following, we are going to consider the equation of motion for β_2, β_3 assuming that it is effectively decoupled from that of all other independent variables (wich will not necessarily coincide with those defined in the Eq. 4). The identification of a set of variables for which this condition holds is a crucial point in our work, as it determines the form of the kinetic energy operator also in the sector involving β_2 and β_3.

5. Quantization of the quadrupole – octupole Hamiltonian

We must now pass from the classical expression T of the kinetic enrgy to the quantum kinetic energy operator $\hat{T}$. This can be done with the Pauli procedure [20] for quantization in a non Cartesian reference.

The classical expression of the kinetic energy has the form

$$T = \frac{1}{2}B_2 \sum_{\mu=-2,2} \left|\dot{\alpha}_\mu^{(2)}\right|^2 + \frac{1}{2}B_3 \sum_{\mu=-3,3} \left|\dot{\alpha}_\mu^{(3)}\right|^2 \Rightarrow \frac{1}{2}\sum_{\mu\nu} \mathcal{G}_{\mu\nu} v_\mu v_\nu \tag{5}$$

where v_ν are the components of the generalized velocity vector $\boldsymbol{v}$. Here, $\boldsymbol{v} = \{\dot{\beta}_2, \dot{\gamma}_2, \dot{\beta}_3, \dot{\gamma}_3, \dot{X}, \dot{Y}, \dot{\xi}_c, \dot{\eta}_c, \dot{\zeta}_c, q_1, q_2, q_3\}$, and q_μ are the components of the angular velocity $\vec{q}$ along the axes of the intrinsic frame.

According to the Pauli recipe, the kinetic energy operator is

$$\hat{T} = -\sum \frac{\hbar^2}{2} G^{-1/2} \frac{\partial}{\partial Q_\mu} [G^{1/2} (\mathcal{G}^{-1})_{\mu\nu} \frac{\partial}{\partial Q_\nu}] \tag{6}$$

where the Q_ν are the dynamical variables and G is the Determinant of the matrix $\mathcal{G}$ defined in the Eq. 5.

If one makes use of the dynamical variables defined in the Eq. 4, the result obviously depends on the choice of the *arbitrary functions* $c_1(\beta_2, \beta_3)$ and $c_2(\beta_2, \beta_3)$. *E.g.*, with the simplest possible choice $c_1 = c_2 = 1$, we would obtain a determinant $G = \mathrm{Det}\mathcal{G} = 1152\beta_2^2\beta_3^2\left(\beta_2^2 + 2\beta_3^2\right)^4\left(\beta_2^2\gamma_2^2 + \beta_3^2\gamma_3^2\right)^2$. In this case, at the limit for $\beta_3 \to 0$, we obtain $G \propto \beta_2^{14}$, while at the same limit the Bohr model gives $G \propto \beta_2^8\ \gamma_2^2$ and in the case of small–amplitude vibrations around a well deformed shape ($\beta \approx \bar{\beta}$ constant) the Frankfurt model [10] gives $G \propto \bar{\beta}_2^4$. Therefore, the choice $c_1 = c_2 = 1$ seems not to be adequate.

Apparently, a better choice could be $c_1 = (\beta_2^2 + 2\beta_3^2)^{1/2}$, $c_2 = (\beta_2^2 + 5\beta_3^2)^{1/2}$, which gives the $\mathcal{G}$ matrix shown in the Table 1. All the non diagonal matrix elements, apart from those of the last line and column, are small at least of the first order in comparison with the corresponding diagonal elements, and it is possible to show [21] that they can be neglected. Those of the last line or column are still small of the first order, but they must be compared with the diagonal element $\mathcal{J}_3$, which is *small of the second order*. Their effect will be discussed in the next Sec. 6.

Now the determinant G has a more reasonable behaviour:

$$G = \mathrm{Det}\mathcal{G} = 1152\beta_2^2\beta_3^2\left(\beta_2^2 + 2\beta_3^2\right)^2\left(\beta_2^2 + 5\beta_3^2\right)^{-1}\left(\beta_2^2\gamma_2 + \sqrt{5}\ \beta_3^2\gamma_3\right)^2 \tag{7}$$

and, when $\beta_3 \to 0$, $G \propto \beta_2^8$ as in the original Bohr model. We can consider

Table 1. The matrix of inertia $\mathcal{G}$: leading terms and relevant first-order terms. Other first-order terms are indicated with the symbol ≈ 0.

	$\dot\beta_2$	$\dot\gamma_2$	$\dot\beta_3$	$\dot\gamma_3$	$\dot X$	$\dot Y$	$\dot\xi$	$\dot\eta$	$\dot\zeta$	q_1	q_2	q_3
$\dot\beta_2$	1	0	0	0	0	0	0	0	0	≈ 0	≈ 0	0
$\dot\gamma_2$	0	β_2^2	0	0	0	0	0	0	0	≈ 0	≈ 0	$\frac{-\sqrt{40}\beta_2\beta_3\xi}{\sqrt{\beta_2^2+5\beta_3^2}}$
$\dot\beta_3$	0	0	1	0	0	0	0	0	0	≈ 0	≈ 0	0
$\dot\gamma_3$	0	0	0	β_3^2	≈ 0	≈ 0	0	≈ 0	≈ 0	≈ 0	≈ 0	$\frac{\sqrt{8}\beta_2\beta_3\xi}{\sqrt{\beta_2^2+5\beta_3^2}}$
$\dot X$	0	0	0	≈ 0	2	0	0	≈ 0	0	≈ 0	≈ 0	$6Y$
$\dot Y$	0	0	0	≈ 0	0	2	0	0	≈ 0	≈ 0	≈ 0	$-6X$
$\dot\xi$	0	0	0	0	0	0	2	0	0	≈ 0	≈ 0	$\frac{\sqrt{8}\beta_2\beta_3\gamma}{\sqrt{\beta_2^2+5\beta_3^2}}$
$\dot\eta$	0	0	0	≈ 0	≈ 0	0	0	2	0	≈ 0	≈ 0	2ζ
$\dot\zeta$	0	0	0	≈ 0	0	≈ 0	0	0	2	≈ 0	≈ 0	-2η
q_1	≈ 0	≈ 0	≈ 0	≈ 0	≈ 0	≈ 0	≈ 0	≈ 0	≈ 0	$\mathcal{J}_1$	0	0
q_2	≈ 0	≈ 0	≈ 0	≈ 0	≈ 0	≈ 0	≈ 0	≈ 0	≈ 0	0	$\mathcal{J}_2$	0
q_3	0	[..]	0	[..]	$6Y$	$-6X$	[..]	2ζ	-2η	0	0	$\mathcal{J}_3$

Note: Here $\gamma = \sqrt{5}\gamma_2 - \gamma_3$, $\mathcal{J}_1 = 3(\beta_2^2 + 2\beta_3^2) + 2\sqrt{3}(\beta_2^2\gamma_2 + \sqrt{5}\beta_3^2\gamma_3)$; $\mathcal{J}_2 = 3(\beta_2^2 + 2\beta_3^2) - 2\sqrt{3}(\beta_2^2\gamma_2 + \sqrt{5}\beta_3^2\gamma_3)$; and $\mathcal{J}_3 = 4(\beta_2^2\gamma_2^2 + \beta_3^2\gamma_3^2) + 18(X^2 + Y^2) + 2(\eta^2 + \zeta^2) + 8\xi^2$.

therefore the present choice of c_1, c_2 and the results given in the Table 1 as a good starting point for our future work.

6. The third component of the angular momentum

Surface vibrations that conserve the axial symmetry correspond to an angular momentum component $L_3 = 0$ along the intrinsic symmetry axis. It is interesting to investigate the relation between the non–axial modes of vibration, described by the dynamical variables γ_2, γ_3, X, Y, ξ, η, ζ and other eigenvalues of L_3

To this purpose, I will follow here a semiclassical approach, simpler and more transparent than the completely consistent one, which can be found in the Ref. [21]. The classical expression for the intrinsic components of the angular momentum is

$$L_\kappa = \partial T / \partial q_\kappa \tag{8}$$

While, at the leading order, the components L_1, L_2 have the usual form $L_\kappa = \mathcal{J}_\kappa q_\kappa$, $\kappa = 1,\ 2$, the third component has a very complicated expression, due to the effect of non diagonal terms of the last line and column:

$$L_3 = \mathcal{J}_3 q_3 + \Big[\frac{\sqrt{8}\beta_2\beta_3}{\sqrt{\beta_2^2 + 5\beta_3^2}} (\gamma\dot\xi - \xi\dot\gamma) + 6(Y\dot X - X\dot Y) + 2(\zeta\dot\eta - \eta\dot\zeta) \Big] \tag{9}$$

where we have put

$$\gamma = \sqrt{5}\gamma_2 - \gamma_3 \ . \tag{10}$$

At this point, it will be convenient to express the variables γ_2 and γ_3 as linear combinations of two new variables, one of which is, obviously, $\gamma = \sqrt{5}\gamma_2 - \gamma_3$. The other one, γ_0, can be chosen proportional to the linear combination which enters in the expression of the determinant G,

$$\gamma_0 = c_0 \left(\beta_2^2 \gamma_2 + \sqrt{5}\ \beta_3^2 \gamma_3 \right) . \tag{11}$$

where, again, the factor c_0 will be considered as an arbitrary function of β_2, β_3, with the only condition that $c_0 \to \beta_2^2$ when $\beta_3 \to 0$.

The expression of L_3 can be substantially simplified with the introduction of a new set of variables v, ϑ, u, φ, w, χ, u_0 and the corresponding conjugate moments p_v, p_ϑ, p_u, p_φ, p_w, p_χ, p_{u_0}. These new variables are related to the old ones by the expressions

$$\begin{aligned}
X &= w \sin\vartheta \\
Y &= w \cos\vartheta \\
\eta &= v \sin\varphi \\
\zeta &= v \cos\varphi \\
\xi &= u \sin\chi \ . \\
\gamma &\equiv \sqrt{5}\gamma_2 - \gamma_3 = \sqrt{2} \left(\sqrt{\beta_2^2 + 5\beta_3^2} \ / \ \beta_2\beta_3 \right) u \cos\chi \\
\frac{\gamma_0}{c_0} &\equiv \beta_2^2\gamma_2 + \sqrt{5}\ \beta_3^2\gamma_3 = \sqrt{\beta_2^2 + 5\beta_3^2}\ u_0
\end{aligned} \tag{12}$$

The expression of the kinetic energy, in terms of the new variables and of their time derivatives, is

$$\begin{aligned}
T = \frac{1}{2} \Big\{ & \dot\beta_2^2 + \dot\beta_3^2 + \dot u_0^2 + 2(\dot v^2 + v^2\dot\varphi^2) + 2(\dot u^2 + u^2\dot\chi^2) + 2(\dot w^2 + w^2\dot\vartheta^2) \\
& + 2q_3 \left[2v^2\dot\varphi + 4u^2\dot\chi + 6w^2\dot\vartheta \right] + \mathcal{J}_1 q_1^2 + \mathcal{J}_2 q_2^2 + \mathcal{J}_3 q_3^2 \Big\}
\end{aligned} \tag{13}$$

with $\mathcal{J}_1 \approx \mathcal{J}_2 \approx 3(\beta_2^2 + 2\beta_3^2)$, $\mathcal{J}_3 = 4u_0^2 + 2v^2 + 8u^2 + 18w^2$. The expression of L_3 takes the simpler form

$$\begin{aligned}
L_3 &= \mathcal{J}_3 q_3 + \left[2v^2\dot\varphi + 4u^2\dot\chi + 6w^2\dot\vartheta \right] \\
&= 2v^2(q_3 + \dot\varphi) + 4u^2(2q_3 + \dot\chi) + 6w^2(3q_3 + \dot\vartheta) + 4u_0^2 q_3 \ .
\end{aligned} \tag{14}$$

We also evaluate the *conjugate moments* of the new variables φ, χ, ϑ and invert this system of equations to obtain their time derivatives in terms of the conjugate moments and L_3:

$$\begin{aligned} p_\varphi &= 2v^2(\dot\varphi + q_3) & \dot\varphi &= \frac{p_\varphi}{2v^2} - \frac{1}{u_0^2}(L_3 - p_\varphi - 2p_\chi - 3p_\vartheta) \\ p_\chi &= 2u^2(\dot\chi + 2q_3) & \dot\chi &= \frac{p_\chi}{2u^2} - \frac{2}{u_0^2}(L_3 - p_\varphi - 2p_\chi - 3p_\vartheta) \\ p_\vartheta &= 2w^2(\dot\vartheta + 3q_3) & \dot\vartheta &= \frac{p_\vartheta}{2\chi_0^2} - \frac{3}{u_0^2}(L_3 - p_\varphi - 2p_\chi - 3p_\vartheta). \end{aligned} \tag{15}$$

and

$$q_3 = \frac{1}{u_0}(L_3 - p_\varphi - 2p_\chi - 3p_\vartheta) \tag{16}$$

This last relation shows that, when $u_0 \to 0$, then $q_3 \to \infty$ unless $L_3 = \Omega \equiv p_\varphi - 2p_\chi - 3p_\vartheta$. This relation has a very simple meaning if the potential does not depend on the variables φ, χ or ϑ. In such a case (a sort of model φ-χ-ϑ–instable, in the sense of the γ–instable model by Wilets and Jean [22]) the conjugate moments of these three angular variables are constants of the motion, with integer eigenvalues n_φ,n_χ and n_ϑ (in units of $\hbar$), and the operator L_3 is diagonal, with eigenvalues $K = n_\varphi + 2n_\chi + 3n_\vartheta$. Therefore, the three degrees of freedom corresponding to the pairs of variables v, φ, u, χ and w, ϑ can be associated with non-axial excitation modes with K =1, 2 and 3, respectively. By using their definitions (Eq. 12), it is easy to verify that they also carry *negative parity.*

It remains to discuss the role of u_0. The variable u_0 measures the triaxiality of the overall tensor of inertia. In this sense, it plays a role similar to that of a_2 in the pure–quadrupole case and, like a_2 in the case of small triaxial deformation, it can be eventually replaced by an angular variable times a proper combinations of β_2 and β_3. If we assume that the differential equation for u_0 can be decoupled from the others, this equation is

$$\left\{\frac{1}{u_0}\frac{\partial}{\partial u_0}\left[u_0\frac{\partial}{\partial u_0}\right] + \frac{2}{\hbar^2}\left[E_{u_0} - U(u_0)\right] - \frac{1}{u_0^2}\left[\frac{K_0}{2}\right]^2\right\}\phi(u_0) = 0 \tag{17}$$

where we have put $K_0 = L_3 - \Omega$. The variable u_0 can take positive as well as negative values. The condition of continuity for the wavefunction $\phi(u_0)$ and its derivative at $u_0 = 0$ imposes that $K_0 = 2n_{u_0}$, with n_{u_0} integer. We can conclude that the degree of freedom associated to the variable u_0 carries two units of angular momentum along the intrinsic axis 3, and it is possible to show that it also carries positive parity.

With this choice of dynamical variables, the determinant G is

$$G \equiv \mathrm{Det}\mathcal{G} = 2304\, u_0^2\, v^2 u^2 w^2 \left(\beta_2^2 + 2\beta_3^2\right)^2 \tag{18}$$

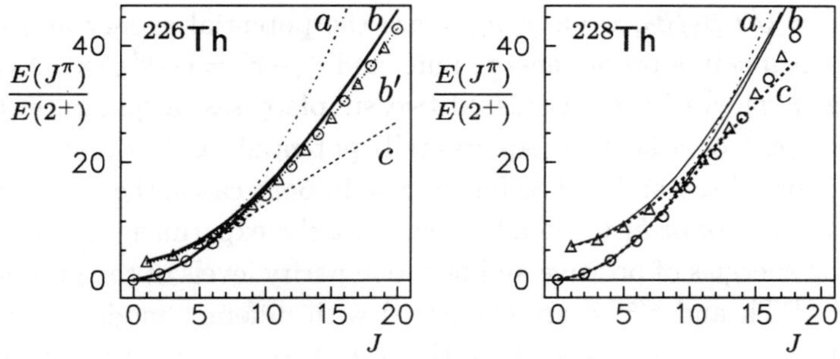

Fig. 1. (From Ref. [21]). Excitation energies for states of positive parity (circles) and negative parity (triangles), in units of $E(2^+)$, for the ground-state band of ^{226}Th and ^{228}Th. Theoretical curves: a – rigid rotor; b (b') – present model with critical potential, fitted on the 1^- state (on high-spin states); c – present model with harmonic potential.

and the inverse of the matrix $\mathcal{G}$ turns out to be diagonal (at the relevant order) in the space of momenta conjugate to the variables defined in the Eq. 15 and of the angular momentum components L_1, L_2 and $K_0 = L_3 - \Omega$. A more formal derivation of these results, involving the derivatives of the Euler angles, can be found in the Appendix C of Ref. [21].

7. Axial octupole mode with stable quadrupole deformation

This is the simplest case in which the properties of octupole excitation can be followed from the limit of harmonic oscillations around the reflection symmetric core to the opposite limit of stable octupole deformation. A detailed discussion of this subject can be found in the Ref. [21]. Here, we only summarize these results.

A preliminary comment is in order. The properties of the quadrupole vibrations around an axially deformed core are better described [10] with respect to the intrinsic parameters $a_2^{(2)}$, $\delta a_0^{(2)} = a_0^{(2)} - \bar{a}_0^{(2)}$ than in terms of the Bohr parameters β_2 and γ_2. We have seen that the parameter u_0 defined in the Eq. 12 plays, in our treatment, a role analogous to that of $a_2^{(2)}$ in the pure quadrupole Hamiltonian. It appears reasonable, therefore, to use it as a dynamical variable instead of defining an angle variable similar to the γ_2 of the quadrupole case. Therefore, we can use the expression of G given in Eq. 18, to derive with the Pauli recipe the differential equation for β_3 (in doing this, we assume decoupling of the β_3 motion from the small–amplitude oscillations in all other degrees of freedom). One obtains

$$\frac{d^2\psi(x)}{dx^2} + \frac{2x}{1+x^2}\frac{d\psi(x)}{dx} + \left[\epsilon - \frac{J(J+1)}{6(1+x^2)} - v(x)\right]\psi(x) = 0 \qquad (19)$$

where $x = \sqrt{2}\ \beta_3/\bar{\beta}_2$, while $v(x)$, ϵ are the potential energy and the energy eigenvalue in a proper energy unit, and $\psi(-x) = (-)^J\psi(x)$. As for the potential $v(x)$, we have considered two simple cases: a quadratic expression $v = \frac{1}{2}cx^2$ or a critical (square-well) potential, as in the X(5) model: $v(x) = 0$ for $|x| < b$ and $= +\infty$ for $|x| > b$. In both cases, the model has one free parameter (c or b) to be adjusted to fit the experimental data. In the Fig. 1 the energies of positive and negative parity levels of the ground-state band of ^{226}Th and ^{228}Th are compared with different model predictions. The former turns out to be close to the results we obtain for a critical-point potential, while the latter is closer to those obtained with a quadratic potential. Relative values of experimental transition strengths, B(E1) and B(E2) have also been calculated and compared with existing experimental data. The results can be found in the Ref. [21]. The agreement is satisfactory, within the (admittedly large) experimental errors.

8. Going close to the quadrupole critical point

If we want to consider the case where the dependence of the potential energy on β_2 is that of a square well extending from $\beta_2 = 0$ to some finite limit $\beta_2^{\rm w}$, we need that the results of our model converge to those of the Bohr model in the limit of small octupole deformation. This is not the case for the Hamiltonian we have used in the previous Section, where u_0 has been used as independent dynamical variable. Since in the Bohr model the variable γ_2 is used in the place of $a_2^{(2)}$, it is now necessary to replace u_0 with a proper adimensional variable γ_0 which would reduce to γ_2 for $\beta_3 \to 0$. A necessary condition we have to fulfill by means of this substitution is that the limit of the determinant G for $\beta_3 \to 0$ assume the correct dependence on β_2^8, as in the Bohr model. However, this is not enough to ensure that, at this limit, the model Hamiltonian converge to that of Bohr.

In fact, if we assume that the complete differential equation obtained with the Pauli quantization rule can be effectively separated in one part depending only on β_2, β_3 and another containing all other dynamical variables, for the former we obtain

$$\left\{\frac{1}{g}\frac{\partial}{\partial\beta_2}\left[g\frac{\partial}{\partial\beta_2}\right]+\frac{1}{g}\frac{\partial}{\partial\beta_3}\left[g\frac{\partial}{\partial\beta_3}\right]+\epsilon-V-\frac{A_J}{\beta^2+2\beta_3^2}\right\}\Psi(\beta_2,\beta_3)=0 \qquad (20)$$

where $g \propto G^{1/2}$, $\epsilon = 2E/\hbar^2$, $V = V(\beta_2, \beta_3)$ and $A_J = J(J+1)/3$. It is convenient to eliminate in the Eq. 20 the first–derivative terms, with the substitution $\Psi(\beta_2,\beta_3) = g^{-1/2}\ \Psi_0(\beta_2,\beta_3)$, to obtain

$$\left\{\frac{\partial^2}{\partial\beta_2^2} + \frac{\partial^2}{\partial\beta_3^2} + \epsilon - V(\beta_2,\beta_3) + V_g(\beta_2,\beta_3) - \frac{A_J}{\beta^2 + 2\beta_3^2}\right\}\Psi_0 = 0 \quad (21)$$

where

$$V_g(\beta_2,\beta_3) = \frac{1}{4g^2}\left[\left(\frac{\partial g}{\partial\beta_2}\right)^2 + \left(\frac{\partial g}{\partial\beta_3}\right)^2\right] - \frac{1}{2g}\left[\frac{\partial^2 g}{\partial\beta_2^2} + \frac{\partial^2 g}{\partial\beta_3^2}\right]. \quad (22)$$

We need, therefore, that also V_g takes the correct value at the limit $\beta_3 \ll \beta_2$. To this purpose, it is sufficient [23] that the first and second derivatives of G with respect to β_3 vanish when $\beta_3 \to 0$. A possible (perhaps, non unique) choice leading to this result corresponds to assuming $c_0 = 1/\sqrt{(\beta_2^2+\beta_3^2)(\beta_2^2+2\beta_3^2)}$ in the definition of γ_0 (Eq. 11). In this case one obtains

$$G \propto \frac{(\beta_2^2+\beta_3^2)^2(\beta_2^2+2\beta_3^2)^4}{(\beta_2^2+5\beta_3^2)^2} \Rightarrow \beta_2^8\left[1 + 16\,(\beta_3/\ \beta_2)^4 + ...\right]$$

and the first and second derivative of g with respect to β_3 vanish for $\beta_3 \to 0$.

9. Specific models for quadrupole–octupole oscillations

We now consider the case of simultaneous quadrupole–octupole oscillations, and in particular the quadrupole motion corresponding to the critical point of phase transition described by the X(5) model. We can start from what seems to be the simplest case: β_3 oscillations of very small amplitude compared to those in β_2, and, therefore, large excitation energy of the negative-parity levels. Experimentally, this happens in ^{150}Nd and ^{152}Sm.

The situation is very much simplified in this case, as far as $\beta_3 \ll \beta_2$ and we can neglect β_3^2 in comparison to β_2^2. One could express the quadrupole and octupole amplitudes in terms of two new variables β and δ,

$$\beta_2 = \beta\cos\delta \qquad\qquad \beta_3 = \beta\sin\delta \quad (23)$$

and confine δ to very small values by a proper potential term. It is clear that also the amplitudes v, u, w must be small compared to β, as well as γ_0 compared to 1. At the moment, however, we will forget their presence and only discuss the Schrödinger equation involving the variables β, δ and the Euler angles. With this *ansatz*, and assuming that the potential energy has the form $(\hbar^2/2)\ V(\beta,\delta)$, for $K = 0$ we obtain

$$\left\{\frac{1}{g}\frac{\partial}{\partial\beta}\left[g\frac{\partial}{\partial\beta}\right] + \frac{1}{g}\frac{\partial}{\partial\delta}\left[\frac{1}{\beta^2}g\frac{\partial}{\partial\delta}\right] + \epsilon - V - \frac{A_J}{\beta^2(1+\sin^2\delta)}\right\}\Psi(\beta,\delta) = 0 \quad (24)$$

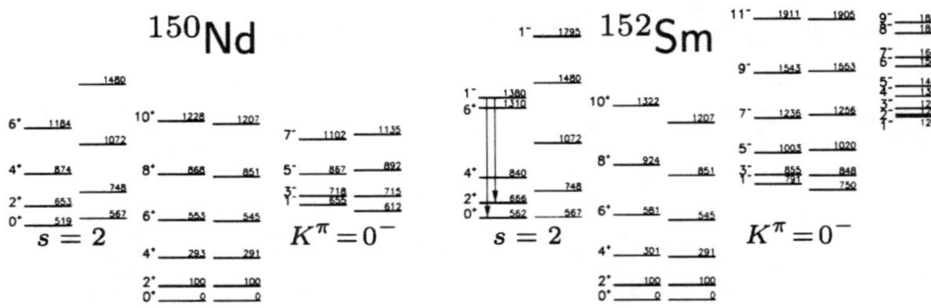

Fig. 2. The experimental (left) and theoretical energies (normalized to that of the first excited state) of the $K = 0$ bands in ^{150}Nd and ^{152}Sm. In the latter, the experimental values for the $K^\pi = 1^-$ band are also shown. Theoretical values for even J and parity are those of the X(5) model, for odd J and parity are obtained with $\Delta_1 = 15$ for ^{150}Nd and with $\Delta_1 = 20$ for ^{152}Sm.

where $g \propto G^{1/2} \propto \beta^5 \; [(1+\sin^2\delta)^2/\;(1+4\sin^2\delta)] \; \gamma_0 \; v \; u \; w$, $\epsilon = 2E/\hbar^2$, and $A_J = J(J+1)/3$. Again, it is convenient to eliminate in the Eq. 24 the first–derivative terms, with the substitution $\Psi(\beta,\delta) = g^{-1/2} \; \Psi_0(\beta,\delta)$ giving, at the limit $\delta \ll 1$,

$$\left\{ \frac{\partial^2}{\partial\beta^2} + \frac{1}{\beta^2}\frac{\partial^2}{\partial\delta^2} + \epsilon - V(\beta,\delta) - \frac{A_J - 7/4}{\beta^2} \right\} \Psi_0 = 0 \; . \tag{25}$$

The Eq. 25 has a structure very similar to that of the Bohr equation for pure quadrupole motion at the limit close to the axial symmetry, with our parameter δ in the place of γ_2. We could assume, also here, that the potential $V(\beta,\delta)$ is the sum of two independent terms, $V = V_\beta(\beta) + V_\delta(\delta)$, and try – as in the X(5) model [24] – an approximate separation of the variables, substituting the factor $1/\beta^2$ with a proper average value in the differential equation for δ. The result would be a level spacing in the excited band very similar to the ground–state band, and, at least in ^{152}Sm and ^{150}Nd, this is the case for the γ band but not for the negative-parity ones [25].

Now we want to explore some alternative procedure which could better account for the experimental data. If, *e.g.* one assumes for δ a square-well potential $V_\delta(\delta) = 0$ for $\delta < \delta_{\rm w}$ and $= +\infty$ elsewhere, the equation becomes exactly separable, with $\Psi_0 = \psi(\beta)\phi(\delta)$:

$$\frac{{\rm d}^2\phi_{\rm k}(\delta)}{{\rm d}\delta^2} + \left[A'_k - V_\delta(\delta)\right]\phi_k(\delta) = 0 \tag{26}$$

$$\frac{{\rm d}^2\psi_{\rm skJ}(\beta)}{{\rm d}\beta^2} + \left[\epsilon_{\rm skJ} - \tilde{V}_\beta(\beta) - \frac{1}{\beta^2}\left(A_J + 2 + \Delta_k\right)\right]\psi_{\rm skJ}(\beta) = 0$$

where $\tilde{V}_\beta = V_\beta(\beta) + (1/\beta^2)[A'_0 - 1/4]$ and $\Delta_k = A'_k - A'_0$. We now consider in particular the case where $\tilde{V}_\beta$ is a critical–point potential, $\tilde{V}_\beta = 0$ when

Table 2. Experimental and calculates B(E1) strengths in ^{152}Sm

Transition	E_i [keV]	$E\gamma$ [keV]	B(E1) [10^{-2}W.u.] experimental	calculated
$1^- \to 0^+$	963	963	0.42 (4)	0.42 (norm.)
$1^- \to 2^+$	963	841	0.77 (7)	0.97
$3^- \to 2^+$	1041	919	0.81 (16)	0.62
$3^- \to 4^+$	1041	675	0.82 (16)	0.99
$1^- \to 0_2^+$	963	279	weak	0.37
$1^- \to 2_2^+$	963	153	0.013 (4)	0.61

Note: Calculated strengths are normalized to the experimental one for the $1^- \to 0^+$ transition.

β is in the interval $0 - \beta_w$ and $= +\infty$ elsewhere*. In this case, for $k = 0$ one obtains the X(5) solution. For the first excited band of negative parity ($k = 1$), the term $\Delta_1 = A_1' - A_0'$ will be considered as an adjustable parameter. In this case, the spectrum of eigenvalues is given by

$$\epsilon(s, J, k) - \epsilon_0 = C\ [x_\nu(s, J, k)]^2 \tag{27}$$

with C constant, $x_\nu(s, J, k)$ the sth zero of the Bessel function $J_\nu(x)$ and $\nu = \sqrt{J(J+1)/3 + 9/4 + \Delta_k}$. The Fig. 2 shows a partial level scheme of ^{152}Sm and ^{150}Nd, normalized to the energy of the first excited state, and compared with the values derived from Eq. 27 (which, for positive–parity states, coincide with those of the X(5) model). The agreement is fairly good in both cases. In ^{152}Sm, the comparison can be extended to the lowest negative–parity state of the $s = 2$ band, if the level reported as $1^{(-)}$ really belongs to this band as it would be suggested by its decay. If it is so, its energy is significantly lower than the model prediction, but this happens also for all the $s = 2$ excited states of even parity and spin.

The B(E1) strength of a few transitions is known, but the theoretical analysis of this information is not easy. In fact, if we remain strictly in the frame of the hydrodynamical model (with a fluid of constant charge density) all E1 transitions are predicted to vanish (at least, in the long wavelength limit). An extension of the model, assuming a constant charge polarizability of the fluid, gives an Electric Dipole operator in the form [26]

$$D_\mu^{(1)} = c\ \beta_2 \beta_3 Y_\mu^{(1)} \tag{28}$$

*With a separation constant $A_0' = 1/4$ one would obtain exactly $\tilde{V}_\beta \equiv V_\beta$. Actually, it is probably unnecessary to assume that this relation holds. In fact, if a phenomenological potential is used for the active variables, this potential should already include the zero–point energies of all other degrees of freedom not explicitly taken into account.

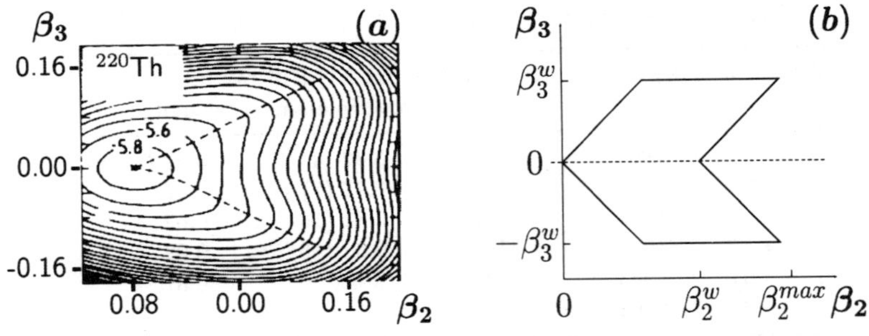

Fig. 3. Part *a*: Potential energy $V(\beta_2, \beta_3)$ calculated by Nazarwicz et al. [29] for ^{220}Th. From this situation of *correlated* quadrupole–octupole vibrations, moving towards a stable quadrupole–octupole deformation, one should cross a critical point approximately described by the potential shown in the part *b*, with $V = 0$ inside the marked area and $= +\infty$ outside. The dotted lines inside this area show the shape of the lattice used for the numerical integration.

with c dependent on the nuclear polarizability.

It would be important to check the validity of Eq. 28 in real nuclei by means of microscopic calculations. Results obtained with a Skyrme Hartree–Foch approach for well deformed nuclei of the Ra – Th region [27], show that the nuclear polarizability c is almost constant in a given nucleus, but changes drastically (even in sign) from one isotope to the next. It is not clear what can happen outside the region of stable quadrupole deformation.

In the Table 2, the values of B(E1) calculated [28] according to Eq. 28 are compared with experimental data. The strengths of E1 transitions inside the $s = 1$ sector of the level scheme are reasonably reproduced, but the experimental values of B(E1) for transitions to the $s = 2$ band are much smaller than the calculated values. Apparently, a selection rule exists, that the model is not able to reproduce.

10. Critical-point behaviour for ^{224}Ra and ^{224}Th ?

When the amplitude of the octupole vibrations is not negligible compared with the quadrupole, *i.e.* when the excitation energy of negative-parity levels is comparable with that of positive-parity ones, the above approximations are no longer valid and, at the moment, the only practicable way seems to be a numerical approach. The phase transition could take place, in this case, between a situation of quadrupole–octupole vibration around a spherical shape (with a potential like in Fig. 3*a*) directly to a permanent quadrupole–octupole deformation. The flat potential at the critical point has been schematized as shown in Fig. 3*b*. With the substitution

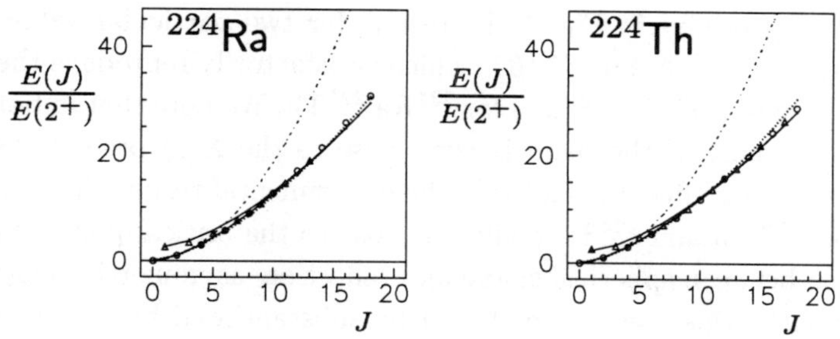

Fig. 4. Experimental excitation energies of the positive parity levels (circles) and of the negative parity ones (triangles) for ^{224}Ra and ^{224}Th, compared with the results of the present model (full line) with the follwing values of the parameters: for ^{224}Ra, $\beta_3^w/\beta_2^w = 0.915$, $\beta_2^{max}/\beta_2^w = 1.333$; for ^{224}Th, $\beta_3^w/\beta_2^w = 0.800$, $\beta_2^{max}/\beta_2^w = 1.360$ (note that the assumed values are not the result of a real optimization, still to be done). The predictions of the X(5) model (dotted lines) and of a rigid-rotor model (dashed-dotted) are shown for comparison.

$\Psi = g^{-1/2}\Psi_0$, the differential equation to be solved takes the form

$$\left[\frac{\partial^2}{\partial\beta_2^2} + \frac{\partial^2}{\partial\beta_3^2} + \epsilon - V(\beta_2, \beta_3) - V_g(\beta_2, \beta_3) - \right] \Psi_0(\beta_2, \beta_3) = 0 \qquad (29)$$

with V_g given in the Eq. 22 and $\Psi_0 = 0$ on the contour of Fig. 3*b*.

The numerical integration has been performed with the finite difference method. Namely, the space is discretized on a rectangular lattice and values of Ψ_0 at the lattice nodes are taken as independent variables. In the place of second derivatives, the ratios of finite differences are used: *e.g.*,

$$\left(\frac{\partial^2\Psi_0}{\partial\beta_2^2}\right)_{x,y} \Rightarrow \frac{\Psi_0(x+\Delta_x, y) - 2\,\Psi_0(x,y) + \Psi_0(x-\Delta_x, y)}{\Delta_x^2}$$

As $\Psi_0(\beta_2, \beta_3) = \pm\Psi_0(\beta_2, -\beta_3)$ for J even/odd, it is enough to consider only the region $\beta_3 > 0$. The lattice centers are chosen as

$$\beta_3 = (k_3 + 1/2)\Delta_y \qquad \beta_2 = (k_2 + k_3 + 1/2)\Delta_x \qquad (30)$$

with $k_2 = 1...n_2$, $k_3 = 1...n_3$, and $\Delta_x = \beta_2^w/(n_2+1)$, $\Delta_y = 2\beta_3^w/(2n_3+1)$. The ratio Δ_y/Δ_x determines the slope of the lateral borders. With this choice, all borders contain a line of nodes, on which we assume $\Psi_0 = 0$. The number of nodes internal to the integration region – and therefore the number of variables – is now $N = n_2 \cdot n_3$, and we obtain a finite dimensional $N \times N$ Hamiltonian matrix. This Hamiltonian has been diagonalized with the Implicitly Restarted Arnoldi – Lanczos method, using the ARPACK package [30]. This kind of analysis is still in progress. As an example, we give in Fig. 4 the results for two particular values of the parameters $\beta_2^{\max}/\beta_2^{\rm w}$ and $\beta_3^{\rm w}/\beta_2^{\rm w}$, which satisfactorily reproduce the experimental values of $E(J^\pi)/E(2^+)$ for ^{224}Ra ^{224}Th. We note that in both cases the positive part of the band is very close to the X(5) predictions, while the negative part fits satisfactorily the experimental results. This fact suggests that ^{224}Ra and ^{224}Th really lye close to the critical point of a phase transition between spherical shape and reflection–asymmetric deformation.

Obviously, this conclusion should be substantiated by further calculations, in particular for the transition amplitudes. This work is in program.

References

1. J. Engel and F. Iachello, *Phys. Rev. Lett.* **54**, p. 1126 (1985).
2. C. Alonso *et al.*, *Nuclear Phys. A* **586**, p. 100 (1995).
3. N. Zamfir and D. Kusnezov, *Phys. Rev. C* **63**, p. 054306 (2001).
4. N. Zamfir and D. Kusnezov, *Phys. Rev. C* **67**, p. 014305 (2003).
5. A. Raduta and D. Ionescu, *Phys. Rev. C* **67**, p. 044312 (2003).
6. T. Shneidman *et al.*, *Phys. Letters B* **526**, p. 322 (2002).
7. T. Shneidman *et al.*, *Phys. Rev. C* **67**, p. 014313 (2003).
8. W. Donner and W. Greiner, *Z. Phys.* **197**, p. 460 (1966).
9. S. Rohozinski, *Rep. Progr. Phys.* **51**, p. 541 (1988).
10. J. Eisenberg and W. Greiner, *Nuclear Theory*, 3rd edn. (Amsterdam, 1987).
11. P. Butler and W. Nazarewicz, *Rev. Mod. Phys.* **68**, p. 349 (1996).
12. P. Lipas and J. Davidson, *Nucl. Phys.* **26**, p. 80 (1961).
13. V. Denisov and A. Dzyublik, *Nucl. Phys. A* **589**, p. 17 (1995).
14. R. Jolos and P. von Brentano, *Phys. Rev. C* **60**, p. 064317 (1999).
15. N. Minkov *et al.*, *Phys. Rev. C* **63**, p. 044305 (2001).
16. D. Bonatsos *et al.*, *Phys. Rev. C* **71**, p. 064309 (2005).
17. N. Minkov *et al.*, *Phys. Rev. C* **73**, p. 044315 (2006).
18. C. Wexler and G. Dussel, *Phys. Rev. C* **60**, p. 014305 (1999).
19. D. A. Bohr, *Dan. Mat. Phys. Medd.* **26** (1952).
20. W. Pauli, in *Handbook der Physik*, ed. A. Smekal (Springer, Berlin, 1933)
21. P. Bizzeti and A. Bizzeti-Sona, *Phys. Rev. C* **70**, p. 064319 (2004).
22. L. Wilets and M. Jean, *Phys. Rev.* **102**, p. 788 (1956).
23. P. Bizzeti and A. Bizzeti-Sona, in *Nuclear Theory 05*, ed. S. Dimitrova (Heron Press, Sofia, 2005).
24. F. Iachello, *Phys. Rev. Lett.* **87**, p. 052502 (2001).
25. P. Bizzeti and A. Bizzeti-Sona, *Phys. Rev. C* **66**, p. R031301 (2002).
26. P. Lipas, *Nucl. Phys.* **40**, p. 629 (1963).
27. A. Tsvenkov, J. Kvasil and R. Nazmitdinov, *J. Phys. G* **28**, p. 2187 (2002).
28. P. Bizzeti and A. Bizzeti-Sona, in *Symmetries and low–energy phase transition in nuclear–structure physics*, eds. G. L. Bianco and D. Balabanski (University of Camerino, Italy, 2005).
29. W. Nazarewicz and P. Olanders, *Nucl. Phys. A* **441**, p. 420 (1985).
30. *http://www.caam.rice.edu/software/ARPACK/.*

SIMULTANEOUS DESCRIPTION OF FOUR POSITIVE AND FOUR NEGATIVE PARITY BANDS

A. A. RADUTA[a),b)] and C. M. RADUTA [b)]

[a)] *Department of Theoretical Physics and Mathematics, Bucharest University, POBox MG11, Romania*

[b)] *Department of Theoretical Physics, Institute of Physics and Nuclear Engineering, Bucharest, POBox MG6, Romania*

A unified description of four positive and four negative parity rotational bands for nuclei with and without octupole deformation is presented within an extended version of the coherent state model. Signatures for octupole deformation in the ground as well as in the excited bands are pointed out. Specific features of octupole deformed nuclei related with the electric and magnetic transition probabilities are presented. Comments concerning the project of enlarging the quadrupole and octupole boson system by coupling a set of particles leading to a final system with a chiral symmetry are added. To save the space, only part of the applications performed in the last decade were reviewed.

Keywords: parity, space reflection symmetry, static octupole deformation, nuclear phase, transition probabilities, quadrupole and octupole bosons

1. Introduction

Low-lying negative parity states were first observed in Ra and Th nuclei by high-resolution alpha spectroscopy measurements by Asaro et al. in 1953 [1] and Stephen et al. in 1955 [2]. Indeed, by angular correlations and gamma coincidence a spin sequence $1, 3, 5, ...$ of negative parity has been identified. The states were interpreted as describing vibrations around a spherical equilibrium shape. Microscopically, such vibrations are caused by particle-hole (ph) excitations induced by octupole-octupole two body interaction. Long time people thought that the negative parity states are similar with the vibrational states of positive parity, the only difference being the parity. The interest for negative parity states increased dramatically when two theoretical works predicted that some nuclei might have a static octupole deformation. Indeed, in Ref.3 Chasman predicted parity doublets for several odd mass isotopes of Ac, Th, and Pa. The doublet members have the same angular momentum, different parities, large connecting E3 transition and almost the same energy. The parity doublet of lowest energy plays the role of a degenerate ground state with a broken parity reflection-symmetry. Advancing the parity doublet hypothesis, Chasman was able to describe consistently the data in the nuclei mentioned above. In the second paper [4], Moller and Nix showed that the binding energy in the mass region of 224 gains about 1.5 MeV when the octupole deformation is included in the mean field of the potential energy.

By contrast to the case of nuclei having only quadrupole deformation, a nuclear surface with a static octupole deformation does not exhibit a space reflection symmetry. On the other hand breaking a symmetry leads to setting on a new nuclear phase, with specific properties. Therefore, one expects that an octupole deformed nucleus has properties which are not met in nuclei with good space reflection symmetry. Due to this feature the study of the rotational bands in octupole deformed nuclei is of a paramount importance. However, it is difficult to identify the nuclei with a static octupole deformation The reason is that there is no measurable observable which may quantitatively describe the octupole deformation. Therefore, information about it should be indirectly obtained from energy levels.

In the framework of microscopic theories the onset of octupole deformation is caused by the octupole interaction of orbits lying close to the Fermi sea and characterized by $\Delta j = \Delta l = 3$. In this context the octupole deformed nuclei should have the Fermi level close to the intruder state. Low lying negative parity states (*nps*) are compatible only with a potential energy exhibiting a flat deformed minimum. This means that a nucleus with a low lying negative parity state may be suspected to have a static octupole deformation. Along the time several signatures for octupole deformation have been pointed out. Here is the list of properties considered to be signatures for a static octupole deformation.

- In the even-even pear shape nuclei the negative and positive parity states (belonging to the ground band) form an alternating parity spectrum with a parity doublet structure.
- The two interleaved bands show an identical J(J+1) pattern, i.e. rotations in the two bands are characterized by identical moments of inertia.
- The negative parity members of doublets have enhanced E1 rates and moments.
- The experimental data show that the hindrance factor for an alpha

decay from a state of a given parity to the doublet member of opposite parity is enhanced by 2-3 orders of magnitude.

Contributions to the description of various properties of the octupole deformed nuclei have been reviewed in several papers [5,6]. Due to the space restriction we confine our review to the contributions performed with a single formalism, namely the one which is able to describe in a consistent manner a large volume of data.

Thus the aim of our lecture is to investigate some properties of octupole bands within a formalism which represents an extension [7–11,13] of the coherent state model (CSM) [14]. Our intentions regard the following features: i) The interleaved structure in the g, β and γ bands for some even-even isotopes of Ra, Th, U, Pu, Gd, Yb. ii) The low excitation energy for the state 1^- in ^{218}Ra. iii)The description of the E1 branching ratios. iv) Are there signatures for static octupole deformation in the excited bands? v) What are the specific properties for the dipole bands. vi) Could an octupole deformed nucleus have a chiral symmetry?

2. Brief review of CSM

The intrinsic deformed states modeling the ground, beta and gamma bands must satisfy a set of restrictions suggested by the experimental data. These restrictions have been formulated in Ref.14 and in brief they are: a) the states are deformed functions of quadrupole bosons, $b^{\dagger}_{2\mu}$; b) the states in the laboratory frame are obtained through an angular momentum projection procedure; c) the states are orthogonal before and after angular momentum projection; d) functions depend on a real parameter simulating the quadrupole deformation; e) in the vibrational region the functions describe a degenerate multi-boson state while in the large deformation regime they are proportional to a Wigner function of a definite K; f) the link between the vibrational and rotational functions is achieved in full agreement with the Sheline-Sakai scheme [15,16]; g) the projected states span a restricted collective space where an effective Hamiltonian is constructed. In Ref.14 we found a solution for the deformed states obeying these criteria.

The projected states defining the ground, beta and gamma bands , are:

$$\varphi^{(i)}_{JM} = N^{(i)}_J P^J_{MK_i}\Phi_i, i = g, \beta, \gamma \tag{2.1}$$

where the projection operator is defined by:

$$P^J_{MK} = \frac{2J+1}{8\pi^2}\int d\Omega D^{J^*}_{MK}(\Omega)R(\Omega), \tag{2.2}$$

and acts on intrinsic mutually orthogonal states of the form

$$\Phi_g = exp\left[d\left(b_{20}^{+} - b_{20}\right)\right] |0>_{(2)} \ ,$$
$$\Phi_\beta = \Omega_\beta^\dagger \Phi_g$$
$$\equiv \left[\left(b_2^\dagger b_2^\dagger b_2^\dagger\right)_{00} + \frac{3}{\sqrt{14}}\, d \left(b_2^\dagger b_2^\dagger\right)_{00} - \frac{d^3}{\sqrt{70}}\right] \Phi_g \ ,$$
$$\Phi_\gamma = \Omega_\gamma^\dagger \Phi_g \equiv \left[\left(b_2^\dagger b_2^\dagger\right)_{22} + \sqrt{\frac{2}{7}}\, d\, b_{22}^\dagger\right] \Phi_g \ . \quad (2.3)$$

It can be checked that these projected states are orthogonal. Also the un-projected states have this property. The parameter d is a real number and simulates the nuclear quadrupole deformation.

An effective Hamiltonian is then constructed such that a maximal decoupling of the projected states is achieved

$$\hat{H}_2 = A_1\left(22\hat{N}_2 + 5\Omega_{\beta'}^\dagger \Omega_{\beta'}\right) + A_2 \hat{J}_2^2 + A_3 \Omega_\beta^\dagger \Omega_\beta \ , \quad (2.4)$$

$\hat{J}_2$ is the total angular momentum, $\hat{N}_2$ is the quadrupole boson number operator and

$$\Omega_{\beta'}^\dagger = (b_2^\dagger b_2^\dagger)_0 - \frac{d^2}{\sqrt{5}}. \quad (2.5)$$

With this Hamiltonian all states of the beta band and those of the gamma band with odd J are completely decoupled. Only the states of the gamma band with even J are coupled to the ground band states of similar angular momentum. In this case, for a given J the energies are computed by diagonalizing the boson Hamiltonian in a two-dimensional space. The coupling is in any case vanishing in the vibrational and rotational limits and is small in the transitional regime. The CSM was successfully used to describe the g, β and γ states, including those of high spin, in the Pt, rare earth and actinide regions.

In order to enlarge the range of applicability we proposed several independent extensions of CSM: a) By adding the qp degrees of freedom the back-bending phenomena in even-even and even-odd nuclei could be studied [17,18]. b) A distinction between protons and neutrons was made in order to allow for a description of the low-lying orbital $M1$ mode [19]. c) Three negative parity bands with $K^\pi = 0^-, 1^-, 2^-$ were described by exciting, with an octupole boson, the g, β, γ bands given by the CSM model [20]. In this way the vibrational-negative parity bands were described.

To stress on the virtues of CSM, it has been extensively compared with others phenomenological models like: i) the liquid drop model (LD); ii) the rotation-vibration model (RV); iii) the Interacting Boson Approximation (IBA). This latter study clearly shows that the two models differ in their basic conceptual assumptions and their range of applicability. Indeed, while IBA is not working for deformed nuclei as well as for the high spin states, the CSM works especially well under the mentioned circumstances due to its semiclassical character.

3. CSM extension to the negative parity states

We suppose that the ground state exhibits both quadrupole and octupole deformations and is described by a product of two coherent functions for quadrupole and octupole bosons, respectively [7]:

$$\Psi_g = e^{f(b_{30}^+ - b_{30})} e^{d(b_{20}^+ - b_{20})} |0\rangle_2 |0\rangle_3, \tag{3.1}$$

d and f are real parameters which simulates the quadrupole and octupole nuclear deformation. This function is a sum of two components with different parities $\Psi^{\pm}$ which define through projection two sets of states, respectively:

$$\varphi_{JM}^{(g,k)} = N_J^{(g,k)} P_{M0}^J \Psi_g^{(k)}, \quad K = \pm \tag{3.2}$$

$$J = \delta_{k,+}(even) + \delta_{k,-}(odd). \tag{3.3}$$

The normalization factors have the expressions:

$$\left(N_J^{(g,k)}\right)^{-2} = e^{-\frac{1}{2}(2y_3+3y_2)}(2J+1)\mathcal{I}_{(g,J)}^{(k)}(y_2, y_3), \; y_2 = d^2, y_3 = f^2.$$

where the overlap integrals are given by:

$$\mathcal{I}_{(g,k)}^J = (3y_3)^J \sum_{p=p_{min}}^{\infty} \sum_{l=0}^{l_{Max}} \left(\frac{y_3^2}{4}\right)^{p-l} (9)^p \left(\frac{5}{27}\right)^l (-)^{l+J}$$
$$\times \frac{(p+J)!(2p+J)!F(-l, 2p-3l+J+1; \frac{9}{10}y_2)}{l!p!(2p-3l+J)!(2p+2J+1)!}.$$

Alternatively, the projected states can be written in a tensorial form as a product of projected states:

$$\varphi_{JM}^{(g,k)} = \mathcal{N}_J^{(g,k)} \sum_{J_2,J_3} (N_{J_3}^{(k)} N_{J_2}^{(g)})^{-1} C_{0\ 0\ 0}^{J_3\ J_2\ J} [\Psi_{J_3}^{(k)} \varphi_{J_2}^{(g)}]_{JM} \; ; \quad k = \pm\,.$$

In this case all matrix elements for quadrupole operators can be expressed as simple functions of overlap integrals:

$$I_J^{(0)} = \int_0^1 P_J(x) e^{d^2 P_2(x)} dx, \tag{3.4}$$

and $I_J^{(1)}$, which is the first derivative of $I_J^{(0)}$ with respect to $y_2(=d^2)$. If the intrinsic ground state has not a good reflection symmetry it sounds reasonable to assume that intrinsic gamma and beta bands have also this property. Then instead of using Φ_γ and Φ_β as model states of the two bands we propose now to choose the following functions:

$$\Psi_\gamma = \Omega_\gamma^+ \Psi_g, \quad \Psi_\beta = \Omega_\beta^+ \Psi_g.$$

Projecting first the parity and then the angular momentum one obtains four bands, two of positive and two of negative parity:

$$\varphi_{JM}^{(i,k)} = \mathcal{N}_J^{(i,k)} P_{MK_i}^J \Psi_i^{(k)} \quad ; i = \beta, \gamma \ ; \ k = \pm \ ,$$

with

$$\Psi_i^{(\pm)} = \Omega_i^+ e^{-\frac{y_3+y_2}{2}} \begin{pmatrix} cosh(fb_{30}^+) \\ sinh(fb_{30}^+) \end{pmatrix} e^{db_{20}^+} \mid 0\rangle_3 |0\rangle_2 \ ;$$

$$i = \beta, \gamma \ ; \ K_i = 2\delta_{i,\gamma}, \ y_2 = d^2, y_3 = f^2.$$

These functions can be also written as superposition of products of quadrupole and octupole projected states.

$$\varphi_{JM}^{(i,k)} = \mathcal{N}_J^{(g,k)} \sum_{J_2,J_3} (N_{J_3}^{(k)} N_{J_2}^{(g)})^{-1} C_{0\ 0\ 0}^{J_3\ J_2\ J} [\Psi_{J_3}^{(k)} \varphi_{J_2}^{(i)}]_{JM}. \qquad (3.5)$$

Since for large values of the deformation parameter d, the projected states $\varphi_{JM}^{(i)}$ with $i = g, \beta, \gamma$ behave like a Wigner function with a definite quantum number "K", they describe rotational bands with $K = 0, 0, 2$, respectively. Analogously, the octupole states $\Psi_{JM}^{(\pm)}$ describe rotational bands having $K = 0$. Consequently, four of the six bands defined above have $K = 0$, while the remaining two are $K = 2$ bands. In order to stress on the parity partnership, we use the suggestive notations $g^\pm, \beta^\pm, \gamma^\pm$ for the six bands. Each pair of bands is expected to give rise to an alternating parity sequence as it happens in the case of ground and $K^\pi = 0^-$ bands, i.e. the $g^\pm$ pair. The set $\{\varphi_{JM}^{(i,k)}\}_{i,k;J,M}$ with $i = g, \beta, \gamma$ and $k = \pm$ is orthogonal. Note that for $f = 0$ only the positive parity states $\varphi_{J_3M_3}^{(+)}$ are well defined. However the limits for "f" going to zero exist both for $k = +$ and $k = -$, and the following relation holds:

$$\lim_{f \to 0} \varphi_{JM}^{(i,+)} = \varphi_{JM}^{(i)}.$$

Thus, the formalism proposed yields the CSM in the limit of $f \to 0$. Following the CSM, we define a model Hamiltonian which is effective in the model space of projected states. The effectiveness criteria is satisfied by:

$$\hat{H} = \hat{H}_2' + \mathcal{B}_1\hat{N}_3(22\hat{N}_2 + 5\Omega^{+}_{\beta'}\Omega_{\beta'}) + \mathcal{B}_2\hat{N}_3\Omega^{+}_{\beta}\Omega_{\beta}$$
$$+ \mathcal{B}_3\hat{N}_3 + \mathcal{A}_{(J23)}\,\vec{J}_2\vec{J}_3 + \mathcal{A}_{(J)}\hat{J}^2,$$

where $\hat{H}_2'$ is obtained from $\hat{H}_2$ by subtracting the rotational term $A_2\hat{J}_2^2$.

The formalism described above is called the extended CSM (ECSM).

4. Extension to the $K^{\pi} = 1^{\pm}$ bands

Here, the ECSM will be further extended by considering the dipole parity partner bands [12,13]. The difficulty encountered when the restricted collective space is enlarged consists in finding an intrinsic state which is orthogonal on the previously defined model states, before as well as after angular momentum projection. The second step is to correct the model Hamiltonian by a term so that the resulting Hamiltonian is effective in the extended space of projected states. A possible solution for the intrinsic state generating the dipole bands is:

$$\Psi^{(1,\pm)} = \Omega_3^{\dagger} b_{31}^{\dagger} \Psi_g^{(\pm)}, \; \Omega_3^{\dagger} = [b_3^{\dagger} b_3^{\dagger}]_0 + \frac{f^2}{\sqrt{7}}. \qquad (4.1)$$

From these states, two sets of angular momentum projected states are obtained, which are hereafter denoted by $\phi^{1,\pm}_{JM}$. These states are weakly coupled to the states of other bands by the $\mathcal{B}_1$ and $\mathcal{B}_3$ terms. Moreover, these terms give large contribution to the diagonal matrix elements involving the projected dipole states. Aiming at describing quantitatively the properties of the dipole states, two terms are added to the model Hamiltonian

$$\Delta H = \mathcal{C}_1\Omega_3^{\dagger}\Omega_3 + \mathcal{C}_2\Omega_3^{\dagger}\hat{N}_2\Omega_3. \qquad (4.2)$$

The new terms affect only the diagonal m.e. of the dipole states. $\mathcal{C}_2$ is determined so that the corresponding contribution to a particular state (say 2^-) cancels the one coming from the $\mathcal{B}_1$ term. $\mathcal{C}_1$ is determined so that the measured excitation energy of the state 1^- is reproduced. The contribution of the $\mathcal{B}_1$ and $\mathcal{B}_3$ terms to the off-diagonal matrix elements characterizing the dipole states amounts to few keV.

Thus, the final Hamiltonian to be used for describing simultaneously four positive and four negative bands, is:

$$H = \mathcal{A}_1(22\hat{N}_2 + 5\Omega^{\dagger}_{\beta'}\Omega_{\beta'}) + \mathcal{A}_2\Omega^{\dagger}_{\beta}\Omega_{\beta} + \mathcal{A}_J\vec{J}^2$$
$$+ \mathcal{B}_3\hat{N}_3 + \mathcal{B}_1\hat{N}_3(22\hat{N}_2 + 5\Omega^{\dagger}_{\beta'}\Omega_{\beta'}) + \mathcal{A}_{(J23)}\vec{J}_2\vec{J}_3$$
$$+ \mathcal{C}_1\Omega_3^{\dagger}\Omega_3 + \mathcal{C}_2\Omega_3^{\dagger}\hat{N}_2\Omega_3. \qquad (4.3)$$

The quantities which are to be calculated are:

- The excitation energies

$$\epsilon_J^{(k)} = E_J^{(k)} - E_0^{(+)} \ .$$

- The dynamic moment of inertia, as function of angular frequency.

$$\hbar\omega = \frac{dE}{dI} \approx \frac{1}{2}(E_I - E_{I-2}), \tag{4.4}$$
$$\mathcal{J}^{(2)}/\hbar^2 = \left(\frac{d\omega}{dI}\right)^{-1} \approx 2/(\omega_I - \omega_{I-2}).$$

- The energy displacement functions

$$\begin{aligned} \delta E(J^-) &= E(J^-) \\ &\quad - \frac{(J+1)E((J-1)^+) + JE((J+1)^+)}{2J+1}, \\ \Delta E_{1,\gamma}(I) &= \frac{1}{16}[6E_{1,\gamma}(I) - 4E_{1,\gamma}(I-1) - 4E_{1,\gamma}(I+1) \\ &\quad + E_{1,\gamma}(I-2) + E_{1,\gamma}(I+2)], \\ E_{1,\gamma}(I) &= E(I+1) - E(I). \end{aligned}$$

- The E1 transition operator is defined as:

$$\begin{aligned} T_{1\mu} = q_1 \sum_{\mu_2,\mu_3} C^{3\ \ 2\ \ 1}_{\mu_3\ \mu_2\ \mu} \quad & (b^+_{3\mu_3} + (-)^{\mu_3} b_{3,-\mu_3}) \\ & \times (b^+_{2\mu_2} + (-)^{\mu_2} b_{2,-\mu_2}) \end{aligned}$$

- Using this transition operator we calculated the ratio:

$$R_1(I) = \frac{B(E1; I^- \to (I+1)^+)}{B(E1; I^- \to (I-1)^+)} \tag{4.5}$$

- Also, the E2 and E3 transition probabilities have been calculated:

$$T_{\lambda\mu} = q_\lambda \left(b^\dagger_{\lambda\mu} + (-)^\mu b_{\lambda,-\mu}\right), \lambda = 2, 3.$$

- The angle between the angular momenta $\vec{J}_2$ and $\vec{J}_3$.

$$cos\varphi = \frac{\langle\phi^{(k)}_{JM}|\vec{J}_2.\vec{J}_3|\phi^{(k)}_{JM}\rangle}{\sqrt{\langle\phi^{(k)}_{JM}|\hat{J}^2_2|\phi^{(k)}_{JM}\rangle\langle\phi^{(k)}_{JM}|\hat{J}^2_3|\phi^{(k)}_{JM}\rangle}}, k = 1, \pm. \tag{4.6}$$

5. Numerical results

Numerical calculations have been performed for: ^{158}Ga, ^{172}Yb, ^{218}Ra, ^{220}Ra, ^{226}Ra, ^{228}Th, ^{232}Th, ^{236}U, ^{238}U, ^{238}Pu. Due to the lack of space we do not present here all results, but only those which are necessary for underlying the relevant ideas. There are several parameters involved in the game, which are to be determined through a fitting procedure: $\mathcal{A}_1, \mathcal{A}_2, \mathcal{A}_J, \mathcal{B}_1, \mathcal{B}_3, \mathcal{A}_{(J23)}, \mathcal{C}_1, \mathcal{C}_2, d, f$.

The structure coefficients $\mathcal{A}$ and $\mathcal{B}$ and the deformation parameters d and f were determined by the least square procedure of the excitation energies in the bands $g^{\pm}, \beta^{\pm}, \gamma^{\pm}$. The remaining two parameters $\mathcal{C}_1$ and $\mathcal{C}_2$ are determined as explained before.

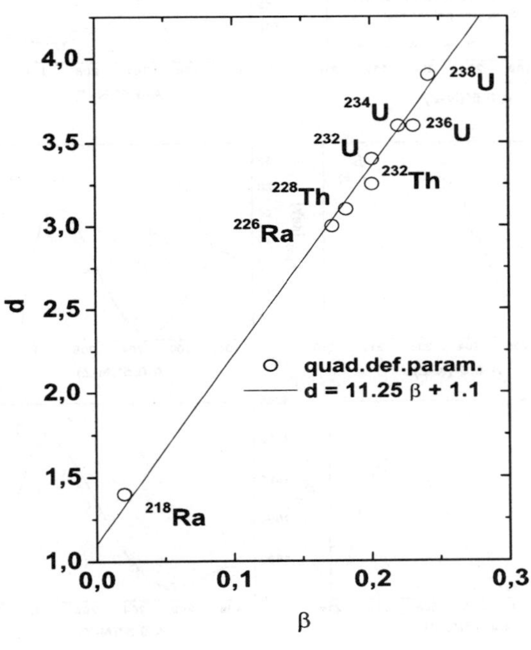

Fig. 1. The deformation parameter d, is plotted as function of the nuclear deformation.

In Fig. 1 the obtained values for the deformation parameter d are plotted as function of the nuclear deformation β, for the actinide isotopes considered. The results for the rare earth isotopes lye on a straight line parallel to the one from Fig . 1. The values of d corresponding to ^{158}Gd and ^{172}Yb are equal to 3 and 3.68, respectively. It is remarkable the linear dependence of the two variables, d and β. Unfortunately, a similar plot for the octupole deformation parameter is not possible due to the lack of experimental data for the nuclear octupole deformation. The calculated values of f are equal to 0.3 except those for ^{226}Ra and ^{238}U which are 0.8 and 0.6, respectively.

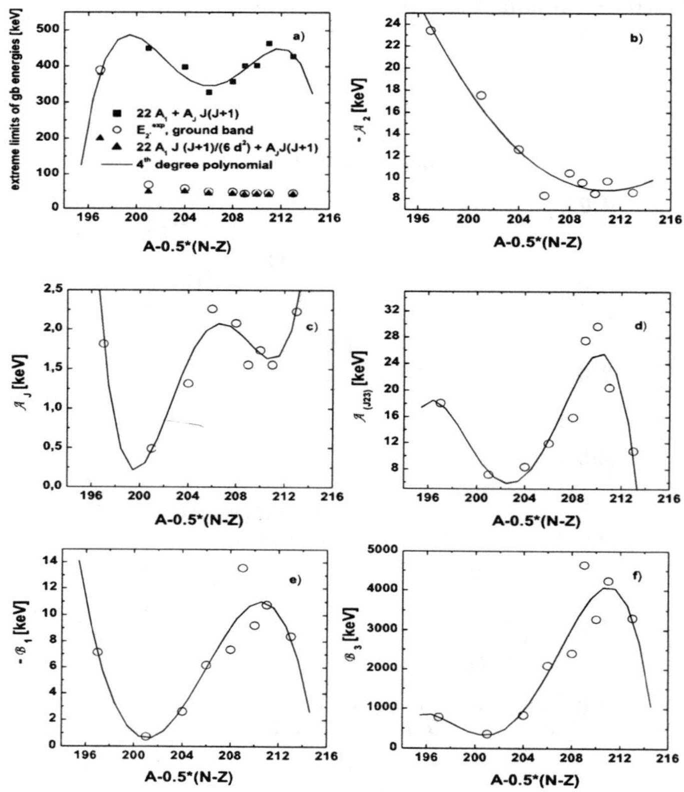

Fig. 2. The coefficients involved in the model Hamiltonian are plotted as function of $A - 0.5 * (N - Z)$.

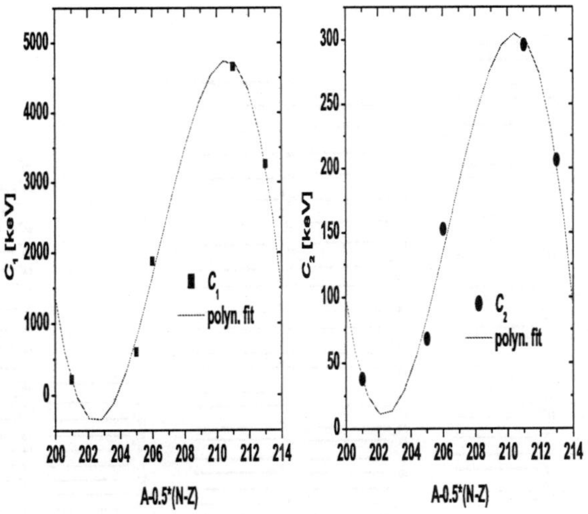

Fig. 3. The structure coefficient C_1 (black square) and C_2, determined as explained in the text, is represented as function of $A - 0.5 * (N - Z)$ (black square). The obtained values are interpolated by a third order polynomial (full line curve).

The structure coefficients are represented as function of $A - 0.5(N - Z)$ in Figs. 2 and 3. The calculated values were interpolated by a smooth curve of a polynomial type. Energies for the partner bands g^+ and g^-, in three isotopes of Ra are represented in Fig. 4. One notes the doublet as well as the interleaved structure of positive and negative parity states. In the formalism described here, the doublets are caused by the small octupole deformation. If the octupole deformation were vanishing, the doublet members would be degenerate. For ^{218}Ra the doublet structure persists also in the region of high spin, while for ^{226}Ra the spectrum in the high angular momentum area becomes equidistant.

Let us now turn our attention to the low position of the $K^\pi = 0^-$ state 1^-. This is caused by the term $\vec{J}_2\vec{J}_3$ which is attractive in the state 1^- and repulsive in other states.

The calculated energies for three pairs of parity partner bands, $g^\pm$, $\beta^\pm$ and $\gamma^\pm$, are compared with the available experimental data in Fig.5. There we give also the parabola $aJ(J + 1) + b$ which fits the energies of the first and the last states in the respective band. In general, the J dependence of

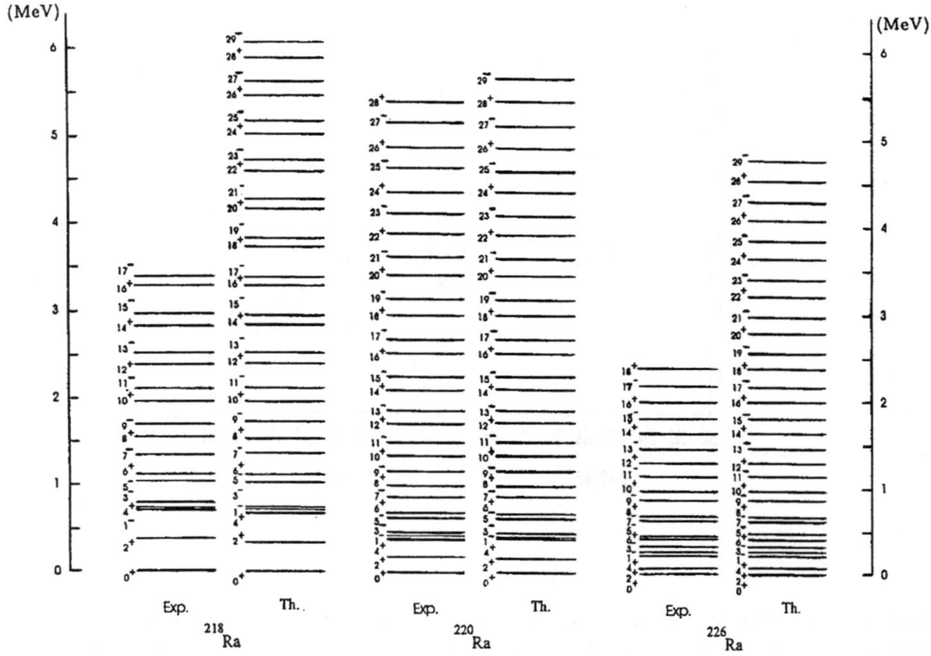

Fig. 4. Experimental and calculated energies for the $g^{\pm}$ bands in ^{218,220,226}Ra (see Ref. [7]).

energies is different from the J(J+1) pattern. The quality of the agreement between the theoretical results and the experimental energies can be best judged by plotting the dynamic moment of inertia which is a very sensitive function of the rotational frequency. Such a plot is shown in Fig. 6 for the pairs of bands $g^{\pm}, \beta^{\pm}, \gamma^{\pm}$. This graph indicates ^{226}Ra as the best candidate for a static octupole deformation in all three bands.

Since in ^{218}Ra the spectrum in the bands $g^{\pm}$ is almost equidistant, the dynamic moment of inertia is very large. Due to this feature, for this case we give, instead, the graph representing the angular momentum as function of the rotational frequency. It is worth noticing that the back and forward bending seen for the experimental energies characterizing the band g^{+}, are nicely reproduced by our calculations. The microscopic interpretation of the first bending is the crossing of a collective band with a two quasiparticle band, while the forward bending is caused by the intersection of the later band with a four quasiparticle band. In our phenomenological description the two bending are caused by the interaction between the quadrupole and octupole degrees of freedom.

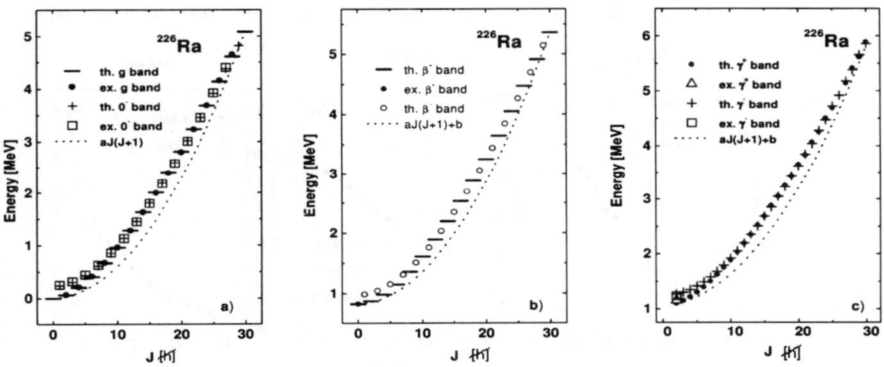

Fig. 5. Experimental and calculated energies for the bands $g^{\pm}, \beta^{\pm}, \gamma^{\pm}$ for ^{226}Ra.

Suppose now that the nuclear system exhibits a static octupole deformation and therefore is described, in the intrinsic frame, by a function with both quadrupole and octupole deformations. If the octupole deformation is small the projected states J^{+}, $(J+1)^{-}$ are close in energies. Since the projected states originates from the same intrinsic state, they are characterized by a single moment of inertia. If the energies of the mentioned states depend linearly on $J(J+1)$ then the first order energy displacement function vanishes for the angular momentum equal to J. Reversely, if the energy displacement function is vanishing at a certain angular momentum, one says that in the corresponding state, the static octupole deformation is set on. However, in many cases the excitation energies deviate drastically from the $J(J+1)$ law. If energies depend quadratically on $J(J+1)$, the vanishing of the second order energy displacement function indicates that the second order derivative for energy with respect to $J(J+1)$ is the same for the two parity partner bands. Therefore, in order to decide whether a state of a certain angular momentum exhibits a static octupole deformation or not, we must analyze simultaneously the first and the second order energy

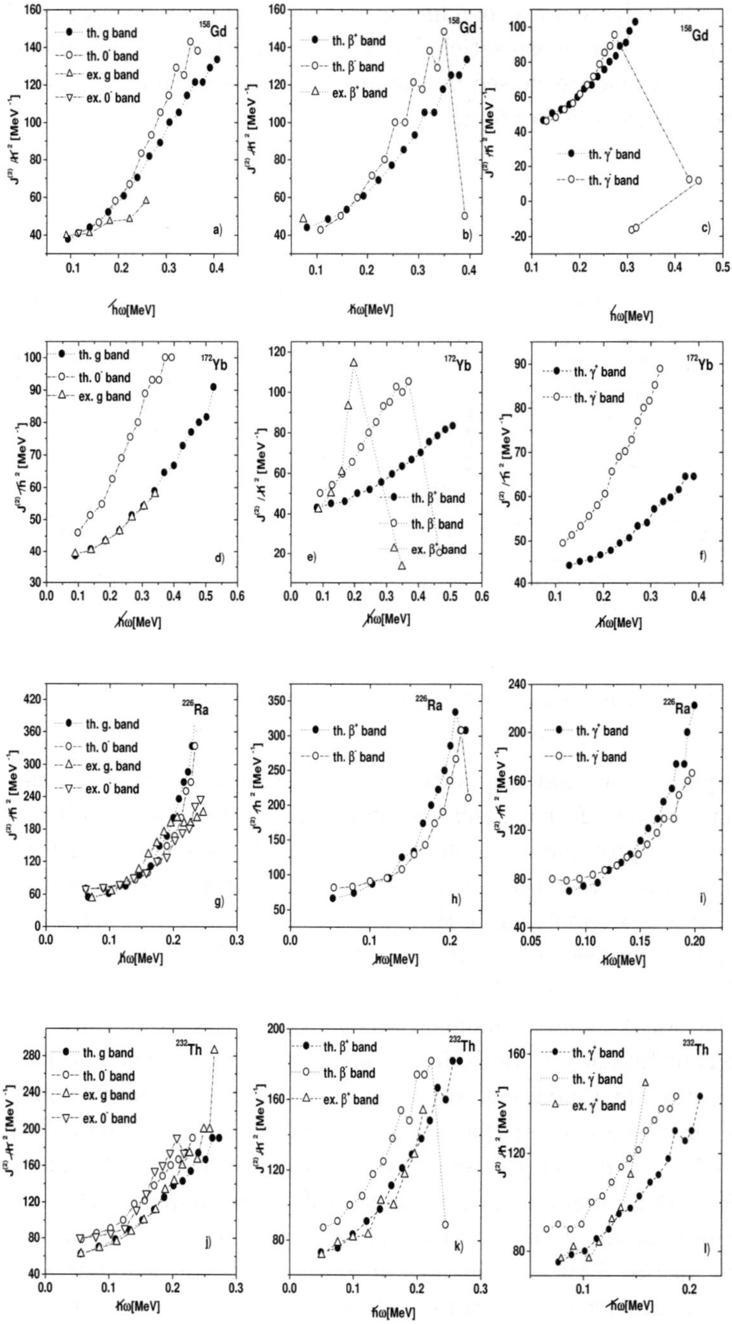

Fig. 6. The dynamic moment of inertia characterizing the bands $g^{\pm}, \beta^{\pm}, \gamma^{\pm}$, is plotted as function of the rotational frequency.

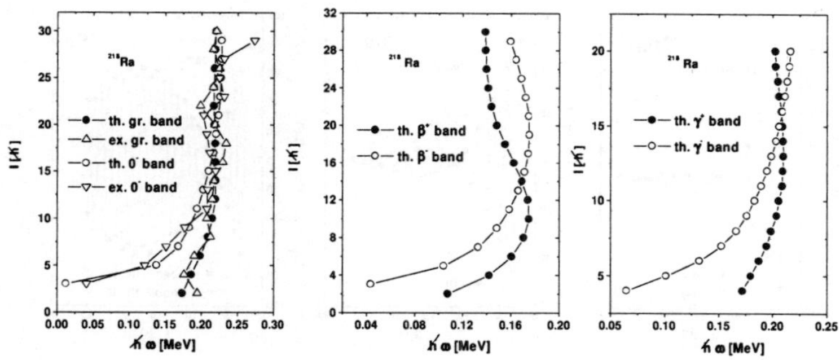

Fig. 7. The angular momentum is plotted as function of the rotational frequency for three parity partner bands, $g^{\pm}, \beta^{\pm}, \gamma^{\pm}$, in ^{218}Ra.

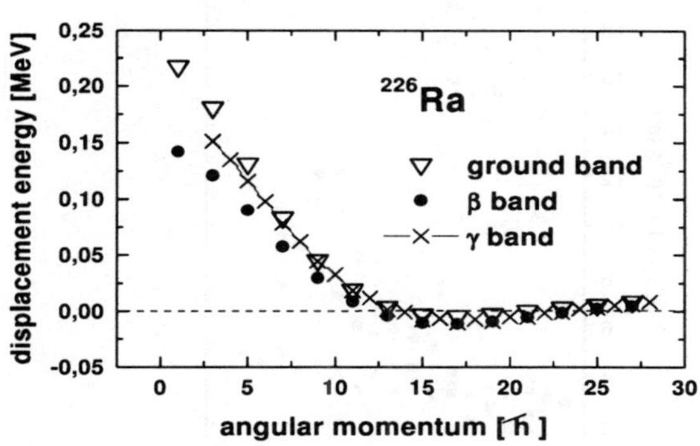

Fig. 8. The first order energy displacement function for ^{226}Ra .

displacement functions. The two functions are given for ^{226}Ra in Figs. 8 and 9, respectively. We note that at least for this isotope the octupole deformation is settled, according to the behavior of the δE function, simultaneously in the three pairs of bands, $g^{\pm}, \beta^{\pm}, \gamma^{\pm}$. The second order energy displacement shows that ground and beta bands get octupole deformation

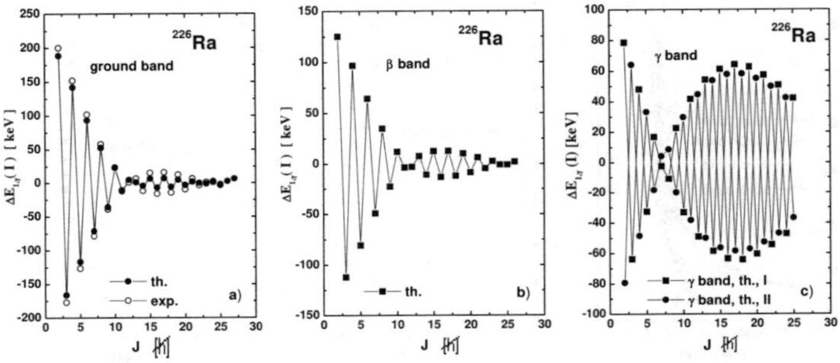

Fig. 9. The second order energy displacement function for ^{226}Ra .

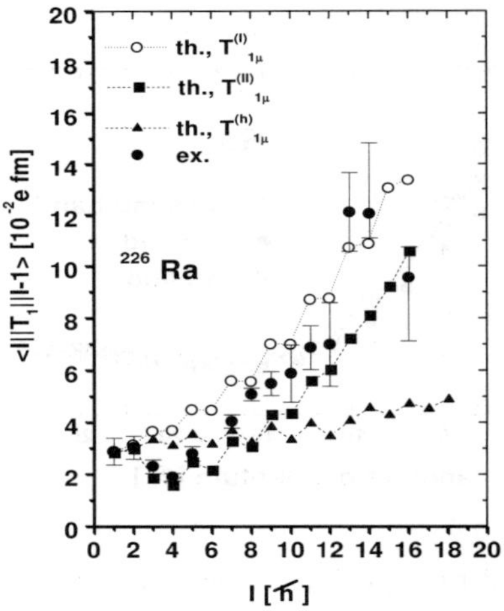

Fig. 10. The matrix elements for the transition $I \rightarrow (I-1)$ is plotted as function of the angular momentum for ^{226}Ra. Data are from Ref. 23. Calculations correspond to three different expressions for the transition operator as explained in Ref.11 .

for the same angular momentum while in the gamma bands the octupole deformation is earlier settled. A systematic analysis of the displacement

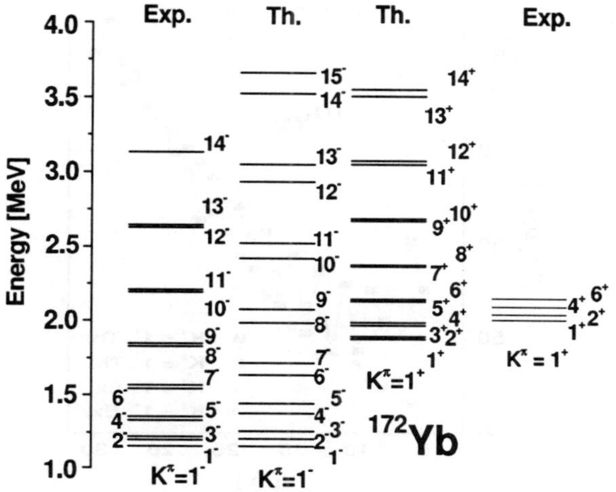

Fig. 11. Theoretical (Th.) and available experimental (Exp.) excitation energies for the $K^\pi = 1^-$ and $K^\pi = 1^+$ in ^{172}Yb.

functions for a large number of nuclei may be found in Refs. 10,21. Therein we identified several distinct situations: a) octupole deformation shows up in all three pairs of bands; b) octupole deformation appears in the bands $g^\pm$ but not in the other bands; c) octupole deformation is settled in the $\gamma^\pm$ bands but not in the others.

Due to the rod effect saying that the charge density is maximum in the region where the surface curvature is maximum, a system having octupole deformation may exhibit a non-vanishing dipole moment. Consequently, interacting with an electromagnetic field such a system can be driven in a state characterized by large E1 rates. In this context one expects that the B(E1) value exhibits a jump at the angular momentum where the octupole deformation is set on. This feature is illustrated in Fig.10 where the reduced matrix element for the transition $I \to (I-1)$ is represented as function of angular momentum. One notes a fairly good agreement between theoretical and experimental data.

The results for the $K^\pi = 1^-$ band energies are presented in Fig.11 for ^{172}Yb, where relevant data are available [22].

In Fig. 12 the dynamic moment of inertia is plotted vs. the angular momentum. From this figure one notices that the results corresponding to even and those corresponding to odd angular momenta are lying on separate smooth curves as if these sets of states belonged to two distinct bands. The remark is valid for both the positive and negative parity bands.

In order to see whether there are signatures of octupole deformation in

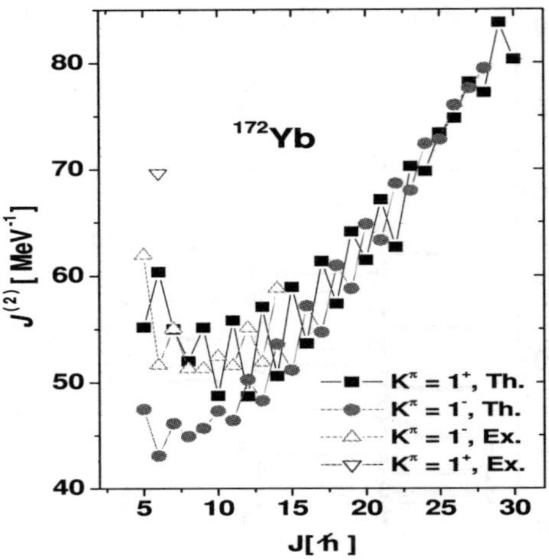

Fig. 12. The dynamic moment of inertia for the dipole bands of positive and negative parity corresponding to the calculated and experimental energies respectively, is plotted as function of the angular momentum

the dipole bands, we show in Fig. 13 the energy displacement functions for the two dipole bands with $K^\pi = 1^\pm$.

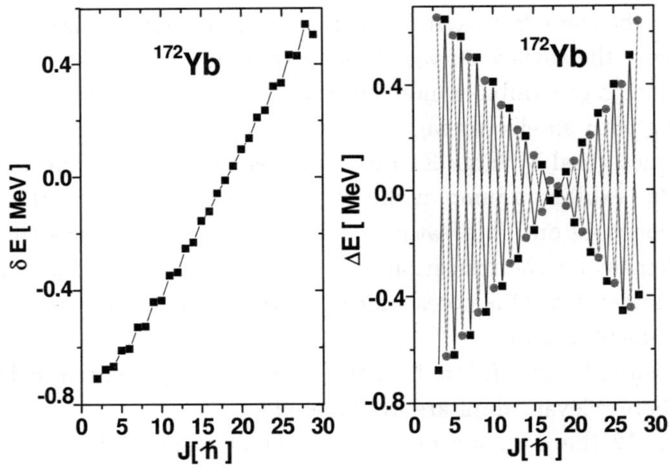

Fig. 13. The energy displacement functions δE (left panel) and ΔE (right panel), given in the text, are plotted as functions of J.

According to Fig.13 , the states of angular momentum equal to 18,19 may have static octupole deformation. To obtain a definite conclusion about the static octupole deformation we have analyzed the E1 and M1 properties of these bands. The relative magnitude of branching ratios for the bands with $K^\pi = 1^+$ and $K^\pi = 1^-$ indicate that the magnetic transitions are stronger for the positive parity states while the E1 transitions prevail for negative parity states. Due to this fact we call the band $K^\pi = 1^+$ as the magnetic band while the negative parity band as the electric band. The branching ratios of the dipole states calculated within the formalism presented above are compared with the corresponding data in Fig.14. In contrast to the case of $K^\pi = 0^-$ band, for the $K^\pi = 1^-$ band there is no jump in the behavior of the B(E1) value. However, the M1 branching ratio from the $K^\pi = 1^+$ to $K^\pi = 0^+$ get a jump for $J = 18, 19$, which are in fact the angular momenta where the energy displacement functions vanish. Due to this feature we consider the big value of the mentioned M1 branching ratio as a signature for the octupole deformation in the dipole bands.

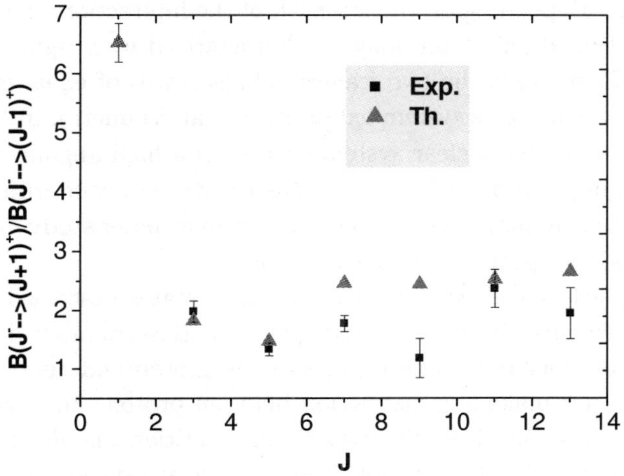

Fig. 14. The branching ratios characterizing the transitions of $K^\pi = 1^-$ states to the ground band states (triangle), are compared with the corresponding experimental data (square). The transition operator used is $T_{1\mu} = T^h_{1\mu} + T^{anh}_{1\mu}$ with the harmonic term defined in the text and $T^{anh}_{1\mu} = q_{anh}\{\left[b_3^\dagger\left(\hat{J}_3\hat{J}_2\right)_2\right]_{1\mu} + \left[\left(\hat{J}_3\hat{J}_2\right)_2 b_{\bar{3}}\right]_{1\mu}\}$. All ratios correspond to the relative effective charge q_{anh}/q_1=-1.722, where q_1 denotes the strength of the harmonic term.

Within ECSM, one can calculate the angle between the angular momenta carried by the quadrupole ($\vec{J}_2$) and octupole ($\vec{J}_3$) bosons respectively, for a state of total angular momentum $\vec{J}$. This angle is shown in Fig. 15 as function of the angular momentum for the states belonging to the four pairs of bands under study. Apart from small details, the features shown in Fig. 15 for ^{226}Ra are common to all nuclei studied by our group. The angle has a saw-tooth structure for the dipole bands. Here the angle characterizing the even and odd angular momenta stay on separate smooth curves suggesting once again that the two sets of states might form different bands. For the bands $g^{\pm}, \beta^{\pm}, \gamma^{\pm}$ the angle is decreasing up to a critical value after which is slightly increasing reaching a plateau at $\varphi = \pi/2$. The interpretation of this result is as follows. If the quadrupole bosons describes an ellipsoidal shape having the axis OZ as symmetry axis, the angular momentum $\vec{J}_2$ is oriented along an axis in the plane XOY, say OX, to which the maximum moment of inertia is associated. The octupole bosons describe a shape for which the moment of inertia corresponding to the axis OZ, is maximum.

Suppose now that a term describing a set of particles and a term describing the interaction between the two sub-systems are added to the model Hamiltonian. Depending on the strength of the interaction, the eigenstates of the resulting Hamiltonian may be characterized by a right or left triad $(\vec{j}, \vec{J}_2, \vec{J}_3)$. In the case the two frames defines states of equal energies one says that the composite system exhibits a chiral symmetry. In this context we may say that the nuclear system excited in a high angular momentum state belonging to either of the six bands $g^{\pm}, \beta^{\pm}, \gamma^{\pm}$, constitutes a precursor of a chiral symmetry system. Such a system is under study in our group, and we hope to report the results very soon.

We may ask ourself whether the magnetic states described in this lecture is related with the scissors mode [24]. The scissors mode describes the angular oscillations of symmetry axes of the proton and neutron systems. Here, we do not make any distinction between protons and neutrons, but we could say that we deal with two distinct entities, one described by the quadrupole and other by octupole bosons. The two systems rotate around axes which make an angle which was just described. By contrast to the scissors mode, where the angle between the symmetry axes is small, here the angle is large. Therefore, we could name the magnetic states described in the present lecture as shares states.

6. Conclusions

The results presented above may be summarized as follows: States of four pairs of partner bands $g^{\pm}, \beta^{\pm}, \gamma^{\pm}, 1^{\pm}$, are projected from four orthogonal states having both quadrupole and octupole deformation. The interleaved

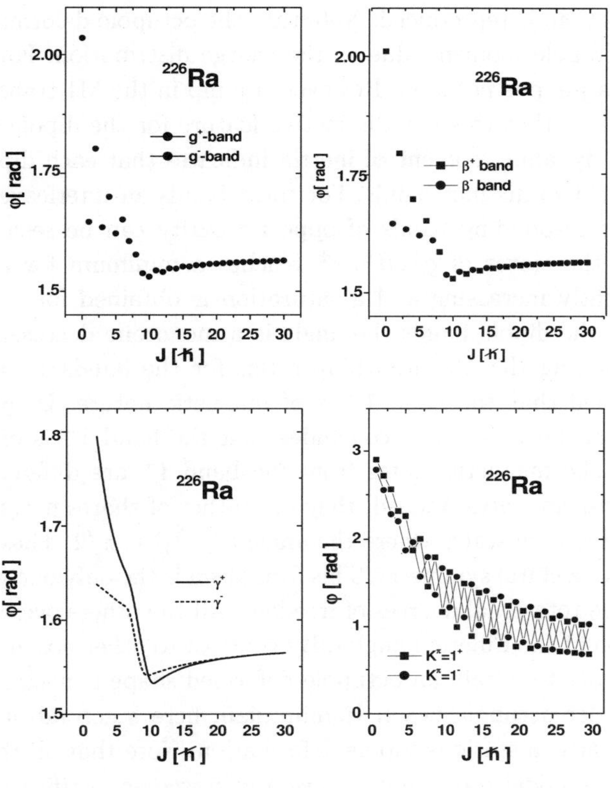

Fig. 15. The angle between the angular momenta carried by the quadrupole and octupole bosons respectively, in the states of $g^{\pm}$ (upper left), $\beta^{\pm}$ (upper right), $\gamma^{\pm}$ (bottom left) and dipole (bottom right) bands, vs. angular momentum.

structure of positive and negative parity states, which have been seen in some nuclei, is well reproduced. The low position for the state 1^- in ^{218}Ra and ^{220}Ra is caused by the $\vec{J}_2\vec{J}_3$ interaction. The back bending of the angular momentum represented as function of the rotational frequency, seen in ^{218}Ra, is nicely reproduced. From the analysis of energy displacement functions (e.d.f.) it results that the settlement of the octupole static deformation in the excited bands, takes place for different angular momenta. Moreover, there are several distinct situations: a) The vanishing of the e.d.f. takes place only for the gb. b) the vanishing takes place in g, β and γ bands but at different angular momenta. c) the vanishing appears in γ band but not in other bands. The jump in the E1 transition probability seen in ^{226}Ra for the $g^{\pm}$ bands, for the angular momenta were the static octupole deformation is set on, is reproduced. Note that the octupole deformation causes an electric dipole moment, due to the charge distribution. For the dipole band such a jump is not seen. However, a jump in the M1 transition shows up. We believe that this is a distinctive feature for the dipole bands. The plot of the dynamic moment of inertia indicates that each dipole band is a reunion of two distinct bands. For these bands an interleaved structure with the corresponding bands of opposite parity can be seen. The angle $(\vec{J}_2, \vec{J}_3)$ in the states of $g^{\pm}, \beta^{\pm}, \gamma^{\pm}$ reaches a minimum for a certain J, then is slightly increasing and a saturation is obtained for $\varphi = \pi/2$. By contrast in the dipole bands the angle is a monotone decreasing function of J. Comparing the M1 branching ratios for the bands 1^+ and 1^- one has concluded that the band 1^+ is of magnetic nature. Doing the same with E1 branching ratio one concludes that the band 1^- is of an electric character. The magnetic states from the band 1^+ are different from the so called scissors states. Indeed, they are rather of shares nature. We have seen that there are states where the angle $(\vec{J}_2, \vec{J}_3)$ is $\pi/2$. These states are precursors of a chiral symmetry. This formalism is the only one which treats correctly the rotational degrees of freedom. All the others overestimate the contribution of the Eulerian angles. By contrast to other boson formalisms, where in order to obtain an octupole deformed shape is necessary to have a fourth order octupole boson Hamiltonian, here a second order term is enough to cause a static octupole deformation. Note that all the terms involved in the model Hamiltonian have a microscopic justification within a boson expansion formalism applied to a two body quadrupole-quadrupole plus octupole-octupole interaction [25].

Some authors believe that the bands of non-vanishing K cannot be of collective nature [26]. Thus, the dipole states of ^{172}Yb are interpreted as two quasi-neutron states [22]. As shown in our lecture, we don't share this opinion, this interpretation being not unique. Here we give three examples which lead to a different conclusion. Indeed, in Ref.27, based on microscopic studies with surface delta interaction, the authors concluded that the $K^{\pi} = 1^-, 2^-$ bands in some actinides, are of collective nature. The branching ratios of the $K^{\pi} = 0^-, 1^-$ were realistically described by a IBA-sdf formalism in Ref.28. In the papers reviewed in our lecture, we provide a consistent description of four bands of non-vanishing K within a collective boson formalism. The anharmonic as well as the quadrupole-octupole coupling terms are simulating the effects interpreted in the quoted papers, as being caused by the single particle motion. In this context we mention again the double bending of angular momentum curve for ^{218}Ra.

What is the predictive power of our formalism? We should recall the fact that the parameters have a smooth dependence on A and Z. Hence, for a nucleus not included in the list presented here, the parameters could be taken from Figs. 1, 2 and 3. For these cases there is no free parameter left. For nuclei considered here a large volume of data could be fairly well described with a relatively small number of parameters. To give an example, for ^{232}Th there are about 65 energy levels known and all of them are described with a deviation of about 20 keV. Also for ^{226}Ra a large number of energy levels and many data concerning the E1 transitions are available. All of them are quite well described.

ECSM achieves an unified description of octupole-vibrational bands in spherical and transitional, and the negative parity bands in octupole deformed nuclei.

References

1. F. Asaro, F. S. Stephen, I. Perlman, Phys. Rev. **92** (1953) 1495.
2. F. S. Stephen, Jr., F. Asaro and I. Perlman, Phys. Rev. **100** (1955) 1543.
3. R. R. Chasman, Phys. Rev. Lett. **42**, 630 (1979); Phys. Lett **B 96**, 7 (1980).
4. P. Möller and J. R. Nix, Nucl. Phys. **A 361**, 117 (1981).
5. S. G. Rohozinski, Rep. Prog. Phys. **51**, 541 (1988).
6. P. A. Butler and W. Nazarewicz, Rev. Mod. Phys. **68**, 349 (1996).
7. A. A. Raduta, Al. H. Raduta, A. Faessler, Phys. Rev. **C 55**, 1747 (1997).
8. A. A. Raduta, Al. H. Raduta, A. Faessler, J. Phys. **G 23**, 149 (1997).
9. A. A. Raduta, A.Faessler, R. K. Sheline, Phys. Rev. **C 57**, 1512 (1998).
10. A. A. Raduta, D. Ionescu, A. Faessler, Phys. Rev. **C 65**, 064322 (2002).
11. A.A. Raduta and D. Ionescu, Phys. Rev. **C67**,044312 (2003).
12. A. A. Raduta, C. M. Raduta, A. Faessler, Phys. Lett. **B 635**, 80 (2006).
13. A. A. Raduta, C. M. Raduta, Nucl. Phys. **A 768**, 170 (2006).
14. A. A. Raduta, *et al.*, Nucl. Phys. **A 381**, 253 (1982).
15. R.K.Sheline, Rev. Mod. Phys. **32**, 1 (1960).
16. M.Sakai, Nucl. Phys. **A104** ; Nucl. Data Tables **10**, 511 (1972).
17. A.A.Raduta, C. Lima and A. Faessler, Z. Phys. **A313**, 69 (1983).
18. A.A.Raduta, S.Stoica, Z. Phys. **A327**, 275 (1987).
19. A. A. Raduta, A. Faessler, V. Ceausescu, Phys. Rev. **36**, 2111 (1987).
20. A. A. Raduta, N. Lo Iudice and I. I. Ursu, Nucl. Phys. A 608, 11 (1996).
21. D. Ionescu, PhD Thesis, IFIN-HH, Bucharest, 2003, unpublished.
22. P.M.Walker *et al.*, Phys. Lett. **87 B**, 339 (1979).
23. H. J. Wollershein *et al.*, Nucl. Phys. **A556**, 261 (1993).
24. N. Lo Iudice and F. Palumbo, Phys. Rev. Lett. **41**, 1532 (1978).
25. A.A. Raduta et *al.*, Phys. Rev. **C 8**, 1525 (1973).
26. K. Neergard, P. Vogel, Nucl. Phys.**A 145**, 33 (1970).
27. A. Faessler, A. Plastino, Z. f. Physik, **203**, 333 (1967).
28. P. von Brentano, N. V. Zamfir, A. Zigles, Phys. Lett. **B 278**, 221 (1992).

A VARIATIONAL METHOD FOR EQUILIBRIUM NUCLEAR SHAPE

D. N. POENARU*

Horia Hulubei National Institute of Physics and Nuclear Engineering (IFIN-HH), RO 077125 Bucharest-Magurele, Romania

** E-mail: Dorin.Poenaru@nipne.ro*

www.theory.nipne.ro

W. GREINER

Frankfurt Institute for Advanced Studies, Johann Wolfgang Goethe University, D-60438 Frankfurt am Main, Germany

E-mail: greiner@fias.uni-frankfurt.de

A method for finding reflection asymmetric or symmetric saddle-point nuclear shapes with axial symmetry is presented. The shape is a solution of an Euler-Lagrange equation, derived by solving the variational problem of minimization of the deformation energy. By introducing phenomenological shell corrections one obtains minima of deformation energy at the saddle-point for binary fission of ^{230,234,238}U nuclei at a non-zero mass asymmetry. Ternary, quaternary, and multicluster fission is also discussed.

Keywords: Cold binary fission; Ternary fission; Quaternary fission; Multicluster fission; Saddle point shapes; Variational method.

1. Introduction

One of the earliest features observed in binary nuclear fission was the preference for breakup into two fragments of unequal mass. Many theoretical attempts to find an explanation in the framework of the liquid drop model (LDM) failed. Only in the late sixties, by adding shell corrections within Strutinsky's macroscopic-microscopic method [1], it was shown that the outer barrier for asymmetric shapes is lower than for symmetric ones.

We consider at the beginning a binary process of cold fission

$$^{A}Z \rightarrow {}^{A_1}Z_1 + {}^{A_2}Z_2 \tag{1}$$

in which a parent nucleus, with Z protons and N neutrons ($A = Z + N$), is split into two fragments $^{A_1}Z_1$ and $^{A_2}Z_2$. The equilibrium nuclear shapes are usually obtained [2] by minimizing the deformation energy for a given surface equation. In the present work we shall describe a method [3–5] allowing to find a saddle-point shape of a multifragment fission (number of fragments $n \geq 2$) as a solution of an integro-differential equation; no *apriori* surface equation has to be given. Applications for ternary and quaternary fission will be illustrated.

In the approach based on a pure LDM [6], saddle-point shapes are always reflection symmetric: the deformation energy increases with the mass-asymmetry parameter $\eta = (A_1 - A_2)/(A_1 + A_2)$. By adding the shell corrections δE, to the LDM deformation energy, $E_{def} = E_{LDM} + \delta E$, we can obtain the minima at a finite value of the mass asymmetry for binary fission. The phenomenological shell correction δE is inspired from the Ref. 7. Results are presented for reflection asymmetric saddle point shapes of uranium even-mass isotopes with A = 230-238.

There are many parametrizations of the nuclear surface described in the literature. The surface equation, determined by a set of deformation coordinates (see various chapters of the books [8–10] and References therein), is frequently used to calculate the potential energy surfaces (PES) which in turn can be applied to find the nuclear deformations and fission barriers, to explain shape isomers, to obtain indications about mass-asymmetry in fission, to calculate half-lives against various decay modes or to study multidimensional tunneling [11], to extend the nuclear chart in the region of superheavy nuclei, etc. The shapes during the fission process have been intensively studied either statically (looking for the minimimum of potential energy [2,6]) or dynamically (by choosing a path with the smallest value of action integral [12,13]). The statical approach using a given parametrization of the surface [2,6,14–16] shows the importance of taking into account a large number of deformation coordinates (al least 5 coordinates are frequently needed). The parametrization of Legendre polynomial expansion with even order deformation parameters α_{2n} up to $n = 18$ was employed [2] to describe various saddle point shapes.

By using the two center shell model [17] to describe the single-particle states, one can follow the shell structure all the way from the original nucleus, over the potential barriers, up to the final stage of separated fragments. Particularly important points on a potential energy surface are those corresponding to the ground-state, saddle-point(s) and scission point [18–21].

The unified approach of cold binary fission, cluster radioactivity, and α-decay [9,22,23] was extended to cold ternary [24] and to multicluster fission including quaternary (two-particle accompanied) fission [25,26]. We stressed the expected enhanced yield of two alpha accompanied fission compared to other combinations of two light particles; it was indeed experimentally confirmed [27–30]. In a cold binary fission the involved nuclei are neither excited nor strongly deformed, hence no neutron is evaporated from the fragments or from the compound nucleus; the total kinetic energy equals the released energy. In a more complex than binary *cold fission* (ternary, quaternary, etc), neutrons could still be emitted from the neck, because the Q-value is positive. In this case their kinetic energy added to those of the fragments should exhaust the total released energy.

The most advanced asymmetric two center shell model [31] was improved [32,33] and applied to calculate potential energy surfaces (PES) for cluster emitters (^{222}Ra, ^{232}U, ^{236}Pu, ^{242}Cm [34]) as well as for ^{228}Th [35] and for light (^{106}Te and ^{212}Po) and superheavy (^{294}118) alpha emitters [36]. Other applications concerns the sub-barrier synthesis of $Z = 118$ isotopes [37] and the study of input channels to produce ^{286,290,298}114 [38,39]. The dynamical calculations have been performed in a multidimensional hyperspace of deformation coordinates followed by minimization of the action integral for all possible charge and mass asymmetries.

The Werner-Wheeler approximation [40] was employed to obtain the nuclear inertia tensor. The pairing correction energy calculated within the BCS approximation [34] was observed to give an important contribution to the deformation energy by lowering the barrier heights and smoothing the shell effects. The strong shell effect associated with the doubly magic character of the daughter ^{208}Pb, which was seen in the systematic analysis of experimental results, comes from a valley present on the PESs of cluster emitters at a relatively high value of the asymmetry parameter.

The potential barrier shape of heavy ion radioactivity obtained for the first time by use of the macroscopic-microscopic method provides further support for the particular choice of the barrier within the superasymmetric fission (ASAF) model, which was very successful in predicting the half-lives.

The BCS pairing has an essential contribution to the cranking inertia tensor [41] which may be expressed with an analytical relationship for a particular choice of the system Hamiltonian of a spheroidal harmonic oscillator without spin-orbit interaction. If the crossing terms P_{ij} with $i \neq j$ are not taken into account, an important error could be induced into the half-life value given by the WKB approximation. Examples for ^{240}Pu [41] illustrate the conclusions.

The α-decay life-times of superheavies and lighter emitters have been calculated [36,42] within our ASAF model, the universal formula, and the semiempirical relationship including shell effects.

In the following we shall present the variational method and some results concerning cold binary, ternary, quaternary, and multicluster fission.

2. Minimization of deformation energy

For axially symmetric shapes around z axis and the tips z_1 and z_2 we are looking for a function $\rho = \rho(z)$ expressing in cylindrical coordinates the surface equation. The dependence of deformation energy on the neutron and proton numbers is contained in the surface energy of a spherical nucleus, E_s^0, the fissility parameter, $X = E_C^0/(2E_s^0)$, as well as in the shell correction of the spherical nucleus δE^0. E_C^0 is the Coulomb energy of the spherical shape for which the radius is $R_0 = r_0 A^{1/3}$. The radius constant is $r_0 = 1.2249$ fm, and $e^2 = 1.44$ MeV·fm is the square of electron charge. The lengths are given in units of the radius, R_0, and the Coulomb potential at the nuclear surface, $V_s = (R_0/Ze)\phi_s$, in units of Ze/R_0. The surface tension and the charge density are denoted by σ and ρ_e respectively. The nuclear surface equation we are looking for should minimize the functional of potential energy of deformation $E_s + E_C$, where

$$E_s = 2\pi\sigma R_0^2 \int_{z_1}^{z_2} \rho(z)\sqrt{1+\rho'^2}dz \tag{2}$$

$$E_C = \frac{2\pi R_0^2 Ze\rho_e}{5} \int_{z_1}^{z_2} \left(\rho^2 - \frac{z}{2}\frac{\partial\rho^2}{\partial z}\right) V_s dz \tag{3}$$

with two constraints: volume conservation,

$$V = \pi R_0^3 \int_{z_1}^{z_2} \rho^2(z)dz = \frac{4\pi R_0^3}{3} \tag{4}$$

and a given deformation parameter,

$$\alpha = \frac{\pi R_0^3}{V} \int_{z_1}^{z_2} F(z,\rho)\rho^2 dz \tag{5}$$

assumed to be an adiabatic variable. By choosing the deformation coordinate as the distance between the centers of mass of the left and right fragments, $\alpha = |z_L^c| + |z_R^c|$, one can reach all intermediate stages of deformation from one parent nucleus to two fragments by a continuos variation of its value. Also a possible dynamical study, for which the center of mass treatment is very important [43], may conveniently use this definition.

We denote with F_1, F_2, F_3, F_4, the corresponding integrands one needs to write the Euler-Lagrange equation:

$$\sigma F_1 = \sigma\rho\left(1+\rho'^2\right)^{1/2} \quad ; \quad \frac{R_0\rho_e\phi_s}{5}F_2 = \frac{R_0\rho_e\phi_s}{5}\left(\rho^2 - z\rho\rho'\right) \qquad (6)$$

$$F_3 = \rho^2 \;\; ; \;\; F_4 = \rho^2 F \qquad (7)$$

The derivatives are easily obtained

$$\frac{\partial F_1}{\partial \rho} = \left(1+\rho'^2\right)^{1/2} \qquad (8)$$

$$\frac{d}{dz}\frac{\partial F_1}{\partial \rho'} = \frac{d}{dz}\left[\frac{\rho\rho'}{(1+\rho'^2)^{1/2}}\right] = \frac{\rho'^2+\rho\rho''}{(1+\rho'^2)^{1/2}} - \frac{\rho\rho'^2\rho''}{(1+\rho'^2)^{3/2}} \qquad (9)$$

$$\frac{\partial F_2}{\partial \rho} = 2\rho - z\rho' \;\; ; \;\; \frac{d}{dz}\frac{\partial F_2}{\partial \rho'} = \frac{d}{dz}(-z\rho) = -\rho - z\rho' \qquad (10)$$

$$\frac{\partial F_3}{\partial \rho} = 2\rho \;\; ; \;\; \frac{\partial F_4}{\partial \rho} = 2\rho\left(F + \frac{\rho}{2}\frac{\partial F}{\partial \rho}\right) \qquad (11)$$

According to the calculus of variations the function $\rho(z)$ minimizing the energy with two constraints should satisfy the Euler-Lagrange equation

$$\sigma\left(\frac{\partial F_1}{\partial \rho} - \frac{d}{dz}\frac{\partial F_1}{\partial \rho'}\right) + \frac{R_0\rho_e\phi_s}{5}\left(\frac{\partial F_2}{\partial \rho} - \frac{d}{dz}\frac{\partial F_2}{\partial \rho'}\right) + \rho\,(2\lambda_1'' + 2\lambda_2'' f) = 0 \qquad (12)$$

leading to

$$\rho\rho'' - \rho'^2 - (\lambda_1 + \lambda_2|z| + 6XV_s)\,\rho(1+\rho'^2)^{3/2} - 1 = 0 \qquad (13)$$

if we choose $F = |z|$ (hence $f = |z|$) and express $3R_0\rho_e/(5\sigma)$ as $6X$ because the Coulomb and surface energy of a spherical nucleus within LDM are given by $E_C^0 = (3Z^2e^2)/(5R_0)$ and $E_s^0 = 4\pi R_0^2\sigma$, respectively.

Alternatively one can obtain from this equation the equivalent relationship

$$2\sigma K + 3\rho_e\phi_s/5 + \lambda_1' + \lambda_2' f = 0 \;\; ; \;\; f(z,\rho) = F(z,\rho) + \frac{\rho}{2}\frac{\partial F(z,\rho)}{\partial \rho} \qquad (14)$$

in which λ_1' and λ_2' are Lagrange multipliers and K is the mean curvature:

$$K = (\mathcal{R}_1^{-1} + \mathcal{R}_2^{-1})/2 \qquad (15)$$

with $\mathcal{R}_1$ and $\mathcal{R}_2$ the principal radii of curvature given by

$$\mathcal{R}_1 = R_0\tau\rho \;\; ; \;\; \mathcal{R}_2 = -R_0\tau^3/\rho'' \;\; ; \;\; \tau^2 = 1 + \rho'^2 \qquad (16)$$

where $\rho' = d\rho/dz$ and $\rho'' = d^2\rho/dz^2$. It is interesting to mention that in the absence of an electric charge, the condition of stable equilibrium at the surface of a fluid [44,45] is given by Laplace formula equating the difference of pressures with the product $2\sigma K$ present in eq. (14).

The position of separation plane between fragments, $z = 0$, is given by the condition $(d\rho/dz)_{z=0} = 0$, which defines the median plane for a usual spherical, ellipsoidal, or "diamond" shape in the ground state, or the middle of the neck for an elongated reflection symmetrical shape on the fission path. For this choice of the function $F(z, \rho)$ one has $f = |z|$. At the left hand side and right hand side tips on the symmetry axis one can write

$$\rho(z_1) = \rho(z_2) = 0 \tag{17}$$

and the transversality conditions

$$\frac{d\rho(z_1)}{dz} = \infty \ ; \quad \frac{d\rho(z_2)}{dz} = -\infty \tag{18}$$

The equation is solved numerically by an iterative procedure checking the minimization of the deformation energy with a given accuracy. The phenomenological shell corrections to the LDM deformation energy are used to obtain reflection asymmetric saddle point shapes. One can develop the computer code for just one of the "fragments" (for example for the right hand one extended from $z = 0$ to $z = z_2$) and then write the result for the other fragment. For symmetrical shapes we have $z_2 = z_p = -z_1$. It is convenient to make a change of the function and variable defined by:

$$u(v) = \Lambda^2 \rho^2[z(v)] \ ; \quad z(v) = z_p - v/\Lambda \tag{19}$$

By substituting into equation (13) one has

$$u'' - 2 - \frac{1}{u}\left[u'^2 + \left(\frac{3XV_s}{2\Lambda} + \frac{\lambda_1 + \lambda_2 z_p}{4\Lambda} - \frac{\lambda_2 v}{4\Lambda^2}\right)(4u + u'^2)^{3/2}\right] = 0 \tag{20}$$

A linear function of v is introduced by adding and subtracting $a + bv$ to $3XV_s/2\Lambda$. The quantity V_{sd} is defined as the deviation of Coulomb potential at the nuclear surface from a linear function of v

$$V_{sd} = \frac{3X}{2\Lambda}V_s - a - vb \tag{21}$$

where the constant

$$a = \frac{3X}{2\Lambda}V_s(v = 0) \tag{22}$$

is chosen to give $V_{sd}(v = 0) = 0$, and

$$b = \left[\frac{3X}{2\Lambda}V_s(v = v_p) - a\right]/v_p \tag{23}$$

in which $v_p = \Lambda z_p$. By equating with 1 the coefficient of v in the new eq. one can establish the following link between Λ and λ_2

$$\Lambda^2 = \lambda_2/4(b-1) \tag{24}$$

In this way $u(v)$ is to be determined by the equation

$$u'' - 2 - \frac{1}{u}[u'^2 + (v - d + V_{sd})(4u + u'^2)^{3/2}] = 0 \tag{25}$$

where the role of a Lagrange multiplier is played by the quantity d which is taken to be constant instead of α. To the tip $z = z_p$, at which $\rho(z_p) = 0$, corresponds $v = 0$, hence $u(0) = \Lambda^2\rho^2(z_p) = 0$. By multiplying with u the equation (25), introducing $v = 0$, and using the relationship $V_{sd}(v = 0) = 0$, it follows that $u'(0) = 1/d$. Consequently the boundary conditions for $u(v)$ are:

$$u(0) = 0, \quad u'(0) = 1/d \tag{26}$$

To $z = 0$, at which $\rho'(0) = 0$ (the middle of the neck for elongated shapes), corresponds $v_p = \Lambda z_p$ and $u'(v_p) = -2\Lambda\rho(0)\rho'(0) = 0$. The point $v = v_p$ in which $u'(v_{pn}) = 0$ is determined by interpolation from two consecutive values of v_p leading to opposite signs of $u'(v)$. The number n of changes of signs is equal to the number of necks plus one given in advance, e.g. for a single neck (binary fission) $n = 2$ and for two necks (ternary fission) $n = 3$, etc.

Although the quantity Λ is not present in eq. (25) we have to know it in order to obtain the shape function $u(v)$. By changing the function and the variable in the eq (4) one has

$$\Lambda = \left\{ \frac{3}{2} \int_0^{v_{pn}} u(v)dv \right\}^{1/3} \tag{27}$$

and the deformation coordinate, $\alpha = z_L^c + z_R^c$, may also be determined. From the dependence $\alpha(d)$, one can obtain the inverse function $d = d(\alpha)$.

In order to find the shape function $u(v)$ we solve eq (25) with boundary conditions written above. One starts with given values of the constants d and n. For reflection symmetric shapes $d_L = d_R$ and $n_L = n_R$. In the first iteration one obtains the solution for a Coulomb potential at the nuclear surface assumed to be a linear function of v, i.e. for $V_s = 0$. Then one calculates the parameters Λ, a, and b, which depend on the Coulomb potential and its deviation V_{sd} from a linear function, and the deformation energy corresponding to the nuclear shape [46,47]. The quantity V_{sd} determined in such a way is introduced in eq (25) and the whole procedure is repeated until the deformation energy is obtained with the desired accuracy. In every iteration the equation is solved numerically with the Runge-Kutta method. One can calculate for different values of deformation α the deformation en-

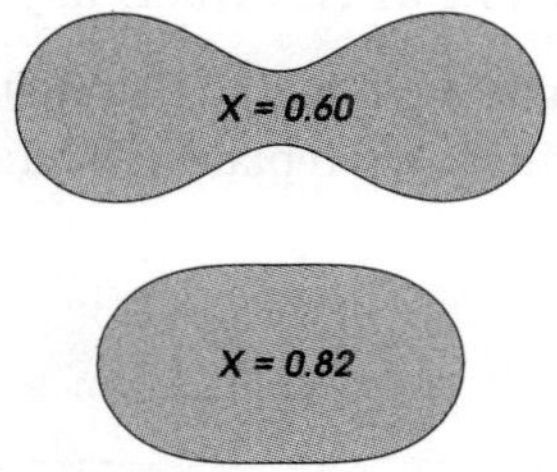

Fig. 1. Saddle point shapes during binary fission of nuclei with fissility $X = 0.60$ and 0.82.

ergy $E_{def}(\alpha)$. The particular value α_s for which $dE_{def}(\alpha_s)/d\alpha = 0$ corresponds to the extremum, i.e. the shape function describes the saddle point, and the unconditional extremum of the energy is the fission barrier. The other surfaces (for $\alpha \neq \alpha_s$) are extrema only with condition $\alpha =$ constant. In this way one can compute the deformation energy versus $d_L = d_R$. The saddle point corresponds to the maximum of this deformation energy.

For *reflection asymmetrical shapes* we need to introduce another constraint: the asymmetry parameter, η, defined by

$$\eta = \frac{M_L - M_R}{M_L + M_R} = \frac{A_1 - A_2}{A_1 + A_2} \tag{28}$$

It should remain constant during variation of the shape function $u(v)$. Consequently eq (25) should be written differently for left hand side and right hand side [5]. There is an almost linear dependence of η from the difference $d_L - d_R$.

3. Saddle point shapes with reflection symmetry

One can test the method by comparing some nuclear shapes within LDM to the standard results for medium and heavy nuclei. A comparison between nuclear shapes at the saddle point for nuclei with fissilities $X = 0.60$ and 0.82 (corresponding to ^{170}Yb and ^{252}Cf nuclei lying on the line of beta-stability) is presented in Figure 1. One can see how the necking-in and the elongation are decreasing ($\alpha = 2.304$ and 1.165) when fissility increases from $X = 0.60$ to $X = 0.82$, in agreement with [2]. In the limit $X = 1$ the saddle point shape is spherical. The method proved its capability by reproducing the well known LDM saddle point shapes.

4. Qualitative explanation of the mass asymmetry in fission

Within LDM a nonzero mass asymmetry parameter leads to a deformation energy which increases with η. We replace η by an almost linear dependent quantity $(d_L - d_R)$.

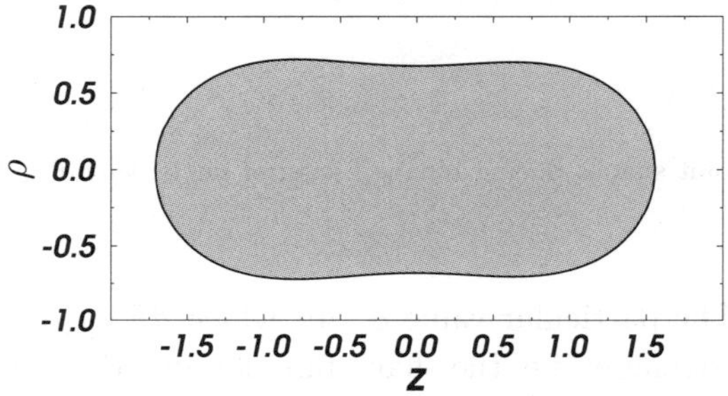

Fig. 2. Mass asymmetric saddle point shape of ^{232}U. Shell effects taken into account.

When the shell effects are taken into account a saddle point solution of the integro-differential equation with reflection asymmetry is obtained (see fig. 2). We use [5] a phenomenological shell correction adapted after Myers and Swiatecki [7].

Results for binary cold fission of parent nuclei $^{230-238}$U are presented in figure 3. The minima of the saddle point energy occur at nonzero mass asymmetry parameters $d_L - d_R$ in the range $0.04, 0.08$. They correspond to η of $0.050, 0.095$ which leads to $A_1 \simeq 125$ in all cases. For experimentally determined mass asymmetry [48,49] the maximum of the fission fragment mass distributions is centered on $A_1 = 140$ in a broad range of mass numbers of parent nuclei. In the figure 3, from the saddle point energies E_{SP} of every nucleus we subtracted its minimum value E_{SP}^{min}. The minimum of the E_{SP} is produced by the negative values of the shell corrections $\delta E - \delta E^0$.

As mentioned by Wilkins et al. [50], calculations of PES for fissioning nuclei "qualitatively account for an asymmetric division of mass". From the qualitative point of view the results displayed in Figure 3 proove the capability of the method to deal with fission mass and charge asymmetry. The experimentally determined mass number of the most probable heavy

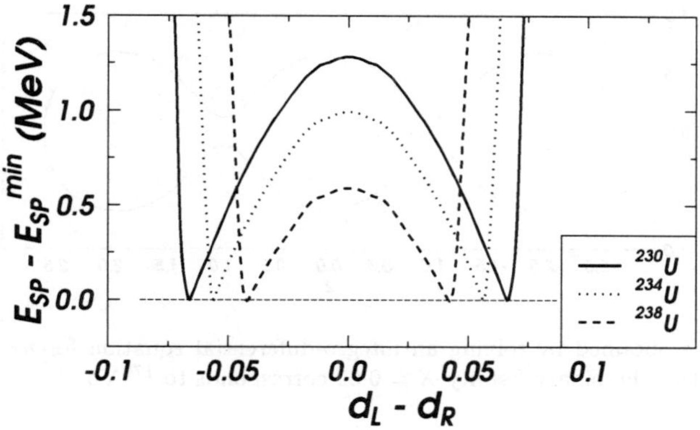

Fig. 3. Difference between the saddle point deformation energy E_{SP} and its minimum value E_{SP}^{min} vs mass asymmetry parameter $(d_L - d_R)$ for cold binary fission of U isotopes in the presence of shell corrections.

fragment [51] for U isotopes ranges from 134 to 140. The corresponding values at the displayed minima in Figure 3 are very close to 125, which means a discrepancy between 6.7 % and 10.7 % for A_H. Only the contribution of shell effects can produce a minimum of the barrier height at a finite value of the mass asymmetry. One may hope to obtain a better agreement with experimental data by using a more realistic shell correction model, based on the recently developed two center shell model [31].

5. Ternary Fission

The particle-accompanied fission (or ternary fission) was observed both in neutron-induced and spontaneous fission since 1946. Several such processes, in which the charged particle is a proton, deuteron, triton, $^{3-8}$He, $^{6-11}$Li, $^{7-14}$Be, $^{10-17}$B, $^{13-18}$C, $^{15-20}$N, $^{15-22}$O, have been detected [52]. Many other heavier isotopes of F, Ne, Na, Mg, Al, Si, P, S, Cl, Ar, and even Ca were also mentioned.

The elongated shape for ternary fission with $d_L = d_R = 7.00$ is shown in figure 4. The configuration with $E/E_s^0 = 0.134$ is not far from a *"true ternary-fission"* [53] in which the three fragments are almost identical: $^{170}_{70}$Yb$\rightarrow^{56}_{23}$V + $^{56}_{23}$V + $^{58}_{24}$Cr and the Q-value is 83.639 MeV. One may compare the above E/E_s^0 value with the touching-point energy of these spherical fragments $(E_t - Q)/E_s^0 = 0.239$. It is larger, as expected, because of the finite neck of the shapes in figure 4. For α-accompanied fission of ^{170}Yb with two $^{83}_{34}$Se fragments $Q = 87.484$ MeV is larger and the touching point energy

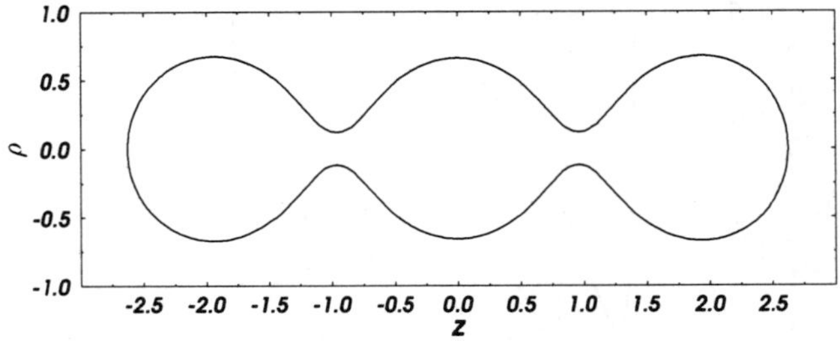

Fig. 4. Shape obtained by solving an integro-differential equation for $n_L = n_R = 3$, $d_L = d_R = 7.00$. The binary fissility $X = 0.60$ corresponds to ^{170}Yb.

$(E_t - Q)/E_s^0 = 0.103$ is lower. A lower $Q = 70.859$ MeV and higher energy barrier $(E_t - Q)/E_s^0 = 0.147$ is obtained for ^{10}Be accompanied fission of ^{170}Yb with $^{80}_{33}$As fission fragments.

Systematic calculations [54] have shown a clear correlation between the Q-values and the measured yield of different isotopes for one cluster accompanied fission. For example, among the He isotopes with mass numbers 4, 6, and 8, ^{4}He leads to the maximum Q-value. The maximum yield was indeed experimentally observed [52] for α accompanied fission. Similarly, among ^{6,8,10,12}Be, the clusters ^{8}Be and ^{10}Be give the maximum Q-values. As ^{8}Be spontaneously breaks into 2α particles it is not easy to measure ^{8}Be accompanied fission yield; consequently ^{10}Be has been most frequently identified. *By detecting, in coincidence, these two alpha particles, the ^{8}Be accompanied fission with a larger yield compared to that of the ^{10}Be one, could be observed in the future.* From ^{12,14,16,18}C the favoured is ^{14}C, and all ^{16,18,20,22}O isotopes have comparable Q-values when they are emitted in a cold binary fission of ^{252}Cf. Nevertheless, ^{20}O is slightly upper than the others. As a rule, if the Q-value is larger the barrier height is smaller, and the quantum tunneling becomes more probable. The stronger emission of ^{14}C compared to ^{12}C has the same explanation as for the ^{14}C radioactivity; the Q-value is larger because the heavy fragment is doubly magic.

We should stress again that if one is interested to estimate the yield in various fission processes, one has to compare the potential barriers and not the Q-values. Our results are in agreement with preceding calculations [55] showing also preference for prolate over oblate shapes. Theoretically it was pointed out by Present in 1941 that Uranium tripartition would release about 20 MeV more energy than the binary one. In spite of having quite large Q values [54], this "true ternary fission" [53,56] is a rather weak process; the strongest phenomenon remains the α-particle-accompanied fission.

By performing dynamical calculations, Hill arrived in his thesis and in [12] at elongated shapes with pronounced necks looking more encouraging for particle-accompanied fission. It would be rewarding to perform successful experiments with nowadays very much improved experimental techniques, despite the previous rather pessimistic conclusion that "true" ternary spontaneous fission is an extremely rare phenomenon.

6. Quaternary and multicluster fission

A shape with four fragments and three necks ($n_L = n_R = 4$, $d_L = d_R = 4.00$) can be seen in figure 5. The shape with $E/E_s^0 = 0.214$ approaches a

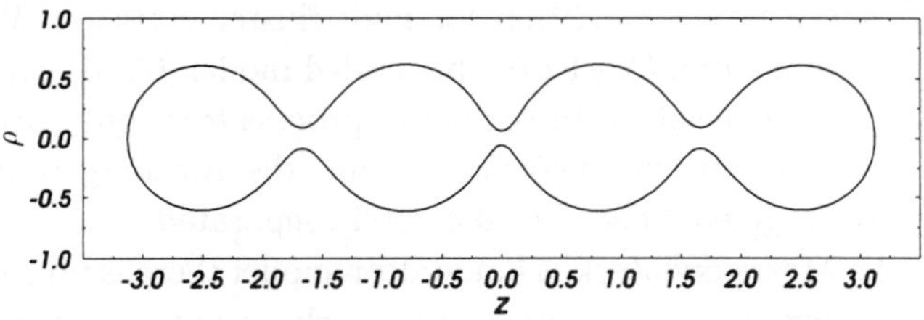

Fig. 5. Nuclear shapes during quaternary fission of a nucleus with fissility $X = 0.60$ for $d_L = d_R = 4.00$.

fission into almost identical four fragments ${}^{170}_{70}\mathrm{Yb} \rightarrow {}^{42}_{17}\mathrm{Cl} + {}^{42}_{17}\mathrm{Cl} + {}^{43}_{18}\mathrm{Ar} + {}^{43}_{18}\mathrm{Ar}$. Again the configuration with aligned spherical fragments in touch is higher in energy: $(E_t - Q)/E_s^0 = 0.324$. Even more complex shapes can be obtained by further increasing the values of $n_L = n_R$.

The best chance to be experimentally observed has a quaternary fission in which two light particles are emitted from a neck formed between two heavy fragments [25,57]. The successful experiment [27,28] on 2α-accompanied fission observed in cold neutron induced fission of ^{233,235}U, confirmed our expectations.

The possibility of a whole family of new decay modes, *the multicluster accompanied fission*, was envisaged [25,57–59]. Besides the fission into two or three fragments, a heavy or superheavy nucleus spontaneously breaks into four, five or six nuclei of which two are asymmetric or symmetric heavy fragments and the others are light clusters, e.g. α-particles, ^{10}Be, ^{14}C, ^{20}O, or combinations of them. Examples were presented for the two-, three- and four cluster accompanied cold fission of ^{252}Cf and ^{262}Rf, in which the emitted clusters are: 2α, $\alpha+^{6}$He, $\alpha+^{10}$Be, $\alpha+^{14}$C, 3α, $\alpha+^{6}$He + ^{10}Be, $2\alpha+^{6}$He, $2\alpha+^{8}$Be, $2\alpha+^{14}$C, and 4α. A comparison was made with the recently observed ^{252}Cf cold binary fission, and cold ternary (accompanied by α particle or by ^{10}Be cluster). The strong shell effect corresponding to the doubly magic heavy fragment ^{132}Sn is emphasized. From the analysis of different configurations of fragments in touch, we conclude that the most favorable mechanism of such a decay mode should be the cluster emission from an elongated neck formed between the two heavy fragments.

In a first approximation, one can obtain an order of magnitude of the potential barrier height by assuming spherical shapes of all the participant nuclei. This assumption is realistic if the fragments are magic nuclei. For deformed fragments it leads to an overestimation of the barrier. By taking into account the prolate deformations, one can get smaller potential barrier height, hence better condition for multicluster emission. We use the Yukawa-plus-exponential (Y+E) double folded model [60,61] extended [47] for different charge densities. In the decay process from one parent to several fragments, the nucleus deforms, reaches the touching configuration, and finally the fragments became completely separated.

Within the Myers-Swiatecki's liquid drop model there is no contribution of the surface energy to the interaction of the separated fragments; the deformation energy has a maximum at the touching point configuration. The proximity forces acting at small separation distances (within the range of strong interactions) give rise in the Y+EM to a term expressed as folllows

$$E_{Yij} = -4\left(\frac{a}{r_0}\right)^2 \sqrt{a_{2i}a_{2j}}\left[g_i g_j\left(4+\frac{R_{ij}}{a}\right) - g_j f_i - g_i f_j\right]\frac{\exp(-R_{ij}/a)}{R_{ij}/a} \tag{29}$$

where

$$g_k = \frac{R_k}{a}\cosh\left(\frac{R_k}{a}\right) - \sinh\left(\frac{R_k}{a}\right) \ ; \ f_k = \left(\frac{R_k}{a}\right)^2 \sinh\left(\frac{R_k}{a}\right) \tag{30}$$

in which R_k is the radius of the nucleus A_kZ_k, $a = 0.68$ is the diffusivity parameter, and a_{2i}, a_{2j} are expressed in terms of the model constants a_s, κ and the nuclear composition parameters I_i and I_j, $a_2 = a_s(1-\kappa I^2)$, $a_s = 21.13$ MeV, $\kappa = 2.3$, $I = (N-Z)/A$, $R_0 = r_0 A^{1/3}$, $r_0 = 1.16$ fm is the radius constant, and e is the electron charge, $e^2 \simeq 1.44$ MeV·fm.

The investigated pairs [62] are the following for the binary fission: ${}^{102,104}_{40}$Zr–${}^{150,148}_{58}$Ce ($N_L = 62, 64$), ${}^{104-108}_{42}$Mo–${}^{148-144}_{56}$Ba ($N_L = 62-66$), ${}^{110}_{44}$Ru–${}^{142}_{54}$Xe ($N_L = 66$), and ${}^{116}_{46}$Pd–${}^{136}_{52}$Te ($N_L = 70$). For cold α accompanied fission [63] one has: ${}^{92}_{36}$Kr–${}^{156}_{60}$Nd ($N_L = 56$), ${}^{96-101}_{38}$Sr–${}^{152-147}_{58}$Ce

$(N_L = 58-63)$, ${}^{100-104}_{40}$Zr–${}^{148-144}_{56}$Ba $(N_L = 60-64)$, ${}^{106-108}_{42}$Mo–${}^{142-140}_{54}$Xe $(N_L = 64-66)$, ${}^{112}_{44}$Ru–${}^{136}_{52}$Te $(N_L = 68)$, and ${}^{116}_{46}$Pd–${}^{132}_{50}$Sn $(N_L = 70)$. There is also one example of detected cold ^{10}Be accompanied fission of ^{252}Cf, namely ${}^{96}_{38}$Sr–${}^{146}_{56}$Ba $(N_L = 58)$.

The new decay modes which have a good chance to be detected are 2α-, 3α-, and 4α accompanied fission. The corresponding Q-values are not smaller compared to what has been already measured, which looks very promising for the possibilty of detecting the 2α-, 3α-, and 4α accompanied fission decay modes. In fact by taking into account the mass-values of the participants, one can see that the Q-value for the 2α accompanied fission may be obtained by translation with +0.091 MeV from the Q-value of the ^{8}Be accompanied fission. A similar translation with -7.275 MeV should be made from the ^{12}C accompanied fission in order to obtain the Q-values of the 3α accompanied fission, etc. Less promising looks the combination of three cluster, α+^{6}He + ^{10}Be accompanied cold fission of ^{252}Cf. As mentioned above, the 2α accompanied fission was already observed.

Different kinds of aligned and compact configurations of fragments in touch may be assumed. The potential barrier for the "polar emission" is much higher than that of the emission from the neck, which explains the experimentally determined low yield of the polar emission compared to the "equatorial" one. As it should be, the compact configuration posses the maximum total interaction energy, hence it has the lowest chance to be observed. The same is true for the quaternary fission when the two clusters are formed in the neck. An important conclusion can be drawn, by generalizing this result, namely: *the multiple clusters should be formed in a configuration of the nuclear system in which there is a relatively long neck between the light $(n-1)$ and heavy (n) fragment.* Such shapes with long necks in fission have been considered [12] as early as 1958. For the "true" ternary fission, in two ^{84}As plus ^{84}Ge, $E_t = 98$ MeV! Despite the larger Q-value (266 MeV), the very large barrier height explains why this split has a low chance to be observed.

The energies of the optimum configuration of fragments in touch, for the 2α-, 3α-, and 4α accompanied cold fission of ^{252}Cf are not much higher than what has been already measured. When the parent nucleus is heavier, the multicluster emission is stronger as we observed by performing calculations for nuclei like ^{252,254}Es, ^{255,256}Fm, ^{258,260}Md, ^{254,256}No, ^{262}Lr, ^{261,262}Rf, etc.

While the minimum energy of the most favorable aligned configuration of fragments in touch, when at least one cluster is not an alpha particle, becomes higher and higher with increasing complexity of the partners, the same quantity for multi alphas remains favorable. In conclusion, we suggested since 1998 experimental searches for the multicluster 2α accompanied fission, for ^{8}Be-, ^{14}C- and ^{20}O accompanied fission. Also, the contribution of the single- and multi-neutron accompanied cold fission mechanism to the prompt neutron emission has to be determined.

7. Conclusions

The method of finding the axially-symmetric shape at the saddle point without introducing *apriori* a parametrization, by solving an integro-differential equation was tested for binary, ternary, and quaternary fission processes within a pure liquid drop model. The well known LDM saddle point shapes are well reproduced. The method proved its practical capability in what concerns fission into two, three, or four identical fragments, for which fission barriers given by shapes with rounded necks are, as expected, lower than those of aligned spherical fragments in touch.

In the absence of any shell corection it is not possible to reproduce the experimental data, or to give results for particle-accompanied fission. By adding (phenomenological) shell corrections we succeded to obtain minima at a finite value of mass asymmetry for the binary fission of $^{230-238}$U nuclei.

Fission barriers for ternary and quaternary fission into identical fragments are lower than for aligned spherical fragments in touch. Our expectations concerning the possibilty to detect quaternary fission as 2α-accompanied fission were experimentally confirmed.

Acknowledgments

This work was partly supported by a grant of the Deutsche Forschungsgemeinschaft, and by Ministry of Education and Research, Bucharest.

References

1. V. M. Strutinsky, *Nucl. Phys. A* **95**, 420 (1967).
2. S. Cohen and W. J. Swiatecki, *Annals of Physics (N. Y.)* **22**, p. 406 (1963).
3. D. N. Poenaru, W. Greiner, Y. Nagame and R. A. Gherghescu, *Journal of Nuclear and Radiochemical Sciences, Japan* **3**, 43 (2002).
4. D. N. Poenaru and W. Greiner, *Europhysics Letters* **64**, 164 (2003).
5. D. N. Poenaru, R. A. Gherghescu and W. Greiner, *Nuclear Physics A* **747**, 182 (2005).
6. V. M. Strutinski, *Soviet Physics JETP* **15**, 1091 (1962), JETF 42 (1962) 1571-1581.

7. W. D. Myers and W. J. Swiatecki, *Nuclear Physics, A* **81**, 1 (1966).
8. J. M. Eisenberg and W. Greiner, *Nuclear Theory*, 3rd edn. (North-Holland, Amsterdam, 1987).
9. D. N. Poenaru and W. Greiner, Theories of cluster radioactivities, in *Nuclear Decay Modes* (Institute of Physics Publishing, Bristol, UK, 1996) pp. 275–336.
10. D. Poenaru and W. Greiner (eds.), *Handbook of Nuclear Properties* (Clarendon Press, Oxford, 1996).
11. A. Iwamoto, *Zeitschrift für Physik, A* **349**, 265 (1994).
12. D. L. Hill, The dynamics of nuclear fission, Proc. of the Second U. N. Int. Conf. on the Peaceful Uses of Atomic Energy, Geneva, 1-13 Sept, 1958, (United Nations, Geneva, 1958), p. 244-247.
13. M. Brack, J. Damgaard, A. S. Jensen, H. C. Pauli, V. M. Strutinsky and C. Y. Wong, *Rev. Mod. Phys.* **44**, 320 (1972).
14. R. W. Hasse and W. D. Myers, *Geometrical relationships of macroscopic nuclear physics* (Springer, Berlin, 1988).
15. R. Smolańczuk, H. V. Klapdor-Kleingrothaus and A. Sobiczewski, *Acta Physica Polonica, B* **24**, 685 (1993).
16. P. Möller, D. G. Madland, A. J. Sierk and A. Iwamoto, *Nature* **409**, 785 (2001).
17. W. Greiner and J. A. Maruhn, *Nuclear Models* (Springer, Berlin, 1996).
18. U. Brosa, S. Grossmann and A. Müller, *Physics Reports* **197**, 167 (1990).
19. Y. Nagame *et al.*, *Physics Letters, B* **387**, 26 (1996).
20. I. Nishinaka, Y. Nagame, K. Tsukada, H. Ikezoe, K. Sueki, H. Nakahara, M. Tanikawa and T. Ohtsuki, *Physical Review, C* **56**, p. 891 (1997).
21. Y. L. Zhao *et al.*, *Physical Review Letters* **82**, 3408 (1999).
22. W. Greiner and D. N. Poenaru, Radioactivity, in *Encyclopedia of Condensed Matter Physics, Vol. 5*, eds. F. Bassani, G. L. Liedl and P. Wyder (Elsevier, Oxford, 2005) pp. 106–116.
23. D. N. Poenaru, Y. Nagame, R. A. Gherghescu and W. Greiner, *Physical Review C* **65**, 054308 (2002), Erratum: C66.049902.
24. D. N. Poenaru, B. Dobrescu, W. Greiner, J. H. Hamilton and A. V. Ramayya, *Journal of Physics G: Nuclear and Particle Physics* **26**, L97 (2000).
25. D. N. Poenaru, W. Greiner, J. H. Hamilton, A. V. Ramayya, E. Hourany and R. A. Gherghescu, *Physical Review, C* **59**, 3457 (1999).
26. D. N. Poenaru, Ternary and multicluster cold fission. In *Nuclei Far from Stability and Astrophysics (Proc. of the NATO Advanced Study Institute, Predeal)*, eds. D. N. Poenaru, H. Rebel and J. Wentz, Series II: Mathematics, Physics and Chemistry, Vol. 17 (Kluwer Academic Publishers, Dordrecht, 2001), pp. 151-162.
27. F. Gönnenwein, P. Jesinger, M. Mutterer, A. M. Gagarski, G. Petrov, W. H. Trzaska, V. Nesvizhevski and O. Zimmer, Quaternary fission, in *Nuclear Physics at Border Lines (Proc. Internat. Conf. Lipari)*, eds. G. Fazio, G. Giardina, F. Hanappe, G. Immè and N. Rowley (World Scientific, Singapore, 2002), pp. 107-111.
28. F. Gönnenwein, P. Jesinger, M. Mutterer, W. H. Trzaska, G. Petrov, A. M.

Gagarski, V. Nesvizhevski and P. Geltenbort, *Heavy Ion Physics* **18**, 419 (2003).
29. D. V. Kamanin *et al.*, Rare fission modes: study of multi-cluster decays of actinide nuclei, in *Proc. of the International Conference on Dynamical Aspects of Nuclear Fission, Smolenice Castle, Slovakia, 2006)"*, in print.
30. Y. Pyatkov *et al.*, Exotic decay modes of ^{242}Pu* from the reaction ^{238}U +^{4}He (40 MeV), in *Proc. of the International Conference on Dynamical Aspects of Nuclear Fission, Smolenice Castle, Slovakia, 2006)"*, in print.
31. R. A. Gherghescu, *Physical Review* **C 67**, 014309 (2003).
32. R. A. Gherghescu and W. Greiner, *Physical Review* **C 68**, 044314 (2003).
33. R. A. Gherghescu, W. Greiner and G. Münzenberg, *Physical Review* **C 68**, 054314 (2003).
34. D. N. Poenaru, R. A. Gherghescu and W.Greiner, *Physical Review, C* **73**, 014608 (2006).
35. D. N. Poenaru, R. A. Gherghescu, I. H. Plonski and W.Greiner, *International Journal of Modern Physics, E* **15**, in print (2006).
36. D. N. Poenaru, I. H. Plonski, R. A. Gherghescu and W. Greiner, *Journal of Physics G: Nuclear and Particle Physics* **32**, 1223 (2006).
37. R. A. Gherghescu, W. Greiner and S. Hofmann, *European Physical Journal* **A 27**, 23 (2006).
38. R. A. Gherghescu, D. N. Poenaru, W. Greiner and Y. Nagame, *Journal of Physics G: Nuclear and Particle Physics* **32**, L73 (2006).
39. R. A. Gherghescu and W. Greiner, *Journal of Physics G: Nuclear and Particle Physics* **A 27**, 23 (2006).
40. R. A. Gherghescu and D. N. Poenaru, *Physical Review, C* **72**, 027602 (2005).
41. D. N. Poenaru, R. A. Gherghescu and W. Greiner, *The European Physical Journal A* **24**, 355 (2005).
42. D. N. Poenaru, I. H. Plonski and W.Greiner, *Physical Review, C* **74**, 014312 (2006).
43. D. N. Poenaru, J. A. Maruhn, W. Greiner, M. Ivaşcu, D. Mazilu and I. Ivaşcu, *Zeitschrift für Physik, A* **333**, 291 (1989).
44. H. Lamb, *Hydrodinamics* (Cambridge University Press, New York, 1932).
45. L. D. Landau and E. M. Lifshitz, *Fluid Mechanics* (Pergamon, London, 1959).
46. D. N. Poenaru and M. Ivaşcu, *Computer Physics Communications* **16**, 85 (1978).
47. D. N. Poenaru, M. Ivaşcu and D. Mazilu, *Computer Physics Communications* **19**, 205 (1980).
48. H. R. von Gunten, *Actinides Rev.* **1**, p. 275 (1969).
49. E. A. C. Croach, *Atomic Data Nucl. Data Tables* **19**, p. 417 (1977).
50. B. D. Wilkins, E. P. Steinberg and R. R. Chasman, *Physical Review, C* **14**, 1832 (1976).
51. A. C. Wahl, *Atomic Data and Nuclear Data Tables* **39**, 1 (1988).
52. M. Mutterer and J. P. Theobald, Particle-accompanied fission, in *Nuclear Decay Modes* (Institute of Physics Publishing, Bristol, Bristol, England, 1996) pp. 487–522.
53. D. N. Poenaru, R. A. Gherghescu and W. Greiner, *Acta Physica Hungarica:*

Heavy Ion Physics **18**, 409 (2003).
54. D. N. Poenaru, W. Greiner and R. A. Gherghescu, *Atomic Data and Nuclear Data Tables* **68**, 91 (1998).
55. H. Diehl and W. Greiner, *Nuclear Physics, A* **229**, 29 (1974).
56. D. N. Poenaru, R. A. Gherghescu, W. Greiner, Y. Nagame, J. H. Hamilton and A. V. Ramayya, *Romanian Reports in Physics* **55**, 781 (2003).
57. D. N. Poenaru, W. Greiner, J. H. Hamilton and A. V. Ramayya, *Journal of Physics G: Nuclear and Particle Physics* **27**, L19 (2001).
58. D. N. Poenaru and W. Greiner, *Journal of Physics G: Nuclear and Particle Physics* **25**, L7 (1999).
59. D. N. Poenaru, W. Greiner, J. Hamilton, E. Hourany and R. A. Gherghescu, Multi-fragment fission. most probable configurations, in *Perspectives in Nuclear Physics (Proc. International Symposium, Atlantis Resort, Nassau, Bahamas, 14-16 November, 1998)*, eds. J. H. Hamilton, H. K. Carter and R. B. Piercey (World Scientific, Singapore, 1999, ISBN 981-02-3774-X).
60. W. Scheid and W. Greiner, *Zeitschrift für Physik, A* **226**, 364 (1969).
61. H. J. Krappe, J. R. Nix and A. J. Sierk, *Phys. Rev. C* **20**, 992 (1979).
62. J. H. Hamilton, A. V. Ramaya *et al.*, *Journal of Physics G: Nuclear and Particle Physics* **20**, L85 (1994).
63. A. V. Ramayya, J. H. Hamilton *et al.*, *Physical Review Letters* **81**, 947 (1998).

SYMMETRIC THREE CENTER SHELL MODEL

R. A. GHERGHESCU*
Horia Hulubei National Institute for Nuclear Physics and Engineering,
Bucharest-Magurele, Romania
** E-mail: radu@radu.nipne.ro*

WALTER GREINER
Frankfurt Institute for Advanced Studies,
Frankfurt am Main, Germany

The symmetric three center shell model has been constructed in order to account for the transition of the level scheme of one parent nucleus towards three equal spherical fragments. The model calculate the energy levels using a three center oscillator potential for partially overlapped nuclei. The oscillator levels and the spin-orbit and l^2 terms are dependent on the distance between the centers of the side fragments, thus on the elongation of the ternary configuration. At the end the independent level schemes of three separated spherical nuclei are obtained. Calculations are applyied to the splitting of ^{144}Nd into three ^{48}Ca spherical systems.

Keywords: shell model; ternary fission; level scheme

1. Introduction

This work addresses to such particular phenomena as the tripartition configuration and ternary fission, reported 50 years ago [1]. In 1958 it was shown [2], on the basis of a macroscopic liquid drop model that for heavy nuclei it is possible to obtain an energetically more favourable division into three or even in four fragments than in the case of binary fission. The progress in understanding the connection between scission configuration and the binary fission mechanism and mass yield is still recent. The reaction energy must be positive [3], when calculated with experimental masses [4]. It is accepted that for ternary and multicluster fission the aligned configurations of fragments in touch, possesing an axial symmetry are more probable the non-axial compact configurations which lead to a larger potential energy. Alpha-accompanied fission remains up to now the most probbale mode of tripartition. Other configurations are possible however, and with recent, more performant detectors, like the GAMMASPHERE in United States, it is now possbile to verify these hypothesis. The ability to predict probable ternary configurations and the most favourable combinations between parent nucleus and the three fragment partition can be supported only using a very specialized model able to describe the microscopic transition from a unique energy level scheme, going through three partially overlapped schemes and ending to the totally separated three potential wells, corresponding to the separated fragments. Such a model has not been done before and is about to be described in this work. The model shall yield the necessary proton and neutron energy levels during the evolution of the nuclear shape from a single system towards three superposed fragment configuration. The proton and neurton level schemes are further used for the calculation of the shell corrections along the ternary phenomenon evolution.

2. Ternary configuration shapes

The shapes which are described by the present model are presented in figure 1 and consist in three intersected spheres corresponding to three symmetrical fragments $A/3$, resulted from the fragmentation of an initial nucleus A. The fragments are colinear, hence one has axial symmetry. $-z_1, 0$ and, z_1 are the centers of the fragments. $-z_{01}$ and z_{01} are the matching points between the intersected fragments. Once the mass and atomic number are given, the only free parameter which is needed to describe a certain point in the ternary shape evolution is the distance between the centers of the side fragments, R. When R increases, the shape goes from one sphere to three.

A typical evolution of the tripartition configuration is obtained when the distance between centers varies from zero (initial nucleus) up to the sum of the radius values for three touching final fragments. The starting point is the initial, spherical parent nucleus. At this moment the three centers completely overlap. When R starts to increase, the three fragments begin to emerge. The nuclear volume is not equally distributed. At the beginning, for small R, most of the volume is divided between the side fragments, whereas the middle fragment has only a small part of it. As R increases, the three fragments shape themselves more clearly, the middle fragment increases in volume and the side ones decrease. Consequently the radii are evolving towards the final values of the separated fragments. This important feature influences the value of the microscopic potential. At the end the three fragments become equal in volume when the touching point

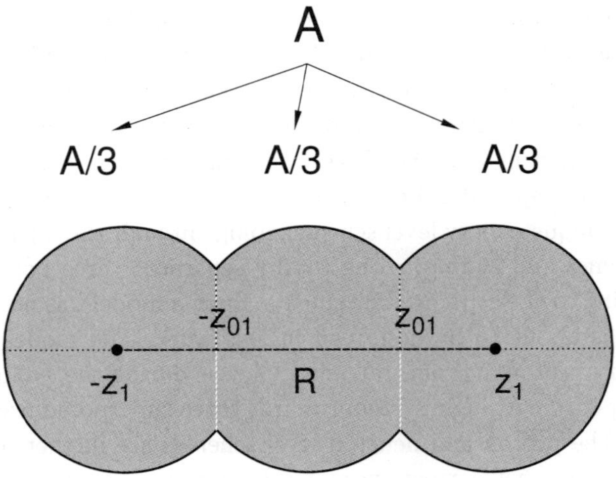

Fig. 1. Typical ternary colinear symmetric configuration for the splitting of an initial A-parent nucleus. The main independent variable, R, is the distance between the side centers. The pint $-z_1, 0, z_1$ are the centers of the fragments.

is reached. An example of the shape evolution is given in figure 2 for such symmetrical splitting.

3. Three center potential

The three center model potential is based on three oscillator type wells which are partially overlapping. It has a ρ-part, perpendicular on the symmetry axis, and a z-part, along the symmetry axis. The total oscillator potential reads:

$$V(\rho, z, \phi) = V(\rho) + V(z) \tag{1}$$

as the sum of the two direction potentials. The ρ-part depends on the perpendicular frequency ω_ρ and the ρ-coordinate:

$$V(\rho) = \frac{1}{2} m_0 \omega_\rho^2 \rho^2 \tag{2}$$

Since the three fragments are equal, the frequency is the same. The potential has to describe the nuclear surface, thus it has to be centered in the middle

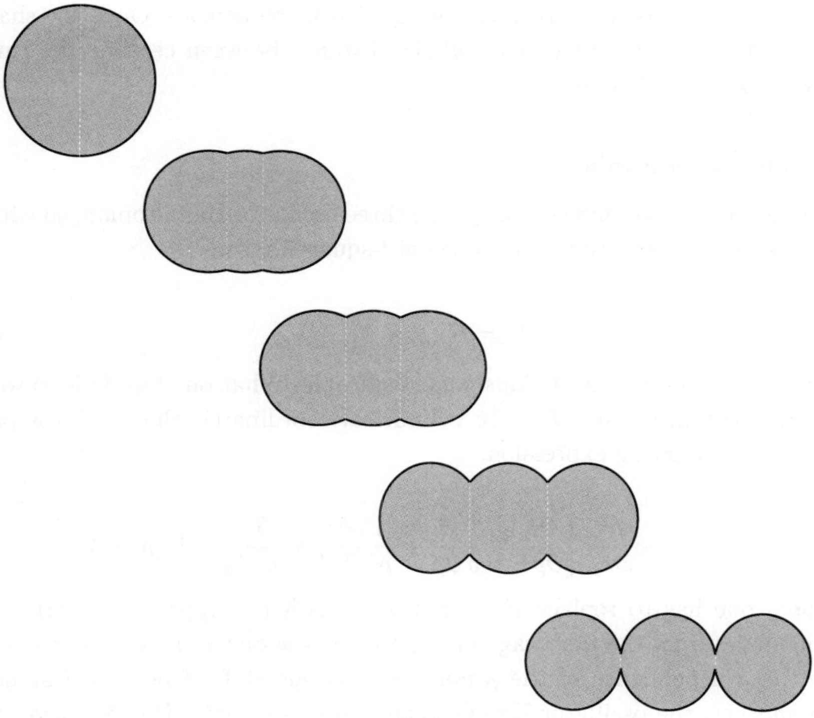

Fig. 2. Ternary shape evolution from a spherical parent nucleus towards three symmetrical spherical fragments, with increasing distance between centers.

of each fragment. This request is fulfilled by the expression of the z-part of the potential $V(z)$, which reads:

$$V(z) = \begin{cases} \frac{1}{2} m_o \omega_z^2 (z - z_1)^2 & , z > z_{01} \\ \frac{1}{2} m_o \omega_z^2 z^2 & , -z_{01} < z < z_{01} \\ \frac{1}{2} m_o \omega_z^2 (z + z_1)^2 & , z < -z_{01} \end{cases} \quad (3)$$

The potential $V(z)$ has three expressions, each of them being active within the corresponding fragment region, where $-z_1, 0$ and z_1 are the colinear centers. Since the volume of the nuclear shape depends on the mass number if incompressibility is assumed, the geometric parameters are directly related to the potential through the oscillator frequencies. Once the shape is given by the mass number and the distance between centers, the three center potential is determined.

4. The Hamiltonian

The total Hamiltonian comprises the three oscillator Hamiltonian, to which one adds the spin-orbit and the usual l-squared term:

$$H = H_{3osc} + V_{\hat{l}\hat{s}} + V_{\hat{l}^2} \tag{4}$$

Such a Hamiltonian is obviously not separable. What one can do is to work on the oscillator part H_{3osc} In cylindrical coordinates, the oscillator part reads the following expression:

$$H_{3osc} = -\frac{\hbar^2}{2m_0}\left[\frac{\partial^2}{\partial\rho^2} + \frac{1}{\rho}\frac{\partial}{\partial\rho} + \frac{1}{\rho^2}\frac{\partial^2}{\partial\phi^2} + \frac{\partial^2}{\partial z^2}\right] + V(\rho) + V(z) \tag{5}$$

where one has to replace $V(\rho)$ and $V(z)$ with the appropriate terms, as one moves from the first fragment to the middle one and the last one. Due to the z-dependence of the ρ-potential, as one shifts from one fragment to another, the oscillator Hamiltonian is not separable. But imposing the condition of equal ρ-frequencies: $\omega_{\rho_1} = \omega_{\rho_2} = \omega_{\rho_3}$ one has the same form for $V(\rho)$. In this situation one can choose the total wave function as the solution of the oscillator Schroedinger equation to be a product of three one-dimensional functions:

$$\Psi(\rho, z, \phi) = \Phi_m(\phi) R_{n_\rho}^{|m|}(\rho) Z_\nu(z) \tag{6}$$

In this case one obtains three one-dimensional eqautions. The Φ and ρ equations are imediately solved. The angular function and the weighted Laguerre polynomial equations produce the solutions along two out of three coordinates:

$$\Phi_m(\phi) = \frac{1}{\sqrt{2\pi}} \exp(im\phi)$$

$$R_{n_\rho}^{|m|}(\rho) = \left(\frac{2\Gamma(n_\rho+1)\alpha_1^2}{\Gamma(n_\rho+|m|+1)}\right)^{\frac{1}{2}} \exp\left(-\frac{\alpha_1^2\rho^2}{2}\right)(\alpha_1^2\rho^2)^{\frac{|m|}{2}} L_{n_\rho}^{|m|}(\alpha_1^2\rho^2)$$

Here $L_{n_\rho}^{|m|}$ is the Laguerre polynomial and Γ is the Gamma function. A small part of the problem is solved, since one has two out of three necessary quantum numbers and two out of the three wave functions. The total three-oscillator energy levels have two parts:

$$E_{3osc} = E_{\rho,\phi} + E_z \tag{7}$$

The (ρ, ϕ) part is already solved by the previous two equations. Once one has the n_ρ and m quantum numbers, the (ρ, ϕ) energy is determined. For the z-part of the energy, one has to solve the corresponding z-axis Schroedinger equation:

$$\left[\frac{\partial^2}{\partial z^2} + \frac{2m_0 E_z}{\hbar^2} - \frac{2m_0}{\hbar^2} V(z)\right] Z(z) = 0 \tag{8}$$

After a series of simple calculations, one obtains the Hermite function typical equations, with two independent solutions:

$$Z_\nu(z) = \exp\left(-z^2/2\right)\mathcal{H}_\nu(\pm z) \tag{9}$$

where $\mathcal{H}_\nu(\pm z)$ is the Hermite function of non-integer indices ν. One observes at this point that ν depends on the geometrical configuration through the potential $V(z)$.

If one replaces the z-potential with each of the three-center expressions, one obtains the solution along the symmetry axis:

$$Z_\nu(z) = \begin{cases} C_{1n} \exp\left[-\frac{\alpha^2(z-z_1)^2}{2}\right]\mathcal{H}_\nu[\alpha(z-z_1)] & , \quad z \geq z_{01} \\ C_{0n} \exp\left(-\frac{\alpha^2 z^2}{2}\right)[\mathcal{H}_\nu(z) + (-1)^n \mathcal{H}_\nu(-z)] & , -z_{01} < z < z_{01} \\ (-1)^n C_{1n} \exp\left[-\frac{\alpha^2(z+z_1)^2}{2}\right]\mathcal{H}_\nu[-\alpha(z+z_1)] \, , & \quad z \leq -z_{01} \end{cases}$$

In this expression α is the frequency-dependent parameter. There are three unknown quantities: two normalization constants, c_{1n} and c_{0n} (due to symmetry the wave function for the two side fragments have the same normalization constant, c_{1n}) and the z-quantum number, ν. From the continuity of the wave function Z_ν and its derivative Z'_ν:

$$Z'_\nu(z) = \begin{cases} C_{1n} \exp\left[-\frac{\alpha^2(z-z_1)^2}{2}\right] \alpha\{-(z-z_1)\mathcal{H}_\nu[\alpha(z-z_1)] \\ \quad +2\nu\mathcal{H}_{\nu-1}[\alpha(z-z_1)]\} \\ \quad z \geq z_{01} \\ C_{0n} \exp - \frac{\alpha^2 z^2}{2} \alpha\{-z[\mathcal{H}_\nu(z)+ \\ (-1)^n \mathcal{H}_\nu(-z)] 2\nu[[\mathcal{H}_{\nu-1}(z) + (-1)^{n+1}\mathcal{H}_{\nu-1}(-z)]\} \\ \quad -z_{01} < z < z_{01} \\ (-1)^n C_{1n} \exp\left[-\frac{\alpha^2(z+z_1)^2}{2}\right] \alpha\{(z+z_1)\mathcal{H}_\nu[-\alpha(z+z_1)] \\ \quad +2\nu\mathcal{H}_{\nu-1}[-\alpha(z+z_1)]\} \\ \quad z \leq -z_{01} \end{cases}$$

at the matching points $\pm z_{01}$ of the z-potential, one obtains two equations. The third is acquired from the normalization condition:

$$C_{1n}^2 j(\nu,\nu;-z_{01},\infty) + C_{0n}^2[j(\nu,\nu;-z_{01},z_{01}) + (-1)^n j(\bar{\nu},\nu;-z_{01},z_{01})] = \frac{\alpha}{2} \tag{10}$$

where $j(\nu,\nu;x_1,x_2)$ are the z-integrals along the symmetry axis between the limits x_1 and x_2. The integrals are performed numerically over the range of each corresponding fragment. The system is solved numerically and the solution of two constants and the z-quantum number is determined for each step of the ternary configuration. With solving this system, the complete function basis, specific to tripartition fragmentation, is obtained.

At this point the first ternary signature of the process evolution is obtained as the variation of the z-quantum numbers ν. The variation is displayed in figure 3.

First calculations have been applied to the symmetric splitting of neodymium in three ^{48}Ca fragments. The starting points are the integers corresponding to a unique center. Then the z-quantum numbers decrease through non-integer values. At the end of the process, the numbers merge three by three into the final integers, specific for each ^{48}Ca totally separated fragment. The three oscillator part is solved. A first set of calculations has as a result the three-oscillator energy levels as a function of the distance between centers. The starting values are the usual one oscillator level sequence, which is obtained here by the total superposition of the three centers and is presented in figure 4. As the distance R increases, the levels are mixing. Towards the end of the process, the levels converge in three identical oscillator schemes, particular for the three ^{48}Ca.

One observes the increment of the space between the shells, as the mass

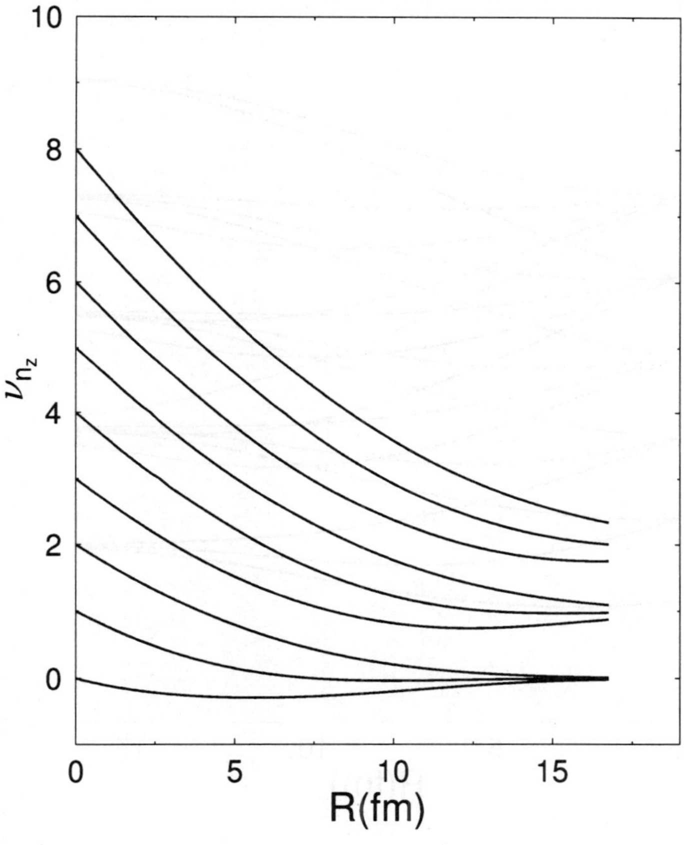

Fig. 3. Variation of the z-quantum numebers ν with the distance between centers R.

number decreases. This is due to the fact that associated nuclear frequency is conversely proportional with the nuclear mass.

5. The spin-orbit and l^2 - terms

In order to complete the energy of the ternary system, one has to add the spin-orbit and l^2 interactions. Due to the fact that spin-orbit intensities κ

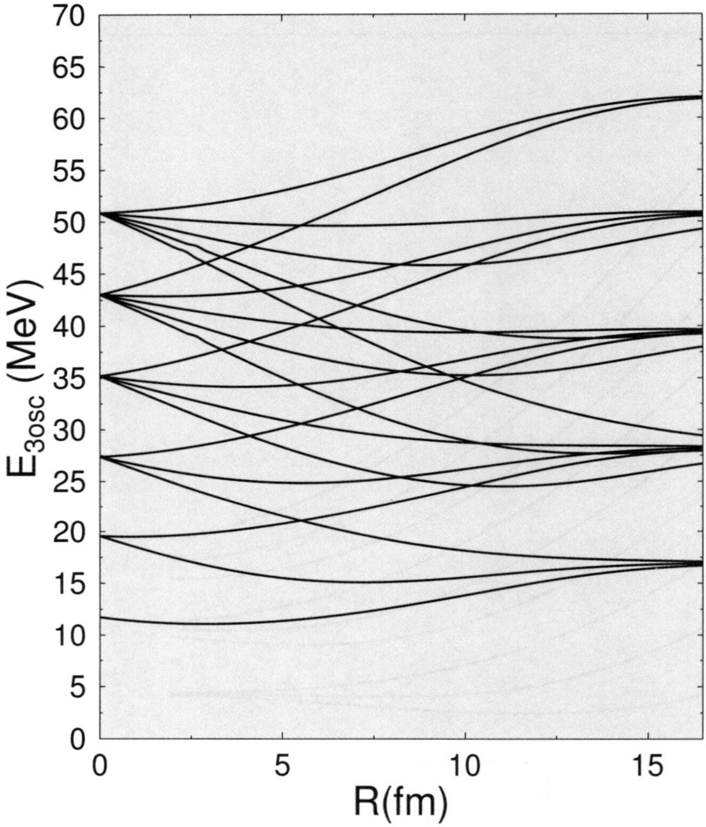

Fig. 4. Three center oscillator level scheme against the distance between centers for the symmetric splitting of ^{144}Nd into three ^{48}Ca fragments.

and μ are nuclear mass dependent these quantities can change when one passes from one fragment to another, within the tripartition shape. This fact makes the intensities z-dependent. Since the spin-orbit operators contain derivatives, the usual expression is not hermitian. For this reason, one shall replace the two potentials with anti-commutators between the strengths as (ρ, z) functions instead of constants, and the angular momentum dependent operators, as one can read:

$$V_{ls} = \begin{cases} -\left\{ \dfrac{\hbar}{m_0\omega_{01}} \kappa_1(\rho, z), (\nabla V(z > z_{01}) \times p)s \right\} & , A_1 - region \\ -\left\{ \dfrac{\hbar}{m_0\omega_{02}} \kappa_2(\rho, z), (\nabla V(-z_{01} < z < z_{01}) \times p)s \right\} & , A_2 - region \\ -\left\{ \dfrac{\hbar}{m_0\omega_{03}} \kappa_3(\rho, z), (\nabla V(z < -z01) \times p)s \right\} & , A_3 - region \end{cases}$$

and the corresponding l^2 dependent potential has a similar expression:

$$V_{l^2} = \begin{cases} -\left\{ \dfrac{\hbar}{m_0^2\omega_{01}^3} \kappa_1\mu_1(\rho, z), (\nabla V(z > z_{01}) \times p)^2 \right\} & , A_1 - region \\ -\left\{ \dfrac{\hbar}{m_0^2\omega_{02}^3} \kappa_2\mu_2(\rho, z), (\nabla V(-z_{01} < z < z_{01}) \times p)^2 \right\} & , A_2 - region \\ -\left\{ \dfrac{\hbar}{m_0^2\omega_{03}^3} \kappa_3\mu_3(\rho, z), (\nabla V(z < -z_{01}) \times p)^2 \right\} & , A_3 - region \end{cases}$$

Each of the operators has three expressions, as one moves from one fragment region A_1 to the next A_2 and so on. The potential is also replaced successively by one of the above three expressions.

The general expression of the spin-orbit operator is constructed using the creation and annihilation operators:

$$ls \rightarrow \tfrac{1}{2}(\mathbf{\Omega}^+\mathbf{s}^- + \mathbf{\Omega}^-\mathbf{s}^+) + \mathbf{\Omega}_z\mathbf{s}_z \tag{11}$$

With these new operators one can construct now the total spin-orbit operator from the creation and annihilation parts. To comply to the anti-commutator rule, three combinations of Heaviside functions are employed. Each of them ensures the action of the specific operator only within the region where the corresponding fragment is active. These regions are bordered by surfaces which pass through the matching points of the ternary configuration. The last step of the spin-orbit part is to obtain the dependence of the operators on the specific ternary configuration at a given geometry. This is fulfilled by the use of different corresponding potentials for each region. The final three expressions for the creation operator read:

$$\langle\prime|\boldsymbol{\Omega}^+|\rangle =$$

$$\langle\prime|\{\boldsymbol{\Omega}^+(z>z_{01}),\Theta(z-z_{01}]\}|\rangle \qquad \rightarrow A_1, \qquad z>z_{01}$$

$$\langle\prime|\{\boldsymbol{\Omega}^+(-z_{01}<z<z_{01}),[\Theta(z+z_{01})-\Theta(z-z_{01})]\}|\rangle \rightarrow A_2, -z_{01}\leq z\leq z_{01}$$

$$\langle\prime|\{\boldsymbol{\Omega}^+(z<-z_{01}),[1-\Theta(z+z_{01})]\}|\rangle \qquad \rightarrow A_3, \qquad z<-z_{01} \tag{12}$$

and the same is available for Ω^- and $\boldsymbol{\Omega}_z$.

The final expressions for the creation operators are frequency and geometry dependent:

$$\boldsymbol{\Omega}^+(z>z_{01}) = -e^{i\varphi}\left[\frac{\partial V(\rho,z>z_{01})(\rho,z)}{\partial\rho}\frac{\partial}{\partial z}-\frac{\partial V(\rho,z>z_{01}))}{\partial z}\frac{\partial}{\partial\rho}-\frac{i}{\rho}\frac{\partial V(\rho,z>z_{01})}{\partial z}\frac{\partial}{\partial\varphi}\right]$$

$$= -e^{i\varphi}\left[m_0\omega_\rho^2\rho\frac{\partial}{\partial z}-m_0\omega_z^2(z-z_1)\frac{\partial}{\partial\rho}-\frac{i}{\rho}m_0\omega_z^2(z-z_1)\frac{\partial}{\partial\varphi}\right]$$

$$\boldsymbol{\Omega}^+(-z_{01}<z<z_{01}) =$$
$$= -e^{i\varphi}\left[\frac{\partial V(\rho,-z_{01}<z<z_{01})}{\partial\rho}\frac{\partial}{\partial z}-\frac{\partial V(\rho,-z_{01}<z<z_{01})}{\partial z}\frac{\partial}{\partial\rho}+\frac{i}{\rho}\frac{\partial V(\rho,-z_{01}<z<z_{01})}{\partial z}\frac{\partial}{\partial\varphi}\right]$$

$$= -e^{i\varphi}\left[m_0\omega_\rho^2\rho\frac{\partial}{\partial z}-m_0\omega_z^2 z\frac{\partial}{\partial\rho}+\frac{i}{\rho}m_0\omega_z^2 z\frac{\partial}{\partial\varphi}\right]$$

$$\boldsymbol{\Omega}^+(z<-z_{01}) = -e^{i\varphi}\left[\frac{\partial V(\rho,z<-z_{01})(\rho,z)}{\partial\rho}\frac{\partial}{\partial z}-\frac{\partial V(\rho,z<-z_{01}))}{\partial z}\frac{\partial}{\partial\rho}-\frac{i}{\rho}\frac{\partial V(\rho,z<-z_{01})}{\partial z}\frac{\partial}{\partial\varphi}\right]$$

$$= -e^{i\varphi}\left[m_0\omega_\rho^2\rho\frac{\partial}{\partial z}-m_0\omega_z^2(z+z_1)\frac{\partial}{\partial\rho}-\frac{i}{\rho}m_0\omega_z^2(z+z_1)\frac{\partial}{\partial\varphi}\right]$$

In this final form one can observe that each operator is centered in the middle of the fragment it represents. The total spin-orbit interaction is the sum of the the three regions:

$$\begin{aligned}V_{\boldsymbol{\Omega}\mathbf{s}} = &-\frac{\hbar}{m_0\omega_{01}}\kappa_1\{\boldsymbol{\Omega}\mathbf{s}(A_1),(A_1)\}\\ &-\frac{\hbar}{m_0\omega_{02}}\kappa_2\{\boldsymbol{\Omega}\mathbf{s}(A_2),(A_2)\}\\ &-\frac{\hbar}{m_0\omega_{01}}\kappa_3\{\boldsymbol{\Omega}\mathbf{s}(A_3),(A_3)\}\end{aligned} \tag{13}$$

Finally the matrix of the total Hamiltonian for the three center shell model can be constructed as the sum of the three superposed oscillators and the angular momentum dependent terms:

$$\langle i|3CSM|j\rangle = E_{3osc}(n_\rho, |m|, \nu) + \langle i|V_{\mathbf{\Omega s}}|j\rangle + \langle i|V_{\mathbf{\Omega}^2}|j\rangle \tag{14}$$

A total number of 14 shells has been used in computation, which yieds 220 levels. Each set is performed separately for protons and neutrons.

6. Results

As the result of diagonalization of the total matrix one obtains the level scheme for colinear ternary fragmentation of a given system. The first set of calculations have been applied to the symmetric splitting of ^{144}Nd into three ^{48}Ca. For zero distance between centers the three fragments completely overlap and one obtains the initial level scheme of the parent nucleus. Then, with increasing R the shape becomes more elongated and the three fragments begin to form. The levels are mixing now and one can observe the existence of energy gaps for certain geometries. At the end, the levels converge towards the typical shells of three separated ^{48}Ca.

Calculations are performed separately for neutrons and protons, since the the spin-orbit strength is different. The final level scheme is used as an input data for computing the shell corrections E_{sh}. This part has been fulfilled by using the Strutinsky method. In figure 5 the results for ^{144}Nd are presented.

A peculiarity in the figure is the deep negative final value of the total shell correction energy. This is due to the double-magicity of the ^{48}Ca fragments. The Z=20 and N=28 proton and neutron magic numbers produce three times their negative Strutinsky shell corrections within the ternary system. A first minimum is observable for a small distance between centers, at the beginning of the process. The shell corrections are added to the macroscopic liquid drop part of the energy. This first minimum in shell corections can produce a small potential pocket, which is in fact due to the initial deformation of neodymium. A maximum is followed by a second minimum in the very deformed, elongated region. At this point the fragments are only partially overlapped so that now one has a ternary effect. At the end of the process, when the three fragments reach the touching point, the individuality of ^{48}Ca is manifested. The magicity of the already formed proton and neutron level schemes produce the deep minimum in the shell corrections. Here one has three times the neutron and proton negative shell

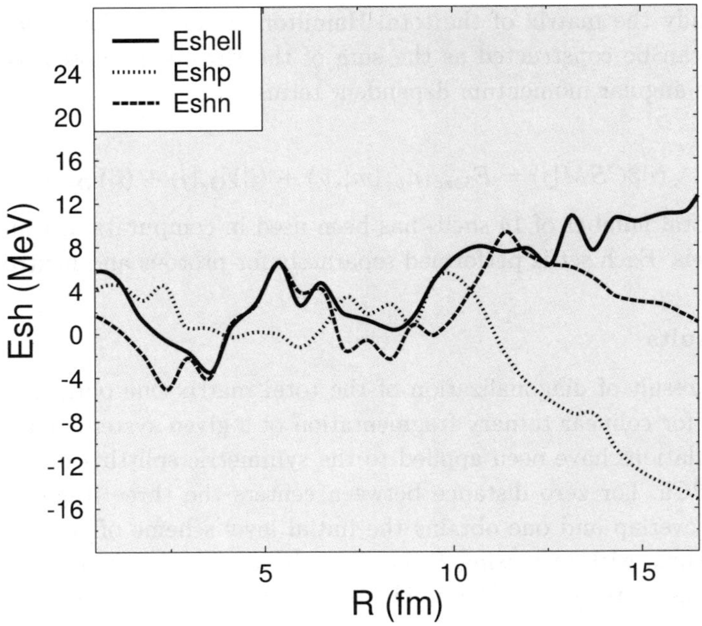

Fig. 5. Calculated shell corrections for protons, neutrons and total value for the symmetric splitting of ^{144}Nd into three ^{48}Ca.

corrections, which correspond to 28 neutron and 20 proton magic numbers closures.

An isobaric reaction has been chosen for the second set of calculations. The parent nucleus is chosen to be ^{144}Dy and is symmetrically divided in three ^{48}Ti fragments. The magicity of the fragments is lost now and the shell corrections are positive. The proton and neutron values for the microscopic Strutinsky corrections are displayed in figure 6.

At the end of the process the total shell corections comprises three times the proton and the neutron fragment values.

The last example has been taken as an even more neutron rich system, with the parent nucleus ^{144}Hf splitting into three ^{48}Cr fragments. One has to mention a deep minimum at a large distance between centers, which implies the possibility of an isomeric quasistable state for an elongated configuration.

In order to have a final result about such a speculation, one has to calculate the macroscopic part of the deformation energy to be added and

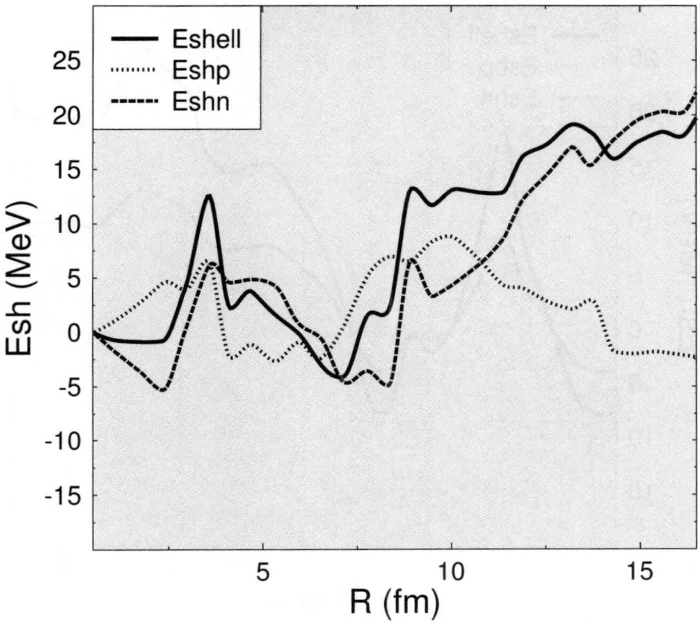

Fig. 6. Calculated shell corrections for protons, neutrons and total value for the symmetric splitting of ^{144}Dy into three ^{48}Ti.

see whether the minimum still exists. The last results are displayed in figure 7 for the ternary fission of ^{144}Hf. One has to mention that the proton and neutron values are identical, since for light fragments the strengths of the spin-orbit and l^2 potentials are equal, and the isospin is zero in this case. The first minimum is noticeable also suggesting a deformed ground state for ^{144}Hf. All calculations have been performed under the supposition of spherical parent nucleus.

7. Conclusions

The three center shell model which has been constructed describes the transition of the parent neutron and proton level scheme to the three partially overlapped and finally separated fragment level schemes. Spin-orbit interaction operators are geometry dependent and generate the appropiate matrix elements influenced by the ternary character of the process. The minima calculated in the shell corrections along the tripartition splitting can lower the macroscopic barrier and decide which parent nucleus can be chosen as

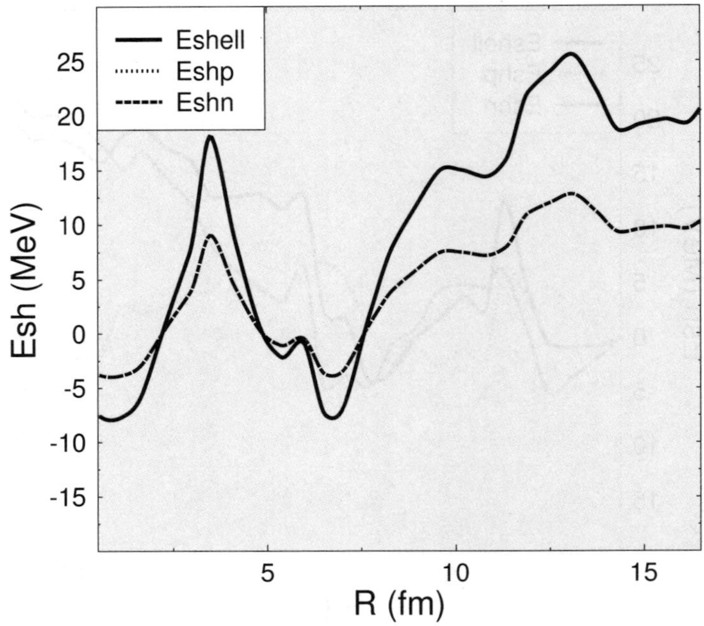

Fig. 7. Calculated shell corrections for protons, neutrons and total value for the symmetric splitting of ^{144}Hf into three ^{48}Cr.

favourable for ternary fission studies. Also minima in the shell corrections obtained with the three center shell model level scheme could influence the stability of an elongated, linear three-body type system. The model also is able to be applied to the study of alpha-chain type of nuclei, as the lightest possible ternary configuration.

Acknowledgments

The present work was partly supported by Deutsche Forschungsgemeinschaft, Bonn and by Ministry of Education and Research, Bucharest.

References

1. L. W. Alvarez, as reported by G. Farwell, E. Segre and C. Wiegand, Phys. Rev. 71 (1947) 327.
2. W. J. Swiatecki, in *Second U. N. Int. Conf. on the peaceful uses of atomic energy*, Geneva (1958) p.248.

3. D. N. Poenaru, W. Greiner and R. A. Gherghescu, Atomic Data Nucl. Data Tables 68 (1998) 91.
4. G. Audi and A. H. Wapstra, Nucl. Phys. A595 (1995) 409.

Precision Measurements with Ion Traps

ARI JOKINEN

Department of Physics, University of Jyväskylä
P.O.Box 35, FIN-40014, Jyväskylä, Finland
E-mail: ari.jokinen@phys.jyu.fi

Ion traps have become an essential tools for nuclear studies. They have been applied widely for measurements of atomic masses with unprecedented precision, but recently they have also been developed for spectroscopic studies, where they can provide clear benefits compared to conventional techniques. In these lecture notes, recent results from Penning trap projects for radioactive ions are discussed with special emphasis on JYFLTRAP project in the Department of Physics, University of Jyväskylä.

1. Introduction

The mass of the ground state of a nucleus is a result of the high order of symmetry of a complex quantum system. Accurate measurements of the ground state masses can therefore provide insight into the underlying features of the nucleonic system, such as charge symmetry, shell effects, coexisting structures and so forth. The accuracy necessary to obtain relevant information ranges from 1 keV, when investigating Coulomb energy differences between mirror nuclei or isospin multiplets - up to 100 keV, when global effects, such as deformation, are studied.

In nuclear astrophysics, the binding energies are one of the most important ingredients for reliable calculations. They affect the rates of the relevant reactions and they influence the time-scale and energy production of nucleosynthesis. In high temperature conditions, they adjust the balance, which defines the process paths. Precision measurements provide important data for fundamental studies of the weak interaction. Of particular interest are measurements related to super-allowed beta decays, which test the conserved vector current (CVC) hypothesis and the unitarity of the Cabibbo-Kobayashi-Maskawa (CKM) guark mixing matrix [1–3]. Nuclear masses and binding data in general can be obtained in various indirect and direct methods. Traditionally the most precise information comes from re-

Fig. 1. Three eigenmotions of the charged particle in the Penning trap. For details, see text. Magnetic filed lines are perpendicular to the radial plain of the ion motion.

action studies with stable beams and targets. Although this method can be accurate, it is applicable only close to the line of the stability.

Binding information for exotic nuclei has mainly been obtained via radioactive decays linking different isotopes and states to stable isotopes. In some cases, a relatively good accuracy could be obtained. However, in many cases such a data can be erroneous due to the limited knowledge decay properties affecting the analysis of experimental data.

The very first direct methods of mass measurements were relying on the mass-dependent curvature of the ions in the dipole magnetic field. Nowadays, a combination of the ISOL-method and Penning trap technology allows direct mass measurements of radioactive isotopes, as demonstrated by the ISOLTRAP experiment at CERN [4]. In JYFL, Penning trap technology is combined with the IGISOL-technique. Thus precision studies of atomic masses can be extended to short-lived exotic isotopes without target-ion source chemistry restrictions [1].

In a Penning trap ion are confined by a combination of strong magnetic field and static cylindrical quadrupole electric field. The first one confines ions radially to two circular eigen motions and axially ions oscillate in a harmonic potential. Three eigen motions are illustrated in the fig. (1). Slow drift at large radius around the magnetic field is called magnetron motion and fast rotation is reduced cyclotron motion. Angular frequencies of these two motions combined to true cyclotron frequency, which is mass-dependent in a constant field according to equation $\omega_c = \omega_- + \omega_+ = \frac{qm}{B}$, where f_c is the cyclotron frequency of an ion with a charge-to-mass ratio of q/m oscillating in the external magnetic field B.

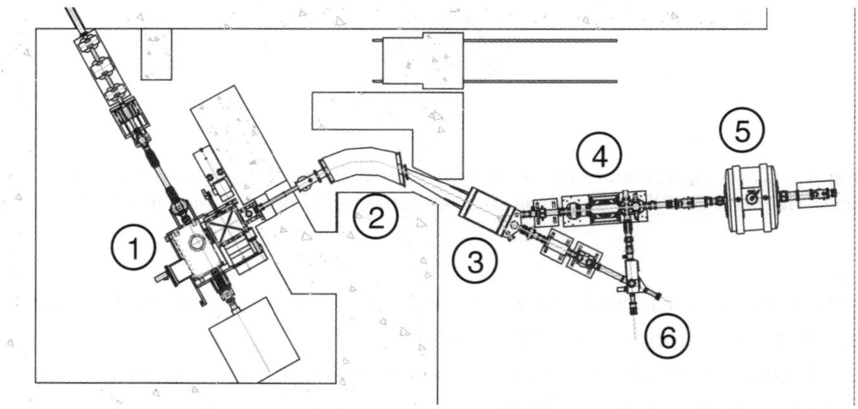

Fig. 2. Layout of the IGISOL-facility. Numbers in the figure have the following meaning: 1 - ion guide, 2 - dipole magnet, 3 - switchyard, 4 - RFQ cooler and buncher, 5 - superconducting solenoid housing two Penning traps, 6 - spectroscopy station in the end of the beam line and connection to the second floor for collinear laser spectroscopy.

2. JYFLTRAP project

Penning trap project in the Department of Physics, University of Jyväskylä was initiated in mid-90's. Trapping facility was constructed in one of the beam lines served by the ion guide isotope separator IGISOL. The layout of the of the IGISOL-facility is shown in figure 2.

IGISOL-technique was developed in Jyväskylä early 80's as a complementary approach to produce low energy ion beams of mass separated radioactive ions [5]. Compared to the conventional ion source based ISOL-facilities, IGISOL is rather insensitive for the chemical and physical properties making it universal source for all elements. It is also very fast allowing extraction of exotic species without decay losses.

The IGISOL technique is based on the stopping of the recoil ions in a buffer gas resulting in a fast reset to the 1+ ionic charge state. With this technique ion beams can be made of short-lived ($T_{1/2} \geq 0.1ms$) radio isotopes of all elements, including the most refractory ones. Another variant of the method is based on selective laser ionization of short-lived radioactive species thermalized in gas as neutral atoms [6].

2.1. *IGISOL*

At IGISOL the projectile beam hits a thin target and product nuclei recoil out as highly charged ions into a fast-flowing buffer gas, usually helium. As the ions slow down and thermalize their charge state changes continuously via charge exchange processes with the gas atoms. A significant fraction retain a 1+ charge state and are guided out of the ion source with the gas flow, whereby they are injected into the mass separator via stages of differential pumping. After acceleration to between 30- and 40 kV depending on the experimental requirements the beam is mass separated by a dipole magnet, allowing separation of nuclei with a typical mass resolving power of the order of 250-500 depending on the operational parameters of the ion guide and the front-end of the separator. The attractive features of this technique are the fast (sub-millisecond) release, and chemical non-selectivity making it possible to produce even the most refractory of elements. In connection with nuclear fission, the IGISOL method has led to the production of neutron-rich refractory isotopes such as Nb, Mo, Tc, Ru and Rh, with beam intensities approximately 10^5 ions/s. The typical transverse emittance of an extracted ion beam is 12 $\pi \cdot mm \cdot mrad$ and the energy spread is relatively large, up to 50-100 eV. In order to reduce these physical parameters an additional cooling is required as will be explained

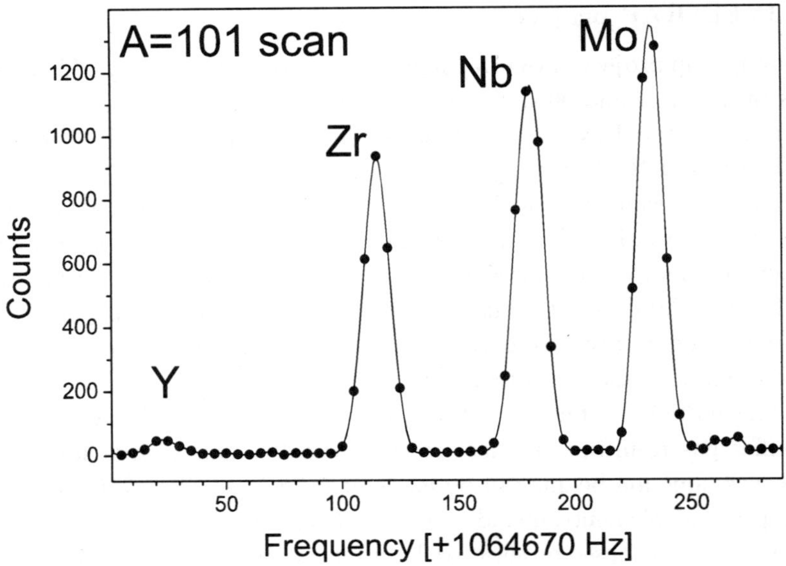

Fig. 3. Mass scan for A=101 isotopes performed in the first trap, purification trap. Different isotopes belonging to the same mass chain are clearly separated.

in the following sections.

2.2. *Linear Paul trap: RFQ cooler and buncher at IGISOL*

The first element of the JYFLTRAP device is a gas-filled radio frequency quadrupole (RFQ), which prepares ion ensemble for the injection into the Penning trap. In an RFQ ions are confined with a transverse time-dependent electric field and energy loss is obtained in ion and buffer gas atom interactions. A small electric potential of about 5 V over the length of the RFQ draws the ions through the cooler, resulting in a transit time of the order of 1 ms. Due to the simultaneous energy loss in ion and buffer gas atom interactions, the energy spread of the ions in an extraction of the device is typically less than 1 eV. With an aid of an axial potential dwell it is also possible to accumulate ions into the RFQ and release them as ion bunches with a duration of a few microseconds. The typical transmission efficiency of 60-70 % for ions with $A \geq 40$ has been obtained. With these properties strongly improved conditions for collinear laser spectroscopy are encountered and the demands for the injection conditions of a Penning trap system are fulfilled. For more details on ion coolers and bunchers, see [7].

2.3. *JYLFTRAP: tandem Penning trap device*

In the purification trap the bunched ion cloud from the RFQ is captured to an axial potential dwell. After a short cooling period, ions are radially excited to large radii and subsequently centered mass selectively by quadrupole excitation with the cyclotron frequency. As a result, only those ions in the resonance with an excitation frequency are centered and others a lost. In a series of test experiments it could be shown that, depending on the buffer gas pressure used in the system, mass-resolving powers between 20000 and 150000 can be obtained. Figure 3 exemplifies the output ion current from the first trap as a function of the resonance frequency. As can be seen, mass resolving power of the method is capable independently center isotopes belonging to the same mass chain [8].

Although the main purpose of the first trap is to prepare the ion cloud for the injection to the second trap, it can also be used for various spectroscopic studies. Combined with a transmission of approximately 30 % and a capture efficiency of 60 % this will allow for strongly enhanced conditions for decay spectroscopy. Mass purification has been applied for example, in the decay studies of neutron-rich Zr isotopes [9,10] and for the determination of the relative isotopic distributions in fission reactions. Another interesting application is charged-particle spectroscopy from a massless source. Isobarically purified sample is transported through narrow channel to the second trap for precision mass measurements. Determination of the cyclotron frequency of the ion of interest is based on the TOF-technique described in [13]. More details on the second trap of the JYFLTRAP can be found in ref. [1]

3. Atomic Masses measurements at JYFLTRAP

A start-up of the JYFLTRAP mass measurements programs has been very succesfull due to an access to variety of low-energy radioactive species. The nuclide chart in a figure 4 summarizes the mass measurement programs in Jyväskylä. Details of the mass measurements with connections to different physics issues are discussed in the following sections.

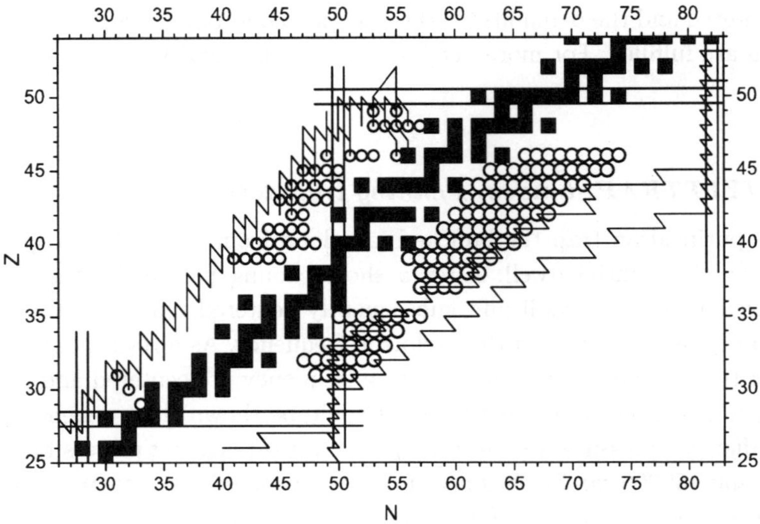

Fig. 4. A part of the chart of the nuclei showing the isotopes studied recently with JYFLTRAP. The chart shows also examples of rp and r process paths. Rp-process path is calculated assuming steady state burning [14]. For r-process path, the relevant parameters are: temperature 1.5 GK and neutron density $N_n = 10^{24}/cm^3$.

3.1. *Binding energies and nuclear structure*

Mass predictions have been widely used to extrapolate masses far from stability, where the lack of experimental data is a severe. Various theoretical approaches have been developed. It is customary to divide them to local and global predictions. The first set of predictions can provide reasonably good results in the limited range of nuclei, as their parameter sets are adjusted to region of interest. Contradictory to these, global predictions aim for the best possible average agreement over the large area, often the whole nuclide chart.

We have recently studied more than 100 masses ranging from Ga to Pd [1,11,12,15] as shown in figure 4. This data provides an important set of mass values for comparison with the recent atomic mass evaluations [16] and modern mass predictions. An example of the comparison of the JYFLTRAP values with the HFB-8 predictions is shown in figure 5. HFB-8 model is one of the recent models used in astrophysical predictions [17]. In this case the average discrepancy between the predicted and the experimental mass excesses is 80 keV. A more thorough analysis, as shown in Fig. 5, implies neutron number dependent structure in the measured mass excesses.

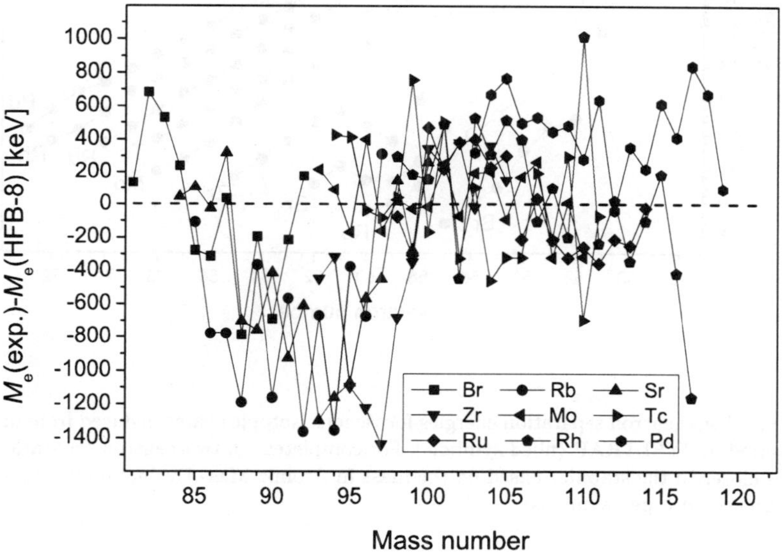

Fig. 5. Comparison of measured mass-excess and those predicted by HFB-8 model [17] for neutron-rich nuclei from Br to Pd.

In addition to the masses itself, the obtained data can be used to determined various binding observables, like pairing energies, particle separation energies, etc. Figure 6 displays the two-neutron separation energies for neutron-rich medium mass isotopes. Apart from the sudden drop at the neutron number N=50 and smooth decrease of the two-neutron separation energy as a function of neutron number, one can observe an irregular behavior between neutron numbers 56 and 60. This effect is most prominent for the Zr-isotopes, and vanishes within a couple of units in proton number.

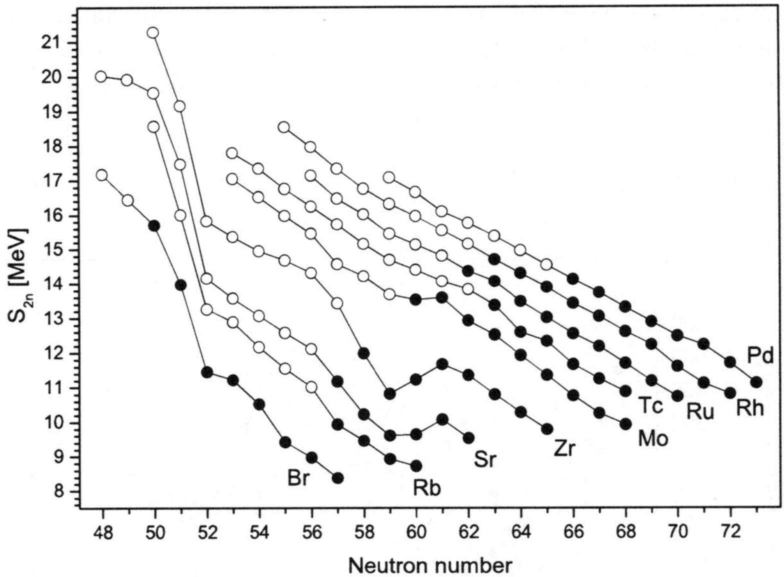

Fig. 6. Two-neutron separation energies for various isotopic chains deduced from masses measured at JYFLTRAP (filled symbols). For completeness, two-neutron separation energies closer to the stability based on the mass in Atomic Mass Evaluation 2003 [16] are also presented (open symbols).

3.2. *Pairing and shell gap energies*

While two neutron separation energies may reveal some information of the structural changes one may also get interesting information by calculating other type of binding energy differences. For, example it is possible to determine the shell gap energy by using equation

$$\delta(N_0)_{2n} = S_{2n}(N_0, Z) - S_{2n}(N_0 + 2, Z) \tag{1}$$

Figure 7 shows extracted shell gaps for closed neutron shell N=50 and subshell closure N=56. It is worth of noticing that shell gap at N=50 tends to get narrower while proton number is decreased. This results is in contradictory to numerous recent articles, where gamma-spectroscopic data and theoretical predictions are compared. A data shown in figure 7 and our new, unpublished data extending down to Ga-isotopes, clearly shows a reduction of the shell gap at the closed neutron shell N=50.

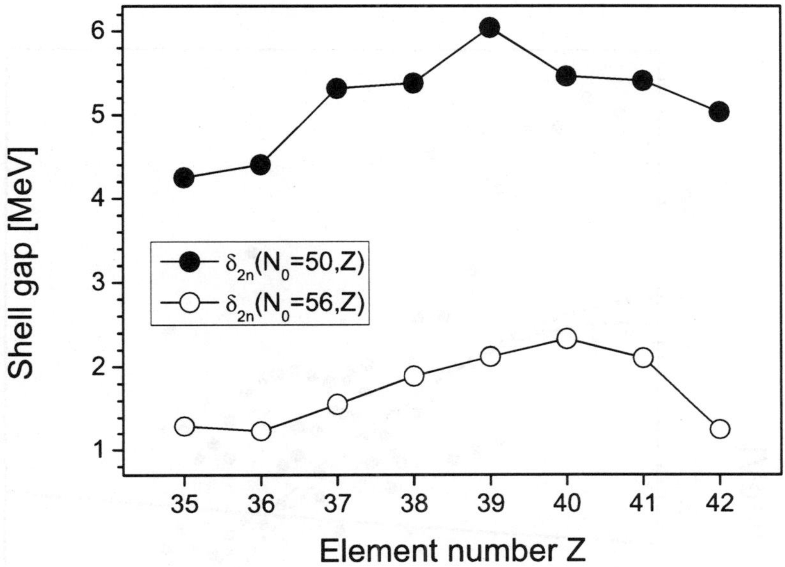

Fig. 7. Shell gap energies for neutron number N=50 and N=56 based on the calculation as given in Eq. 1 with experimental two neutron separation energies from JYFLTRAP.

Pairing energy can be obtained from binding energies by applying 3-point formula

$$P_3(N) = \frac{(-1)^N}{4}[2 \cdot S_n(N) - S_n(N+1)S_n(N-1)] \tag{2}$$

where masses are taken from the experimental data. Figure 8 collects experimental pairing energies in one systematical plot. Selected axes of the plot are useful for the compilation of large set of the data. All in all, a general trend of reduced pairing as a function asymmetry term I=(N-Z)/A is observed.

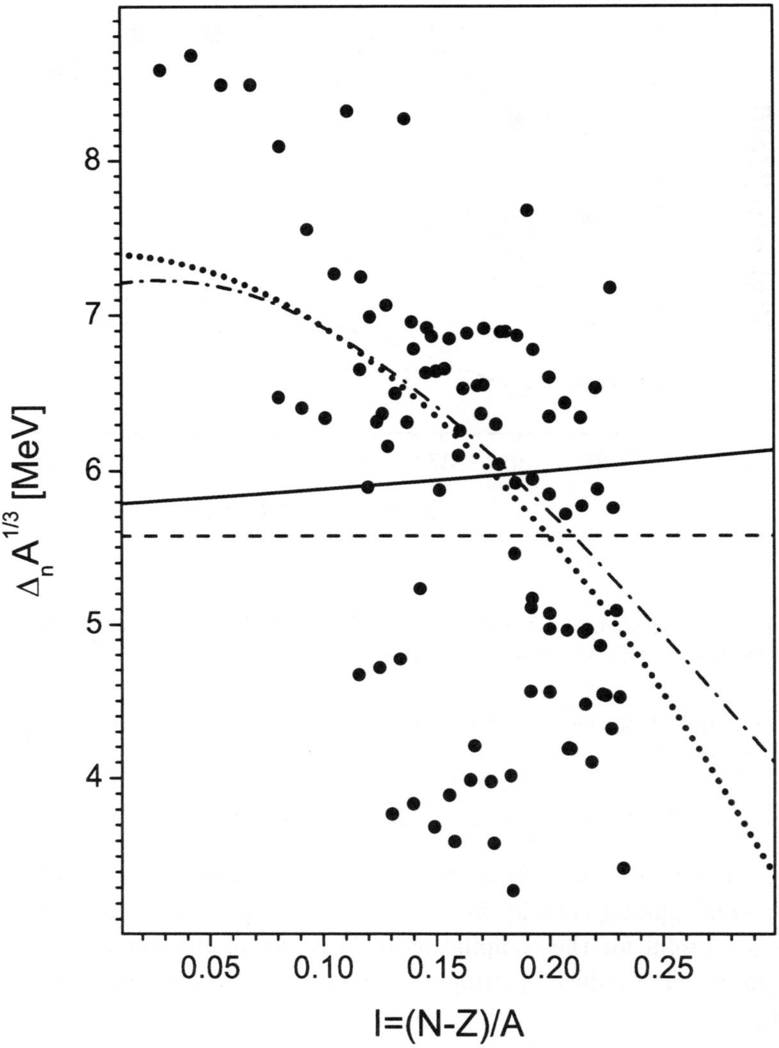

Fig. 8. Neutron pairing energies for neutron-rich even-Z isotopes from Se to Pd calculated from the recent high-precision Penning trap data from JYFLTRAP (and from ISOLTRAP for Kr-isotopes [18]). Solid and dashed lines are based on the simple $\Delta_n = \frac{12}{A^{1/2}}$ formula for fixed Z and A, respectively. Dash-dotted and dotted lines are global pairing fits according to refs. [19,20]

If pairing energies are plotted directly as a function of neutron numbers, as shown in Fig. 9, some regular patterns can be observed. There is an increase of the pairing energy while moving from highly deformed Zr (Z=40) isotopes to more spherical Pd (Z=46) isotopes. Neutron sub shell closure at N=56 seems to increase the pairing energy, especially for deformed nuclei, where pairing energies are lower than global formulas predict.

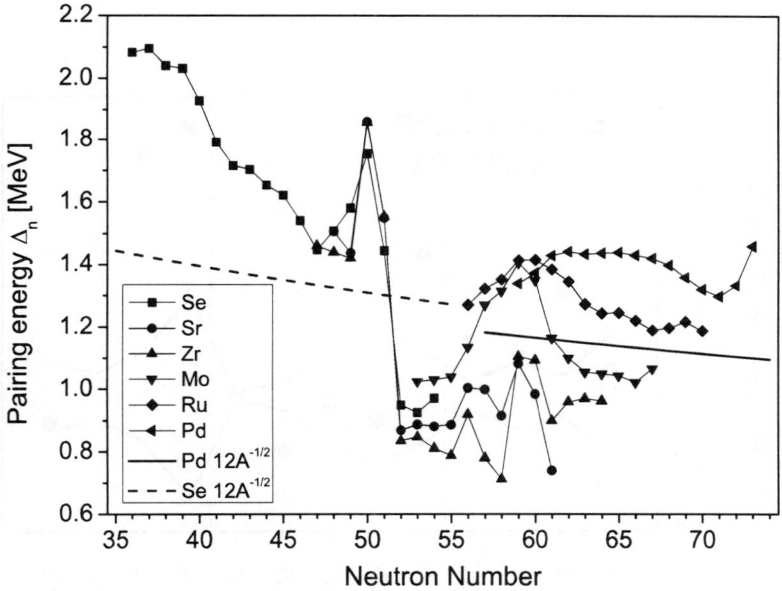

Fig. 9. Neutron pairing energies as a function of neutron number for even-Z isotopes from Se to Pd.

3.3. *Atomic Masses and the Nucleosynthesis*

Proton and neutron separation energies are in a central role in nuclear astrophysics modeling for rapid neutron capture process in neutron-rich nuclei and for rapid proton capture process in neutron-deficient nuclei. We have initiated a project to measure the binding energies of nuclei located in the expected region of the rp-process and p-process paths. These measurements include $^{80-83}$Y, $^{83-86,88}$Zr and $^{85-88}$Nb isotopes, of which ^{84}Zr has been measured for the first time. The obtained data has improved considerably S_p and Q_{EC} values for astrophysically important nuclei [21]. Figure 10 shows a comparison of the experimental and compiled proton separation energies. Rather good agreement with the recent AME compilation has been obtained for Y and Zr isotopes. In case of Nb-isotopes, our results disagree with compiled values. It is obvious that compiled estimates for proton separation energies has to be revised, especially of the lightest Nb-isotopes which could impact the estimated position of the proton dripline [21].

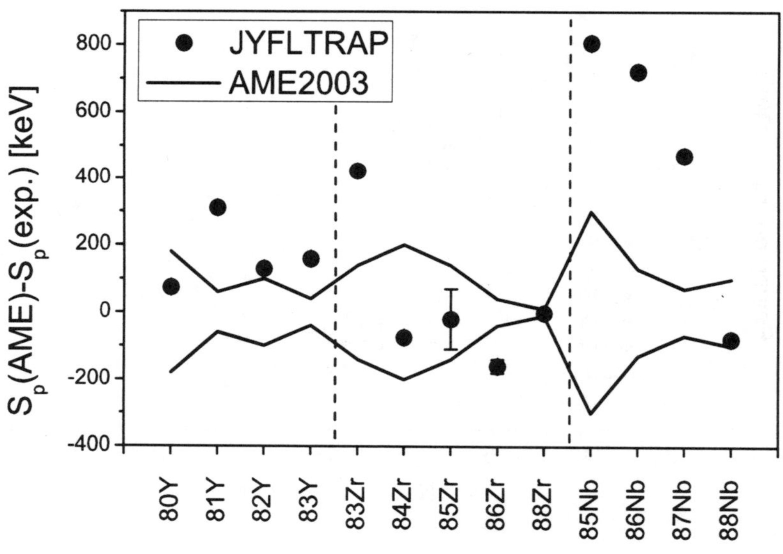

Fig. 10. Proton separation energies for neutron-deficient nuclei. Difference between compiled values and experimental values are given. Solid lines corresponds to uncertainties of compiled values.

4. Precision Q_{EC} Determination for Superallowed Beta Decay

Assuming that the half-lives and branching ratios are well known, a high accuracy of the Q_{EC}-value of super allowed beta-decay is needed for a precise determination of the ft-value resulting in an accurate measurement of the vector coupling constant G_d. This in turn allows to determine V_{ud}, the up-down matrix element of the Cabbibo-Kobayashi-Maskawa Matrix (CKM). The obtained information tests the Conserved Vector Current (CVC) hypothesis and the unitarity of the CKM and physics beyond the standard model. So far four superallowed beta emitters have been studied

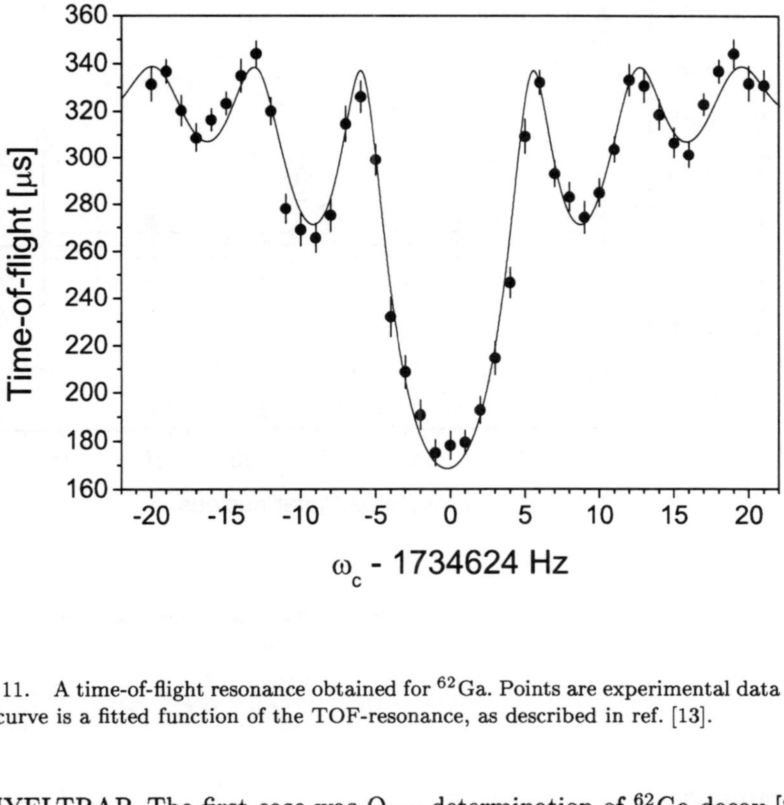

Fig. 11. A time-of-flight resonance obtained for ^{62}Ga. Points are experimental data and the curve is a fitted function of the TOF-resonance, as described in ref. [13].

at JYFLTRAP. The first case was Q_{EC} determination of ^{62}Ga decay [22], where Q_{EC} value was determined with the precision of 540 eV. An example of the time-of flight resonances for ^{62}Ga is shown in figure 11. By combining value with the recent high-precision branching ratio measurement from TRIUMF [23], an experimental uncertainty was reduced to the level that the dominant uncertainty originates from theoretical corrections.

Triggered by the anomalous Q_{EC}-value claimed for the beta decay of ^{46}V [24], an independent measurement was performed at JYFLTRAP. In connection to this experiment, Q_{EC} values for $^{26}Al^m$ and ^{42}Sc were also re-checked [25]. JYFLTRAP results confirmed Q_{EC} value of ^{46}V obtained in ref. [24] as well as compiled values [26] for $^{26}Al^m$ and ^{42}Sc. An updated collection of precisely measured comparative half-lives is presented in figure 12.

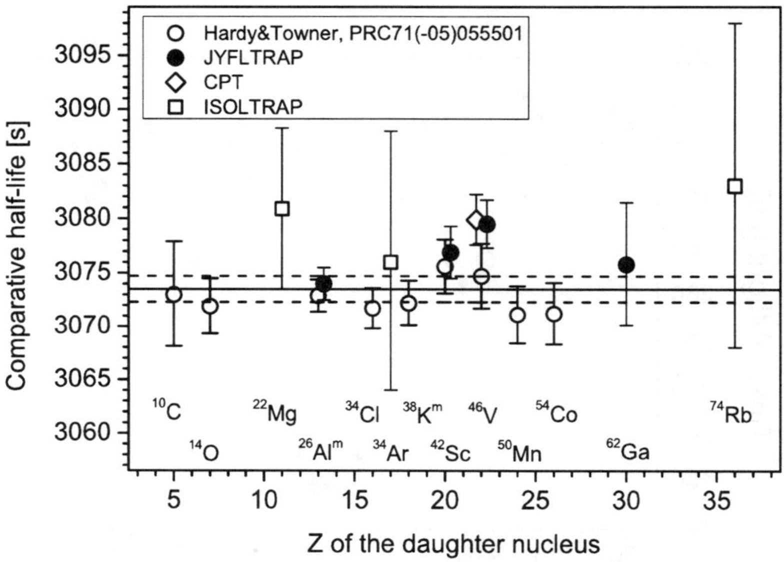

Fig. 12. Summary of comparative half-lives for nine well studied cases. In addition, recent new values from Penning trap projects are also included.

5. Summary

Atomic mass data can contribute significantly or the better understanding of the nuclear structure and the binding energies are also of importance for the nuclear astrophysics. Precise atomic masses of super allowed beta emitter parent and daughter contribute to the low-energy CVC tests and for the unitarity test of CKM-matrix. Penning traps have become the most accurate tools for direct mass determination of radioactive isotopes. A continuous development of trapping techniques has increased and will increase the accuracy, sensitivity and efficiency further from the present values.

Acknowledgments

This work has been supported by the EU 6th Framework Programme, Integrated Infrastructure Initiative - Joint Research Project Activities Contract number: 506065 (EURONS, JRA's TRAPSPEC. LASER and DLEP) and by the Academy of Finland under the Finnish Centre of Excellence Programmes 2000-2005 (Project No. 44875, Nuclear and Condenced Matter Physics Programme) and 2006- 2011 (Nuclear and Accelerator Based Physics Programme at JYFL). A. Jokinen is is indebted to financial support from the Academy of Finland (Project numbers 46351).

References

1. A. Jokinen et al. *Int. J. Mass Spectr.* **251**, 204 (2006).
2. K. Blaum *Phys. Rep.* **425**, 1 (2006).
3. D. Lunney et al. *Rev. Mod. Phys.* **75**, 1021 (2003).
4. G. Bollen et al. *Nucl. Instr. Meth. A* **368**, 675 (1996).
5. J. Äystö *Nucl. Phys. A* **693**, 477 (2001).
6. I. Moore et al. *J. Phys. G* **31**, S1499 (2005).
7. J. Äystö and A. Jokinen *J. Phys. B Atom. Mol. Opt. Phys.* **36**, 573 (2003).
8. V. Kolhinen et al. *Nucl. Instr. Meth. A* **528**, 776 (2004).
9. S. Rinta-Antila et al. *Phys. Rev. C* **70**, 011301R (2004).
10. S. Rinta-Antila et al. *Eur. Phys. J.* **A**, submitted (2006).
11. U. Hager et al. *Phys. Rev. Lett.* **96**, 052504 (2006).
12. S. Rahaman et al. *Eur. Phys. J.* **A**, submitted (2006).
13. M. König et al. *Int. J. Mass. Spectr. Ion Process.* **142**, 95 (1995).
14. H. Schatz et al. *Phys. Rev. Lett.* **86**, 3471 (2001).
15. U. Hager et al. *Phys. Rev. C* , submitted (2006).
16. G. Audi et al., *Nucl. Phys. A* **729**, 3 (2003).
17. M. Samyn et al., *Phys. Rev. C* **70** 044309 (2004).
18. P. Delahaye et al. *Phys. Rev. C* **74**, 034331 (2006).
19. D.G. Madland and J.R. Nix *Nucl. Phys. A* **476**, 1 (1988).
20. P. Vogel et al. *Phys. Lett. B* **139**, 227 (1984).
21. A Kankainen et al. *Eur. Phys. J. A* **29**, 271 (2006).
22. T. Eronen et al. *Phys. Lett. B* **636**, 191 (2006).
23. B. Hyland et al. *Phys. Rev. Lett.* **97**, 102501 (2006).
24. G. Savard et al. *Phys. Rev. Lett.* **95**, 102501 (2006).
25. T. Eronen et al. *Phys. Rev. Lett.* , in press (2006).
26. J. Hardy and I. Towner *Phys. Rev. C* **71**, 055501 (2005).

Connecting critical point symmetries to the shape/phase transition region of the Interacting Boson Model

E. A. McCUTCHAN

Wright Nuclear Structure Laboratory, Yale University, New Haven, Connecticut 06520-8124, USA

DENNIS BONATSOS

Institute of Nuclear Physics, National Centre for Scientific Research "Demokritos", GR-15310 Aghia Paraskevi, Attiki, Greece

N. V. ZAMFIR

National Institute of Physics and Nuclear Engineering, Bucharest-Magurele, Romania

The parameter independent (up to overall scale factors) predictions of the X(5)-β^2, X(5)-β^4, and X(3) models, which are variants of the X(5) critical point symmetry developed within the framework of the geometric collective model, are compared to two-parameter calculations in the framework of the interacting boson approximation (IBA) model. The results show that these geometric models coincide with IBA parameters consistent with the phase/shape transition region of the IBA for boson numbers of physical interest (close to 10). ^{186}Pt and ^{172}Os are identified as good examples of X(3), while ^{146}Ce, ^{174}Os and ^{158}Er, ^{176}Os are identified as good examples of X(5)-β^2 and X(5)-β^4 behavior respectively.

1. Introduction

Critical point symmetries [1,2], describing nuclei at points of shape/phase transitions between different limiting symmetries, have recently attracted considerable attention, since they lead to parameter independent (up to overall scale factors) predictions which are found to be in good agreement with experiment [3–6]. The X(5) critical point symmetry [2], was developed to describe analytically the structure of nuclei at the critical point of the transition from vibrational [U(5)] to prolate axially symmetric [SU(3)] shapes. The solution involves a five-dimensional infinite square well potential in the β collective variable and a harmonic oscillator potential in the

γ variable. The success of the X(5) model in describing the properties of some nuclei with parameter free (except for scale) predictions has led to considerable interest in such simple models to describe transitional nuclei. Since its development, numerous extensions involving either no free parameters or a single free parameter have been developed. Those approaches which involve a single parameter include replacing the infinite square well potential with a sloped well potential [7], exact decoupling of the β and γ degrees of freedom [8], and displacement of the infinite square well potential, or the confined β-soft model [9]. Parameter free variants of the X(5) model include the X(5)-β^2 and X(5)-β^4 models [10], in which the infinite square well potential is replaced by a β^2 and a β^4 potential respectively, as well as the X(3) model [11], in which the γ degree of freedom is frozen to $\gamma = 0$, resulting in a three-dimensional Hamiltonian, in which an infinite square well potential in β is used.

Prior to these simple geometric models, shape/phase transitions were investigated [12] within the interacting boson approximation (IBA) model [13] by constructing the classical limit of the model, using the coherent state formalism [14,15]. Using this method it was shown [12,15] that the shape/phase transition between the U(5) and SU(3) limiting symmetries is of first order, while the transition between the U(5) and O(6) (γ-unstable) limiting symmetries is of second order. Furthermore, the region of phase coexistence within the symmetry triangle [16] of the IBA has been studied [17–19] and its borders have been determined [20,21], while a similar structural triangle for the geometric collective model has been constructed [22].

It is certainly of interest to examine the extent to which the parameter free (up to overall scale factors) predictions of the various critical point symmetries and related models, built within the geometric collective model, are related to the shape/phase transition region of the IBA. It has already been found [23] that the X(5) predictions cannot be exactly reproduced by any point in the two-parameter space of the IBA, while best agreement is obtained for parameters corresponding to a point close to, but outside the shape/phase transition region of the IBA. In the present work we examine the extent to which the predictions of the X(5)-β^2, X(5)-β^4, and X(3) models can be reproduced by two-parameter IBA calculations using boson numbers of physical interest (close to 10) and the relation of these geometrical models to the shape/phase transition region of the IBA. Even-even nuclei corresponding to reasonable experimental examples of the manifestation of the X(3), X(5)-β^2, and X(5)-β^4 models are also identified.

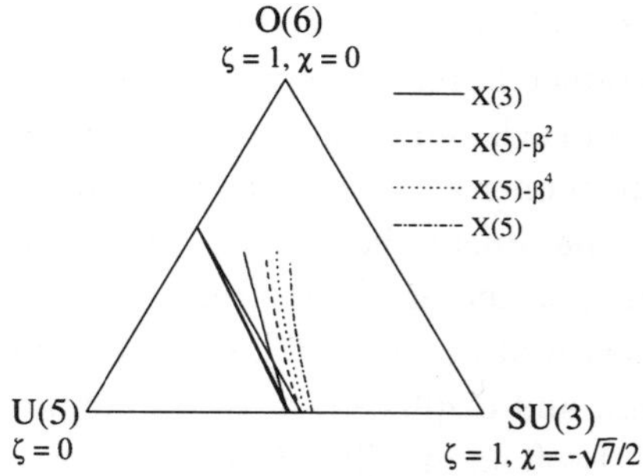

Fig. 1. IBA symmetry triangle illustrating the dynamical symmetry limits and their corresponding parameters. The phase transition region of the IBA, bordered by ζ^* on the left and by ζ^{**} on the right, as well as the loci of parameters which reproduce the $R_{4/2}$ ratios of X(3) (2.44), X(5)-β^2 (2.65), X(5)-β^4 (2.77), and X(5) (2.90) are shown for $N_B = 10$. The line defined by $\zeta_{\rm crit}$ is also shown, lying to the right of the left border and almost coinciding with it.

2. The IBA Hamiltonian and symmetry triangle

The study of shape/phase transitions in the IBA is facilitated by writing the IBA Hamiltonian in the form [18,20]

$$H(\zeta,\chi) = c\left[(1-\zeta)\hat{n}_d - \frac{\zeta}{4N_B}\hat{Q}^\chi\cdot\hat{Q}^\chi\right], \tag{1}$$

where $\hat{n}_d = d^\dagger\cdot\tilde{d}$, $\hat{Q}^\chi = (s^\dagger\tilde{d} + d^\dagger s) + \chi(d^\dagger\tilde{d})^{(2)}$, N_B is the number of valence bosons, and c is a scaling factor. The above Hamiltonian contains two parameters, ζ and χ, with the parameter ζ ranging from 0 to 1, and the parameter χ ranging from 0 to $-\sqrt{7}/2 = -1.32$. With this parameterization, the entire symmetry triangle of the IBA, shown in Fig. 1, can be described, along with each of the three dynamical symmetry limits of the IBA. The parameters (ζ,χ) can be plotted in the symmetry triangle by converting them into polar coordinates [24]

$$\rho = \frac{\sqrt{3}\zeta}{\sqrt{3}\cos\theta_\chi - \sin\theta_\chi}, \qquad \theta = \frac{\pi}{3} + \theta_\chi, \tag{2}$$

where $\theta_\chi = (2/\sqrt{7})\chi(\pi/3)$.

Using the coherent state formalism of the IBA [13–15] one can obtain the scaled total energy, $E(\beta,\gamma)/(cN_B)$, in the form [19]

$$\varepsilon(\beta,\gamma) = \frac{\beta^2}{1+\beta^2}\left[(1-\zeta) - (\chi^2+1)\frac{\zeta}{4N_B}\right] - \frac{5\zeta}{4N_B(1+\beta^2)}$$

$$-\frac{\zeta(N_B-1)}{4N_B(1+\beta^2)^2}\left[4\beta^2-4\sqrt{\frac{2}{7}}\chi\beta^3\cos 3\gamma+\frac{2}{7}\chi^2\beta^4\right], \tag{3}$$

where β and γ are the two classical coordinates, related [13] to the Bohr geometrical variables [25].

As a function of ζ, a shape/phase coexistence region [17] begins when a deformed minimum, determined from the condition $\frac{\partial^2\varepsilon}{\partial\beta^2}|_{\beta_0\neq 0}=0$, appears in addition to the spherical minimum, and ends when only the deformed minimum remains. The latter is achieved when $\varepsilon(\beta,\gamma)$ becomes flat at $\beta=0$, fulfilling the condition [20] $\frac{\partial^2\varepsilon}{\partial\beta^2}|_{\beta=0}=0$, which is satisfied for

$$\zeta^{**}=\frac{4N_B}{8N_B+\chi^2-8}. \tag{4}$$

The former, ζ^*, can be derived from the results of Ref. [26]. For $\chi=-\sqrt{7}/2$ this point is given by

$$\zeta^*=\frac{(896\sqrt{2}+656R)N_B}{-1144\sqrt{2}+123R+(1536\sqrt{2}+164R)N_B} \tag{5}$$

where

$$R=\sqrt{\frac{35456}{15129}+\frac{32\ 6^{2/3}}{41}}-\sqrt{\frac{70912}{15129}-\frac{32\ 6^{2/3}}{41}+\frac{3602816}{15129\sqrt{1108+369\ 6^{2/3}}}} \tag{6}$$

In between there is a point, $\zeta_{\rm crit}$, where the two minima are equal and the first derivative of ε_{min}, $\partial\varepsilon_{min}/\partial\zeta$, is discontinuous, indicating a first-order phase transition. For $\chi=-\sqrt{7}/2$ this point is [21]

$$\zeta_{\rm crit}=\frac{16N_B}{34N_B-27}. \tag{7}$$

Expressions for ζ^* and ζ_{crit} involving the parameter χ can also be deduced using the results of Ref. [26].

The range of ζ corresponding to the region of shape/phase coexistence shrinks with decreasing $|\chi|$ and converges to a single point for $\chi=0$, which is the point of a second-order phase transition between U(5) and O(6), located on the U(5)–O(6) leg of the symmetry triangle (which is characterized by $\chi=0$) at $\zeta=N_B/(2N_B-2)$, as seen from Eq. (4). The phase transition region of the IBA is included in Fig. 1. For $N_B=10$, it is clear that the left border of the phase transition region, defined by ζ^*, and the line defined by $\zeta_{\rm crit}$ nearly coincide. For $\chi=-1.32$, in particular, one

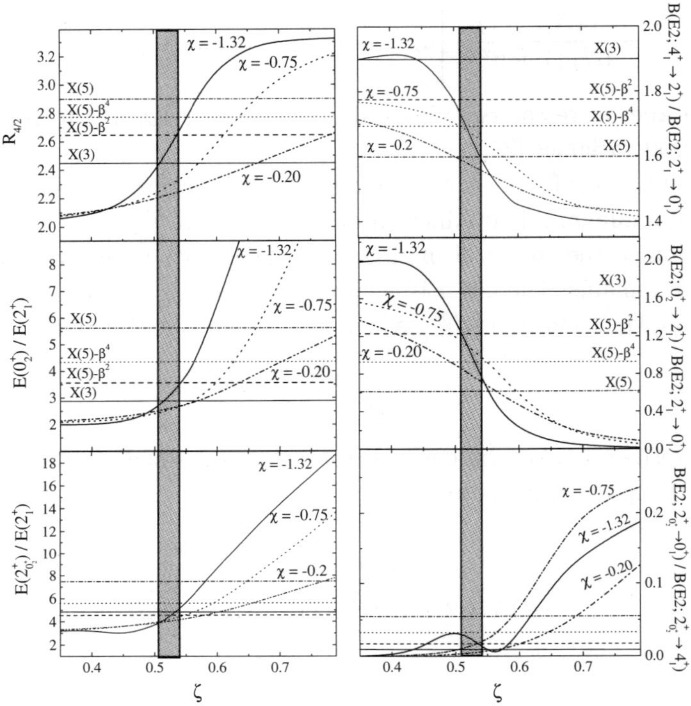

Fig. 2. Evolution of some key structural observables with the IBA parameters ζ and χ for $N_B = 10$. The predictions of X(3), X(5)-β^2, X(5)-β^4, and X(5) are indicated in each panel by a horizontal line. The values of ζ corresponding to the phase transition region of the IBA, bordered by $\zeta_{\rm crit}$ (approximately equal to ζ^*) on the left and by ζ^{**} on the right, are marked by the shaded area. The small dependence of the phase transition region on χ is not shown.

has $\zeta^* = 0.507$ and $\zeta_{\rm crit} = 0.511$. Therefore in what follows we shall use $\zeta_{\rm crit}$ as the approximate left border of the phase transition region.

3. Comparison of X(3), X(5)-β^2, and X(5)-β^4 predictions to the IBA

The most basic structural signature of the geometrical models, X(3), X(5)-β^2, and X(5)-β^4 is the yrast band energy ratio $R_{4/2} \equiv E(4_1^+)/E(2_1^+)$. Since this is a simple and often experimentally well known observable, we use the $R_{4/2}$ ratio as a starting point for these calculations. A constant value of the $R_{4/2}$ ratio can be obtained in the IBA for a small range of ζ values (since both provide a measure of the quadrupole deformation) and a wider range of χ values. Figure 1 gives the loci of the parameters which reproduce the $R_{4/2}$ ratios of X(3) (2.44), X(5)-β^2 (2.65), X(5)-β^4 (2.77), and X(5) (2.90), for $N_B = 10$. As expected from the varying $R_{4/2}$ ratios, there is a smooth evolution of the lines from X(3) up through X(5), corresponding to an increase in the average ζ value. Relating to the phase/shape transition region of the IBA, the X(3) locus begins on the U(5)-SU(3) leg of the triangle close to the left border of the phase/shape transition region and then crosses it as the absolute value of χ decreases. The X(5)-β^2 locus starts within the right border of the phase/shape transition region on the bottom leg of the triangle, then diverges slightly away from it. The X(5)-β^4 and X(5) loci lie just beyond the phase/shape transition region on the U(5)-SU(3) leg of the triangle, then move away from it. This evolution can be understood by considering the potentials used in these solutions. X(5)-β^2 uses a harmonic oscillator potential while X(5)-β^4 involves a potential intermediate between the β^2 potential and the infinite square well potential of X(5). Note that each of these modified versions of the X(5) solution are at some point closer to the phase/shape transition region of the IBA than X(5) itself.

To investigate the agreement between these different models and the IBA further, Fig. 2 illustrates some key structural observables as a function of the parameter ζ for different values of the parameter χ. The shape/phase transition region of the IBA, bordered by ζ^* (which almost coincides with $\zeta_{\rm crit}$, as discussed in relation to Fig. 1) on the left and by ζ^{**} on the right, is marked by the shaded area. Note that the shaded area corresponds to the phase/shape transition region for $\chi = -1.32$. As $\chi \to 0$, this region becomes increasingly narrower. This χ dependence is not shown in Fig. 2, since it is small. The parameter-independent predictions of X(3), X(5)-β^2, X(5)-β^4, and X(5) are shown as horizontal lines. In Fig. 2 and the following discussion, the notation $E(2^+_{0^+_2})$ refers to the energy of 2^+ state belonging to the 0^+_2 band. For the observables involving energies (left column of Fig. 2), the X(3) and X(5)-β^2 models coincide exactly with the predictions of the IBA in the phase transition region for χ values close to -1.32. The predictions for X(3) (solid line) intersect the IBA predictions for $\chi = -1.32$ on, or near, the left border of the phase transition region. For the energy ratios given in Fig. 2, $R_{4/2}$, $E(0^+_2)/E(2^+_1)$, and $E(2^+_{0^+_2})/E(2^+_1)$, the IBA predictions for ζ_{crit} and $\chi = -1.32$ are 2.44, 2.65, and 4.02, respectively. These are in close agreement with the X(3) predictions of 2.44, 2.87, and 4.83, respectively. On the other hand, the predictions for X(5)-β^2 (dashed line) for all three energy ratios intersect with the IBA predictions for $\chi = -1.32$ on, or very close to, the right border of the phase/shape transition region.

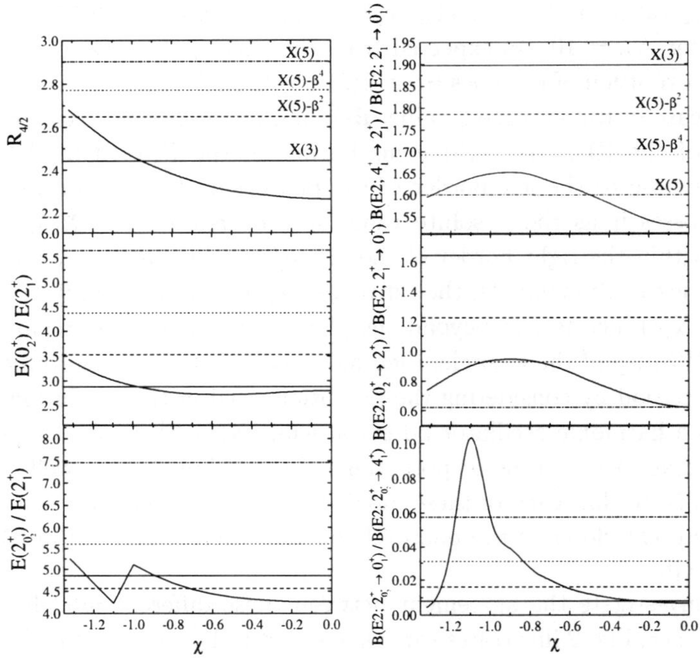

Fig. 3. The same structural observables shown in Fig. 2 presented as functions of the IBA parameter χ for ζ corresponding to the right border of the phase transition region of the IBA (i.e. with ζ fixed at the ζ^{**} value) for $N_B = 10$.

The IBA predictions at ζ^{**} and $\chi = -1.32$ for $R_{4/2}$, $E(0_2^+)/E(2_1^+)$, and $E(2_{0_2^+}^+)/E(2_1^+)$ are 2.68, 3.41, and 5.23, respectively. These are very similar to the X(5)-β^2 predictions of 2.65, 3.56, and 4.56, respectively. The X(5)-β^4 and X(5) predictions for energies do not coincide with the predictions of the IBA in the phase/shape transition region for any set of parameter values. Overall, the intersection of the IBA calculations with the X(5)-β^4 and X(5) predictions lies closest to the phase transition region of the IBA for $\chi = -1.32$ and moves further away for decreasing $|\chi|$. In summary, for the observables involving the energy ratios in Fig. 2, the X(3) solution is quite similar to the IBA predictions for $\chi = -1.32$ at ζ_{crit}, the X(5)-β^2 solution corresponds closely to the IBA predictions for $\chi = -1.32$ at ζ^{**}, and the X(5)-β^4 and X(5) solutions do not match the IBA predictions in the phase/shape transition region for any value of χ.

For the observables involving electromagnetic transition strengths (right column of Fig. 2), the correspondence between the different geometrical models and the IBA predictions in the phase/shape transition region changes somewhat. For the $B(E2)$ ratios, $B_{4/2} \equiv B(E2;\, 4_1^+ \to 2_1^+)/B(E2;\, 2_1^+ \to 0_1^+)$ and $B_{0/2} \equiv B(E2;\, 0_2^+ \to 2_1^+)/B(E2;\, 2_1^+ \to 0_1^+)$, the X(3) predictions do not coincide with the predictions of the IBA in the phase/shape transition region for any value of χ, with the X(3) predictions being larger than those of the IBA in the phase/shape coexistence region. The X(5)-β^2 solution shows a better correspondence with the IBA in the phase/shape transition region for the $B_{4/2}$ and $B_{0/2}$ ratios. The X(5)-β^2 predictions for $B_{4/2}$ and $B_{0/2}$ of 1.77 and 1.21, respectively, are very close to the IBA predictions at ζ_{crit} and $\chi = -1.32$ of 1.73 and 1.23, respectively. The X(5)-β^4 solution also intersects the IBA predictions for $B_{4/2}$ and $B_{0/2}$ for ζ values within the phase/shape transition region. For the final $B(E2)$ ratio, $B(E2;\, 2^+_{0_2^+} \to 0_1^+)/B(E2;\, 2^+_{0_2^+} \to 4_1^+)$, all three geometrical models predict values < 0.04, which is consistent with the predictions of the IBA in within the phase/shape transition region for all values of χ.

Motivated by the above findings, we show in Fig. 3 the same structural observables studied in Fig. 2, but now as a function of the IBA parameter χ and with ζ fixed to the value ζ^{**}, corresponding to the right border of the phase transition region, for $N_B = 10$. The predictions for each of the different geometrical models are again indicated by horizontal lines. Each of the geometrical models intersects the IBA predictions for ζ^{**} for one or more observables, although the results do not coincide exactly for the complete set of observables. For example, the X(3) predictions for the observables $R_{4/2}$, $R_{0/2} \equiv E(0_2^+)/E(2_1^+)$, and $E(2^+_{0_2^+})/E(2_1^+)$ coincide exactly with the phase transition region of the IBA for $\chi \sim -0.9$. However, this same agreement is not obtained for the $B(E2)$ strengths. Considering all the observables given in Fig. 3, the X(5)-β^2 model perhaps provides the closest level of agreement with the predictions of the phase/shape transition region in the IBA. For a χ value of ~ -1.2, the energy ratio predictions of X(5)-β^2 are very well reproduced by the IBA calculations for ζ^{**}. The $B(E2)$ ratios show less agreement but are reproduced within an order of magnitude or better.

In view of the above results, we compare in Fig. 4(top) the parameter independent level scheme of X(3) to the level scheme of the IBA with $N_B = 10$, $\chi = -1.32$, and $\zeta = \zeta_{\rm crit} = 0.51$. Similarly in Fig. 4 (middle) the parameter independent level scheme of X(5)-β^2 is compared to the level scheme of the IBA with $N_B = 10$, $\chi = -1.32$, and $\zeta = \zeta^{**} = 0.54$, and in Fig. 4 (bottom) the level scheme of X(5)-β^4 is compared to the level scheme of the IBA with $N_B = 10$, $\chi = -1.32$, and $\zeta = 0.55$. The energy levels in the ground state band and the excited K $= 0_2^+$ band are reproduced quite

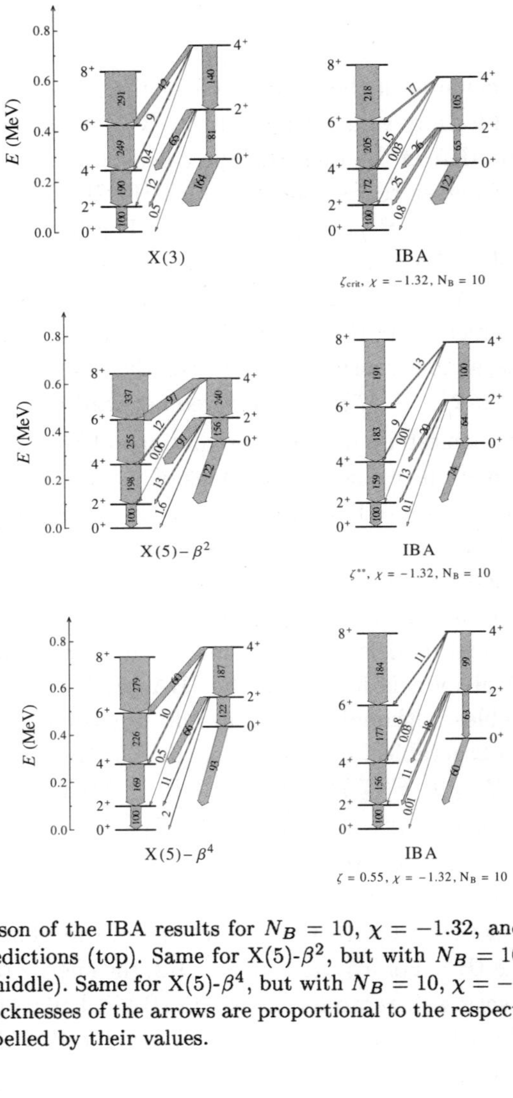

Fig. 4. Comparison of the IBA results for $N_B = 10$, $\chi = -1.32$, and $\zeta = \zeta_{\text{crit}} = 0.51$ with the X(3) predictions (top). Same for X(5)-β^2, but with $N_B = 10$, $\chi = -1.32$, and $\zeta = \zeta^{**} = 0.54$ (middle). Same for X(5)-β^4, but with $N_B = 10$, $\chi = -1.32$, and $\zeta = 0.55$ (bottom). The thicknesses of the arrows are proportional to the respective $B(E2)$ values, which are also labelled by their values.

well. Both the intra- and inter-band $B(E2)$ strengths are consistently lower in the IBA compared with the predictions of the geometrical models. The decays from the 2^+ and 4^+ members of the $\mathrm{K} = 0_2^+$ sequence in each of the geometrical models exhibit a pattern where spin-ascending branches are enhanced and spin-descending branches are highly suppressed. While the spin-descending branches are consistently highly suppressed in the IBA predictions, the spin-ascending branches are not as strong as given in the geometrical models. Varying χ away from -1.32 can increase the strength of the spin-ascending branches, as will be seen in the discussion of Section 4.

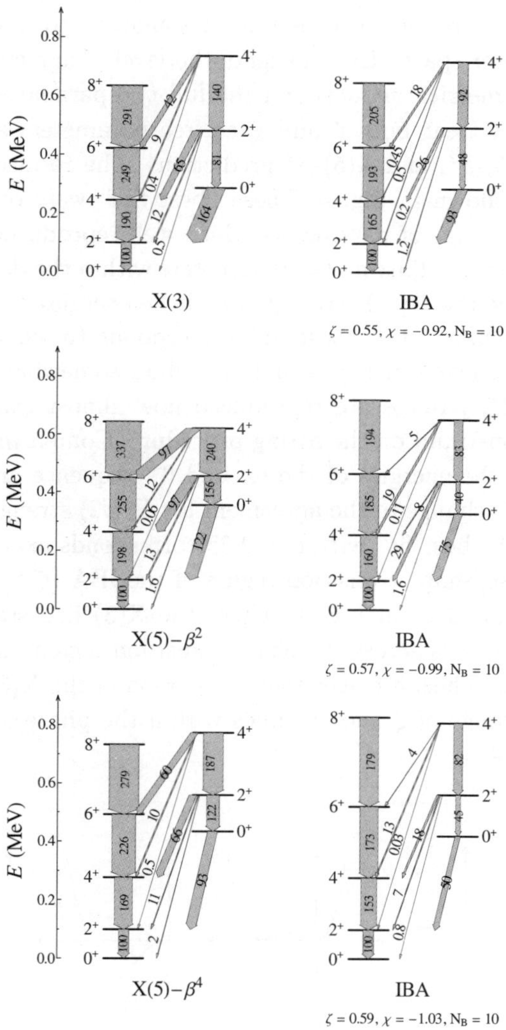

Fig. 5. IBA "best fits" to the X(3), X(5)-β^2, and X(5)-β^4 predictions, produced by treating ζ and χ as free parameters. The resulting parameters are included in the figure. The thicknesses of the arrows are proportional to the respective $B(E2)$ values, which are also labelled by their values.

Till now, our focus has been on the correspondence between the predictions of the X(3), X(5)-β^2, and X(5)-β^4 models and the IBA predictions within the phase/shape coexistence region and near the bottom leg of the triangle. While indeed we find reasonable agreement between the geometrical models and the IBA predictions within the phase/shape coexistence region, there is no reason to constrain the analysis to a single region of the IBA parameter space. To investigate the level of agreement obtainable between the geometrical models and the full two-parameter space of IBA for $N_B = 10$, we treat both ζ and χ as free parameters and attempt to fit the X(3), X(5)-β^2, and X(5)-β^4 predictions. The resulting spectra and parameters are shown in Fig. 5. These "best fits" were obtained by first determining the range of parameters which well reproduced the $R_{4/2}$ and $R_{0/2}$ ratios and then adjusting the parameters within the determined range to best reproduce the $B(E2)$ strengths and the energies of the higher spin states. For each model the "best fit" corresponds to values of ζ slightly larger than those given in Fig. 4 and $|\chi|$ values somewhat less than 1.32. The $R_{4/2}$ and $R_{0/2}$ ratios are reproduced now almost exactly, since this was an initial constraint on the fitting procedure. Comparing to the results given in Fig. 4, the energies of the excited 0^+ sequence are better reproduced, with little change to the agreement for $B(E2)$ strengths. In the case of X(3), the IBA "best fit" with $\zeta = 0.55$ corresponds exactly to the right edge of the phase/shape transition region of the IBA (ζ^{**}) for $\chi = -0.92$. This result is hinted at in Fig. 1, where the X(3) line starts at ζ_{crit} for $\chi = -1.32$ then moves across the phase transition region, intersecting with ζ^{**} for $\chi \sim -0.9$. Thus, a reasonable description of the X(3) model can be obtained for a range of ζ and χ values within the phase/shape transition region of the IBA.

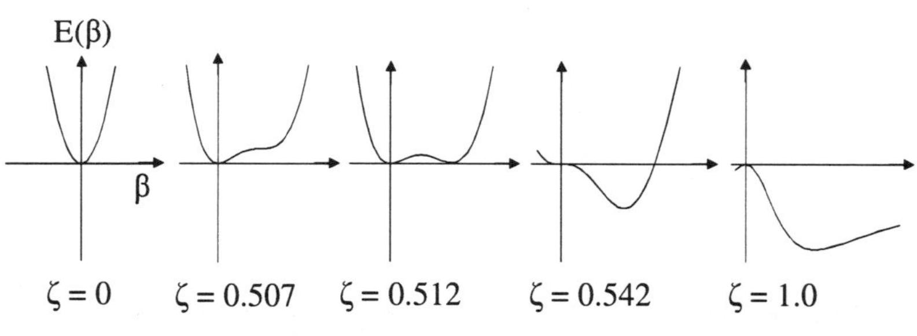

Fig. 6. Evolution with ζ of IBA total energy curves for $N_B = 10$ and $\chi = -1.32$.

Since we have found that the X(3), X(5)-β^2, and X(5)-β^4 predictions are best reproduced by IBA Hamiltonians with $\chi = -1.32$, or close to it, it is instructive to study [21] the evolution with increasing ζ of the IBA total energy curves for $N_B = 10$ and $\chi = -1.32$, shown in Fig. 6. It is clear that at $\zeta = \zeta_{\text{crit}}$, where the two minima are equal, the total energy curve can be quite well approximated by an infinite square well, which is the potential used in X(3), best reproduced with $\zeta = \zeta_{\text{crit}}$. In contrast, at $\zeta = \zeta^{**}$ a deeper minimum at positive β starts to develop. The total energy curves at and beyond ζ^{**} are quite similar to the β^2 and β^4 potentials, when the latter are supplemented by a $L(L+1)/(3\beta^2)$ centrifugal term [8], found recently through the use of novel techniques allowing for the exact numerical diagonalization of the Bohr Hamiltonian [27–29].

4. Comparison of X(3), X(5)-β^2, and X(5)-β^4 predictions to experiment

Several nuclei in the rare-earth region with $N = 90$ have been identified [5,6] as candidates for the X(5) critical point model. Therefore one obvious region to look for candidates for the X(3), X(5)-β^2, and X(5)-β^4 models is in the neighbors to these nuclei. In addition, within the framework of the IBA, detailed fits [30,31] to Os and Pt isotopes have identified nuclei which lie close to the shape/phase transition region of the IBA. Candidates can be identified by considering the trajectories of different isotopic chains in the IBA symmetry triangle and the lines corresponding to the X(3), X(5)-β^2, and X(5)-β^4 models, seen in Fig. 1, and in addition, the best agreement for $R_{0/2}$. We identify ^{186}Pt and ^{172}Os as candidates for the X(3) model, ^{146}Ce and ^{174}Os as candidates for the X(5)-β^2 model, and ^{158}Er and ^{176}Os as candidates for the X(5)-β^4 model. The experimental level schemes of these nuclei are compared to the relevant geometrical model predictions as well as IBA calculations in Figs. 7-9.

Considering that these nuclei were essentially chosen on the basis of only their $R_{4/2}$ ratio and $R_{0/2}$ ratio, the level of agreement for the other experimental observables is quite impressive. Overall the spacings in the $K = 0_2^+$ excited sequence are well reproduced by both the geometrical models and the IBA calculations. The one exception is in ^{172}Os, where experimentally the first two levels of the $K = 0_2^+$ band are lying too close, resulting in $R_{4/2} \sim 7$, which cannot be correct, indicating that the experimental information on these levels should be reconsidered. The X(3) model shows excellent agreement with the yrast band $B(E2)$ values in ^{186}Pt and ^{172}Os, whereas the IBA significantly underpredicts them. Identical results are found for the yrast band $B(E2)$ strengths in ^{158}Er (compared with the X(5)-β^4) model). In ^{174}Os, the yrast $B(E2)$ strengths are overestimated by the X(5)-β^2 model, while the IBA calculations provide a reasonable descrip-

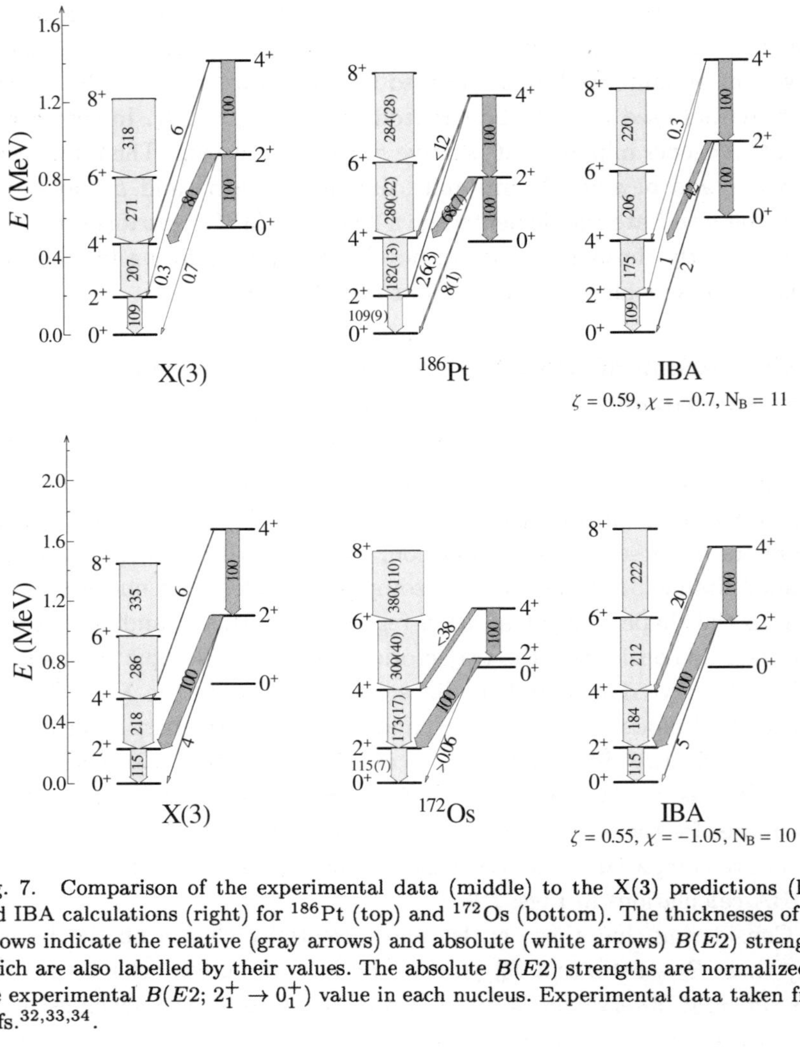

Fig. 7. Comparison of the experimental data (middle) to the X(3) predictions (left) and IBA calculations (right) for ^{186}Pt (top) and ^{172}Os (bottom). The thicknesses of the arrows indicate the relative (gray arrows) and absolute (white arrows) $B(E2)$ strengths which are also labelled by their values. The absolute $B(E2)$ strengths are normalized to the experimental $B(E2;\, 2_1^+ \to 0_1^+)$ value in each nucleus. Experimental data taken from Refs.[32,33,34].

tion. The branching ratios from excited states in the $\mathrm{K} = 0_2^+$ sequence are also well reproduced by the geometrical models. Overall, each of these candidate nuclei present an enhanced decay to the spin-ascending branch and a suppression to the spin-descending branch, in agreement with the predictions of X(3), X(5)-β^2, and X(5)-β^4. The IBA calculations also generally follow this pattern, with the exception of the predictions for ^{174,176}Os.

Given the narrowness of the shape/phase transition region of the IBA, it is not surprising that good experimental examples of X(3), X(5)-β^2, and

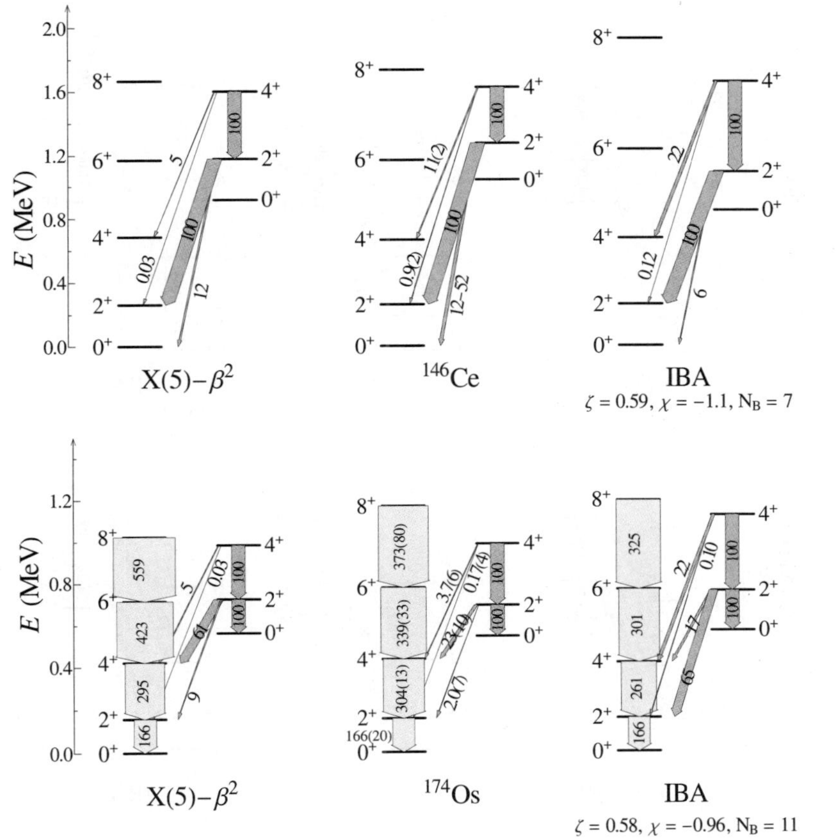

Fig. 8. Same as Fig. 7, but for comparison of the experimental data (middle) to the X(5)-β^2 predictions (left) and IBA calculations (right) for ^{146}Ce (top) and ^{174}Os (bottom). Experimental data taken from Refs.[35,36].

X(5)-β^4 are provided by neighboring even-even nuclei (^{172}Os–^{174}Os–^{176}Os). The IBA total energy curves for these nuclei, obtained from Eq. (3) and the parameters given in the captions of Figs. 7-9, are shown in Fig. 10. With increasing neutron number, the total energy curves evolve from a shallow deformed minimum in ^{172}Os to more pronounced single deformed minima in ^{174,176}Os. Qualitatively speaking, the evolution of these potentials resembles the evolution of the potentials one would obtain in moving from X(3) to X(5)-β^2, to X(5)-β^4, namely a flat bottomed potential in X(3) followed by a potential where the single deformed minimum becomes larger, as in X(5)-β^2 and X(5)-β^4. More specifically, the slight preference for a deformed

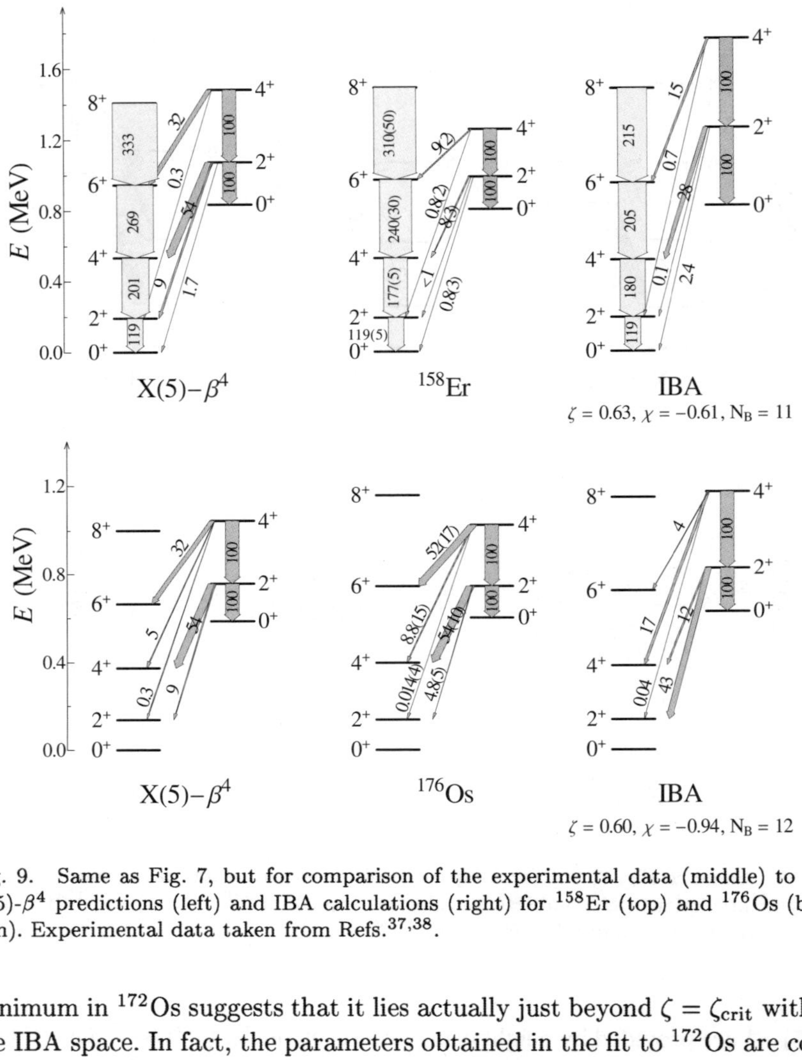

Fig. 9. Same as Fig. 7, but for comparison of the experimental data (middle) to the X(5)-β^4 predictions (left) and IBA calculations (right) for ^{158}Er (top) and ^{176}Os (bottom). Experimental data taken from Refs.[37,38].

minimum in ^{172}Os suggests that it lies actually just beyond $\zeta = \zeta_{\rm crit}$ within the IBA space. In fact, the parameters obtained in the fit to ^{172}Os are consistent with the parameters obtained for the "best fit" of the X(3) solution corresponding to ζ^{**}.

5. Discussion

In the present work the parameter independent (up to overal scale factors) predictions of the X(5)-β^2, X(5)-β^4, and X(3) models, which are variants of the X(5) critical point symmetry developed within the framework of the geometric collective model, are compared to the results of two-parameter

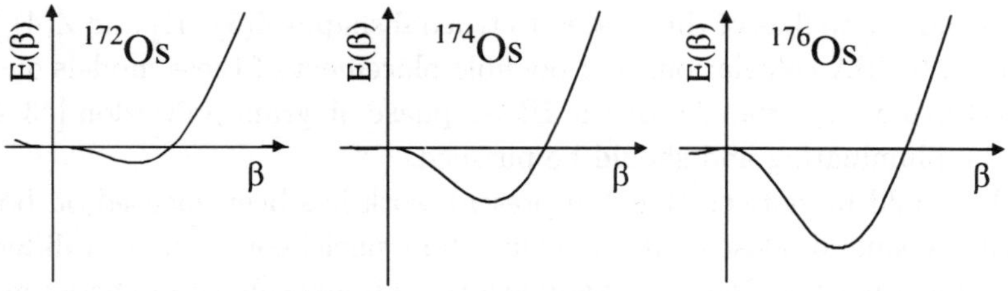

Fig. 10. IBA total energy curves for ^{172}Os (left), ^{174}Os (middle), and ^{176}Os (right) obtained from Eq. (3) using the parameter sets quoted in the captions of Figs. 7-9.

interacting boson approximation (IBA) model calculations, with the aim of establishing a connection between these two approaches. The study is focused on boson numbers of physical interest (around 10). It turns out that both X(3) and X(5)-β^2 lie close to the U(5)–SU(3) leg of the IBA symmetry triangle and within the narrow shape/phase transition region of the IBA. In particular, for $\chi = -1.32$, X(3) lies close to ζ_{crit}, the left border of the shaded shape/phase transition region of the IBA, corresponding to IBA total energy curves with two equal minima, while X(5)-β^2 lies near the right border of the shape/phase transition region, ζ^{**}, corresponding to IBA total energy curves with a single deformed minimum. A set of neighboring even-even nuclei exhibiting the X(3), X(5)-β^2, and X(5)-β^4 behaviors have been identified (^{172}Os-^{174}Os-^{176}Os). Additional examples for X(3), X(5)-β^2, and X(5)-β^4 are found in ^{186}Pt, ^{146}Ce, and ^{158}Er, respectively. The level of agreement of these parameter free, geometrical models with these candidate nuclei is found to be similar to the predictions of the two-parameter IBA calculations.

It is intriguing that the X(3) model, which corresponds to an *exactly separable* γ-rigid (with $\gamma = 0$) solution of the Bohr collective Hamiltonian, is found to be related to the IBA results at ζ_{crit}, which corresponds to the critical case of two degenerate minima in the IBA total energy curve, approximated by an infinite square well potential in the model. It is also remarkable that the X(5)-β^2 model, which uses *the same approximate separation of variables* as the X(5) critical point symmetry, is found to correspond to the right border (ζ^{**}) of the shape/phase transition region, related to the onset of total energy curves with a single deformed minimum, comparable in shape with the β^2 potential used in the model in the presence of a $L(L+1)/(3\beta^2)$ centrifugal term [8].

Comparisons in the same spirit of the parameter independent predictions of the E(5) critical point symmetry [1] and related E(5)-β^{2n} models [39,40], as well as of the related to triaxial shapes Z(5) [41] and Z(4) [42] models, to IBA calculations and possible placement of these models on the IBA-1 symmetry triangle (or the IBA-2 phase diagram polyerdon [43–45]) can be illuminating and should be pursued.

It should be noticed that the present work has been focused on boson numbers equal or close to 10, to which many nuclei correspond. A different but interesting question is to examine if there is any connection between the X(3), X(5)-β^2, and/or X(5)-β^4 models and the IBA for large boson numbers. This is particularly interesting especially since it has been established (initially for $N = 1,000$ [39], recently corroborated for $N = 10,000$ [46]) that the IBA critical point of the U(5)-O(6) transition for large N corresponds to the E(5)-β^4 model, i.e. to the E(5) model employing a β^4 potential in the place of the infinite well potential.

References

1. F. Iachello, Phys. Rev. Lett. **85**, 3580 (2000).
2. F. Iachello, Phys. Rev. Lett. **87**, 052502 (2001).
3. R. F. Casten and N. V. Zamfir, Phys. Rev. Lett. **85**, 3584 (2000).
4. R. M. Clark et al., Phys. Rev. **C69**, 064322 (2004).
5. R. F. Casten and N. V. Zamfir, Phys. Rev. Lett. **87**, 052503 (2001).
6. R. M. Clark et al., Phys. Rev. **C68**, 037301 (2003).
7. M.A. Caprio, Phys. Rev. **C69**, 044307 (2004).
8. M.A. Caprio, Phys. Rev. **C72**, 054323 (2005).
9. N. Pietralla and O.M. Gorbachenko, Phys. Rev. **C70**, 011304(R) (2004).
10. D. Bonatsos, D. Lenis, N. Minkov, P. P. Raychev, and P. A. Terziev, Phys. Rev. **C69**, 014302 (2004).
11. D. Bonatsos, D. Lenis, D. Petrellis, P. A. Terziev, and I. Yigitoglu, Phys. Lett. **B632**, 238 (2006).
12. D. H. Feng, R. Gilmore, and S. R. Deans, Phys. Rev. **C23**, 1254 (1981).
13. F. Iachello and A. Arima, The Interacting Boson Model, Cambridge University Press, Cambridge, 1987.
14. J. N. Ginocchio and M. W. Kirson, Phys. Rev. Lett. **44**, 1744 (1980).
15. A. E. L. Dieperink, O. Scholten, and F. Iachello, Phys. Rev. Lett. **44**, 1747 (1980).
16. R. F. Casten, Nuclear Structure from a Simple Perspective, Oxford University Press, Oxford, 1990.
17. F. Iachello, N. V. Zamfir, and R. F. Casten, Phys. Rev. Lett. **81**, 1191 (1998).
18. N. V. Zamfir, P. von Brentano, R. F. Casten, and J. Jolie, Phys. Rev. **C66**, 021304 (2002).
19. F. Iachello and N. V. Zamfir, Phys. Rev. Lett. **92**, 212501 (2004).
20. V. Werner, P. von Brentano, R. F. Casten, and J. Jolie, Phys. Lett. **B527**, 55 (2002).
21. N. V. Zamfir and G. E. Fernandes, in Proceedings of the Eleventh Interna-

tional Symposium on Capture Gamma Ray Spectroscopy and Related Topics (Prohonice, 2002), edited by J. Kvasil, P. Cejnar, and M. Krticka, World Scientific, Singapore, 2003.
22. J. Y. Zhang, R. F. Casten, and N. V. Zamfir, Phys. Lett. **B407**, 201 (1997).
23. E. A. McCutchan, N. V. Zamfir, and R. F. Casten, Phys. Rev. **C71**, 034309 (2005).
24. E. A. McCutchan, N. V. Zamfir, and R. F. Casten, Phys. Rev. **C69**, 064306 (2004).
25. A. Bohr, Mat. Fys. Medd. K. Dan. Vidensk. Selsk. **26**, no. 14 (1952).
26. E. López-Moreno and O. Castaños, Phys. Rev. **C54**, 2374 (1996).
27. D. J. Rowe, Nucl. Phys. **A735**, 372 (2004).
28. D. J. Rowe, P. S. Turner, and J. Repka, J. Math. Phys. **45**, 2761 (2004).
29. D. J. Rowe and P. S. Turner, Nucl. Phys. **A753**, 94 (2005).
30. E. A. McCutchan and N. V. Zamfir, Phys. Rev. **C71**, 054306 (2005).
31. E. A. McCutchan, R. F. Casten, and N. V. Zamfir, Phys. Rev. **C71**, 061301(R) (2005).
32. C. M. Baglin, Nucl. Data Sheets **99**, 1 (2003).
33. J. C. Walpe, PhD. Thesis, University of Notre Dame, 1999.
34. B. Singh, Nucl. Data Sheets **75**, 199 (1995).
35. L. K. Peker and J. K. Tuli, Nucl. Data Sheets **82**, 187 (1997).
36. E. Browne and Huo Junde, Nucl. Data Sheets **87**, 15 (1999).
37. R. G. Helmer, Nucl. Data Sheets **101**, 325 (2004).
38. M.A. Basunia, Nucl. Data Sheets **107**, 791 (2006).
39. J. M. Arias, C. E. Alonso, A. Vitturi, J. E. García-Ramos, J. Dukelsky, and A. Frank, Phys. Rev. **C68**, 041302 (2003).
40. D. Bonatsos, D. Lenis, N. Minkov, P. P. Raychev, and P. A. Terziev, Phys. Rev. **C69**, 044316 (2004).
41. D. Bonatsos, D. Lenis, D. Petrellis, and P. A. Terziev, Phys. Lett. **B588**, 172 (2004).
42. D. Bonatsos, D. Lenis, D. Petrellis, P. A. Terziev, and I. Yigitoglu, Phys. Lett. **B621**, 102 (2005).
43. J. M. Arias, J. E. García-Ramos, and J. Dukelsky, Phys. Rev. Lett. **93**, 212501 (2004).
44. M. A. Caprio and F. Iachello, Phys. Rev. Lett. **93**, 242502 (2004).
45. M. A. Caprio and F. Iachello, Ann. Phys. (NY) **318**, 454 (2005).
46. J. E. García-Ramos, J. Dukelsky, and J. M. Arias, Phys. Rev. **C72**, 037301 (2005).

X(3): an exactly separable γ-rigid version of the X(5) critical point symmetry

DENNIS BONATSOS, D. LENIS, D. PETRELLIS

Institute of Nuclear Physics,
National Centre for Scientific Research "Demokritos",
GR-15310 Aghia Paraskevi, Attiki, Greece

P. A. TERZIEV

Institute for Nuclear Research and Nuclear Energy,
Bulgarian Academy of Sciences,
72 Tzarigrad Road, BG-1784 Sofia, Bulgaria

I. YIGITOGLU

Hasan Ali Yucel Faculty of Education, Istanbul University,
TR-34470 Beyazit, Istanbul, Turkey

A γ-rigid version (with $\gamma = 0$) of the X(5) critical point symmetry is constructed. The model, to be called X(3) since it is proved to contain three degrees of freedom, utilizes an infinite well potential, is based on exact separation of variables, and leads to parameter free (up to overall scale factors) predictions for spectra and $B(E2)$ transition rates, which are in good agreement with existing experimental data for ^{172}Os and ^{186}Pt. An unexpected similarity of the β_1-bands of the X(5) nuclei ^{150}Nd, ^{152}Sm, ^{154}Gd, and ^{156}Dy to the X(3) predictions is observed.

1. Introduction

Critical point symmetries [1,2], describing nuclei at points of shape phase transitions between different limiting symmetries, have recently attracted considerable attention, since they lead to parameter independent (up to overall scale factors) predictions which are found to be in good agreement with experiment [3–6]. The X(5) critical point symmetry [2], in particular, is supposed to correspond to the transition from vibrational [U(5)] to prolate axially symmetric [SU(3)] nuclei, materialized in the $N = 90$ isotones ^{150}Nd [7], ^{152}Sm [5], ^{154}Gd [8,9], and ^{156}Dy [9,10].

On the other hand, it is known that in the framework of the nuclear

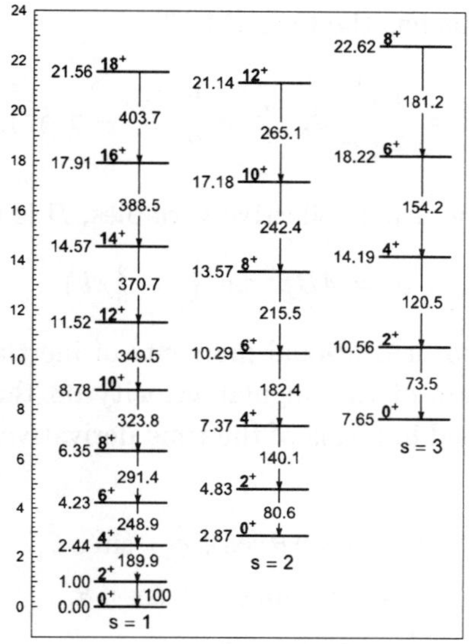

Fig. 1. Energy levels of the ground state ($s = 1$), β_1 ($s = 2$), and β_2 ($s = 3$) bands of X(3), normalized to the energy of the lowest excited state, 2_1^+, together with intra-band $B(E2)$ transition rates, normalized to the transition between the two lowest states, $B(E2; 2_1^+ \to 0_1^+)$. Interband transitions are listed in Table 1. See Section 3 for further discussion.

collective model [11], which involves the collective variables β and γ, interesting special cases occur by "freezing" the γ variable [12] to a constant value.

In the present work we constuct a version of the X(5) model in which the γ variable is "frozen" to $\gamma = 0$, instead of varying around the $\gamma = 0$ value within a harmonic oscillator potential, as in the X(5) case. It turns out that only three variables are involved in the present model, which is therefore called X(3). Exact separation of the β variable from the angles is possible. Experimental realizations of X(3) appear to occur in ^{172}Os and ^{186}Pt, while an unexpected agreement of the β_1-bands of the X(5) nuclei ^{150}Nd, ^{152}Sm, ^{154}Gd, and ^{156}Dy to the X(3) predictions is observed.

In Section 2 the X(3) model is constructed, while numerical results and comparisons to experiment are given in Section 3, and a discussion of the present results and plans for further work in Section 4.

2. The X(3) model

In the collective model of Bohr [11] the classical expression of the kinetic energy corresponding to β and γ vibrations of the nuclear surface plus rotation of the nucleus has the form [11,13]

$$T = \frac{1}{2}\sum_{k=1}^{3} J_k\, \omega_k'^2 + \frac{B}{2}\,(\dot\beta^2 + \beta^2\dot\gamma^2), \tag{1}$$

where β and γ are the usual collective variables, B is the mass parameter,

$$J_k = 4B\beta^2 \sin^2\left(\gamma - \tfrac{2}{3}\pi k\right) \tag{2}$$

are the three principal irrotational moments of inertia, and ω_k' ($k = 1$, 2, 3) are the components of the angular velocity on the body-fixed k-axes, which can be expressed in terms of the time derivatives of the Euler angles $\dot\phi, \dot\theta, \dot\psi$ [13,14]

$$\begin{aligned} \omega_1' &= -\sin\theta\cos\psi\,\dot\phi + \sin\psi\,\dot\theta, \\ \omega_2' &= \sin\theta\sin\psi\,\dot\phi + \cos\psi\,\dot\theta, \\ \omega_3' &= \cos\theta\,\dot\phi + \dot\psi. \end{aligned} \tag{3}$$

Assuming the nucleus to be γ-rigid (i.e. $\dot\gamma = 0$), as in the Davydov and Chaban approach [12], and considering in particular the axially symmetric prolate case of $\gamma = 0$, we see that the third irrotational moment of inertia J_3 vanishes, while the other two become equal $J_1 = J_2 = 3B\beta^2$, the kinetic energy of Eq. (1) reaching the form [13,15]

$$T = \frac{1}{2}3B\beta^2(\omega_1'^2 + \omega_2'^2) + \frac{B}{2}\,\dot\beta^2 = \frac{B}{2}\left[3\beta^2(\sin^2\theta\,\dot\phi^2 + \dot\theta^2) + \dot\beta^2\right]. \tag{4}$$

It is clear that in this case the motion is characterized by three degrees of freedom. Introducing the generalized coordinates $q_1 = \phi$, $q_2 = \theta$, and $q_3 = \beta$, the kinetic energy becomes a quadratic form of the time derivatives of the generalized coordinates [13,16]

$$T = \frac{B}{2}\sum_{i,j=1}^{3} g_{ij}\;\dot q_i \dot q_j, \tag{5}$$

with the matrix g_{ij} having a diagonal form

$$g_{ij} = \begin{pmatrix} 3\beta^2\sin^2\theta & 0 & 0 \\ 0 & 3\beta^2 & 0 \\ 0 & 0 & 1 \end{pmatrix}. \tag{6}$$

(In the case of the full Bohr Hamiltonian [11] the square matrix g_{ij} is 5-dimensional and non-diagonal [13,16].) Following the general procedure of quantization in curvilinear coordinates one obtains the Hamiltonian operator [13,16]

$$H = -\frac{\hbar^2}{2B}\Delta + U(\beta) = -\frac{\hbar^2}{2B}\left[\frac{1}{\beta^2}\frac{\partial}{\partial\beta}\beta^2\frac{\partial}{\partial\beta} + \frac{1}{3\beta^2}\Delta_\Omega\right] + U(\beta), \quad (7)$$

where Δ_Ω is the angular part of the Laplace operator

$$\Delta_\Omega = \frac{1}{\sin\theta}\frac{\partial}{\partial\theta}\sin\theta\frac{\partial}{\partial\theta} + \frac{1}{\sin^2\theta}\frac{\partial^2}{\partial\phi^2}. \quad (8)$$

The Schrödinger equation can be solved by the factorization

$$\Psi(\beta,\theta,\phi) = F(\beta)\, Y_{LM}(\theta,\phi), \quad (9)$$

where $Y_{LM}(\theta,\phi)$ are the spherical harmonics. Then the angular part leads to the equation

$$-\Delta_\Omega Y_{LM}(\theta,\phi) = L(L+1)Y_{LM}(\theta,\phi), \quad (10)$$

where L is the angular momentum quantum number, while for the radial part $F(\beta)$ one obtains

$$\left[\frac{1}{\beta^2}\frac{d}{d\beta}\beta^2\frac{d}{d\beta} - \frac{L(L+1)}{3\beta^2} + \frac{2B}{\hbar^2}\Big(E - U(\beta)\Big)\right]F(\beta) = 0. \quad (11)$$

As in the case of X(5) [2], the potential in β is taken to be an infinite square well

$$U(\beta) = \begin{cases} 0, & 0 \leq \beta \leq \beta_W \\ \infty, & \beta > \beta_W \end{cases}, \quad (12)$$

where β_W is the width of the well. In this case $F(\beta)$ is a solution of the equation

$$\left[\frac{d^2}{d\beta^2} + \frac{2}{\beta}\frac{d}{d\beta} + \left(k^2 - \frac{L(L+1)}{3\beta^2}\right)\right]F(\beta) = 0 \quad (13)$$

in the interval $0 \leq \beta \leq \beta_W$, where reduced energies $\varepsilon = k^2 = 2BE/\hbar^2$ [2] have been introduced, while it vanishes outside. Substituting $F(\beta) = \beta^{-1/2}f(\beta)$ one obtains the Bessel equation

$$\left[\frac{d^2}{d\beta^2} + \frac{1}{\beta}\frac{d}{d\beta} + \left(k^2 - \frac{\nu^2}{\beta^2}\right)\right]f(\beta) = 0, \quad (14)$$

where

$$\nu = \sqrt{\frac{L(L+1)}{3} + \frac{1}{4}}, \quad (15)$$

the boundary condition being $f(\beta_W) = 0$. The solution of (13), which is finite at $\beta = 0$, is then

$$F(\beta) = F_{sL}(\beta) = \frac{1}{\sqrt{c}}\,\beta^{-1/2} J_\nu(k_{s,\nu}\beta), \tag{16}$$

with $k_{s,\nu} = x_{s,\nu}/\beta_W$ and $\varepsilon_{s,\nu} = k^2_{s,\nu}$, where $x_{s,\nu}$ is the s-th zero of the Bessel function of the first kind $J_\nu(k_{s,\nu}\beta_W)$ and the normalization constant $c = \beta_W^2\, J^2_{\nu+1}(x_{s,\nu})/2$ is obtained from the condition $\int_0^{\beta_W} F^2_{sL}(\beta)\,\beta^2 d\beta = 1$. The corresponding spectrum is then

$$E_{s,L} = \frac{\hbar^2}{2B}\,k^2_{s,\nu} = \frac{\hbar^2}{2B\beta_W^2}\,x^2_{s,\nu}. \tag{17}$$

It should be noticed that in the X(5) case [2] the same Eq. (14) occurs, but with $\nu = \sqrt{\frac{L(L+1)}{3} + \frac{9}{4}}$, while in the E(3) Euclidean algebra in 3 dimensions, which is the semidirect sum of the T_3 algebra of translations in 3 dimensions and the SO(3) algebra of rotations in 3 dimensions [17], the eigenvalue equation of the square of the total momentum, which is a second-order Casimir operator of the algebra, also leads [17,18] to Eq. (14), but with $\nu = L + \frac{1}{2}$.

From the symmetry of the wave function of Eq. (9) with respect to the plane which is orthogonal to the symmetry axis of the nucleus and goes through its center, follows that the angular momentum L can take only even nonnegative values. Therefore no γ-bands appear in the model, as expected, since the γ degree of freedom has been frozen.

In the general case the quadrupole operator is

$$T^{(E2)}_\mu = t\,\beta\Big[D^{2\,*}_{\mu,0}(\Omega)\cos\gamma + \frac{1}{\sqrt{2}}[D^{2\,*}_{\mu,2}(\Omega) + D^{2\,*}_{\mu,-2}(\Omega)]\sin\gamma\Big], \tag{18}$$

where Ω denotes the Euler angles and t is a scale factor. For $\gamma = 0$ the quadrupole operator becomes

$$T^{(E2)}_\mu = t\,\beta\sqrt{\frac{4\pi}{5}}\,Y_{2\mu}(\theta,\phi). \tag{19}$$

$B(E2)$ transition rates

$$B(E2; sL \to s'L') = \frac{1}{2L+1}\left|\langle s'L'||T^{(E2)}||sL\rangle\right|^2 \tag{20}$$

are calculated using the wave functions of Eq. (9) and the volume element $d\tau = \beta^2 \sin\theta\, d\beta d\theta d\phi$, the final result being

$$B(E2; sL \to s'L') = t^2\left(C^{L'\,0}_{L\,0,\,2\,0}\right)^2 I^2_{sL;s'L'}, \tag{21}$$

Table 1. Interband $B(E2; L_i \to L_f)$ transition rates for the X(3) model, normalized to the one between the two lowest states, $B(E2; 2_1^+ \to 0_1^+)$.

$L_i \to L_f$	X(3)	$L_i \to L_f$	X(3)	$L_i \to L_f$	X(3)
$0_2 \to 2_1$	164.0				
$2_2 \to 4_1$	64.5	$2_2 \to 2_1$	12.4	$2_2 \to 0_1$	0.54
$4_2 \to 6_1$	42.2	$4_2 \to 4_1$	8.6	$4_2 \to 2_1$	0.43
$6_2 \to 8_1$	31.1	$6_2 \to 6_1$	6.7	$6_2 \to 4_1$	0.51
$8_2 \to 10_1$	24.4	$8_2 \to 8_1$	5.5	$8_2 \to 6_1$	0.56
$10_2 \to 12_1$	19.9	$10_2 \to 10_1$	4.7	$10_2 \to 8_1$	0.59
$12_2 \to 14_1$	16.6	$12_2 \to 12_1$	4.0	$12_2 \to 10_1$	0.60
$14_2 \to 16_1$	14.2	$14_2 \to 14_1$	3.5	$14_2 \to 12_1$	0.60
$16_2 \to 18_1$	12.3	$16_2 \to 16_1$	3.1	$16_2 \to 14_1$	0.60
$18_2 \to 20_1$	10.9	$18_2 \to 18_1$	2.8	$18_2 \to 16_1$	0.59
$20_2 \to 22_1$	9.7	$20_2 \to 20_1$	2.5	$20_2 \to 18_1$	0.58
$0_3 \to 2_2$	209.1				
$2_3 \to 4_2$	92.0	$2_3 \to 2_2$	16.2	$2_3 \to 0_2$	0.67
$4_3 \to 6_2$	65.3	$4_3 \to 4_2$	12.2	$4_3 \to 2_2$	0.47
$6_3 \to 8_2$	50.9	$6_3 \to 6_2$	10.1	$6_3 \to 4_2$	0.52
$8_3 \to 10_2$	41.6	$8_3 \to 8_2$	8.6	$8_3 \to 6_2$	0.57
$10_3 \to 12_2$	35.0	$10_3 \to 10_2$	7.5	$10_3 \to 8_2$	0.61
$12_3 \to 14_2$	30.1	$12_3 \to 12_2$	6.6	$12_3 \to 10_2$	0.63
$14_3 \to 16_2$	26.3	$14_3 \to 14_2$	5.9	$14_3 \to 12_2$	0.65
$16_3 \to 18_2$	23.3	$16_3 \to 16_2$	5.4	$16_3 \to 14_2$	0.66
$18_3 \to 20_2$	20.8	$18_3 \to 18_2$	4.9	$18_3 \to 16_2$	0.66
$20_3 \to 22_2$	18.8	$20_3 \to 20_2$	4.5	$20_3 \to 18_2$	0.66

where $C^{L'0}_{L0,20}$ are Clebsch–Gordan coefficients and the integrals over β are

$$I_{sL;s'L'} = \int_0^{\beta_W} \beta\, F_{sL}(\beta)\, F_{s'L'}(\beta)\, \beta^2\, d\beta. \tag{22}$$

The following remarks are now in place.

1) In both the X(3) and X(5) [2] models, $\gamma = 0$ is considered, the difference being that in the former case γ is treated as a parameter, while in the latter as a variable. As a consequence, separation of variables in X(3) is exact (because of the lack of the γ variable), while in X(5) it is approximate.

2) In both the X(3) and E(5) [1] models a potential depending only on β is considered and exact separation of variables is achieved, the difference being that in the E(5) model the γ variable remains active, while in the X(3) case it is frozen. As a consequence, in the E(5) case the equation involving the angles results in the solutions given by Bès [19], while in the X(3) case the usual spherical harmonics occur.

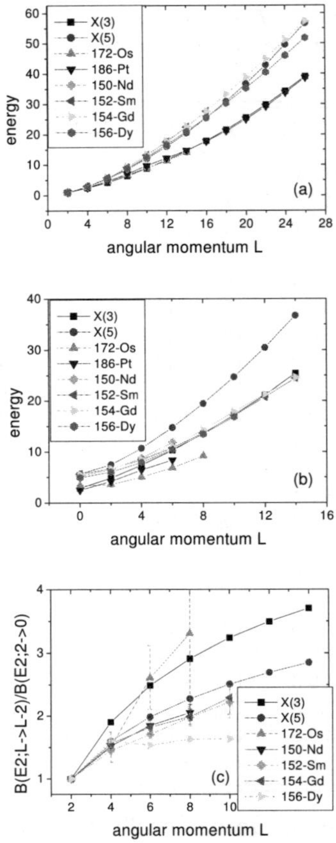

Fig. 2. (a) Energy levels of the ground state bands of the X(3) and X(5) [2] models, compared to experimental data for ^{172}Os [20], ^{186}Pt [21], ^{150}Nd [22], ^{152}Sm [23], ^{154}Gd [24], and ^{156}Dy [25]. The levels of each band are normalized to the 2_1^+ state. (b) Same for the β_1-bands, also normalized to the 2_1^+ state. (c) Same for existing intraband $B(E2)$ transition rates within the ground state band, normalized to the $B(E2; 2_1^+ \to 0_1^+)$ rate. The data for ^{156}Dy are taken from Ref. [9]. See Section 3 for further discussion.

3. Numerical results and comparison to experiment

The energy levels of the ground state band ($s = 1$), as well as of the β_1 ($s = 2$) and β_2 ($s = 3$) bands, normalized to the energy of the lowest excited state, 2_1^+, are shown in Fig. 1, together with intraband $B(E2)$ transition rates, normalized to the transition between the two lowest states, $B(E2; 2_1^+ \to 0_1^+)$, while interband transitions are listed in Table 1.

The energy levels of the ground state band of X(3) are also shown in

Table 2. Relative $B(E2)$ branching ratios for the X(3) model compared to existing experimental data [27] for ^{186}Pt.

$L_i \to L_f$	exp.	X(3)	$L_i \to L_f$	exp.	X(3)
$2_2 \to 0_2$	100	100	$4_2 \to 2_2$	100	100
$2_2 \to 0_1$	8(1)	0.7	$4_2 \to 2_1$	2.6(3)	0.3
$2_2 \to 4_1$	68(7)	80	$4_2 \to 4_1$	< 12	6

Fig. 2(a), where they are compared to the experimental data for ^{172}Os [20] (up to the point of bandcrossing) and ^{186}Pt [21]. In the same figure the ground state band of X(5), along with the experimental data for the $N = 90$ isotones ^{150}Nd [22], ^{152}Sm [23], ^{154}Gd [24], and ^{156}Dy [25], which are considered as the best realizations of X(5) [5,7–10], are shown for comparison. The energy levels of the β_1-band for the same models and nuclei are shown in Fig. 2(b), while existing intraband $B(E2)$ transition rates for the ground state band are shown in Fig. 2(c). The following comments are now in place.

1) The ground state bands of ^{172}Os and ^{186}Pt are in very good agreement with the X(3) predictions, while the β_1-bands are a little lower. Similarly, the ground state bands of ^{150}Nd, ^{152}Sm, ^{154}Gd, and ^{156}Dy are in very good agreement with the X(5) predictions, while the β_1 bands beyond $L = 4$ are much lower. This discrepancy is known to be fixed by considering [26] a potential with linear sloped walls instead of an infinite well potential. What occured rather unexpectedly is the fact that the β_1 bands of the $N = 90$ isotones [the best experimental examples of X(5)] from $L = 4$ upwards agree very well with the X(3) predictions. This could be interpreted as indication that the bandhead of the β_1 band is influenced by the presence of the γ degree if freedom, but the excited levels of this band beyond $L = 4$ are not influenced by it. Detailed measurements of intraband $B(E2)$ transition rates within the β_1-bands of these $N = 90$ isotones could clarify this point.

2) Existing intraband $B(E2)$ transition rates for the ground state band of ^{172}Os (below the region influenced by the bandcrossing) are in good agreement with X(3), being quite higher than the ^{150}Nd, ^{152}Sm, and ^{154}Gd rates, as they should. [The $B(E2)$ rates of ^{156}Dy are known [9] to be in less good agreement with X(5), as also seen in Fig. 2(c).] However, more intraband and interband transitions (and with smaller error bars) are needed before final conclusions could be drawn. The same holds for ^{186}Pt, for which experimental information on B(E2)s is missing [21,27]. The relative branching ratios known in ^{186}Pt [27] are given in Table 2, being in good agreement with the X(3) predictions.

The placement of the above mentioned nuclei in the symmetry triangle

[28] of the Interacting Boson Model (IBM) [29] can be illuminating. All of the above mentioned N=90 isotones lie close to the phase coexistence and shape phase transition region of the IBM, with ^{152}Sm being located on the U(5)-SU(3) side of the triangle [30], while ^{154}Gd and ^{156}Dy gradually move towards the center of the triangle [31]. ^{172}Os [32] and ^{186}Pt [27] also appear near the center of the symmetry triangle and close to the transition region of the IBM.

It should be noticed that the critical character of ^{186}Pt is also supported by the criteria posed in Ref. [33]. In particular, a relatively abrupt change of the $R_4 = E(4_1^+)/E(2_1^+)$ ratio occurs between ^{186}Pt and ^{184}Pt, as seen in the systematics presented in Ref. [32], while 0_2^+ shows a minimum at ^{186}Pt, as seen in the systematics presented in Ref. [27], especially if the 0_2^+ energies are normalized with respect to the 2_1^+ state of each Pt isotope. Furthermore, ^{186}Pt is located at the point where the crossover of 0_2^+ and 2_γ^+ occurs, as seen in the systematics presented in Ref. [27].

4. Discussion

In summary, a γ-rigid (with $\gamma = 0$) version of the X(5) model is constructed. The model is called X(3), since it is proved that only three variables occur in this case, the separation of variables being exact, while in the X(5) case approximate separation of the five variables occuring there is performed. The parameter free (up to overall scale factors) predictions of X(3) are found to be in good agreement with existing experimental data of ^{172}Os and ^{186}Pt, while a rather unexpected agreement of the β_1-bands of the X(5) nuclei ^{150}Nd, ^{152}Sm, ^{154}Gd, and ^{156}Dy to the X(3) predictions is observed. The need for further $B(E2)$ measurements in all of the above-mentioned nuclei is emphasized.

Acknowledgements

One of the authors (IY) is thankful to the Turkish Atomic Energy Authority (TAEK) for support under project number 04K120100-4.

References

1. F. Iachello, Phys. Rev. Lett. **85**, 3580 (2000).
2. F. Iachello, Phys. Rev. Lett. **87**, 052502 (2001).
3. R. F. Casten and N. V. Zamfir, Phys. Rev. Lett. **85**, 3584 (2000).
4. R. M. Clark et al., Phys. Rev. C **69**, 064322 (2004).
5. R. F. Casten and N. V. Zamfir, Phys. Rev. Lett. **87**, 052503 (2001).

6. R. M. Clark et al., Phys. Rev. C **68**, 037301 (2003).
7. R. Krücken et al., Phys. Rev. Lett. **88**, 232501 (2002).
8. D. Tonev et al., Phys. Rev. C **69**, 034334 (2004).
9. A. Dewald et al., Eur. Phys. J. A **20**, 173 (2004).
10. M. A. Caprio et al., Phys. Rev. C **66**, 054310 (2002).
11. A. Bohr, Mat. Fys. Medd. K. Dan. Vidensk. Selsk. **26**, no. 14 (1952).
12. A. S. Davydov and A. A. Chaban, Nucl. Phys. **20**, 499 (1960).
13. A. G. Sitenko and V. K. Tartakovskii, Lectures on the Theory of the Nucleus, Atomizdat, Moscow, 1972 (in Russian).
14. R. N. Zare, Angular Momentum, Wiley, New York, 1988.
15. A. S. Davydov, Theory of the Atomic Nucleus, Fizmatgiz, Moscow, 1958.
16. J. M. Eisenberg and W. Greiner, Nuclear Theory, Vol. I: Nuclear Models, North-Holland, Amsterdam, 1970.
17. A. O. Barut and R. Raczka, Theory of Group Representations and Applications, World Scientific, Singapore, 1986.
18. D. Bonatsos, D. Lenis, N. Minkov, P. P. Raychev, and P. A. Terziev, Phys. Rev. C **69**, 044316 (2004).
19. D. R. Bès, Nucl. Phys. **10**, 373 (1959).
20. B. Singh, Nucl. Data Sheets **75**, 199 (1995).
21. C. M. Baglin, Nucl. Data Sheets **99**, 1 (2003).
22. E. der Mateosian and J. K. Tuli, Nucl. Data Sheets **75**, 827 (1995).
23. A. Artna-Cohen, Nucl. Data Sheets **79**, 1 (1996).
24. C. W. Reich and R. G. Helmer, Nucl. Data Sheets **85**, 171 (1998).
25. C. W. Reich, Nucl. Data Sheets **99**, 753 (2003).
26. M. A. Caprio, Phys. Rev. C **69**, 044307 (2004).
27. E. A. McCutchan, R. F. Casten, and N. V. Zamfir, Phys. Rev. C **71**, 061301 (2005).
28. R. F. Casten, Nuclear Structure from a Simple Perspective, Oxford University Press, Oxford, 1990.
29. F. Iachello and A. Arima, The Interacting Boson Model, Cambridge University Press, Cambridge, 1987.
30. N. V. Zamfir, E. A. McCutchan, and R. F. Casten, Yad. Fiz. **67**, 1856 (2004) [Phys. At. Nucl. **67**, 1829 (2004)].
31. E. A. McCutchan, N. V. Zamfir, and R. F. Casten, Phys. Rev. C **69**, 063406 (2004).
32. E. A. McCutchan and N. V. Zamfir, Phys. Rev. C **71**, 054306 (2005).
33. N. V. Zamfir, E. A. McCutchan, and R. F. Casten, in Nuclear Physics, Large and Small: International Conference on Microscopic Studies of Collective Phenomena, edited by R. Bijker, R. F. Casten, and A. Frank, AIP Conf. Proc. **726**, 187 (2004).

Chaotic Behavior of Nuclear Systems

J. M. G. Gómez, L Muñoz, and J. Retamosa
Departamento de Física Atómica, Molecular y Nuclear, Universidad Complutense de Madrid, E-28040 Madrid, Spain

R. A. Molina and A. Relaño
Instituto de Estructura de la Materia, CSIC, E-28006 Madrid, Spain

E. Faleiro
Departamento de Física Aplicada, E. U. I. T. Industrial, Universidad Politécnica de Madrid, E-28012 Madrid, Spain

A survey of chaotic dynamics in atomic nuclei is presented, using on the one hand standard statistics of quantum chaos studies, and on the other a new approach based on time series analysis methods. First we emphasize the energy and isospin dependence of nuclear chaoticity, based on shell-model energy spectra fluctuations in Ca, Sc and Ti isotopes, which are analyzed using standard statistics such as the nearest level spacing distribution and the Dyson-Mehta Δ_3 statistic. For all the Ca isotopes, in the ground state region the energy level fluctuations show strong deviations from GOE predictions. When one or two neutrons are replaced by protons, Sc is closer to GOE and Ti is even more chaotic. Thus we find a clear isospin dependence in the degree of nuclear chaoticity. Afterward, we discuss chaos in nuclei using a new approach based on the analogy between the sequence of energy levels and a discrete time series. Considering the energy spectrum fluctuations of quantum systems as a discrete time series, we suggest the following conjecture: *The energy spectra of chaotic quantum systems are characterized by* $1/f$ *noise.* Moreover, we show that the spectra of integrable quantum systems exhibit $1/f^2$ noise. Although an exact proof of this conjecture is not available yet, we use random matrix theory to derive theoretical expressions that reproduce to a good approximation the power laws of type $1/f$ and $1/f^2$ characteristic of chaotic and integrable systems, respectively. We note that $1/f$ noise is a very ubiquitous property, since many complex systems in nature and in human society exhibit the same kind of fluctuations as chaotic quantum systems.

1. Introduction

Quantum chaos has made important progress during the two last decades. As far as stationary systems are concerned, the amount of compiled results shows a clear relationship between the fluctuations properties of the energy levels of a system and the large time scale behavior of its classical analog. The pioneering work of Berry, Bohigas and others [1,2] lead to an important and concise statement: the spectral properties of simple systems known to be ergodic in the classical limit, follow very closely those of the classical ensembles of random matrices. In this case the distribution of the nearest neighbor spacings is the Wigner distribution [3,4]. It corresponds to the existence of linear repulsion between adjacent levels. On the contrary, integrable systems lead to level fluctuations that are well described by the Poisson distribution, i.e., levels behave as if they were uncorrelated.

The history of random matrix theory (RMT) started in the 1950's, almost twenty years before the birth of quantum chaos. An interesting description of the origin of RMT has recently been presented by O. Bohigas [5]. After the end of World War II, military as well as civilian purposes lead to an impressive effort in connection with nuclear fission. For instance, it was important to understand the properties of large sequences of compound nucleus resonances observed in slow neutron scattering by heavy nuclei. Wigner studied the nearest neighbor spacing distribution $P(s)$ of these high level density sequences, where s is the spacing between two consecutive levels. He assumed that the Hamiltonian that governs the behavior of a complicated system is a matrix with random matrix elements and no particular properties, except for its symmetric nature. Assuming that this distribution was universal and independent of the details of the Hamiltonian of the system, he proposed an analytical expression for the $P(s)$ distribution, known as the Wigner surmise,

$$P(s) = \frac{\pi}{2} s \exp\left(-\frac{\pi}{4} s^2\right). \tag{1}$$

Later, it was shown that this shape is close to the asymptotic result for $N \times N$ random matrices.

In this talk we shall constantly compare the fluctuation properties of nuclear and RMT spectra. In Section 2 we give an outline of chaos in nuclei, emphasizing the different behavior of nuclear motion in the ground state region and in highly excited states. In Section 3 we present an statistical analysis of the spectral fluctuations of Ca, Sc and Ti isotopes, and show that there is a strong energy and isospin dependence of nuclear chaoticity.

In Section 4 a new approach to quantum chaos based on time series methods for the analysis of spectral fluctuations is presented. We introduce the $1/f$ noise conjecture for the power spectrum of chaotic quantum systems in Section 6 and in Section 7 we show that this conjecture is in agreement with an analytical expression obtained from RMT. Finally, the main conclusions are summarized in Section 8.

2. Chaos in nuclei

Real and complex quantum systems are usually not fully chaotic or integrable, and many questions on their chaotic and regular motion are still open. In this context, the atomic nucleus can be considered as a small laboratory where the principles obtained in schematic systems may be tested. The information on regular and chaotic nuclear motion available from experimental data is rather limited, because the analysis of energy levels requires the knowledge of sufficiently large pure sequences, i.e. consecutive level samples all with the same quantum numbers (J, π, T) in a given nucleus. The situation is rather clear above the one-nucleon emission threshold, where a large number of neutron and proton $J^{\pi} = 1/2^{+}$ resonances are identified. As Fig. 1 shows, the agreement between the spectral fluctuations of this Nuclear Data Ensemble (NDE) [6] and the GOE predictions is excellent. In the low energy domain, however, it is rather difficult (if not impossible) to get large enough pure sequences. For this reason the conclusions are less clear, although there is some evidence that, at low energy, the spectral fluctuations are close to GOE predictions in spherical nuclei whilst they deviate towards the Poisson distribution in deformed nuclei [7,8].

In order to get a deeper understanding of what happens in the low energy region we can use the shell model with configuration mixing. Realistic shell-model calculations have shown that the *sd* shell nuclei exhibit a strong chaotic behavior [9,10]. In the *pf* shell nuclei, however, the situation seems to be more interesting. The spectral properties of the Ca isotopes and the A=46 isobars were recently studied [11–13] and it was found that in the low energy domain the Ca isotopes did not follow the GOE predictions. It was further found that the simple substitution of a neutron by a proton lead to spectral fluctuations compatible with the GOE. Thus in this region of the nuclear chart it becomes possible to study with realistic shell-model calculations how nuclei evolve from quasiregular to chaotic motion as a function of isospin and excitation energy. Moreover, the large shell model

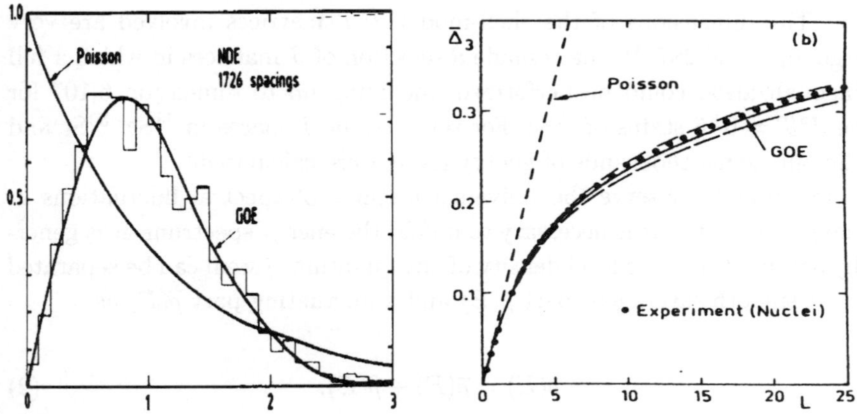

Fig. 1. Comparison of Poisson and GOE predictions with the nuclear data ensemble (NDE) results for the nearest neighbor spacing distribution $P(s)$ (left panel), and the Δ_3 statistic (right panel).

spaces involved, with full diagonalization of matrices with dimensions up to 6107, provide reliable values for the usual statistics.

In the following section we present a detailed study of energy level statistics of ^{46}Ti and the Ca and Sc isotopes from $A = 46$ to $A = 52$. We shall see how an order to chaos transition takes place as the excitation energy increases in Ca isotopes. A similar, but more abrupt transition takes place as one or two neutrons are replaced by protons in these nuclei, showing the existence of a clear isospin dependence in the degree of nuclear chaoticity.

3. Chaoticity of Ca, Sc and Ti isotopes

To study the T dependence of nuclear chaos, we have performed a detailed comparative analysis of spectral properties of the $A = 46, 48, 50$ and 52 Ca and Sc isotopes and ^{46}Ti [13]. The $T = T_z$ states are considered in all the cases.

We follow the shell-model procedure to obtain the low energy levels for a given nucleus. The most bound particles are assumed to form an inert core whilst the remaining nucleons move in a few single-particle orbits, the so-called valence space. For the lower part of the pf shell it is usual to take a core of ^{40}Ca and a valence space made of the $f_{7/2}$, $p_{3/2}$, $f_{5/2}$ and $p_{1/2}$ shells. We extract the single-particle energies from experiment and the residual two-body interaction is a monopole improved version of the Kuo-Brown interaction called KB3 [14]. The construction and subsequent diagonalization of the JT matrices was carried out using the code NATHAN

[15]. The dimensions of the shell-model $J^\pi T$ matrices involved are very large, up to 36,287. We have made a selection of J matrices in which a full diagonalization could be performed, including up to dimension 6,107 for the $J^\pi T = 0^+5$ states of ^{52}Sc. For some larger J spaces in ^{48}Sc, ^{50}Sc and ^{52}Sc only some thousands of energy levels were calculated.

In order to observe the universal features of spectral fluctuations in quantum systems, it is necessary to unfold the energy spectrum. It is generally accepted that the level density of any quantum system can be separated into a smooth (averaged) part $\overline{\rho}(E)$ and a fluctuating part $\widetilde{\rho}(E)$ as

$$\rho(E) = \overline{\rho}(E) + \widetilde{\rho}(E). \tag{2}$$

Since level fluctuation amplitudes are modulated by the value of the mean level density $\bar{\rho}(E)$, to compare the fluctuations of different systems we have to remove the smooth behavior of the level density. The unfolding consists in locally mapping the real spectrum into another one with mean level density equal to one. The actual energy levels E_i are mapped into new dimensionless levels ϵ_i,

$$E_i \longrightarrow \epsilon_i = \bar{N}(E_i), \;\; i = 1, \cdots, N, \tag{3}$$

where N is the dimension of the spectrum and $\bar{N}(E)$ is given by

$$\bar{N}(E) = \int_{-\infty}^{E} dE' \bar{\rho}(E'). \tag{4}$$

This function is a smooth approximation to the step function $N(E)$ that gives the true number of levels up to energy E. The form of the function $\bar{\rho}(E)$ can be determined by a best fit of $\bar{N}(E)$ to $N(E)$.

The nearest neighbor spacing sequence is defined by

$$s_i = \epsilon_{i+1} - \epsilon_i, \;\; i = 1, \cdots, N-1. \tag{5}$$

For the unfolded levels, the mean level density is equal to 1 and $\langle s \rangle = 1$. In practical cases the unfolding procedure can be a difficult task for systems where there is no analytical expression for the mean level density [16].

Generally, two suitable statistics are used to study the fluctuation properties of the unfolded spectrum. The nearest neighbor spacing distribution $P(s)$ gives information on the short range correlations among the energy levels. The $\Delta_3(L)$ statistic makes it possible to study correlations of length L: as we change the L value we obtain information on the level correlations at all scales.

The first neighbor spacing distribution $P(s)$ has been studied including all the levels up to 5 MeV and 10 MeV above the yrast line, and without any cutoff. The unfolding is performed for each $J^\pi, T = T_z$ pure sequence separately and then the unfolded spacings are gathered into a single set for each nucleus to get better statistics. Table 1 shows the different J matrices considered for each nucleus and also the total number of spacings included, depending on the energy limit.

Nucleus	^{46}Ca	^{48}Ca	^{50}Ca	^{52}Ca	^{46}Sc	^{48}Sc	^{50}Sc	^{52}Sc	^{46}Ti
J	0-12	0-12	0-12	0-12	0-12	0-4,7-12	0-1,9-12	0,11,12	0-5,8-12
E<5MeV	96	128	140	133	155	173	120	55	77
E<10MeV	441	654	818	658	1050	1328	950	405	553
All E	3937	11981	17203	11981	25498	14207	8031	11493	4508

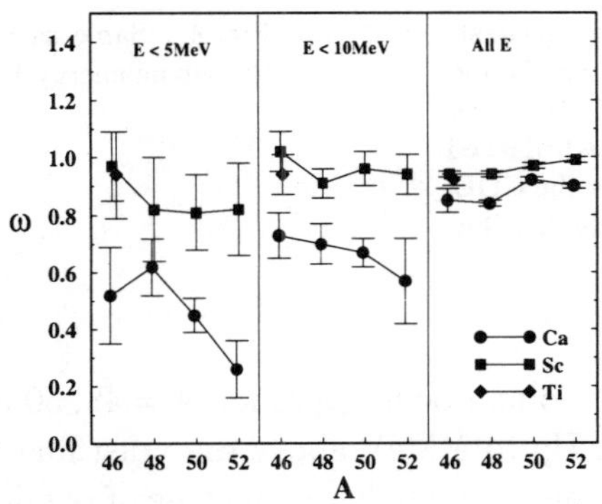

Fig. 2. Brody parameter ω for the $A = 46$, 48, 50 and 52 Ca and Sc isotopes and ^{46}Ti, using all the energy levels up to 5 MeV and 10 MeV over the yrast line, and the full spectrum.

We use the Brody parameter [9] ω to measure the degree of chaoticity. The value $\omega = 1$ corresponds to the GOE limit and $\omega = 0$ to Poisson statistics. The values of the Brody parameter are displayed in Fig. 2 separated in three sub-panels according to the energy cutoff. Figs. 3 and 4 compare the

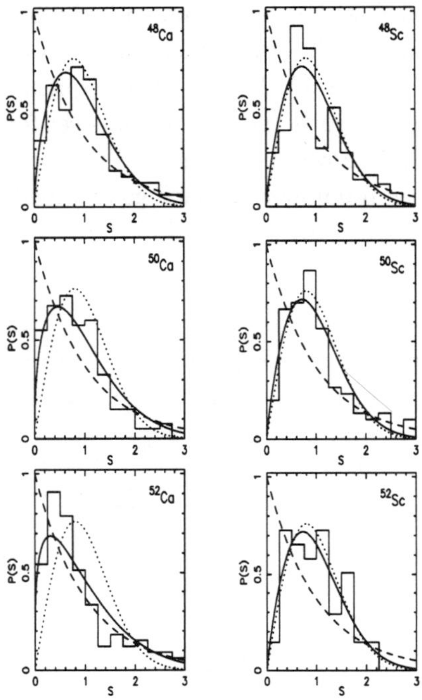

Fig. 3. Distribution of nearest-neighbor spacings $P(s)$ for Ca and Sc energy levels up to $5\,MeV$ over the yrast line. The dotted, dashed and solid curves stand for the GOE, Poisson and best fit Brody distributions, respectively.

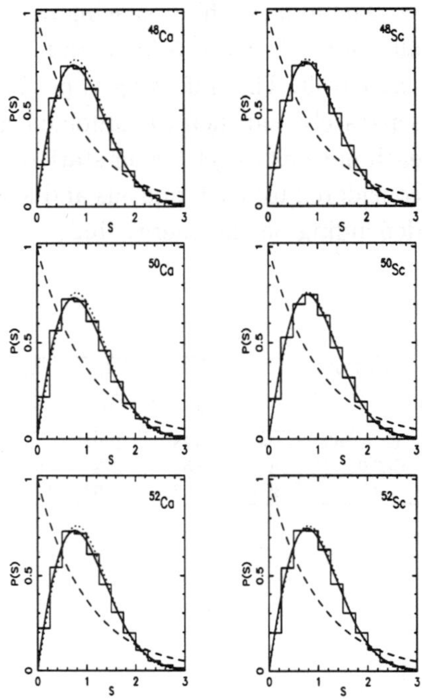

Fig. 4. Same as Fig. 3, but using the whole energy level spectrum.

$P(s)$ distributions of Ca and Sc isotopes for $A = 48$, 50 and 52, depending on the energy cutoff. Up to 5 MeV above the yrast line, Ca isotopes show spectral fluctuations intermediate between those of regular and chaotic systems, except ^{52}Ca which essentially is a regular system. On the contrary, all the Sc isotopes and ^{46}Ti are very close to GOE fluctuations. For a given A, the big differences between Sc and Ca isotopes must be due to the residual two-body interaction, because the single particle energies are the same in both cases. It was argued [12] that the neutron-neutron interaction is much weaker that the neutron-proton interaction and thus the central field motion is less affected by the former interaction.

Another interesting feature observed in Fig. 2 is that the ω parameter for the Ca isotopes shows a strong fall from $A = 48$ to $A = 52$, where $\omega = 0.25$.

It can also be observed looking at Fig. 3. As we go down from the top left to the bottom left panel, the Brody line moves away from the Wigner curve and progressively approaches the Poisson line. This astonishing result means that the two-body interaction is almost unable to perturb the single-particle motion in the low energy levels of ^{52}Ca.

As we go up in the energy spectrum, the new states become more and more complex combinations of many different configurations. Finally, when the whole energy spectrum is taken into account, the $P(s)$ distribution is dominated by the high density central part of the spectrum. Fig. 4 shows that when the whole energy spectrum is analyzed, the Ca isotopes become almost fully chaotic systems, with $P(s)$ very close to the Wigner surmise. But nevertheless, it can be seen that in Ca isotopes it is not quite as close to the chaotic limit as in Sc isotopes.

The ω value for ^{46}Ti has been included in Fig. 2 to show that, for all the energies, the Brody parameter already reaches its maximum for ^{46}Sc. Therefore, according to short range level correlation behavior, replacing a single neutron by a proton in Ca isotopes causes a transition from a quasi-regular to a chaotic regime. It is remarkable that this transition takes place abruptly at all excitation energies in all the isotopes. A second replacement of a neutron by a proton does not seem to produce appreciable effects.

In order to confirm the previous results, we have computed the $\Delta_3(L)$ statistic for some $J^\pi, T = T_z$ sequences. To obtain the $\Delta_3(L)$ value for each L, we take its average value over many overlapping intervals of L unfolded spacings along the whole spectrum. Therefore, the results given below concern the full spectrum and not only the low energy region. Fig. 5 shows $\Delta_3(L)$ values for $L \leq 50$, using the $J^\pi = 0^+$, $T = T_z$ levels of ^{46}Ca, ^{46}Sc and ^{46}Ti. Of the three nuclei, only ^{46}Ti follows the GOE line, at least until $L = 50$. For ^{46}Sc the Δ_3 is close to GOE predictions up to a certain separation value, $L_{sep} \simeq 30$ where it upbends from the GOE curve. In ^{46}Ca the upbending starts at a smaller value $L_{sep} \simeq 10$. In a chaotic system, Wigner level repulsion gives rise to a rigid spectrum, where long range correlations are suppressed in comparison with a random sequence of levels. For such a system the Δ_3 increases logarithmically following the GOE prediction. The upbending from the GOE curve and a linear growth of this statistic reveals a departure from the chaotic regime. This kind of behavior of the Δ_3 statistic was first seen in one-body systems like billiards [17] and then in shell-model calculations by several authors [18]. The values of L_{sep} differ from one case to another depending on the valence space or the effective interaction for reasons which are not fully understood as yet. But for the same interaction and configuration space, the Δ_3 behavior clearly shows a strong isospin dependence in the $A = 46$ nuclei, with chaoticity increasing as T decreases. This happens not only from Ca to Sc, but also from Sc to Ti. The same phenomenon is observed for other J values.

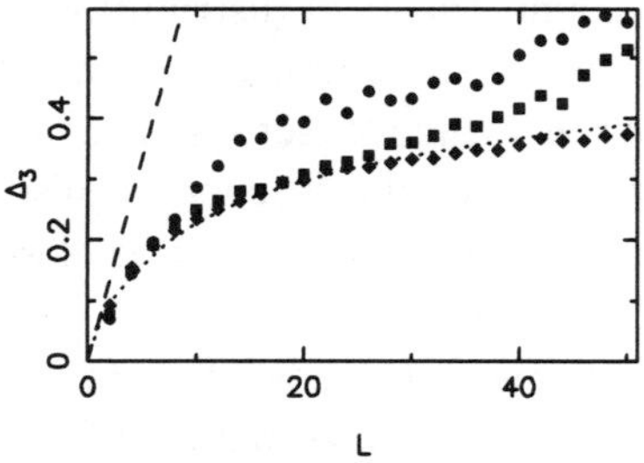

Fig. 5. Average Δ_3 for all the $J^\pi = 0^+, T = T_z$ levels of ^{46}Ca (dots), ^{46}Sc (squares) and ^{46}Ti (diamonds). The dotted and dashed curves represent the GOE and Poisson Δ_3 values, respectively.

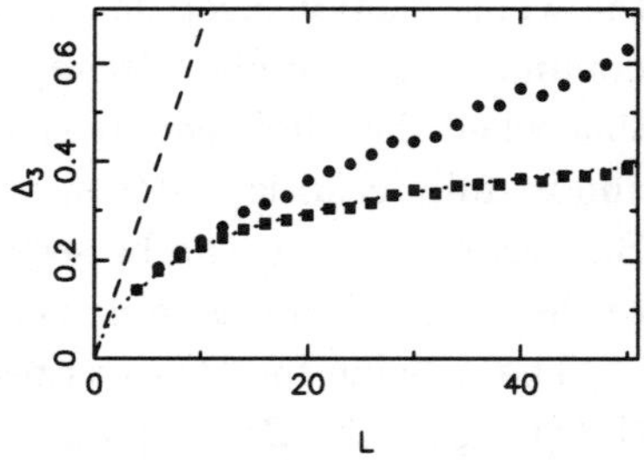

Fig. 6. Same as Fig. 5 for all the $J^\pi = 5^+, T = T_z$ levels of ^{52}Ca and ^{52}Sc.

As another example, we compare in Fig. 6 the Δ_3 values for the $J^\pi = 0^+$, $T = T_z$ spectra of ^{52}Ca and ^{52}Sc. Here we see again that ^{52}Sc is clearly more chaotic than ^{52}Ca. Notice that in this case $L_{sep} > 50$ for ^{52}Sc, and comparing with Fig. 5 it seems to be more chaotic than ^{46}Sc. The main reason may be that there are more proton-neutron interactions in ^{52}Sc.

Summarizing the analysis of the spectral fluctuations, there exists a clear excitation energy and isospin dependence in the chaoticity degree of nuclear motion. It increases from Ca to Sc and from Sc to Ti. It is observed not only in the ground state region, but along the whole spectrum. When the full spectrum is taken into account, the $P(s)$ distribution is not very sensitive to the isospin dependence, but the effect is clearly seen in the Δ_3 statistic.

4. Time series, chaos and 1/f noise

We present now a recently discovered, very different approach to quantum chaos [19], which is based on traditional methods of time series analysis.

The essential feature of chaotic energy spectra in quantum systems is the existence of level repulsion and correlations. To study these correlations, we can consider the energy spectrum as a discrete signal, and the sequence of energy levels as a time series where the energy plays the role of time. We shall see that examination of the power spectrum of energy level fluctuations reveals very accurate power laws for completely regular or completely chaotic Hamiltonian quantum systems. It turns out that chaotic systems have $1/f$ noise, in contrast to the $1/f^2$ Brown noise of regular systems.

We characterize the spectral fluctuations by the statistic δ_n [20] defined by

$$\delta_n = \sum_{i=1}^{n} (s_i - \langle s \rangle) = \sum_{i=1}^{n} w_i, \tag{6}$$

where the index n runs from 1 to $N-1$. The quantity w_i gives the fluctuation of the i-th spacing from its mean value $< s >= 1$.

The statistic δ_n represents the deviation of the unfolded excitation energy from its mean value n, and it is closely related to the accumulated level density fluctuations given by $\widetilde{N}(E)$. Indeed, we note that we can also write that $\delta_n = -\widetilde{N}(E_{n+1})$, if we appropriately shift the ground state energy; thus, it represents the accumulated level density fluctuations at $E = E_{n+1}$.

The function δ_n has a formal similarity with a time series. For example, we may compare the energy level spectrum with the diffusion process of a particle. The analogy is clear if the index i of the nearest level spacings is considered as a discrete time, and the spacing fluctuation w_i as the analogue of the particle displacement d_i from the collision at time i to the next collision. Certainly, there are some differences between the two systems, but the essential point is that the function δ_n is the analogue of the particle total displacement at time n.

Our aim is to study the δ_n signal of chaotic quantum systems. We can analyze their spectral statistics with numerical techniques normally used in the study of complex systems and try to relate the emerging properties with some universal features that appear in many other branches of physics. One of those techniques is the calculation of the power spectrum $S(k)$ of a discrete and finite series δ_n given by

$$S(k) = \left|\hat{\delta}_k\right|^2, \quad \hat{\delta}_k = \frac{1}{N} \sum_n \delta_n \exp\left(\frac{2\pi i k n}{N}\right), \tag{7}$$

where $\hat{\delta}_k$ is the Fourier transform of δ_n, and N is the size of the series.

5. Time series analysis of chaotic quantum systems

5.1. *Atomic nuclei*

As an example of a very chaotic system, we take the atomic nucleus at high excitation energy, where the level density is very large. To obtain the energy spectrum, shell-model calculations for selected nuclei are performed, using realistic interactions that reproduce well experimental data of nuclei in the appropriate mass region. The Hamiltonian matrices for different angular momenta, parity and isospin are fully diagonalized using the shell-model code Nathan [15], and careful global unfolding is performed. Then, sets of 256 consecutive levels of the same $J^{\pi}T$, from the high level density region, are selected. To characterize the statistical properties of the δ_n signal, we calculate an ensemble average of its power spectrum, in order to reduce statistical fluctuations and clarify its main trend. The average $\langle S(k)\rangle$ is calculated with 25 sets.

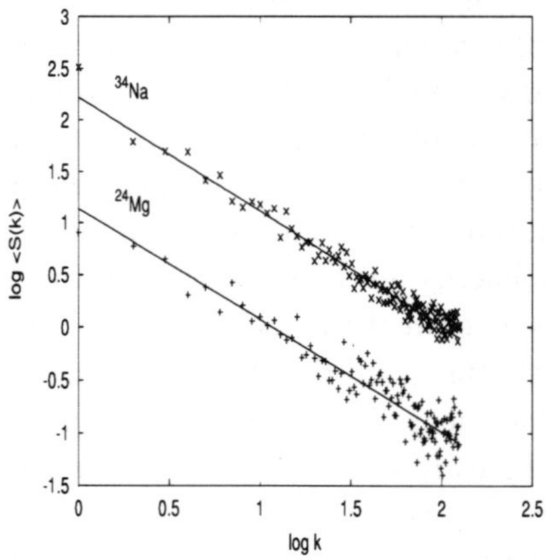

Fig. 7. Average power spectrum of the δ_n function for ^{24}Mg and ^{34}Na, using 25 sets of 256 levels from the high level density region. The plots are displaced to avoid overlapping.

Fig. 7 shows the results for a typical stable *sd* shell nucleus, ^{24}Mg, with matrix dimensionalities up to about 2000, using the effective W interaction [21]; and for a very exotic nucleus, ^{34}Na, with dimensions up to about 5000, in the *sd* proton and *pf* neutron shells, using a realistic interaction [22]. Clearly, the power spectrum of δ_n follows closely a power law. We may assume the simple functional form

$$\langle S(k)\rangle \sim \frac{1}{k^{\alpha}}. \tag{8}$$

A least squares fit to the data of Fig. 7 gives $\alpha = 1.11 \pm 0.03$ for ^{34}Na, and $\alpha = 1.06 \pm 0.05$ for ^{24}Mg. These results rise the question of whether there is a general relationship between quantum chaos and the power spectrum of the δ_n fluctuations of the system.

5.2. *RMT ensembles*

Probably, the simplest and most reliable way to clarify this issue is to compare δ_n and $\langle S(k)\rangle$ for Poisson energy levels and random matrix spectra. Random matrix theory plays a predominant role in the description of chaotic quantum systems [23]. It deals with three basic Hamiltonian matrix ensembles: The Gaussian orthogonal ensemble (GOE) of N-dimensional matrices, the Gaussian unitary ensemble (GUE), and the Gaussian symplectic ensemble (GSE), which apply to different systems, depending on the integer or half-integer spin, the time-reversal and the rotational symmetries of the system. As an additional collectivity, we introduce here the ensemble of diagonal matrices whose elements are random Gaussian variables, and call it the *Gaussian diagonal ensemble* (GDE).

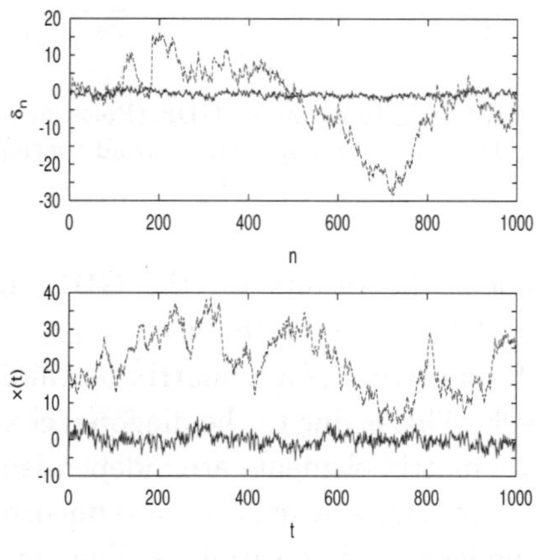

Fig. 8. Comparison of the δ_n function for Poisson (dashed line) and GOE spectra (solid line), with a standard time series $x(t)$ with $1/k^\alpha$ power spectrum, for $\alpha = 2$ (dashed line) and $\alpha = 1$ (solid line).

Fig. 8 shows the signal δ_n for a GDE (Poisson) and a GOE spectrum of dimension 1000. Clearly the two signals are very different. It is also illustrative to compare those signals with a discrete time series $x(t)$, with $1/k$ and $1/k^2$ power laws, generated with the random-phase approximation procedure [24]. The similarity of the $\alpha = 2$ time series with the Poisson spectrum, and the $\alpha = 1$ time series with the GOE spectrum is obvious.

To compute the average $\langle S(k) \rangle$, we generate 30 different matrices of dimension 1000 for each type of random matrix ensemble. Fig. 9 shows the results of these calculations in a decimal log-log scale. In all the cases the main trend is essentially linear, except for very high frequencies, where some deviation is observed, probably due to finite size effects.

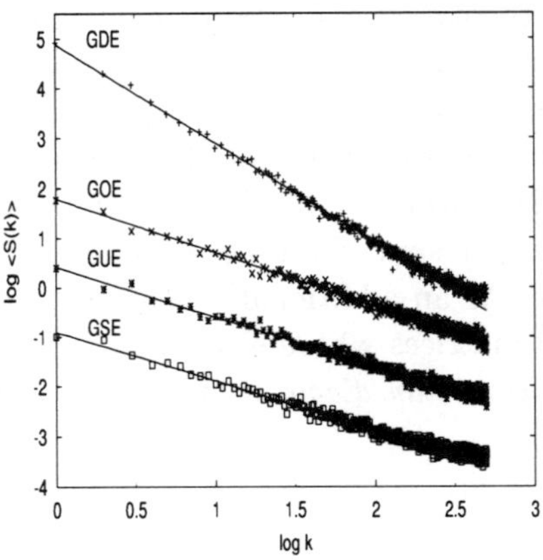

Fig. 9. Power spectrum of the δ_n function for GDE (Poisson) energy levels, compared to GOE, GUE and GSE. The plots are displaced to avoid overlapping.

Let us first comment the results for the GDE. Ignoring frequencies greater than $\log k = 2.2$, the fit to (8) gives $\alpha_{\mathrm{GDE}} = 1.99$ with an uncertainty near 2%. The spectrum of any matrix pertaining to GDE consists of N uncorrelated levels. This is due to the diagonal character of the matrix and to the fact that its matrix elements are independent random variables. Consequently, the nearest level spacings are also uncorrelated and δ_n is just a sum of $N-1$ independent random variables. The power spectrum of such a signal is well known to present $1/k^2$ behavior, and that is in full agreement with our numerical value for α in the Poisson spectrum. Furthermore, Berry and Tabor [1] showed that in a semiclassical integrable system, the spacings s_i are random independent variables for $i >> 1$. As a consequence their δ_n power spectrum behaves as $1/k^2$. However this behavior may be modified by the levels of the ground state region.

By contrast, the spectrum of any GOE member of large dimension is generally considered the paradigm of chaotic quantum spectra. It presents level correlations at all scales. The same applies to GUE and GSE, in increasing order of level repulsion. As is well known, the nearest neighbor spacing distribution for these three ensembles behaves as $P(s) \sim s^\beta$ for small s, where β is known as the level repulsion parameter. For our diagonal ensemble with Poisson statistics, $\beta = 0$, while $\beta = 1, 2, 4$ for GOE, GUE and GSE, respectively [23]. The power spectrum of δ_n for the three latter ensembles is also displayed in Fig. 9. The fit of $\langle S(k) \rangle$ to the power law (8) is excellent. For the exponents we obtain $\alpha_{\text{GOE}} = 1.08$, $\alpha_{\text{GUE}} = 1.02$ and $\alpha_{\text{GSE}} = 1.00$. In all the cases the error of the linear regression is about 2%. The three ensembles yield the same power law, with $\alpha \simeq 1$. Clearly, the the power spectrum $\langle S(k) \rangle$ behaves as $1/k^\alpha$ both in regular and chaotic energy spectra, but level correlations decrease the exponent from the $\alpha = 2$ limit for uncorrelated spectra to apparently a minimum value $\alpha = 1$ for chaotic quantum systems.

6. The power spectrum conjecture

The concept of quantum chaos has no precise definition as yet. Quantum systems with classical analogues are considered chaotic when their classical analogues are chaotic. Quantum systems without classical analogues may be called chaotic if they show the same kind of fluctuations as chaotic quantum systems with classical analogues. In practice, the Bohigas-Gianoni-Schmit RMT conjecture [2] is generally used as a criterion. But the results obtained above for the power spectrum of the δ_n statistic suggest a new conjecture: *The energy spectra of chaotic quantum systems are characterized by* $1/f$ *noise* [19].

This conjecture has several appealing features. It is a property characterizing the chaotic spectrum by itself, without any reference to the properties of other systems like GOE. It is universal for all kinds of chaotic quantum systems, either time reversal invariant or not, either of integer or half-integer spin. Furthermore, the $1/f$ noise characterization of quantum chaos includes these physical systems into a widely spread kind of systems appearing in many fields of science, which display $1/f$ fluctuations. Thus, the energy spectrum of chaotic quantum systems exhibits the same kind of fluctuations as many other complex systems. However, there is no indication that $1/f$ spectral fluctuations in a quantum system implies $1/f$ noise in its classical analogue. Neither have we found any relationship with $1/f$ noise in classical chaotic phenomena like intermittency [25].

7. RMT derivation of the $1/f$ noise in quantum chaos

Finding the origin of the ubiquitous $1/f$ noise in the time fluctuations of complex systems is an important open problem. In the case of quantum systems, an exact and complete proof of the $1/f$ noise conjecture in the power spectrum of the δ_n statistic seems to be extremely difficult. However, it can be theoretically studied in semiclassical systems or random matrix ensembles, where the mathematical tractability of these systems may help to understand the origin of the $1/f$ noise in quantum systems. In this contribution we present an explicit expression of the average value of $S(k)$ for fully chaotic and integrable systems, obtained in the framework of RMT.

Except for integrable systems, one of the main features of quantum spectra is that successive level spacings are not independent, but correlated quantities. This property makes exceedingly difficult to work directly with the discrete δ_n sequence. The statistical properties of the fluctuating part of the density can be measured in terms of the spectral form factor, defined as

$$K(\tau) = \left\langle \lim_{L\to\infty} \frac{1}{2L} \left| \int_{-L}^{L} d\epsilon \widetilde{\rho}(\epsilon) e^{-2\pi i \epsilon \tau} \right|^2 \right\rangle . \tag{9}$$

that is, as the power spectrum of the fluctuating part of the energy level density.

We have recently shown that the power spectrum of δ_n for fully chaotic or integrable systems can be written in terms of $K(\tau)$. The derivation of this result involves cumbersome calculations which are out of the scope of this paper. For a more detailed account of the derivation, The interested reader can see ref [26]. The final result is

$$\begin{aligned} \langle S(k) \rangle_\beta = \frac{N^2}{4\pi^2} \left\{ \frac{K\left(\frac{k}{N}\right) - 1}{k^2} + \frac{K\left(1 - \frac{k}{N}\right) - 1}{(N-k)^2} \right\} \\ + \frac{1}{4 \sin^2 \left(\frac{\pi k}{N}\right)} + \Delta, \quad k = 1, 2, \cdots, N-1, \quad N \gg 1. \end{aligned} \tag{10}$$

where β is the repulsion parameter of RMT ensembles and takes the values $\beta = 1$ for GOE, $\beta = 2$ for GUE, and $\beta = 4$ for GSE [20]. Here $\Delta = 0$ for integrable systems and $\Delta = -1/12$ for chaotic systems. This equation, together with the appropriate values of $K(\tau)$, gives explicit expressions of $\langle S(k) \rangle$ for specific ensembles or systems.

When $k \ll N$ the first term of eq. (10) becomes dominant and we can write

$$\langle P_k^\delta \rangle_\beta = \begin{cases} \dfrac{N}{2\beta\pi^2 k}, & \text{for chaotic systems,} \\ \dfrac{N^2}{4\pi^2 k^2}, & \text{for integrable systems.} \end{cases} \tag{11}$$

These expressions show that, for small frequencies, the excitation energy fluctuations exhibit $1/f$ noise in chaotic systems and $1/f^2$ noise in integrable systems. As we shall see below, these power laws are also approximately valid through almost the whole frequency domain, due to partial cancellation of higher order terms. Only near $k = N/2$ the effect of these terms becomes appreciable.

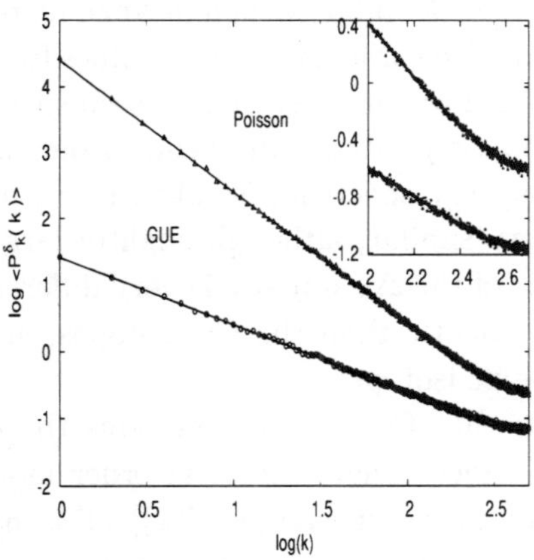

Fig. 10. Theoretical power spectrum of the δ_q function for GUE and integrable systems (solid lines), compared to numerical averages calculated using 30 GUE matrices of dimension $N = 10^3$ (circles) and 30 Poisson level sequences of length $N = 10^3$ (triangles).

To test all these theoretical expressions we have compared their predictions to numerical results obtained for two different ensembles. Fig. 10 displays the theoretical values of $\langle S(k) \rangle$ for GUE and integrable systems, as given by (10), together with the numerical average values for 500 GUE matrices and 500 Poisson level sequences. In order to enlarge the high frequency region, where the numerical results show small deviations from the $1/f^\alpha$ power law behavior, an upper right panel is added to the figure. The agreement between the theoretical and numerical results is excellent at all frequencies (note that there are no free parameters in the analytical result).

8. Conclusions

We have studied the fluctuation properties of nuclear energy levels in terms of the $P(s)$ and Δ_3 statistics. The short range correlations of Ca energy levels up to 5 MeV above the yrast line are far from the GOE limit. The Brody parameter ω falls from 0.6 in ^{48}Ca to 0.2 in ^{52}Ca. Thus there is a strong dependence on the number of active neutrons. The $P(s)$ analysis has also been performed for energy levels up to 10 MeV above the yrast line and for the whole energy spectrum. As we move up in energy, the chaoticity in Ca isotopes increases smoothly until $\omega \simeq 0.9$ for the whole energy spectrum.

In order to study if there is an isospin dependence, we made similar calculations for the $T = T_z$ states of Sc isotopes and ^{46}Ti. For all these nuclei and for all the excitation energy regions the short range energy level correlations are close to GOE limit, although some increase with excitation energy is also observed. The variation of the ω values from Ca to Sc is always quite large, especially at low energies. Thus it seems that replacing a single neutron by a proton in Ca produces an abrupt order to chaos transition.

When the whole energy spectrum is taken into account, the ω values for Ca and Sc are very similar, although slightly larger for Sc for all A. However, the behavior of the Δ_3 statistic is very different showing that the Sc isotopes are more chaotic than the Ca isotopes, and also that ^{46}Ti is more chaotic than the Sc isotopes.

We have also seen that for quantum systems the δ_n function can be considered as a time series, where the level order index n plays the role of a discrete time. The power spectrum $\langle S(k)\rangle$ of δ_n has been studied for representative energy spectra of regular and chaotic quantum systems. Neat power laws $\langle S(k)\rangle \sim 1/k^\alpha$ have been found in all cases. For Poisson spectra, we get $\alpha = 2$, as expected for independent random variables. For spectra of atomic nuclei at higher energies, in regions of high level density, and for the GOE, GUE and GSE ensembles, we obtain $\alpha = 1$. Moreover, we have derived an analytical expression for the power spectrum of δ_n in the framework of RMT. The agreement between the parameter free analytical results and the numerical random matrix calculations is excellent.

All these results suggest the conjecture that chaotic quantum systems are characterized by $1/f$ noise in the energy spectrum fluctuations. We would like to emphasize that the power spectrum of δ_n is not a mere new statistic to measure the chaoticity of the system. It provides an intrinsic characterization of quantum chaotic systems without any reference to the properties of other systems, like RMT ensembles. As is well known $1/f$

noise is quite ubiquitous in complex systems. It characterizes, for example, sunspot activity, music, and the beats of a healthy heart [27]. And we believe that it characterizes quantum chaos as well.

Finally, let us mention that the transition from order to chaos has been recently studied in semiclassical systems, like the Robnik billiard [28], the Kicked-top, and the coupled quartic oscillator [29], using the power spectrum approach. The remarkable result is that a power law of type $1/f^\alpha$ is found at all stages through the transition. The parameter α changes smoothly from 2 to 1 as the system evolves from fully regular to fully chaotic motion.

This work is supported in part by Spanish Government grants BFM2003-04147-C02 and FTN2003-08337-C04-04.

References

1. M. V. Berry and M. Tabor, Proc. R. Soc. London A **356**, 375 (1977).
2. O. Bohigas, M. J. Giannoni, and C. Schmidt, Phys. Rev. Lett. **52**,1 (1984).
3. E. P. Wigner, Interational Conference on the Neutron interactions with the Nucleus. Columbia University, 1957. Columbia University Report CU-175, p.49.
4. L. D. Landau, and Ya. Smorodinsky, Nuclear Reactions (Statistical Theory), Ch. 7 in Lectures in Nuclear Theory p. 69 (Plenum, New York)
5. O. Bohigas, in *Recent Perspectives in Random Matrix Theory and Number*, F. Mezzadri and N. C. Snaith, Eds., Cambridge University Press (2005), p. 147.
6. R. U. Haq, A. Pandey and O. Bohigas, Phys. Rev. Lett. **48**, 1086 (1982).
7. J. F. Shriner jr., G. E. Mitchell and T. von Egidy, Z. Phys. A**338**, 309 (1991).
8. J. D: Garret, J. R. German, L. Courtney and J. M. Espino, in Future Directions in Nuclear Physics, eds. J. Dudek and B. Haas (A. I. P., N.Y., 1992) p. 345.
9. T. A. Brody, J. Flores, J. B. French, P. A. Mello, A. Pandey, S. S. M. Wong, Rev. Mod. Phys. **53**, 385, (1981).
10. M. Horoi, V. Zelevinsky, and B. A: Brown, Phys. Rev. Lett. **74**, 5194 (1995).
11. E. Caurier, J. M. G. Gómez, V. R. Manfredi and L. Salasnich, Phys. Lett. B**365**,7 (1996).
12. J. M. G. Gómez, V. R. Manfredi L. Salasnich and E. Caurier, Phys. Rev. C**58**,2108 (1998).
13. R. A. Molina, J. M. G. Gómez and J. Retamosa, Phys. Rev. C**63**, 014311 (2000).
14. A. Poves and A. Zuker, Phys. Rep. 70,4 (1981).
15. E. Caurier, G. Martínez-Pinedo, F. Nowacki, A. Poves, J. Retamosa, and A. P. Zuker, Phys. Rev. C **59**, 2033 (1999).
16. J. M. G. Gómez, R. A. Molina, A. Relaño and J. Retamosa, Phys. Rev. E **66**, 036209 (2002).

17. H.D. Gräf, H.L. Harney, H. Lengler, C.H. Lewenkopff, C. Rangacharyulu, A. Righter, P. Schardt and H.A. Weindenmüler, Phys. Rev. Lett. **69**, 1296 (1992).
18. V. Zelevinsky, B.A. Brown, N. Frazier and M. Horoi, Phys. Rep. 276, 35 (1996).
19. A. Relaño, J. M. G. Gómez, R. A. Molina, J. Retamosa and E. Faleiro, Phys. Rev. Lett. **89**, 244102 (2002).
20. M. L. Mehta, *Random Matrices*, (Academic Press, 1991)
21. B. H. Wildenthal, in *Progress in Particle and Nuclear Physics*, Ed. D. H. Wilkinson, Vol. **11** (Pergamon, Oxford 1984).
22. E. Caurier, F. Nowacki, A. Poves, and J. Retamosa, Phys. Rev. C **58**, 2033 (1998).
23. T. Guhr, A. Müller-Groeling and H. A. Weidenmüller, Phys. Rep. **299**, 189 (1998).
24. N. P. Greis and H. S. Greenside, Phys. Rev. A **44**, 2324 (1991).
25. H. G. Schuster, *Deterministic Chaos: an introduction* (Weinheim VCH, 1995).
26. E. Faleiro, J.M.G. Gómez, R. A. Molina, L. Muñoz, A. Relaño, J. Retamosa, Phys. Rev. Lett. **93**, 244101 (2004).
27. B. B. Mandelbrot, *Multifractals and 1/f noise* (Springer, New York, 1999).
28. J. M. G. Gómez, A. Relaño, J. Retamosa, E. Faleiro, M. Vranicar, and M. Robnik. Phys. Rev. Lett. **94**, 084101 (2005).
29. M. S. Santhanam, J. N. Bandyopadhyay, Phys. Rev. Lett. **95**, 114101 (2005).

Nuclear physics for astrophysics with radioactive nuclear beams: indirect methods*

Livius TRACHE
Cyclotron Institute
Texas A&M University
College Station, TX 77843-3366, USA
E-mail: livius_trache@tamu.edu

The lecture describes a number of indirect methods in nuclear astrophysics using radioactive beams: Coulomb dissociation, transfer reactions (the ANC method), breakup of loosely bound nuclei at intermediate energies, other spectroscopic measurements. The examples chosen are be drawn from the experiments the author was involved together with his group at Texas A&M University. One example discussed in particular is that of the reactions used to determine the S_{17} astrophysical factor for the $^7Be(p,\gamma)^8B$ reaction, crucial for the understanding the solar neutrino problem. We discuss also the case of the proton drip line nucleus ^{23}Al. From its study we extract data to determine the stellar reaction rates for ^{22}Mg(p,γ)^{23}Al and ^{22}Na(p,γ)^{23}Mg. Both breakup and beta-decay methods were used in this study.

1. Introduction

Let me start by saying that I am very glad to be here again. I have been at these schools many times, beginning in the mid-seventies, in all three positions: student, lecturer or organizer. And I always learned a lot, particularly as a student. I am, therefore, very grateful to the organizers of the current edition, firstly to Apolodor, for giving me the opportunity to be in Predeal this year.

Given the title of this reunion, a school, no matter what the composition of the actual audience, I concieved my presentation as a lecture addressed to students, making it more general and simplified, rather than a report at a conference presenting latest personal results to a field of specialists. The specificity will stem from that that, as illustrative examples, I'll use only experiments or studies to which I participated directly. Also, given the singularity of the subject at this school, I'll start with a broader introduction.

2. Introduction, some vocabulary

All chemical elements in the Universe as we know it were produced in processes that we call generically nucleosynthesis. Nucleosynthesis occured in various stages of the evolution of the Universe, in various places and in different types of events: Big Bang nucleosynthesis or later stellar evolution, far away or around us, explosive or steady burning. And we have firm evidence collected in the latest decades that nucleosynthesis happens today, even in our own galaxy, close to where we live. We also know today that the nuclear processes occuring in stars are not only the source of energy for cosmic processes, but also that nucleosynthesis gives us unique and undelible fingerprints of these processes. Many nucleosynthesis scenarios exist today, some were formulated for some time (Big Bang Nucleosynthesis, Inhomogeneous Big Bang Nucleosynthesis, the s-process, the r-process, the rp-process, etc.), some are newer proposals. The possibilities to check the detailed predictions of specific models occured only recently, with the availability of more nuclear data, of advances in understanding the dynamics of non-equilibrium processes, and increased computing power. It turns out that an important component of all these nucleosynthesis model calculations is represented by the data for the nuclear processes involved. Only good nuclear physics data permit to make definite, quantitative predictions that can be checked against the ever increasing observational data sought and obtained by astrophysicists. This is the object of the nuclear physics for astrophysics, a subject that we most often call ***nuclear astrophysics.*** The present lecture will not deal at all with specificities of the dynamics of different stellar processes, but only with the nuclear reactions involved, in particular with how we obtain these data from indirect measurements with radioactive beams.

There are many nuclear reactions and nuclear processes that occur in stars. For example, one important class is that of radiative capture reactions X(p,γ), X(n,γ), X(α,γ), etc. There are many problems we encounter in a nuclear astrophysics laboratory when we do direct measurements (= exactly the reactions that occur in stars, at stellar energies), but two main categories occur when we want to study the relevant reactions. One stems from the fact that the energies involved in stars are small (tens or hundreds of keV/u) and consequently the cross sections are extremly small in reactions involving charged particles, due to the Coulomb repulsion. This presents experimental challenges, related to beam intensities and background. As these reactions (radiative proton capture, radiative alpha capture, (α,p), (α,n) ...) have cross sections of the order of nanobarns, picobarns, and even smaller at stellar energies, we can only make studies at somewhat larger energies (MeV/u, some few hundreds keV/u) and than extrapolate our measurements down to the relevant energy window (called Gamow peak). The other is that most of the reactions in stars involve short lived, unstable nuclei produced in the preceding step of the stellar reaction chain. Therefore, many of the reactions cannot be measured with the stable beams and targets available in laboratory. These two major problems lead nuclear physicists to seek indirect methods for nuclear astrophysics and to the use of radioactive beams. Our subject today.

A number of specific notions are introduced in nuclear astrophysics: the astrophysical S-factor (to separate the Coulomb barrier penetration), the reaction rate (integrated over the maxwellian distribution), the Gamow peak (the region in the energy maxwellian distribution that contributes to the reaction rate). I'll not include them in this typescript, for their precise definitions consult a book on nuclear astrophysics, like the very popular one of Rolfs and Rodney [1].

2.1. *The $^7Be(p,\gamma)^8B$ reaction in the Sun*

The search for solar neutrinos and later for a solution of the solar neutrino problem was a major and extremely fruitful quest of the last few decades (see a good review in Ref. [2]). It involved major experimental and theoretical efforts and lead to major breakthroughs, such as the confirmation of the neutrino oscillations ([3–6] and the references therein). We knew for sometime that only nuclear reactions could be the source of Sun's energy and that those reaction have to take place deep inside the core of the Sun, at temperatures much larger than those of the corona, the region from which the light comes to us. The energy is produced in the pp chains and in the CNO cycle, both with the same result of transforming 4 protons into a ^{4}He nucleus, 2 positrons, 2 neutrinos and energy. The neutrinos produced in the nuclear reactions are the only signal that can carry information about those reactions from the interior of the Sun all the way to our terrestrial detectors. Essentially, only when solar neutrinos were detected by Davis [3], we could say we had experimental evidence that nuclear reactions are indeed fueling the Sun, and probably the other stars too. Solar neutrinos are produced at different stages in the pp burning processes. In one of them, the ppIII chain, which only accounts for about 0.1% of the energy balance, ^{7}Be nuclei are produced. Further, they capture another proton through the reaction ^{7}Be(p,γ)^{8}B. ^{8}B is not stable, it beta-decays into a short lived ^{8}Be, a positron and a neutrino that can have a maximum energy of 15 MeV, the largest energy of all neutrinos produced in solar processes. However, most of the neutrinos detected by SuperKamiokande and SNO detectors are ^{8}B neutrinos, and therefore, for reliable calculations of the neutrino flux in the standard solar model [6] we need good data for the reaction rates involved, in particular for the radiative proton capture reaction ^{7}Be(p,γ)^{8}B. The magnitude of this reaction cross section is measured by what we call the S_{17} astrophysical factor. The precise determination of S_{17} was, therefore, the subject of a large number of experimental and theoretical efforts ([7–15],to name only a few of the most recent), summarized in many recent publications [16–18]. In spite of these efforts there is no clear consensus on the value of $S_{17}(0)$ at the desired 5% precision, with apparent discrepancies between values given by some direct measurements and some from indirect methods, like Coulomb dissociation [13,16,19], (^{7}Be,^{8}B) proton transfer reactions [9,22] and breakup of ^{8}B [11,23–25]. Consequently, several new experiments are under way or planned.

These all make the above reaction one of the most important ones in nuclear astrophysics, and I'll use it as case study for the indirect methods treated here. The average value obtained by Cyburt et al. [18] $S_{17}(0) = 20.8 \pm 0.6(stat) \pm 1.0(syst)$ eV b using all radiative capture data (direct measurements) is dominated by a single measurement that claims a very good precision [17] $S_{17}(0) = 22.1 \pm 0.6(stat) \pm 0.6(theor)$ eV b.

3. Indirect methods in nuclear astrophysics

The principle behind the use of indirect methods is simple: we (1) do experiments at higher energies (tens of MeV/u or higher), typically above the Coulomb barrier, where cross sections are larger, in order to obtain nuclear information that we consequently (2) transform in reaction cross sections and reaction rates for astrophysically relevant energies. I want to stress here that in both steps above, calculations are needed, and therefore good knowledge of the dynamics of the processes involved is crucial. Typically more information from other experiments is needed or valuable. Another point I want to stress is that it is important to chose, and therefore seek at step (1), the nuclear information (typically nuclear structure) that is most relevant in obtaining the reaction rates sought at step (2) (see the discussion about ANCs vs. spectroscopic factors in Sect. 3.2 and that about the spin of the g.s. of ^{23}Al in Sect. 3.3).

There are only a few indirect methods applied in nuclear astrophysics:

a) Coulomb dissociation

b) transfer reactions (the ANC method)
c) breakup at intermediate energies
d) Trojan horse method
e) other spectroscopic methods, in particular the study of resonances.

I will discuss each of them in what follows, except d), the Trojan horse method, with which I personally only brushed with briefly (and because of time constrains). See Refs. for this subject [26,27]. Methods a)-c) will be presented in connection with the study of the ^{7}Be(p,γ)^{8}B reaction, crucial for the solar neutrino production, and I will present more extensively examples for cases b), c) and e).

3.1. *The Coulomb dissociation*

The Coulomb dissociation is a method specifically introduced for nuclear astrophyiscs two decades ago [28,29], and the Predeal school was one of the first places where prof. Rebel talked about it. It can be used to determine cross sections (or equivalently, astrophysical S-factors) for radiative capture reactions with charged particles (protons or alphas). Schematically, it works as follows! Instead of studying the radiative capture reaction X(p,γ)Y at a definite center-of-mass energy E_p, process in which a gamma-ray of energy $E_\gamma = E_p + S_p$ is emitted (S_p=binding energy of the proton in nucleus Y), we could measure the inverse process: photodissociation. There a photon of energy E$_\gamma$ interacts with nucleus Y producing the dissociation $Y + \gamma \to X + p$, in which a proton-core system of energy $E_p = E_\gamma - S_p$ results. Then the Fermi golden rule of detailed equilibrium can be used to relate the cross section of the two processes. Baur, Bertulani and Rebel proposed to replace the real photons needed in photodissociation with virtual photons. A fast moving projectile Y in the strong Coulomb field of a high Z target senses a field of virtual photons that induces the dissociation of the projectile $Y \to X + p$. The resulting cross section for Coulomb dissociation is a product between the photodissociation cross section and the number of virtual photons of the particular multipolarity and energy needed:

$$\frac{d^2\sigma}{dE_\gamma d\Omega}(E_\gamma,\theta) = \frac{1}{E_\gamma}\left[\frac{dN(E1,E_\gamma)}{d\Omega}\sigma_{E1}^{photo}(E_\gamma) + \frac{dN(E2,E_\gamma)}{d\Omega}\sigma_{E2}^{photo}(E_\gamma)\right] \tag{1}$$

The photodissociation cross section is directly related to the radiative capture cross section sought in nuclear astrophysics. In reality problems arise from the need of relatively large projectile incident energies to produce enough virtual protons of the large E$_\gamma$ energy needed to produce photodissociation. Another problem stems from the fact that different multipoles can contribute in different proportions in Coulomb dissociation and in radiative capture (see eq. above). Therefore a disentangling of different multipole contributions from angular distribution measurements in Coulomb dissociation is needed before translating the results into astrophysical S-factors for radiative capture. That may be experimentally very demanding. Also, it is difficult, if not impossible, to separate the contribution of the nuclear field from that of the Coulomb field in dissociation at large energies. This is done requiring that dissociation happens at large impact parameters, which translates into measurements very close to zero degrees, experimentally a difficult task. However a number of very good Coulomb experiments have been done to obtain astrophysical data, too many to cite them all, and the method is considered rather well established. One important conceptual advantage of the method is that from Coulomb dissociation the energy dependence $S(E)$ can be experimentally extracted (E = relative p-core X energy). Currently the method is considered very appropriate for use with proton rich radioactive beams at intermediate energies obtained through projectile fragmentation. Notable contributions were brought by the Coulomb dissociation studies of $^{8}B \rightarrow {}^{7}Be+p$ to the determination of the S_{17} factor for solar neutrinos [13,16,19]. Initially somewhat lower values were obtained from Coulomb dissociation data than from direct measurements $S_{17}(0) = 18.6 \pm 0.4(exp) \pm 1.1(syst)$ eV b [16], a latter re-analysis of the GSI2 set of data gets even closer [20]. The largest impact of these experiments is that they gave essentially the same result for S_{17} with a totally different approach, and with totally different sources of errors or inaccuracies than the direct measurements, putting the nuclear physics part of the solar neutrinos on a firm basis.

3.2. *The ANC method*

In the past years we used the proton transfer reactions $^{10}B(^{7}Be,^{8}B)^{9}Be$ [9] and $^{14}N(^{7}Be,^{8}B)^{13}C$ [21,22] and what became known as the Asymptotic Normalization Coefficient method [31] to determine the amplitude of the tail of the overlap integral of the ground state wave function of ^{8}B onto the two-body channel $^{7}Be+p$. It is known for long time [32] that this normalization of the wave function of the last proton at large distances that determines entirely the astrophysical S-factors for charged particle radiative capture. The method consists in determining this normalization constant from transfer reactions at 10-15 MeV/u, which are also peripheral, but happen at smaller distances and, therefore, have much larger cross sections. Typically, from nucleon transfer reactions, spectroscopic factors

$S^A_{n_A l_A j_A}$ are extracted, we extract the asymptotic normalizations $C^2_{n_A l_A j_A}$ of the overlap integrals.

$$\frac{d\sigma}{d\Omega} = \sum S^A_{n_A l_A j_A} S^B_{n_B l_B j_B} \sigma_{DWBA}(\theta) = \sum C^2_{n_A l_A j_A} C^2_{n_B l_B j_B} \frac{\sigma_{DWBA}(\theta)}{b^2_{n_A l_A j_A} b^2_{n_B l_B j_B}} \quad (2)$$

One advantage of extracting the ANCs, versus the spectroscopic factors, is that the former are insensitive to the choice of the geometry of the nucleon binding potential (reflected in the single particle ANCs b^2_{nlj}) that we chose in the DWBA calculations (see Fig. 1).

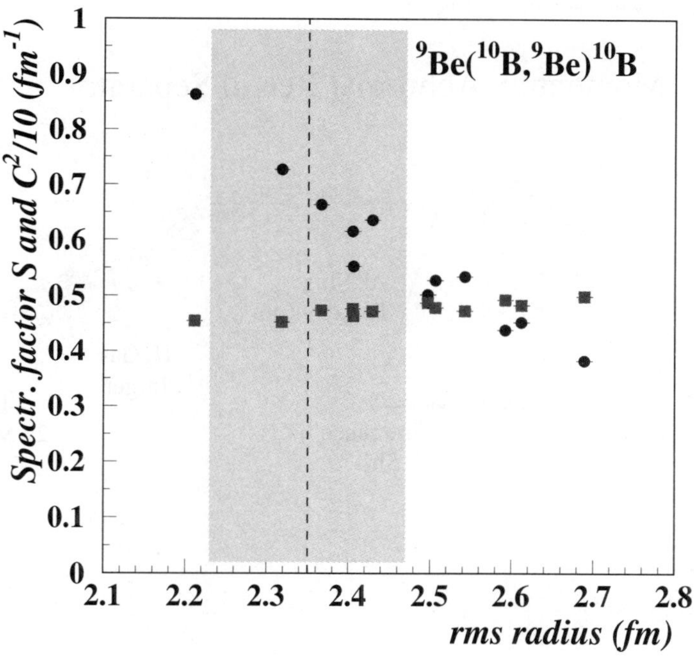

Fig. 1. Spectroscopic factors (circles) and ANCs (squares) are extracted from experimental data using proton binding potentials of different geometries. The ANC values are insensitive to the geometry chosen. For each geometry (r_0, a) used, the rms radius of the ^{10}B charge distribution was calculated and is used as ordinate in the graph. The vertical bar shows the experimental radius with its error (shaded area).

The ANC method was proved to reproduce within 9% the results from direct measurements in the case of the proton radioactive capture $^{16}O(p,\gamma)^{17}F$ [33] and was successfully used for a number of cases. The two (^{7}Be,^{8}B) experiments cited above lead to mutually consistent ANC values. To analyze the transfer reaction X(^{7}Be,^{8}B)Y using DWBA calculations to extract the ANC, the optical model potentials (OMP) are needed in both the incoming ^{7}Be+X ^{14}N channel and in the outgoing ^{8}B+Y ^{13}C channel. In both measurements the radioactive beam ^{7}Be at 12 MeV/u was obtained by bombarding a H_2 cryogenic gas target with ^{7}Li primary beam from the Texas A&M University superconducting cyclotron K500. Using the Momentum Achromat Recoil Separator (MARS), secondary beams of about 10^5 pps, high purity (above 99%) and sizes of about 4 mm diameter on target, were selected (Fig. 2).

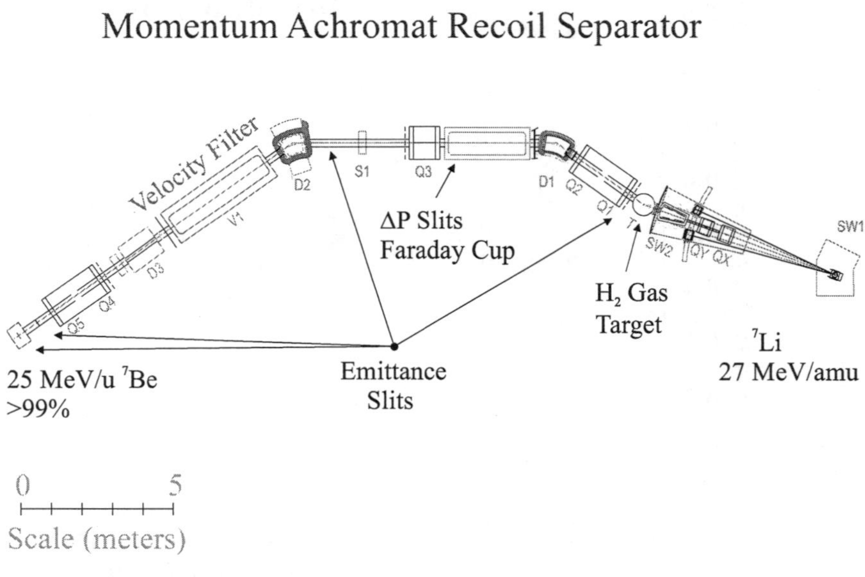

Fig. 2. The production of radioactive beams in MARS.

They bombarded secondary ^{10}B or melamine (composite $C_3N_6H_6$) targets, and 2 to 4 Si detector telescopes were positioned at angles covering angles 8 up to 30 deg. to measure simultaneously elastic scattering and transfer reaction products. The measurement of the elastic scattering was made in order to get the optical potential used in the incoming channels in DWBA calculations for the proton transfer reactions (^{7}Be,^{8}B). In the latter case, we also measured the elastic scattering of ^{8}B beam on a C target with the aim of checking directly, for the first time, the optical model we used in the outgoing channel ^{8}B+^{13}C. For more details, see Refs. cited above. We also use the mixing ratio between the $1p_{1/2}$ and $1p_{3/2}$ determined in the (^{7}Li,^{8}Li) experiment [34]. The measurements permited us to extract the Asymptotic Normalization Coefficient (ANC) of ^{8}B from the proton transfer (^{7}Be,^{8}B) reaction: $C^2(^8B, p_{3/2}) = 0.418 \pm 0.040$ fm^{-1}, resulting in $C^2_{tot}(^8B) = C^2_{p_{3/2}} + C^2_{p_{1/2}} = 0.471 \pm 0.044$ fm^{-1}. This lead to an average value for the astrophysical S-factor for the ^{7}Be(p,γ)^{8}B reaction $S_{17}(0) = 18.2 \pm 1.7$ eV b, in reasonable agreeement with the results of the other methods.

The astrophysical S-factors for the radiative proton capture reactions ^{11}C(p,γ)^{12}N, ^{13}N(p,γ)^{14}O and ^{12}N(p,γ)^{13}O were also obtained from the measurement of ANCs from proton transfer reactions at 12 MeV/u with radioactive beams separated with MARS: ^{14}N(^{11}C,^{12}N)^{13}C [35], ^{14}N(^{13}N,^{14}O)^{13}C [36] and ^{14}N(^{12}N,^{13}O)^{13}C [37]. A large number of elastic scattering and transfer reaction studies with loosely bound but stable beams were made to establish and test the method, as well as to establish a successful double folding procedure to predict OMP at around 10 MeV/u [38,48].

3.3. *Breakup of loosely bound nuclei at intermediate energies*

A few years back we proposed to extract astrophysical S-factors from one-nucleon-removal (or breakup) reactions of loosely bound nuclei at intermediate energies or later [11,30,34]. The method is based on data showing that the structure of loosely bound nuclei is dominated by one or two nucleons orbiting a core. Consequently, we use the fact that their breakup is essentially a peripheral process, and therefore, the breakup cross-sections can give information about the wave function of the last nucleon at large distances from the core. More precisely, asymptotic normalization coefficients (ANCs) can be determined. We show that there exists a favorable kinematical window (30-150 MeV/u) in which breakup reactions are highly peripheral and are dominated by the external part of the wave function and, therefore, the ANC is the better quantity to be extracted. The approach offers an alternative and complementary technique to extracting ANCs from transfer reactions explained above, with the advantage that beams of much poorer quality and intensity are sufficient.

In the breakup of loosely bound nuclei at intermediate energies, a nucleus $B = (Ap)$, where B is a bound state of the core A and the nucleon p, is produced by fragmentation from a primary beam, separated and then used to bombard a secondary target. In measurements, the core A is detected, measuring its parallel and transverse momenta and eventually the gamma-rays emitted from its deexcitation. Spectroscopic information can be extracted from these experiments, such as the orbital momentum of the relative motion of the nucleon and the contribution of different core states, typically comparing the measured momentum distributions with those calculated with Glauber models. The integrated cross sections can be used to extract absolute spectroscopic factors [39] or the ANC [11]. The latter approach has, again, the advantage that it is independent of the geometry of the proton binding potential. Here first we use the well studied case of ^{8}B breakup as a benchmark to demonstrate the usefulness of the method and show the possibilities of the Glauber reaction model used (details of the model in refs. [40,41]). The model calculations were extensively tested against existing experimental data: parallel and transversal momentum distributions on a wide range of targets, from Be to Pb, and energies. We have shown that existing experimental data at energies between 30 and 1000 MeV/nucleon [24,42–45] on a range of light and heavy targets translate into consistent values of the ANC, which is then used to determine the astrophysical factor S_{17}. Two approaches were used in calculations. The first is a potential approach. To obtain the folded potentials needed in the S-matrix calculations we used the JLM effective nucleon-nucleon interaction [46], using the procedure and the renormalizations of ref. [48]. We applied this technique for energies below 285 MeV/nucleon only and on all targets. In a second approach, the Glauber model in the optical limit was used. Calculations were done using different effective nucleon-nucleon interactions with different ranges. No new parameters were adjusted. The contribution of the ^{7}Be core excitation was calculated for each target and at each energy using the data from an experiment which disentangle it [45], and corrected for in all cases. For details on the procedure see [25]. In Fig. 3 we show that from the widely varying breakup cross sections (upper panel) on all targets and at so different energies, we extract ANCs which are consistent with a constant value (lower panel). However, we see that a certain

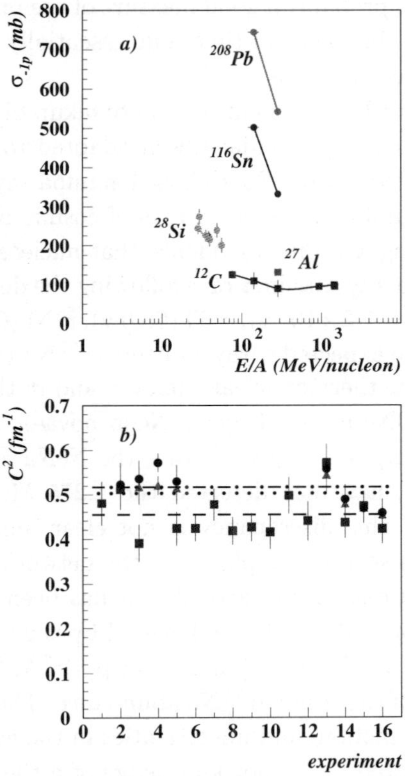

Fig. 3. a) The one-proton-removal cross sections on C, Al, Sn and Pb targets, depending on energy. b) The ANCs determined from the breakup of ^{8}B at 28-1000 MeV/nucleon using the data above and various effective interactions: JLM (squares), "standard" (circles) and "Ray" (triangles). The dashed, dotted and dash-dotted lines are the averages of the three interactions above, in that order. List of experiments in Ref. [25].

dependence on the used effective NN interaction exists, which may point to the limitations of our present knowledge of the effective nucleon-nucleon interactions. Taking the unweighted average of all 31 determinations we found an ANC $C^2_{tot}(JLM) = 0.483 \pm 0.050$ fm^{-1} (fig. 3). The value is in agreement with that determined using the (^{7}Be,^{8}B) proton transfer reactions at 12 MeV/u [9,22]. The two values agree well, in spite of the differences in the energy ranges and in the reaction mechanisms involved. The ANC extracted leads to the astrophysical factor $S_{17}(0) = 18.7 \pm 1.9$ eV$\cdot$ b for the key reaction for solar neutrino production ^{7}Be(p,γ)^{8}B. The uncertainties quoted are only the standard deviation of the individual values around the average, involving therefore the experimental and theoretical uncertainties. This 10% error bar is probably a good measure of the precision we can claim from the method at this point in time, due essentially to the uncertainties in the cross section calculations.

Further, a proposed experiment for the breakup of ^{23}Al is discussed to show that the method is particularly well adapted to rare isotope beams produced using fragmentation. Space-based gamma-ray telescopes have for some time had the ability to detect γ-rays of cosmic origin. They have already provided strong and direct evidence that nucleosynthesis is an ongoing process in our Galaxy. Gamma-rays following the decay of long-lived isotopes like ^{26}Al ($t_{1/2} = 7.2\times10^5$ y), ^{44}Ti (60.0 y), ^{56}Ni (6.1 d), etc. have been observed. Among the expected γ-ray emitters is ^{22}Na ($t_{1/2}$=2.6 y), thought to be produced in the thermonuclear runaway and in the high-temperature phase of so-called ONe novae (Oxygen-Neon novae) through the reaction chain ^{20}Ne(p,γ)^{21}Na(p,γ)^{22}Mg($\beta\gamma$)^{22}Na – the NeNa cycle [1,52–54]. Measurements, however, have not detected the 1.275 MeV gamma-ray from ^{22}Na. The origin of this discrepancy is not clear, but a poor knowledge of the reaction cross sections employed in the network calculations for the rp-process may be a reason. In particular, it has been proposed that ^{22}Na itself or its precursor ^{22}Mg could be depleted by the radiative proton capture reactions ^{22}Na(p,γ)^{23}Mg [55] and ^{22}Mg(p,γ)^{23}Al [56], both leading to a serious reduction of the residual ^{22}Na abundance. The former is the prime candidate for this depletion, and uncertainties in the evaluation of its reaction rate are dominated by a poor knowledge of a the relevant resonances (position and resonance strengths). For the second reaction a further complication appeared when some measurements of the reaction cross sections for N=10 isotones and Z=13 isotopes around 30 MeV/nucleon on a ^{12}C target found a remarkable enhancement for ^{23}Al, which led the authors to the conclusion that it is one of the rare proton halo nuclei [59]. This was explained with a presumed level inversion between the $2s_{1/2}$ and $1d_{5/2}$ orbitals. The inversion was further supported by several microscopic nuclear structure calculations [60,61]. (NNDC [47] gives $J^\pi=3/2^+$). If the above mentioned inversion is correct, it will affect the radiative capture cross section much more strongly than any other uncertainties: we recalculate the astrophysical S-factor (fig. 3a) and the stellar reaction rate (fig. 3b) for the ^{22}Mg(p,γ)^{23}Al reaction and find an increase of 30 to 50 times over the current estimate of the rate for the temperature range $T_9 = 0.1 - 0.3$. We set to determine this spin by measuring the momentum distributions from the breakup of ^{23}Al in an experiment next month in GANIL (already successfully carried on at the time of typescript submission). The momentum distribution for the $2s_{1/2}$ case should be considerably narrower than the one for the $1d_{5/2}$ case. Simultaneously we have addressed the problem of spin determination in experiments at TAMU, as explained in the next section.

4. Other spectroscopic methods

There are a number of spectroscopic data that are either needed or useful in assessing astrophysically important reaction cross sections. I'll only discuss here two examples, both from the same study [62] of beta-decay of the proton rich nucleus ^{23}Al. Pure samples of ^{23}Al were obtained with a 48-MeV/nucleon ^{24}Mg beam bombarding a H_2 cryogenic gas target, 2.5 mg/cm^2 thick. Recoiling ^{23}Al nuclei from the p(^{24}Mg,^{23}Al)2n reaction were separated in MARS. The secondary beam then passed through a 50-μm-thick Kapton foil into air, then through a plastic scintillator foil, which counted the ions, through a set of Al attenuators, and finally stopped in the 76-μm-thick aluminized mylar tape of the tape-transport system. Beta singles and β-γ coincidences were measured with a plastic scintillator and a well efficiency calibrated HpGe detector. They allowed us to determine the absolute branching ratios and ft-values for transitions to final states in the daughter nucleus ^{23}Mg, including some with already known spins and parities. From the latter we unambiguously determined $J^\pi = 5/2^+$ for the ^{23}Al ground state. This contradicts the earlier suggestion that $J^\pi = 1/2^+$. This decreases the possibility of a simple explanation for the non-observation of the 1.275 MeV line from ^{22}Na decay in spectra taken by space-based gamma-ray telescopes. There is no evidence that its precursor, ^{22}Mg, can be depleted in ONe novae explosions by the reaction ^{22}Mg(p,γ)^{23}Al.

We have also found two states in ^{23}Mg with small ft-values at 7803(2) keV and 7787(2) keV, and identified them to be the isobaric analog state of the ^{23}Al ground state and a $J^\pi = (7/2)^+$ state, which likely dominates the proton-decay spectrum, respectively. Both are resonances contributing to the depletion reaction, ^{22}Na(p,γ)^{23}Mg. For the latter resonance at $E_{res} =$ 207(2) keV we find its resonance strength to be $\omega\gamma = 2.6(9)$ meV, making it the dominant contributor to the reaction rate at the temperatures of explosive H burning in ONe novae. This resonant strength determination, based on data from several indirect experimental data is in good agreement, but totally independent, from a determination from a direct measurement ^{22}Na(p,γ)^{23}Mg involving a radioactive ^{22}Na target [63].

5. Conclusions

The above S_{17}values extracted from Coulomb dissociation, from proton transfer reactions (^{7}Be,^{8}B) and from the breakup of ^{8}B are in agreement with all the values obtained from other indirect methods and with most of those from direct measurements, but one. Our experience leads us to believe that one cannot expect precisions under 10% from this type of indirect determinations, and this is sufficient for most of the cases encountered in nuclear astrophysics.

In the last Section I have shown how diverse spectroscopic data can be used to determine astrophysical reaction rates. The example used was that of the $\beta-$decay of the proton rich drip-line nucleus ^{23}Al.

6. Acknowledgments

This paper is based mostly on work done with the people of the MARS group at the Cyclotron Institute, Texas A&M University: RE Tribble, CA Gagliardi, JC Hardy, AM Mukhamedzhanov, A Azhari, X Tang, VE Iacob, N. Nica, G. Tabacaru, A Banu and many students. F. Carstoiu of IFIN-HH Bucharest very beneficially collaborated with us over many years. I thank them all. The original articles are cited throughout this paper. The work was supported in part by the U.S. Department of Energy under Grant No. DE-FG03-93ER40773.

References

1. C. Rolfs and W. Rodney, Cauldrons in the Cosmos, Univ. of Chicago Press, Chicago 1988.
2. W.C. Haxton, P.D. Parker, C.E. Rolfs, Nucl.Phys. **A777**, 226 (2006).
3. R. Davis, D.S. Harmer and K. C. Hoffman, Phys. Rev. Lett. **20**, 1205 (1968).
4. SuperKamiokande collaboration, S. Fukuda *et al.*, Phys. Rev. Lett. **86**, 5651 (2001).
5. SNO collaboration, S. N. Ahmed *et al.*, Phys Rev. Lett. **87**, 071301 (2003).
6. J. N. Bahcall, M. H. Pinsoneault, Phys. Rev. Lett. **92,** 121301 (2004).
7. F. Hammache *et al.*, Phys. Rev. Lett. **86,** 3985 (2001).
8. F. Strieder *et al.*, Nucl. Phys. **A696**, 219 (2001).
9. A. Azhari *et al.*, Phys. Rev. Lett. **82**, 3960 (1999); Phys. Rev. C **60,** 055803 (1999).
10. B. Davids *et al.*, Phys. Rev. Lett. **86**, (2001).
11. L. Trache, F. Carstoiu, C. A. Gagliardi and R.E. Tribble, Phys. Rev. Lett. **87,** 271102 (2001).
12. A. R. Junghans *et al.*, Phys. Rev. Lett. **88,** 041101 (2002).
13. F. Schumann *et al.*, Phys. Rev. Lett. **90,** 232501 (2003).

14. L. T. Baby *et al.*, Phys. Rev. C **67**, 065805 (2003).
15. P. Descouvemont, Phys. Rev. C 70, 065802 (2004).
16. B. Davids and S. Typel, Phys. Rev. C **68,** 045802 (2003).
17. A. R. Junghans *et al.*, Phys. Rev. C **68**, (2003).
18. R. H. Cyburt, B. Davids, and B.K. Jennings, Phys. Rev. C **70**, 045501 (2004).
19. Kikuchi et al., EuroPhys. J. **A 3**, 213 (1998).
20. F. Schumann *et al.*, Phys. Rev. C **73**, 015806 (2006).
21. A. Azhari *et al.*, Phys. Rev. C **60,** 055803 (1999).
22. G. Tabacaru *et al.*, Phys. Rev. C **73**, 025808 (2006).
23. A. B. Brown *et al.*, Phys. Rev. C **65**, (2002).
24. J. Enders *et al.*, Phys. Rev. C **67,** 064301 (2003).
25. L. Trache, F. Carstoiu, C. A. Gagliardi and R.E. Tribble, Phys. Rev. C **69**, (2004).
26. G. Baur, Phys. Lett. **178B**, 135 (1986).
27. C. Spitaleri, in S. Stoica, L. Trache and R.E. Tribble (eds.), "*Exotic Nuclei and Nuclear/Particle Astrophysics*", p. 316, World Scientific, Singapore, 2006.
28. G. Baur, C.A. Bertulani and H. Rebel, Nucl. Phys. **A458**, 188 (1986).
29. G. Baur and H. Rebel, Ann. Rev. Nucl. Sci. **46**, 321 (1996).
30. L. Trache, F. Carstoiu, M. A. Mukhamedzhanov and R.E. Tribble, Phys. Rev. C **66,** 035801 (2002).
31. A.M. Mukhamedzhanov, C.A. Gagliardi and R.E. Tribble, Phys. Rev. C **63**, 024612 (2001).
32. R.F. Christy and I. Duck, Nucl. Phys. **24**, 89 (1963).
33. C.A. Gagliardi *et al.*, Phys. Rev. C **59**, (1999).
34. L. Trache *et al.*, Phys. Rev. C **67**, 062801(R) (2003).
35. X. Tang *et al.*, Phys. Rev. C **67**, 015804 (2003).
36. X. Tang *et al.*, Phys. Rev. C **69**, 055807 (2004).
37. A. Banu *et al.*, to be publ. in Phys. Rev. C.
38. F. Carstoiu L. Trache, C.A. Gagliardi and R.E. Tribble, Phys. Rev. C **70**, 054610 (2004).
39. P. G. Hansen and B. M. Sherrill, Nucl. Phys. **A693**, 133 (2001).
40. E. Sauvan *et al.*, Phys. Rev. C **69**, (2004).
41. K. Hencken, G. Bertsch and H. Esbensen, Phys. Rev. C **54**, 3043 (1996).
42. F. Negoita *et al.*, Phys. Rev. C **54,** 1787 (1996).
43. B. Blank *et al.*, Nucl. Phys. **A624,** 242 (1997).
44. R. E. Warner *et al.*, Phys. Rev. C **74**, 014605 (2006).
45. D. Cortina-Gil *et al.*, Nucl. Phys. **A720,** 3 (2003).
46. J. P. Jeukenne, A. Lejeune and C. Mahaux, Phys. Rev. C **16,** 80 (1977).
47. National Nuclear Data Center, Brookhaven National Laboratory Online Data Service.
48. L. Trache *et al.* Phys. Rev. C **61**, 024612 (2000).
49. B. Davids *et al.*, Phys. Rev. C **63**, 065806 (2001).
50. C. Bertulani and G. Baur, Nucl. Phys. **A480**, 615 (1988).
51. L. Ray, Phys. Rev. C **20,** 1857 (1979).
52. S. Starrfield, J.W. Truran, M. Wiescher and W.M. Sparks, Mon. Not. R.

Astron. Soc. **296**, 502 (1998).
53. J. Jose, A. Coc and M. Hernanz, Astroph. J. **520**, 347 (1999).
54. S. Wanajo, M. Hashimoto and K. Homono, Astrophys. J. **523**, 409 (1999).
55. M. Arnould and W. Beelen, Astron. Astroph. **33**, 215 (1974); M. Arnould and H. Norgaard, Astron. Astroph. **64**, 195 (1978).
56. M. Wiescher *et al.*, Nucl. Phys. **A484**, 90 (1988).
57. I. Iyundin *et al.*, Astron. Astroph. **300**, 422 (1995).
58. R. Diehl, Nucl. Phys. **A718**, 52c (2003) and refs therein.
59. X. Z. Cai *et al*, Phys. Rev. C **65**, 024610 (2002).
60. H.-Y. Zhang *et al.*, Chin. Phys. Lett. **19**, 1599 (2002); ibidem **20**, 46 (2003), 1234.
61. H.Y. Zhang *et al.*, Nucl. Phys. **A722**, 518c (2003).
62. V.E. Iacob *et al.*, Phys. Rev. C **74**,045810 (2006).
63. F. Stegmuller *et al.*, Nucl. Phys. **A601**, 168 (1996).

I.2 Microscopic Formalisms

NEW MICROSCOPIC APPROACHES TO THE NUCLEAR EIGENVALUE PROBLEM

N. LO IUDICE+, F. ANDREOZZI AND A. PORRINO

Dipartimento di Scienze Fisiche, Universitá di Napoli "Federico II"
and Istituto Nazionale di Fisica Nucleare,
Monte S Angelo, Via Cintia I-80126 Napoli, Italy
+E-mail: loiudice@na.infn.it
www.unina.it

F. KNAPP* and J. KVASIL

Institute for Particle and Nuclear Physics MFF, Charles University,
V Holesovickach 2, 180 00 Praha 8, Czech Republic
kvasil@ipnp.troja.mff.cuni.cz

We outline two methods for facing the nuclear eigenvalue problem, one suitable for low-lying spectroscopy and the other for describing collective modes. The first is an an important sampling algorithm which generates a subset of exact eigensolutions of the nuclear shell model Hamiltonian within a truncated space. The other is an equation of motion method which generates iteratively a microscopic multiphonon basis and is therefore especially suitable for the investigation of collective modes. Numerical tests of both methods are presented.

Keywords: Nuclear many-body; Eigenvalue problem; New methods.

1. Introduction

The spectacular progress in computer power registered in the last two decades has stimulated the development of sophisticated techniques for facing the nuclear eigenvalue problem. These new approaches either face directly the diagonalization of the nuclear shell model (SM) Hamiltonian or circumvent the problem by resorting to stochastic methods. To the first category belongs an iterative diagonalization algorithm we have developed recently. This will be illustrated here in some more detail.

SM is, in general, not suitable for studying nuclear collective proper-

*Also at Nuclear Physics Institute, 250 68 Řež, Czech Republic.

ties. These are studied traditionally in mean field approaches, like Tamm-Dancoff (TDA) or random-phase approximation (RPA), which select the configurations building up the collective states. TDA and RPA, however, are harmonic approximations and, therefore, are not suitable for describing the anharmonic features of the nuclear spectra. As we shall see, there is growing evidence of multiphonon excitations and anharmonicities in nuclei and a raising interest in developing many-body approaches capable of accounting for these new features. We will briefly discuss some of them with special attention at an iterative equation of motion method which we have proposed for an exact study of multiphonon excitations in a microscopic multiphonon space.

2. Large scale shell model approaches

The stochastic techniques are basically Monte Carlo methods. They have proved to be extremely powerful in accounting for the ground states properties and computing nuclear responses [1]. The direct diagonalization approaches are mostly based on Lanczos [2] algorithm. A bridge between the two approaches is represented by a method which generates stochastically a truncated basis for the direct diagonalization of the many-body Hamiltonian [3]. As we shall see, each method has remarkable virtues but also specific limitations.

2.1. *Direct diagonalization: Lanczos method*

Standard diagonalization methods are not suitable for large scale shell model calculations. They require in fact N^3 operations, N being the dimension of the matrix [4]. On the other hand, one is interested only in few nuclear eigenstates. Efficient iterative methods are now available for this purpose. A very successful one is the Lanczos method. The goal of Lanczos method is to construct an orthogonal basis which brings the Hamiltonian matrix H to tridiagonal form.

The action of H on a starting (pivot) vector $\mid 0>$ decomposes the space into two orthogonal pieces

$$H \mid 0 >= \alpha_0 \mid 0 > + \mid w_1 >, \qquad\qquad < 0 \mid w_1 >= 0, \qquad (1)$$

where $\alpha_0 =< 0 \mid H \mid 0 >$ and the state $\mid w_1 >$ is unnormalized. We can then construct the normalized vector $\mid 1 >=\mid w_1 > / < w_1 \mid w_1 >^{1/2}$ and obtain $\beta_1 =< 0 \mid H \mid 1 >=< w_1 \mid w_1 >^{1/2}$. The state $\mid 1 >$ is thereby determined. We now iterate the procedure. By operating on the generic state $\mid k >$ we obtain the following orthogonal decomposition

$$H \mid k >= \beta_{k-1} \mid k-1 > +\alpha_k \mid k > +\beta_{k+1} \mid k+1 >, \tag{2}$$

where $\mid k >=\mid w_k > / < w_k \mid w_k >^{1/2}$ and

$$\alpha_k =< k \mid H \mid k >, \qquad \beta_k =< k-1 \mid H \mid k >=< w_k \mid w_k >^{1/2}. \tag{3}$$

The tridiagonal matrix so obtained is to be diagonalized at each iterative step k until convergence to a selected sets of eigenvalues is reached.

The Lanczos procedure is related to the number of iterations, rather than the dimensions of the matrix. Moreover, when used in the $m-$scheme, takes full advantage of the sparse nature of the nuclear Hamiltonian matrix. It is therefore a fast and efficient tool for determining the shell model states [5]. Its numerical implementation, however, deserves special care. In fact, the generated states are mathematically but not numerically orthogonal. The error so induced propagates rapidly with the number of iterations and may yield "ghost" or spurious states, like the multiple appearance of the same state or the intrusion of states of unwanted angular momenta. Several codes based on Lanczos are now available. Among them, *Antoine* [6], which works in the $m-$scheme, is widely adopted.

2.2. *Shell Model Monte Carlo*

Alternative approaches, which circumvent the direct diagonalization of H, are the shell model Monte Carlo (SMMC) methods [7,8]. The central quantity of these methods is the imaginary-time evolution operator, U = exp($-\beta H$), where $\beta = 1/T$ is interpreted as the inverse of the temperature T. In SMMC one computes expectation values

$$< O >= \frac{< \Phi \mid e^{-\beta/2\ H} O e^{-\beta/2\ H} \mid \Phi >}{< \Phi \mid e^{-\beta\ H} \mid \Phi >} = \frac{Tr\ e^{-\beta H} O}{Tr\ e^{-\beta H}}, \tag{4}$$

where the trace $TrX = \sum_i < i \mid X \mid i >$ sums over all many-body states of the system. In the zero temperature limit, one gets the ground state expectation value

$$< O >_0 = lim_{\beta \to 0} \frac{< \Phi \mid e^{-\beta/2\ H} O e^{-\beta/2\ H} \mid \Phi >}{< \Phi \mid e^{-\beta\ H} \mid \Phi >} = \frac{< \Psi \mid O \mid \Psi >}{< \Psi \mid \Psi >}. \tag{5}$$

Next step is to write the Hamiltonian in the form

$$H = \sum_\alpha \epsilon_\alpha b_\alpha + \frac{1}{2} \sum_\alpha V_\alpha b_\alpha^\dagger b_\alpha, \tag{6}$$

where b_α is a density operator bilinear in the Fermionic creation and annihilation operators $a^\dagger$ and a. In the case of a single density operator b (no sums over α), the difficulty of computing the evolution operator comes only from the two body term. One can, however, linearize the Hamiltonian by means of the Hubbard -Stratonovich transformation

$$e^{-\beta H} = \sqrt{\frac{\beta \mid V \mid}{2\pi}} \int d\sigma e^{-\frac{1}{2}\beta|V|\sigma^2} e^{-\beta h}, \tag{7}$$

where σ is a $c-$number field and $h = \epsilon b + sV\sigma b$ is now a one body Hamiltonian with $s = 1$ for $V < 0$ and $s = i$ for $V > 0$.

For a realistic Hamiltonian of the form (6), since $[b_\alpha, b_\beta] \neq 0$, one must split the interval β into N_t time slices of length $\Delta\beta = \beta/N_t$ so that the evolution operator can be written as

$$e^{-\beta H} = [e^{-\Delta\beta H}]^{N_t}. \tag{8}$$

One then linearizes for each time slice $n = 1, \ldots N_t$ using auxiliary fields $\sigma_{\alpha n}$ obtaining

$$< O >= \frac{\int D_\sigma W_\sigma O_\sigma}{\int D\sigma W_\sigma}, \tag{9}$$

where

$$W_\sigma = e^{-\Delta\beta \sum_{\alpha n} |V_\alpha||\sigma_{\alpha n}|^2} TrU_\sigma, \qquad O_\sigma = \frac{TrU_\sigma O}{TrU_\sigma}$$
$$D\sigma = \Pi_{\alpha n} d\sigma_{\alpha n} d\sigma *_{\alpha n} \left(\frac{\Delta\beta \mid V_\alpha \mid}{2\pi}\right) \tag{10}$$

and

$$U_\sigma = U_{N_t} \ldots U_2 U_1, \qquad U_n = e^{-\Delta\beta h_n}$$
$$h_n = \sum_\alpha (\epsilon_\alpha + s_\alpha V_\alpha \sigma *_\alpha) b_\alpha + c.c. \tag{11}$$

For a high accuracy, the number of time slices N_t must be very large. Since there is a variable for each operator at each time slice, the dimension D of these integrals, being of the order $N_s N_t$, is also very large and might exceed 10^5. On the other hand, Eq. (9) can be written as

$$< O >= \int D_\sigma P_\sigma O_\sigma \tag{12}$$

where

$$P_\sigma = \frac{W_\sigma}{\int D\sigma W_\sigma} \tag{13}$$

can be considered a probability density, since $P_\sigma \geq 0$ and $\int D\sigma P_\sigma = 1$. Thus, $< O >$ comes out to be the average of O_σ weighted by P_σ. One may choose randomly a set S of configuration σ_s with probability P_σ and approximate $< O >$ with

$$< O >= \int D_\sigma P_\sigma O_\sigma \approx \frac{1}{S} \sum_{s=1,S} O_s, \tag{14}$$

where O_s is the value of O_σ at the field configuration σ_s. The same $< O >$, depending on the random choice of the field configurations, is a random variable. In virtue of the central limit theorem, its average value is the required value with an uncertainty

$$\sigma^2_{<O>} = \frac{1}{S} \int D_\sigma P_\sigma (O_\sigma - < O >)^2 \approx \frac{1}{S^2} \sum_s (O_\sigma - < O >)^2. \tag{15}$$

It remains now to use a method for generating the field configurations. Generally one adopts the Metropolis, Rosenbluth, Rosenbluth, Teller and Teller algorithm.

The method adopted requires that the weight function W_σ be real and non negative. Unfortunately, this occurs only for schematic Hamiltonian such as pairing plus quadrupole. In general, W_σ may be negative. This is the famous ***sign problem*** of MC methods which has not found yet a clear-cut solution.

Even with this limitation, SMMC has proved to be a powerful and efficient tool for studying ground-state and thermal properties of medium-mass nuclei as well as electroweak nuclear properties such as Gamow Teller strength distributions and the dipole giant resonance.

2.3. *Quantum Monte Carlo Diagonalization*

SMMC evaluates expectation values and strength functions, but does not give explicit eigenvectors. These can be obtained only by explicit diagonalization of H. The quantum Monte Carlo diagonalization (QMCD) method combines the two approaches [9]. It is based on direct diagonalization, but the basis states are generated stochastically by MC. The QMCD exploits the fact the the imaginary-time evolution operator behaves as a filter which yields only the ground state for $\beta \to \infty$. Such a state can then be determined stochastically as

$$| \Psi_g >= \sum_{MC;\sigma} | \Phi_\sigma >, \tag{16}$$

where

$$| \Phi_\sigma > \propto \Pi_\sigma e^{-\Delta\beta h(\sigma)} | \Psi_0 >, \tag{17}$$

Ψ_0 being a pivot state.

The main idea of QMCD is that the $| \Phi_\sigma >$ can be interpreted as basis vectors and, thus, the Hamiltonian H can be diagonalized in the space spanned by these vectors. This idea has then lead to the following procedure for generating the basis. Suppose that a set of k basis vectors has been generated stochastically from a pivot state. Then a new vector $| \Phi_\sigma >$ is generated stochastically and added to the set. We diagonalize the Hamiltonian in this enlarged space. If the energy of the ground state is lowered appreciably, $| \Phi_\sigma >$ is included in the basis. It is discarded otherwise. The iteration proceeds until the energy eigenvalue converges reasonably well.

It is clear from this brief outline, that QMCD selects the basis states stochastically, but the energies and wave functions are determined by the diagonalization. Once a basis has been selected for the ground state, one may implement the method to select a basis for a second excited state and so on. In this way the low energy states are determined. The random sampling of the QMCD allows to determine each eigensolution by diagonalizing the Hamiltonian in a restricted shell model space. On the other hand, the basis states so generated are not orthogonal and form in general a redundant set. Moreover, they do not have the spin as good quantum number. Specific procedures have been developed to obviate at these shortcomings.

QMCD has been adopted extensively and with success for systematics in the (s, d) and (pf) nuclear regions.

3. A new shell model algorithm

The approach we have developed [10,11] faces directly the diagonalization of the Hamiltonian. It is an iterative algorithm which generates a selected set of eigenvectors of a large matrix and is extremely simple to be implemented. Is is also endowed with an importance sampling which allows for a drastic reduction of the space and offers other important advantages. For illustrative purposes we will show how the method applies to typical nuclei in different regions of the periodic table.

3.1. *The algorithm*

For the sake of simplicity, we consider a symmetric matrix

$$A = \{a_{ij}\} = \{\langle i \mid \hat{A} \mid j \rangle\} \tag{18}$$

representing a self-adjoint operator $\hat{A}$ in an orthonormal basis $\{| 1\rangle, | 2\rangle, \ldots, | N\rangle\}$. The algorithm goes through several iteration loops. The first loop consists of the following steps:
1a) Diagonalize the two-dimensional matrix (a_{ij}) (i,j=1,2),
1b) select the lowest eigenvalue λ_2 and the corresponding eigenvector

$$| \phi_2\rangle c_1^{(2)} \mid 1\rangle + c_2^{(2)} \mid 2\rangle, \tag{19}$$

1c) for $j = 3, \ldots, N$, diagonalize the matrix

$$\begin{pmatrix} \lambda_{j-1} & b_j \\ b_j & a_{jj} \end{pmatrix} \tag{20}$$

where $b_j \langle \phi_{j-1} \mid \hat{A} \mid j\rangle$ and select the lowest eigenvalue λ_j together with the corresponding eigenvector $| \phi_j\rangle$. This zero approximation loop yields the approximate eigenvalue and eigenvector

$$E^{(1)} \equiv \lambda_N, \qquad | \psi^{(1)}\rangle \equiv | \phi_N\rangle \sum_{i=1}^{N} c_i^{(N)} \mid i\rangle. \tag{21}$$

With these new entries we start an iterative procedure which goes through $n = 2, 3, \ldots$ refinement loops, consisting of the same steps with the following modification. At each step $j = 1, 2, \ldots, N$ of the n-th loop ($n > 1$) we have to solve an eigenvalue problem of general form, since the states $| \phi_{j-1}\rangle$ and $| j\rangle$ are no longer orthogonal. The eigenvalue $E^{(n)} \equiv \lambda_N$ and eigenvector $| \psi^{(n)}\rangle \equiv | \phi_N\rangle$ obtained after the n-th loop are proved to converge to the exact eigenvalue E and eigenvector $| \psi\rangle$ respectively [10].

The algorithm has been shown to be completely equivalent to the method of optimal relaxation [12] and has therefore a variational foundation. Because of its matrix formulation, however, it can be generalized with minimal changes so as to generate at once an arbitrary number n_v of eigensolutions.

Indeed, we have to replace the two dimensional matrix (20) with a multidimensional one

$$\begin{pmatrix} \lambda_{j-1} & b_j \\ b_j & a_{jj} \end{pmatrix} \Longrightarrow \begin{pmatrix} \Lambda_k & B_k \\ B_k^T & A_k \end{pmatrix}, \tag{22}$$

where Λ_k is a n_v-dimensional diagonal matrix whose non-zero entries are the eigenvalues $\lambda_1^{(k-1)}, \lambda_2^{(k-1)}, \ldots, \lambda_{n_v}^{(k-1)}$, $A_k = \{a_{ij}\}$ is a p-dimensional submatrix, B_k and its transpose are matrices composed of the matrix elements $b_{ij}^{(k)} = \langle \phi_i^{(k-1)} \mid \hat{A} \mid j\rangle$. A loop procedure similar, though more general, to the one adopted in the one-dimensional case, yields a set of n_v eigenvalues $E_1, \ldots, E_{n_v}$ and corresponding eigenvectors $\psi_1, \ldots, \psi_{n_v}$.

3.2. *Importance sampling*

The just outlined algorithm, though of simple implementation, becomes inadequate when the dimensions of the Hamiltonian matrix become prohibitively large. In these cases, one must rely on some importance sampling which allows for a truncation of the space by selecting only the basis states relevant to the exact eigensolutions. Our sampling goes through the following steps:

1a) Bring the v-dimensional principal submatrix $\{a_{ij}\}\,(i,j = 1,v)$ through the diagonal form Λ_v with eigenvalues $\lambda_1, \lambda_2, \ldots, \lambda_v$.

1b) For $j = v+1,\ldots,N$, diagonalize the $v+1$-dimensional matrix

$$A = \begin{pmatrix} \Lambda_v & \vec{b}_j \\ \vec{b}_j^T & a_{jj} \end{pmatrix}, \tag{23}$$

where $\vec{b}_j = \{b_{1j}, b_{2j}, \cdot, b_{vj}\}$.

1c) Select the lowest v eigenvalues λ'_i, $(i = 1,v)$ and accept the new state only if

$$\sum_{i=1,v} \mid \lambda'_i - \lambda_i \mid > \epsilon \tag{24}$$

Otherwise restart from point 1b) with a new j.

The outcome of this procedure is that the selected states span a $n_s(< N)$ dimensional space, so that the subsequent refinement loops iterate only on the smaller set of n_s basis vectors. The number of operations is thereby reduced by a factor N/n_s. The sampling procedure has also the important virtue of generating a scaling law for the eigenvalues. Indeed the sampling parameter ϵ scales with n according to

$$\epsilon = b\frac{N}{n^2} \exp\left[-c\frac{N}{n}\right]. \tag{25}$$

This induces for the energies the following scaling law

$$E = E_0 + b\frac{N}{n} \exp\left[-c\frac{N}{n}\right], \tag{26}$$

where b, c, and E_0 are constants specific of each state and the full dimension N provides the scale. This law, which is somewhat different from the one proposed in Refs. [13], is valid for all states and nuclei examined and follows directly from the sampling [11].

3.3. *Multipartitioning method*

The extent of truncation induced by the sampling is maximal when the eigenvectors are highly localized. This condition is fulfilled in most physical problems. Even when this is not the case, we can approach the above condition by using a correlated basis obtained by a multipartitioning method [14]. This goes through the following prescriptions:
i) Partition the shell model space for N valence nucleons into orthogonal subspaces, P_1 and P_2 according to

$$P = P_1 + P_2, \tag{27}$$

ii) distribute N_1 and N_2 nucleons ($N_1 + N_2 = N$) among these subspaces in all possible ways,
iii) decompose the Hamiltonian H into

$$H = H_1 + H_2 + H_{12}, \tag{28}$$

iv) solve the eigenvalue equations

$$H_i \mid \alpha_i N_i\rangle = E_{\alpha_i} \mid \alpha_i N_i\rangle \tag{29}$$

obtaining the eigenstates $\mid \alpha_1 N_1\rangle$ and $\mid \alpha_2 N_2\rangle$ of H_1 and H_2 respectively in P_1 and P_2. Once this is done, it is possible to replace the standard shell model basis with one composed of the states

$$\mid \alpha N\rangle = \mid \alpha_1 N_1 \alpha_2 N_2\rangle. \tag{30}$$

We use the above basis to diagonalize the residual term H_{12} of the SM Hamiltonian. The new basis is in general highly correlated and, therefore, highly localized in the Fock space, a feature which enhances considerably the efficiency of the method.

3.4. *Selective numerical tests*

We applied the sampling algorithm to the semi-magic ^{108}Sn, the N=Z even-even ^{48}Cr and the $N > Z$ odd-even ^{133}Xe. The model spaces are:
1) $P \equiv \{2d5/2, 1g7/2, 2d3/2, 3s1/2, 1h11/2\}$ for the 8 valence neutrons of ^{108}Sn and for the 4 valence protons and 3 valence neutron holes of ^{133}Xe,
2)$P \equiv \{1f7/2, 1f5/2, 2p3/2, 2p1/2\}$ for the 4 valence protons and neutrons of ^{48}Cr.

We adopted a realistic effective interaction deduced from the Bonn-A potential [15] for ^{108}Sn and ^{133}Xe, and used the KB3 interaction [16] for ^{48}Cr.

The large energy gap ($\sim$ 2 MeV) between the two groups of subshells has suggested the following partitioning for ^{108}Sn

$$P \equiv \{2d5/2, 1g7/2, 2d3/2, 3s1/2, 1h11/2\} \begin{array}{l} \nearrow P_1 \equiv \quad \{2d5/2, 1g7/2\} \\ \qquad\qquad\qquad + \\ \searrow P_2 \equiv \{2d3/2, 3s1/2, 1h11/2\}. \end{array} \tag{31}$$

For ^{48}Cr and ^{133}Xe, we simply decompose the space into a proton and neutron subspace $P = P_p + P_n$.

The convergence rate of the low-lying levels is illustrated in Fig. 1 for the semi-magic ^{108}Sn and the N=Z even-even ^{48}Cr.

In all nuclei and for all states, the eigenvalues decrease monotonically and smoothly with n. Only in few cases, the energies undergo a jump from an upper to a lower curve, a signal of energy crossing. The subsequent behavior, however, is smooth as for the other states. It follows that, in all cases, starting from a sufficiently small ϵ, the energies scale with the dimensions n according to the law (26). This allows to extrapolate to asymptotic eigenvalues which differ from the exact ones in the second or third decimal digit.

The convergence to the exact values is quite rapid. The curves reach a plateau of practically constant energies starting from a n value which is smaller than the full dimension N by more than one order of magnitude in ^{108}Sn, ^{48}Cr, and by more than two in ^{133}Xe.

An equally fast convergence is reached for the eigenfunctions of the n-dimensional truncated Hamiltonian matrix

$$\psi_n = \sum_{i=1}^{n} c_i^{(n)} \mid i\rangle, \tag{32}$$

where $\mid i\rangle$ are the correlated basis states obtained by the partitioning method.

As shown in Fig. 2, the overlap of ψ_n with the exact eigenvector ψ for the first five $J^\pi = 2^+$ of ^{108}Sn and $J^\pi = 0^+$ of ^{48}Cr converges fast to unity, even if, in some cases, the overlap is very small at small n. Small fluctuations are noticeable at small n. They reflect the interference between the components of different wave functions in correspondence of partial energy crossings. The above two features represent a further proof of the *robustness* of the iterative algorithm.

To complete the analysis we studied the convergence of the strengths of the $E2$ transitions. In all cases, the strengths have a smooth behavior and reach soon a plateau. Their smooth behavior allows for an extrapolation to asymptotic values through a formula having the same structure as the

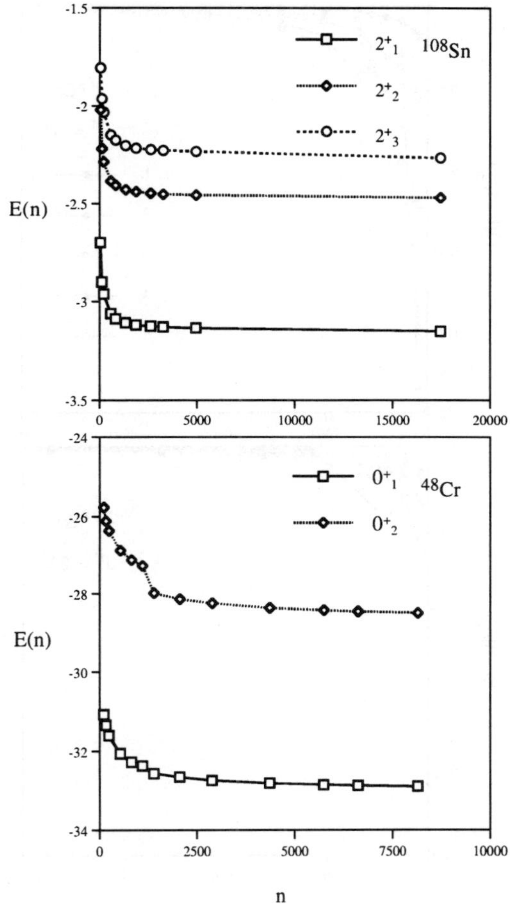

Fig. 1. Eigenvalues versus the dimensions n of the truncated matrices resulting from the sampling in ^{108}Sn and ^{48}Cr.

scaling law adopted for the energies (Eq. 26). The rapid convergence of the $E2$ strengths is quite significant in view of the extreme sensitivity of the transition strengths to even very small components of the wave function.

3.5. *Remarks*

The importance sampling algorithm is simple, easy to be implemented and allows to reduce the sizes of the Hamiltonian matrix by at least an order of magnitude with no detriment to the accuracy. Moreover, it generates extrapolation laws to asymptotic eigenvalues and $E2$ transition probabilities

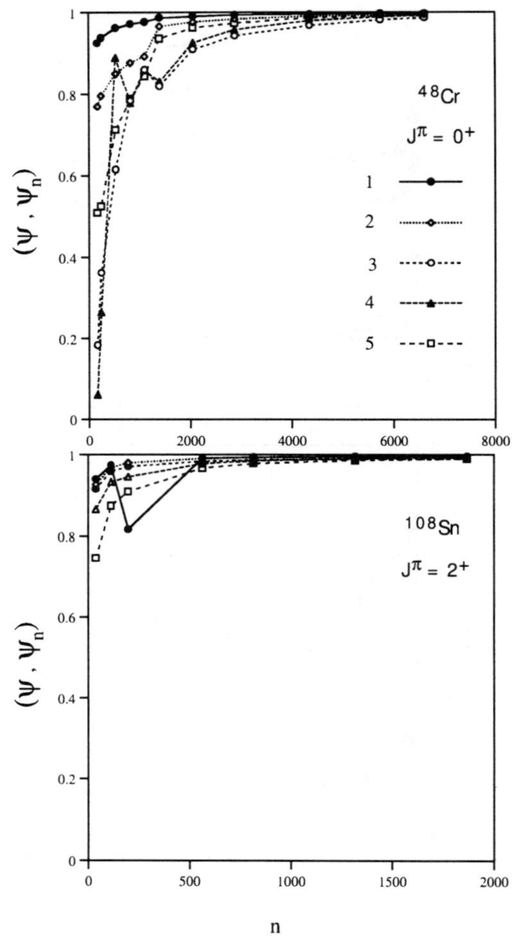

Fig. 2. Overlap of sampled wave functions with the corresponding exact ones

which coincide practically with the exact corresponding quantities.

The method is especially effective when applied to ^{133}Xe, having a neutron excess. We feel therefore confident that the sampling will enable us to face successfully the eigenvalue problem in heavier nuclei, all having a neutron excess.

Finally, it may be worth to stress the analogy of the present algorithm with the real space renormalization group [17]. In the latter approach one goes through the following steps: 1) Break a one dimensional chain of spins, into finite blocks; 2) select a block B and the corresponding Hamiltonian

H_B represented by an $m \times m$ matrix and finds the eigensolutions; 3) consider two-joined blocks with corresponding Hamiltonian H_{BB} represented by $m^2 \times m^2$ dimensional matrix; 4) diagonalize the new Hamiltonian and extract the lowest m eigensolutions; 5) use these to construct the block B'; 6) replace B with B' and repeat the procedure starting from 3). The correspondence with the steps of our method is quite clear!

The real space renormalization group came out to be not enough accurate for describing systems like the Heisenberg and Hubbard models. The difficulties reside in the inconsistency of the boundary conditions of the single blocks B with the ones of the joined blocks BB. This problem was circumvented by a new reformulation known as density matrix renormalization group [18]. This new reformulation has been adapted to the nuclear system [19]. In its latest version, this method could compute the low energy spectra of ^{48}Cr [20]. It is interesting to see the future developments. We can state, however, that the boundary condition problem, the main limit of the real space renormalization group in extended systems, is absent in finite nuclei. There is therefore no a priori reason for privileging the density matrix over the real space renormalization group in the nuclear many-body problem.

4. Collective modes and anharmonicities in nuclei

The evidence of multiphonon collective modes in nuclei, predicted already within the Bohr-Mottelson model, [21] has grown rapidly in recent years. At low-energy, resonance fluorescence scattering experiments have detected double-quadrupole, double-octupole and mixed quadrupole-octupole multiplets. [22] A combination of $\gamma-$ray spectroscopy techniques have allowed to identify unambiguously and fully characterize a class of multiphonon quadrupole states with proton-neutron (F-spin) mixed symmetry. [23–25] Evidence of three-phonon excitations of quadrupole nature has also been gained. [23–25] At high energy, the double giant dipole resonance has been observed in a number of different reactions. [26–28]

These discoveries have triggered a series of theoretical investigations of phenomenological as well as microscopic nature. The interacting-boson model (IBM) [29] was adopted with success for a systematic study of the low-energy multiphonon modes. More detailed investigations were carried out within microscopic schemes, which extend RPA in various ways. Most of these extensions are based on or inspired by the Fermion-Boson (FB) mapping technique. [30–32] The IBM itself is to be considered a phenomenological realization of the Fermion-Boson mapping. [29] A microscopic approach, explicitly based on the FB mapping, has been developed for studying the double giant dipole resonance. [33]

Among the approaches not explicitly based on the FB mapping, it is worth mentioning the nuclear field theory, [34] especially suitable for characterizing the anharmonicities of the vibrational spectra and the spreading widths of the giant resonances, and the quasiparticle-phonon model (QPM). [35] In the QPM, a Hamiltonian of generalized separable form is expressed in terms of RPA quasi-boson operators and then diagonalized in a severely truncated space which includes a selected set of two and three RPA phonons. The method has been extensively adopted to describe both low and high energy multiphonon excitations, like the mixed-symmetry states [36] and the double giant dipole resonance. [37] Other microscopic methods have tried to go beyond the quasi-boson approximation but with limited success. [38,39].

Here, we present an iterative equation of motion method [40] which generates a basis of TDA multiphonon states for the exact solution of the nuclear eigenvalue problem.

4.1. *An equations of motion approach for generating a multiphonon basis*

Our goal is to generate a set of multiphonon states $\mid n;\alpha\rangle$ which diagonalize the Hamiltonian H within each separate subspace spanned by states with n phonons, so that

$$< n;\beta|H|n;\alpha >= E_{\alpha}^{(n)})\delta_{\alpha\beta}, \tag{33}$$

where $\mid n;\alpha\rangle =\mid \nu_1 \dots \nu_n\rangle$ and the labels ν_i denote the quantum numbers of the i_{th} phonon. Under this request, we obtain

$$< n;\beta|\,[H,b^{\dagger}_{ph}]\,|n-1;\alpha > = (E_{\beta}^{(n)} - E_{\alpha}^{(n-1)}) < n;\beta|b^{\dagger}_{ph}|n-1;\alpha >, \tag{34}$$

where $b^{\dagger}_{ph} = a^{\dagger}_p a_h$ is a bilinear form in the operators $a^{\dagger}_p$ and a_h which create respectively a particle (p) and a hole (h) with respect to the unperturbed ground state (ph vacuum).

We then write the Hamiltonian in second quantized form and expand the commutator $[H, a^{\dagger}_p a_h]$ on the left-hand side of the equation. After a linearization procedure, we obtain for the n-phonon subspace the eigenvalue equation

$$\sum_{\gamma p' h'} A_{\alpha\gamma}^{(n)}(ph\ ;\ p'h')\ X_{\gamma\beta}^{(n)}(p'h') = E_{\beta}^{(n)}\ X_{\alpha\beta}^{(n)}(ph), \tag{35}$$

where

$$X^{(n)}_{\alpha\beta}(ph) \equiv < n; \beta | b^{\dagger}_{ph} | n-1; \alpha > \tag{36}$$

and

$$A^{(n)}_{\alpha\gamma}(ph;p'h') = \delta_{hh'}\delta_{pp'}\delta^{(n-1)}_{\alpha\gamma}\left[(\epsilon_p - \epsilon_h) + E^{(n-1)}_{\alpha}\right] +$$
$$\sum_{h_1} V_{p'h_1h'p}\,\rho^{(n-1)}_{\alpha\gamma}(h_1h) - \sum_{p_1} V_{p'hh'p_1}\rho^{(n-1)}_{\alpha\gamma}(pp_1) +$$
$$\frac{1}{2}\left[\delta_{hh'}\sum_{p_1p_2} V_{p'p_1pp_2}\rho^{(n-1)}_{\alpha\gamma}(p_1p_2) - \delta_{pp'}\sum_{h_1h_2} V_{hh_1h'h_2}\rho^{(n-1)}_{\alpha\gamma}(h_1h_2)\right]. \tag{37}$$

The symbols ϵ_p (ϵ_h) are single particle (hole) energies, V_{ijkl} the matrix elements of the two-body potential, and

$$\rho^{(n)}_{\alpha\gamma}(kl) = < n;\ \gamma | a^{\dagger}_k a_l | n;\ \alpha > \tag{38}$$

defines the density matrix with $a^{\dagger}_k a_l$ written in normal order with respect to the ph vacuum.

For $n = 1$, the density matrices appearing in Eq. (37) take the values

$$\rho^{(0)}_{\alpha\gamma}(pp') = \delta_{\alpha,0}\delta_{\gamma,0} < 0 \mid a^{\dagger}_p a_{p'} \mid 0 >= 0$$
$$\rho^{(0)}_{\alpha\gamma}(hh') = \delta_{\alpha,0}\delta_{\gamma,0} < 0 \mid a^{\dagger}_h a_{h'} \mid 0 >= \delta_{h,h'}$$

so that, Eqs. (35,37) yield the standard Tamm-Dancoff equations. Our method is, therefore, nothing but the extension of Tamm-Dancoff method to multiphonon spaces.

4.2. *Redundancy-free multiphonon basis*

It is easy to infer from the expression (36) of the vector amplitudes that the eigenvectors generated from solving such a system of equations are linear combinations of N_r states $b^{\dagger}_{ph}|n-1;\ \alpha>$. These are linearly dependent and, therefore, form an overcomplete set.

In order to extract a linear independent basis, we expand $|n;\ \beta>$ in terms of the redundant N_r states

$$|n;\ \beta >= \sum_{\alpha\, ph} C^{(n)}_{\alpha\beta}(ph)\, b^{\dagger}_{ph}\ |n-1;\ \alpha > . \tag{39}$$

Upon insertion in Eqs. (36) and (37), we get

$$X = \mathcal{D}C \tag{40}$$
$$A\mathcal{D}C = E\mathcal{D}C, \tag{41}$$

where $\mathcal{D}$ is the overlap or metric matrix

$$d_{\alpha\beta}^{(n-1)}(ph;p'h') = \langle n-1;\beta \mid b_{p'h'} b_{ph}^{\dagger} \mid n-1;\alpha\rangle. \tag{42}$$

Eq. (41) defines an eigenvalue equation of general form. It is, however, ill-defined. The matrix $\mathcal{D}$ is singular, since its determinant vanishes.

The traditional methods adopted to overcome this problem are based on the direct diagonalization of $\mathcal{D}$ [41], which is time consuming. Moreover, the calculation of the metric matrix is a highly non trivial task, requiring elaborated diagrammatic techniques and complex iterative procedures. [38, 39,42,43]

In our approach, the metric matrix is given by the simple formula

$$d_{\alpha\beta}^{(n-1)}(ph;p'h') = \sum_{\gamma} \left[\delta_{pp'}\delta_{\gamma\beta} - \rho_{\gamma\beta}^{(n-1)}(pp')\right] \rho_{\alpha\gamma}^{(n-1)}(hh'), \tag{43}$$

where the matrix densities are computed by using the recursive relations

$$\rho_{\alpha\beta}^{(n)}(h_1h_2) = \sum_{ph\gamma\delta} C_{\alpha\gamma}^{(n)}(ph)\left[\delta_{hh_1}\delta_{\gamma\delta}X_{\gamma\beta}^{(n)}(ph_2) - X_{\delta\beta}^{(n)}(ph)\,\rho_{\gamma\delta}^{(n-1)}(h_1h_2)\right]$$

$$\rho_{\alpha\beta}^{(n)}(p_1p_2) = \sum_{ph\gamma\delta} C_{\alpha\gamma}^{(n)}(ph)\,X_{\delta\beta}^{(n)}(p_1h)\left[\delta_{pp_2}\delta_{\gamma\delta} - \rho_{\gamma\delta}^{(n-1)}(pp_2)\right]. \tag{44}$$

Moreover, we have avoided the direct diagonalization of $\mathcal{D}$ by adopting an alternative method based on the Choleski decomposition to extract a set of linear independent states and, thereby, generate the n-phonon basis.

Once this is done, we evaluate the amplitudes $X_{\alpha\beta}^{(n)}(ph)$ and the density matrix $\rho_{\alpha\beta}^{(n)}(kl)$ by making use of Eqs. (40) and (44) respectively. $X_{\alpha\beta}^{(n)}(ph)$ and $\rho_{\alpha\beta}^{(n)}(kl)$ are the new entries for the equations of motion in the $(n+1)$-phonon subspace.

The iterative procedure is clearly outlined at this stage. To implement it, we have just to start with the lowest trivial 0-phonon subspace, the *ph* vacuum, and, then, solve the equations of motions step by step up to a convenient n-phonon subspace.

The multiphonon basis is, thereby, generated. In such a basis, the Hamiltonian gets so decomposed

$$H = \sum_{n\alpha} E_{\alpha}^{(n)} \mid n;\alpha\rangle\langle n;\alpha \mid + \sum_{n\alpha n'\beta} \mid n';\beta\rangle\langle n';\beta \mid \; H \; \mid n;\alpha\rangle\langle n;\alpha \mid, \tag{45}$$

where $n' = n\pm 1, n\pm 2$ and the off diagonal terms are given by simple recursive formulas.

The diagonalization of the Hamiltonian yields exact eigenvalues. The corresponding eigenvectors have the phonon structure

$$| \Psi_\nu \rangle = \sum_{n,\alpha_n} C^{(\nu)}_{\alpha_n} \mid n;\alpha_n\rangle = \sum_{\nu_1 \dots \nu_n} C^{(\nu)}_{\nu_1 \dots \nu_n} \mid \nu_1 \dots \nu_i \dots \nu_n \rangle, \qquad (46)$$

which provides simple recursive relations for the transition amplitudes.

4.3. *A numerical illustrative application of the method:* ^{16}O

We choose as testing ground for our method ^{16}O, whose low-lying excitations are known to have a highly complex *ph* structure. [44] The low-energy positive parity spectrum was studied in a shell model calculation which included up to $4p-4h$ and $4\hbar\omega$ configurations [45] and, more recently, within a no-core and an algebraic symplectic shell model [46] up to $6\hbar\omega$.

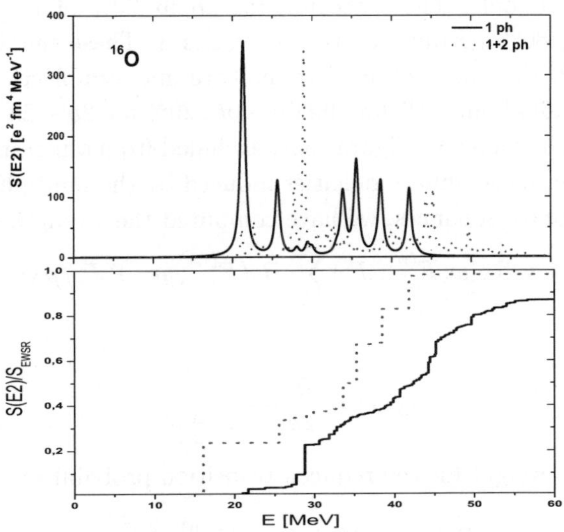

Fig. 3. $E2$ strength distribution and running sum in ^{16}O.

For our illustrative purposes, we have included all *ph* configurations up to $n = 3$ and $3\hbar\omega$, a space considerably smaller than the one adopted in shell model. On the other hand, our method generates at once the whole spectrum of positive and negative parity states and, therefore, allows to study the high energy spectroscopic properties. We used a Hamiltonian composed of a Nilsson unperturbed piece plus a bare G-matrix deduced from the Bonn-A potential. [15]

We achieve a complete separation of the intrinsic from the center of mass motion by resorting to the method of Palumbo [47], applied to standard shell model by Glockner and Lawson [48] and, since then, widely adopted in nuclear structure studies. It consists of adding an Harmonic Oscillator Hamiltonian in the center of mass coordinated multiplied by a coupling constant. If all configurations up to $N\hbar\omega$ are included, as in our case, each eigenfunction of the full Hamiltonian gets factorized into an intrinsic and a center of mass components. For a large enough coupling constant, the center of mass excited states are pushed high up in energy, leaving at the low physical energies only the intrinsic states, namely the eigenfunctions with the center of mass in the ground state.

Being our space confined to $3-\hbar\omega$, the ground state contains correlations up to 2-phonons only. These account for about 20% of the state, while the remaining 80% pertains to the ph vacuum. These numbers may be compared with the ones obtained by no-core and symplectic shell model calculations, [46] about 60% for the $0p-0h$, 20% for $2p-2h$ and 20% for the other more complex configurations, excluded from our restricted space.

To investigate the anharmonicities induced by the multiphonon configurations on giant resonances, we have computed the strength function

$$\mathcal{S}(\omega; F_\lambda^{(n)}, J^\pi) \approx \mathcal{S}_\Delta(\omega; F_\lambda^{(n)}, J^\pi) \sum_\nu B_\nu(F_\lambda^{(n)}; gr \to J^\pi)\, \rho_\Delta(\omega-\omega_\nu), \quad (47)$$

where

$$\rho_\Delta(x) = \frac{\Delta}{2\pi} \frac{1}{x^2 + (\frac{\Delta}{2})^2} \quad (48)$$

is a Lorentzian weight for the reduced transition probability

$$B_\nu(F_\lambda^{(n)}, J^\pi) \left| \langle \Phi_{\nu J^\pi} \| F_\lambda^{(n)} \| 0 \rangle \right|^2 . \quad (49)$$

The field $F_\lambda^{(n)}$ is

$$\mathcal{M}(E\lambda, \mu) = \frac{e}{2} \begin{pmatrix} 1 \\ \tau_3 \end{pmatrix} r^n Y_{\lambda\mu}(\hat{r}). \quad (50)$$

for the quadrupole and the isovector dipole transitions and

$$F_{\lambda=1\mu}^{(n=3)} = r^3 Y_{1\mu}(\hat{r}) \quad (51)$$

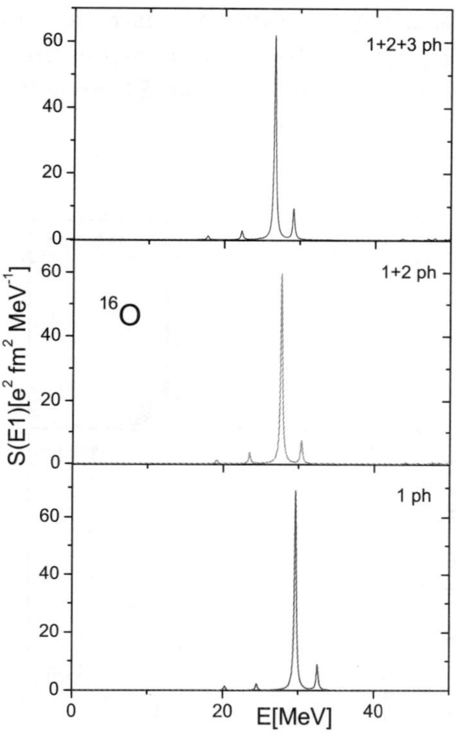

Fig. 4. Isovector $E1$ strength distributions in ^{16}O.

for the isoscalar dipole excitations (squeezed dipole mode). It is important to notice the absence of any corrective term, generally included in order to eliminate the spurious contribution due to the center of mass excitation. Such a term is not necessary in our approach which guarantees a complete separation of the center of mass from the intrinsic motion.

The two-phonon configurations have a damping and spreading effect on the $E2$ strength (Fig. 3). Because of the two-phonon coupling, some strength is pushed too high in energy (second panel) so as to deplete the EWSR. This anomaly is an indication that the phonon space considered here is too restricted.

The multiphonon configurations affect little the isovector $E1$ response (Fig. 4), but have a dramatic spreading effect on the isoscalar $E1$ strength

(Fig. 5). Such a spreading was expected, since the energy range of the ISGDR, around $3\hbar\omega$, is accessible to ph as well as $2p-2h$ and $3p-3h$ configurations. The isovector and isoscalar $E1$ strengths are both at too

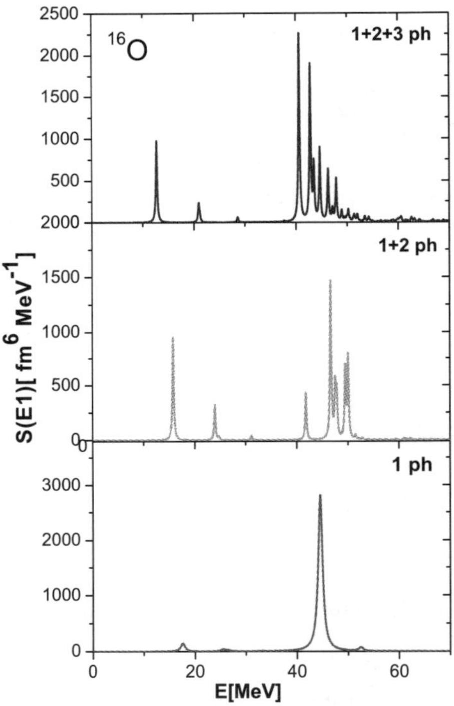

Fig. 5. Isoscalar $E1$ strength distributions in ^{16}O.

high energies with respect to experiments or mean field estimates. The peak of the IVGDR is about 5 MeV above the experimental one, while the ISGDR is pushed by about 6 MeV with respect to experiments [49]. These upward shifts are due, to a large extent, to the Nilsson potential which induces a too large gap between major shells.

The phonon space adopted here is sufficient for our illustrative purposes, but too restricted to describe exhaustively and faithfully all spectroscopic properties of ^{16}O. Extending the calculation to a larger space is not straightforward. Indeed, the number of density matrices to be computed increases so rapidly with the number of phonons as to render the procedure unbearably slow. The method, however, generates a basis of correlated states. It is therefore conceivable that most of them are non collective and unnecessary. The selection of the relevant basis states may be done efficiently by the importance sampling algorithm outlined in the first part [11], which allows a severe truncation while monitoring the accuracy of the solutions.

As for the high energy excitations, however, we believe that enlarging further the space will affect modestly the $E1$ giant resonances investigated here. Indeed, since our phonon Hamiltonian couples states differing by two-phonons, at most, the anharmonicities on the $E1$ one phonon states come almost entirely from the coupling with two and three phonons, accounted here. It remains, therefore, to exploit the sensitivity of the $E1$ response to the single particle energies.

5. Concluding remarks

The two methods outlined here are in many ways complementary. One is a shell model algorithm and, as such, covers the low-energy spectroscopy. The other aims at the description of collective modes, generally out of reach in standard shell model calculations.

The shell model algorithm has been shown to be fast, *robust*, yielding always stable numerical solutions, free of *ghost* states, and extremely simple to be implemented. Moreover, it is naturally endowed with an importance sampling, which allows for a drastic truncation of the matrices, while keeping the accuracy of the solutions under strict control. The importance sampling provides also scaling laws which allow to extrapolate to the exact eigenvalues. Since the truncation is far more effective in nuclei with neutron excess [11], we feel confident that the sampling may be successfully applied to heavy nuclei.

The equation of motion method proposed for treating multiphonon excitations leads to eigenvalue equations of simple structure in any n-phonon subspace. It is, therefore, not only exact but also of easy implementation for a Hamiltonian of general form. Moreover, it generates at once the whole nuclear spectrum. It is therefore suitable for studying the low-lying spectroscopic properties as well as the high energy giant resonances.

The method can be extended in several ways. It can be reformulated so as to include RPA phonons. This extension, however, might be unnecessary since the method, already in its present TDA formulation, yields an explicitly correlated ground state.

A formulation in terms of quasi-particles rather than particle-hole states is also straightforward and especially suitable for studying anharmonicities and multiphonon excitations in open shell nuclei not easily accessible to shell model methods.

Once a reliable importance sampling method will be implemented so as to reduce drastically the dimensions of the phonon subspaces, it should be possible to use the method for a reliable study of anharmonicities and multiphonon excitations in heavy spherical as well as deformed nuclei.

Acknowledgments

Work supported in part by the Italian Ministero della Istruzione Universitá e Ricerca (MIUR) and by the research plan MSM 0021620834 and GAUK 222/2006/B-FYZ/MFF of Czech Republic.

References

1. See for instance J.A. White, S.E. Koonin, and D.J. Dean, *Phys. Rev. C* **61**, 034303 (2000).
2. See for instance G. H. Golub and C. F. Van Loan, *Matrix Computations*, (John Hopkins University Press, Baltimore 1996).
3. T. Otsuka, M. Honma, and T. Mizusaki, *Phys. Rev. Lett.* **81**, 1588 (1998).
4. J.H. Wilkinson, *The Algebraic Eigenvalue Problem*, (Clarendom Press, Oxford, 1965).
5. E. Caurier *et al*, *Rev. Mod. Phys.* **77**, 427 (2005)
6. E. Caurier and F. Nowacki, *Acta Physica Polonica* **30**, 705 (1999)
7. S.E. Koonin, and D.J. Dean, and K. Langanke, *Phys. Rep.* **278**, 1 (1997).
8. S.E. Koonin, and D.J. Dean, and K. Langanke, *Annu. Rev. Nucl. Part. Sc.* **47**, 463 (1997).
9. T. Otsuka *et al*, *Prog. Part. Nucl. Phys.* **47**, 319 (2001).
10. F. Andreozzi, A. Porrino, and N. Lo Iudice, *J. Phys. A: Math. Gen.* **35**, L61 (2002).
11. F. Andreozzi, N. Lo Iudice, and A. Porrino, *J. Phys. : Nucl. Part. Phys.* **29**, 2319 (2003).
12. I. Shavitt, C. F. Bender, A. Pipano, and R. P. Hosteny, *J. Computational Phys.* **11**, 90 (1973).
13. M. Horoi, A. Volya, and V. Zelevinsky, *Phys. Rev. Lett.* **82**, 2064 (1999).
14. F. Andreozzi and A. Porrino, *J. Phys. G: Nucl. Part. Phys.* **27**, 845 (2001).
15. R. Machleidt, *Adv. Nucl. Phys.* **19**, 189 (1989).
16. E. Caurier, A.P. Zuker, A. Poves, and G. Martinez-Pinedo, *Phys. Rev. C* **50**, 225 (1994).
17. K. G. Wilson, *Rev. Mod. Phys.* **47**, 773 (1975).
18. S. R. White, *Phys. Rev. B* **48**, 10345 (1993).
19. J. Dukesky and S. Pittel, *Phys. Rev. C* **63**, 061303 (2001)
20. S. Pittel and N. Sandulesku, *Phys. Rev. C* **73**, 014301 (2006)

21. A. Bohr and B. R. Mottelson *Nuclear Structure* Vol. II (Benjamin, New York, 1975).
22. M. Kneissl, H. H. Pitz and A. Zilges, *Prog. Part. Nucl. Phys.* **37**, 439 (1996) and references therein.
23. N. Pietralla *et al.*, *Phys. Rev. Lett.* **83**, 1303 (1999).
24. C. Fransen *et al.*, *Phys. Rev. C* **71**, 054304 (2005).
25. M. Kneissl, N. Pietralla and A. Zilges, *J. Phys. G: Nucl. Part. Phys.* **32**, R217 (2006) and references therein.
26. N. Frascaria, *Nucl. Phys.* **A482**, 245c (1988) and references therein.
27. T. Auman, P. F. Bortignon, H. Hemling, *Annu. Rev. Nucl. Part. Sci.* **48**, 351 (1998) and references therein.
28. C. A. Bertulani and V. Yu. Ponomarev, *Phys. Rep.* **321**, 139 (1999) and references therein.
29. For a review see A. Arima and F. Iachello, *Adv. Nucl. Phys.* **13**, 139 (1984).
30. S. T. Belyaev and V. G. Zelevinsky, *Nucl. Phys.* **39**, 582 (1962).
31. T. Marumori, M. Yamamura, and A. Tokunaga, *Progr. Thor. Phys.* **31**, 1009 (1964).
32. For a review see A. Klein and E. R. Marshalek, *Rev. Mod. Phys.* **63**, 375 (1991).
33. F. Catara, P. Chomaz, and N. Van Giai, *Phys. Lett.* **B 233** (1989) 6; *Phys. Lett.* **B 277**, 1 (1992).
34. P. F. Bortignon, R. A. Broglia, D. R. Bes, and R. Liotta, *Phys. Rep.* **30**, 305 (1977).
35. V. G. Soloviev, *Theory of Atomic Nuclei: Quasiparticles and Phonons* Institute of Physics, Bristol, 1992.
36. N. Lo Iudice and Ch. Stoyanov, *Phys. Rev. C* **62** (2000) 047302; *Phys. Rev. C* **65**, 064304 (2002).
37. V. Yu. Ponomarev, P. F. Bortignon, R. A. Broglia, and V. V. Voronov, *Phys. Rev. Lett.* **85**, 1400 (2000).
38. C. Pomar, J. Blomqvist, R. J. Liotta, and A. Insolia, *Nucl. Phys.* **A515**, 381 (1990).
39. M. Grinberg, R. Piepenbring, K. V. Protasov, B. Silvestre-Brac, *Nucl. Phys.* **A597**, 355 (1996).
40. F. Andreozzi, N. Lo Iudice, and A. Porrino, F. Knapp, J. Kvasil, *Phys. Rev. C.* to be submitted.
41. D. J. Rowe, *J. Math. Phys.* **10**, 1774 (1969).
42. R. J. Liotta and C. Pomar, *Nucl. Phys.* **A382**, 1 (1982).
43. K. V. Protasov, B. Silvestre-Brac, R. Piepenbring, and M. Grinberg, *Phys. Rev. C* **53**, 1646 (1996).
44. G. E. Brown, A. M. Green, *Nucl. Phys.* **75**, 401 (1966).
45. W. C. Haxton and C. J. Johnson, *Phys. Rev. Lett.* **65**, 1325 (1990).
46. J. P. Draayer, private communication.
47. F. Palumbo, *Nuc. Phys.* **99**, 100 (1967).
48. D. H. Glockner and R. D. Lawson, *Phys. Lett.* **B 53**, 313 (1974).
49. Y.-W. Lui, H. L. Clark, and D. H. Youngblood, *Phys. Rev. C* **64**, 064308 (2001).

NUCLEAR SYMMETRIES AND ANOMALIES

LARRY ZAMICK* and ALBERTO ESCUDEROS**

Department of Physics & Astronomy, Rutgers University, 136 Frelinghuysen Road Piscataway, New Jersey 08854-8019, United States of America
** E-mail: lzamick@physics.rutgers.edu*
*** E-mail: aescuder@physics.rutgers.edu*
www.physics.rutgers.edu

First in a single-j-shell calculation ($j = f_{7/2}$) we discuss various symmetries, e.g., two to one in ^{44}Ti vs. ^{43}Ti. We note that the wave function amplitudes for T(higher) states are coefficients of fractional parentage, and that orthogonality of T(higher) and T(lower) states leads to useful results. Then we consider what happens if $T = 0$ two-body matrix elements are set equal to zero. We find a partial dynamical symmetry with several interesting degeneracies. It is noted that some formulae developed for identical particles also apply to different (companion) problems involving mixed systems of protons and neutrons. In the $g_{9/2}$ shell, where one can have for the first time seniority violation for identical particles, we find some interesting yet unproven results. Finally, we discuss shell-model calculations for the magnetic moments of different nuclei, such as ^{52}Ti, even-even Ca isotopes, and $N = Z$ nuclei.

Keywords: Nuclear structure, shell model, magnetic moments.

1. Single j Shell Calculations—$f_{7/2}$

We use as basis states for even-even Ti isotopes

$$\left[(f^p_{7/2})^{J_P}(f^n_{7/2})^{J_N}\right]^I \equiv [J_P J_N]^I \ , \tag{1}$$

where p and n are the number of valence protons and neutrons; J_P and J_N are the angular momenta of the protons and the neutrons; and I is the total angular momentum [1].

We write the wave function as

$$\Psi^{I\alpha} = \sum_{J_P J_N} D^{I\alpha}(J_P J_N)[J_P J_N]^I \ . \tag{2}$$

Here $D^{I\alpha}(J_P J_N)$ is the probability amplitude that protons couple to J_P and neutrons to J_N.

We have

$$\sum_{J_P J_N} D^{I\alpha}(J_P J_N) D^{I\alpha'}(J_P J_N) = \delta_{\alpha\alpha'} \qquad \text{(orthonormality)}$$

$$\sum_{\alpha} D^{I\alpha}(J_P J_N) D^{I\alpha}(J'_P J'_N) = \delta_{J_P J'_P}\delta_{J_N J'_N} \qquad \text{(completeness)} \qquad (3)$$

To get the D's, we need an interaction. We can take matrix elements from experiment, e.g., the spectrum of the "two-particle system" ^{42}Sc. Table 1 gives the excitation energies $E^*(J)$, which we use to identify $\langle (f^2_{7/2})^J |V| (f^2_{7/2})^J \rangle = E^*(J)$.

Table 1. Experimental excitation energy of the lowest state of angular momentum J in ^{42}Sc.

$T = 1$		$T = 0$	
$J = 0$	0.0000	$J = 1$	0.6111
$J = 2$	1.5863	$J = 3$	1.4904
$J = 4$	2.8153	$J = 5$	1.5101
$J = 6$	3.2420	$J = 7$	0.6163

Note that even-J states have isospin 1, i.e., they are isotriplets. The wave functions are space-spin antisymmetric and can occur in ^{42}Ca, ^{42}Sc, and ^{42}Ti. Assuming charge independence, the spectra of even-J states in these three nuclei would be identical. The odd-J states can only be present in ^{42}Sc, the neutron-proton system.

For the calcium isotopes in the single-j model, the allowed angular momenta and seniorities are [2]

^{42}Ca, ^{46}Ca (2 holes)	$J = 0, 2, 4, 6$	
^{43}Ca, ^{45}Ca (3 holes)	$J = 3/2, 5/2, 7/2, 9/2, 11/2, 15/2$	
^{44}Ca	$J = 0$	(seniority $v = 0$)
	$J = 2, 4, 6$	(seniority $v = 2$)
	$J = 2, 4, 5, 8$	(seniority $v = 4$)

In this simple model, the spectra would be the same for ^{43}Ca and ^{45}Ca (hole-hole = particle-particle).

The spectra of $v = 2$ states would be the same for ^{42}Ca, ^{44}Ca, and ^{46}Ca, but ^{44}Ca would also have added $v = 4$ states. In Ca, for a given pair (v, J), there is only one state. So the wave function is completely determined by the Pauli principle, not the interaction. The excitation energies are, of course, affected by the interaction.

Table 2. Wave functions for a system of p protons coupled to J_P and n neutrons coupled to J_N with total angular momentum I.

TITANIUM 43:						
			$I = 3.5$			
		Energy	0.00000	2.79305	4.14201	4.39375
J_p	J_n				$T = 3/2$	
3.5	0.0		0.78776	−0.35240	−0.50000	0.07248
3.5	2.0		0.56165	0.73700	0.37268	0.04988
3.5	4.0		0.22082	−0.37028	0.50000	−0.75109
3.5	6.0		0.12340	−0.44219	0.60093	0.65431
TITANIUM 44:						
			$I = 0.0$			
		Energy	0.00000	5.58610	8.28402	8.78750
J_p	J_n				$T = 2$	
0.0	0.0		0.78776	−0.35240	−0.50000	0.07248
2.0	2.0		0.56165	0.73700	0.37268	0.04988
4.0	4.0		0.22082	−0.37028	0.50000	−0.75109
6.0	6.0		0.12340	−0.44219	0.60093	0.65431

We now proceed to mixed systems of protons and neutrons. In Table 2 we show the wave functions for ^{43}Ti and ^{44}Ti represented as column vectors [1].

There are several comments to be made. Let us first focus on $I = 0$ states of ^{44}Ti. For the ground state, the dominant amplitudes are $D(00) = 0.78776$ and $D(22) = 0.56165$. This is reminiscent of the Interacting Boson Model II, where s and d bosons are dominant [3,4]. Of course, as one looks at states with higher excitation energy, this is no longer true.

Let us now look at the $J = j$ states of ^{43}Ti (also ^{43}Sc). We assign isospin labels only to states of higher isospin. Thus, three states have $T = 1/2$ and one has $T = 3/2$. The $T = 3/2$ state is an analog state of a corresponding state in ^{43}Ca. In ^{43}Ca all states of the j^3 configuration must have $T = 3/2$ because $T \geq |T_3|$ and $T_3 = -3/2$.

Now in ^{43}Ca we make a fractional parentage decomposition

$$\Psi(j^3)^{Iv} = \sum_{J_A} (j^2 J_A j|\}j^3 Iv)[(j^2)^{J_A} j]^I \ . \tag{4}$$

Each term on the right is not antisymmetric in all three particles, but the total sum is. The cfp's are useful in order to single out one particle from the rest.

Now an important point to be made is that in ^{43}Ti the $D(J_P J_N)$'s for the $T = 3/2$ state are precisely coefficients of fractional parentage (do not worry about phases). They do not depend on the interaction. This is intuitively obvious. We are splitting up a three-particle wave function into one particle with $J_P = j$ and two particles with J_N.

1.1. *Two to one symmetry*

We purposely showed ^{43}Ti ($I = j$) and ^{44}Ti ($I = 0$) next to each other. Please note that the wave function amplitudes for corresponding states are identical and that the excitation energies in ^{44}Ti are twice those in ^{43}Ti. We call this a two to one symmetry. When one realizes that such a symmetry exists [1], it is not too difficult to prove why this is so. This was done by Zamick and Devi [5].

As a consequence of this symmetry, the spectroscopic factor for the ground-state transition in the reaction ^{44}Ti(p,d)^{43}Ti is 2 and no strength remains for excited states. Also a similar two to one relationship holds for ^{48}Ti and ^{49}Ti (and its cross conjugate ^{47}Sc). Hence, the reaction ^{48}Ti(d,p)^{49}Ti will only populate the $J = 7/2^-$ ground state, again with a spectroscopic strength of 2 [1].

1.2. *Orthogonalizing T-greater states with T-lower states*

We now give a modern twist to an old-fashioned calculation. Since the higher isospin wave function amplitudes are fractional parentage coefficients, independent of the nuclear interaction, we can use these to set constraints on the T lower states.

For the $I = 0$ states in ^{44}Ti, the possible isospins are $T = 0$ and $T = 2$ (only one). The $T = 2$ state must be orthogonal to the $T = 0$ states. Besides, for $T = 2$, $D^{I\alpha}(J_P J_P) = (j^2 J_P j^2 J_P | \} j^4 0)$, i.e., a two-particle cfp. But Devi and Zamick [5] showed the following relation: $(j^2 J_P j^2 J_P | \} j^4 0) = (j^2 J_P j | \} j^3 j)$. Hence,

$$\sum_{J_P} D^I(J_P J_P)(j^2 J_P | \} j^3 j) = 0 \qquad \text{for } T = 0$$

$$= 1 \qquad \text{for } T = 2 \tag{5}$$

Using a Redmond relation [6–8] for $n = 2$

$$2 \begin{Bmatrix} j & j & J_1 \\ I_s & j & J_0 \end{Bmatrix} (-1)^{J_0 + J_1} \sqrt{(2J_0+1)(2J_1+1)} =$$

$$= -\delta_{J_0 J_1} \delta_{v_0 v_1} + 3 \sum_{v_s} [j^2(v_0 J_0) j I_s | \} j^3 v_s I_s][j^2(v_1 J_1) j I_s | \} j^3 v_s I_s] \ , \tag{6}$$

we get

$$2\sum_{J_P}\sqrt{(2J_P+1)(2J_{12}+1)}\begin{Bmatrix} j & j & J_P \\ j & j & J_{12}\end{Bmatrix} D(J_PJ_P) = -D(J_{12}J_{12}) \quad \text{for } T=0$$
$$= 2D(J_{12}J_{12}) \quad \text{for } T=2. \tag{7}$$

These relations were used by us to simplify the expression for the number of $J = 0$ pairs in even-even Ti isotopes [9].

The original expression is

$$N_{np}(J_{12}) = 2n\left|\sum_J D(J,Jv)[j^{n-1}(J_0)j|\}j^nJv]\right.$$
$$\left.\times\sqrt{(2J+1)(2J_{12}+1)}\begin{Bmatrix} j & j & J \\ j & J_0 & J_{12}\end{Bmatrix}\right|^2 \tag{8}$$

We get rid of annoying Racah coefficients and find the number of np pairs in $^{40+n}$Ti with angular momentum zero is

$$\frac{2|D(00)|^2}{n} \qquad \text{for } T=0$$

and

$$\frac{2n(2j+1-n)}{(2j+1)(n+1)} \qquad \text{for } T=2 \tag{9}$$

Rosensteel and Rowe [10] use the same equation (7) to count the number of states of three identical particles (e.g., electrons) with a given I. Instead of isospin $T = 0$ and $T = 2$, they have quasispin $S = 0$ and $S = 2$, and instead of the amplitudes $D(JJ)$, they have operators $Z(J) = \frac{1}{2}\left[[a^+a^+]^J[\tilde{a}\tilde{a}]^J\right]^0$. The $T = 2$ version of Eq. (7) appeared in an earlier work of Schwartz and de Shalit [11], but in the context of three identical particles.

The problem of the number of states of a given angular momentum has been vigorously persued by several other groups [12–15]. From these works, one learns, for example, that the number of states for three particles in a single j shell with total angular momentum $I = j$ is equal to

$$\frac{1}{3}\left(\frac{2j+1}{2} + 2\sum_{J_0\ \text{even}}(2J_0+1)\begin{Bmatrix} j & j & J_0 \\ j & j & J_0\end{Bmatrix}\right), \tag{10}$$

which, in turn, has been shown by Ginocchio and Haxton [12] to be equal to $[(2j+3)/6]$, where the square brackets mean the integer part of what is inside. But this is quite different from the companion problem also using Eq. (7) concerning the number of $J = 0$ pairs in the even-even Ti isotopes.

1.3. *Application–The primitive origin of the Interacting Boson Model*

Here we follow the work of Robinson and Zamick [16]. Consider the ground state wave function of ^{44}Ti shown in Table 2. Suppose for a T lower ground state we keep only the $(L_P = 0, L_N = 0)$ and $(L_P = 2, L_N = 2)$ terms, loosely speaking only s and d bosons:

$$\Psi = D^0(00)[0\ 0]^0 + D^0(22)[2\ 2]^0 \,. \tag{11}$$

We have the normalization condition $D^0(00)^2 + D^0(22)^2 = 1$, plus the fact that this state is orthogonal to the $T_{\min} + 2$ state. This **completely** determines the wave function. There is no freedom in the amount of "d boson" admixture in the ground state.

We obtain the following values of $(D^0(00), D^0(22))$

$$\begin{aligned} {}^{44}\mathrm{Ti}:&\quad 1/\sqrt{14}(\sqrt{5}, 3)\\ {}^{46}\mathrm{Ti}:&\quad 1/\sqrt{8}(\sqrt{5}, \sqrt{3})\\ {}^{48}\mathrm{Ti}:&\quad 1/\sqrt{6}(\sqrt{5}, 1)\end{aligned}$$

Compared with a full calculation, there is too much $(L_P = 2, L_N = 2)$ admixture in ^{44}Ti, but the results for ^{46}Ti and ^{48}Ti are surprisingly good.

This is part of a more general expression of Zamick, Mekjian, and Lee for $I = 0$ states of Ti isotopes [17]:

$$\sqrt{\frac{2j+1-n}{(n+1)(2j+1)}}D^0(00) - M\sqrt{\frac{2n}{(n+1)(2j+1)(2j-1)}} \begin{aligned} &= 0\,,\ T = T_{\min}\\ &= 1\,,\ T = T_{\min}+2 \end{aligned} \tag{12}$$

where $M = \sum_{J\geq 2} D^0(JJ)\sqrt{2J+1}$, with J even.

1.4. *Other single j symmetries*

For a nucleus in which there are the same number of neutron holes as there are proton particles, we find that one can assign a signature quantum number to the wave functions. For example, ^{48}Ti consists of two protons and two neutron holes. The wave function amplitudes are such that

$$D^I(L_P L_N) = (-1)^{I+s} D^I(L_N L_P) \,. \tag{13}$$

For some wave functions, $s = +1$ (positive signature), and for others, $s = -1$ (negative signature).

One consequence of this symmetry is that the double beta decay matrix element of the operator $[\sigma(1)\times\sigma(2)]^{J=2}t_+(1)t_+(2)$ vanishes from the ground state of ^{48}Ca to the 2_1^+ state of ^{48}Ti in the single j shell approximation [18].

Related to the above for midshell, i.e., ^{48}Cr, one has that $(-1)^{(v_P+v_N)/2}$ is a good quantum number, where v_P (v_N) is the seniority of the protons (neutrons).

2. Partial Dynamical Symmetry When $T = 0$ Two-Body Matrix Elements are Set Equal to Zero

In a model, we set all isospin $T = 0$ two-body matrix elements to zero but keep the $T = 1$ matrix elements as before [19]. We find in a single-j-shell calculation that for certain angular momenta (but not for others), several states become degenerate and the wave functions have good (J_P, J_N) as dual quantum numbers, where J_P is the angular momentum of the protons and J_N that of the neutrons.

For example, in ^{43}Sc, the $J = 1/2^-$ and $J = 13/2^-$ states with isospin $T = 1/2$ are degenerate. The configuration of the $J = 1/2$ state is $[j, J_N = 4]^{I=1/2}$. For the $13/2^-$ state, there are two configurations: $[j\ 4]^{I=1/2}$ and $[j\ 6]^{I=1/2}$. In general, the wave function is of the form $a[j\ 4] + b[j\ 6]$, but when we set the $T = 0$ matrix elements to zero, we get separate eigenstates $[j\ 4]$ and $[j\ 6]$. Likewise, in ^{44}Ti the $I = 3_2^+, 7_2^+, 9_1^+$, and 10_1^+ states are degenerate, all with configurations $[J_P = 4, J_N = 6] \pm [J_P = 6, J_N = 4]$.

We clearly need two conditions: a degeneracy condition and a condition to make (J_P, J_N) good dual quantum numbers. For the $I = 13/2$ state, the latter is ($j = 7/2$)

$$\begin{Bmatrix} j & j & 4 \\ j & 13/2 & 6 \end{Bmatrix} = 0\ , \tag{14}$$

which generalizes to

$$\begin{Bmatrix} j & j & (2j-3) \\ j & (3j-4) & (2j-1) \end{Bmatrix} = 0\ . \tag{15}$$

And in ^{44}Ti

$$\begin{Bmatrix} j & j & I_0 \\ j & j & I_0 \\ 6 & 4 & I \end{Bmatrix} = 0\ , \tag{16}$$

with $I_0 = 4$ for $I = 3$ and 7, and $I_0 = 6$ for $I = 9$ and 10.

There are also degeneracy conditions. For ^{43}Sc, to explain the degeneracies of $13/2^-, 17/2^-$, and $19/2^-$, all with the configuration $[j\ 6]$, we note

$$\begin{Bmatrix} j & j & (2j-1) \\ j & I & (2j-1) \end{Bmatrix} = \frac{-1}{8j-2} \tag{17}$$

for $I = 13/2, 17/2$, and $19/2$ (but not for $I = 15/2$).

For ^{44}Ti, we get two degeneracy conditions:

$$\begin{Bmatrix} j & j & (2j-3) \\ j & j & (2j-1) \\ (2j-3) & (2j-1) & I \end{Bmatrix} = \frac{1}{4(4j-5)(4j-1)}, \tag{18}$$

$$\begin{Bmatrix} j & j & (2j-1) \\ j & j & (2j-1) \\ (2j-1) & (2j-1) & I \end{Bmatrix} = \frac{1}{2(4j-1)^2}. \tag{19}$$

Arima and Zhao also derived these conditions using a J-pairing Hamiltonian [13].

But where do these conditions come from? They seem mysterious. The common thread for the angular momenta is not a priori obvious, but it turns out there is one. The angular momenta involved in the partial dynamical symmetry are those that **cannot** occur for a system of identical particles, e.g., the Ca isotopes.

Consider a basic state of two protons and one neutron

$$\left[[j(1)j(2)]^{J_P} j(3)\right]^I \qquad (J_P \text{ is even}). \tag{20}$$

Let us try to antisymmetrize this:

$$\begin{aligned}(1 - P_{13} - P_{23})& \left[[j(1)j(2)]^{J_P} j(3)\right]^I \\ &= \sum_{J_A} \left(\delta_{J_A J_P} + 2\sqrt{(2J_P+1)(2J_A+1)} \begin{Bmatrix} j & j & J_P \\ j & I & J_A \end{Bmatrix} \right) \\ &\quad \times \left[[j(1)j(2)]^{J_A} j(3)\right]^I .\end{aligned} \tag{21}$$

But we know that not all I are possible for three identical particles, e.g., $j = f_{7/2}$, $I = 1/2, 13/2, 17/2$, and $19/2$. Therefore, we must have for each J_A

$$\delta_{J_A J_P} + 2\sqrt{(2J_P+1)(2J_A+1)} \begin{Bmatrix} j & j & J_P \\ j & I & J_A \end{Bmatrix} = 0 . \tag{22}$$

Hence, $\left\{\begin{matrix} 7/2 & 7/2 & 4 \\ 7/2 & 13/2 & 6 \end{matrix}\right\} = 0$. But this is precisely the condition that (J_P, J_N) are good dual quantum numbers for a system of two protons and one neutron with isospin $T = 1/2$ and $I = 13/2$.

Also, for $J_P = J_A$, we get $\left\{\begin{matrix} j & j & J_P \\ j & I & J_P \end{matrix}\right\}(2J_P + 1) = -1/2$ for all I for which there are no states for three indentical particles, e.g., $I = 1/2, 13/2, 17/2$, and $19/2$. But this is the condition needed to explain the degeneracy of the states $J_P = 4$ for $I = 1/2$ and $13/2$, as well as the degeneracy of the states with $J_P = 8$, $I = 13/2, 17/2$, and $19/2$ when the $T = 0$ interaction is set equal to zero.

We now consider four particles. Start with a basis state of ^{44}Ti with two protons and two neutrons

$$\left[(j_1 j_2)^{J_P}(j_3 j_4)^{J_N}\right]^I . \tag{23}$$

Try to antisymmetrize it

$$\begin{aligned}&(1 - P_{13} - P_{14} - P_{23} - P_{24})\left[(jj)^{J_P}(jj)^{J_N}\right]^I \\ &= \sum_{\substack{J_A J_B \\ \text{even}}} \left[\delta_{J_P J_A}\delta_{J_N J_B} - 4\sqrt{(2J_P+1)(2J_N+1)(2J_A+1)(2J_B+1)}\left\{\begin{matrix} j & j & J_P \\ j & j & J_N \\ J_A & J_B & I \end{matrix}\right\}\right] \\ &\quad\times \left[(jj)^{J_A}(jj)^{J_B}\right]^I . \end{aligned} \tag{24}$$

If there is no state of four identical particles with total angular momentum I, we must have

$$\delta_{J_P J_A}\delta_{J_N J_B} - 4\sqrt{(2J_P+1)(2J_N+1)(2J_A+1)(2J_B+1)}\left\{\begin{matrix} j & j & J_P \\ j & j & J_N \\ J_A & J_B & I \end{matrix}\right\} = 0 . \tag{25}$$

These are exactly the conditions needed to make (J_P, J_N) good dual quantum numbers in a companion problem in ^{44}Ti when the two-body $T = 0$ interaction is set equal to zero. Furthermore, the energies of (J_P, J_N) states are independent of I for those I that cannot occur for four identical particles, e.g., $I = 3^+, 7^+, 9^+$, and 10^+.

Table 3 shows what happens to the wave functions when the two-body $T = 0$ matrix elements are set to zero.

We show in Fig. 1 the spectrum of ^{44}Ti in a single-j-shell calculation. In the first column, we have V(^{42}Sc); in the second, T0V(^{42}Sc), where the $T = 0$ matrix elements are set to zero. One sees the degeneracy of $3_2^+, 7_2^+, 9_1^+$, and 10_1^+. Lastly, we have experiment.

Table 3. Change in the wave functions when the two-body $T = 0$ matrix elements are set to zero.

SCANDIUM 43:									
		$I = 6.5$							
		3.50013	4.95078						
J_p	J_n								
3.5	4.0	0.98921	−0.14647		$\longrightarrow$	1	0		
3.5	6.0	0.14647	0.98921			0	1		
		$I = 7.5$							
		3.51123	7.29248						
J_p	J_n		$T = 3/2$						
3.5	4.0	0.87905	−0.47673		$\longrightarrow$	Unchanged			
3.5	6.0	0.47673	0.87905						
TITANIUM 44:									
		$I = 10.0$							
		7.38394	8.90568	10.02992					
J_p	J_n			$T = 1$					
4.0	6.0	0.70089	−0.09358	0.70711			$1/\sqrt{2}$	0	$1/\sqrt{2}$
6.0	4.0	0.70089	−0.09358	−0.70711	$\longrightarrow$		$1/\sqrt{2}$	0	$-1/\sqrt{2}$
6.0	6.0	0.13234	0.99120	0.00000			0	1	0

We now show selected figures for full fp calculations in which we give results for the full FPD6 interaction and for T0FPD6, the latter being an interaction in which **all** the $T = 0$ two-body matrix elements have been set equal to zero. By removing and then restoring the $T = 0$ two-body matrix elements, we can get an idea of the importance of the neutron-proton interaction in the $T = 0$ channel.

We show in Fig. 2 the even-J states of ^{50}Cr obtained in a full fp calculation [20]. One notes that one gets reasonable results for T0FPD6, even though we made a severe change. This shows that it is difficult to tell from yrast spectra alone the importance of the correlations brought about by the neutron-proton interaction in $T = 0$ states. Also, the fact that the spectra are so close shows that this study is of more than academic interest.

We show in Fig. 3 a more challenging case—a full fp shell calculation of the $T = 0$ states of ^{46}V, where we have adjusted the $J = 0$ ground states for the two interactions to coincide (otherwise, the entire $T = 0$ spectrum for T0FPD6 is shifted up by about 2.5 MeV).

There are some differences, of course. The spectra for all $J \geq 8$ are shifted down for T0FPD6 by about 1 MeV, give or take, relative to FPD6. But the overall agreement with experiment is about just as good—the experimental levels are also below the predictions of FPD6 (although the

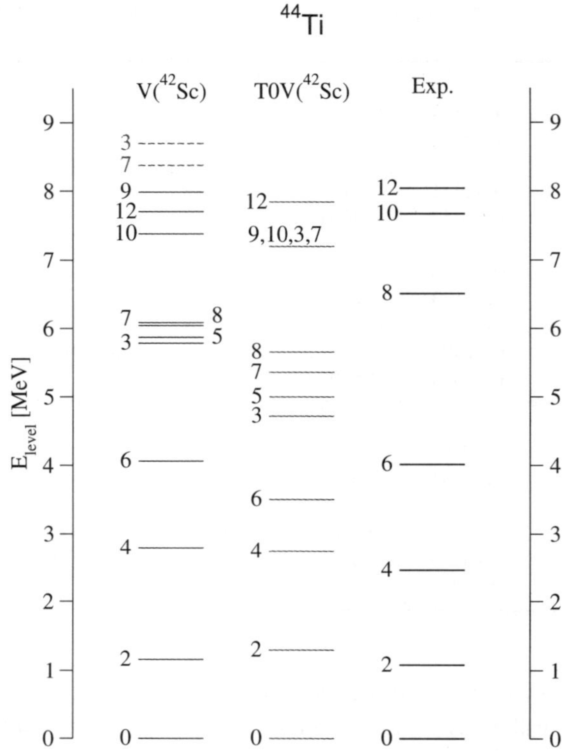

Fig. 1. Single-j-shell calculation of ^{44}Ti.

$J = 16^+$ assignment is still tentative).

To find evidence of n-p correlations, one has to look elsewhere, e.g., Gamow-Teller transitions, scissors mode excitations, etc., but we will not discuss these in detail here.

In Table 4 we show results for energy splittings that would be zero in a single-j-shell calculation in which the $T = 0$ two-body matrix elements are set to zero, i.e., if T0V(^{42}Sc) were used [21,22]. In a full fp calculation, it is no longer true that the splitting would be zero when the $T = 0$ matrix elements are set equal to zero. However, in general, the splittings should be smaller. We show in this table the results of FPD6 and T0FPD6 in a full fp calculation. With the exception of the 3_2^+–7_2^+ splitting in ^{44}Ti, the T0FPD6 splittings **are** much smaller than those of FPD6. In most of these cases, then, the splittings come dominantly from the FPD6 interaction. For example, in ^{47}V the $(29/2)_1^-$–$(31/2)_1^-$ splitting with T0FPD6 is only

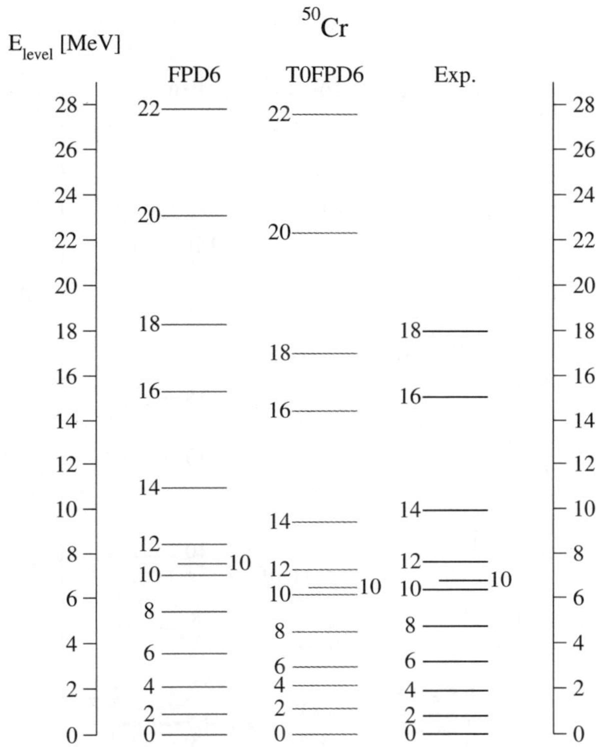

Fig. 2. Full fp shell calculations with (FPD6) and without (T0FPD6) the $T = 0$ two-body matrix elements compared with experiment for the even-J states of ^{50}Cr.

0.072 MeV, while with the full FPD6 it is 0.780 MeV and the experimental value is 0.765 MeV. In ^{45}Ti, the respective calculated splittings for $(25/2)_1^-$–$(27/2)_1^-$ are 0.105 MeV and 0.955 MeV. Unfortunately, the energy of the $25/2^-$ state is not known.

3. A Brief Discussion of the $g_{9/2}$ Shell

We have found some interesting results in our brief excursion into the $g_{9/2}$ shell. Consider three identical particles in the $g_{9/2}$ shell, e.g., three proton holes (^{97}Ag) or three particles (^{93}Tc). We define:

$$\Delta E_3 = E_3(I_{\text{max}}) - E_3(I_{\text{min}}) = E_3(21/2) - E_3(3/2) \,. \tag{26}$$

Now, for five identical particles and a seniority-conserving interaction, we have: $\Delta E_5 = \Delta E_3$. And with the $Q \cdot Q$ interaction (which does not

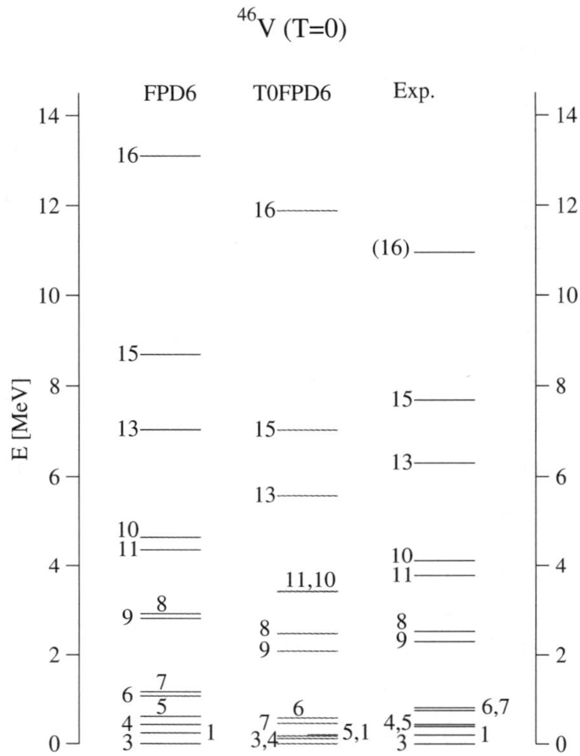

Fig. 3. Full fp shell calculations with (FPD6) and without (T0FPD6) the $T = 0$ two-body matrix elements compared with experiment for the $T = 0$ states of ^{46}V.

conserve seniority): $\Delta E_5 = -\Delta E_3$. This is a very interesting result yet to be proved (see Fig. 4). In the $f_{7/2}$ shell, the result is $\Delta E_5 = \Delta E_3$ even with $Q \cdot Q$. We note that $3/2^-$ states have not yet been identified experimentally for proton holes in $g_{9/2}$ relative to $N = 50, Z = 50$ or for $g_{9/2}$ neutrons relative to $N = 40, Z = 40$. We urge experimentalists to look for these states so as to ascertain the relative importance of $Q \cdot Q$. From our calculations, we do not expect the $3/2^-$ to be higher than $21/2^-$ for the five-particle system in ^{95}Rh. We are not sure what will happen for the five-neutron system ^{85}Zr.

Consider now four $g_{9/2}$ neutrons with $I = 4$. There is one state of seniority $v = 2$ and two states with $v = 4$. Even with a seniority-violating interaction, one state remains a pure $v = 4$ state. The other $v = 4$ state and the $v = 2$ state mix. The pure $v = 4$ state has the following structure:

Table 4. Splitting in energies (MeV) for states that were degenerate in the single j shell with $T0E(J)$. All experimental energies are taken from the National Nuclear Data Center (`http://www.nndc.bnl.gov/`).

	ΔE	Single j	FPD6	T0FPD6	Exp.
^{43}Sc (^{43}Ti)	$(1/2)^-_1 - (13/2)^-_1$	0.816	−1.866	−0.126	
	$(13/2)^-_2 - (17/2)^-_1$	0.653	0.399	−0.023	
	$(17/2)^-_1 - (19/2)^-_1$	0.653	1.020	0.090	1.237
^{44}Ti	$3^+_2 - 7^+_2$	0.320	−1.031	−1.104	
	$7^+_2 - 9^+_1$	0.391	0.379	0.181	
	$9^+_1 - 10^+_1$	0.600	1.214	0.094	
	$10^+_2 - 12^+_1$	1.203	1.617	0.133	0.945
^{45}Ti	$(25/2)^-_1 - (27/2)^-_1$	0.580	0.955	0.105	
^{46}V	$12^+_1 - 13^+_1$	0.863	0.500	0.170	1.163
	$13^+_2 - 15^+_1$	0.809	1.394	0.176	
^{47}V	$(29/2)^-_1 - (31/2)^-_1$	0.229	0.780	0.072	0.765

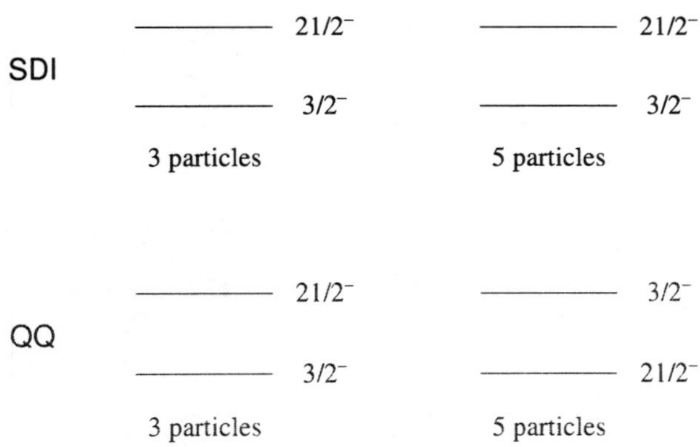

Fig. 4. Energy splittings for three and five identical particles in the $g_{9/2}$ shell with a seniority-conserving interaction (SDI or Surface Delta) and a seniority-violating interaction ($Q \cdot Q$).

$I = 4$:

J_0	v_0	$v = 4$
1.5	3	0.1222
2.5	3	0.0548
3.5	3	0.6170
4.5	1	0.0000
4.5	3	0.0000
5.5	3	−0.4043
6.5	3	−0.6148
7.5	3	−0.1597
8.5	3	0.1853

This result also has to be proved.

In Fig. 5 we show the experimental spectra of odd-even nuclei consisting of proton holes relative to $N = 50$, $Z = 50$. The nuclei in question are ^{97}Ag, ^{95}Rh, and ^{93}Tc. The spectra show reasonable shell model behaviour. We can get pretty good agreement in a single-j-shell calculation ($g_{9/2}$) using the spectrum of the two-hole system ^{98}Cd to get the hole-hole interaction. Rather than showing the results, we will show how one can get an idea of the goodness of the shell model from some general properties.

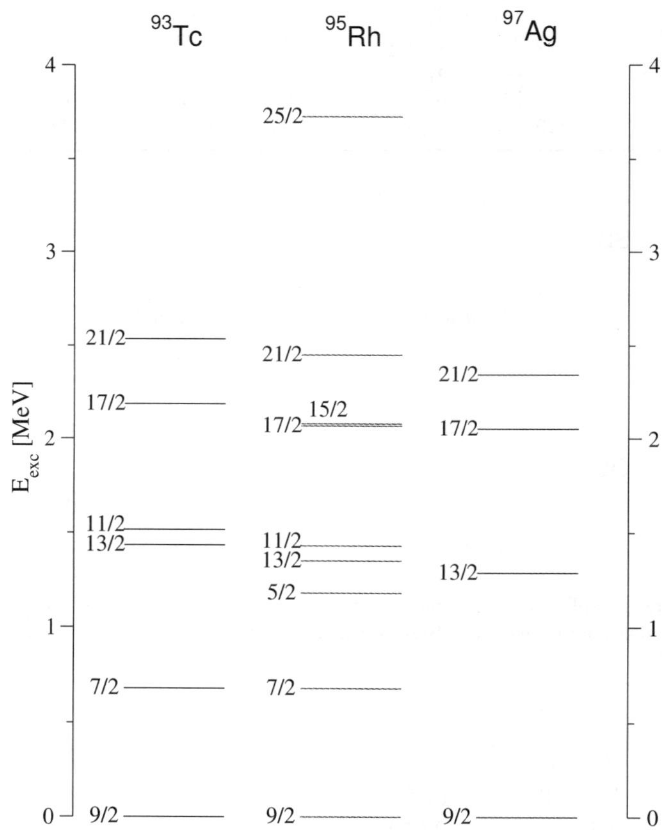

Fig. 5. Experimental spectra of ^{93}Tc, ^{95}Rh, and ^{97}Ag.

For example, the hole spectrum in the single j shell should be the same as the particle spectrum. By looking at ^{93}Tc and ^{97}Ag, we see that the spectra are reasonably close.

For the five-particle (or five-hole) nucleus ^{95}Rh, we have states which are admixtures of basis states of seniority 1, 3, and 5. If we have a seniority-conserving interaction, then the spectra of seniority-3 states in the five-particle (or five-hole) system would be the same as that for three particles (or holes). Comparing ^{95}Rh with ^{93}Tc, the spectra are rather similar, indication that seniority-violating interactions such as $Q \cdot Q$ are not very important.

Furthermore, we checked, and indeed the calculated $21/2^{-}$–$3/2^{-}$ splitting in ^{95}Rh is about the same as in ^{97}Ag. Recall that with pure $Q \cdot Q$, the splittings would be equal but opposite for the two nuclei in question.

4. Nuclear Magnetic Moments

Magnetic moments are very sensitive to nuclear structure effects. We are finding that often what are thought to be very good effective interactions are giving bad results for magnetic moments, or g factors ($g = \mu/I$). We here discuss calculations performed with Shadow Robinson and Yitzhak Sharon, in collaboration with experimental groups at Rutgers and Bonn, lead by Noemie Koller and Karl-Heinz Speidel, respectively.

For example, in Table 5 we can see the comparison of experiment and a full fp calculation for ^{52}Ti [23]. The measured value for the 2^{+} state, 0.83, is more than a factor of two larger than the calculations. The best fit to the 4^{+} state is with the FPD6 interaction, which is generally considered inferior to KB3 and GXPF1. This is but one example of many.

Table 5. Comparison of g factors between experiment and different full fp calculations for ^{52}Ti.

	Experiment	KB3	GXPF1	FPD6
$g(2^{+})$	0.83(19)	0.350	0.314	0.375
$g(4^{+})$	0.46(15)	1.002	1.045	0.580

4.1. *The even-even Ca isotopes*

In the Ca isotopes [24,25], there are many surprises and one must go beyond the fp model space to explain them. Consider the g factors of the lowest 2^{+} states in ^{42}Ca, ^{44}Ca, and ^{46}Ca. In the single-j-shell model for identical particles, all g factors are the same. For $j = L + 1/2$ neutrons, the g factor is $-1.913/j$, which for $f_{7/2}$ yields a value of -0.56. However, the measured

g factors of ^{42}Ca and ^{44}Ca are positive. To explain this, we have to add intruder state admixtures a la Gerace and Green (also known as highly deformed states). We write the wave functions as

$$C\Psi(fp)^n + D\Psi(\text{intruder}) \; . \tag{27}$$

We show the measured g factors and the values of C and D required to fit them in Table 6. In ^{44}Ca our admixtures are consistent with values obtained from static quadrupole moments ($C^2 = 0.43$ and $D^2 = 0.57$) and from stripping reactions ($C^2 = 0.47$ and $D^2 = 0.53$).

Table 6. Experimental g factors and values of C^2 and D^2 (see text) for several Ca isotopes.

	g measured	C^2	D^2
^{42}Ca	+0.04(6)	0.45(6)	0.55(6)
^{44}Ca	+0.16(3)	0.35(4)	0.65(4)
^{46}Ca	−0.25(5)	0.92(7)	0.08(7)

4.2. *Isoscalar moments*

We list in Table 7 the g factors for 2^+ states of $N = Z$ nuclei [25]. They all have isospin $T = 0$. All the measured and calculated g factors are close to +0.5. Although this is consistent with the rotational value $g_{\text{rot}} = Z/A$, one cannot conclude from this alone that the nuclei are deformed. The point is that one can also get these g factors close to 0.5 in a single-j-shell calculation. There is a remarkable insensitivity to the amount of configuration mixing is put in.

Table 7. Comparison of g factors between experiment and calculation for different $N = Z$ nuclei.

	Experiment	Calculation
^{32}S	+0.44(10)	0.50
^{36}Ar	+0.52(18)	0.488
^{44}Ti	+0.52(15)	0.514

Acknowledgements

We thank Shadow Robinson, Noemie Koller, Gerfried Kumbartzki, Yitzhak Sharon, Karl-Heinz Speidel, Igal Talmi, and Ben Bayman for their help. A.E. acknowledges support from the Secretaría de Estado de Educación y Universidades (Spain) and the European Social Fund.

References

1. J. D. McCullen, B. F. Bayman, and L. Zamick, *Phys. Rev.* **134** (1964) B515; A. Escuderos, L. Zamick, and B. F. Bayman, LANL nucl-th/0506050.
2. A. de Shalit and I. Talmi, *Nuclear Shell Theory* (Academic Press, New York, 1963); I. Talmi, *Simple Models of Complex Nuclei* (Harwood Academic, Reading, UK, 1993).
3. A. Arima, T. Otsuka, F. Iachello, and I. Talmi, *Phys. Rev. Lett.* **68B** (1977) 205.
4. J. A. Evans, J. P. Elliott, and S. Szpikowski, *Nucl. Phys.* **A435** (1985) 317.
5. L. Zamick and Y. D. Devi, *Phys. Rev.* **C60** (1999) 054317.
6. P. J. Redmond, *Proc. R. Society London* **A222** (1954) 84.
7. B. R. Judd, *Second Quantization and Atomic Spectroscopy* (John Hopkins Press, 1967).
8. L. Zamick and A. Escuderos, *Annals of Physics* **321** (2006) 987.
9. L. Zamick, A. Escuderos, S. J. Lee, A. Z. Mekjian, E. Moya de Guerra, A. A. Raduta, and P. Sarriguren, *Phys. Rev.* **C71** (2005) 034317.
10. G. Rosensteel and D. J. Rowe, *Phys. Rev.* **C67** (2003) 014303.
11. C. Schwartz and A. de Shalit, *Phys. Rev.* **94** (1954) 1257.
12. J. N. Ginocchio and W. C. Haxton, in *Symmetries in Science VI*, ed. B. Gruber and M. Ramek (Plenum, New York, 1993), pp. 263–273.
13. Y. M. Zhao and A. Arima, *Phys. Rev.* **C68** (2003) 044310; **C70** (2004) 034306; **C71** (2005) 047304; **C72** (2005) 054307.
14. I. Talmi, *Phys. Rev.* **C72** (2005) 037302.
15. L. Zamick and A. Escuderos, *Phys. Rev.* **C71** (2005) 014315; **C72** (2005) 044317.
16. S. J. Q. Robinson and L. Zamick, *Phys. Rev.* **C70** (2004) 057301.
17. L. Zamick, A. Z. Mekjian, and S. J. Lee, *J. Kor. Phys. Soc.* **47** (2005) 18.
18. L. Zamick and E. Moya de Guerra, *Phys. Rev.* **C34** (1986) 290.
19. S. J. Q. Robinson and L. Zamick, *Phys. Rev.* **C64** (2001) 057302.
20. S. J. Q. Robinson, A. Escuderos, and L. Zamick, *Phys. Rev.* **C72** (2005) 034314.
21. A. Escuderos, B. F. Bayman, L. Zamick, and S. J. Q. Robinson, *Phys. Rev.* **C72** (2005) 054301.
22. A. Escuderos, S. J. Q. Robinson, and L. Zamick, *Phys. Rev.* **C73** (2006) 027301.
23. K.-H. Speidel et al., *Phys. Lett.* **B633** (2006) 219.
24. M. J. Taylor et al., *Phys. Lett.* **B559** (2003) 187; **B605** (2005) 265.
25. S. Schielke et al., *Phys. Lett.* **B571** (2003) 29.

MODERN SHELL-MODEL CALCULATIONS

A. COVELLO

Dipartimento di Scienze Fisiche, Università di Napoli Federico II,
and Istituto Nazionale di Fisica Nucleare,
Complesso Universitario di Monte S. Angelo, Via Cintia I-80126 Napoli, Italy
E-mail: covello@na.infn.it

The present paper is comprised of two parts. First, we give a brief survey of the theoretical framework for microscopic shell-model calculations starting from the free nucleon-nucleon potential. In this context, we discuss the use of the low-momentum nucleon-nucleon (NN) interaction $V_{\rm low-k}$ in the derivation of the shell-model effective interaction and emphasize its practical value as an alternative to the Brueckner G-matrix method. Then, we present some results of recent studies of nuclei near doubly magic ^{132}Sn, which have been obtained starting from the CD-Bonn potential renormalized by use of the $V_{\rm low-k}$ approach. The comparison with experiment shows how shell-model effective interactions derived from modern NN potentials are able to provide an accurate description of nuclear structure properties.

1. Introduction

A fundamental problem of nuclear physics is to understand the properties of nuclei starting from the forces among nucleons. Within the framework of the shell model, which is the basic approach to nuclear structure calculations in terms of nucleons, this problem implies the derivation of the model-space effective interaction from the free nucleon-nucleon (NN) potential. Although efforts in this direction started some forty years ago, [1,2] for a long time there was a widespread skepticism about the practical value of what had become known as "realistic shell-model calculations" (see, e.g., Ref. 3). This was mainly related to the highly complicated nature of the nucleon-nucleon force, in particular the presence of a very strong repulsion at short distances which, in turn, made very difficult solving the nuclear many-body problem. As a consequence, in most of the shell-model calculations through the mid 1990s either empirical effective interactions containing several adjustable parameters have been used or the two body matrix elements have been treated as free parameters.

From the late 1970s on, however, there has been substantial progress toward a microscopic approach to nuclear structure calculations starting from the free NN potential V_{NN}. This has concerned both the two basic ingredients which come into play in this approach, namely the NN potential and the many-body methods for deriving the model-space effective interaction, $V_{\rm eff}$.

These improvements brought about a revival of interest in realistic shell-model calculations. This started in the early 1990s and continued to increase during the following years. The main aim of the initial studies was to give an answer to the key question of whether calculations of this kind were able to provide an accurate description of nuclear structure properties. By the end of the 1990s it became clear (see Ref. 4) that shell-model calculations employing effective interactions derived from realistic NN potentials can provide, with no adjustable parameters, a quantitative description of nuclear structure properties. As a consequence, in the past few years the use of these interactions has been rapidly gaining ground, opening new perspectives to nuclear structure theory.

As mentioned above, a main difficulty encountered in the derivation of $V_{\rm eff}$ from the free NN potential is the existence of a strong repulsive core. As is well known, the traditional way to overcome this difficulty is the Brueckner G-matrix method. Recently, a new approach has been proposed [5,6] which consists in deriving from V_{NN} a renormalized low-momentum potential, $V_{\rm low-k}$, that preserves the physics of the original potential up to a certain cutoff momentum Λ. This is a smooth potential which can be used directly to derive $V_{\rm eff}$. As we shall discuss in more detail in Sect. 4, we have shown [6–8] that this approach provides an advantageous alternative to the use of the G matrix.

The purpose of these lectures is to give a short overview of the theoretical framework for realistic shell-model calculations and to present, by way of illustration, some results of recent calculations employing the $V_{\rm low-k}$ approach to the renormalization of the bare NN interaction.

The outline of the lectures is as follows. In Sec. 2 a brief pedagogical review of the NN interaction is given, which is mainly aimed at highlighting the considerable progress made in this field over a period of about 50 years. The derivation of the shell-model effective interaction is discussed in Sec. 3, while the low-momentum NN potential $V_{\rm low-k}$ is introduced in Sec 4. Selected results of calculations for nuclei around doubly magic ^{132}Sn are reported and compared with experiment in Sec. 5. The last section, Sec. 6, provides a brief summary and concluding remarks.

2. The nucleon-nucleon potential

The nucleon-nucleon interaction has been extensively studied since the discovery of the neutron and in the course of time there have been a number of review papers marking the advances in the understanding of its nature. A review of the major progress of the 1990s including references through 2000 can be found in Ref. 9. Here, I shall only give a brief historical account and a survey of the main aspects relevant to nuclear structure, the former serving the purpose to look back and recall how hard it has been going from nucleon-nucleon interaction to nuclear structure.

Let us start from the end of the 1950s. At that time the state of the art was summarized by M. L. Goldberger [10] in the following way: "*There are few problems in modern theoretical physics which have attracted more attention than that of trying to determine the fundamental interaction between two nucleons. It is also true that scarcely ever has the world of physics owed so little to so many. In general, in surveying the field one is oppressed by the unbelievable confusion and conflict that exists. It is hard to believe that many of the authors are talking about the same problem, or, in fact, that they know what the problem is*".

In the next decade, however, quantitative one-boson-exchange potentials were developed, following the experimental discovery of heavy mesons in the early 1960s. This brought about a more optimistic view of the field. Quoting from the Summary [11] of the 1967 NN Interaction Conference at the University of Florida in Gainesville: "*It would appear that our view has improved considerably from the bleak picture of 1960. Indeed several relatively simple and accurate descriptions of the nucleon-nucleon interaction based upon meson field theory have emerged. While the formalisms used differ greatly, it appears now that these theories have the same physical substance and that the various authors are not only talking about the same problem but that the correspondences between the various languages are being established*".

By the early 1980s the main questions concerning the NN interaction had been clearly emerged. In the words of R. Vinh Mau [12] at the International Conference on Nuclear Physics held in Florence in 1983: "*i) Do we have at our disposal a model of NN interaction based on sound theoretical grounds which at the same time can fit quantitatively the vast wealth of existing NN data?" ii) If such a free NN interaction exists, is it usable in predicting properties of complex nuclei? How the predictions compare with data?*". In the conclusions of his talk Vinh Mau answers to these questions essentially in the affirmative.

The knowledge of the NN interaction around 1990 may be summarized by the statement [13] of R. Vinh Mau at the International Conference on Nuclear Physics held in São Paulo in 1989: "*As time elapses, there is more and more evidence, thanks to the new high precision experimental data, that the description of the long-range and medium-range NN interaction in terms of hadronic (nucleons, mesons, isobars) degrees of freedom is quantitatively very successful*".

The above statement was well justified by the advances made in the previous decade, during which the Nijmegen78 [14], Paris [15], and Bonn [16] potentials, all based on meson theory, were constructed. These potentials fitted the NN scattering data below 300 MeV available at that time with χ^2/datum = 5.12, 3.71, and 1.90, respectively [17]. A detailed discussion of these three potentials can be found in Ref. 17. To make it suitable for application in nuclear structure, an energy-independent one-boson parametrization of the full Bonn potential was also developed [18], which has become known as Bonn-A potential. The phase-shift predictions by this potential are very similar to the ones by the Bonn full model with a χ^2/datum of about 2.

Over the last ten years or so both the Paris and Bonn-A potentials have been used in nuclear structure calculations. In some cases comparisons between the results given by these two potentials have been made. From our own calculations for several medium-heavy nuclei [19–21], it has turned out that Bonn-A leads to the best agreement with experiment for all the nuclei considered.

From the early 1990s on there has been much progress in the field of nuclear forces. In the first place, the NN phase-shift analysis was greatly improved by the Nijmegen group [22]. Then, based upon this analysis, a new generation of high-quality NN potentials has come into play which fit the Nijmegen database (this contains 1787 *pp* and 2514 *np* data below 350 MeV) with a χ^2/datum $\approx$ 1. These are the potentials constructed by the Nijmegen group, Nijm-I, NijmII and Reid93 [23], the Argonne V_{18} potential [24], and the CD-Bonn potential [25]. The latter is essentially a new version of the one-boson-exchange potential including the π, ρ and ω mesons plus two effective scalar-isoscalar σ bosons, the parameters of which are partial-wave dependent. This additional fit freedom produces a χ^2/datum of 1.02 for the 4301 data of the Nijmegen database, the total number of free parameters being 43. In this connection, it may be mentioned that since 1992 the number of NN data has considerably increased. This has produced the "1999 database" [9,25] which contains 5990 *pp* and *np* data. The χ^2/datum for the CD-Bonn potential in regard to the latter database remains 1.02. [25].

All the high-precision NN potentials mentioned above have a large number of free parameters, say about 45, which is the price one has to pay to achieve a very accurate fit of the world NN data. This makes it clear that, to date, high-quality potentials with an excellent χ^2/datum ≈ 1 can only be obtained within the framework of a substantially phenomenological approach. Since these potentials fit almost equally well the NN data up to the inelastic threshold, their on-shell properties are essentially identical, namely they are phase-shift equivalent. In addition, they all predict almost identical deuteron observables (quadrupole moment and D/S-state ratio) [9].

While they have also in common the inclusion of the one-pion exchange contribution, their off-shell behavior may be quite different. A detailed comparison between their predictions is given in Ref. 9. I only mention here that the predicted D-state probability of the deuteron ranges from 4.85% for CD-Bonn to 5.76% for V_{18}. In this context, the question arises of how much nuclear structure results may depend on the NN potential one starts with. We shall consider this important point in Sec. 4.

The brief review of the NN interaction given above has been mainly aimed at highlighting the progress made in this field over a period of about 50 years. As already pointed out in the Introduction, and as we shall see in Sec. 5, this has been instrumental in paving the way to a more fundamental approach to nuclear structure calculations than the traditional, empirical one. It is clear, however, that from a first-principle point of view a substantial theoretical progress in the field of NN interaction is still in demand. This is not likely to be achieved along the lines of the traditional meson theory. Indeed, in the past few years efforts in this direction have been made within the framework of the chiral effective field theory. The literature on this subject, which is still actively pursued, is by now very extensive and even a brief summary is outside the limits of these lectures. Thus I shall only give here a bare outline of some aspects which are relevant to my presentation. Comprehensive reviews may be found in [28–30].

The approach to the NN interaction based upon chiral effective field theory was started by Weinberg [26,27] some fifteen years ago and then developed by several authors. The basic idea [26] is to derive the NN potential starting from the most general chiral Lagrangian for low-energy pions and nucleons, consistent with the symmetries of quantum chromodynamics, in particular the spontaneously broken chiral symmetry. The chiral Lagrangian provides a perturbative framework for the derivation of the nucleon-nucleon potential. In fact, it was shown by Weinberg [27] that a systematic expansion of the nuclear potential exists in powers of the small parameter Q/Λ_χ, where Q denotes a generic low-momentum and $\Lambda_\chi \approx 1$ GeV is the chiral symmetry breaking scale. This perturbative low-energy theory is called chiral perturbation theory. The contribution of any diagram to the perturbation expansion is characterized by the power ν of the momentum Q, and the expansion is organized by counting powers of Q. This procedure [27] is referred to as power counting.

In the decade following the initial work by Weinberg, where only the lowest order NN potential was obtained, the effective chiral potential was extended to order $(Q/\Lambda_\chi)^3$ [next-to-next-to-leading order (NNLO), ν=3)] by various authors (see Ref. 30 for a comprehensive list of references through 2002). An accurate NNLO potential, called Idaho potential, was constructed by Entem and Machleidt [31,32]. With 46 parameters, the NN data below 210 MeV were reproduced with a χ^2/datum = 0.98. [33]. Shell-model calculations using this chiral potential yielded very good results for nuclei with two valence particles in various mass region. [8,34].

More recently, chiral potentials at the next-to-next-to-next-to-leading order (N^3LO, fourth order) have been constructed [35,36]. The potential developed in the work of Ref. 35, dubbed Idaho N^3LO, includes 24 contact terms (24 parameters) which contribute to the partial waves with $L \leq 2$. With 29 parameters in all, it gives a χ^2/datum for the reproduction of the 1999 *np* and *pp* databases below 290 MeV of 1.10 and 1.50, respectively. A brief survey of the current status of the chiral potentials as well as a list of references to recent nuclear structure studies employing the Idaho N^3LO potential may be found in Ref. 37. We only mention here that the use of this potential by our own group [38–40] has produced very promising results.

The foregoing discussion has all been focused on the two-nucleon force. As is well known, the role of three-nucleon interactions in light nuclei has been, and is currently, actively investigated. However, to touch upon this topic is clearly beyond the scope of these lectures. Here, I may only mention that in recent years the Green's function Monte Carlo (GFMC) method has proved to be a valuable tool for calculations of properties of light nuclei using realistic two-nucleon and three-nucleon potentials. [41,42] In particular, the combination of the Argonne V_{18} potential and Illinois-2 three-nucleon potential has yielded good results for energies of nuclei up to ^{12}C [43].

3. The shell-model effective interaction

The shell-model effective interaction V_{eff} is defined, as usual, in the following way. In principle, one should solve a nuclear many-body Schrödinger equation of the form

$$H\Psi_i = E_i\Psi_i, \tag{1}$$

with $H = T + V_{NN}$, where T denotes the kinetic energy. This full-space many-body problem is reduced to a smaller model-space problem of the form

$$PH_{\text{eff}}P\Psi_i = P(H_0 + V_{\text{eff}})P\Psi_i = E_iP\Psi_i. \tag{2}$$

Here $H_0 = T + U$ is the unperturbed Hamiltonian, U being an auxiliary potential introduced to define a convenient single-particle basis, and P denotes the projection operator onto the chosen model space,

$$P = \sum_{i=1}^{d} |\psi_i\rangle\langle\psi_i|, \tag{3}$$

d being the dimension of the model space and $|\psi_i\rangle$ the eigenfunctions of H_0. The effective interaction V_{eff} operates only within the model space P. In operator form it can be schematically written [46,47] as

$$V_{\text{eff}} = \hat{Q} - \hat{Q}'\int\hat{Q} + \hat{Q}'\int\hat{Q}\int\hat{Q} - \hat{Q}'\int\hat{Q}\int\hat{Q}\int\hat{Q} + \ \dots\ , \tag{4}$$

where $\hat{Q}$, usually referred to as the $\hat{Q}$-box, is a vertex function composed of irreducible linked diagrams, and the integral sign represents a generalized folding operation. $\hat{Q}'$ is obtained from $\hat{Q}$ by removing terms of first order in the interaction. Once the $\hat{Q}$-box is calculated, the folded-diagram series of Eq. (4) can be summed up to all orders by iteration methods.

A main difficulty encountered in the derivation of V_{eff} from any modern NNpotential is the existence of a strong repulsive core which prevents its direct use in nuclear structure calculations. This difficulty is usually overcome by resorting to the well known Brueckner G-matrix method. The G-matrix is obtained from the bare NN potential V_{NN} by solving the Bethe-Goldstone equation

$$G(\omega) = V_{NN} + V_{NN}Q_2\frac{1}{\omega - Q_2TQ_2}Q_2G(\omega), \tag{5}$$

where T is the two-nucleon kinetic energy and ω is an energy variable, commonly referred to as starting energy. The operator Q_2 is the Pauli exclusion operator for two interacting nucleons, to make sure that the intermediate states of G must not only be above the filled Fermi sea but also outside the model space within which Eq. (5) is to be solved.

As mentioned in the Introduction, the use of the G matrix has long proved to be a valuable tool to overcome the difficulty posed by the strong short-range repulsion contained in the free NN potential. However, the G matrix is model-space dependent as well as energy dependent; these dependences make its actual calculation rather involved. In this context, it may be recalled that an early criticism of the G-matrix method to eliminate effects of the repulsive core in the NN potential dates back to the 1960s [45]. Quoting from the Introduction of Ref. 45: "*To include in a potential a hard core and then remove its catastrophic effect on the independent-particle motion would, if performed correctly, appear to be an impressive but quite pointless feat of mathematical gymnastics*".

Based on the idea of finding a more convenient way to handle this problem, a method was developed for deriving directly from the phase shifts a set of matrix elements of V_{NN} in oscillator wave functions [44,45]. This resulted in the well-known Sussex interaction which has been used in several nuclear structure calculations.

Since then, however, there has been a considerable improvement in the techniques to calculate the G matrix, which has been routinely used in practically all realistic calculations through 2000. Nevertheless, the idea of bypassing the G-matrix approach to the renormalization of the bare NN potential has remained very appealing. Recently, a new approach has been proposed [5,6] that achieves this goal. In the next section, I shall only give a bare outline of it, while a detailed description can be found in Ref. 6.

4. The low-momentum nucleon-nucleon potential $V_{\rm low-k}$

As pointed out in the Introduction, we "smooth out" the strong repulsive core contained in the bare NN potential V_{NN} by constructing a low-momentum potential $V_{\rm low-k}$. This is achieved by integrating out the high-momentum modes of V_{NN} down to a cutoff momentum Λ. This integration is carried out with the requirement that the deuteron binding energy and low-energy phase shifts of V_{NN} are preserved by $V_{\rm low-k}$. This requirement may be satisfied by the following T-matrix equivalence approach. We start from the half-on-shell T matrix for V_{NN}

$$T(k',k,k^2) = V_{NN}(k',k) + \wp \int_0^\infty q^2 dq V_{NN}(k',q) \frac{1}{k^2-q^2} T(q,k,k^2)\,, \quad (6)$$

where $\wp$ denotes the principal value and k, k', and q stand for the relative momenta. The effective low-momentum T matrix is then defined by

$$T_{\rm low-k}(p',p,p^2) = V_{\rm low-k}(p',p) + \wp \int_0^{\Lambda} q^2 dq V_{\rm low-k}(p',q) \frac{1}{p^2-q^2} T_{\rm low-k}(q,p,p^2)\,, \quad (7)$$

where the intermediate state momentum q is integrated from 0 to the momentum space cutoff Λ and $(p',p) \le \Lambda$. The above T matrices are required to satisfy the condition

$$T(p',p,p^2) = T_{\rm low-k}(p',p,p^2)\,; \quad (p',p) \le \Lambda\,. \quad (8)$$

The above equations define the effective low-momentum interaction $V_{\rm low-k}$, and it has been shown [6] that they are satisfied when $V_{\rm low-k}$ is given by the Kuo-Lee-Ratcliff (KLR) folded-diagram expansion, [47,48] originally designed for constructing shell-model effective interactions, see Eq. (4).

In addition to the preservation of the half-on-shell T matrix, which implies preservation of the phase shifts, this $V_{\rm low-k}$ preserves the deuteron binding energy, since eigenvalues are preserved by the KLR effective interaction. For any value of Λ, $V_{\rm low-k}$ can be calculated very accurately using iteration methods. Our calculation is performed by employing the iterative implementation of the Lee-Suzuki method [49] proposed in Ref. 50.

The $V_{\rm low-k}$ given by the T-matrix equivalence approach mentioned above is not Hermitian. Therefore, an additional transformation is needed to make it Hermitian. To this end, we resort to the Hermitization procedure suggested in Ref. 50, which makes use of the Cholesky decomposition of symmetric positive definite matrices.

Once the $V_{\rm low-k}$ is obtained, we use it, plus the Coulomb force for protons, as input interaction for the calculation of the matrix elements of $V_{\rm eff}$. The latter is derived by employing a folded-diagram method (see Sec. 3), which was previously applied to many nuclei [4] using G-matrix interactions. Since $V_{\rm low-k}$ is already a smooth potential, it is no longer necessary to calculate the G matrix. We therefore perform shell-model calculations following the same procedure as described, for instance, in Refs. 20 and 51, except that the G matrix used there is replaced by $V_{\rm low-k}$. More precisely, we first calculate the $\hat{Q}$-box including diagrams up to second order in the two-body interaction. The shell-model effective interaction is then obtained by summing up the $\hat{Q}$-box folded-diagram series using the Lee-Suzuki iteration method [49].

As mentioned in the Introduction, we have assessed the merit of the $V_{\rm low-k}$ approach in practical applications. To this end, we have compared the results of shell-model calculations performed by starting from the CD-Bonn potential and deriving $V_{\rm eff}$ through both the $V_{\rm low-k}$ and G-matrix approaches. In particular, results for ^{18}O are presented in Ref. 6 while the calculations of Ref. 7 concern ^{132}Sn neighbors. A comparison between the G matrix and $V_{\rm low-k}$ spectra for the heavy-mass nucleus ^{210}Po can be found in Ref. 8. In all these calculations the cutoff parameter Λ has been chosen around 2 fm^{-1}, in accord with the criterion given in Ref. 6. It has been a remarkable finding of these studies that the $V_{\rm low-k}$ results are as good or even slightly better than the G-matrix ones.

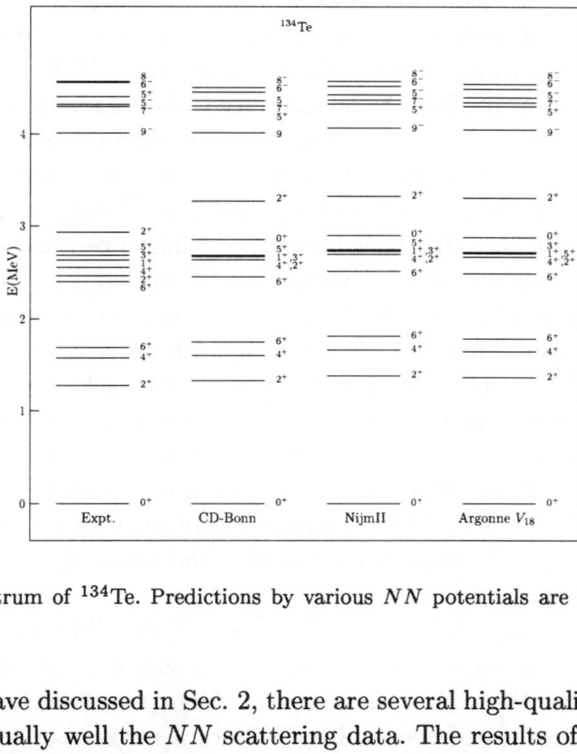

Fig. 1. Spectrum of ^{134}Te. Predictions by various NN potentials are compared with experiment.

As we have discussed in Sec. 2, there are several high-quality potentials which fit equally well the NN scattering data. The results of our realistic shell-model calculations reported in the next section have all been obtained using a input the CD-Bonn potential. This may raise the question of how much they depend on this choice of the NN potential. We have verified that shell-model effective interactions derived from phase-shift equivalent NN potentials through the $V_{\rm low-k}$ approach do not lead to significantly different results. Here, by way of illustration, we present the results obtained for the nucleus ^{134}Te. This nucleus has only two valence protons and thus offers the opportunity to test directly the matrix elements of the various effective interactions. In Fig. 1 we show, together with the experimental spectrum, the spectra obtained by using the CD-Bonn, NijmII, and Argonne V_{18} potentials, all renormalized through the $V_{\rm low-k}$ procedure with a cutoff momentum Λ=2.2 fm^{-1}. From Fig. 1 we see that the calculated spectra are very similar, the differences between the level energies not exceeding 80 keV. It is also seen that the agreement with experiment is very good for all the three potentials.

5. Review of selected results

In this section, we report some selected results of our recent shell-model studies [52–54] of neutron-rich nuclei beyond doubly magic ^{132}Sn.

The study of exotic nuclei around doubly magic ^{132}Sn is a subject of special interest, as it offers the opportunity to explore for possible changes in nuclear structure properties when moving toward the neutron drip line. In this context, great attention is currently focused on nuclei with valence neutrons outside the $N = 82$ shell closure. This is motivated by the fact that some of the data that have become available appear to be at variance with what one might expect by extrapolating the existing results for $N < 82$ nuclei. In particular, some peculiar properties have been recently observed in the two nuclei ^{134}Sn and ^{135}Sb which, with an N/Z ratio of 1.68 and 1.65, respectively, are at present the most exotic nuclei beyond ^{132}Sn for which information exists on excited states. This is the case of the first 2^+ state in ^{134}Sn which, lying at 726 keV excitation energy, is the lowest first-excited 2^+ level observed in a semi-magic even-even nucleus over the whole chart of nuclides. As for ^{135}Sb, there is a significant drop in the energy of the lowest-lying $5/2^+$ state as compared to the values observed for the Sb isotopes with $N \leq 82$.

We consider here the three nuclei ^{134}Sn, ^{135}Sb, ^{134}Sb, the latter being the more appropriate system to study the proton-neutron effective interaction in the ^{132}Sn region. In the calculations for these nuclei ^{132}Sn is assumed to be a closed core and the valence neutrons can occupy the six levels $0h_{9/2}$, $1f_{7/2}$, $1f_{5/2}$, $2p_{3/2}$, $2p_{1/2}$, and $0i_{13/2}$ of the 82-126 shell, while for the odd proton in ^{134,135}Sb the model space includes the five levels $0g_{7/2}$, $1d_{5/2}$, $1d_{3/2}$, $2s_{1/2}$, and $0h_{11/2}$ of the 50-82 shell. The proton and neutron single-particle energies have been taken from the experimental spectra of ^{133}Sb and ^{133}Sn, respectively. The energy of the proton $s_{1/2}$ and neutron $i_{13/2}$ level, which are still missing, are from the studies of Refs. 19 and 55, respectively. All the adopted values are reported in Ref. 52.

The two-body effective interaction is derived by means of the $\hat{Q}$-box folded-diagram method (see Sec. 3) from the CD-Bonn NN potential renormalized through use of the $V_{\rm low-k}$ procedure with a cutoff momentum Λ $=2.2$ fm^{-1}. The computation of the diagrams included in the $\hat{Q}$-box is performed within the harmonic-oscillator basis using intermediate states composed of all possible hole states and particle states restricted to the five shells above the Fermi surface. The oscillator parameter used is $\hbar\omega = 7.88$ MeV.

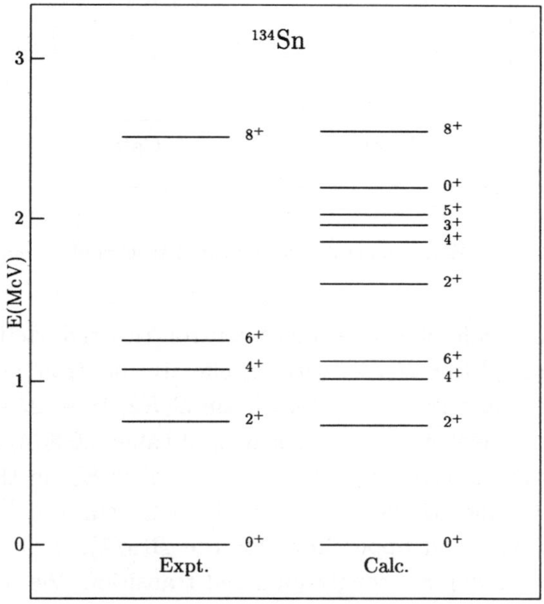

Fig. 2. Experimental and calculated spectra of ^{134}Sn.

The experimental [56–58,60] and calculated spectra of ^{134}Sn and ^{135}Sb are compared in Figs. 2 and 3. ¿From these figures we see that the experimental levels are very well reproduced by the theory. Note that the very low-energy positions of both the first-excited 2^+ and $5/2^+$ states in ^{134}Sn and ^{135}Sb, respectively, are well accounted for. As for the latter, it is shown in Ref. 52 that it is the admixed nature of the $5/2^+$ state that explains its anomalously low position.

Very recently, the $B(E2; 0^+ \rightarrow 2_1^+)$ value in ^{134}Sn has been mea-

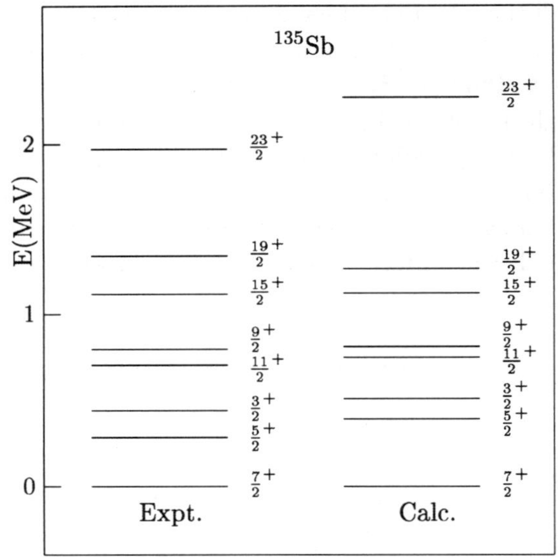

Fig. 3. Experimental and calculated spectra of ^{135}Sb.

sured [59] using Coulomb excitation of neutron-rich radioactive ion beams. We have calculated this $B(E2)$ with an effective neutron charge of $0.70\,e$, according to our early study [55]. We obtain $B(E2; 0^+ \rightarrow 2_1^+) = 0.033\ e^2\text{b}^2$, in excellent agreement with the experimental value $0.029(4)\ e^2\text{b}^2$.

As regards the electromagnetic properties of ^{135}Sb, in the very recent work of Refs. 61 and 62 the lifetime of the $5/2^+$ state in ^{135}Sb has been measured. A very small upper limit for the $B(M1)$, $0.29 \cdot 10^{-3}\,\mu_N^2$, was found, thus evidencing a strongly hindered transition. We have calculated the $B(M1; 5/2^+ \rightarrow 7/2^+)$ making use of an effective $M1$ operator which includes first-order diagrams in $V_{\text{low}-\text{k}}$. Our predicted value is $4.0 \cdot 10^{-3}\,\mu_N^2$. Keeping in mind that in our calculation we do not include any meson-exchange correction, the agreement between the experimental and calculated $B(M1)$ may be considered quite satisfactory.

Let us now come to the one-proton, one-neutron nucleus ^{134}Sb. The calculated energies of the $\pi g_{7/2} \nu f_{7/2}$ and $\pi d_{5/2} \nu f_{7/2}$ multiplets are reported in Fig. 4, where they are compared with the experimental data [58,60]. The first eight calculated states arise from the $\pi g_{7/2} \nu f_{7/2}$ configuration and have their experimental counterpart in the eight lowest-lying experimental states. The wave functions of these states are characterized by very little configuration mixing. As for the $\pi d_{5/2} \nu f_{7/2}$ multiplet, we find that the 1^-, 2^-, 4^-, and 6^- members correspond to the yrare states, while both the other two, with $J^\pi = 3^-$ and 5^-, to the third excited state. As is shown in Fig. 4, only the 1^- and 2^- members of the $\pi d_{5/2}\nu f_{7/2}$ multiplet are known. As regards the structure of the states belonging to the

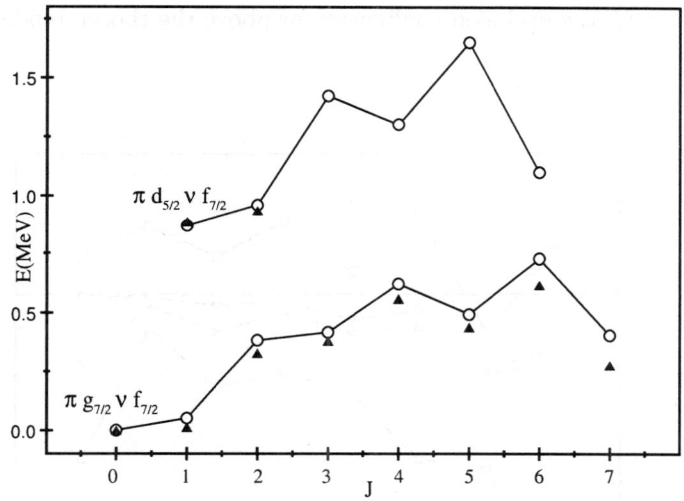

Fig. 4. Proton-neutron $\pi g_{7/2}\nu f_{7/2}$ and $\pi d_{5/2}\nu f_{7/2}$multiplets in ^{134}Sb.

$\pi d_{5/2}\nu f_{7/2}$ multiplet, we find that all members receive significant contributions from configurations other than the dominant one.

From Fig. 4 we see that the agreement between theory and experiment is very good, the discrepancies being in the order of a few tens of keV for most of the states. It is an important outcome of our calculation that we predict almost the right spacing between the 0^- ground state and first excited 1^- state. In fact, the latter has been observed at 13 keV excitation energy, our value being 53 keV.

In this context, it is interesting to try to understand what makes our proton-neutron matrix elements appropriate to the description of the multiplets in ^{134}Sb, in particular the very small energy spacing between the 0^- and the 1^- states. To this end, in Ref. 53 an analysis has been performed of the various contributions to the effective interaction, focusing attention on the $\pi g_{7/2}\nu f_{7/2}$ configuration.

As mentioned above, our effective interaction is calculated within the framework of a $\hat{Q}$-box folded-diagram method. In particular, the $\hat{Q}$ box is composed of first- and second-order diagrams in the $V_{\rm low-k}$ derived from the CD-Bonn potential. In other words, the matrix elements of the effective interaction contain the $V_{\rm low-k}$ plus additional terms which take into account core-polarization effects arising from $1p-1h$ ("bubble" diagram) and $2p-2h$ excitations.They also include the so-called ladder diagrams, which must compensate for the excluded configurations above the chosen model space.

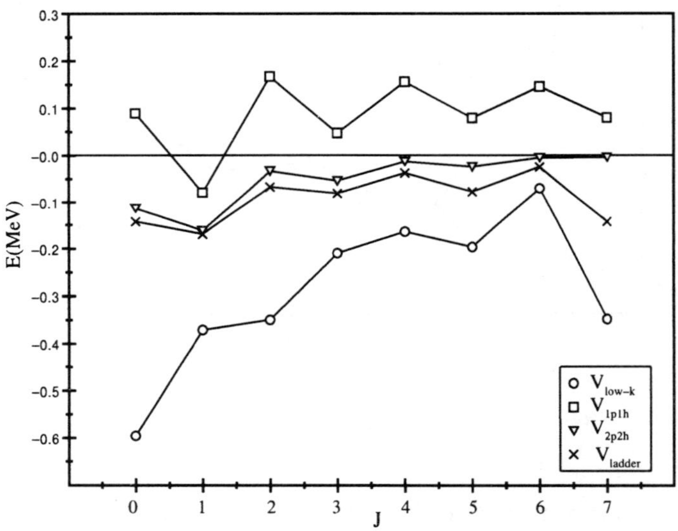

Fig. 5. Diagonal matrix elements of $V_{\rm low-k}$ and contributions from the two-body second-order diagrams for the $\pi g_{7/2}\nu f_{7/2}$ configuration. See text for comments.

In Fig. 5 we show the $\pi g_{7/2}\nu f_{7/2}$ matrix elements of the $V_{\rm low-k}$ as a function of J together with the second-order two-body contributions. ¿From the inspection of this figure we see that the incorrect behavior of the $V_{\rm low-k}$ matrix elements is "healed" by the $V_{\rm 1p1h}$, $V_{\rm 2p2h}$ and $V_{\rm ladder}$ corrections. In particular it appears that a crucial role is played by the bubble diagram, especially as regards the position of the 1^- state. It is worth noting that the folding procedure provides a common attenuation of all matrix elements, which does not affect the overall behavior of the multiplet.

6. Concluding remarks

In these lectures, I have tried to give a self-contained, albeit brief, survey of modern shell-model calculations employing two-body effective interactions derived from the free nucleon-nucleon potential. A main feature of these calculations is that no adjustable parameter appears in the determination of the effective interaction. This removes the uncertainty inherent in the traditional use of empirical interactions, making the shell model a truly microscopic theory of nuclear structure.

I have shown how the $V_{\text{low}-\text{k}}$ approach to the renormalization of the strong short-range repulsion contained in all modern NN potentials is a valuable tool for nuclear structure calculations. This potential may be used directly in shell-model calculations without the need of first calculating the Brueckner G-matrix. In this context, it is worth emphasizing that the $V_{\text{low}-\text{k}}$'s extracted from various phase-shift equivalent potentials give very similar results in shell-model calculations, suggesting the realization of a nearly unique low-momentum NN potential.

In the last part of this paper I have presented, by way of illustration, some selected results of recent calculations for nuclei beyond doubly magic ^{132}Sn. These neutron-rich nuclei, which lie well away from the valley of stability, offer the opportunity for a stringent test of the matrix elements of the effective interaction. The very good agreement with the available experimental data shown in Sec. 5 supports confidence in the predictive power of realistic shell-model calculations in the regions of shell closures off stability, which are of great current interest.

Acknowledgments

This work was supported in part by the Italian Ministero dell'Istruzione, dell'Università e della Ricerca.

References

1. J. F. Dawson, I Talmi, and J. D. Walecka, *Ann. Phys. (NY)* **18**, 339 (1962).
2. T. T. S. Kuo and G. E. Brown, *Nucl. Phys.*, **85**, 40 (1966).
3. J. P. Elliott, in *Shell Model and Nuclear Structure: Where Do We Stand?*, ed. A. Covello (World Scientific, Singapore, 1989), pp. 13-28.
4. A. Covello, L. Coraggio, A. Gargano, and N. Itaco, *Acta Phys. Pol. B* **32**, 871 (2001), and references therein.
5. S. Bogner, T. T. S. Kuo, and L. Coraggio, *Nucl. Phys. A* **684**, 432c (2001).
6. S. Bogner, T. T. S. Kuo, L. Coraggio, A. Covello, and N. Itaco, *Phys. Rev. C* **65**, 051301(R) (2002).

7. A. Covello, L. Coraggio, A. Gargano, N. Itaco, and T. T. S. Kuo, in *Challenges of Nuclear Structure*, ed. A. Covello (World Scientific, Singapore, 2002), pp. 139-146.
8. A. Covello, in *Proc. Int. School of Physics "E. Fermi", Course CLIII*, eds. A. Molinari, L. Riccati, W. M. Alberico and M. Morando (IOS Press, Amsterdam, 2003), pp. 79-91.
9. R. Machleidt and I. Slaus, *J. Phys. G* **27**, R69 (2001).
10. M. L. Goldberger, in *Proc. Midwestern Conference on Theoretical Physics* (Purdue University, Lafayette, Ind., April 1960), pp. 50-63.
11. A. E. S. Green, M. H. MacGregor, and R. Wilson, *Rev. Mod. Phys.* **39**, 497 (1967) .
12. R. Vinh Mau, in *Proc. Int. Conference on Nuclear Physics, Florence, 1983* (Tipografia Compositori, Bologna, 1983), pp. 61-84.
13. R. Vinh Mau, in *Proc. of the 1989 Int. Conference on Nuclear Physics, São Paulo* (World Scientific, Singapore, 1990), pp. 189-209.
14. M. M. Nagels, T. A. Rijken, and J. J. de Swart, *Phys. Rev. D* **17**, 768 (1978).
15. M. Lacombe, B. Loiseau, J. M. Richard, R. Vinh Mau, J. Côtè, P. Pirés, and R. De Tourreil, *Phys. Rev. C* **21**, 861 (1980).
16. R. Machleidt, K. Holinde, and Ch. Elster, *Phys. Rep.* **149**, 1 (1987).
17. R. Machleidt and G. Q. Li, *Phys. Rep.* **242**, 5 (1994).
18. R. Machleidt, *Adv. Nucl. Phys.* **19**, 189 (1989).
19. F. Andreozzi, L. Coraggio, A. Covello, A. Gargano, T. T. S. Kuo, Z. B. Li, and A. Porrino, *Phys. Rev. C* **54**, 1636 (1996).
20. A. Covello, F. Andreozzi, L. Coraggio, A. Gargano, T. T. S. Kuo, and A. Porrino, *Prog. Part. Nucl. Phys.* **38**, 165 (1997).
21. A. Covello, L. Coraggio, A. Gargano, N. Itaco, and T. T. S. Kuo, in *Proc. of the Nuclear Structure 98 Conference, Gatlinburg, Tennessee, 1998, AIP Conf. Proc* **481**, ed. C. Baktash (AIP, New York, 1999), p. 56-65.
22. V. G. J. Stoks, R. A. M. Klomp, M. C. M. Rentmeester and J. J. de Swart, *Phys. Rev. C* **48**, 792 (1993).
23. V. G. J. Stoks, R. A. M. Klomp, C. P. F. Terheggen, J. J. de Swart, *Phys. Rev. C* **49**, 2950 (1994).
24. R. B.Wiringa, V. G. J. Stoks, and R. Schiavilla, *Phys. Rev. C* **51**, 38 (1995).
25. R. Machleidt, *Phys. Rev. C* **63**, 024001 (2001).
26. S. Weinberg, *Phys. Lett. B* **251**, 288 (1990).
27. S. Weinberg, *Nucl. Phys. B* **363**, 3 (1991).
28. U. van Kolck, *Prog. Part. Nucl. Phys.* **43**, 337 (1999).
29. S. R. Beane, P. F. Bedaque, W. C. Haxton, D. R. Phillips, and M.J. Savage, in *At the frontiers of particle physics - Handbook of QCD*, ed. M. Shifman (World Scientific, Singapore, 2001), Vol. 1, p. 133.
30. P. F. Bedaque and U. van Kolck, *Annu. Rev. Nucl. Part. Sci* **52**, 339 (2002).
31. D. R. Entem and R. Machleidt, in *Challenges of Nuclear Structure*, ed. A. Covello (World Scientific, Singapore, 2002), p. 113-127.
32. D. R. Entem and R. Machleidt, *Phys. Lett. B* **524**, 93 (2002).
33. D. R. Entem, R. Machleidt, and H. Witala, *Phys. Rev. C* **65**, 064005 (2002).
34. L. Coraggio, A. Covello, A. Gargano, N. Itaco, T. T. S. Kuo, D. R. Entem,

and R. Machleidt, *Phys. Rev. C* **66**, 021303(R) (2002).
35. D. R. Entem and R. Machleidt, *Phys. Rev. C* **68**, 041001(R) (2003).
36. E. Epelbaum, W. Glöckle, and U.-G. Meissner, *Nucl. Phys. A* **747**, 362 (2005).
37. R. Machleidt, in *Proc. 18th Int. IUPAP Conference on Few-body Problems in Physics*, *Nucl. Phys. A*, to be published.
38. L. Coraggio, N. Itaco, A. Covello, A. Gargano, and T. T. S. Kuo, *Phys. Rev. C* **68**, 034320 (2003).
39. L. Coraggio, A. Covello, A. Gargano, N. Itaco, T. T. S. Kuo, and R. Machleidt, *Phys. Rev. C* **71**, 014307 (2005).
40. L. Coraggio, A. Covello, A. Gargano, N. Itaco, and T. T. S. Kuo, *Phys. Rev. C* **73**, 014304 (2006).
41. S. C. Pieper, V. R. Pandharipande, R. B. Wiringa, and J. Carlson, *Phys. Rev. C* **64**, 014001 (2001).
42. S. C. Pieper, K. Varga, and R. B. Wiringa, *Phys. Rev. C* **66**, 044310 (2002).
43. S. C. Pieper, *Nucl. Phys. A* **751**, 516c (2005).
44. J. P. Elliott, H. A. Mavromatis, and E. A. Sanderson, *Phys. Lett. B* **24**, 358 (1967).
45. J. P. Elliott, A. D. Jackson, H. A. Mavromatis, E. A. Sanderson, and B. Singh, *Nucl. Phys. A* **121**, 241 (1968).
46. T. T. S. Kuo and E. M. Krenciglowa, *Nucl. Phys. A* **342**, 454 (1980).
47. T. T. S. Kuo and E. Osnes, *Lecture Notes in Physics*, Vol. 364 (Springer-Verlag, Berlin, 1990).
48. T. T. S. Kuo, S. Y. Lee, and K. F. Ratcliff , *Nucl. Phys. A* **176**, 65 (1971).
49. K. Suzuki and S. Y. Lee (1980)*Prog. Theor. Phys.* **64**, 2091 (1980).
50. F. Andreozzi *Phys. Rev. C* **54**, 684 (1996).
51. M. F. Jiang, R. Machleidt, D. B. Stout, and T. T. S. Kuo *Phys. Rev. C* **46**, 910 (1992).
52. L. Coraggio, A. Covello, A. Gargano, and N. Itaco, *Phys. Rev. C* **72**, 057302 (2005).
53. L. Coraggio, A. Covello, A. Gargano, and N. Itaco, *Phys. Rev. C* **73**, 031302(R) (2006).
54. A. Covello, L. Coraggio, A. Gargano, and N. Itaco, *Eur. Phys. J. A*, to be published.
55. L. Coraggio, A. Covello, A. Gargano, and N. Itaco, *Phys. Rev. C* **65**, 051306(R) (2002).
56. C. T. Zhang *et al.*, *Z. Phys. A* **358**, 9 (1997).
57. A. Korgul *et al.*, *Eur. Phys. J. A* **7**, 167 (2000).
58. J. Shergur *et al.*, *Phys. Rev. C* **71**, 064321 (2005).
59. J. R. Beene *et al. Nucl. Phys. A* **746**, 471c (2004).
60. Data extracted using the NNDC On-line Data Service from the ENSDF database, version of October 13, 2006.
61. H. Mach *et al.*, in *Key Topics in Nuclear Structure*, ed. A. Covello (World Scientific, Singapore, 2005), pp. 205-211.
62. A. Korgul *et al. Eur. Phys. J. A* **25**, s01, 123 (2005).

Nuclear Superfluidity in Exotic Nuclei and Neutron Stars

Nicolae Sandulescu

Institute of Physics and Nuclear Engineering, 76900 Bucharest, Romania

Nuclear superfludity in exotic nuclei close to the drip lines and in the inner crust matter of neutron stars have common features which can be treated with the same theoretical tools. In the first part of my lecture I discuss how two such tools, namely the HFB approach and the linear response theory can be used to describe the pairing correlations in weakly bound nuclei, in which the unbound part of the energy spectrum becomes important. Then, using the same models, I shall discuss how the nuclear superfluidity can affect the thermal properties of the inner crust of neutron stars.

1. Introduction

The basic features of nuclear superfluidity are the same in finite nuclei and in infinite Fermi systems such as neutron stars. Yet, in atomic nuclei the pairing correlations have special features related to the finite size of the system. The way how the finite size affects the pairing correlations depends on the position of the chemical potential. If the chemical potential is deeply bound, like in stable and heavy nuclei, the finite size influences the pairing correlations mainly through the shell structure induced by the spin-orbit interaction. The situation becomes more complex in weakly bound nuclei close to the drip lines, where the chemical potential approaches the continuum threshold. In this case the inhomogeneity of the pairing field can produce a strong coupling between the bound and the unbound part of the single-particle spectrum. This is an issue which will be discussed in the first part of my lecture. More precisely, I shall discuss how the continuum coupling and the pairing correlations can be treated in the framework of the Hartree-Fock-Bogoliubov (HFB) approach and linear response theory.

The neutron-drip line put a limit to the neutron-rich nuclei which can be produced in the laboratory or in supernova explosions (via rapid neutron-cupture processes). However, this limit can be overpassed in the inner crust of neutron stars since there the driped neutrons are kept together with the neutron-rich nuclei by the gravitational preasure.

The superfluid properties of the inner crust have been considered long ago in connection with post-glitch timing observations and colling processes [1–3]. However, although the neutron star matter superfluidity has been intensively studied in the last decades [4], so far only a few microscopic calculations have been done for the superfluidity of inner crust matter. The most sophisticated microscopic calculations done till now use the framework of HFB approach at finite temperature [5]. They will be discussed in the second part of my lecture. The discussion will be focused on the effects induced by the pairing correlations on the specific heat and on the cooling time of the inner curst of neutron stars.

2. Pairing correlations in exotic nuclei

2.1. *Continuum-HFB approach*

The tool commonly used for treating pairing correlations in exotic nuclei close to the drip lines is the Hartree-Fock-Bogoliubov (HFB) approach [6]. The novel feature of interest here is how within this approach one can treat properly the quasiparticle states belonging to the continuum spectrum [7]. This issue is discussed below.

The HFB equations for local fields and spherical symmetry have the form:

$$\begin{pmatrix} h(r)-\lambda & \Delta(r) \\ \Delta(r) & -h(r)+\lambda \end{pmatrix} \begin{pmatrix} U_i(r) \\ V_i(r) \end{pmatrix} = E_i \begin{pmatrix} U_i(r) \\ V_i(r) \end{pmatrix} , \tag{1}$$

where λ is the chemical potential, $h(r)$ is the mean field hamiltonian and $\Delta(r)$ is the pairing field. The fields depend on particle density $\rho(r)$ and pairing density $\kappa(r)$ given by:

$$\rho(r) = \frac{1}{4\pi} \sum_i (2j_i+1) V_i^*(r) V_i(r), \tag{2}$$

$$\kappa(r) = \frac{1}{4\pi} \sum_i (2j_i+1) U_i^*(r) V_i(r). \tag{3}$$

In the calculations presented here the mean field is described with a Skyrme force and for the pairing interaction it is used a density- dependent force of zero range of the following form [8]:

$$V(\mathbf{r}-\mathbf{r}') = V_0[1-\eta(\frac{\rho}{\rho_0})^\alpha]\delta(\mathbf{r}-\mathbf{r}') \equiv V_{eff}(\rho(r))\delta(\mathbf{r}-\mathbf{r}'). \tag{4}$$

For this force the pairing field is given by $\Delta(r) = \frac{V_{eff}}{2}\kappa(r)$.

The HFB equations have two kind of solutions. Thus, between 0 and $-\lambda$ the quasiparticle spectrum is discrete and both upper and lower components of the radial HFB wave function decay exponentially at infinity. On the other hand, for $E > -\lambda$ the spectrum is continuous and the solutions are:

$$u_{lj}(E,r) = C[cos(\delta_{lj})j_l(\alpha_1 r) - sin(\delta_{lj})n_l(\alpha_1 r)] \ ,$$
$$v_{lj}(E,r) = D_1 h_l^{(+)}(i\beta_1 r) \ , \tag{5}$$

where j_l and n_l are spherical Bessel and Neumann functions repectively and δ_{lj} is the phase shift corresponding to the angular momentum (lj). The phase shift is found by matching the asymptotic form of the wave function written above with the inner radial wave function. From the energy dependence of the phase shift one can determine the energy regions of quasiparticle resonant states. In HFB they are of two types. A first type corresponds to the single-particle resonances of the mean field. A second kind of resonant states is specific to the HFB approach and corresponds to the bound single-particle states which in the absence of pairing correlations have an energy $\epsilon < 2\lambda$. Among all possible resonances, of physical interest are the ones close to the continuum threshold. An exemple of such resonances is shown in section 2.4.

2.2. *Resonant states in the BCS approach*

The low-lying quasiparticle resonances can be also calculated in the framework of the BCS approach. The BCS approximation is obtained by neglecting in the HFB equations the non-diagonal matrix elements of the pairing field. This means that in the BCS limit one neglects the pairing correlations induced by the pairs formed in states which are not time-reversed partners.

The extension of BCS equations for taking into account the effect of resonant states was proposed in Refs. [9,10]. For the case of a general pairing interaction the corresponding resonant-BCS equations read [9]:

$$\Delta_i = \sum_j V_{i\bar{i}j\bar{j}} u_j v_j + \sum_\nu V_{i\bar{i},\nu\epsilon_\nu\overline{\nu\epsilon_\nu}} \int_{I_\nu} g_\nu(\epsilon)u_\nu(\epsilon)v_\nu(\epsilon)d\epsilon \ , \tag{6}$$

$$\Delta_\nu \equiv \sum_j V_{\nu\epsilon_\nu\overline{\nu\epsilon_\nu},j\bar{j}} u_j v_j + \sum_{\nu'} V_{\nu\epsilon_\nu\overline{\nu\epsilon_\nu},\nu'\epsilon_{\nu'}\overline{\nu'\epsilon_{\nu'}}} \int_{I_{\nu'}} g_{\nu'}(\epsilon')u_{\nu'}(\epsilon')v_{\nu'}(\epsilon')d\epsilon' \ , \tag{7}$$

$$N = \sum_i v_i^2 + \sum_\nu \int_{I_\nu} g_\nu(\epsilon) v_\nu^2(\epsilon) d\epsilon \ . \tag{8}$$

Here Δ_i is the gap for the bound state i and Δ_ν is the averaged gap for the resonant state ν. The quantity $g_\nu(\epsilon) = \frac{2j_\nu+1}{\pi}\frac{d\delta_\nu}{d\epsilon}$ is the continuum level density and δ_ν is the phase shift of angular momentum $(l_\nu j_\nu)$. The factor $g_\nu(\epsilon)$ takes into account the variation of the localisation of scattering states in the energy region of a resonance (i.e., the width effect) and becomes a delta function in the limit of a very narrow width. In these equations the interaction matrix elements are calculated with the scattering wave functions at resonance energies and normalised inside the volume where the pairing interaction is active.

The BCS equations written above have been solved with a single particle spectrum corresponding to a HF [9] and a RMF [11] mean field. It was shown that by including in the BCS equations a few relevant resonances close to the continuum threshold one can get results very similar to the ones obtained in the HFB and RHB calculations. One can thus conclude that in nuclei close to the dripline the quasiparticle spectrum is dominated by a few low-lying resonances. In many odd-even nuclei close to the drip lines these low-lying quasiparticle resonances might be the only measurable excited states. Their widths can be calculated from the phase shift behaviour or from the imaginary part of the energies associated to the Gamow states [12].

In even-even nuclei the low-lying quasiparticle resonances could form the main component of unbound collective excitations with a finite time life. How these excitations can be treated in nuclei close to the drip line is discussed in the next section.

2.3. *Linear response theory with pair correlations and continuum coupling*

The collective excitations of atomic nuclei in the presence of pairing correlations is usually described in the Quasiparticle-Random Phase Approximation (QRPA) [6]. In nuclei characterized by a small nucleon separation energy, the excited states are strongly influenced by the coupling with the quasiparticle continuum configurations. Among the configurations of particular interest are the two-quasiparticle states in which one or both quasiparticles are in the continuum. In order to describe such excited states one needs a proper treatment of the continuum coupling, which is missing in the usual QRPA calculations based on a discrete quasiparticle spectrum. In this section we discuss how the pairing and the continuum coupling can be treated in the framework of the linear response theory [13,14].

The response of the nuclear system to an external perturbation is obtained from the time-dependent HFB equations (TDHFB) [6]:

$$i\hbar\frac{\partial\mathcal{R}}{\partial t} = [\mathcal{H}(t) + \mathcal{F}(t), \mathcal{R}(t)], \tag{9}$$

where $\mathcal{R}$ and $\mathcal{H}$ are the time-dependent generalized density and the HFB Hamiltonian, respectively. $\mathcal{F}$ is the external oscillating field :

$$\mathcal{F} = Fe^{-i\omega t} + h.c.. \tag{10}$$

In Eq. (10) F includes both particle-hole and two-particle transfer operators :

$$F = \sum_{ij} F^{11}_{ij} c^\dagger_i c_j + \sum_{ij} (F^{12}_{ij} c^\dagger_i c^\dagger_j + F^{21}_{ij} c_i c_j), \tag{11}$$

and $c^\dagger_i$, c_i are the particle creation and annihilation operators, respectively.

In the small amplitude limit the TDHFB equations become:

$$\hbar\omega\mathcal{R}' = [\mathcal{H}', \mathcal{R}^0] + [\mathcal{H}^0, \mathcal{R}'] + [F, \mathcal{R}^0], \tag{12}$$

where the superscript ' stands for the corresponding perturbed quantity.

The variation of the generalized density $\mathcal{R}$' is expressed in term of 3 quantities, namely ρ', κ' and $\bar{\kappa}'$, which are written as a column vector :

$$\rho' = \begin{pmatrix} \rho' \\ \kappa' \\ \bar{\kappa}' \end{pmatrix}, \tag{13}$$

where $\rho'_{ij} = \left\langle 0|c^\dagger_j c_i|' \right\rangle$ is the variation of the particle density, $\kappa'_{ij} = \langle 0|c_j c_i|'\rangle$ and $\bar{\kappa}'_{ij} = \left\langle 0|c^\dagger_j c^\dagger_i|' \right\rangle$ are the fluctuations of the pairing tensor associated to the pairing vibrations and $|'\rangle$ denotes the change of the ground state wavefunction $|0>$ due to the external field.

The variation of the HFB Hamiltonian is given by:

$$\mathcal{H}' = \mathbf{V}\rho', \tag{14}$$

where $\mathbf{V}$ is the matrix of the residual interaction expressed in terms of the second derivatives of the HFB energy functional, namely:

$$\mathbf{V}^{\alpha\beta}(\mathbf{r}\sigma, \mathbf{r}'\sigma') = \frac{\partial^2\mathcal{E}}{\partial\rho_\beta(\mathbf{r}'\sigma')\partial\rho_{\bar{\alpha}}(\mathbf{r}\sigma)}, \quad \alpha, \beta = 1, 2, 3. \tag{15}$$

In the above equation the notation $\bar{\alpha}$ means that whenever α is 2 or 3 then $\bar{\alpha}$ is 3 or 2.

Introducing for the external field the three dimensional column vector:

$$\mathbf{F} = \begin{pmatrix} F^{11} \\ F^{12} \\ F^{21} \end{pmatrix}, \tag{16}$$

the density changes can be written in the standard form:

$$\rho' = \mathbf{GF} \ , \tag{17}$$

where $\mathbf{G}$ is the QRPA Green's function obeying the Bethe-Salpeter equation:

$$\mathbf{G} = (\mathbf{1} - \mathbf{G_0 V})^{-1} \mathbf{G_0} = \mathbf{G_0} + \mathbf{G_0 V G}. \tag{18}$$

The unperturbed Green's function $\mathbf{G_0}$ has the form:

$$\mathbf{G_0}^{\alpha\beta}(\mathbf{r}\sigma, \mathbf{r}'\sigma'; \omega) = \sum_{ij} \frac{\mathcal{U}_{ij}^{\alpha 1}(\mathbf{r}\sigma)\bar{\mathcal{U}}_{ij}^{*\beta 1}(\mathbf{r}'\sigma')}{\hbar\omega - (E_i + E_j) + i\eta} - \frac{\mathcal{U}_{ij}^{\alpha 2}(\mathbf{r}\sigma)\bar{\mathcal{U}}_{ij}^{*\beta 2}(\mathbf{r}'\sigma')}{\hbar\omega + (E_i + E_j) + i\eta}, \tag{19}$$

where E_i are the qp energies and $\mathcal{U}_{ij}$ are 3 by 2 matrices expressed in term of the two components of the HFB wave functions [13]. The $\sum$ symbol in Eq. (19) indicates that the summation is taken over the discrete and the continuum quasiparticle states.

The QRPA Green's function can be used for calculating the strength function associated with various external perturbations. For instance, the transitions from the ground state to the excited states induced by a particle-hole external field can be described by the strength function:

$$S(\omega) = -\frac{1}{\pi} Im \int F^{11*}(\mathbf{r})\mathbf{G}^{11}(\mathbf{r}, \mathbf{r}'; \omega) F^{11}(\mathbf{r}') d\mathbf{r} \ d\mathbf{r}' \tag{20}$$

where $\mathbf{G}^{11}$ is the (ph,ph) component of the QRPA Green's function. Another process which can be described in the same manner is the two-particle transfer from the ground state of a nucleus with A nucleons to the excited states of a nucleus with A+2 nucleons. For this process the strength function is:

$$S(\omega) = -\frac{1}{\pi} Im \int F^{12*}(\mathbf{r}) \mathbf{G}^{22}(\mathbf{r},\mathbf{r}';\omega) F^{12}(\mathbf{r}') d\mathbf{r}\, d\mathbf{r}' \tag{21}$$

where $\mathbf{G}^{22}$ denotes the (pp,pp) component of the Green's function.

2.4. *Quasiparticle excitations and two-neutron transfer in neutron-rich nuclei*

The formalisms presented above are illustrated here for the case of neutron-rich oxygen isotopes [13,14]. First, we present an exemple of quasiparticle resonances calculated in the framework of the continuum-HFB (cHFB) approach introduced in section 3.1. In the cHFB calculations the mean field quantities are evaluated using the Skyrme interaction SLy4 [17], while for the pairing interaction we take a zero-range density-dependent force. The parameters of the pairing force are given in Ref [13]. The HF single-particle and HFB quasiparticle energies corresponding to the *sd* shell and to the $1f_{7/2}$ state are listed in Table 1. One can notice that in both HF and cHFB calculations the state $1f_{7/2}$ is a wide resonance for $^{18-22}$O nuclei, while the state $1d_{3/2}$ is a narrow resonance. As seen below, these one-quasiparticle resonances determine essentially the two-quasiparticle states which are the stongest populated in even-even oxygen isotopes.

Table 1. HF and HFB energies in oxygen isotopes.

		^{18}O	^{20}O	^{22}O
$1d_{5/2}$	HF	-6.7	-7.0	-7.45
$1d_{5/2}$	cHFB	2.26	2.08	2.30
$2s_{1/2}$	HF	-4.0	-4.2	-4.6
$2s_{1/2}$	cHFB	3.46	2.28	1.05
$1d_{3/2}$	HF	(0.46;0.02)	(0.51;0.03)	(0.42;0.02)
$1d_{3/2}$	cHFB	(7.74;0.12)	(6.60;0.29)	(5.39;0.01)
$1f_{7/2}$	HF	(5.50;1.35)	(5.24;1.24)	(4.86;1.04)
$1f_{7/2}$	cHFB	(12.77;1.13)	(12.14;0.83)	(10.05;0.69)

The two-quasiparticle states are calculated by using the response theory described in section 2.3. In the calculations one includes the full discrete and continuum HFB spectrum up to 50 MeV. These states are used to construct the unperturbed Green's function $\mathbf{G_0}$. After solving the Bethe-Salpeter equation for the QRPA Green function one constructs the strength functions written in section 2.3. The details of the calculations can be found in Ref [14]. Here we show the results obtained for the collective states excited in the two-neutron transfer. The strength function corresponding to a neutron pair transferred to the oxygen isotope ^{22}O is shown in Fig.1.

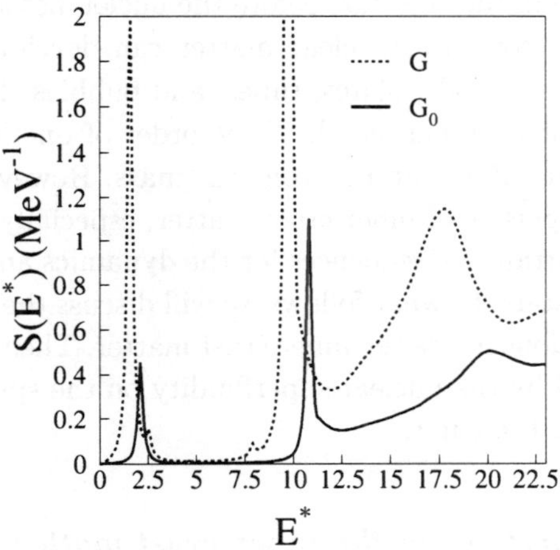

Fig. 1. The response function for the two-neutron transfer on 22O. The unperturbed response is in solid line and the QRPA response in dashed line.

For the isotope ^{22}O the subshell $d_{5/2}$ is essentially blocked for the pair transfer. Therefore in this nucleus we can clearly identify only two peaks below 11 MeV, corresponding to a pair transferred to the states $2s_{1/2}$ and $2d_{3/2}$. The strength function shown in Fig. 1 shows also a broad peak around 20 MeV. This peak is built mainly upon the single-particle resonance $1f_{7/2}$ and its cross section is much larger than the one associated to the lower energy transfer modes. Since this high energy transfer mode is formed mainly by single-particle states above the valence shell, this mode is similar to the giant pairing vibration mode suggested long ago [15]. Although such a mode has not been detected yet, the pair transfer reactions involving exotic loosely bound nuclei may offer a better chance for this undertaking [16].

3. Superfluid and thermal properties of neutron stars crust

In this section we analyse the superfluid properties of a nuclear system in which the limit of the neutron-drip line is overpassed: the inner crust of neutron stars. The inner crust consists of a lattice of neutron-rich nuclei immersed in a sea of unbound neutrons and relativistic electrons. Down to the inner edge of the crust, the crystal lattice is most probably formed by spherical nuclei. More inside the star, before the nuclei dissolve completely into the liquid of the core, the nuclear matter can develop other exotic configurations as well, i.e, rods, plates, tubes, and bubbles [18]. The thickness of the inner crust is rather small, of the order of one kilometer, and its mass is only about 1% of the neutron star mass. However, in spite of its small size, the properties of inner crust matter, especially its superfluid properties, have important consequences for the dynamics and the thermodynamics of neutron stars. In what follows we will discuss the main features of the pairing correlations inside the inner crust matter. Then we shall focus on the effects induced by the nuclear superfluidity on the specific heat and on the cooling time of the curst.

3.1. *Superfluid properties of the inner crust matter*

The superfluid properties of the inner crust matter discussed here [5] are based on the finite-temperature HFB (FT-HFB) approach. For zero range forces and spherical symmetry the radial FT-HFB have a similar form as the HFB equations at zero temperature, ie.,

$$\begin{pmatrix} h_T(r) - \lambda & \Delta_T(r) \\ \Delta_T(r) & -h_T(r) + \lambda \end{pmatrix} \begin{pmatrix} U_i(r) \\ V_i(r) \end{pmatrix} = E_i \begin{pmatrix} U_i(r) \\ V_i(r) \end{pmatrix} , \tag{22}$$

where E_i is the quasiparticle energy, U_i, V_i are the components of the radial FT-HFB wave function and λ is the chemical potential. The quantity $h_T(r)$ is the thermal averaged mean field hamiltonian and $\Delta_T(r)$ is the thermal averaged pairing field. The latter depends on the average pairing density κ_T. In a self-consistent calculation based on a Skyrme-type force , $h_T(r)$ is expressed in terms of thermal averaged densities, i.e., kinetic energy density $\tau_T(r)$, particle density $\rho_T(r)$ and spin density $J_T(r)$, in the same way as in the Skyrme-HF approach. The thermal averaged densities mentioned above are given by [5]:

$$\rho_T(r) = \frac{1}{4\pi}\sum_i (2j_i + 1)[V_i^*(r)V_i(r)(1 - f_i) + U_i^*(r)U_i(r)f_i]$$

$$\kappa_T(r) = \frac{1}{4\pi}\sum_i (2j_i+1)U_i^*(r)V_i(r)(1-2f_i)$$

$$J_T(r) = \frac{1}{4\pi}\sum_i (2j_i+1)[j_i(j_i+1) - l_i(l_i+1) - \frac{3}{4}]$$
$$\times\{V_i^2(1-f_i) + U_i^2 f_i\}$$

$$\tau_T(r) = \frac{1}{4\pi}\sum_i (2j_i+1)\{[(\frac{dV_i}{dr} - \frac{V_i}{r})^2 + \frac{l_i(l_i+1)}{r^2}V_i^2]$$
$$\times(1-f_i) + [(\frac{dU_i}{dr} - \frac{U_i}{r})^2 + \frac{l_i(l_i+1)}{r^2}U_i^2]f_i\},$$
(23)

where $f_i = [1 + exp(E_i/k_B T)]^{-1}$ is the Fermi distribution, k_B is the Boltzmann constant and T is the temperature. The summations in the equations above are over the whole quasiparticle spectrum, including the unbound states.

The FT-HFB equations are solved for the spherical Wigner-Seitz cells determined in Ref. [20]. To generate far from the nucleus a constant density corresponding to the neutron gas, the FT-HFB equations are solved by imposing Dirichlet-Neumann boundary conditions at the edge of the cell [20], i.e., all wave functions of even parity vanish and the derivatives of odd-parity wave functions vanish. In the FT-HFB calculations we use for the particle-hole channel the Skyrme effective interaction SLy4 [17], which has been adjusted to describe properly the mean field properties of neutron-rich nuclei and infinite neutron matter. In the particle-particle channel we employ a density dependent zero range force. Since the magnitude of pairing correlations in neutron matter is still a subject of debate, the parameter of the pairing force are chosen so as to describe two different scenarios for the neutron matter superfluidity. Thus, for the first calculation we use the parameters: V_0=-430 MeV fm^3, η=0.7, and α=0.45. With these parameters and with a cut-off energy for the quasiparticle spectrum equal to 60 MeV one obtains approximately the pairing gap given by the Gogny force in nuclear matter [8]. In the second calculation we reduce the strength of the force to the value V_0=-330 MeV fm^3. With this value of the strength we simulate the second scenario for the neutron matter superfluidity, in which the screening effects would reduce the maximum gap in neutron matter to a value around 1 MeV [21].

The FT-HFB results are shown here for two representative Wigner-Seitz cells chosen from Ref. [20]. These cells contain Z=50 protons and have rather different baryonic densities, i.e., 0.0204 fm^{-3} and 0.00373 fm^{-3}. The cells, which contain N=1750 and N=900 neutrons, respectively, are denoted below as a nucleus with Z protons and N neutrons, i.e., ^{1800}Sn and ^{950}Sn. The FT-HFB calculations are done up to a maximum temperature of T=0.5 MeV, which is covering the temperature range of physical interest [22].

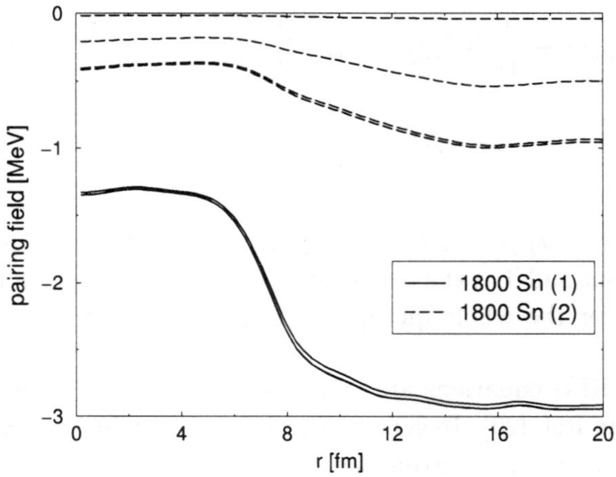

Fig. 2. Neutron pairing fields for the cell ^{1800}Sn calculated at various temperatures. The numbers 1 and 2 which follow the cell symbol (see the inset) indicate the variant of the pairing force used in the calculations. The full and the dashed lines corresponds (from bottom upwards) to the set of temperatures T=$\{0.0, 0.5\}$ MeV and T=$\{0.0, 0.1, 0.3, 0.5\}$MeV, respectively.

The temperature dependence of the pairing fields in the two cells presented above is shown in Figs.2-3. First, we notice that for all temperatures the nuclear clusters modify significantly the profile of the pairing field. One can also see that for most of the cases the temperature dependence of the pairing field is significant. This is clearly seen in the low-density cell ^{950}Sn.

3.2. *Specific heat of the inner crust baryonic matter*

The superfluid properties of the neutrons discussed in the previous section have a strong influence on the specific heat of the inner crust matter [5]. The specific heat of a given cell of volume V is defined by:

$$C_V = \frac{1}{V}\frac{\partial \mathcal{E}(T)}{\partial T}, \tag{24}$$

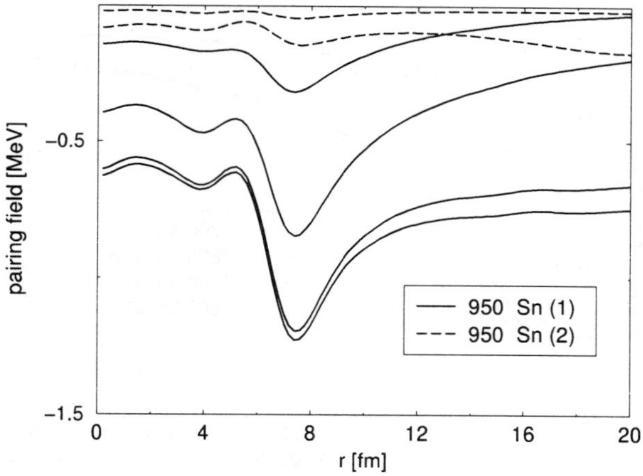

Fig. 3. The same as in Fig.1, but for the cell ^{950}Sn. The full and the dashed lines corresponds (from bottom upwards) to the set of temperatures T={0.0, 0.1, 0.3, 0.5} and T={0.0, 0.1}MeV, respectively

where $\mathcal{E}(T)$ is the total energy of the baryonic matter inside the cell, i.e.,

$$\mathcal{E}(T) = \sum_i f_i E_i. \tag{25}$$

Due to the energy gap in the excitation spectrum, the specific heat of a superfluid system is dramatically reduced compared to its value in the normal phase. Since the specific heat depends exponentially on the energy gap, its value for a Wigner-Seitz cell is very sensitive to the local variations of the pairing field induced by the nuclear clusters. This can be clearly seen in Fig.4, where the specific heat is plotted for the cell ^{1800}Sn and for the neutrons uniformly distributed in the same cell. One can notice that at T=0.1 MeV and for the first pairing force the presence of the cluster increases the specific heat by about 6 times compared to the value for the uniform neutron gas. However, the most striking fact seen in Fig.4 is the huge difference between the predictions of the two pairing forces. Thus, for T=0.1 MeV this difference amounts to about 7 orders of magnitude.

The behaviour of the specific heat for the low- density cell ^{950}Sn is shown in Fig.5. For the first pairing force we can also see that at T=0.1 MeV the cluster increases the specific heat by about the same factor as in the cell ^{1800}Sn. However, for the second pairing force the situation is opposite: the presence of the nucleus decreases the specific heat instead of increasing it.

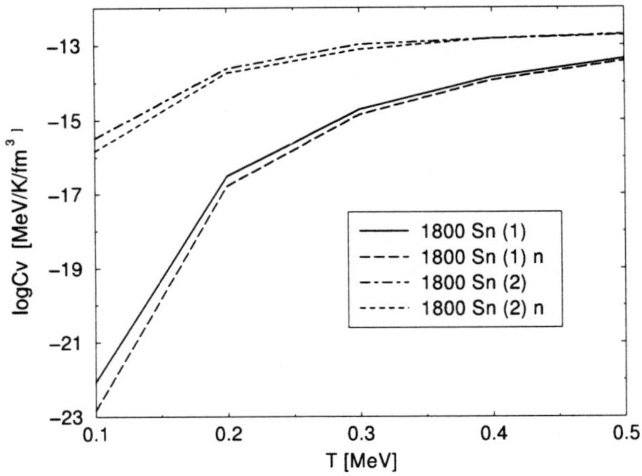

Fig. 4. Specific heat for the cell ^{1800}Sn as a function of temperature. The notations used in the inset and the representation of the calculated values are the same as in Figs.1-3.

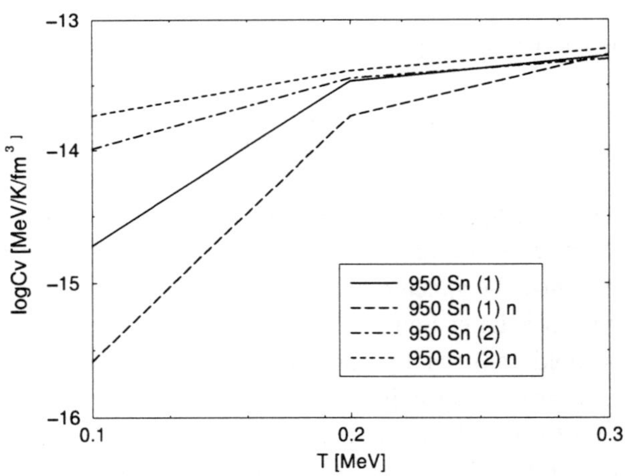

Fig. 5. The same as in Fig.4, but for the cell ^{950}Sn.

3.3. *Collective modes in the inner crust matter*

In the calculations presented in the previous section the specific heat of inner crust matter was evaluated by considering only non-interacting quasiparticles states. However, the specific heat can be also strongly affected by the collective modes created by the residual interaction between the quasiparticles, especially if these modes appear at low-excitation energies.

The collective modes in the inner crust matter were calculated in Ref. [23] in the framework of linear response theory discussed in section 2.3. The most important result of these calculations is the apparence of very collective modes at low energies, of the order of the pairing gap. An example of such mode is seen in Fig.6, were is shown the quadrupole response for the cell ^{1800}Sn. As can be clearly seen from Fig.6 , when the residual interaction is introduced among the quasiparticles the unperturbed spectrum, distributed over a large energy region, is gathered almost entirely in the peak located at about 3 MeV. This peak collects more than 99% of the total quadrupole strength and is extremely collective. An indication of the extreme collectivity of this low-energy mode can be also seen from its reduced transition probability, B(E2), which is equal to 25.10^3 Weisskopf units. This value of B(E2) is two orders of magnitude higher than in standard nuclei. This underlines the fact that this WS cell cannot be simply considered as a giant nucleus. The reason is that in this cell the collective dynamics of the neutron gas dominates over the cluster contribution.

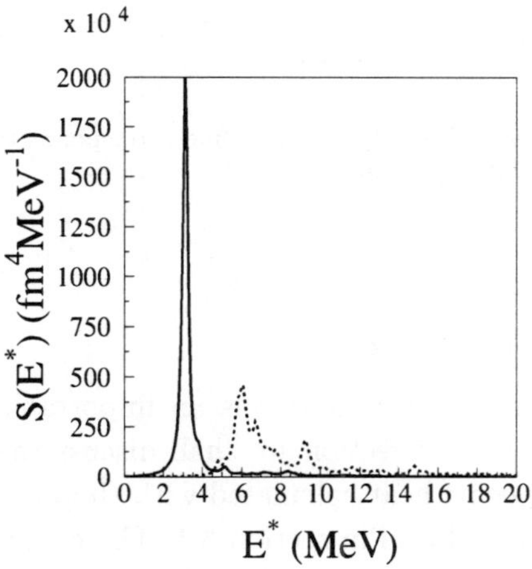

Fig. 6. Quadrupole strength distribution of neutrons for the cell ^{1800}Sn. The full curve represents the QRPA strength, and the dashed line is the HFB unperturbed strength.

The collective excitations located at low energies can affect significantly the specific heat of the inner crust baryonic matter. This can be seen in Fig.7, where the specific heat corresponding to the collective modes (of multipolarity L=0,1,2,3) is shown. We notice that for T=0.1 MeV the specific heat given by the collective modes is of the same order of magnitute as the one corresponding to HFB spectrum. Therefore one expects that the collective modes of the inner crust matter could affect strongly the thermal behaviour of the crust.

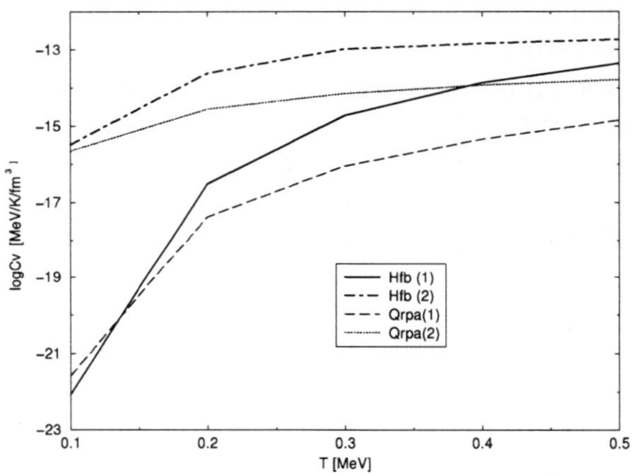

Fig. 7. Specific heat in the cell ^{1800}Sn. The dashed line corresponds to the collective modes and the full line to the HFB spectrum

3.4. *Cooling time of the inner crust*

The specific heat of the inner crust matter is an important quantity for cooling time calculations. In this section we shall discuss the sensivity of the cooling time on the specific heat calculated with the two scenarios for the nuclear superfluidity introduced in section 3.1. The cooling process we analyse here corresponds to a fast cooling mechanism (e.g., induced by direct Urca reactions). In this case the interior of the star cools much faster than the crust. The cooling time is defined as the time needed for the cooling wave to traverse the crust and to arrive at the surface of the star. According to numerical simulations [24], the cooling time is proportional to the square of the crust size and to the specific heat. This result was used by Pizzochero et al [25] for estimating the cooling time in a simple model, which we have also employed in our calculations [26]. Thus, the crust is devided in shells of thickness R_i for which the thermal difusivity D_i could be considered as constant. Then the cooling time is obtained by summing the contribution of each shell, i.e.,

$$t_{cool} = \frac{1}{\gamma} \sum_i \frac{R_i^2}{D(i,T)} \tag{26}$$

The thermal diffusivity is given by $D = \frac{\kappa}{C_V}$, where κ is the thermal conductibility and C_V is the specific heat. The thermal conductivity is mainly determined by the electrons and in our calculations we have used the values reported by Lattimer et al. [24]. The specific heat C_V has major contributions from the electrons, which can be easily calculated, and from the neutrons of the inner crust. As we have seen above, the specific heat of the neutrons depends strongly on pairing correlations. In order to see if this dependence has observationally consequences for the cooling time, we have performed two calculations, corresponding to the strong and the weak pairing forces introduced in section 3.1. The calculations show that the cooling time is increasing with more than 80% if for calculating the pairing correlations we use a weak pairing force instead of a strong pairing force. This result indicates that the nuclear superfluidity of the inner crust matter plays a crucial role for the cooling time calculation of neutron stars.

4. Conclusions

In this lecture we have shown how the HFB approach and the linear response theory can be used to describe the pairing correlations in exotic nuclei and in the inner crust of neutron stars. Thus, for the nuclei close to the drip lines we have discussed how one can incorporate in the two approaches mentioned above the effects of the continuum coupling on pairing correlations. Then, using the same models, we have analysed what are the effects of pairing correlations on the specific heat and on the cooling time of neutron stars. We thus found that the cooling time depends very strongly on the intensity of pairing correlations in nuclear matter. Because the intensity of pairing correlations in nuclear matter is still unclear, at present one can only estimate the limits in which the cooling time of the inner crust can vary. Since the pairing correlations in nuclear matter and in nuclei are in fact correlated, one hopes that these limits could be reduced by systematic studies of pairing in both infinite and finite nuclear systems.

References

1. D. Pines and M. Ali Alpar, Nature (London) **316**, 27 (1985)
2. J. A. Sauls, in *Timing Neutron Stars*, ed. by H. Ogelman and E. P. J. van den Heuvel (Dordrecht, Kluwer, 1989) pp. 457
3. M. Prakash, Phys. Rep. **242** 387 (1994);
4. U. Lombardo and H-J. Schulze, in *Physics of Neutron Star Interiors*, ed. by D.Blaschke et al (Springer,2001) pp.30
5. N. Sandulescu, Phys. Rev. C **70** 025801
6. P. Ring, P. Schuck, *The nuclear many-body problem, Springer-Verlag* (1980).
7. M. Grasso, N. Sandulescu, Nguyen Van Giai, and R. J. Liotta, Phys. Rev. **C64** 064321 (2001)
8. G. F. Bertsch and H. Esbensen, Ann. Phys. (N.Y.) **209** 327 (1991)
9. N. Sandulescu, N. Van Giai, and R.J. Liotta, Phys. Rev. **C 61** 061301(R) (2000)
10. N. Sandulescu, R. J. Liotta and R. Wyss, Phys. Lett. **B394** 6 (1997)
11. N. Sandulescu, L. S. Geng, H. Toki, and G. C. Hillhouse, Phys. Rev. C **68** 054323 (2003)
12. R. Id. Betan, N. Sandulescu, T. Vertse, Nucl. Phys. **A771** 93 (2006)
13. E. Khan, N. Sandulescu, M. Grasso, Nguyen Van Giai, *Phys. Rev* **C66** (2002) 024309.
14. E. Khan, N. Sandulescu, Nguyen Van Giai, M. Grasso, *Phys. Rev.* **C69** (2004) 014314
15. M. W. Herzog, R. J. Liotta and T. Vertse *Phys. Lett.* **B165** (1985) 35.
16. L. Fortunato, W. von Oertzen, H. M. Sofia and A. Vitturi, *Eur. Phys. J.* **A14** (2002) 37.
17. E. Chabanat, P. Bonche, P. Haensel, J. Meyer, R. Schaeffer, Nucl. Phys. **A635** 231 (1998)
18. C. J. Pethick and D. G. Ravenhall, Annu. Rev. Nucl. Part. Sci. **45** 429 (1995)
19. N. Sandulescu, Nguyen Van Giai, and R. J. Liotta, Phys. Rev. **C69** 045802 (2004)
20. J. W. Negele and D. Vautherin, Nucl. Phys. **A207** 298 (1973)
21. C. Shen, U. Lombardo, P. Schuck, W. Zuo, and N. Sandulescu, Phys. Rev. **C67** 061302R (2003)
22. K. A. van Riper, Ap.J. **75** 449 (1991)
23. E. Khan, N. Sandulescu and Nguyen Van Giai, Phys. Rev. **C71** 042801 (2005)
24. J. M. Lattimer et al, ApJ **425** 802 (1994)
25. P. M. Pizzochero, F. Barranco, E. Vigetti, and R. A. Broglia, ApJ **569** (2002) 381
26. J. Margueron, C. Monrozeau, N. Sandulescu, in preparation

BEYOND MEAN FIELD APPROACHES AND EXOTIC NUCLEAR STRUCTURE PHENOMENA

A. PETROVICI*

Horia Hulubei National Institute for Physics and Nuclear Engineering, R-077125 Bucharest-Magurele, Romania
** E-mail: spetro@ifin.nipne.ro*

We present the microscopic description of some exotic nuclear structure phenomena and decays identified experimentally at low, intermediate and high spins in the A≃ 70 mass region within the *complex* VAMPIR approaches. Special emphasis is put on the very large many-nucleon model spaces required in order to determine the structure of the wave functions.

Keywords: VAMPIR models; Shape-coexistence and -mixing; Isospin mixing and superallowed Fermi beta decay; Magnetic bands.

1. Introduction

The nuclei near the N=Z line in the A≃ 70 mass region display some interesting nuclear structure effects like shape-coexistence and -mixing [1]. Since in N≃Z nuclei neutrons and protons fill the same single-particle orbits one expects a strong competition of the neutron and proton alignment as well as significant effects from the neutron-proton pairing correlations [2]. Rapid changes in structure with particle number, angular momentum and excitation energy are expected to be revealed [3–5]. Furthermore, these nuclei play an important role in nuclear astrophysics, since their weak decay determines details of the nucleosynthesis. Also the investigation of superallowed Fermi β decays between analog states, which provide a test of the validity of the conserved vector current hypotesis and the unitarity of the Cabibbo-Kobayashi-Maskawa matrix is of particular interest in nuclei with A≥62, where the charge induced isospin-mixing is expected to be large [6,7]. Magnetic rotation, a new kind of nuclear rotation, has attracted great interest in the recent years for medium mass nuclei, too [8,9].

Obviously, microscopic nuclear structure calculations for such medium heavy nuclei are extremely involved. The adequate model spaces are far

too large to allow for a complete diagonalisation of an appropriate effective many-body Hamiltonian and thus one has to rely on suitable approximate methods. Furthermore, the appropriate effective Hamiltonian itself is not known a priori and can only be determined by an iterative process of many time-consuming calculations. Both, the limitation of the particular approximate method used as well as the insufficient knowledge of the appropriate Hamiltonian, will leave some uncertainties in the quantitative results especially if small effects are to be investigated.
The *complex* VAMPIR approaches allowing to use rather large single-particle basis systems as well as general two-body interactions provide the possibility to accomplish really *large-scale* nuclear structure studies going far beyond the possibilities of the shell-model configuration-mixing approach.
These models are based on chains of variational calculations using symmetry-projected Hartree-Fock-Bogoliubov (HFB) vacua. The use of essentially *complex* HFB transformations in these approaches allows to account for neutron-proton pairing and unnatural-parity correlations.
In Section 2 we shall briefly summarize the essential features of two of these approaches and will give some details on the effective interaction and the model space. In Section 3 results on the shape coexistence effects in ^{78}Kr will be discussed. Section 4 presents results on the effect of isospin-mixing on superallowed Fermi beta decay in the A=70 and A=74 isovector triplets of nuclei. In Section 5 then some details on the *complex* EXCITED VAMPIR description of the mechanism responsible for the magnetic rotation in this mass region will be presented. Conclusions are drawn in Section 6.

2. Theoretical Framework

We shall first summarize the theoretical tools which are needed for the symmetry-projected variational approaches discussed later on.
Let $\{|i\rangle \equiv |\tau nljm\rangle\}$ be a complete set of eigenstates of some spherical single particle potential, e.g., the harmonic oscillator. Here τ denotes the isospin projection, and n the radial quantum number, l is the orbital, j the total angular momentum and m the 3–projection of the latter. The corresponding creation and annihilation operators are denoted by $\{c_i^\dagger, c_k^\dagger, ...\}$ and $\{c_i, c_k, ...\}$, respectively. They fulfill the usual anticommutation relations for Fermion field operators.
The model space is then defined by a finite, M-dimensional subset of these basis states $\{|i\rangle, |k\rangle, ...\}_M$. We assume furthermore that the effective many-nucleon Hamiltonian appropriate for this model space is known and can be

written in the chosen representation as a sum of only one- and two-body terms

$$\hat{H} = \sum_{i,k=1}^{M} t(ik)c_i^\dagger c_k + \frac{1}{4}\sum_{i,k,r,s=1}^{M} v(ikrs)c_i^\dagger c_k^\dagger c_s c_r\,. \tag{1}$$

Here the $t(ik)$ are the matrix elements of the kinetic energy operator, which have to be replaced by some suitably chosen single-particle energies $\varepsilon(i)\delta_{ik}$ if an inert core is used, and $v(ikrs) \equiv \langle ik|\hat{v}|rs - sr\rangle$ denote the antisymmetrized matrix elements of the effective two body interaction.

We then introduce quasi-particle creators and annihilators via the unitary $(2M \times 2M)$ HFB transformation

$$\begin{pmatrix} a^\dagger \\ a \end{pmatrix} = F\begin{pmatrix} c^\dagger \\ c \end{pmatrix} = \begin{pmatrix} A^T & B^T \\ B^\dagger & A^\dagger \end{pmatrix}\begin{pmatrix} c^\dagger \\ c \end{pmatrix} \tag{2}$$

The corresponding quasi-particle vacuum can be represented as

$$|F\rangle = \prod_{\alpha=1}^{M'} a_\alpha(F)|0\rangle \text{ where } a_\alpha(F)|0\rangle = \begin{cases} \neq 0 & \text{for} \alpha = 1, ..., M' \leq M \\ = 0 & \text{else} \end{cases} \tag{3}$$

Here $|0\rangle$ is the usual particle vacuum. Configurations of this type serve in the following as basic building blocks of the theory. In general the HFB transformations and therefore the vacua, too, violate all the symmetries required by the many body Hamiltonian, except for the "number parity". To obtain physical configurations one has therefore first to restore these broken symmetries $s \equiv N, Z, I, \pi$. This is done via the projection operator

$$\hat{\vartheta}^s_M \equiv \sum_{K=-I}^{+I} \hat{\Theta}^s_{MK}\, f_K = \sum_{K=-I}^{+I} \hat{P}^I_{MK}\, \hat{Q}(N)\hat{Q}(Z)\hat{p}(\pi)\, f_K \tag{4}$$

which besides the parity projection, the projection onto the desired proton and neutron numbers and the restoration of the total angular momentum and its 3-component, involves dynamical variables f_K which, in principle, like the underlying transformation F should be determined by variation. Requiring time-reversal invariance as well as axial symmetry for the HFB transformation, however, only vacua with $K = 0$ do remain and f_0 is given by the normalisation. The resulting symmetry-projected configurations

$$|\phi_i; sM\rangle \equiv \frac{\hat{\Theta}^s_{M0}|F^s_i\rangle}{\sqrt{\langle F^s_i|\hat{\Theta}^s_{00}|F^s_i\rangle}} \tag{5}$$

are used as trial configurations in the following. Though time reversal and axial symmetry are imposed, allowing for parity and neutron-proton mixing as well as admitting *essentially complex* HFB transformations, the vacua contain all posssible two-nucleon couplings. Using the Bloch-Messiah theorem the HFB vacuum can be written as

$$|F>=\left\{\prod_{m=1/2}^{m_{max}}\left(\prod_{\alpha}^{(m)}[u_\alpha+v_\alpha b_\alpha^\dagger b_{\bar{\alpha}}^\dagger]\right)\right\}|0> \tag{6}$$

where the $b_\alpha^\dagger$ are given by the unitary first Bloch–Messiah transformation

$$b_\alpha^\dagger = \sum_{\tau_i,n_i,l_i,j_i}^{(m_\alpha>0)} D_{i\alpha}^* c_i^\dagger \,. \tag{7}$$

The basic building block of the HFB vacuum can now be written as

$$b_\alpha^\dagger b_{\bar{\alpha}}^\dagger = \sum_{\tau=p,n}\sum_{\underline{i}\leq\underline{k}}^{(m_\alpha\tau)} [1+\delta(\underline{i},\underline{k})]^{-1}\sum_I(-)^{j_k+l_k-m_\alpha}(j_ij_kI|m_\alpha-m_\alpha 0)\times$$

$$\left\{\left[Re(D_{i_\tau\alpha}^*D_{k_\tau\alpha})[1+(-)^{l_i+l_k+I}]+iIm(D_{i_\tau\alpha}^*D_{k_\tau\alpha})[1-(-)^{l_i+l_k+I}]\right]\left[c_{\underline{i}}^\dagger c_{\underline{k}}^\dagger\right]_{1T_z}^{I0}\right\}$$

$$+\sum_{\underline{i}}^{(m_\alpha p)}\sum_{\underline{k}}^{(m_\alpha n)}\sum_{IT}(1/2 1/2 T|-1/2 1/2 0)(-)^{j_k+l_k-m_\alpha}(j_ij_kI|m_\alpha-m_\alpha 0)\times$$

$$\left\{\left[Re(D_{i_p\alpha}^*D_{k_n\alpha})[1+(-)^{l_i+l_k+I}]+iIm(D_{i_p\alpha}^*D_{k_n\alpha})[1-(-)^{l_i+l_k+I}]\right]\left[c_{\underline{i}}^\dagger c_{\underline{k}}^\dagger\right]_{T0}^{I0}\right\} \tag{8}$$

where the operator

$$[c_{\underline{i}}^\dagger c_{\underline{k}}^\dagger]_{TT_z}^{IM} \equiv \sum_{m_im_k\tau_i\tau_k}(j_ij_kI|m_im_kM)(\frac{1}{2}\frac{1}{2}T|\tau_i\tau_kT_z)c_i^\dagger c_k^\dagger \tag{9}$$

creates a pair of nucleons coupled to the angular momentum and isospin quantum numbers I, M and T, T_z, respectively. The time-reversal invariant basic building block of the vacuum contains $T=1$ neutron-neutron and proton-proton pairing as well as $T=1$ and $T=0$ proton-neutron pairs, which all may be coupled to arbitrary I^π. Time-reversal even pairs have purely real, time-reversal odd ones purely imaginary coefficients.

In the VAMPIR approximation now for each spin-parity I^π in a considered nucleus N, Z only one single trial wave function of the type (5) is admitted. Minimisation of the energy functional

$$E_1^s[F_1^s] = \langle\phi_1; sM|\hat{H}|\phi_1; sM\rangle \tag{10}$$

with respect to arbitrary variations of the underlying transformation F_1^s yields then the optimal representation for the energetically lowest (yrast) state of a particular spin-parity by a single symmetry-projected HFB determinant. Excited states can then be constructed via exactly the same procedure : after the VAMPIR yrast solution has been obtained, it is eliminated from the model space via Gram-Schmidt orthogonalisation, and the optimal one determinant representation for the first excited state with the same symmetry quantum numbers is determined. Eliminating this state from the variational space, too, the second excited state can be constructed and so on. The trial wave function for the i-th excited state has finally the form

$$|\Phi_i; sM\rangle \equiv \hat{T}_{i-1}\hat{\Theta}^s_{M0}|F_i^s\rangle \tag{11}$$

and differs from (5) only by the Gram-Schmidt projection operator $\hat{T}_{i-1}$ which eliminates the $i-1$ lower states already obtained in the previous steps of the above described procedure and takes care of the normalisation. After i states of the type (11) have been constructed, the residual interaction in between them is diagonalized and we obtain the EXCITED VAMPIR wave functions as

$$|\Psi_\alpha^{(i)}; sM\rangle = \sum_{j=1}^{i} |\Phi_j; sM\rangle g_{j\alpha}^{(i)} \qquad \text{for} \qquad \alpha = 1, ..., i \tag{12}$$

Left to be defined are the model space and the effective interaction. For nuclei in the $A \simeq 70$ mass region a ^{40}Ca core is used and the $1p_{1/2}$, $1p_{3/2}$, $0f_{5/2}$, $0f_{7/2}$, $1d_{5/2}$ and $0g_{9/2}$ oscillator orbits for both protons and neutrons are taken as single-particle basis states. The effective two-body interaction is constructed from a nuclear matter G-matrix based on the Bonn One-Boson-Exchange potential (Bonn A). It is modified by short range Gaussians for the isospin $T = 1$ and $T = 0$ in order to increase the pairing correlations in the corresponding channels. In addition the isoscalar interaction was modified by monopole shifts for all the diagonal matrix elements of the form $\langle 0g_{9/2}0f; IT = 0|\hat{G}|0g_{9/2}0f; IT = 0\rangle$ with $0f$ denoting either the $0f_{5/2}$ or the $0f_{7/2}$ orbit, and in the $\langle 1p1d_{5/2}; IT = 0|\hat{G}|1p1d_{5/2}; IT = 0\rangle$ matrix elements, where 1p denotes either the $1p_{1/2}$ or the $1p_{3/2}$ orbit. These shifts have been introduced in order to influence the onset of deformation.

3. Shape Coexistence Effects in ^{78}Kr

Recently the experimental level scheme of ^{78}Kr could be extended up to spin 26^+ [10] and also lifetime, g-factor and spectroscopic quadrupole moment measurements have been reported [11–13]. The experimental information available for this nucleus suggests shape coexistence and mixing at low, intermediate as well as high spins where a forking was identified at spin 24^+ [10].

We investigated the lowest few even spin positive parity states up to spin 26^+ in ^{78}Kr [14]. First VAMPIR calculations for each yrast state under consideration were performed. In each case starting wave functions with different intrinsic prolate (p) or oblate (o) deformations were used and then always the energetically lowest resulting projected solution was taken. The excited states with the same quantum numbers were then constructed by independent variational calculations imposing always orthogonality with respect to all the solutions already obtained. Finally, within the *complex* EXCITED VAMPIR approach the residual interaction between all the solutions for the same quantum numbers were diagonalized. We considered the lowest 15 up to 17 EXCITED VAMPIR configurations for the states with spin $\leq 16^+$. For the higher spins a considerably higher density of the basis states was found and thus the EXCITED VAMPIR basis has been extended here up to a maximum of 24 configurations for spin 24^+. Strong mixing was obtained for particular states as can be seen from Table 1. In this table we present for some of the calculated states the amount of prolate and oblate mixing for the configurations contributing more than 3% to the final wave functions. We obtained a large variety of structures including pure oblate or prolate states built essentially by a single projected configuration, as well as strong mixing of configurations of the same oblate or prolate nature, and also states dominated by a variable mixing of differently deformed oblate and prolate configurations.

The high density of states of a given spin and the strong mixing of the states result in a very complicated decay pattern as illustrated in Fig. 1. The states have been grouped in "bands" solely on the basis of the B(E2) values connecting them. In Table 2 the B(E2;$I \rightarrow I-2$) values for all significant decay branches of the bands displayed in Fig. 1 are presented. In Table 3 furthermore the significant B(E2;$I \rightarrow I-2$) strengths for the transitions feeding the $po(p)_1$, $op(op)_2$ and $op(o)_5$ bands in Fig. 1 from different 22^+, 24^+ and 26^+ are given. In agreement with the experimental findings the EXCITED VAMPIR results indicate that each state of spin 22^+ and 24^+ is fed by a few branches of comparable strength. Also each 26^+ state is decaying by a few significant branches. The high density of strongly mixed configurations at spin 24^+ and 26^+ illustrated in Table 1 explains the experimentally identified forking at spin 24^+.

Table 1. The amount of mixing for the states in ^{78}Kr presented in Fig. 1.

$I[\hbar]$	o-mixing (%)	p-mixing (%)	$I[\hbar]$	o-mixing (%)	p-mixing (%)
0_1^+	13(4)(3)	74(4)	16_1^+	7	56(27)(6)
0_2^+	51(8)(3)	24(11)	16_2^+	97	
0_3^+	22	75	16_3^+		81(17)
0_4^+	18(3)	61(10)(6)	16_4^+	3	43(36)(10)(3)
			16_5^+	83(5)	6
2_1^+	5	88(3)			
2_2^+	82(4)(3)	8	18_1^+		70(12)(7)(3)(3)(3)
2_3^+	5	91	18_2^+	94(5)	
2_4^+	4	72(13)(7)	18_3^+	6	66(11)(7)(3)
			18_5^+	84(4)(3)	3
4_1^+	3	88(7)			
4_2^+	91	5	20_1^+	3	80(5)(3)
4_3^+		88(6)	20_2^+	85(9)	
4_4^+		74(9)(7)(6)	20_3^+		68(19)(6)(4)
			20_5^+	79(7)(3)	6
6_1^+		64(31)	$20_{\beta_1}^+$	9(4)(4)	70(8)
6_2^+	92(3)	3			
6_3^+		51(27)(16)(3)	22_1^+	4	51(20)(12)(7)
6_4^+		60(13)(13)(5)(4)	22_2^+	85(5)(4)(3)	
6_5^+	53	21(14)	22_3^+		96
			22_5^+	88(3)	6
8_1^+		50(42)(5)	$22_{\beta_1}^+$		38(26)(15)(12)(5)(3)
8_2^+	71(14)(8)(5)		$22_{\beta_2}^+$		40(31)(12)(8)(4)
8_3^+		48(31)(11)(7)			
8_4^+		75(8)(7)(4)	24_2^+	54(15)(2)	17(6)
8_5^+	71(19)(4)		$24_{\alpha_1}^+$	20(5)	67
			$24_{\alpha_2}^+$	25	47(19)(3)
10_1^+		82(7)(5)	$24_{\alpha_3}^+$	73(15)(4)	
10_2^+	82(7)(4)(4)		$24_{\beta_1}^+$		46(28)(22)
10_3^+		87(9)	$24_{\beta_2}^+$		47(42)(9)
10_4^+		79(10)(4)(3)			
10_5^+	77(10)(9)(4)		$26_{\alpha_1}^+$	40(5)%	48(3)%
			$26_{\alpha_2}^+$	47(7)(7)(3)%	30%
12_1^+		82(15)	$26_{\alpha_3}^+$		93%
12_2^+	94(5)		$26_{\alpha_4}^+$	79(12)%	4%
12_3^+		82(14)	$26_{\beta_1}^+$		94(4)%
12_4^+		72(22)			
12_5^+	96(3)				
14_1^+		70(20)(9)			
14_2^+	95(3)				
14_3^+		69(18)(12)			
14_4^+	3	74(15)(4)(3)			
14_5^+	93				

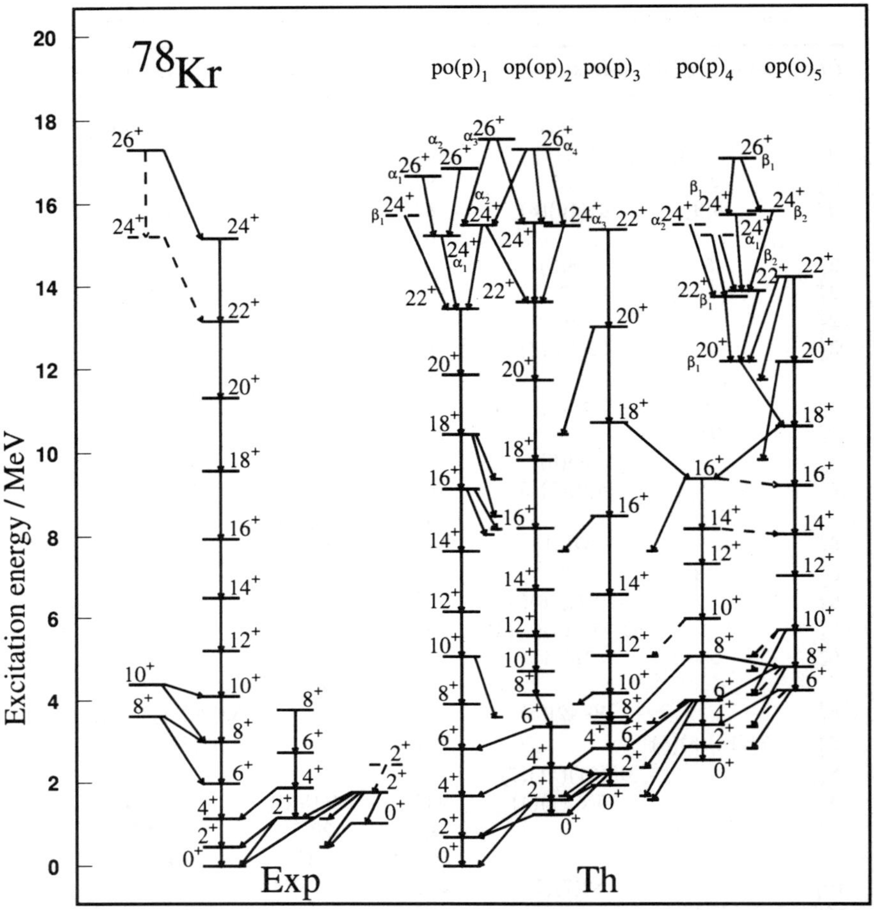

Fig. 1. The theoretical spectrum of ^{78}Kr for even spin positive parity states calculated within the *complex* EXCITED VAMPIR approximation is compared to the experimental results [10,11].

In order to illustrate the collectivity of the considered states, the B(E2) strengths for branches which are not shown in Fig. 1, are presented, too (in the square brackets). Due to the very strong mixing of some states with intermediate angular momentum, either the B(E2) strength for the decay of the corresponding states is strongly fragmented, or the fastest decay path is represented by some M1, ΔI=0 branches. To give some hints about the evolution of different static and dynamic properties of the states building the “possible bands” in nuclei dominated by shape coexistence we illustrate them for few other observables. The spectroscopic quadrupole moments for

Table 2. $B(E2; I \to I-2)$ values (in e^2fm^4) for some states of the nucleus ^{78}Kr presented in Fig. 1. The labels of the secondary branches indicate the end point of the transitions. In Fig. 1 the transitions corresponding to the strengths given in brackets are not shown. The available data are presented for comparison in the second column.

$I[\hbar]$	Exp	$po(p)_1$	$op(op)_2$	$po(p)_3$	$po(p)_4$	$op(o)_5$
2^+	1210 (80)	1170 58op(op)$_2$	1066 118po(p)$_3$ 47po(p)$_1$	1201 70op(op)$_2$	1186 [192]	
4^+	1740 (140)	1806 27po(p)$_3$	1849 38po(p)$_3$	1722 [19]op(op)$_2$	1907	
6^+	2020 (340)	1908 [97]po(p)$_3$	1945 [123]	1095 [462][79] [48]po(p)$_4$	832 549po(p)$_3$ 60po(p)$_1$ 46op(op)$_2$ [277]	480po(p)$_4$ 123po(p)$_3$ 44op(op) [630] [291][126][89]
8^+	1940 (300)	1881 124po(p)$_4$	1228 [440]op(o)$_5$ [252]	1506 [162]po(p)$_4$	1413 194po(p)$_3$ [331]	798 912op(op)$_2$ 90po(p)$_4$ [288]
10^+	1400 (300)	1728 163po(p)$_3$ [71]po(p)$_4$	1774 [157]	1543 131po(p)$_1$	1320 65po(p)$_1$ [171][144]	1368 254op(op)$_2$ [80]
12^+	1920 (1300)	1714	2025	1632	640 [650][150][135]	2007
14^+		402 [1352]	2104	1550	1572	1800
16^+		1009 168op(o)$_5$ 124po(p)$_4$ [162]	2056	1135 265po(p)$_1$ [170]	1152 140po(p)$_1$ [115]	1695
18^+		809 214po(p)$_4$ 131po(p)$_3$	2066	1038 226po(p)$_4$ [252]		1169 182po(p)$_4$ [180]
20^+		976	1726	644 91po(p)$_1$ [354][225][103]		1268 200op(op)$_2$ [135]
22^+		692 [115][97]	994 [364][101]	1056 [347]		1201 384op(op)$_2$ 142β_1
24^+			714 [97][50]			

the states building the selected bands reveal the influence of the variable mixing of configurations with different intrinsic quadrupole deformations, too. The results for the lowest two 2^+ states, $Q(2_1^+) = -61.38efm^2$ and

Table 3. $B(E2; I \to I-2)$ values (in e^2fm^4) for some states of the nucleus ^{78}Kr presented in Fig. 1. The labels of the branches indicate the end point of the transitions. In Fig. 1 the transitions corresponding to the strengths given in brackets are not shown.

$I[\hbar]$	α_1	α_2	α_3	α_4	β_1	β_2
22^+					$562\beta_1$ [267][114][74]	$328\beta_1$ [541][93][74]
24^+	$163po(p)_1$ $137\beta_2$ $118\beta_1$ [378][60]	$287\beta_1$ $263op(op)_2$ $182po(p)_1$ [101]	$163op(op)_2$ [113] [78]		$334\beta_2$ $188po(p)_1$ [366][90] [50]	$324\beta_2$ [519] [71][71] [58][53]
26^+	$552\alpha_1$ [155][112] [86]	$367\alpha_1$ [316][68]	$474\alpha_2$ $288op(op)_2$ [99][37]	$405op(op)_2$ $314\alpha_2$ $153\alpha_3$	$501\beta_2$ $394\beta_1$	

$Q(2_2^+) = +55.23efm^2$ are in good agreement with the exeprimental values $Q(2_1^+) = -61(3)efm^2$ and $Q(2_2^+) = +44(6)efm^2$ [11]. The alignment plot giving the angular momentum contribution of the $g_{9/2}$ neutrons and protons in the direction of total angular momentum is presented in Fig. 2. These results can be connected with the proton and neutron occupations of the $g_{9/2}$ orbital and also reflect the trends in the evolution of the g-factors with increasing spin. A very good agreement between the theoretical and the experimental results is obtained for the measured g-factors at low spins. The reported experimental results [13] are the following: g(2_1^+) = 0.43(3), g(4_1^+) = 0.46(7), g(2_2^+) = 0.54(10), while the corresponding calculated values are: g(2_1^+) = 0.45, g(4_1^+) = 0.50, g(2_2^+) = 0.50. The calculated g-factors are very small for the intermediate spins and even negative values are obtained for the spins 8^+ and 10^+ within the op(op)$_2$ and op(o)$_5$ bands. From the effective g-factor measurement of the yrast 8^+ a neutron alignment is expected [12].

Concerning the excited 0^+ states we predict the existence of the second 0^+ state based mainly on oblate configurations which would correspond to an yet unobserved 0^+ state in ^{78}Kr attributed to the oblate band. This prediction gets support from the experimental information concerning the value of the electric monopole strength $\rho^2(E0; 0_2^+ \to 0_1^+) = 0.047(13)$ [15]. Our results indicate $\rho^2(E0; 0_2^+ \to 0_1^+) = 0.0007$ while $\rho^2(E0; 0_3^+ \to 0_1^+) = 0.017$.

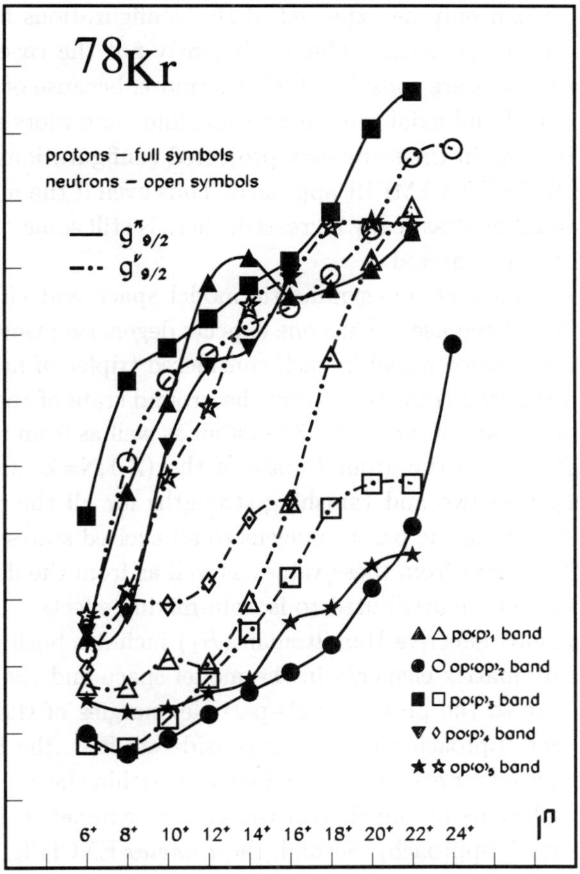

Fig. 2. The alignment plot for some states investigated in ^{78}Kr nucleus.

4. Coulomb-Induced Isospin-Mixing Effects on the Superallowed Fermi Beta Decay

In order to investigate the superallowed Fermi β decay between analog states we calculated the lowest 0^+ states in ^{70}Se, ^{70}Br, ^{70}Kr, ^{74}Kr, ^{74}Rb, and ^{74}Sr. The energetically lowest 13 $I^{\pi}=0^+$ *complex* EXCITED VAMPIR configurations have been considered for each nucleus of the A=70 isovector triplet, while the lowest 18 configurations have been taken into account for the A=74 triplet of nuclei.

Isospin is not considered as a good quantum number in the VAMPIR approaches. Consequently, even if a charge symmetric interaction is used, good

total isospin can only be expected, if the configurations form a complete set under isospin-rotations. This is obviously not the case, if only the 13 or 18 lowest states are considered. Furthermore, because of the assumption of time-reversal and axial symmetry, some four- and more-nucleon correlations are missing in the symmetry-projected configurations created by the *complex* EXCITED VAMPIR approach. Thus even if the number of configurations would be drastically increased, there is still some "spurious isospin impurity" to be expected.
In the first step a charge-symmetric model space and effective Hamiltonian, H_0, have been used. Thus one expects degenerate isovector excitation spectra for the three nuclei in each considered triplet of nuclei and for the superallowed Fermi transitions from the ground state of the (Z+1,N=Z-1)- to the ground state of the (Z,N=Z)-system as well as from the ground state of the (Z,N=Z)- to the ground state of the (Z-1,N=Z+1)-system both a total strength of two and vanishing strengths for all the transitions from the ground state of the parent nucleus to all excited states of the daughter nucleus. Deviations from these values as well as from the degeneracy of the spectra can then be attributed to isospin-mixing effects.
The "Coulomb" effective Hamiltonian (H_1) includes both the corresponding two-body matrix elements in the model space and the contribution of the ^{40}Ca core to the proton single-particle energies of the valence orbits. Two different approaches have been considered: first, the Hamiltonian H_1 was diagonalized for each considered nucleus within the *complex* EXCITED VAMPIR solutions obtained with the charge symmetric Hamiltonian H_0 ("perturbative" approach). Second, the *complex* EXCITED VAMPIR procedure was repeated with H_1 right from the beginning ("variational" approach). It should be mentioned that the Coulomb interaction removes the degeneracy and nicely reproduces the correct energy differences between the ground states of the three nuclei in each considered triplet. For each triplet of nuclei we calculated the total strength (S_T) of the transitions from the parent ground state to all the calculated 0^+ states of the daughter nucleus and the strength for the ground to ground transition (S_{g-g}). A sumarry of the results is presented in Table 4 and Table 5 for the mass number 70 and 74, respectively.
Estimating the error by summing the missing strengths between all and the analog transitions for the charge symmetric case ($\epsilon_1 = (S_T(H_0) - S_{g-g}(H_0))/2$) and the missing total strengths between the variational calculations performed with H_0 and H_1 ($\epsilon_2 = (S_T(H_0) - S_T(H_1))/2$)one may conclude that in the A=70 triplet the upper limit for the isospin mixing

Table 4. The total (S_T) and analog (S_{g-g}) Fermi β decay strengths of selected A=70 nuclei for the "no-Coulomb" (H_0) and "Coulomb" (H_1) effective Hamiltonian. The "perturbative" (p) and the "variational" (v) approaches are described in the text.

Parent nucleus	H_0 S_T	H_0 S_{g-g}	H_1^p S_T	H_1^p S_{g-g}	H_1^v S_T	H_1^v S_{g-g}
^{70}Kr	1.975	1.967	1.970	1.935	1.946	1.917
^{70}Br	1.977	1.967	1.979	1.967	1.959	1.951

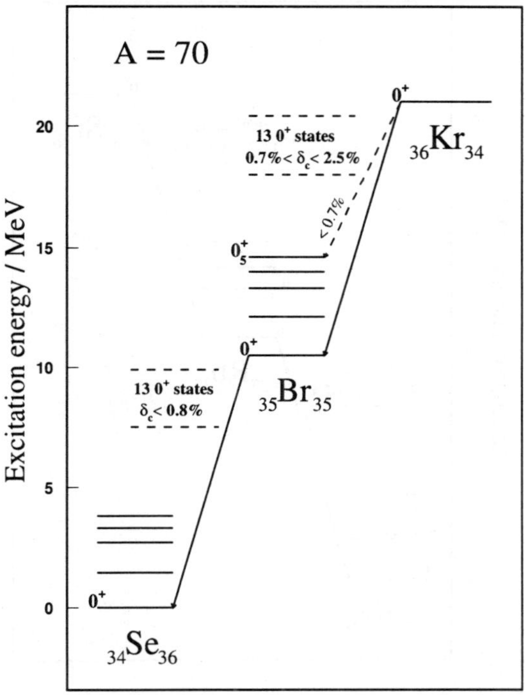

Fig. 3. The isospin-mixing correction (δ_c) for the A=70 isovector triplet.

effect on the Br to Se ground to ground transition is about 0.8 percent with this strength distributed over many excited states, while for the Kr to Br ground to ground transition a depletion of at least 0.7 and at most 2.5 percent is obtained. In this latter case a non-analog branch feeding the fourth excited 0^+ state in Br with an upper limit of 0.7 percent is obtained.

Table 5. The total (S_T) and analog (S_{g-g}) Fermi β decay strengths of selected A=74 nuclei for the "no-Coulomb" (H_0) and "Coulomb" (H_1) effective Hamiltonian. The "perturbative" (p) and the "variational" (v) approaches are described in the text.

Parent nucleus	H_0 S_T	H_0 S_{g-g}	H_1^p S_T	H_1^p S_{g-g}	H_1^v S_T	H_1^v S_{g-g}
^{74}Sr	1.954	1.947	1.940	1.918	1.932	1.893
^{74}Rb	1.957	1.948	1.948	1.929	1.946	1.924

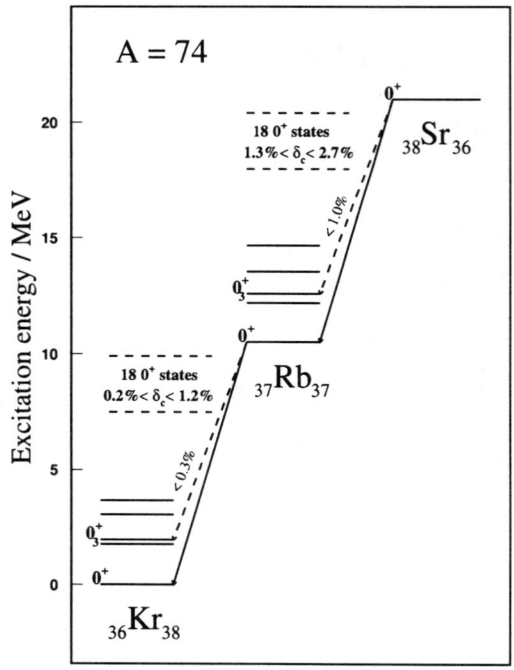

Fig. 4. The isospin-mixing correction (δ_c) for the A=74 isovector triplet.

The results are summarized in Fig. 3.

For the A=74 isovector triplet one obtains a depletion of the ground to ground decay from Sr to Rb in between 1.3 and 2.7 percent of the sum rule strength and a non-analog branch with an upper limit of about 1 percent from the ground to the second excited state. For the Rb to Kr ground to ground decay the depletion is in between 0.2 and 1.2 percent. Here an upper limit of only 0.3 percent is obtained for the branch feeding the second excited state in Kr. Experimentally, the non-analog Fermi-decay to the first exited 0^+ state in Kr has been investigated [16,17] and has turned out to be very weak in agreement with the result of our calculations. The results are summarized in Fig. 4.

The effects of the isospin-mixing on the superallowed Fermi transitions are rather small and thus difficult to describe by any microscopic many-body theory quantitatively. Small changes in the effective interaction and/or the size of the model space could yield considerable changes in the quantitative results. However, we consider that (with a conservative estimation of the errors) we have obtained at least the rough magnitude of the effects to be expected. Furthermore, since it turned out that a large fraction of the depleted strength of the ground to ground transition can be attributed to particular non-analog decay branches at least in some cases (^{74}Sr $\rightarrow$ ^{74}Rb and ^{70}Kr $\rightarrow$ ^{70}Br), there is some hope for experimental detection.

5. Magnetic Bands in ^{82}Rb

The *complex* EXCITED VAMPIR approach was applied to the description of the recently identified magnetic cascade of negative-parity states in the odd-odd ^{82}Rb nucleus, characterized by strong M1, $\Delta I = 1$ transitions as well as rather weak crossover B(E2) strengths [18]. The most important signature of such a magnetic rotational (or shears) band is a sharp decrease in the B(M1) values with increasing angular momentum [8].

We calculated negative-parity states up to spin 18^- in ^{82}Rb including in the many-nucleon bases at least 30 symmetry-projected configurations for the states of spin higher than 9^- [9]. Searching for $\Delta I = 1$ cascades of states connected by strong B(M1) transitions as well as significant B(E2,$\Delta I = 2$) strengths, like the one identified experimentally, we linked the states in bands. In Fig. 5 we present the calculated lowest few states which are based on spherical configurations, as well as the lowest collective band which was found to have oblate character (*o*-band).

The excited even- and odd-spin negative-parity states are characterized by a high level density for spins above 11^-. The most probable candidates for the experimental "magnetic sequence of states" are the states linked in the band labeled m-band in Fig. 5, while the m^*-band is characterized by few significant M1 transitions and relatively weak B(E2) strengths at intermediate spins. A strong mixing of configurations having different oblate and prolate deformations in the intrinsic system dominates the structure of the high spin states belonging to these bands. The agreement of the calculated

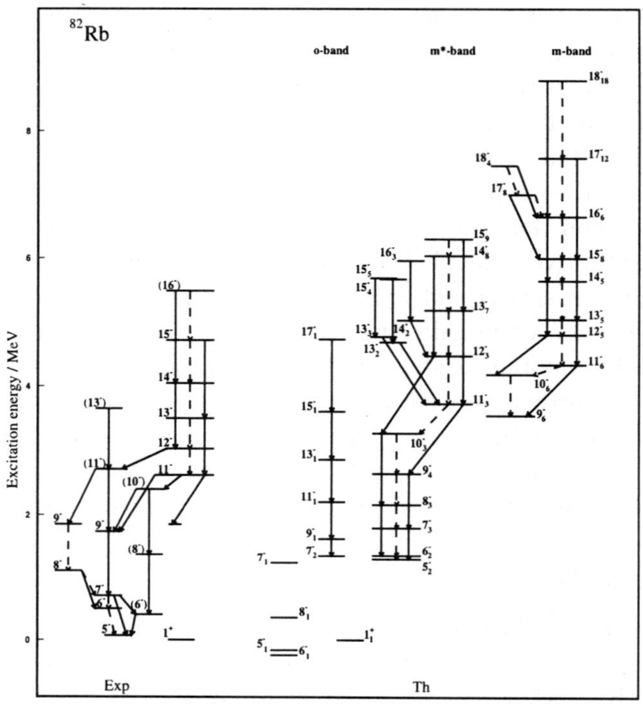

Fig. 5. The spectrum of ^{82}Rb for negative-parity states calculated within the *complex* EXCITED VAMPIR approximation is compared to the experimental results [18].

B(E2,$\Delta I = 2$) values characterizing the m-band with the available data is rather good, while for the m^*-band weaker strengths have been obtained as can be seen in Table 6. Due to the high density of states obtained for spins $I^\pi \geq 12^-$ each state is decaying and also fed by many significant B(E2,$\Delta I = 2$) branches. This aspect is illustrated for the highest calculated spins presenting in Fig. 5 an alternative way to continue the m-band above spin 16^- and the m^*-band above spin 11^-.

The results presented in Table 7 for the B(M1,$\Delta I = 1$) strengths characterizing the m-band show a reasonable agreement with the available data. One obtains the required decrease of the B(M1) values with increasing spin up to the highest calculated members included in the m-band. Both types of proton contributions (orbital and spin) are determined by rearrangements of the particles occupying the $0g_{9/2}$ orbital. The neutron contribution is determined essentially by particles occupying the $1p_{3/2}$ and $1p_{1/2}$ orbitals. The analysis of the alignment for the m^*-band and m-band indicates strong

Table 6. $B(E2; I \to I-2)$ values (in $e^2 fm^4$) for some states of the nucleus ^{82}Rb.

$I^\pi[\hbar]$	*Exp* [18]	*o*-band	m^*-band	*m*-band
7^-			815	
8^-			668	
9^-		562	824	
10^-			828	
11^-		516	828	840
12^-			490	196
13^-	384^{+131}_{-106}	424	273	583
14^-	511^{+193}_{-129}		101	487
15^-	>110	436	205	618
16^-				349
17^-		330		302
18^-				66

Table 7. $B(M1; I \to I-1)$ values (in μ_N^2) for some states of the nucleus ^{82}Rb.

$I^\pi[\hbar]$	*Exp* [18]	m^*-band	*m*-band
7^-		1.66	
8^-		1.36	
9^-		2.09	
10^-		1.00	0.48
11^-		1.99	1.13
12^-	$1.24^{+0.37}_{-0.24}$	0.35	0.45
13^-	$0.77^{+0.20}_{-0.13}$	0.86	0.86
14^-	$0.74^{+0.22}_{-0.14}$	0.06	0.46
15^-	>0.11	0.26	1.04
16^-			0.40
17^-			0.07
18^-			0.33

alignment for the $0g_{9/2}$ protons, but slow and weak alignment for the $0g_{9/2}$ neutrons. Furthermore, the angular momentum contribution of the protons occupying the $0g_{9/2}$ orbital as procentage of the total spin is decreasing with increasing spin. The main property of the magnetic rotational bands, the decrease of the B(M1) strength with increasing spin, can be explained by the increased mixing of configurations underlying the structure of the states building the bands.

6. Conclusions

The VAMPIR approaches provide a good description of the exotic nuclear structure phenomena identified at low, intermediate and high spins in the A$\simeq$ 70 nuclei using rather large model spaces and realistic nucleon-nucleon interactions. Very large many-nucleon model spaces are required in order to determine the structure of the corresponding wave functions.

Acknowledgments

The results presented in these lectures are obtained in collaboration with Prof. K. W. Schmid and Prof. Amand Faessler from the University of Tübingen, Germany, within the DFG Project 436 RUM 113/20/0-2, and PhD student O. Radu from NIPNE Bucharest, Romania.

References

1. A. Petrovici, K.W. Schmid, A. Faessler, *Nucl. Phys. A* **605**, 290 (1996).
2. A. Petrovici, K.W. Schmid, A. Faessler, *Nucl. Phys. A* **647**, 197 (1999).
3. A. Petrovici, K.W. Schmid, A. Faessler, J.H. Hamilton, A.V. Ramayya, *Progr. Part. Nucl. Phys.* **43**, 485 (1999) and references therein.
4. A. Petrovici, K.W. Schmid, A. Faessler, *Nucl. Phys. A* **665**, 333 (2000).
5. A. Petrovici, K.W. Schmid, A. Faessler, *Nucl. Phys. A* **728**, 396 (2003).
6. I. S. Towner and J. C. Hardy, *Phys. Rev. C* **66** 035501 (2002).
7. A. Petrovici, K.W. Schmid, O. Radu, A. Faessler, *Nucl. Phys. A* **747**, 44 (2005).
8. S. Frauendorf, *Z. Phys. A* **358**, 163 (1997).
9. A. Petrovici, K.W. Schmid, O. Radu, A. Faessler, *Eur. Phys. J. A* **28**, 19 (2006).
10. H. Sun et al., *Phys. Rev. C* **59**, 655 (1999).
11. F. Becker et al., *Nucl. Phys. A* **770**, 107 (2006).
12. J. Billowes et al., *Phys. Rev. C* **47**, R917 (1993).
13. T. J. Mertzimekis et al., *Phys. Rev. C* **64**, 024314-1 (2001).
14. A. Petrovici, K.W. Schmid, O. Radu, A. Faessler, *J. Phys. G: Nucl. Part. Phys.* **32**, 583 (2006).
15. A. Giannatiempo et al., *Phys. Rev. C* **52**, 2444 (1995).
16. M. Oinonen, *Phys. Lett. B* **511**, 145 (2001).
17. E. F. Zganjar et al., *Eur. Phys. J. A* **15**, 229 (2002).
18. R. Schwengner et al., *Phys. Rev. C* **66**, 024310 (1993).

SUPERFLUID-NORMAL PHASE TRANSITION IN FINITE SYSTEMS AND ITS EFFECT ON DAMPING OF HOT GIANT RESONANCES*

NGUYEN DINH DANG†

1) Heavy-ion nuclear physics laboratory, Nishina Center for Accelerator-Based Science, RIKEN, 2-1 Hirosawa, Wako city, 351-0198 Saitama, Japan
2) Institute for Nuclear Science and Technique, Hanoi, Vietnam
† *E-mail: dang@riken.jp*

Thermal fluctuations of quasiparticle number are included making use of the secondary Bogolyubov's transformation, which turns quasiparticles operators into modified-quasiparticle ones. This restores the unitarity relation for the generalized single-particle density operator, which is violated within the Hartree-Fock-Bogolyubov (HFB) theory at finite temperature. The resulting theory is called the modified HFB (MHFB) theory, whose limit of a constant pairing interaction yields the modified BCS (MBCS) theory. Within the MBCS theory, the pairing gap never collapses at finite temperature T as it does within the BCS theory, but decreases monotonously with increasing T. It is demonstrated that this non-vanishing thermal pairing is the reason why the width of the giant dipole resonance (GDR) does not increase with T up to $T \sim 1$ MeV. At higher T, when the thermal pairing is small, the GDR width starts to increase with T. The calculations within the phonon-damping model yield the results in good agreement with the most recent experimental systematic for the GDR width as a function of T. A similar effect, which causes a small GDR width at low T, is also seen after thermal pairing is included in the thermal fluctuation model.

1. Introduction

It is well known that infinite systems undergo a sharp phase transition from the superfluid phase to the normal-fluid one at finite temperature T. Marked by a collapse of the pairing correlations (pairing gap), and a near divergence of the heat capacity at a critical temperature T_c, this phase transition is a second-order one. The critical temperature is found to be

*Invited lecture at the Predeal international summer school in nuclear physics on "Collective motion and phase transitions in nuclear systems", 28 August - 9 September, 2006, Predeal, Romania

$T_c \simeq 0.567\Delta(0)$ for infinite systems, where $\Delta(0)$ is the pairing gap at zero temperature $T = 0$ [1].

The application of the BCS theory and its generalization, the Hartree-Fock-Bogolyubov (HFB) theory, to finite Fermi systems paved the way to study the superfluid-normal (SN) phase transition in nuclei at finite temperature [2–4]. Soon it has been realized that the BCS and HFB theories ignore a number of quantal and thermodynamic fluctuations, which become large in small systems because of their finiteness. As a consequence, the unitarity relation for the generalized particle-density matrix R, which requires $R^2 = R$, is violated. In deed, within the HFB theory at $T \neq 0$, one has $Tr[R^2(T) - R(T)] = 2\delta\mathcal{N}^2 \equiv 2\sum_i \delta\mathcal{N}_i^2 = 2\sum_i n_i(1 - n_i) > 0$ where $n_i = [e^{\beta E_i} + 1]^{-1}$ is the occupation number of non-interacting quasiparticles with energy E_i at temperature $T = 1/\beta$ on the i-th orbital [4]. Large thermal fluctuations smooth out the sharp second-order SN phase transition. As the result the pairing gap does not collapse as has been predicted by the BCS theory, but decreases monotonously as the temperature increases, and remains finite even at rather high T [5–7]. So far these fluctuations were taken into account based on the macroscopic Landau theory of phase transitions [5,6]. This concept is close to that of the static-path approximation, which treats thermal fluctuations on all possible static paths around the mean field [7].

It will be shown in the first part of this lecture that the recently proposed modified-BCS (MBCS) theory [8,9], and its generalization, the modified-HFB (MHFB) theory [10] take into account the fluctuations of quasiparticle number in a microscopic way. The MHFB theory restores the unitarity relation by explicitly including the quasiparticle-number fluctuations, making use of a secondary Bogolyubov transformation from quasiparticle operators to modified quasiparticle ones. In the limiting case of a constant pairing interaction G the MHFB equation is reduced to the MBCS one.

The second part of the lecture represents an application of the MBCS theory in the study of the damping of giant dipole resonances (GDR) in hot nuclei, which are formed at high excitation energies E^* in heavy-ion fusion reactions or in the inelastic scattering of light particles (nuclei) on heavy targets. The γ-decay spectra of these compound nuclei show the existence of the GDR, whose peak's energy depends weakly on the excitation energy E^*. The dependence of the GDR on the temperature T has been experimentally extracted when the angular momentum of the compound nucleus is low, as in the case of the light-particle scattering experiments, or when it can be separated out from the excitation energy E^*. These measurements have showed that the GDR width remains almost constant at $T \leq 1$ MeV, but sharply increases with T up to $T \simeq 2$ - 3 MeV, and saturates at higher T [11]. The phonon-damping model (PDM), proposed by the lecturer in collaboration with Arima [12], explains the GDR width's increase and saturation by coupling the GDR to non-collective particle-particle (pp) and hole-hole (hh) configurations, which appear due to the deformation of the Fermi surface at $T \neq 0$. It will be shown that, by including non-vanishing MBCS thermal pairing, the PDM is also able to predict the GDR width at low T.

2. Modified HFB theory at finite temperature and its limit, modified BCS theory

2.1. *HFB theory*

The HFB theory is based on the self-consistent Hartree-Fock (HF) Hamiltonian with two-body interaction

$$H = \sum_{ij} \mathcal{T}_{ij} a_i^\dagger a_j + \frac{1}{4} \sum_{ijkl} v_{ijkl} a_i^\dagger a_j^\dagger a_l a_k \ , \tag{1}$$

where $i, j, ..$ denote the quantum numbers characterizing the single-particle orbitals, $\mathcal{T}_{ij}$ are the kinetic energies, and v_{ijkl} are antisymmetrized matrix elements of the two-body interaction. The HFB theory approximates Hamiltonian (1) by an independent-quasiparticle Hamiltonian $H_{\rm HFB}$

$$H - \mu\hat{N} \approx H_{\rm HFB} = E_0 + \sum_i E_i \alpha_i^\dagger \alpha_i \ , \tag{2}$$

where $\hat{N}$ is the particle-number operator, μ is the chemical potential, E_0 is the energy of the ground-state $|0\rangle$, which is defined as the vacuum of quasiparticles:

$$\alpha_i|0\rangle = 0 \ , \tag{3}$$

and E_i are quasiparticle energies. The quasiparticle creation $\alpha_i^\dagger$ and destruction α_i operators are obtained from the single-particle operators $a_i^\dagger$ and a_i by the Bogolyubov transformation, whose matrix form is

$$\begin{pmatrix} \alpha^\dagger \\ \alpha \end{pmatrix} \begin{pmatrix} U & V \\ V^* & U^* \end{pmatrix} \begin{pmatrix} a^\dagger \\ a \end{pmatrix} \tag{4}$$

with the properties

$$UU^\dagger + VV^\dagger = \mathbf{1} \ , \quad UV^{\rm T} + VU^{\rm T} = 0 \ , \tag{5}$$

where **1** is the unit matrix, and the superscript ^{T} denotes the transposing operation. The quasiparticle energies E_i and matrices U and V are determined as the solutions of the HFB equations, which are usually derived by applying either the variational principle of Ritz or the Wick's theorem.

At finite temperature T the condition for a system to be in thermal equilibrium requires the minimum of its grand potential Ω

$$\Omega = \mathcal{E} - TS - \mu N \ , \tag{6}$$

with the total energy $\mathcal{E}$, the entropy S, and particle number N, namely

$$\delta\Omega = 0 \ . \tag{7}$$

This variation defines the density operator $\mathcal{D}$ with the trace equal to 1

$$\mathrm{Tr}\mathcal{D} = 1 \ , \quad \delta\Omega/\delta\mathcal{D} = 0 \tag{8}$$

in the form

$$\mathcal{D} = Z^{-1}\mathrm{e}^{-\beta(H-\mu\hat{N})}, \quad Z = \mathrm{Tr}[\mathrm{e}^{-\beta(H-\mu\hat{N})}] \ , \quad \beta = T^{-1} \ , \tag{9}$$

where Z is the grand partition function. The expectation value $\prec \hat{\mathcal{O}} \succ$ of any operator $\hat{\mathcal{O}}$ is then given as the average in the grand canonical ensemble

$$\prec \hat{\mathcal{O}} \succ = \mathrm{Tr}(\mathcal{D}\hat{\mathcal{O}}) \ . \tag{10}$$

This defines the total energy $\mathcal{E}$, entropy S, and particle number N as

$$\mathcal{E} = \mathrm{Tr}(\mathcal{D}H) \ , \quad S = -\mathrm{Tr}(\mathcal{D}\ln\mathcal{D}) \ , \quad N = \mathrm{Tr}(\mathcal{D}\hat{N}) \ . \tag{11}$$

The FT-HFB theory replaces the unknown exact density operator $\mathcal{D}$ in Eq. (9) with the approximated one, $\mathcal{D}_{\mathrm{HFB}}$, which is found in Ref. [3] by substituting Eq. (2) in to Eq. (9) as

$$\mathcal{D}_{\mathrm{HFB}} \prod_i [n_i\hat{\mathcal{N}}_i + (1-n_i)(1-\hat{\mathcal{N}}_i)] \ , \tag{12}$$

where $\hat{\mathcal{N}}_i$ is the operator of quasiparticle number on the i-th orbital

$$\hat{\mathcal{N}}_i = \alpha_i^{\dagger}\alpha_i \ , \tag{13}$$

and n_i is the quasiparticle occupation number. Within the FT-HFB theory n_i is defined according to Eq. (10) as

$$n_i = \langle\hat{\mathcal{N}}_i\rangle = \frac{1}{\mathrm{e}^{\beta E_i} + 1} \ , \tag{14}$$

where the symbol $\langle\ldots\rangle$ denotes the average similar to (10), but in which the approximated density operator $\mathcal{D}_{\mathrm{HFB}}$ (12) replaces the exact one, i.e.

$$\langle\hat{\mathcal{O}}\rangle = \mathrm{Tr}(\mathcal{D}_{\mathrm{HFB}}\hat{\mathcal{O}}) \ . \tag{15}$$

The generalized particle-density matrix R is related to the generalized quasiparticle-density matrix Q through the Bogolyubov transformation (4) as

$$R = \mathcal{U}^\dagger Q \mathcal{U} \,, \tag{16}$$

where

$$R = \begin{pmatrix} \rho & \tau \\ -\tau^* & 1-\rho^* \end{pmatrix} , \quad Q = \begin{pmatrix} q & t \\ -t^* & 1-q^* \end{pmatrix} = \begin{pmatrix} n & 0 \\ 0 & 1-n \end{pmatrix} , \tag{17}$$

with

$$\mathcal{U} = \begin{pmatrix} U^* & V^* \\ V & U \end{pmatrix} , \quad \mathcal{U}\mathcal{U}^\dagger = \mathbf{1} \,. \tag{18}$$

The matrix elements of the single-particle matrix ρ and particle pairing tensor τ within the FT-HFB approximation are evaluated as

$$\rho_{ij} = \langle a_j^\dagger a_i \rangle \,, \qquad \tau_{ij} = \langle a_j a_i \rangle \,, \tag{19}$$

while those of the quasiparticle matrix q are given in terms of the quasiparticle occupation number since

$$q_{ij} = \langle \alpha_j^\dagger \alpha_i \rangle = \delta_{ij} n_i \,, \qquad t_{ij} = \langle \alpha_j \alpha_i \rangle = 0 \,, \tag{20}$$

which follow from the HFB approximation (2). Using the inverse transformation of (4), the particle densities are obtained as [3]

$$\rho = U^{\mathrm{T}} n U^* + V^\dagger (1-n) V \,, \qquad \tau = U^{\mathrm{T}} n V^* + V^\dagger (1-n) U \,. \tag{21}$$

By minimizing the grand potential Ω according to Eq. (7), the FT-HFB equations were derived in the following form [3]

$$\begin{pmatrix} \mathcal{H} & \Delta \\ -\Delta^* & -\mathcal{H}^* \end{pmatrix} \begin{pmatrix} U_i \\ V_i \end{pmatrix} = E_i \begin{pmatrix} U_i \\ V_i \end{pmatrix} , \tag{22}$$

where

$$\mathcal{H} = \mathcal{T} - \mu + \Gamma \,, \qquad \Gamma_{ij} = \sum_{kl} v_{ikjl} \rho_{lk} \,, \qquad \Delta_{ij} = \frac{1}{2} \sum_{kl} v_{ijkl} \tau_{kl} \,. \tag{23}$$

The total energy $\mathcal{E}$, entropy S, and particle number N from Eq. (11) are now given within the FT-HFB theory as

$$\mathcal{E} = \mathrm{Tr}[(\mathcal{T} + \frac{1}{2}\Gamma)\rho + \frac{1}{2}\Delta\tau^\dagger] \,, \tag{24}$$

$$S = -\sum_i [n_i \ln n_i + (1-n_i)\ln(1-n_i)] \,, \tag{25}$$

$$N = \mathrm{Tr}\rho \ , \tag{26}$$

from which one can easily calculate the grand potential Ω (6).

At zero temperature ($T = 0$) the quasiparticle occupation number vanishes: n_i =0, and the average (15) reduces to the average in the quasiparticle vacuum (3). The quasiparticle-density matrix Q (17) becomes

$$Q(T=0) \equiv Q_0 = \begin{pmatrix} 0 & 0 \\ 0 & 1 \end{pmatrix} \ , \qquad \text{for which} \quad Q_0^2 = Q_0 \ . \tag{27}$$

Therefore, for the generalized particle-density matrix $R_0 = R(T=0)$ the following unitarity relation holds

$$R_0^2 = R_0 \ , \qquad \text{where} \qquad R_0 = \mathcal{U}^\dagger Q_0 \mathcal{U} \ . \tag{28}$$

However, the idempotent (28) no longer holds at $T \neq 0$. Indeed, from Eqs. (16) and (17) it follows that

$$R - R^2 = \mathcal{U}^\dagger (Q - Q^2) \mathcal{U} \ , \tag{29}$$

which leads to

$$\mathrm{Tr}(R - R^2) = \mathrm{Tr}(Q - Q^2) = 2\sum_i n_i(1 - n_i) \equiv 2(\delta\mathcal{N})^2 \neq 0 \ , \qquad (T \neq 0) \ . \tag{30}$$

The quantity $\delta\mathcal{N}^2 = \sum_i n_i(1 - n_i)$ in Eq. (30) is nothing but the quasiparticle-number fluctuation since

$$\delta\mathcal{N}^2 = \langle\hat{\mathcal{N}}^2\rangle - \langle\hat{\mathcal{N}}\rangle^2 = \langle\sum_i \hat{\mathcal{N}}_i + \sum_{i\neq j} \hat{\mathcal{N}}_i\hat{\mathcal{N}}_j\rangle - \sum_i n_i^2 - \sum_{i\neq j} n_i n_j$$

$$= \sum_i n_i(1 - n_i) = \sum_i \delta\mathcal{N}_i^2 \ , \tag{31}$$

where $\delta\mathcal{N}_i^2 = n_i(1 - n_i)$ is the fluctuation of quasiparticle number on the i-th orbital. Therefore, in order to restore the idempotent of type (28) at $T \neq 0$ a new approximation should be found such that it includes the quasiparticle-number fluctuation in the quasiparticle-density matrix.

2.2. *MHFB theory*

Let us consider, instead of the FT-HFB density operator $\mathcal{D}_{\mathrm{HFB}}$ (12), an improved approximation, $\bar{\mathcal{D}}$, to the density operator $\mathcal{D}$. This approximated density operator $\bar{\mathcal{D}}$ should satisfy two following requirements:

(i) The average

$$\langle\langle\hat{\mathcal{O}}\rangle\rangle \equiv \mathrm{Tr}(\bar{\mathcal{D}}\hat{\mathcal{O}}), \tag{32}$$

in which $\bar{\mathcal{D}}$ is used in place of $\mathcal{D}$ (or $\mathcal{D}_{\rm HFB}$), yields

$$\bar{R} = \mathcal{U}^\dagger \bar{Q} \mathcal{U} \tag{33}$$

for the Bogolyubov transformation $\mathcal{U}$ (18), where one has the modified matrices

$$\bar{R} = \begin{pmatrix} \bar{\rho} & \bar{\tau} \\ -\bar{\tau}^* & 1-\bar{\rho}^* \end{pmatrix} , \quad \bar{Q} = \begin{pmatrix} \bar{q} & t \\ -\bar{t}^* & 1-\bar{q}^* \end{pmatrix} , \tag{34}$$

with

$$\bar{\rho}_{ij} = \langle\langle a_j^\dagger a_i \rangle\rangle , \quad \bar{\tau}_{ij} = \langle\langle a_j a_i \rangle\rangle , \tag{35}$$

$$\bar{q}_{ij} = \langle\langle \alpha_j^\dagger \alpha_i \rangle\rangle \delta_{ij} \bar{n}_i , \quad \bar{t}_{ij} = \langle\langle \alpha_j \alpha_i \rangle\rangle = \mathbf{\Lambda}_{ij} \tag{36}$$

instead of matrices R and Q in Eqs. (17), (19), and (20). The non-zero values of $\bar{t}_{ij}$ in Eq. (36) are caused by the quasiparticle correlations in the thermal equilibrium, which are now included in the average $\langle\langle \ldots \rangle\rangle$ using the density operator $\bar{\mathcal{D}}$.

(ii) The modified quasiparticle-density matrix $\bar{Q}$ satisfies the unitarity relation

$$(\bar{Q})^2 = \bar{Q} . \tag{37}$$

The solution of Eq. (37) immediately yields the matrix $\mathbf{\Lambda}$ in the canonical form

$$\mathbf{\Lambda} = \sqrt{\bar{n}(1-\bar{n})} \equiv \begin{pmatrix} & & & & 0 & -\Lambda_1 \\ & & & & \Lambda_1 & 0 \\ & & 0 & -\Lambda_2 & & \\ & & \Lambda_2 & 0 & & \\ & \cdot & & & & \\ \cdot & & & & & \end{pmatrix} , \quad \Lambda_i = \sqrt{\bar{n}_i(1-\bar{n}_i)} . \tag{38}$$

Comparing this result with Eq. (31), it is clear that tensor $\mathbf{\Lambda}$ consists of the quasiparticle-number fluctuation $\delta\bar{\mathcal{N}}_i = \sqrt{\bar{n}_i(1-\bar{n}_i)}$. From Eq. (33) it is easy to see that the unitarity relation holds for the modified generalized single-particle density matrix $\bar{R}$ since $\bar{R} - \bar{R}^2 = \mathcal{U}^\dagger(\bar{Q} - \bar{Q}^2)\mathcal{U} = 0$ due to Eq. (37) and the unitary matrix $\mathcal{U}$.

Let us define the modified-quasiparticle operators $\bar{\alpha}_i^\dagger$ and $\bar{\alpha}_i$, which behave in the average (32) exactly as the usual quasiparticle operators $\alpha_i^\dagger$ and α_i do in the quasiparticle ground state, namely

$$\langle\langle \bar{\alpha}_i^\dagger \bar{\alpha}_k \rangle\rangle \langle\langle \bar{\alpha}_i^\dagger \bar{\alpha}_k^\dagger \rangle\rangle \langle\langle \bar{\alpha}_k \bar{\alpha}_i \rangle\rangle = 0 . \tag{39}$$

In the same way as for the usual Bogolyubov transformation (4), we search for a transformation between these modified-quasiparticle operators ($\bar{\alpha}_i^\dagger$, $\bar{\alpha}_i$) and the usual quasiparticle ones ($\alpha_i^\dagger$, α_i) in the following form

$$\begin{pmatrix} \bar{\alpha}^\dagger \\ \bar{\alpha} \end{pmatrix} \begin{pmatrix} w & z \\ z^* & w^* \end{pmatrix} \begin{pmatrix} \alpha^\dagger \\ \alpha \end{pmatrix} , \tag{40}$$

with the unitary property similar to Eq. (5) for U and V matrices : $ww^\dagger + zz^\dagger = \mathbf{1}$. Using the inverse transformation of (40) and the requirement (39), we obtain

$$\bar{n}_i = \langle\langle \alpha_i^\dagger \alpha_i \rangle\rangle = \sum_k z_{ik} z_{ik}^* . \tag{41}$$

From this equation and the unitarity condition (28), it follows that $zz^\dagger = \bar{n}$ and $ww^\dagger = \mathbf{1} - \bar{n}$. Since $\mathbf{1} - \bar{n}$ and $\bar{n}$ are real diagonal matrices, the canonical form of matrices w and z is found as

$$w = \begin{pmatrix} w_1 & 0 & & & & \\ 0 & w_1 & & & & \\ & & w_2 & 0 & & \\ & & 0 & w_2 & & \\ & & & & \cdot & \\ & & & & & \cdot \end{pmatrix} , \quad z = \begin{pmatrix} & & & & 0 & -z_1 \\ & & & & z_1 & 0 \\ & & 0 & -z_2 & & \\ & & z_2 & 0 & & \\ & \cdot & & & & \\ \cdot & & & & & \end{pmatrix} , \tag{42}$$

where $w_i = \sqrt{1 - \bar{n}_i}$, $z_i = \sqrt{\bar{n}_i}$.

We now show that we can obtain the idempotent $\bar{R}^2 = \bar{R}$ by applying the secondary Bogolyubov transformation (40), which automatically leads to Eq. (37). Indeed, using the inverse transformation of (40) with matrices w and z given in Eq. (42), we found that the modified quasiparticle-density matrix $\bar{Q}$ can be obtained as

$$\mathcal{W}^\dagger \bar{Q}_0 \mathcal{W} = \begin{pmatrix} \bar{n} & [\sqrt{\bar{n}(1-\bar{n})}]^\dagger \\ \sqrt{\bar{n}(1-\bar{n})} & 1 - \bar{n} \end{pmatrix} \equiv \bar{Q} , \tag{43}$$

where

$$\mathcal{W} = \begin{pmatrix} (\sqrt{1-\bar{n}})^* & (\sqrt{\bar{n}})^* \\ \sqrt{\bar{n}} & \sqrt{1-\bar{n}} \end{pmatrix} , \quad \mathcal{W}\mathcal{W}^\dagger = \mathbf{1} , \tag{44}$$

and

$$\bar{Q}_0 = \begin{pmatrix} \langle\langle \bar{\alpha}^\dagger \bar{\alpha} \rangle\rangle & \langle\langle \bar{\alpha}\bar{\alpha} \rangle\rangle \\ \langle\langle \bar{\alpha}^\dagger \bar{\alpha}^\dagger \rangle\rangle & 1 - \langle\langle \bar{\alpha}^\dagger \bar{\alpha} \rangle\rangle \end{pmatrix} \begin{pmatrix} 0 & 0 \\ 0 & 1 \end{pmatrix} , \quad \bar{Q}_0^2 = \bar{Q}_0 , \tag{45}$$

due to Eq. (39). This result shows another way of deriving the modified quasiparticle-density matrix $\bar{Q}$ (34) from the density matrix $\bar{Q}_0$ of the modified quasiparticles ($\bar{\alpha}_i^\dagger$, $\bar{\alpha}_i$). This matrix $\bar{Q}_0$ is identical to the zero-temperature quasiparticle-density matrix Q_0 (27). Substituting this result into the right-hand side (rhs) of Eq. (33), we obtain

$$\bar{R} = \bar{\mathcal{U}}^\dagger \bar{Q}_0 \bar{\mathcal{U}} \ , \tag{46}$$

where

$$\bar{\mathcal{U}} = \mathcal{W}\mathcal{U}\begin{pmatrix} \bar{U}^* & \bar{V}^* \\ \bar{V} & \bar{U} \end{pmatrix}$$

$$\begin{pmatrix} (\sqrt{1-\bar{n}})^* U^* + (\sqrt{\bar{n}})^* V & (\sqrt{1-\bar{n}})^* V^* + (\sqrt{\bar{n}})^* U \\ \sqrt{1-\bar{n}} V + \sqrt{\bar{n}} U^* & \sqrt{1-\bar{n}} U + \sqrt{\bar{n}} V^* \end{pmatrix} . \tag{47}$$

This equation is the generalized form of the modified Bogolyubov coefficients $\bar{u}_j$ and $\bar{v}_j$ given in Eq. (38) of Ref. [9]. From Eqs. (18), (44), and (47), it follows that $\bar{\mathcal{U}}\bar{\mathcal{U}}^\dagger = \mathbf{1}$, i.e. transformation (46) is unitary. Therefore, from the idempotent (45) it follows that $\bar{R}^2 = \bar{R}$.

Applying the Wick's theorem for the ensemble average, one obtains the expressions for the modified total energy $\bar{\mathcal{E}}$

$$\bar{\mathcal{E}} = \mathrm{Tr}[(\mathcal{T} + \frac{1}{2}\bar{\Gamma})\bar{\rho} + \frac{1}{2}\bar{\Delta}\bar{\tau}^\dagger] \ , \tag{48}$$

where

$$\bar{\Gamma}_{ij} = \sum_{kl} v_{ikjl}\bar{\rho}_{lk} \ , \quad \bar{\Delta}_{ij} = \frac{1}{2}\sum_{kl} v_{ijkl}\bar{\tau}_{kl} \ . \tag{49}$$

From Eq. (46) we obtain the modified single-particle density matrix $\bar{\rho}$ and modified particle-pairing tensor $\bar{\tau}$ in the following form

$$\bar{\rho} = U^{\mathrm{T}}\bar{n}U^* + V^\dagger(1-\bar{n})V + U^{\mathrm{T}}\left[\sqrt{\bar{n}(1-\bar{n})}\right]^\dagger V + V^\dagger\sqrt{\bar{n}(1-\bar{n})}U^* \ , \tag{50}$$

$$\bar{\tau} = U^{\mathrm{T}}\bar{n}V^* + V^\dagger(1-\bar{n})U + U^{\mathrm{T}}\left[\sqrt{\bar{n}(1-\bar{n})}\right]^\dagger U + V^\dagger\sqrt{\bar{n}(1-\bar{n})}V^* \ . \tag{51}$$

As compared to Eq. (21) within the FT-HFB approximation, Eqs. (50) and (51) contain the last two terms $\sim [\sqrt{\bar{n}(1-\bar{n})}]^\dagger$ and $\sim \sqrt{\bar{n}(1-\bar{n})}$, which arise due to quasiparticle-number fluctuation. Also the quasiparticle occupation number is now $\bar{n}$ [See Eq. (36)] instead of n (14).

We derive the MHFB equations following the same variational procedure, which was used to derive the FT-HFB equations in Ref. [3]. According it, we minimize the grand potential $\delta\bar{\Omega} = 0$ by varying U, V, and $\bar{n}$, where

$$\bar{\Omega} = \bar{\mathcal{E}} - T\bar{S} - \bar{\mu}N \ . \tag{52}$$

The MHFB equations formally look like the FT-HFB ones, namely (22)

$$\begin{pmatrix} \bar{\mathcal{H}} & \bar{\Delta} \\ -\bar{\Delta}^* & -\bar{\mathcal{H}}^* \end{pmatrix} \begin{pmatrix} U_i \\ V_i \end{pmatrix} = \bar{E}_i \begin{pmatrix} U_i \\ V_i \end{pmatrix} , \tag{53}$$

where, however

$$\bar{\mathcal{H}} = \mathcal{T} - \bar{\mu} + \bar{\Gamma} \tag{54}$$

with $\bar{\Gamma}$ and $\bar{\Delta}$ given by Eq. (49). The equation for particle number N within the MHFB theory is

$$N = \mathrm{Tr}\bar{\rho} \ . \tag{55}$$

By solving Eq. (53), one obtains the modified quasiparticle energy $\bar{E}_i$, which is different from E_i in Eqs (22) due to the change of the HF and pairing potentials. Hence, the MHFB quasiparticle Hamiltonian H_{MHFB} can be written as

$$H - \bar{\mu}\hat{N} \approx H_{\mathrm{MHFB}} = \bar{E}_0 + \sum_i \bar{E}_i \hat{\mathcal{N}}_i \ , \tag{56}$$

instead of (2). This implies that the approximated density operator $\bar{D}$ (32) within the MHFB theory can be represented in the form similar to (12), namely

$$\bar{\mathcal{D}} \equiv \mathcal{D}_{\mathrm{MHFB}} \prod_i [\bar{n}_i \hat{\mathcal{N}}_i + (1 - \bar{n}_i)(1 - \hat{\mathcal{N}}_i)]. \tag{57}$$

From here it follows that the formal expression for the modified entropy $\bar{S}$ is the same as that given in Eq. (25), i.e.

$$\bar{S} = -\sum_i [\bar{n}_i \ln \bar{n}_i + (1 - \bar{n}_i)\ln(1 - \bar{n}_i)] \ , \tag{58}$$

Using the thermodynamic definition of temperature in terms of entropy $1/T = \delta\bar{S}/\delta\bar{\mathcal{E}}$ and carrying out the variation over $\delta\bar{n}_i$, we find

$$\frac{\delta\bar{\mathcal{E}}}{\delta\bar{n}_i} \equiv \bar{E}_i = T\frac{\delta\bar{S}}{\delta\bar{n}_i} T\ln\left(\frac{1 - \bar{n}_i}{\bar{n}_i}\right) \ . \tag{59}$$

Inverting Eq. (59), we obtain

$$\bar{n}_i = \frac{1}{e^{\beta\bar{E}_i} + 1} \ . \tag{60}$$

This result shows that the functional dependence of quasiparticle occupation number $\bar{n}_i$ on quasiparticle energy and temperature within the MHFB theory is also given by the Fermi-Dirac distribution of noninteracting quasiparticles but with the modified energies $\bar{E}_i$ defined by the MHFB equations (53). Therefore we will omit the bar over $\bar{n}_i$ and use the same Eq. (14) with E_i replaced with $\bar{E}_i$ for the MHFB equations.

2.3. *MBCS theory*

In the limit with equal pairing matrix elements $G_{ij} = G$, neglecting the contribution of G to the HF potential so that $\bar{\Gamma} = 0$, the HF Hamiltonian becomes

$$\bar{\mathcal{H}}_{ij} = (\epsilon_i - \bar{\mu})\delta_{ij} \ . \tag{61}$$

The pairing potential (49) takes now the simple form

$$\bar{\Delta} = -G \sum_{k>0} \bar{\tau}_{k\tilde{k}} \ . \tag{62}$$

The Bogolyubov transformation (4) for spherical nuclei reduces to

$$\alpha^{\dagger}_{jm} = u_j a^{\dagger}_{jm} + v_j (-)^{j+m} a_{j-m} \ ,$$

$$(-)^{j+m}\alpha_{j-m} = u_j (-)^{j+m} a_{j-m} - v_j a^{\dagger}_{jm} \ , \tag{63}$$

while the secondary Bogolyubov transformation (40) becomes [9]

$$\bar{\alpha}^{\dagger}_{jm} = \sqrt{1-n_j}\,\alpha^{\dagger}_{jm} - \sqrt{n_j}(-)^{j+m}\alpha_{j-m} \ ,$$

$$(-)^{j+m}\bar{\alpha}_{j-m} = \sqrt{1-n_j}(-)^{j+m}\alpha_{j-m} + \sqrt{n_j}\,\alpha^{\dagger}_{jm} \ . \tag{64}$$

The U, V, $1-n$, n, and $\sqrt{n(1-n)}$ matrices are now block diagonal in each two-dimensional subspace spanned by the quasiparticle state $|j\rangle$ and its time-reversal partner $|\widetilde{j}\rangle = (-)^{j+m}|j-m\rangle$

$$U = \begin{pmatrix} u_j & 0 \\ 0 & u_j \end{pmatrix} , \quad V = \begin{pmatrix} 0 & v_j \\ -v_j & 0 \end{pmatrix} , \tag{65}$$

$$1 - n = \begin{pmatrix} 1-n_j & 0 \\ 0 & 1-n_j \end{pmatrix} , \quad n = \begin{pmatrix} n_j & 0 \\ 0 & n_j \end{pmatrix} ,$$

$$\sqrt{n(1-n)} = \begin{pmatrix} 0 & -\sqrt{n_j(1-n_j)} \\ \sqrt{n_j(1-n_j)} & 0 \end{pmatrix} , \tag{66}$$

Substituting these matrices into the rhs of Eqs. (50) and (51), we find

$$\bar{\rho}_{j\,\tilde{j}} = (1-2n_j)v_j^2 + n_j - 2\sqrt{n_j(1-n_j)}\,u_j v_j \ , \tag{67}$$

$$\bar{\tau}_{j\,\tilde{j}} = -(1-2n_j)u_j v_j + \sqrt{n_j(1-n_j)}\,(u_j^2 - v_j^2) \ . \tag{68}$$

Substituting now Eqs. (68) and (67) into the rhs of Eqs. (62) and (55), respectively, we obtain the MBCS equations for spherical nuclei in the following form:

$$\bar{\Delta} = G\sum_j \Omega_j[(1-2n_j)u_jv_j - \sqrt{n_j(1-n_j)}(u_j^2 - v_j^2)] \ , \tag{69}$$

$$N = 2\sum_j \Omega_j[(1-2n_j)v_j^2 + n_j - 2\sqrt{n_j(1-n_j)}u_jv_j] \ . \tag{70}$$

Comparing the conventional FT-BCS equations, we see that the MBCS equations explicitly include the effect of quasiparticle-number fluctuation $\sim \delta\mathcal{N}_j$ in the last terms at their rhs, which are the thermal gap $-G\sum_j \Omega_j\sqrt{n_j(1-n_j)}(u_j^2 - v_j^2)$, and the thermal-fluctuation of particle number $\delta N = \sum_j \delta\mathcal{N}_j - 4\sum_j \Omega_j u_j v_j(\delta\mathcal{N}_j)$ in Eq. (70). These terms are ignored within the FT-BCS theory. Hence Eqs. (69) and (70) show for the first time how the effect of statistical fluctuations is included in the MBCS (MHFB) theory at finite temperature on a microscopic ground. So far this effect was treated only within the framework of the macroscopic Landau theory of phase transition [5].

3. Phonon-damping model in quasiparticle representation

The quasiparticle representation of the PDM Hamiltonian [13] is obtained by adding the superfluid pairing interaction and expressing the particle (p) and hole (h) creation and destruction operators, $a_s^\dagger$ and a_s ($s = p, h$), in terms of the quasiparticle operators, $\alpha_s^\dagger$ and α_s, using the Bogolyubov's canonical transformation. As a result, the PDM Hamiltonian for the description of Eλ excitations can be written in spherical basis as

$$H = \sum_{jm} E_j\alpha_{jm}^\dagger\alpha_{jm} + \sum_{\lambda\mu i}\omega_{\lambda i}b_{\lambda\mu i}^\dagger b_{\lambda\mu i} +$$

$$\frac{1}{2}\sum_{\lambda\mu i}\frac{(-)^{\lambda-\mu}}{\hat{\lambda}}\sum_{jj'} f_{jj'}^{(\lambda)}\left\{u_{jj'}^{(+)}\left[A_{jj'}^\dagger(\lambda\mu) + A_{jj'}(\lambda\tilde{\mu})\right] + \right. \tag{71}$$

$$\left. v_{jj'}^{(-)}\left[B_{jj'}^\dagger(\lambda\mu) + B_{jj'}(\lambda\tilde{\mu})\right]\right\}\left(b_{\lambda\mu i}^\dagger + b_{\lambda\tilde{\mu}i}\right) \ ,$$

where $\hat{\lambda} = \sqrt{2\lambda+1}$. The first term at the rhs of Hamiltonian (71) corresponds to the independent-quasiparticle field. The second term stands for the phonon field described by phonon operators, $b^{\dagger}_{\lambda\mu i}$ and $b_{\lambda\mu i}$, with multipolarity λ, which generate the harmonic collective vibrations such as GDR. Phonons are ideal bosons within the PDM, i.e. they have no fermion structure. The last term is the coupling between quasiparticle and phonon fields, which is responsible for the microscopic damping of collective excitations.

In Eq. (71) the following standard notations are used

$$A^{\dagger}_{jj'}(\lambda\mu) = \sum_{mm'} \langle jmj'm'|\lambda\mu\rangle \alpha^{\dagger}_{jm}\alpha^{\dagger}_{j'm'} , \tag{72}$$

$$B^{\dagger}_{jj'}(\lambda\mu) = -\sum_{mm'} (-)^{j'-m'} \langle jmj' - m'|\lambda\mu\rangle \alpha^{\dagger}_{jm}\alpha_{j'm'} , \tag{73}$$

with $(\lambda\tilde{\mu}) \longleftrightarrow (-)^{\lambda-\mu}(\lambda-\mu)$. Functions $u^{(+)}_{jj'} \equiv u_j v_{j'} + v_j u_{j'}$ and $v^{(-)}_{jj'} \equiv u_j u_{j'} - v_j v_{j'}$ are combinations of Bogolyubov's u and v coefficients. The quasiparticle energy E_j is calculated from the single-particle energy ϵ_j as

$$E_j = \sqrt{(\epsilon'_j - \epsilon_{\rm F})^2 + \Delta^2} , \qquad \epsilon'_j \equiv \epsilon_j - G v_j^2, \tag{74}$$

where the pairing gap Δ and the Fermi energy $\epsilon_{\rm F}$ are defined as solutions of the BCS equations. At $T \neq 0$ the thermal pairing gap $\Delta(T)$ (or $\bar{\Delta}(T)$) is defined from the finite-temperature BCS (or MBCS) equations.

The equation for the propagation of the GDR phonon, which is damped due to coupling to the quasiparticle field, is derived making use of the double-time Green's function method (introduced by Bogolyubov and Tyablikov, and developed further by Zubarev [14]). Following the standard procedure of deriving the equation for the double-time retarded Green's function with respect to the Hamiltonian (71), one obtains a closed set of equations for the Green's functions for phonon and quasiparticle propagators. Making the Fourier transform into the energy plane E, and expressing all the Green functions in the set in terms of the one-phonon propagation Green function, we obtain the equation for the latter, $G_{\lambda i}(E)$, in the form

$$G_{\lambda i}(E) = \frac{1}{2\pi} \frac{1}{E - \omega_{\lambda i} - P_{\lambda i}(E)} , \tag{75}$$

where the explicit form of the polarization operator $P_{\lambda i}(E)$ is

$$P_{\lambda i}(E) = \frac{1}{\hat{\lambda}^2} \sum_{jj'} [f^{(\lambda)}_{jj'}]^2 \Bigg[\frac{(u^{(+)}_{jj'})^2 (1 - n_j - n_{j'})(\epsilon_j + \epsilon_{j'})}{E^2 - (\epsilon_j + \epsilon_{j'})^2} - \frac{(v^{(-)}_{jj'})^2 (n_j - n_{j'})(\epsilon_j - \epsilon_{j'})}{E^2 - (\epsilon_j - \epsilon_{j'})^2} \Bigg] . \tag{76}$$

The polarization operator (76) appears due to ph – phonon coupling in the last term of the rhs of Hamiltonian (71). The phonon damping $\gamma_{\lambda i}(\omega)$ (ω real) is obtained as the imaginary part of the analytic continuation of the polarization operator $P_{\lambda i}(E)$ into the complex energy plane $E = \omega \pm i\varepsilon$. Its final form is

$$\gamma_{\lambda i}(\omega) = \frac{\pi}{2\hat{\lambda}^2} \sum_{jj'} [f_{jj'}^{(\lambda)}]^2 \times$$

$$\Big\{ (u_{jj'}^{(+)})^2 (1 - n_j - n_{j'}) [\delta(E - E_j - E_{j'}) - \delta(E + E_j + E_{j'})] - \tag{77}$$

$$(v_{jj'}^{(-)})^2 (n_j - n_{j'}) [\delta(E - E_j + E_{j'}) - \delta(E + \epsilon_j - \epsilon_{j'})] \Big\}.$$

The energy $\bar{\omega}$ of giant resonance (damped collective phonon) is found as the solution of the equation: $\bar{\omega} - \omega_{\lambda i} - P_{\lambda i}(\bar{\omega}) = 0$. The width Γ_λ of giant resonance is calculated as twice of the damping $\gamma_\lambda(\omega)$ at $\omega = \bar{\omega}$, where $\lambda = 1$ corresponds to the GDR width $\Gamma_{\rm GDR}$. The latter has the form

$$\Gamma_{\rm GDR} = 2\pi \Big\{ F_1^2 \sum_{ph} [u_{ph}^{(+)}]^2 (1 - n_p - n_h) \delta(E_{\rm GDR} - E_p - E_h) +$$

$$F_2^2 \sum_{s>s'} [v_{ss'}^{(-)}]^2 (n_{s'} - n_s) \delta(E_{\rm GDR} - E_s + E_{s'}) \Big\}, \tag{78}$$

where $(ss') = (pp')$ and (hh') with p and h denoting the orbital angular momenta j_p and j_h for particles and holes, respectively. The first sum at the rhs of Eq. (78) is the quantal width $\Gamma_{\rm Q}$, which comes from the couplings of quasiparticle pairs $[\alpha_p^\dagger \otimes \alpha_h^\dagger]_{LM}$ to the GDR. At zero pairing they correspond to the couplings of ph pairs, $[a_p^\dagger \otimes a_{\tilde{h}}]_{LM}$ to the GDR. The second sum comes from the coupling of $[\alpha_s^\dagger \otimes \alpha_{\tilde{s}'}]_{LM}$ to the GDR, and is called the thermal width $\Gamma_{\rm T}$ as it appears only at $T \neq 0$. At zero pairing they are pp (hh) pairs, $[a_s^\dagger \otimes a_{\tilde{s}'}]_{LM}$ (The tilde $\tilde{}$ denotes the time-reversal operation).

The line shape of the GDR is described by the strength function $S_{\rm GDR}(\omega)$, which is derived from the spectral intensity in the standard way using the analytic continuation of the Green function (75) and by expanding the polarization operator (76) around $\omega = E_{\rm GDR}$. The final form of $S_{\rm GDR}(\omega)$ is [12,13]

$$S_{\rm GDR}(\omega) = \frac{1}{\pi} \frac{\gamma_{\rm GDR}(\omega)}{(\omega - E_{\rm GDR})^2 + \gamma_{\rm GDR}^2(\omega)}. \tag{79}$$

The PDM is based on the following assumptions:

a1) The matrix elements for the coupling of GDR to non-collective *ph* configurations, which causes the quantal width Γ_Q, are all equal to F_1. Those for the coupling of GDR to *pp* (*hh*), which causes the thermal width Γ_T, are all equal to F_2.

a2) It is well established that the microscopic mechanism of the quantal (spreading) width Γ_Q comes from quantal coupling of *ph* configurations to more complicated ones, such as $2p2h$ ones. The calculations performed in Refs. [15] within two independent microscopic models, where such couplings to $2p2h$ configurations were explicitly included, have shown that Γ_Q depends weakly on T. Therefore, in order to avoid complicate numerical calculations, which are not essential for the increase of Γ_{GDR} at $T \neq 0$, such microscopic mechanism is not included within PDM, assuming that Γ_Q at $T = 0$ is known. The model parameters are then chosen so that the calculated Γ_Q and E_{GDR} reproduce the corresponding experimental values at $T = 0$.

Within assumptions (a1) and (a2) the model has only three T-independent parameters, which are the unperturbed phonon energy ω_q, F_1, and F_2. The parameters ω_q and F_1 are chosen so that after the *ph*-GDR coupling is switched on, the calculated GDR energy E_{GDR} and width Γ_{GDR} reproduce the corresponding experimental values for GDR on the ground-state. At $T \neq 0$, the coupling to *pp* and *hh* configurations is activated. The F_2 parameter is then fixed at $T = 0$ so that the GDR energy E_{GDR} does not change appreciably with T.

4. Numerical results

4.1. *Temperature dependence of pairing gap*

Shown in Fig. 1 (a) is the temperature dependence of the neutron pairing gap $\bar{\Delta}_\nu$ for ^{120}Sn, which is obtained from the MBCS equation (69) using the single-particle energies determined within the Woods-Saxon potential at $T = 0$. The pairing parameter G_ν is chosen to be equal to 0.13 MeV, which yields $\bar{\Delta}(T = 0) \equiv \bar{\Delta}(0) \simeq 1.4$ MeV. Contrary to the BCS gap (dotted line), which collapses at $T_c \simeq 0.79$ MeV, the gap $\bar{\Delta}$ (solid line) does not vanish, but decreases monotonously with increasing T at $T \geq 1$ MeV resulting in a long tail up to $T \simeq 5$ MeV. This behavior is caused by the thermal fluctuation of quasiparticle number in the MBCS equations (69). As the result, the heat capacity [Fig. 1 (b)] has no divergence at T_c, which is seen within the BCS theory.

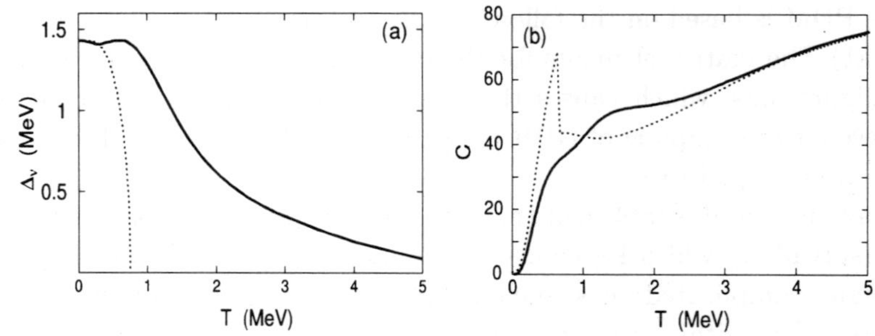

Fig. 1. Neutron pairing gap (a) and heat capacity (b) for ^{120}Sn as functions of T. Solid and dotted lines show the results obtained within MBCS and BCS theories, respectively.

4.2. *Temperature dependence of GDR width*

The GDR widths as a function of T for ^{120}Sn obtained within the PDM are compared in Fig. 2 (a) with the experimental data and the prediction by the thermal fluctuation model (TFM) [16].

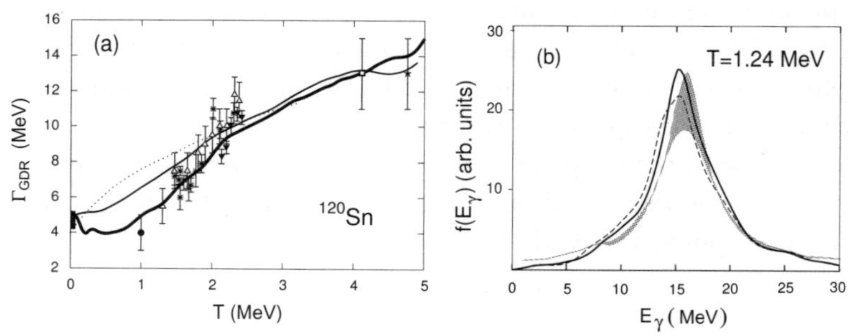

Fig. 2. (b): GDR width $\Gamma_{\rm GDR}$ as a function of T for ^{120}Sn. The thin and thick solid lines show the PDM results obtained neglecting pairing and including the renormalized gap $\widetilde{\Delta} = [1+1/\delta N^2]\bar{\Delta}$, respectively. The gap $\widetilde{\Delta}$ includes the correction $\delta N^2 = \bar{\Delta}(0)^2\sum_j(j+1/2)/[(\epsilon_j - \bar{\epsilon}_{\rm F})^2 + \bar{\Delta}(0)^2]$ due to an approximate number projection. The prediction by the TFM is shown as the dotted line [16]; (b): GDR strength function at $T = 1.24$ MeV. The dashed and solid lines show the results obtained without and including the gap $\widetilde{\Delta}$, while experimental results are shown as the shaded area.

The TFM interprets the broadening of the GDR width via an adiabatic coupling of GDR to quadrupole deformations induced by thermal fluctuations. Even when thermal pairing is neglected the PDM prediction, (the thin solid line) is already better than that given by the TFM, including the region of high T where the width's saturation is reported. The increase of the total width with T is driven by the increase of the thermal width $\Gamma_{\rm T}$, which is caused by coupling to *pp* and *hh* configurations, since the quantal width Γ_Q is found to decrease slightly with increasing T [12]. The inclusion of thermal pairing, which yields a sharper Fermi surface, compensates the smoothing of the Fermi surface with increasing T. This leads to a much weaker T-dependence of the GDR width at low T. As a result, the values of the width predicted by the PDM in this region significantly drop (the thick solid line), recovering the data point at $T = 1$ MeV. The GDR strength function obtained including the MBCS gap is also closer to the experimental data than that obtained neglecting the thermal gap [Fig. 2 (b)].

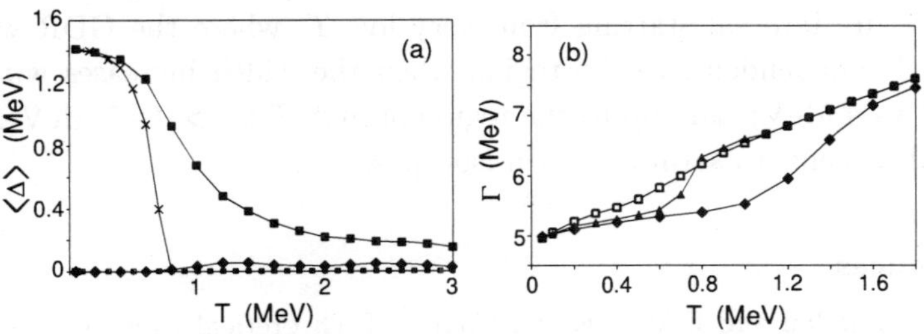

Fig. 3. (a): Pairing gaps for ^{120}Sn averaged over thermal shape fluctuations versus T. Lines with triangles and crosses are the usual BCS proton and neutron pairing gaps, respectively, while those with diamonds and squares denote the corresponding pairing gaps, which also include thermal fluctuations of pairing fields. (b): GDR widths for ^{120}Sn versus T. Open squares, triangles, and diamonds denote the widths obtained without pairing, including BCS pairing, and thermally fluctuating pairing field from (a), respectively.

The results discussed above have also been confirmed by our recent calculations within a macroscopic approach, which takes pairing fluctuations into account along with the thermal shape fluctuations [17]. Here the free energies are calculated using the Nilsson-Strutinsky method at $T \neq 0$, including thermal pairing correlations. The GDR is coupled to the nuclear shapes through a simple anisotropic harmonic oscillator model with a separable dipole-dipole interaction. The observables are averaged over the shape parameters and pairing gap. Our study reveals that the observed quenching of GDR width at low T in ^{120}Sn and ^{148}Au can be understood in terms of simple shape effects caused by pairing correlations. Fluctuations in pairing field lead to a slowly vanishing pairing gap [Fig. 3 (a)], which influences the structural properties even at moderate T (~1 MeV). We found that the low-T structure and hence the GDR width are quite sensitive to the change of the pairing field [Fig. 3 (b)].

5. Conclusions

It has been shown in the present lecture that the MHFB and MBCS theories are microscopic approaches, which take into account thermal fluctuations of quasiparticle number. These large thermal fluctuations smooth out the sharp SN phase transition in finite nuclei. As a result, the thermal pairing gap does not collapse, but decreases monotonously with increasing temperature T, remaining finite even at T as high as 4 - 5 MeV. This non-vanishing thermal pairing gap keeps the width of GDR remain almost constant at low T ($\leq$ 1 MeV for ^{120}Sn) when it is included in the PDM. In this way the PDM becomes a semi-microscopic model that is able to describe the temperature dependence of the GDR width in a consistent way within a large temperature interval starting from very low T, where the GDR width is nearly T-independent, to the region when the width increases with T (1 $\leq T \leq$ 3 - 4 MeV), and up to the region of high T ($T >$ 4 - 5 MeV), where the width seems to saturate in tin isotopes.

References

1. L.D. Landau and E.M. Lifshitz, Course of Theoretical Physics, Vol. 5: Statistical Physics (Moscow, Nauka, 1964) pp. 297, 308.
2. M. Sano and S. Yamazaki, Prog. Theor. Phys. **29**, 397 (1963).
3. A.L. Goodman, Nucl. Phys. **A 352**, 30 (1981).
4. A.L. Goodman, Phys. Rev. C **29**, 1887 (1984).
5. L.G. Moretto, Phys. Lett. B **40**, 1 (1972).
6. N. Dinh Dang, Z. Phys. A **335**, 253 (1990).
7. N.D. Dang, P. Ring, and R. Rossignoli, Phys. Rev. C **47**, 606 (1993).
8. N. Dinh Dang and V. Zelevinsky, Phys. Rev. C **64**, 064319 (2001).
9. N. Dinh Dang and A. Arima, Phys. Rev. C **67**, 014304 (2003).
10. N. Dinh Dang and A. Arima, Phys. Rev. C **68**, 014318 (2003).
11. M.N. Harakeh and A. van der Woude, Giant resonances - Fundamental high-frequency modes of nuclear excitation (Oxford, Clarendon Press, 2001) p. 638.
12. N. Dinh Dang and A. Arima, Phys. Rev. Lett. **80**, 4145 (1998); Nucl. Phys. **A 636**, 427 (1998).
13. N. Dinh Dang N. *et al.*, Phys. Rev. **C 63**, 044302 (2001).
14. N.N. Bogolyubov and S. Tyablikov, *Sov. Phys. Doklady* **4**, 6 (1959); D.N. Zubarev, Nonequilibrium Statistical Thermodynamics (Plenum, NY, 1974);
15. P.F. Bortignon P.F. *et al.*, Nuc. Phys. **A 460**, 149 (1986); N. Dinh Dang, Nucl. Phys. **A 504**, 143 (1989).
16. D. Kuznesov *et al.*, Phys. Rev. Lett. **81**, 542 (1998).
17. P. Arumugam and N. Dinh Dang, RIKEN Accel. Prog. Rep. **39**, 28 (2006).

Analysis of the low-lying collective states using the MAVA

J. KOTILA and J. SUHONEN
Department of Physics, University of Jyväskylä
P.O.Box 35, FIN-40014, Jyväskylä, Finland
E-mail: jenni.kotila@phys.jyu.fi

D.S. DELION
National Institute of Physics and Nuclear Engineering
P.O. Box MG-6, Bucharest Măgurele, Romania

Anharmonic features of the low-lying collective states in cadmium, ruthenium and molybdenum isotopes have been investigated systematically by using the Microscopic Anharmonic Vibrator Approach (MAVA). MAVA is based on a large single-particle valence space and a realistic nuclear Hamiltonian which is used to generate the one-phonon states by the use of the Quasiparticle Random-Phase Approximation (QRPA). The same Hamiltonian is also used to introduce anharmonicities into the description of the low-lying excited states leading to dynamical splitting of the energies of the two-phonon vibrational states. Comparison of the calculated energies and $B(E2)$ values with the available data points to mixing between anharmonic vibrations and deformed intruder degrees of freedom in the case of cadmium isotopes, a shape transition in the case of ruthenium isotopes and the discussed molybdenum isotopes are suggested to be closer to anharmonic vibrators than deformed rotors.

1. Introduction

Spherical and nearly spherical nuclei have usually low-lying collective spectra with a vibrational-type behavior below the pairing gap. These collective states have been investigated in various ways, both experimentally and theoretically, as collective phonons and their multiples. In theoretical analysis, by means of quasiparticle description of the superfluid nuclei, these vibrational phonons have been taken as coherent combinations of two-quasiparticle states. Collective (an)harmonic vibrational states built of two or even three of these phonons (two-phonon and three-phonon states) have been studied systematically by phenomenological analysis along the years. Microscopic description of these multiphonon states involves configuration mixing of two-, four-, six-quasiparticle, etc. degrees of freedom.

Describing the low-energy collective phonons of the medium-heavy and heavy open-shell nuclei within the framework of the QRPA is convenient from the microscopic point of view. The QRPA describes harmonic small-amplitude vibrations around a spherical nuclear shape [1]– [3] leading to collective low-energy solutions of the QRPA equations which can be combined to multiphonon states. In our theoretical framework, MAVA [4], the two-phonon states are built of the QRPA phonons, and the one-phonon and two-phonon states can interact among each other through the H_{31} part of the quasiparticle representation of the residual two-body Hamiltonian. The action of H_{31} and the metric matrix, containing the overlaps between the two-phonon states, breaks the degeneracy of the two-phonon triplet. The Pauli principle is included by diagonalizing first the metric matrix, thus creating a complete orthonormal basis, and then diagonalizing the residual Hamiltonian in this basis.

The selected chains of isotopes are known to contain clear indications of vibrational excitations in their low-energy spectra. Microscopical description of these states and their electromagnetic decays has been a major challenge already for a long time. The experimental data suggests that $^{110-120}$Cd are anharmonic vibrators where the anharmonicities are in a position to push the three-phonon 0^+ state below the two-phonon 0^+ state for the heavier $^{116-120}$Cd. The key in the many discussions concerning Ruthenium isotopes is the phenomena of shape transitions and shape coexistence along the isotopic chain of ruthenium isotopes. Also the structure of Mo isotopes undergoes a change from a spherical nucleus ^{92}Mo to a rotational-like at ^{104}Mo. In addition, the excited 0^+ state observed at an energy near the 2_1^+ state in both ^{98}Mo and ^{100}Mo is a signature of shape coexistence.

2. Theoretical background

2.1. *QRPA*

We will describe collective low-lying excitations in even-even nuclei in terms of single-particle eigenstates in a given spherically symmetric mean field. These states are labeled by spherical single-particle quantum numbers, i.e., isospin, energy eigenvalue, angular momentum, total spin and its projection. To denote them we use the following shorthand notation

$$a^{\dagger}_{\tau\epsilon ljm} \rightarrow a^{\dagger}_{jm}. \tag{1}$$

For the phonon operator describing collective excitations in even-even nuclei within the QRPA we use the restricted presentation written as

$$
\begin{aligned}
Q^{\dagger}_{a_2\alpha_2\mu_2} = \sum_{\tau=\pi,\nu}\sum_{j_1\leq j_2} & [X_\tau(j_1j_2;a_2\alpha_2)\bar{A}^{\dagger}_{\alpha_2\mu_2}(j_1j_2) \\
& - Y_\tau(j_1j_2;a_2\alpha_2)(-)^{\alpha_2-\mu_2}\bar{A}_{\alpha_2-\mu_2}(j_1j_2)],
\end{aligned}
\tag{2}
$$

where $a_2\alpha_2$ denotes the two-particle quantum numbers: the energy eigenvalue, angular momentum (and parity). Now the normalized pair-creation operator is defined by coupling two particle-creation operators to some angular momentum, i.e,

$$
\begin{aligned}
A^{\dagger}_{\alpha_2\mu_2}(j_1j_2) &= \frac{1}{\Delta_{j_1j_2}}\sum_{m_1+m_2=\mu_2}\langle j_1m_1;j_2m_2|\alpha_2\mu_2\rangle a^{\dagger}_{j_1m_1}a^{\dagger}_{j_2m_2}, \\
\Delta_{j_1j_2} &= \sqrt{1+\delta(-)^{\alpha_2}}.
\end{aligned}
\tag{3}
$$

The boson commutation rules for the QRPA phonons lead to the usual orthonormality relations between QRPA amplitudes allowing us to invert Eq. (2) in a standard way. The QRPA equation of motion

$$
[\hat{H}, Q^{\dagger}_{a_2\alpha_2\mu_2}] = E_{a_2\alpha_2}Q^{\dagger}_{a_2\alpha_2\mu_2} \tag{4}
$$

leads to the following matrix equation:

$$
\begin{pmatrix} A & B \\ B & -A \end{pmatrix}\begin{pmatrix} X_\tau \\ Y_\tau \end{pmatrix} = E_{a_2\alpha_2}\begin{pmatrix} X_\tau \\ Y_\tau \end{pmatrix}, \tag{5}
$$

where the matrix elements are given in terms of symmetrized double commutators between the Hamiltonian and basis pair operators. Here we take the BCS state to be the vacuum on which the matrix elements are estimated. The A and B matrices contain the $\pi\pi\pi\pi$, $\pi\pi\nu\nu$ and $\nu\nu\nu\nu$ parts yielding to eigenvectors containing both the proton-proton and neutron-neutron two-quasiparticle amplitudes.

2.2. *MAVA*

In the MAVA the two-phonon states are built in terms of the QRPA degrees of freedom using the equation-of-motion technique of Rowe [2]:

$$
\left[\hat{H}, \Gamma^{\dagger}_{a_4\alpha_4\mu_4}\right] = E_{a_4\alpha_4}\Gamma^{\dagger}_{a_4\alpha_4\mu_4}, \tag{6}
$$

where the ansatz wave function is now taken to be of the form

$$
\begin{aligned}
\Gamma^{\dagger}_{a_4\alpha_4\mu_4} = & \sum_{a_2} Z_1(a_2;a_4\alpha_4)Q^{\dagger}_{a_2\alpha_4\mu_4} \\
& + \sum_{a_2\alpha_2\leq b_2\beta_2} Z_2(a_2\alpha_2b_2\beta_2;a_4\alpha_4)\left(Q^{\dagger}_{a_2\alpha_2}Q^{\dagger}_{b_2\beta_2}\right)_{\alpha_4\mu_4}.
\end{aligned}
\tag{7}
$$

The two-phonon part contains the angular momentum (and parity) α_4 and the z-projection μ_4. The quantum number a_4 indicates the eigenvalue index of the final diagonalized MAVA wave function. Eq.(6) combined with the ansatz (7) leads to a system of equations

$$\begin{pmatrix} E_{a_2\alpha_4}\delta_{a_2a'_2} & H_{12}(a_2;a'_2\alpha'_2b'_2\beta'_2) \\ H_{21}(a_2\alpha_2b_2\beta_2;a'_2) & H_{22}(a_2\alpha_2b_2\beta_2;a'_2\alpha'_2b'_2\beta'_2) \end{pmatrix} \begin{pmatrix} Z_1(a'_2;a_4\alpha_4) \\ Z_2(a'_2\alpha'_2b'_2\beta'_2;a_4\alpha_4) \end{pmatrix} =$$
$$E_{a_4\alpha_4} \begin{pmatrix} \delta_{a_2a'_2} & 0 \\ 0 & I_{\alpha_4}(a_2\alpha_2b_2\beta_2;a'_2\alpha'_2b'_2\beta'_2) \end{pmatrix} \begin{pmatrix} Z_1(a'_2;a_4\alpha_4) \\ Z_2(a'_2\alpha'_2b'_2\beta'_2;a_4\alpha_4) \end{pmatrix} \tag{8}$$

Here the metric matrix,

$$I_{\alpha_4}(a_2\alpha_2b_2\beta_2;a'_2\alpha'_2b'_2\beta'_2) = \langle| \left[(Q_{b_2\beta_2}Q_{a_2\alpha_2})_{\alpha_4}, \left(Q^\dagger_{a'_2\alpha'_2}Q^\dagger_{b'_2\beta'_2} \right)_{\alpha_4} \right] |\rangle \tag{9}$$

is a relevant part in the equations and consists of overlaps between all the two-phonon combinations included in the calculations. The angular-momentum dependence of the metric matrix contributes to the splitting of the two-phonon-like MAVA states. The Pauli principle is preserved in diagonalization of the metric matrix. The matrix element that gives the main contribution to the two-phonon energies can be written as a sum of energies of its one-phonon constituents and the metric matrix:

$$H_{22}(a_2\alpha_2b_2\beta_2;a'_2\alpha'_2b_2\beta'_2) = \langle \left[(Q_{b_2\beta_2}Q_{a_2\alpha_2})_{\alpha_4}, H, \left(Q^\dagger_{a'_2\alpha'_2}Q^\dagger_{b'_2\beta'_2} \right)_{\alpha_4} \right] \rangle =$$
$$\frac{1}{2}(E_{a_2\alpha_2} + E_{b_2\beta_2} + E_{a'_2\alpha'_2} + E_{b'_2\beta'_2}) I_{\alpha_4}(a_2\alpha_2b_2\beta_2;a'_2\alpha'_2b'_2\beta'_2). \tag{10}$$

2.3. *MAVA-2*

Considering properly the fermionic structure of QRPA phonons, i.e. calculating the exact commutation relation between QRPA phonons, one obtains the following relation for $H_{22}(a_2\alpha_2b_2\beta_2;a'_2\alpha'_2b'_2\beta'_2)$:

$$\begin{aligned} H_{22}(a_2\alpha_2b_2\beta_2;a'_2\alpha'_2b'_2\beta'_2) &= \frac{1}{2}(E_{a_2\alpha_2} + E_{b_2\beta_2} + E_{a'_2\alpha'_2} + E_{b'_2\beta'_2}) \\ &\times \left\{ I_{\alpha_4}(a_2\alpha_2b_2\beta_2;a'_2\alpha'_2b'_2\beta'_2) - \langle Q_{b_2\beta_2}\hat{N}_{a_2\alpha_2;a'_2\alpha'_2}Q^\dagger_{b'_2\beta'_2}\rangle_{\alpha_4} \right\} \\ &+ \sum_{c_2} E_{c_2\gamma_2} \langle Q_{b_2\beta_2}\hat{N}_{a_2\alpha_2;c_2\gamma_2}\hat{N}_{c_2\gamma_2;a'_2\alpha'_2}Q^\dagger_{b'_2\beta'_2}\rangle_{\alpha_4}, \end{aligned} \tag{11}$$

which introduces the expectation values of the $\hat{N}$ and $\hat{N}\hat{N}$ operators on the BCS vacuum. This improved version is called MAVA-2. More detailed derivation of MAVA and MAVA-2 equations can be found in Refs. 4 and 5.

2.4. *Electromagnetic transitions*

The reduced matrix element of a two-body transition operator connecting a MAVA or MAVA-2 eigenstate with the ground state is given in terms of the one-phonon components of the eigenstate:

$$\langle a_4\alpha_4||T_{\alpha_4}||0\rangle = \hat{\alpha_4}\sum_{a_2} Z_1(a_2;a_4\alpha_4)\sum_{\tau}\sum_{i\leq j}\xi_{\tau\alpha_4}(ij)[X_\tau(ij;a_2\alpha_4)+Y_\tau(ij;a_2\alpha_4)], \tag{12}$$

where ξ is defined as

$$\xi_{\tau J}(i_\tau j_\tau) = \frac{e_\tau}{\hat{J}\Delta_{ij}}\langle \tau i||T_J||\tau j\rangle(u_i v_j + v_i u_j). \tag{13}$$

Matrix element connecting two eigenstates is a superposition of components containing products between the one-phonon and two-phonon amplitudes multiplied by the metric matrix

$$\begin{aligned}\langle a_4'\alpha_4'||T_{\gamma_2}||a_4\alpha_4\rangle &= \sum_{\tau}\sum_{i\leq j}\sum_{c_2}\xi_{\tau\gamma_2}(ij)[X_\tau(ij;c_2\gamma_2)+Y_\tau(ij;c_2\gamma_2)]\\ &\times\Big\{\hat{\alpha}_4'\sum_{a_2}\sum_{a_2'\alpha_2'\leq b_2'\beta_2'} Z_1(a_2;a_4\alpha_4)Z_2(a_2'\alpha_2'b_2'\beta_2';a_4'\alpha_4')I_{\alpha_4'}(a_2'\alpha_2'b_2'\beta_2';a_2\alpha_4c_2\gamma_2)\\ &+\hat{\alpha}_4\sum_{a_2\alpha_2\leq b_2\beta_2} Z_1(a_2';a_4'\alpha_4')Z_2(a_2\alpha_2b_2\beta_2;a_4\alpha_4)I_{\alpha_4}(a_2'\alpha_4'c_2\gamma_2;a_2\alpha_2b_2\beta_2)\Big\}.\end{aligned} \tag{14}$$

It should be stressed that the metric matrix, through its consideration of the Pauli principle, is the most important ingredient in our approach for description of both eigenstates and electromagnetic transitions.

3. Numerical application

Our microscopic approach starts from a single-particle basis of suitable size. Eigenenergies of the spherical Woods-Saxon nuclear mean field [6], with the Coulomb terms included, are used in the present work. However, the single-particle wave functions are taken to be eigenstates of a spherical harmonic-oscillator with a suitable oscillator constant. This is a good approximation for bound states in nuclei. As the residual two-body interaction the G-matrix elements of the Bonn one-boson-exchange potential [7] were used when calculating the BCS occupation amplitudes and the QRPA eigenstates. Different channels of this interaction are scaled by constants as described in Refs. 8 and 9. The pairing strength for protons and neutrons was adjusted by requiring the calculated pairing gaps to reproduce the empirical ones obtained from the proton and neutron separation energies [10].

The G-matrix elements for the $J^{\pi} = 2^{+}, 4^{+}$ multipoles in the QRPA calculations have been parametrized by two parameters [9], namely the particle-hole parameter, g_{ph}, and the particle-particle parameter, g_{pp}. The value of the particle-hole parameter affects the energy of the lowest 2^{+} and 4^{+} states in the QRPA calculation. These parameters are chosen so that the corresponding experimental energies are correctly reproduced.

The electric decay amplitudes can be used to produce the $B(\mathrm{E2})$ values which can be compared with experimental data. To do this comparison we adopt proton and neutron effective charges, e_{p} and e_{n}, which reproduce the measured $B(E2; 2_1^+ \to 0_1^+)$ value.

4. Results and discussion

4.1. *Cadmium isotopes*

We present our results for the MAVA energies and $B(E2)$ values in Tables 1–3. Table 1 shows the theoretical and experimental energies of the two-phonon type of levels in $^{110-120}$Cd. As one can see for $^{110-116}$Cd the correspondence is rather nice. However, the ordering of the states in not correct and this can be explained by the missing of the three-phonon and/or intruder degrees of freedom. For ^{118,120}Cd the calculated 0^{+} state drops in energy due to increasing collectivity of the lowest 2^{+} phonon of the QRPA. We are also giving two possible experimental values for the energy of the $0^{+}_{2-\mathrm{ph}}$ state in $^{116-120}$Cd isotopes to allow for the discussion of intruder states. In Tables 2 and 3 we summarize the experimental and theoretical

Table 1. Experimental and theoretical energies of two-phonon states in the $^{110-120}$Cd isotopes.

	$0^{+}_{2-\mathrm{ph}}$		$2^{+}_{2-\mathrm{ph}}$		$4^{+}_{2-\mathrm{ph}}$	
A	exp	th	exp	th	exp	th
110	1.4731	1.250	1.4758	1.479	1.5425	1.425
112	1.2245	1.218	1.31241	1.505	1.4156	1.369
114	1.1345	1.034	1.2097	1.361	1.2837	1.236
116	1.2826 1.3803	1.111	1.2130	1.204	1.2194	1.145
118	1.2856 1.6151	0.887	1.2696	1.286	1.1649	1.282
120	1.3889 1.7449	0.995	1.3231	1.373	1.2037	1.347

information about the values of the ratio

$$R_{if}(E2) = \frac{B(E2; J_i^+ \to J_f^+)}{B(E2; 2_1^+ \to 0_+^+)}. \tag{15}$$

Comparison of theory and data suggests that for $^{110-114}$Cd the experimental 0_2^+, 2_2^+ and 4_1^+ states correspond to the calculated triplet of anharmonic two-phonon like states. Judging by the transitions from two-phonon type states to 2_1^+ state one can say that ^{112}Cd is more pure vibrator than theory predicts and ^{114}Cd is less pure vibrator than theory predicts. For ^{116}Cd the choise 0_3^+,2_2^+ and 4_1^+ would best correspond to calculated two-phonon triplet and the experimental 0_2^+ state is explained as deformed intruder state. For the heavier cadmiums there is not enough experimental data to make any definite conclusions. The availability of experimental data al-

Table 2. Experimental and theoretical ratios R_{if} for the $^{110-114}$Cd isotopes.

	^{110}Cd		^{112}Cd		^{114}Cd	
Transition	th	exp [11]	th	exp [12]	th	exp [13]
$2_1^+ \to 0_1^+$	1	1	1	1	1	1
$0_{2-ph}^+ \to 2_1^+$	1.463		1.349	1.69(49)	1.333	0.88(12)
$2_{2-ph}^+ \to 2_1^+$	1.535	1.09(20)	1.551	0.50(11)	1.306	0.92(12)
$4_{2-ph}^+ \to 2_1^+$	1.546	1.69(25)	1.506	2.02(23)	1.590	2.16(3) [14]
$2_{2-ph}^+ \to 0_1^+$	0.025	0.049(8)	0.059	0.020(4)	0.108	0.016(3)
$4_2^+ \to 4_{2-ph}^+$	0.000		0.000		0.010	0.41(14)
$4_2^+ \to 2_{2-ph}^+$	0.006		0.003		0.017	1.03(49)
$4_2^+ \to 2_1^+$	0.169		0.218		0.161	0.013(4)

Table 3. The same as Table 2 for the $^{116-120}$Cd isotopes.

	^{116}Cd		^{118}Cd		^{120}Cd	
Transition	th	exp [15]	th	exp [16]	th	exp
$2_1^+ \to 0_1^+$	1	1	1	1	1	1
$0_{2-ph}^+ \to 2_1^+$	1.429	0.92(36) [17]	1.402		1.402	
$2_{2-ph}^+ \to 2_1^+$	1.097	1.00(7)	1.484	1.00(10)	1.549	
$4_{2-ph}^+ \to 2_1^+$	1.590	1.70(56) [17]	1.651		1.583	
$2_{2-ph}^+ \to 0_1^+$	0.138	0.051(4)	0.083	0.058(6)	0.036	
$4_2^+ \to 4_{2-ph}^+$	0.001	0.0565(4)	0.000	0.09(9)	0.000	
$4_2^+ \to 2_{2-ph}^+$	0.008	1.00(7)	0.003	1.00(10)	0.000	
$4_2^+ \to 2_1^+$	0.035	0.005(1)	0.019	0.01(1)	0.043	

lows us to make some observations about the decay of the theoretical 4_2^+ state. The theoretical 4_2^+ is a typical two-quasiparticle state described by a noncollective QRPA phonon. The agreement between the data and the calculations is much worse for the decay of this state than for the other low-lying states. An explanation to this discrepancy could be that the measured 4_2^+ state is not of two-quasiparticle character but rather of a three-phonon or intruder character. More detailed analysis of the results can be found in Refs. 18 and 19. As a conclusion one can say that the deformed intruder states have a small effect on the $B(E2)$ values but a considerable effect on the energies.

4.2. *Ruthenium isotopes*

The results for $^{98-106}$Ru are presented in Figs. 1 and 2 and Table 4. In the case of ^{98}Ru all the $B(E2)$ values are reproduced extremely well including the very weak transition $2^+_{2-\mathrm{ph}} \to 0^+_{\mathrm{g.s}}$ and the energy correspondence is especially satisfying for the $4^+_{2-\mathrm{ph}}$ state. This means that ^{98}Ru can be considered to be a good anharmonic vibrator of (nearly) spherical shape.

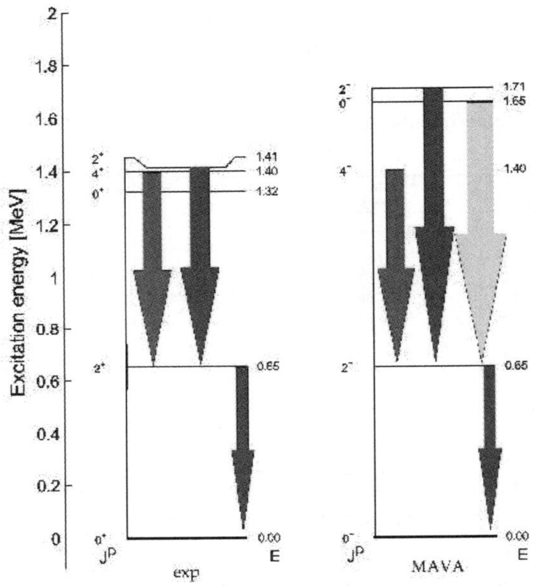

Fig. 1. Experimental [20] and theoretical low-energy spectra of ^{98}Ru.

For the other Rutheniums the situation is worse. Judging by the transitions

Table 4. Experimental and theoretical ratios R_{if} for $^{102-106}$Ru.

TRANSITION	102 MAVA	102 exp [22]	104 MAVA	104 exp [23]	106 MAVA	106 exp
$2^+_1 \to 0^+_1$	1	1	1	1	1	1
$0^+_{2-\mathrm{ph}} \to 2^+_1$	1.432	0.80(9)	1.297	0.42(7)	1.296	
$2^+_{2-\mathrm{ph}} \to 2^+_1$	1.433	0.62(8)	1.317	0.63(11)	1.330	
$4^+_{2-\mathrm{ph}} \to 2^+_1$	1.231	1.5(3)	1.141	1.2(3)	1.174	
$2^+_{2-\mathrm{ph}} \to 0^+_1$	0.050	0.0250(8)	0.068	0.035(9)	0.086	

from the two-phonon-type states to the 2^+_1 one can say that $^{100-106}$Ru seem to be less pure vibrators than theory predicts. It is instructive to look at the evolution of the vibrational excitations to rotational ones within the Sheline-Sakai scheme. In this scheme the two-phonon 0^+, 2^+ and 4^+ states of the anharmonic spherical vibrator evolve to the corresponding rotational states of a well-deformed nucleus. These states are the 0^+ β-band head,

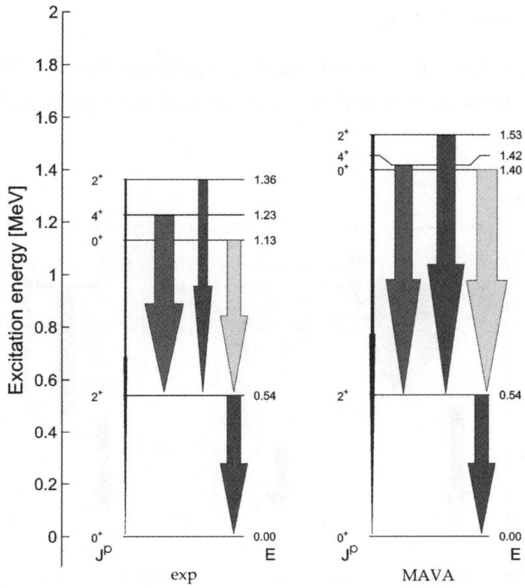

Fig. 2. Experimental [21] and theoretical low-energy spectra of ^{100}Ru.

2^+ γ-band head and 4^+ member of the ground-state band. In the rotational regime the 2^+ member of the ground-state band corresponds to the 2^+_1 vibrational state of the spherical anharmonic vibrator and the γ decay feeding of this state from the 0^+, 2^+ and 4^+ states above can be classified to interband and intraband transitions. The intraband transition $4^+ \to 2^+_1$ is typically strong reproducing the Alaga Rule:

$$\frac{B(E2; 4^+ \to 2^+_1)}{B(E2; 2^+_1 \to 0^+_{g.s})} = 25 \begin{pmatrix} 4 & 2 & 2 \\ 0 & 0 & 0 \end{pmatrix}^2 = 1.43. \tag{16}$$

The Experimental data would point to increasing rotational character in ruthenium nuclei as function of A. This shows up as weakening of the interband type of transitions which can not be reproduced by MAVA, and persistence of the strong intraband transitions. Thus, according to our calculations, ^{100}Ru can be interpreted as transitional nucleus between the anharmonic spherical vibrator ^{98}Ru and the quasirotational heavier $^{102-106}$Ru isotopes. Deeper analysis of the calculations can be found in Refs. 24 and 19.

4.3. *Molybdenum isotopes*

We summarize the relevant information concerning the $^{94-100}$Mo in Figs. 3–6. In the calculations we have used both MAVA and MAVA-2. The overall

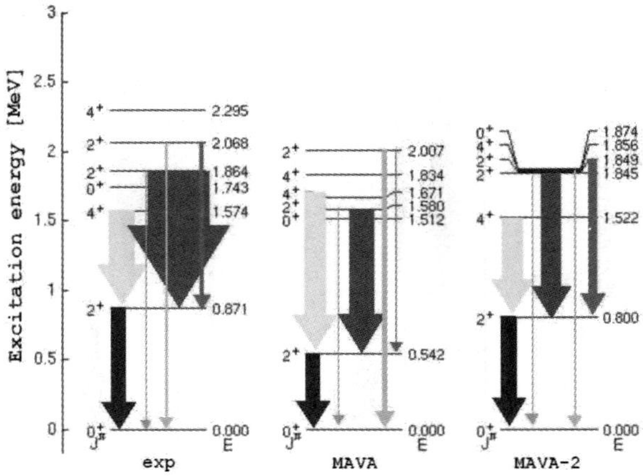

Fig. 3. Experimental [25] and theoretical low-energy spectra of ^{94}Mo.

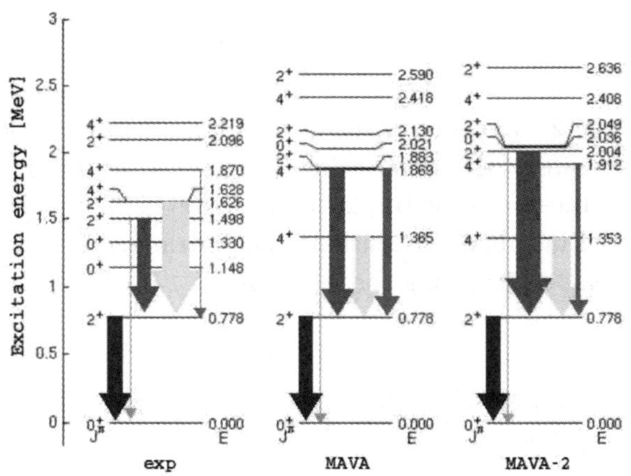

Fig. 4. Experimental [26] and theoretical low-energy spectra of ^{96}Mo.

structure of the energy spectrum is acceptable for ^{94}Mo. Available data corresponds rather well to the calculated $B(E2)$ values and the 2_2^+ and 4_1^+ states are nicely reproduced. One can also say that even though MAVA-2 does not reproduce the correct energy splitting it does reproduce most of the energies more accurately than MAVA as well as the transitions to the ground state. For the ^{96}Mo the case is worse. The first 4^+ state comes too low in energy and other states are pushed too high although the 4_2^+ state is well reproduced. Again MAVA-2 reproduces transitions from 4^+ states

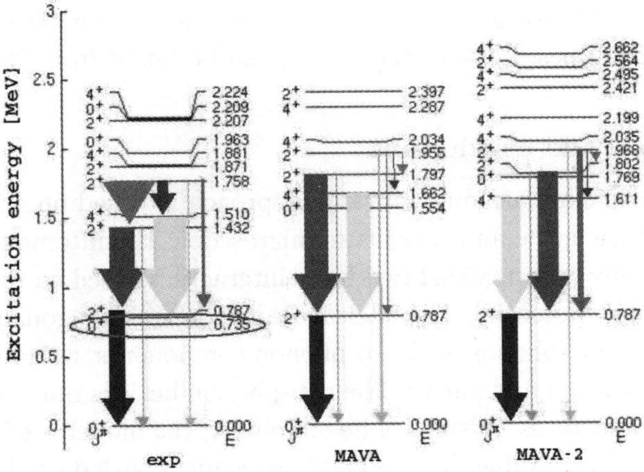

Fig. 5. Experimental [20] and theoretical low-energy spectra of ^{98}Mo.

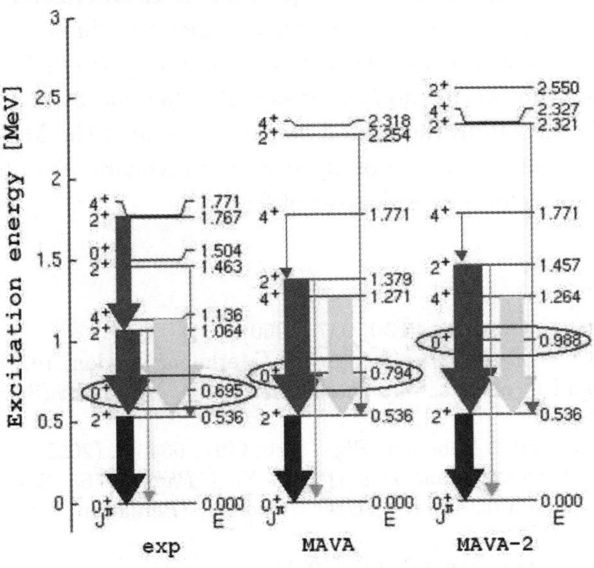

Fig. 6. Experimental [21] and theoretical low-energy spectra of ^{100}Mo.

better. Taking the lowest experimental 0^+ state to be an intruder state leads to a good agreement with the data in the case of ^{98}Mo. MAVA-2 gives a better correspondence with experimental 4_1^+ and 0_2^+ states and reproduces transitions to ground state better. Also the structure of the theoretical spectrum of ^{100}Mo is quite along the lines of the experimental one: the low-lying 0^+ state is quite nicely reproduced and the states are grouped in pairs like in the data. Available data corresponds rather well to the calculated $B(E2)$ values. This suggests that the $^{94-100}$Mo are throughout anharmonic vibrators and no clear sign of shape transition can be seen as in the case of Ruthenium isotopes. More of the results can be found from Ref. 5.

5. Summary and conclusions

The microscopic anharmonic vibrator approach is based on a large single-particle valence space and a realistic microscopic Hamiltonian, using phenomenologically renormalized two-body interaction based on the Bonn one-boson-exchange potential. The nuclear Hamiltonian is diagonalized in a basis containing one-phonon and two-phonon components, coupled to a given angular momentum and parity. The two-phonon basis is built using 2^+ and 4^+ QRPA eigenstates. In spite of its simplicity, the model predicts energies and ratios of $B(E2)$ values in reasonable agreement with data. Furthermore, we can make conclusions about three-phonon states and intruder states and we can see pieces of evidence of shape transitions and shape coexistence. In the near future MAVA will be used to study the properties of other isotopic chains with known two-phonon states as well as beta decay and double-beta decay transitions within the proton-neutron version of the MAVA. Further extensions are the description of alpha decays involving two-phonon states and addition of deformation degrees of freedom.

References

1. M. Baranger, *Phys. Rev.* **120**, 957 (1960).
2. D.J. Rowe, *Nuclear Collective Motion* (Methuen, London, 1970).
3. P. Ring and P. Schuck, *The Nuclear Many-Body Problem* (Springer-Verlag, Berlin, 1980).
4. D.S. Delion and J. Suhonen, *Phys. Rev.* **C67**, 034301 (2003).
5. J. Kotila, J. Suhonen and D.S. Delion, *Nucl. Phys.* **A765**, 354 (2006).
6. A. Bohr, B.R. Mottelson, *Nuclear Structure*, (Benjamin, New York, 1969), Vol. I.
7. K. Holinde, *Phys. Rep.* **68**, 121 (1981).
8. J. Suhonen, T. Taigel and A. Faessler, *Nucl. Phys.* **A486**, 91 (1988).
9. J. Suhonen, *Nucl. Phys.* **A563**, 205 (1993).
10. G. Audi and A.H. Wapstra, *Nucl. Phys.* **A565**, 1 (1993).
11. D. De Frenne and E. Jacobs, *Nucl. Data Sheets* **89**, 534 (2000).
12. D. De Frenne and E. Jacobs, *Nucl. Data Sheets* **79**, 668 (1996).
13. J. Blachot and G. Margueir, *Nucl. Data Sheets* **75**, 750 (1995).
14. R. Julin, *Physica Scripta* **T56**, 151 (1995).
15. Youbao Wang *et. al.*, *Phys. Rev.* **C64**, 054315 (2001).
16. A. Aprahamian, D. Brenner, R.F. Casten, R.F. Gill and A. Piotrowski, *Phys. Rev. Lett.* **59**, 535 (1987).
17. J. Blachot, Nucl. *Data Sheets* **92**, 473 (2001).

18. J. Kotila, J. Suhonen and D.S. Delion, *Phys. Rev.* **C68**, 014307 (2003).
19. J. Kotila, J. Suhonen and D. S. Delion, *Czech. J. Phys.* **56**, 473 (2006).
20. B. Singh, *Nucl. Data Sheets* **84**, 565 (1998).
21. B. Singh, *Nucl. Data Sheets* **81**, 1 (1997).
22. D. De Frenne and E. Jacobs, *Nucl. Data Sheets* **83**, 535 (1998).
23. J. Blachot, *Nucl. Data Sheets* **64**, 1 (1991).
24. J. Kotila, J. Suhonen and D.S. Delion, *Phys. Rev.* **C68**, 054322 (2003).
25. J.K. Tuli, *Nucl. Data Sheets* **66**, 1 (1992).
26. L.K. Peker, *Nucl. Data Sheets* **68**, 165 (1993).

I.3 Relativistic Nuclear Structure

Covariant Density Functional Theory: Description of Rare Nuclei

G. A. Lalazissis

Department of Theoretical Physics, Aristotle University of Thessaloniki
GR-54124, Thessaloniki, Greece
** E-mail: glalazis@auth.gr*

Relativistic Hartree-Bogoliubov (RHB) theory is a powerful tool for the description of the properties of exotic systems. It is described in terms of a covariant density functional and with the use of only a limited number of phenomenological parameters, the theory is able to provide a unified description of nuclear structure properties throughout the periodic table. Here, the covariant density functional theory in nuclei, its various extensions and several applications for nuclei away from stability line, are presented.

1. Introduction

Models based on concepts of nonrenormalizable effective relativistic field theories and density functional theory provide a very interesting theoretical framework for studies of nuclear structure phenomena far from the valley of β-stability. A well known example of an effective theory of nuclear structure is Quantum Hadrodynamics (QHD) [1], a field theoretical framework of Lorentz-covariant, meson-nucleon or point-coupling models of nuclear dynamics. A variety of nuclear phenomena have been described with QHD models: nuclear matter, properties of finite spherical and deformed nuclei, hypernuclei, neutron stars, nucleon-nucleus and electron-nucleus scattering, relativistic heavy-ion collisions. In particular, structure models based on the relativistic mean-field (RMF) approximation have been successfully applied in the description of properties of spherical and deformed β-stable nuclei, and more recently in studies of exotic nuclei far from the valley of beta stability. The RMF models provide a microscopically consistent, and yet simple and economical treatment of the nuclear many-body problem. By adjusting just a few model parameters, coupling constants and effective masses, to global properties of simple, spherical and stable nuclei, it has been possible to describe in detail a variety of nuclear structure phenomena over the whole periodic table, from light nuclei to superheavy elements. Detailed reviews of QHD and its applications can be found in Refs. [1–4]. In the last five or six years, in particular, the relativistic Hartree-Bogoliubov (RHB) model [5] has been employed in analysis of structure phenomena in exotic nuclei far from the valley of β-stability. This model represents a relativistic extension of the conventional Hartree-Fock-Bogoliubov framework, and provides a unified description of mean-field and pairing correlations.

In this article we review some recent applications of the RHB model and relativistic (Q)RPA model in the description of rare nuclei. In section 2 we outline the RHB model. Sections 3 and 4 contain some applications of the RHB model for ground state properties of exotic nuclei close to drips lines, while in section 5 an application of the relativistic RPA model for neutron rich nuclei is presented. Finally, section 6 summarizes our main conclusions.

2. The relativistic Hartree-Bogoliubov model

In the framework of models based on the relativistic mean-field approximation the nucleus is described as a system of Dirac nucleons that interact in a relativistic covariant manner through the exchange of virtual mesons. The isoscalar scalar σ-meson, the isoscalar vector ω-meson, and the isovector vector ρ-meson build the minimal set of meson fields that is necessary for a quantitative description of bulk and single-particle nuclear properties [1–5]. The model is defined by the Lagrangian density

$$\mathcal{L} = \mathcal{L}_N + \mathcal{L}_m + \mathcal{L}_{int}. \tag{1}$$

$\mathcal{L}_N$ denotes the Lagrangian of the free nucleon

$$\mathcal{L}_N = \bar{\psi}\left(i\gamma^\mu \partial_\mu - m\right)\psi, \tag{2}$$

where m is the bare nucleon mass and ψ denotes the Dirac spinor. $\mathcal{L}_m$ is the Lagrangian of the free meson fields and the electromagnetic field

$$\begin{aligned}\mathcal{L}_m = &\frac{1}{2}\partial_\mu\sigma\partial^\mu\sigma - \frac{1}{2}m_\sigma^2\sigma^2 - \frac{1}{4}\Omega_{\mu\nu}\Omega^{\mu\nu} + \frac{1}{2}m_\omega^2\omega_\mu\omega^\mu \\ &-\frac{1}{4}\vec{R}_{\mu\nu}\vec{R}^{\mu\nu} + \frac{1}{2}m_\rho^2\vec{\rho}_\mu\vec{\rho}^\mu - \frac{1}{4}F_{\mu\nu}F^{\mu\nu},\end{aligned} \tag{3}$$

with the corresponding masses m_σ, m_ω, m_ρ, and $\Omega_{\mu\nu}$, $\vec{R}_{\mu\nu}$, $F_{\mu\nu}$ are field tensors

$$\begin{aligned}\Omega_{\mu\nu} &= \partial_\mu\omega_\nu - \partial_\nu\omega_\mu \\ \vec{R}_{\mu\nu} &= \partial_\mu\vec{\rho}_\nu - \partial_\nu\vec{\rho}_\mu \\ F_{\mu\nu} &= \partial_\mu A_\nu - \partial_\nu A_\mu .\end{aligned} \tag{4}$$

The minimal set of interaction terms is contained in $\mathcal{L}_{int}$

$$\mathcal{L}_{int} = -\bar{\psi}\Gamma_\sigma\sigma\psi - \bar{\psi}\Gamma^\mu_\omega\omega_\mu\psi - \bar{\psi}\vec{\Gamma}^\mu_\rho\vec{\rho}_\mu\psi - \bar{\psi}\Gamma^\mu_e A_\mu\psi. \tag{5}$$

The vertices read

$$\Gamma_\sigma = g_\sigma, \quad \Gamma^\mu_\omega = g_\omega\gamma^\mu, \quad \vec{\Gamma}^\mu_\rho = g_\rho\vec{\tau}\gamma^\mu, \quad \Gamma^m_e = e\frac{1-\tau_3}{2}\gamma^\mu, \tag{6}$$

with the coupling constants g_σ, g_ω, g_ρ and e. This simple linear model, however, does not provide a quantitative description of complex nuclear systems. An effective density dependence has been introduced [6] by replacing the quadratic σ-potential $\frac{1}{2}m_\sigma^2\sigma^2$ with a quartic potential

$$U(\sigma) = \frac{1}{2}m_\sigma^2\sigma^2 + \frac{g_2}{3}\sigma^3 + \frac{g_3}{4}\sigma^4 . \tag{7}$$

This potential includes the non-linear σ self-interactions with two additional parameters g_2 and g_3.

From the model Lagrangian density the classical variation principle leads to the equations of motion. The time-dependent Dirac equation for the nucleon reads

$$\left[\gamma^\mu(i\partial_\mu + V_\mu) + m + S\right]\psi = 0. \tag{8}$$

If one neglects retardation effects for the meson fields, a self-consistent solution is obtained when the time-dependent mean-field potentials

$$\begin{aligned} S(\mathbf{r},t) &= g_\sigma\sigma(\mathbf{r},t) \ , \\ V_\mu(\mathbf{r},t) &= g_\omega\omega_\mu(\mathbf{r},t) + g_\rho\vec{\tau}\vec{\rho}_\mu(\mathbf{r},t) + eA_\mu(\mathbf{r},t)\frac{(1-\tau_3)}{2} \ , \end{aligned} \tag{9}$$

are calculated at each step in time from the solution of the stationary Klein-Gordon equations

$$-\Delta\phi_m + U'(\phi_m) = \pm\left\langle\bar{\psi}\Gamma_m\psi\right\rangle , \tag{10}$$

where the (+) sign is for vector fields and the (-) sign for the scalar field. The index m denotes mesons and the photon, i.e. $\phi_m \equiv \sigma, \omega^\mu, \vec{\rho}^\mu, A^\mu$. This approximation is justified by the large meson masses. The corresponding meson exchange forces are of short range and therefore retardation effects can be neglected.

In practical applications to nuclear matter and finite nuclei, the relativistic models are used in the *no-sea* approximation: the Dirac sea of states with negative energies does not contribute to the densities and currents. For a nucleus with A nucleons

$$\langle\bar{\psi}\Gamma_m\psi\rangle = \sum_{i=1}^{A} \bar{\psi}_i(\mathbf{r},t)\Gamma_m\psi_i(\mathbf{r},t) \, , \tag{11}$$

where the summation is performed only over the occupied orbits in the Fermi sea of positive energy states. The set of coupled equations (8) and (10) define the relativistic mean field (RMF) model. In the stationary case they reduce to a nonlinear eigenvalue problem, and in the time-dependent case they describe the nonlinear propagation of the Dirac spinors in time [7].

The lowest order of the quantum field theory is the *mean-field* approximation: the meson field operators are replaced by their expectation values. The A nucleons, described by a Slater determinant $|\Phi\rangle$ of single-particle spinors ψ_i, $(i = 1, 2, ..., A)$, move independently in the classical meson fields. The sources of the meson fields are defined by the nucleon densities and currents. The ground state of a nucleus is described by the stationary self-consistent solution of the coupled system of Dirac and Klein-Gordon equations. The couplings of the meson fields to the nucleon are adjusted to reproduce the properties of nuclear matter and finite nuclei. The σ-meson approximates a large attractive scalar field that is produced by very complicated microscopic processes, such as uncorrelated and correlated two-pion exchange. The ω-meson describes the short range repulsion between the nucleons, and the ρ-meson carries the isospin quantum number. The latter is required by the large empirical asymmetry potential in finite nuclear systems. The basic ingredient of the microscopic nuclear force is the pion. In relativistic mean-field models the pion does not contribute on the Hartree level due to parity conservation.

2.1. *Covariant density functional theory with pairing*

The RHB model can be easily derived within the framework of covariant density functional theory [8]. When pairing correlations are included, the energy functional depends not only on the density matrix $\hat{\rho}$ and the meson fields ϕ_m, but in addition also on the pairing tensor:

$$E_{RHB}[\hat{\rho}, \hat{\kappa}, \phi_m] = E_{RMF}[\hat{\rho}, \phi_m] + E_{pair}[\hat{\kappa}] \, , \tag{12}$$

where $E_{RMF}[\hat{\rho}, \phi]$ is the RMF-functional is defined as:

$$E_{RMF}[\hat{\rho}, \phi_m] = \mathrm{Tr}\left[(\boldsymbol{\alpha}\cdot\boldsymbol{p} + \beta m)\hat{\rho}\right] \pm \int \left[\frac{1}{2}(\nabla\phi_m)^2 + U(\phi_m)\right] d^3r$$
$$+\mathrm{Tr}\left[(\Gamma_m\phi_m)\hat{\rho}\right]. \tag{13}$$

and the pairing energy $E_{pair}[\hat{\kappa}]$ is given by

$$E_{pair}[\hat{\kappa}] = \frac{1}{4}\mathrm{Tr}\left[\hat{\kappa}^* V^{pp}\hat{\kappa}\right]. \tag{14}$$

V^{pp} denotes a general two-body pairing interaction. The equation of motion for the generalized density matrix reads

$$i\partial_t\mathcal{R} = [\mathcal{H}(\mathcal{R}), \mathcal{R}] \ , \tag{15}$$

and the generalized Hamiltonian $\mathcal{H}$ is obtained as a functional derivative of the energy with respect to the generalized density

$$\mathcal{H}_{RHB} \ = \ \frac{\delta E_{RHB}}{\delta\mathcal{R}} \ = \ \begin{pmatrix} \hat{h}_D - m - \lambda & \hat{\Delta} \\ -\hat{\Delta}^* & -\hat{h}_D + m + \lambda \end{pmatrix}. \tag{16}$$

The self-consistent mean field $\hat{h}_D$ is the Dirac Hamiltonian. In the static case with time-reversal symmetry is has the form

$$\hat{h}_D = -i\boldsymbol{\alpha}\boldsymbol{\nabla} \ + \ V_0(\boldsymbol{r}) \ + \ \beta(m + S(\boldsymbol{r})) \tag{17}$$

and the pairing field is an integral operator of the form:

$$\Delta_{ab}(\mathbf{r}, \mathbf{r}') = \frac{1}{2}\sum_{c,d} V^{pp}_{abcd}(\mathbf{r}, \mathbf{r}')\kappa_{cd}(\mathbf{r}, \mathbf{r}'), \tag{18}$$

where a, b, c, d denote quantum numbers that specify the Dirac indices of the spinors, and $V^{pp}_{abcd}(\mathbf{r}, \mathbf{r}')$ are the matrix elements of a general two-body pairing interaction.

The stationary limit of Eq. (15) describes the ground state of an open-shell nucleus [10]. It is determined by the solutions of the Hartree-Bogoliubov equations

$$\begin{pmatrix} \hat{h}_D - m - \lambda & \hat{\Delta} \\ -\hat{\Delta}^* & -\hat{h}_D + m + \lambda \end{pmatrix} \begin{pmatrix} U_k(\mathbf{r}) \\ V_k(\mathbf{r}) \end{pmatrix} = E_k \begin{pmatrix} U_k(\mathbf{r}) \\ V_k(\mathbf{r}) \end{pmatrix}. \tag{19}$$

The chemical potential λ is determined by the particle number subsidiary condition in order that the expectation value of the particle number operator in the ground state equals the number of nucleons. The column vectors denote the quasiparticle wave functions, and E_k are the quasiparticle energies. The dimension of the RHB matrix equation is two times the dimension of the corresponding Dirac equation. For each eigenvector (U_k, V_k) with positive quasiparticle energy $E_k > 0$, there exists an eigenvector (V_k^*, U_k^*) with quasiparticle energy $-E_k$. Since the baryon quasiparticle operators satisfy fermion commutation relations, the levels E_k and $-E_k$ cannot be occupied simultaneously. For the solution that corresponds to a ground state of a nucleus with even particle number, one usually chooses the eigenvectors with positive eigenvalues E_k. The eigensolutions of Eq. (19) form a set of orthogonal (normalized) single quasiparticle states. The corresponding eigenvalues are the single quasiparticle energies. The self-consistent iteration procedure is performed in the basis of quasiparticle states. The self-consistent quasiparticle eigenspectrum is then transformed into the canonical basis of single-nucleon states. The transformation to the canonical basis determines the energies and occupation probabilities of single-nucleon states, which correspond to the self-consistent solution for the ground state of a nucleus. The RHB equations are solved self-consistently, with potentials determined in the mean-field approximation from solutions of static Klein-Gordon equations

$$\left[-\Delta + m_\sigma^2\right] \sigma(\mathbf{r}) = -g_\sigma \, \rho_s(\mathbf{r}) - g_2 \, \sigma^2(\mathbf{r}) - g_3 \, \sigma^3(\mathbf{r}) \tag{20}$$

$$\left[-\Delta + m_\omega^2\right] \omega^0(\mathbf{r}) = g_\omega \, \rho_v(\mathbf{r}) \tag{21}$$

$$\left[-\Delta + m_\rho^2\right] \rho^0(\mathbf{r}) = g_\rho \, \rho_3(\mathbf{r}) \tag{22}$$

$$-\Delta \, A^0(\mathbf{r}) = e \, \rho_p(\mathbf{r}) \tag{23}$$

for the σ-meson, the ω-meson, the $\vec{\rho}$-meson and the photon field, respectively. Because of charge conservation, only the 3-rd component of the isovector ρ-meson contributes. In the ground-state solution for an even-even nucleus there are no currents (time reversal invariance) and the corresponding spatial components $\boldsymbol{\omega}$, $\boldsymbol{\rho}_3$, $\boldsymbol{A}$ of the vector fields vanish. In nuclei with an odd number of protons or neutrons time reversal symmetry is broken, and the resulting spatial components of the meson fields play an essential role in the description of magnetic moments, and of moments of inertia in rotating nuclei. The equation for the isoscalar scalar σ-meson field contains nonlinear terms. As we have already emphasized, the inclusion of nonlinear meson self-interaction terms in meson-exchange RMF models is absolutely necessary for a quantitative description of ground-state properties of spherical and deformed nuclei [4]. The source terms in equations (20) to (23) are sums of bilinear products of baryon amplitudes

$$\rho_s(\mathbf{r}) = \sum_{k>0} V_k^\dagger(\mathbf{r})\gamma^0 V_k(\mathbf{r}) \tag{24}$$

$$\rho_v(\mathbf{r}) = \sum_{k>0} V_k^\dagger(\mathbf{r}) V_k(\mathbf{r}) \tag{25}$$

$$\rho_3(\mathbf{r}) = \sum_{k>0} V_k^\dagger(\mathbf{r})\tau_3 V_k(\mathbf{r}) \tag{26}$$

$$\rho_{\rm em}(\mathbf{r}) = \sum_{k>0} V_k^\dagger(\mathbf{r})\frac{1-\tau_3}{2} V_k(\mathbf{r}) \ , \tag{27}$$

where the sum over positive-energy states ($\sum_{k>0}$) corresponds to the *no-sea* approximation. The self-consistent solution of the Dirac-Hartree-Bogoliubov integro-differential equations and nonlinear Klein-Gordon equations for the meson fields determines the ground state of a nucleus.

Table 1. The parameter set NL3 (from Ref. [9]).

M = 939.000 (MeV)	m_ω = 782.501 (MeV)
m_ρ = 763.000 (MeV)	m_σ = 550.124 (MeV)
g_σ = 10.217	g_ω = 12.868
g_ρ = 4.474	
g_2 = -10.431 (fm^{-1})	g_3 = -28.885

The most successful RMF effective interactions are purely phenomenological, with parameters adjusted to reproduce the nuclear matter equation of state and a set of global properties of spherical closed-shell nuclei. In most applications of the RHB model, in particular, we have used the NL3 effective interaction [9] for the RMF effective Lagrangian. Properties calculated with NL3 indicate that this is probably the best nonlinear meson-exchange effective interaction so far, both for nuclei at and far away from the line of β-stability. In Ref. [10] it was suggested that the pairing part of the well known and very successful Gogny force [11] should be employed in the *pp*-channel:

$$V^{pp}(1,2) = \sum_{i=1,2} e^{-((\mathbf{r}_1-\mathbf{r}_2)/\mu_i)^2} \left(W_i + B_i P^\sigma - H_i P^\tau - M_i P^\sigma P^\tau\right) , \tag{28}$$

with the set D1S [11] for the parameters μ_i, W_i, B_i, H_i, and M_i ($i = 1,2$). This force has been very carefully adjusted to the pairing properties of finite nuclei all over the periodic table. In particular, the basic advantage of the

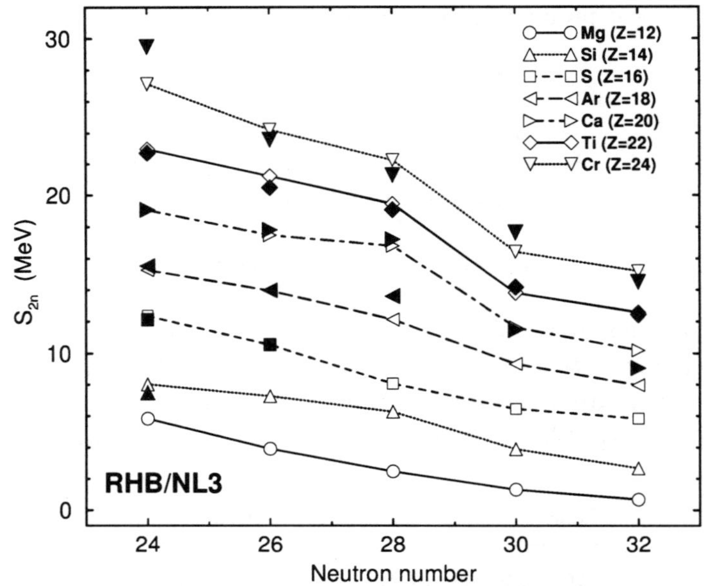

Fig. 1. Two-neutron separation energies in the $N \approx 28$ region calculated in the RHB model and compared with experimental data (filled symbols) from the compilation of G. Audi and A. H. Wapstra.

Gogny force is the finite range, which automatically guarantees a proper cut-off in momentum space.

3. Ground-state properties of neutron-rich nuclei

3.1. *Shape coexistence in the deformed $N = 28$ region*

The region of neutron-rich $N \approx 28$ nuclei exhibits many interesting phenomena: the average nucleonic potential is modified, shell effects are suppressed, large quadrupole deformations are observed as well as shape coexistence, isovector quadrupole deformations are predicted at the drip-lines. The detailed knowledge of the microscopic structure of these nuclei is also essential for the modelling of the nucleosynthesis of the heavy Ca, Ti and Cr isotopes. The structure of exotic neutron rich-nuclei with $12 \leq Z \leq 20$ and, in particular, of the light $N = 28$ nuclei has been analyzed in the RHB model. Especially interesting is the influence of the spherical shell $N = 28$

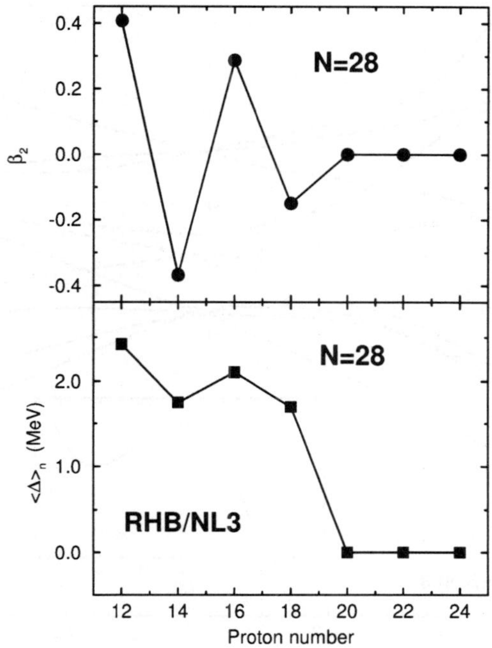

Fig. 2. Self-consistent RHB quadrupole deformations for the ground-states of the $N = 28$ isotones (top). Average neutron pairing gaps $< \Delta_N >$ as function of the proton number (bottom).

on the structure of nuclei below ^{48}Ca, the deformation effects that result from the $1f7/2 \to fp$ core breaking, and the shape coexistence phenomena predicted for these γ-soft nuclei [12].

Fig. 1 shows the two-neutron separation energies for the even-even nuclei $12 \leq Z \leq 24$ and $24 \leq Z \leq 32$. The values that correspond to the self-consistent RHB ground-states (symbols connected by lines) are compared with experimental data and extrapolated values from Ref. [13] (filled symbols). The NL3 effective interaction has been used for the RMF Lagrangian, and the Gogny interaction with the parameter set D1S for the pairing channel. The theoretical values reproduce in detail the experimental separation energies, except for ^{48}Cr. In general, it has been found that the RHB model binding energies are in very good agreement with experimental data when one of the shells (proton or neutron) is closed, or when valence protons and neutrons occupy different major shells (i.e. below and above N and/or $Z = 20$). The differences are more pronounced when both protons

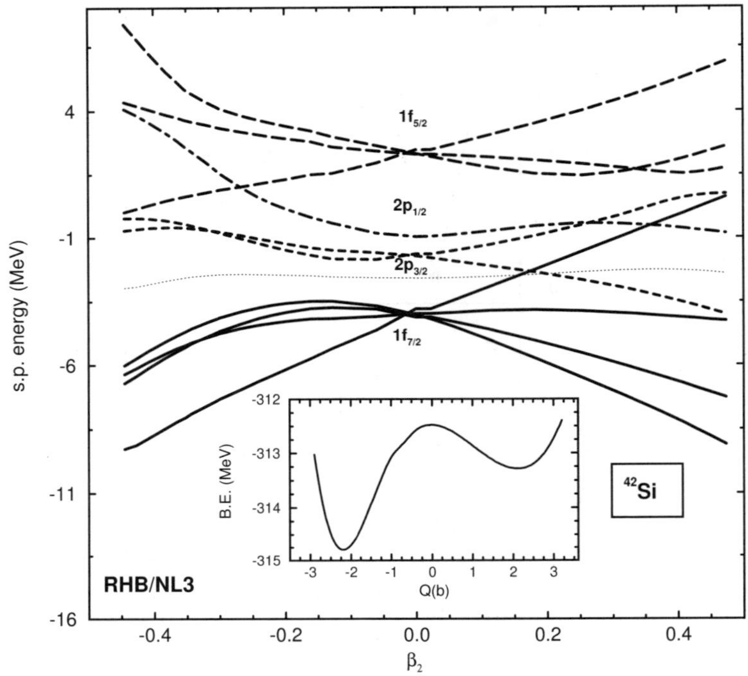

Fig. 3. The neutron single-particle levels for ^{42}Si as function of the quadrupole deformation. The energies in the canonical basis correspond to ground-state RHB solutions with constrained quadrupole deformation. The dotted line denotes the neutron Fermi level. In the insert the corresponding total binding energy curve is shown.

and neutrons occupy the same major shell, and especially for the $N = Z$ nuclei. For these nuclei additional correlations should be taken into account and, in particular, proton-neutron pairing could affect the masses.

The predicted mass quadrupole deformations for the ground states of $N = 28$ nuclei are shown in the upper panel of Fig. 2. The staggering between prolate and oblate configurations indicates that the potential is γ-soft. The absolute value of the deformation decreases towards the $Z = 20$ closed shell. Starting with Ca, the $N = 28$ nuclei are spherical in the ground state. The calculated quadrupole deformations are in agreement with previously reported theoretical results [14] (prolate for $Z = 16$, oblate for $Z = 18$), and with available experimental data: $|\beta_2| = 0.258(36)$ for ^{44}S [15,16], and $|\beta_2| = 0.176(17)$ for ^{46}Ar [17]. In the lower panel of Fig. 2 the average values of the neutron pairing gaps for occupied canonical

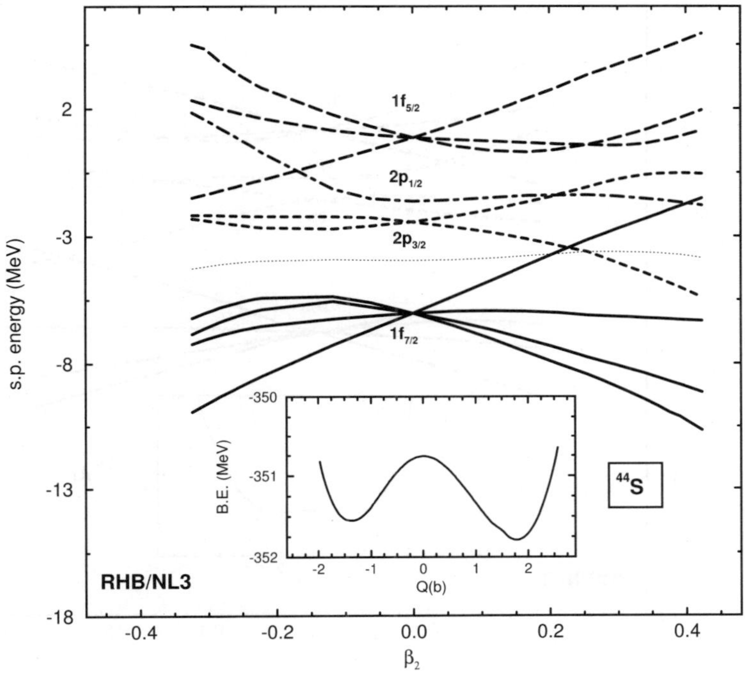

Fig. 4. Same as in Fig. 3, but for ^{44}S.

states are displayed. $< \Delta_N >$ provides an excellent quantitative measure of pairing correlations. The calculated values of $< \Delta_N > \approx 2$ MeV correspond to those found in open-shell Ni and Sn isotopes [18]. The spherical shell closure $N = 28$ is strongly suppressed for nuclei with $Z \leq 18$, and only for $Z \geq 20$ neutron pairing correlations vanish.

The fully self-consistent RHB model calculations provide the details of the single-neutron spectrum, necessary for a microscopic analysis of the formation of minima in the binding energy. Figs. 3-5 display the single-neutron levels in the canonical basis for the $N = 28$ nuclei ^{42}Si, ^{44}S, and ^{46}Ar, respectively. The single-neutron eigenstates of the density matrix are obtained from constrained RHB calculations performed by imposing a quadratic constraint on the quadrupole moment. The canonical states are plotted as function of the quadrupole deformation, and the dotted curve denotes the position of the Fermi level. In the insert the corresponding total binding energy curve is shown as function of the quadrupole moment. For

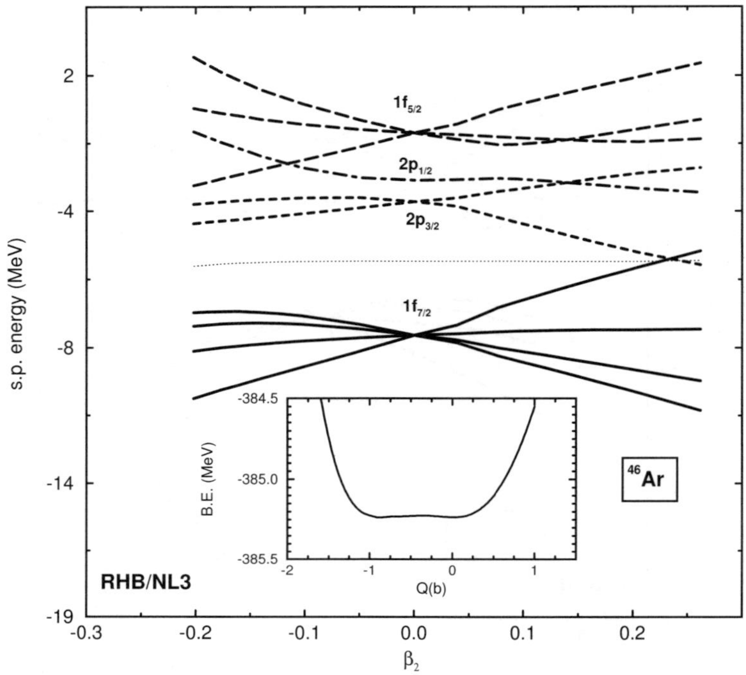

Fig. 5. Same as in Fig. 3, but for ^{46}Ar.

^{42}Si the binding energy displays a deep oblate minimum ($\beta_2 \approx -0.4$). The second, prolate minimum is found at an excitation energy of $\approx$ 1.5 MeV. Shape coexistence is more pronounced for ^{44}S. The ground state is prolate deformed, the calculated deformation is in excellent agreement with experimental data [15,16]. The oblate minimum is found only $\approx$ 200 keV above the ground state. Finally, for the nucleus ^{46}Ar a very flat energy surface is found on the oblate side. The deformation of the ground-state oblate minimum agrees with the experimental data [17], the spherical state is only few keV higher. It is also very interesting to note how the spherical gap between the $1f_{7/2}$ orbital and the $2p_{3/2}$, $2p_{1/2}$ orbitals evolves with proton number. While the gap is really strongly reduced for ^{42}Si and ^{44}S, in the $Z = 18$ isotone ^{46}Ar the spherical gap is already $\approx$ 4 MeV. From ^{48}Ca, of course, the $N = 28$ nuclei become spherical. The single-neutron canonical states in Figs. 3-5 clearly display the disappearance of the spherical $N = 28$ shell closure for neutron-rich nuclei below $Z = 18$.

3.2. *Neutron halo in light nuclei*

In some loosely bound nuclear systems at the drip-lines the neutron density distribution displays an extremely long tail: the neutron halo. The resulting large interaction cross sections have provided the first experimental evidence for halo nuclei [19]. This phenomenon has been analyzed with a variety of theoretical models [20,21]. For very light nuclei, in particular, an approach based on the separation into core plus valence space nucleons (three-body Borromean systems) has been employed. For heavier neutron-rich nuclei one expects that mean-field models should provide a better description of ground-state properties. In a mean-field approach the neutron halo and the stability against nucleon emission can only be explained with the inclusion of pairing correlations. Both the properties of single-particle states near the neutron Fermi level, and the pairing interaction, are important in the formation of the neutron halo.

In Ref. [22] the RHB framework has been applied in the analysis of the possible formation of the neutron halo in Ne isotopes. The NL3 effective interaction has been used for the RMF Lagrangian, and pairing correlations have been described by the Gogny D1S interaction. Fully self-consistent RHB calculations have been performed for the ground-states of neutron-rich Ne nuclei. In Fig. 6 the calculated *rms* radii of the Ne isotopes are plotted as functions of the neutron number. Neutron and proton rms radii are shown, as well as the $N^{1/3}$ curve normalized so that it coincides with the neutron radius in ^{20}Ne. The neutron radii follow the mean-field $N^{1/3}$ curve up to $N \approx 22$. For larger values of N the neutron radii display a sharp increase, while the proton radii stay practically constant. This sudden increase of the neutron rms radii has been interpreted as evidence for the formation of a multi-particle halo. The effect is also illustrated in the plot of proton and neutron density distributions in Fig. 7. The proton density profiles do not change with the number of neutrons, while the neutron density distributions display an abrupt change between ^{30}Ne and ^{32}Ne.

The microscopic origin of the neutron halo has been found in the delicate balance between the self-consistent mean-field and the pairing field. This is shown in Fig. 6, where the neutron single-particle states $1f_{7/2}$, $2p_{3/2}$ and $2p_{1/2}$ in the canonical basis, and the Fermi energy are plotted as function of the neutron number. For $N \leq 22$ the triplet of states is high in the continuum, and the Fermi level uniformly increases toward zero. The triplet approaches zero energy, and a gap is formed between these states and all other states in the continuum. The shell structure dramatically changes at $N \geq 22$. Between $N = 22$ and $N = 32$ the Fermi level is practically

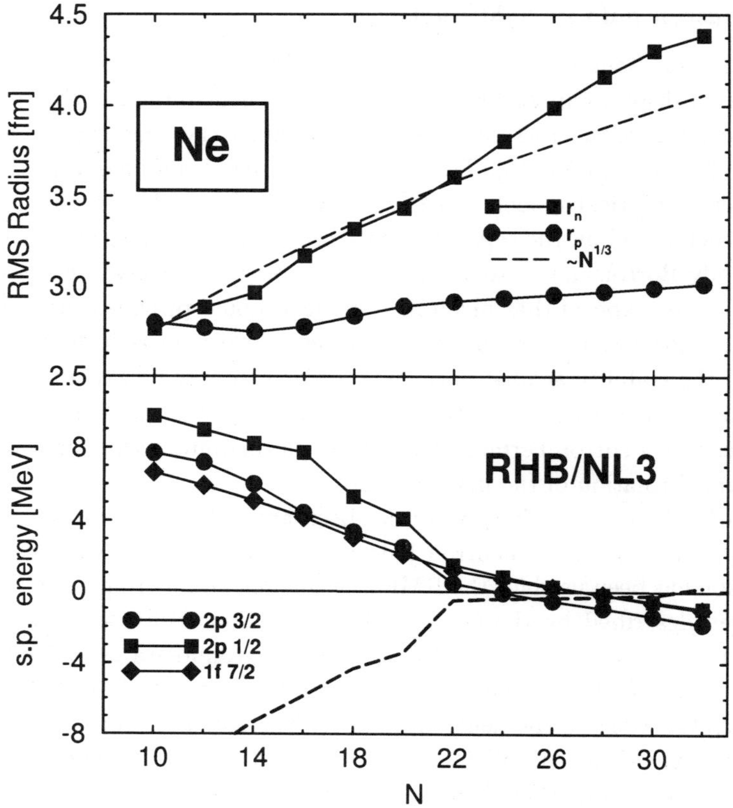

Fig. 6. Proton and neutron *rms* radii for Ne isotopes (top), and the 1f-2p single-particle neutron levels in the canonical basis (bottom), calculated with the NL3 + Gogny D1S effective interaction.

constant and very close to the continuum. The addition of neutrons in this region of the drip does not increase the binding. Only the spatial extension of neutron distribution displays an increase. The formation of the neutron halo is related to the quasi-degeneracy of the triplet of states $1f_{7/2}$, $2p_{3/2}$ and $2p_{1/2}$. The pairing interaction promotes neutrons from the $1f_{7/2}$ orbital to the 2p levels. Since these levels are so close in energy, the total binding energy does not change significantly. Due to their small centrifugal barrier, the $2p_{3/2}$ and $2p_{1/2}$ orbitals form the halo. A similar mechanism has been suggested in Ref. [23] for the experimentally observed halo in the nucleus ^{11}Li. There the formation of the halo is determined by the pair of neutron

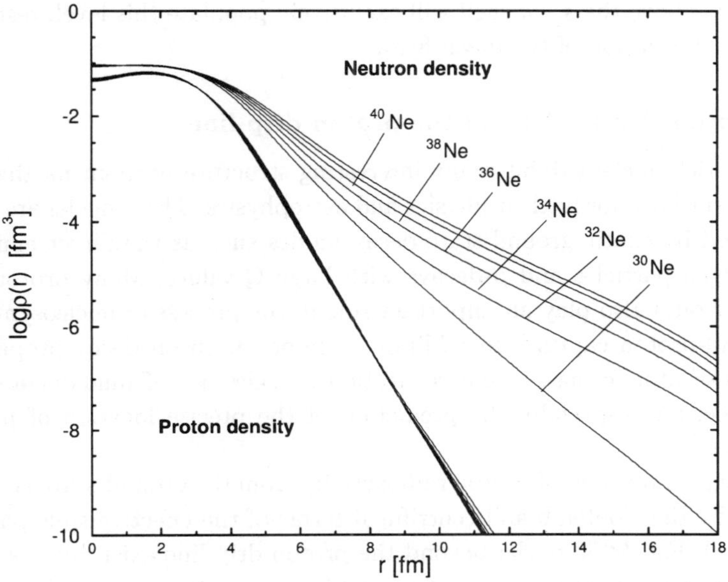

Fig. 7. Proton and neutron density distribution of the Ne isotopes, calculated with the NL3 + Gogny D1S effective interaction.

levels $1p_{1/2}$ and $2s_{1/2}$.

The RHB calculations performed in Ref. [24] have shown that the triplet of single-particle states near the neutron Fermi level: $1f_{7/2}$, $2p_{3/2}$ and $2p_{1/2}$, as well as the neutron pairing interaction, determine the location of the neutron drip-line, the formation of the neutron skin, or eventually of the neutron halo in light nuclei. For C, N, O and F the triplet is still high in the continuum at $N = 20$, and the pairing interaction is to weak to promote pairs of neutrons into these levels. All mean-field effective interactions predict similar results, and the neutron drip is found at $N = 18$ or $N = 20$. For Ne, Na, and Mg the states $1f_{7/2}$, $2p_{3/2}$ and $2p_{1/2}$ are much lower in energy, and for $N \geq 20$ the neutrons populate these levels. The neutron drip can change by as much as twelve neutrons by adding just one or two protons. The model predicts the formation of neutron skin, and eventually neutron halo in Ne and Na. This is due to the fact that the triplet of states is almost degenerate in energy for $N \geq 20$. For Mg the $1f_{7/2}$ lies deeper and neutrons above the $s-d$ shell will exclusively populate this level, resulting in the deformation of the mean field.

4. Proton-rich nuclei and the proton drip-line

Proton-rich nuclei exhibit many interesting structure phenomena that are important both for nuclear physics and astrophysics. These nuclei are characterized by exotic ground-state decay modes such as the direct emission of charged particles and β-decays with large Q-values. Many proton-rich nuclei should also play an important role in the process of nucleosynthesis by rapid-proton capture. In addition to information on decay properties, studies of atomic masses and separation energies are of fundamental importance, and especially the prediction of the precise location of proton drip-line [25].

The phenomenon of proton radioactivity from the ground-state is determined by the Coulomb and centrifugal terms of the effective sinle-particle potential. For Z≤50 nuclei beyond the proton drip-line exist only as short lived resonances, and ground-state single-proton decay probably cannot be observed directly. On the other hand, the relatively high potential energy barrier enables the observation of ground-state proton emission from medium-heavy and heavy nuclei. At the drip-lines proton emission competes with β^+ decay, for heavy nuclei also fission or α decay can be favored. Experimental studies of ground-state proton radioactivity in odd-Z nuclei 51≤Z≤55 and 69≤Z≤ 83, have shown that the systematics of spectroscopic factors is consistent with half-lives calculated in the spherical WKB or DWBA approximations. Reported ground-state proton decay rates in region of light rare-earth nuclei, on the other hand, confirm the occurrence of strongly deformed systems at the drip-lines. The lifetimes of deformed proton emitters provide direct information on the Nilsson configuration occupied by the odd proton and, therefore, on the shape of the nucleus. In Refs. [26–28] we have employed the RHB model in a detailed analysis of deformed odd-Z proton-rich nuclei with 53≤Z≤69. Ground-state properties of isotopes at the proton drip-line have been calculated with the NL3 + D1S combination of effective interactions in the *ph* and *pp* channels, respectively. In Figs. 8 and 9 the one-proton separation energies are displayed for the odd-Z nuclei $53 \leq Z \leq 69$, as functions of the number of neutrons. The model predicts the drip-line nuclei: ^{110}I, ^{115}Cs, ^{118}La, ^{124}Pr, ^{129}Pm, ^{134}Eu, ^{139}Tb, ^{146}Ho, and ^{152}Tm.

For the ground-state proton emission to occur, the valence proton must penetrate the wide Coulomb and centrifugal potential barriers, and this

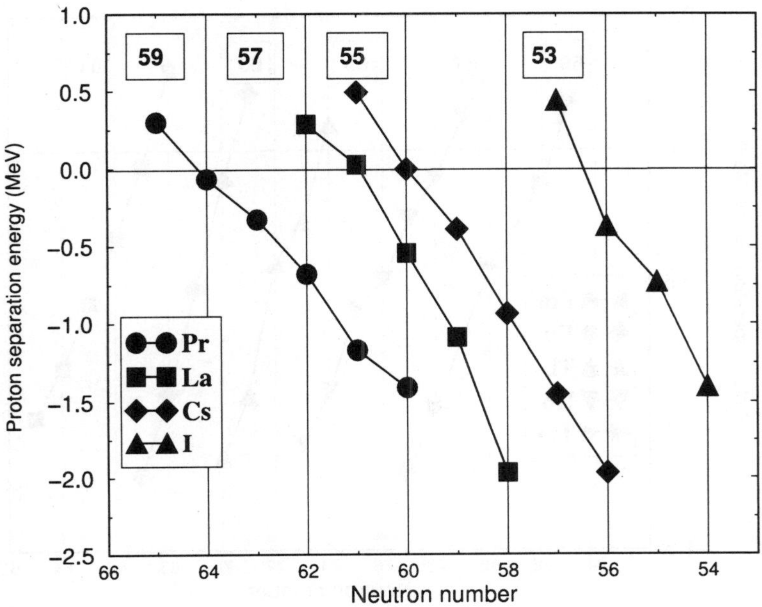

Fig. 8. Calculated one-proton separation energies for odd-Z nuclei $53 \leq Z \leq 59$ at and beyond the drip line.

process competes with the β^+ decay. The half-life of the decay strongly depends on the energy of the odd proton and on its angular momentum. For a typical rare-earth nucleus the energy window in which ground-state proton decay can be directly observed is about 0.8 – 1.7 MeV. For the most probable proton emitters, in Table 2 we enclose the ground-state properties calculated in the RHB model. For each nucleus we include the one-proton separation energy S_p, the quadrupole deformation β_2, the deformed single-particle orbital occupied by the odd valence proton, and the corresponding theoretical spectroscopic factor u^2. The spectroscopic factor of the deformed odd-proton orbital is defined as the probability that this state is found empty in the daughter nucleus with even number of protons. The results of RHB calculations are compared with the predictions of the finite-range droplet mass (FRDM) model: the projection of the odd-proton angular momentum on the symmetry axis and the parity of the odd-proton state Ω_p^π [29], the one-proton separation energy [29], and the ground-state quadrupole deformation [30].

The theoretical separation energies are also compared with experimen-

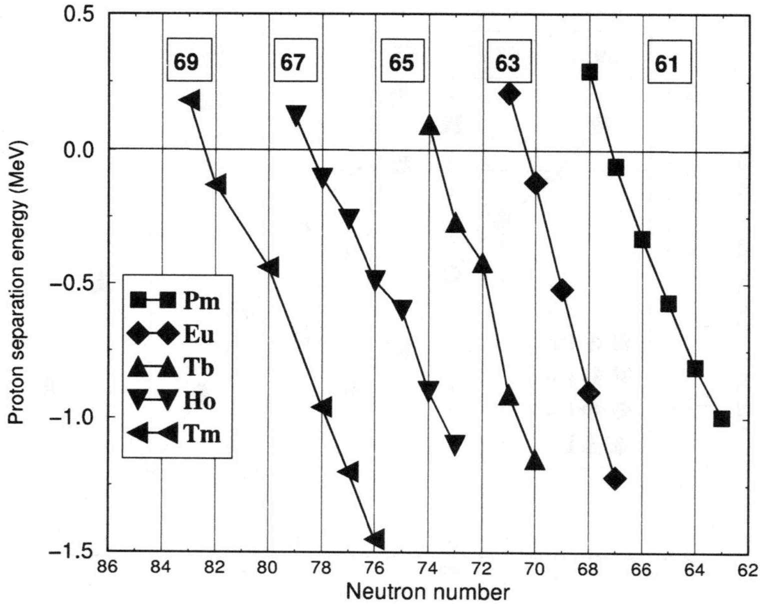

Fig. 9. Calculated one-proton separation energies for odd-Z isotopes $61 \leq Z \leq 69$ at and beyond the drip line.

Table 2. Odd-Z ground-state proton emitters in the region of nuclei with $53 \leq Z \leq 69$.

	S_p	β_2	p-orbital	u^2	Ω_p^π [29]	S_p [29]	β_2 [30]	E_p exp.
^{108}I	-0.73	0.16	$3/2^+$[422]	0.79	$1/2^+$	-1.12	0.15	
^{109}I	-0.37	0.16	$3/2^+$[422]	0.81	$1/2^+$	-0.95	0.16	0.8126(40)
^{112}Cs	-1.46	0.20	$1/2^+$[420]	0.74	$3/2^+$	-0.76	0.21	0.807(7)
^{113}Cs	-0.94	0.21	$1/2^+$[420]	0.73	$3/2^+$	-0.76	0.21	0.9593(37)
^{116}La	-1.09	0.30	$3/2^-$[541]	0.73	$3/2^+$	-0.67	0.28	
^{120}Pr	-1.17	0.33	$3/2^-$[541]	0.33	$3/2^-$	-0.66	0.32	
^{124}Pm	-1.00	0.35	$5/2^-$[532]	0.72	$5/2^-$	-1.34	0.33	
^{125}Pm	-0.81	0.35	$5/2^-$[532]	0.74	$5/2^-$	-1.24	0.33	
^{130}Eu	-1.22	0.34	$5/2^-$[532]	0.44	$3/2^+$	-1.17	0.33	
^{131}Eu	-0.90	0.35	$5/2^+$[413]	0.44	$3/2^+$	-1.01	0.33	0.950(8)
^{135}Tb	-1.15	0.34	$3/2^+$[411]	0.62	$3/2^+$	-1.15	0.33	
^{136}Tb	-0.90	0.32	$3/2^+$[411]	0.65	$3/2^+$	-0.55	0.31	
^{140}Ho	-1.10	0.31	$7/2^-$[523]	0.61	$7/2^-$	-0.81	0.30	
^{141}Ho	-0.90	0.32	$7/2^-$[523]	0.64	$7/2^-$	-0.89	0.29	1.169(8)
^{145}Tm	-1.43	0.23	$7/2^-$[523]	0.47	$1/2^+$	-1.0	0.25	1.728(10)
^{146}Tm	-1.20	-0.21	$7/2^-$[523]	0.50	$7/2^-$	-0.60	-0.20	1.120(10)
^{147}Tm	-0.96	-0.19	$7/2^-$[523]	0.55	$7/2^-$	-0.56	-0.19	1.054(19)

tal data on ground-state proton radioactivity from ^{109}I [31], ^{112}Cs [32], ^{113}Cs [33], ^{131}Eu, ^{141}Ho [34], ^{145}Tm [33], ^{146}Tm [36], and ^{147}Tm [31]. The model does not reproduce the observed anomaly in the one proton separation energies of ^{112}Cs and ^{113}Cs. The ^{131}Eu transition has an energy $E_p = 0.950(8)$ MeV and a half-life 26(6) ms, consistent with decay from either $3/2^+[411]$ or $5/2^+[413]$ Nilsson orbital. For ^{141}Ho the transition energy is $E_p = 1.169(8)$ MeV, and the half-life 4.2(4) ms is assigned to the decay of the $7/2^-[523]$ orbital. The calculated RHB proton separation energy, both for ^{131}Eu and ^{141}Ho, is -0.9 MeV. In the RHB calculation for ^{131}Eu the odd proton occupies the $5/2^+[413]$ orbital, while the ground state of ^{141}Ho corresponds to the $7/2^-[523]$ proton orbital. This orbital is also occupied by the odd proton in the calculated ground states of ^{145}Tm, ^{146}Tm and ^{147}Tm. For the proton separation energies we obtain: -1.43 MeV in ^{145}Tm, -1.20 MeV in ^{146}Tm, and -0.96 MeV in ^{147}Tm. These are compared with the experimental values for transition energies: $E_p = 1.728(10)$ MeV in ^{145}Tm, $E_p = 1.120(10)$ MeV in ^{146}Tm, and $E_p = 1.054(19)$ MeV in ^{147}Tm. Calculations also predict possible proton emitters ^{136}Tb and ^{135}Tb with separation energies -0.90 MeV and -1.15 MeV, respectively. In both isotopes the predicted ground-state proton configuration is $3/2^+[411]$. Another possible proton emitter is ^{130}Eu with separation energy -1.22 MeV and the last occupied proton orbital $5/2^-[532]$ or $5/2^+[413]$.

The structure of nuclei at the proton drip line in the mass region $60 < A < 100$ is important for the process of nucleosynthesis during explosive hydrogen burning. The exact location of the proton drip line determines a possible path of the rapid proton capture process. The path of the rp-process lies between the line of β-stability and the drip line, and it is a very complicated function of the physical conditions, temperature and density, governing the explosion. The input for rp-process nuclear reaction network calculations includes the nuclear masses, or proton separation energies of the neutron deficient isotopes, the proton capture rates, their inverse photodisintegration rates, the β-decay and electron capture rates. In addition to its importance for astrophysical processes, the information about the exact location of the drip line, as well as the proton separation energies beyond the drip line, are essential for studies of ground-state proton radioactivity. No examples of ground-state single-proton emitters below $Z = 50$ have been reported so far, and therefore theoretical studies might provide important information for future experiments in this region.

In Fig. 10 we display the section of the chart of the nuclides along the proton drip line in the region $31 \leq Z \leq 49$ [37]. The RHB (NL3+D1S)

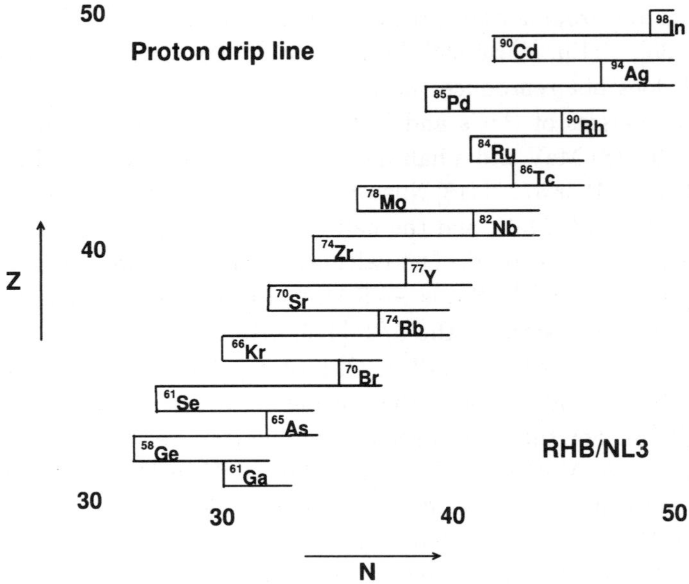

Fig. 10. The proton drip line in the region $31 \leq Z \leq 49$. On this section of the chart of the nuclides the last bound isotopes for each element are indicated. Nuclei to the left are predicted to be proton unstable by the RHB (NL3+D1S) calculation.

calculation predicts the last bound isotopes for each element. Nuclei to the left are proton unstable. For the odd-Z nuclei the proton drip line can be compared with experimental results. For $Z = 31$ and $Z = 33$ the calculated drip line nuclei ^{61}Ga and ^{65}As, respectively, are in agreement with experimental data reported in Refs. [38,39]. These two nuclei are on the rp-process path suggested by Champagne and Wiescher [40]. For $Z = 35$ the RHB calculation predicts that the last proton bound isotope is ^{70}Br. The isotope ^{69}Br is calculated to be proton unbound in most mass models, and no evidence for this isotope was found in experiments. For $Z = 37$ experimental data reported in Refs. [38,41] confirm that ^{74}Rb is the last proton bound nucleus, in agreement with the result of the present calculation. For $Z = 39, 41, 43$ the lightest isotopes observed in experiment [42] are ^{78}Y, ^{82}Nb and ^{86}Tc, respectively. While for Nb and Tc these results correspond to the drip line as calculated in the present work, for Y the RHB model predicts that the last proton bound nucleus is ^{77}Y. This isotope would then be the heaviest $T_z = -\frac{1}{2}$ nucleus. The calculated odd-Z drip line nuclei ^{90}Rh and ^{94}Ag were observed in the experiment reported in Ref. [43], and experimental evidence for ^{98}In was reported in Ref. [44].

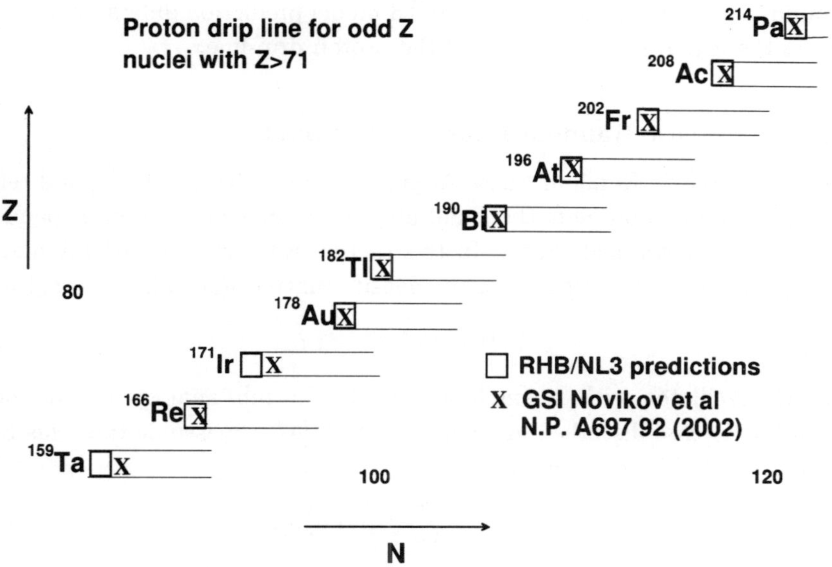

Fig. 11. The proton drip line in the region $73 \leq Z \leq 91$.

For the even-Z nuclei in this region it is not possible to compare the calculated proton drip line with experimental data. While for the odd-Z elements most of the last proton bound nuclei lie on the $N = Z$ line, with just few $T_z = -\frac{1}{2}$ nuclei, the proton drip line for even-Z elements is calculated to be at $T_z = -3$, or even at $T_z = -\frac{7}{2}$. The only exception is the drip line nucleus ^{84}Ru with $T_z = -2$. Nuclei with such extreme values of T_z are virtually impossible to produce in experiments, and since they lie so far away from the rp-process path, the even-Z proton drip line nuclei in this mass region play no role in the process of nucleosynthesis during explosive hydrogen burning.

An important topic of future experiments on proton-rich nuclei will be the possible observation of ground state proton emission in the suburanium region. In Fig. 11 we display the map of the proton drip-line for odd-Z nuclei with $73 \leq Z \leq 91$, calculated in the RHB model with the NL3+D1S effective interaction [45]. The RHB prediction for the last proton-bound isotope of each element is compared with experimental data on the proton drip-line [46]. An excellent agreement between theory and experiment is found, and only for Ta and Ir the RHB model prediction differs by one unit from the experimental position of the proton drip-line.

5. Relativistic Random Phase Approximation

The Relativistic Random Phase Approximation (RRPA) (see [5] and references therein) represents the small amplitude limit of the time-dependent relativistic mean-field theory. In this section we derive the RRPA matrix equations from the response of the density matrix $\hat{\rho}(t)$ to an external field

$$\hat{F}(t) = \hat{F}\mathrm{e}^{-i\omega t} + h.c. \;, \tag{29}$$

which oscillates with a small amplitude. In full analogy to the non-relativistic case discussed in Chap. 8 of Ref. [47] we assume that this field is represented by the operator

$$\hat{f}(t) = \sum_{kl} f_{kl}(t)\, \hat{a}_k^\dagger \hat{a}_l . \tag{30}$$

The equation of motion for the density operator reads

$$i\partial_t \hat{\rho} = \left[\hat{h}(\hat{\rho}) + \hat{f}(t), \hat{\rho}\right] \;, \tag{31}$$

In the small amplitude limit the density matrix is expanded to linear order

$$\hat{\rho}(t) = \hat{\rho}^{(0)} + \delta\hat{\rho}(t) \;, \tag{32}$$

where $\hat{\rho}^{(0)}$ is the stationary ground-state density.

The RRPA equations form read [48]

$$\begin{pmatrix} A^J & B^J \\ B^{*J} & A^{*J} \end{pmatrix} \begin{pmatrix} X_{\tilde{p}h}^{\nu,JM} \\ Y_{\tilde{p}h}^{\nu,JM} \end{pmatrix} = \omega_\nu \begin{pmatrix} 1 & 0 \\ 0 & -1 \end{pmatrix} \begin{pmatrix} X_{\tilde{p}h}^{\nu,JM} \\ Y_{\tilde{p}h}^{\nu,JM} \end{pmatrix} \tag{33}$$

The matrix elements contain the single-particle energies and the two-body interaction

$$A^J_{j_{\tilde{p}} j_h, j_{\tilde{q}} j_i} = \left(\epsilon_{j_{\tilde{p}}} - \epsilon_{j_h}\right) \delta_{j_{\tilde{p}} j_{\tilde{q}}} \delta_{j_h j_i} + V^J_{j_{\tilde{p}} j_i j_h j_{\tilde{q}}} \tag{34}$$

$$B^J_{j_{\tilde{p}} j_h, j_{\tilde{q}} j_i} = (-1)^{j_{\tilde{q}} - j_i + J} V^J_{j_{\tilde{p}} j_{\tilde{q}} j_h j_i} . \tag{35}$$

$\tilde{p}$ denotes both particle and antiparticle states, h denotes states in the Fermi sea.

5.1. *Isovector dipole response in nuclei with a large neutron excess*

The multipole response of unstable nuclei far from the line of β-stability presents a very active field of research, both experimental and theoretical. These nuclei are characterized by unique structure properties: the weak binding of the outermost nucleons and the effects of the coupling between bound states and the particle continuum. On the neutron rich side, in particular, the modification of the effective nuclear potential leads to the formation of nuclei with very diffuse neutron densities, to the occurrence of the neutron skin and halo structures. These phenomena also affect the multipole response of unstable nuclei, in particular the electric dipole and quadrupole excitations, and new modes of excitations might arise in nuclei near the drip line

In Refs. [53,54] we have applied the relativistic random phase approximation (RRPA) in the analysis of the evolution of the isovector dipole response in nuclei with a large neutron excess. Studies of low-energy collective isovector modes, in particular, might provide important information on the isospin and density-dependent parts of the effective interactions used in nuclear structure models. We have analyzed the isovector dipole strength distributions, the transition densities, the neutron and proton *ph* excitations which determine the structure of the RRPA transition amplitudes, the transition currents and the velocity distributions.

The solutions of the RRPA equations are used to evaluate the electric dipole response

$$B(EJ,\omega_\nu) = \left| \sum_{j_{\bar{p}} j_h} X^{\nu,J0}_{j_{\bar{p}} j_h} \langle j_{\bar{p}} || \hat{Q}_J || j_h \rangle + (-)^{j_{\bar{p}} - j_h + J} Y^{\nu,J0}_{j_{\bar{p}} j_h} \langle j_h || \hat{Q}_J || j_{\bar{p}} \rangle \right|^2 \tag{36}$$

for the isovector dipole operator

$$\hat{Q}^{T=1}_{1\mu} = \frac{N}{N+Z} \sum_{p=1}^{Z} r_p Y_{1\mu} - \frac{Z}{N+Z} \sum_{n=1}^{N} r_n Y_{1\mu} \,. \tag{37}$$

The discrete spectra are averaged with the Lorentzian distribution

$$R(E) = \sum_i B(E1, 1_i \to 0_f) \frac{\Gamma^2}{4\left(E - E_i\right)^2 - \Gamma^2}, \tag{38}$$

with $\Gamma = 0.5$ MeV as an arbitrary choice for the width of the Lorentzian.

The energy of the resonance is defined as the centroid energy

$$\bar{E} = \frac{m_1}{m_0} , \tag{39}$$

with the energy weighted moments for discrete spectra

$$m_k = \sum_i B(E1, 1_i \rightarrow 0_f) E_i^k . \tag{40}$$

For $k = 1$ this equation defines the energy weighted sum rule (EWSR).

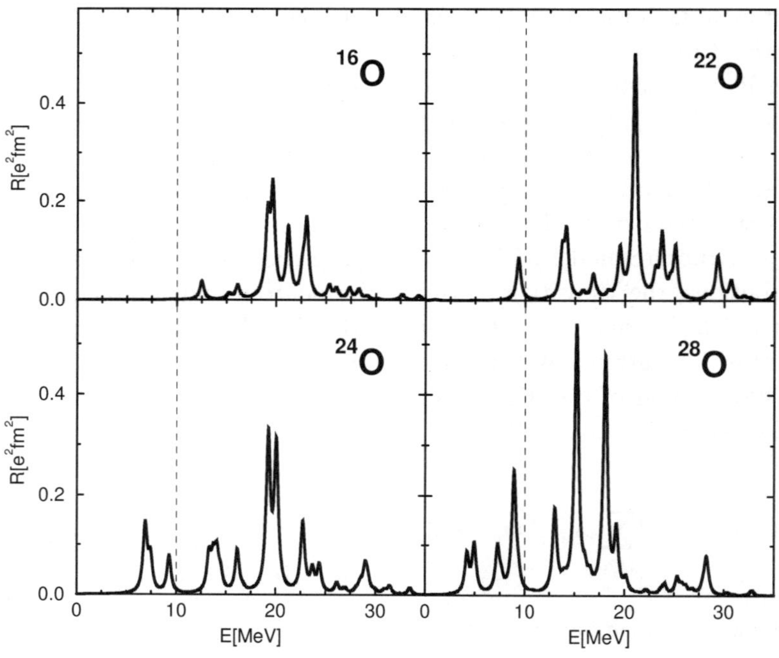

Fig. 12. The RRPA isovector dipole strength distributions in oxygen isotopes calculated with the NL3 effective interaction. The thin dashed line separates the region of giant resonances from the low-energy region below 10 MeV.

The dipole response in nuclei with large neutron to proton ratio is characterized by the fragmentation of the strength distribution and its spreading into the low-energy region, and by the mixing of isoscalar and isovector modes. In light nuclei the onset of dipole strength in the low-energy region is due to single particle excitations of the loosely bound neutrons. In Fig. 12

we display the isovector dipole strength distributions (38) for ^{16}O, ^{22}O, ^{24}O and the hypothetical nucleus ^{28}O. Although ^{24}O is the last bound oxygen isotope, in many mean-field calculations, including the present with the NL3 effective interaction, the neutron drip line is located at ^{28}O. Already for ^{16}O the isovector dipole strength distribution is strongly fragmented with the centroid energy at $\bar{E}$=21.8 MeV. The thin dashed line tentatively separates the region of giant resonances from the low-energy region below 10 MeV. By increasing the number of neutrons, two main effects are observed: a) an increased fragmentation of the dipole strength, and b) the appearance of low lying strength below 10 MeV. The relative contribution of the low-energy region increases with the neutron excess.

In recent experimental studies [49,50] the evolution of the dipole strength with neutron excess was measured for the neutron-rich oxygen isotopes with $A = 17 - 22$. It was found that for all neutron-rich isotopes the dipole strength appears to be strongly fragmented, with a considerable fraction observed well below the giant dipole resonance. In the present RRPA calculation we find 2.5%, 7.0% and 8.6% of the EWSR in the energy region below 10 MeV for ^{22}O, ^{24}O and ^{28}O, respectively. In comparison, the large scale shell model calculation of Ref. [51] predicts that the low-lying dipole strength below 15 MeV exhausts 10% of the classical sum rule in ^{22}O, and 8.6% in ^{24}O. What is the nature of these isovector dipole states? The question whether the soft, i.e. low-lying dipole excitations are collective or single-particle has been addressed, for example, in Ref. [52] for the light neutron halo nuclei ^{11}Li and ^{11}Be. It has been shown that the soft modes, which result from the large spatial extension of the bound single-particle states, represent a new type of non-resonant independent single-particle excitations. The narrow width and the large transition strength, which characterize these excitations, are not caused by a coherent superposition of particle-hole (*ph*) configurations like in collective states.

In the RRPA calculation of the oxygen isotopes we have analyzed in more detail the structure of the main peaks in the low-energy region of the isovector dipole strength distribution (Fig. 12). For a state at energy ω_ν, the contribution of a particular proton or neutron *ph* configuration is determined by the RRPA amplitude

$$\xi_{\tilde{p}h} = \left|X^{\nu}_{\tilde{p}h}\right|^2 - \left|Y^{\nu}_{\tilde{p}h}\right|^2 \;, \tag{41}$$

with X and Y defined by the RRPA equation (33), and the normalization condition

$$\sum_{\tilde{p}h} \xi_{\tilde{p}h} = 1 \,. \tag{42}$$

For ^{22}O we find only one strong peak below 10 MeV: the state at 9.3 MeV exhausts 2.5% of the EWSR, and its wave function is very simple: single neutron excitations (93% $1d_{5/2} \to 2p_{3/2}$) and (3% $1d_{5/2} \to 1f_{7/2}$). The neutron *ph* excitations determine also the main peaks in the low-energy region of ^{24}O and ^{28}O. For ^{24}O we find three strong peaks at 6.9 MeV (3.1% EWSR), 7.4 MeV (1.6% EWSR) and 9.3 MeV (2.3% EWSR). These states correspond to the neutron *ph* excitations: (93% $2s_{1/2} \to 2p_{3/2}$), (96% $2s_{1/2} \to 2p_{1/2}$) and (94% $1d_{5/2} \to 2p_{3/2}$), respectively. A similar structure of the RRPA amplitudes is also found for the more fragmented strength function of ^{28}O. In all neutron-rich oxygen isotopes the isovector dipole response in the low-energy region below 10 MeV is characterized by single particle transitions, in contrast to the coherent superposition of many *ph* configurations, which characterizes the excitations in the region of giant resonances.

In heavier nuclei, on the other hand, low-lying dipole states appear which are characterized by a more distributed structure of the RRPA amplitude, exhausting approximately 2% of the EWSR. A very interesting sequence of medium-heavy neutron-rich nuclei is the chain of Sn isotopes with $50 \leq N \leq 82$. In Fig. 13 we display the isovector dipole strength distributions for ^{100}Sn, ^{114}Sn, ^{120}Sn and ^{132}Sn. With the increase of the number of neutrons, the onset of low-lying strength below 10 MeV is observed. The dipole states in this energy region exhibit a structure similar to that observed in other neutron-rich medium-heavy and heavy nuclei: among several peaks characterized by single particle transitions, between 7 MeV and 9 MeV a state is found with a more distributed structure of the RRPA amplitude, exhausting approximately 2% of the EWSR. In ^{132}Sn, for example, this state is calculated at 8.6 MeV and it exhausts 1.4% of the EWSR. Nine neutron *ph* configurations contribute with more than 0.1% to the total RRPA intensity. The total contribution of proton *ph* excitations is only 10.4%, well below the ratio Z/N expected for a GDR state.

The transition densities to the two states at 8.6 MeV and 14.8 MeV in ^{132}Sn are plotted in Fig. 14. In the upper panel the proton, neutron, isoscalar and isovector components are displayed. Although the isoscalar B(E1) to all states must vanish identically, the corresponding isoscalar transition densities to individual states need not be identically zero. The transition densities for the state at 14.8 MeV display a radial dependence

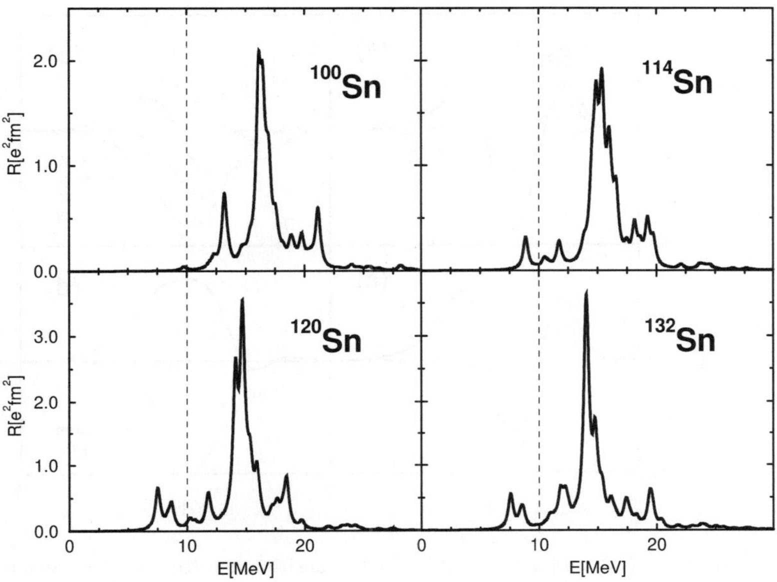

Fig. 13. The RRPA isovector dipole strength distributions in Sn isotopes calculated with the NL3 effective interaction. The thin dashed line separates the region of giant resonances from the low-energy region below 10 MeV.

characteristic for the isovector giant dipole resonance: the proton and neutron densities oscillate with opposite phases; the amplitude of the isovector transition density is much larger than that of the isoscalar component; and at large radii both the isovector and isoscalar transition densities have a similar radial dependence. The transition densities for the state at 8.6 MeV exhibit a rather different radial dependence: the proton and neutron densities in the interior region are not out of phase; there is almost no contribution from the protons in the surface region; the isoscalar transition density dominates over the isovector one in the interior; the neutron transition density displays a long tail in the radial coordinate. In the lower panel, for the pygmy state at 8.6 MeV (c) and for the GDR state at 14.8 MeV (d), for the pygmy state at 8.6 MeV (c) and for the GDR state at 14.8 MeV (d), the contributions of the excess neutrons ($50 < N \leq 82$) (solid), and of the proton-neutron core ($Z, N \leq 50$) (dashed) are displayed separately. By comparing with the transition densities shown in the upper panel, we notice that there is practically no contribution from the core neutrons ($N \leq 50$).

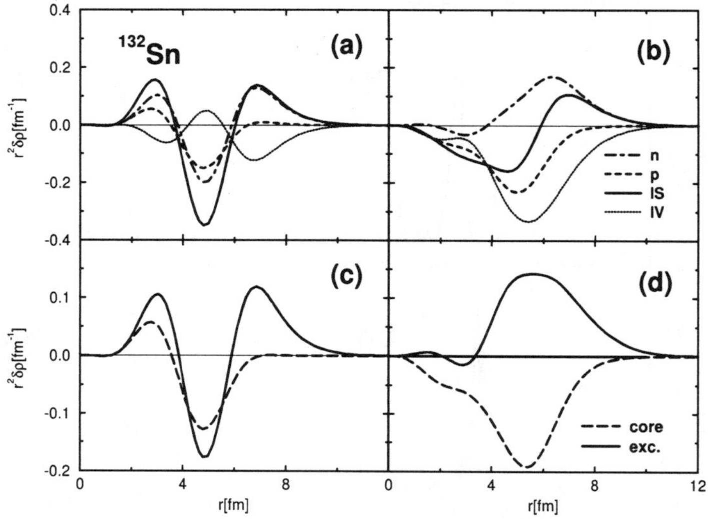

Fig. 14. Isovector (IV) and isoscalar (IS) dipole transition densities for the states at 8.6 MeV (a) and 14.8 MeV (b) in ^{132}Sn. In the lower part of the figure the contributions of the excess neutrons ($50 < N \leq 82$) (solid), and of the proton-neutron core ($Z, N \leq 50$) (dashed) are displayed separately for the state at 8.6 MeV (c), and 14.8 MeV (d). The transition densities are multiplied by r^2.

The *ph* excitations of core neutrons are, of course, at much higher energies. For the GDR state, therefore, the transition densities of the core nucleons and of the excess neutrons have opposite phases (isovector mode). The absolute radial dependence is similar, with the amplitude strongly peaked in the surface region. The two transition densities have the same sign for the pygmy state at 8.6 MeV. The core contribution, however, vanishes for large r and only oscillations of the excess neutrons are observed on the surface of ^{132}Sn.

A similar result has also been obtained in our RRPA calculation of the isovector dipole response in ^{208}Pb [53]. In the left panel of Fig. 15 we display the isovector dipole strength distribution in ^{208}Pb, calculated with the NL3 effective interaction. In addition to the GDR at 12.95 MeV and several low-lying single *ph* states, a collective pygmy dipole state is identified at 7.29 MeV, with proton *ph* excitations contributing only 14% to the total RRPA intensity. The calculated energy of the main peak at $E_p = 12.95$ MeV should be compared with the experimental value of the excitation energy of the isovector giant dipole resonance: 13.3 ± 0.1 MeV [55]. The transi-

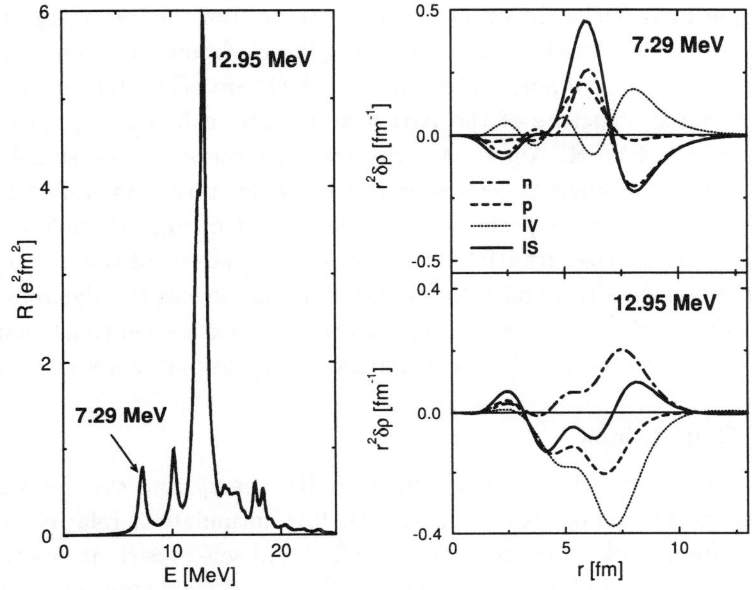

Fig. 15. The isovector dipole strength distribution in ^{208}Pb (left panel), and the transition densities for the two peaks at 7.29 MeV and 12.95 MeV (right panel). Both the isoscalar and the isovector transition densities are displayed, as well as the separate proton and neutron contributions. All transition densities are multiplied by r^2

tion densities to the states at 7.29 MeV and at 12.95 MeV are displayed in the right panel of Fig. 15. The proton and neutron contributions are shown separately; the dotted line denotes the isovector transition density and the solid line is used for the isoscalar transition density. The transition densities for the main peak at 12.95 MeV display the usual radial dependence of isovector giant dipole resonances. The state at 7.29 MeV, on the other hand, corresponds to the collective pygmy mode characterized by the coherent superposition of neutron particle-hole excitations. The transition densities to the pronounced peak at 10.1 MeV display a radial behavior intermediate between those shown in Fig. 15 but closer to the isovector giant dipole at 12.95 MeV, i.e. this state belongs to the tail of the IVGDR. It is interesting to note that our theoretical analysis has prompted an experimental study (high-resolution (γ, γ')) of the electric dipole response in ^{208}Pb [56]. A pygmy dipole resonance has been observed with a centroid energy right at the neutron emission threshold $E_t = 7.37$ MeV. Summarizing, in light nuclei the onset of dipole strength in the low-energy region is due to single particle excitations of the loosely bound neutrons. In heavier nuclei low-lying dipole states appear which are characterized by a more distributed structure of the RRPA amplitude, exhausting approximately 2% of the EWSR. Among several peaks characterized by single particle transitions, a single collective isovector dipole state is identified below 10 MeV. A coherent superposition of many neutron particle-hole configurations characterizes its RRPA amplitude. An analysis of the corresponding transition densities and velocity distributions reveals the dynamics of the dipole pygmy resonance: the vibration of the excess neutrons against the inert core composed of equal numbers of protons and neutrons.

6. Conclusions

The relativistic Hartree-Bogoliubov (RHB) theory, and the (quasiparticle) random phase-approximation (R(Q)RPA) formulate a relativistic mean-field framework which provides a unified and self-consistent description of ground states and properties of excited states in stable nuclei, exotic nuclei far from β-stability, and in nuclear systems at the nucleon drip-lines. In the present work, we have reviewed the recent development and applications of the RHB model and the RRPA/RQRPA in the description of exotic nuclear structure. Particularly, we have focused on the description of neutron halos in light nuclei, the suppression of shell effects and the onset of deformation and shape coexistence in neutron-rich nuclei. In the proton rich side we have investigated the location of the proton drip-line and ground-state proton emitters in the region of deformed rare-earth nuclei. Finally, the relativistic RRPA/RQRPA model has been applied to an analysis of the low-energy multipole response of unstable weakly bound neutron-rich nuclei.

Acknowledgments

This work has been supported by the Programme Pythagoras II of the Greek MoE and RA and of the European Union under project 80661.

References

1. B. D. Serot and J. D. Walecka, Int. J. Mod. Phys. **E6**, 515 (1997).
2. P.-G. Reinhard, Rep. Prog. Phys. **52**, 439 (1989).
3. B. D. Serot, Rep. Prog. Phys. **55**, 1855 (1992).
4. P. Ring, Progr. Part. Nucl. Phys. **37**, 193 (1996).
5. D. Vretenar, A.V. Afanasjev, G.A. Lalazissis, P. Ring Physics Reports 409, 101 (2005).

6. J. Boguta and A. R. Bodmer, Nucl. Phys. **A292**, 413 (1977). (2000).
7. D. Vretenar, H. Berghammer, and P. Ring, Nucl. Phys. **A581**, 679 (1995).
8. G. A. Lalazissis, P. Ring, and D. Vretenar (Eds.), *Extended Density Functionals in Nuclear Structure Physics*, Lecture Notes in Physics 641, (Springer, Berlin Heidelberg 2004bite).
9. G. A. Lalazissis, J. König, and P. Ring, Phys. Rev. **C55**, 540 (1997).
10. T. Gonzales-Llarena, J. L. Egido, G. A. Lalazissis, and P. Ring, Phys. Lett. **B379**, 13 (1996).
11. J. F. Berger, M. Girod, and D. Gogny, Nucl. Phys. **A428**, 32c (1984).
12. G. A. Lalazissis, D. Vretenar, P. Ring, M. Stoitsov, and L. Robledo, Phys. Rev. **C60**, 014310 (1999).
13. G. Audi and A. H. Wapstra, Nucl. Phys. **A595**, 409 (1995).
14. T. R. Werner, J. A. Sheikh, M. Misu, W. Nazarewicz, J. Rikovska, K. Heeger, A. S. Umar, and M. R. Strayer, Nucl. Phys. **A597**, 327 (1996).
15. T. Glasmacher *et al.*, Phys. Lett. **B395**, 163 (1997).
16. T. Glasmacher *et al.*, Nucl. Phys. **A630**, 278c (1998).
17. H. Scheit *et al.*, Phys. Rev. Lett. **77**, 3967 (1996).
18. G. A. Lalazissis, D. Vretenar, and P. Ring, Phys. Rev. **C57**, 2294 (1998)
19. I. Tanihata *et al.*, Phys. Rev. Lett. **55**, 2676 (1985).
20. I. Tanihata, Prog. Part. Nucl. Phys. **35**, 505 (1995).
21. P. Hansen, A. S. Jensen, and B. Jonson, Annu. Rev. Nucl. Part. Phys. **45**, 591 (1995).
22. W. Pöschl, D. Vretenar, G. A. Lalazissis, and P. Ring, Phys. Rev. Lett. **79**, 3841 (1997).
23. J. Meng and P. Ring, Phys. Rev. Lett. **77**, 3963 (1996).
24. G. A. Lalazissis, D. Vretenar, W. Pöschl, and P. Ring, Nucl. Phys. **A632**, 363 (1998).
25. P. J. Woods and C. N. Davids, Annu. Rev. Nucl. Part. Sci. **47**, 541 (1997).
26. D. Vretenar, G. A. Lalazissis, and P. Ring, Phys. Rev. Lett. **82**, 4595 (1999).
27. G. A. Lalazissis, D. Vretenar, and P. Ring, Nucl. Phys. **A650**, 133
28. G.A. Lalazissis, D. Vretenar, and P. Ring, Phys. Rev. C **60**, 051302 (1999).
29. P. Möller, J. R. Nix, and K. L. Kratz, At. Data Nucl. Data Tables **66**, 131 (1997).
30. P. Möller, J. R. Nix, W. D. Myers, and W. J. Swiatecki, At. Data Nucl. Data Tables **59**, 185 (1995).
31. P. J. Sellin *et al.*, Phys. Rev. **C47**, 1933 (1993).
32. R. D. Page *et al.*, Phys. Rev. Lett. **72**, 1798 (1994).
33. J. C. Batchelder *et al.*, Phys. Rev. **C57**, R1042 (1998).
34. C. N. Davids *et al.*, Phys. Rev. Lett. **80**, 1849 (1998).
35. K. Rykaczewski *et al.*, Phys. Rev. **C60**, 011301 (1999).
36. K. Livingston *et al.*, Phys. Lett. **B312**, 46 (1993).
37. G. A. Lalazissis, D. Vretenar, and P. Ring, Nucl. Phys. **A679**, 481 (2001).
38. M. F. Mohar *et al.*, Phys. Rev. Lett. **66**, 1571 (1991).
39. J. A. Winger *et al.*, Phys. Lett. **B299**, 214 (1993).
40. A. E. Champagne and M. Wiescher, Annu. Rev. Nucl. Part. Sci. **42**, 39 (1992).

41. S. Jokinen *et al.*, Z. Phys. **A355**, 227 (1996).
42. S. J. Yennello *et al.*, Phys. Rev. **C46**, 2620 (1992).
43. M. Hencheck *et al.*, Phys. Rev. **C50**, 2219 (1994).
44. K. Rykaczewski *et al.*, Phys. Rev. **C52**, R2310 (1995).
45. G. Lalazissis, D. Vretenar, and P. Ring, Phys. Rev. **C69**, 017301 (2004).
46. Y. N. Novikov *et al.*, Nucl. Phys. **A697**, 92 (2002).
47. P. Ring and P. Schuck, *The Nuclear Many-Body Problem*, Springer Verlag, New York 1980.
48. F. Dawson and R. J. Furnstahl, Phys. Rev. C **42** (1990) 2009.
49. T. Aumann *et al.*, Nucl. Phys. **A649**, 297c (1999).
50. A. Leistenschneider *et al.*, Phys. Rev. Lett. **86**, 5442 (2001).
51. H. Sagawa and T. Suzuki, Phys. Rev. **C59**, 3116 (1999).
52. H. Sagawa, N. Van Giai, N. Takigawa, M. Ishihara, and K. Yazaki, Z. Phys. **351**, 385 (1995).
53. D. Vretenar, N. Paar, P. Ring, and G. A. Lalazissis, Phys. Rev. **C63**, 047301 (2001).
54. D. Vretenar, N. Paar, P. Ring, and G. A. Lalazissis, Nucl. Phys. **A692**, 496 (2001).
55. J. Ritman *et al.*, Phys. Rev. Lett. **70**, 533 (1993).
56. N. Ryezayeva *et al.*, Phys. Rev. Lett. **89**, 272502 (2002).

Mean-field description of nuclei

Thomas J. Bürvenich

Frankfurt Institute for Advanced Studies,
Johann Wolfgang Goethe University
Max-von-Laue-Str. 1
60438 Frankfurt am Main
Germany
E-mail: buervenich@fias.uni-frankfurt.de

The conceptual framework of self-consistent mean-field models is presented. After discussing the parameter adjustment of these models, their overall predictive power is investigated, and, as a special application, predictions for fission barriers of superheavy nuclei are studied.

1. Introduction

The description of atomic nuclei still provides fascinating challenges. These systems are femto-laboratories of interacting quantum-mechanical systems. Their number of constituents places them in a regime where exact few-body methods are not applicable anymore. On the other hand, they have too few constituents that allows one to resort to statistical methods. Nuclei provide insight into fundamental questions such as structure formation of quantum-mechanical many-body systems, tunneling phenomena of many-body system, and many more. Also, their structure and dynamics are interesting topics for themselves.

There are many different models and approaches on the market that aim at the descriptrion of atomic nuclei. In this contribution, we will describe the approach of *self-consistent mean-field models.* The goal is to develop and progressively enhance a universal energy functional for nuclear ground states. Universal in this context means that, once the coupling constants have been fixed, the model can be applied throughout the nuclear chart, with very light systems (A<16) being an exception. Self-consistent mean-field models provide thus far the only framework that can be extrapolated to very heavy and superheavy nuclei.

2. The framework

Our task is to approximately describe the nuclear many body-problem [1]. In order to start, we can take advantage of the fact that solutions of the Schrödinger equation

$$i\hbar\frac{\partial}{\partial t}\psi = \hat{H}\psi \tag{1}$$

can be constructed with the help of the superposition principle and expanded into any complete basis set $|\phi_a\rangle$. Thus, in most general terms, we can write down the solution for Eq. (1) as

$$\psi = \sum_a c_a|\phi_a\rangle \tag{2}$$

Similarly, in the many-body case, the wave-function of the interacting system obeys a Schrödinger equation, and its exact solution $|\Psi\rangle$ can be expanded into a complete basis of many-body wave-functions $|\Phi_a\rangle$, i.e.

$$\Psi = \sum_a C_a|\Phi_a\rangle \tag{3}$$

The basis states $|\Phi_a\rangle$ can be chosen to be Slater determinants, corresponding to independent particles.

Mean-field models now take the extreme stand-point of choosing only one Slater-determinant for the description of the many-body state [1]. The single-particle states of this Slater determinant, however, are being optimized by means of a variational principle, i.e.,

$$\delta\langle\Psi|\hat{H}\Psi\rangle = 0. \tag{4}$$

yielding the equations of motions for the single-particle states which are coupled self-consistently by a mean potential or *mean field.*

The description of the nucleus in the intrinsic frame leads to violation of several symmetries, e.g., translational invariance, rotational symmetry, exact particle number (upon the inclusion of pairing) [1]. This breaking of symmetries, however, enlarges the Hilbert space, but in a somewhat uncontrolled way. For example, violating good angular momentum allows the description of deformed nuclei. Broken symmetries can be restored by projection schemes, either before or after the variation.

Self-consistent models for the nucleus (for a comprehensive review see Ref. [2]) have recently experienced a reinterpretation in terms of *effective field theory* and *density functional theory* [3], which partly explains their success in accurately describing nuclear ground-state properties. As effective field theories for nucleonic degrees of freedom, they are constructed to incorporate explicitly physics below a certain energy scale λ, which is of $\mathcal{O}(1\ \text{GeV})$. Short-distance physics and correlations, vacuum polarization and all effects at higher energies are being absorbed into the various interaction terms and coupling constants.

The *Hohenberg-Kohn theorem* [4,5] states that the energy of a many-body state is a unique functional of its density. Thus, *in principle*, it is possible to achieve an exact treatment of such a system if the right energy functional is found. In such a description, all correlations are present in the energy and the density distribution (not in the wave-function, however, which is a Slater-determinant of Kohn-Sham orbitals). For the description of a self-bound nuclear system in the intrinsic frame, generalizations of this theorem are necessary and available [6]. Unfortunatley, these theorems are non-constructive and provide no handle for the development of such functionals. Thus, current efforts involve ways to systematically construct these functionals and subsequently improve them.

An important feature of nuclear structure models is to provide the strong spin-orbit force in nuclei. In non-relativistic models it is taken care of by an extra term, while in relativistic approaches it is automatically included in the model structure and yields no additional adjustable parameters.

We can now summarize the conceptual background of self-consistent mean-field models:

- Construct an effective interation or energy functional for point-like nucleons
- Absorb non-resolved physics (short-distance physics, vacuum physics, ...) into terms and parameters
- Introduce many-body approximations and solve the reduced problem numerically
- Adjust the coupling constants introduced through the interaction (6 to 10, 2 for pairing): a *force* is born (biased, thus there are many forces)
- Predict nuclear ground-state observables throughout the nuclear chart and extrapolate to unknown regions

As an example, we present the relativistic mean-field model with point-like interactions [2,10].

$$\mathcal{L} = \mathcal{L}^{\text{free}} + \mathcal{L}^{\text{4f}} + \mathcal{L}^{\text{hot}} + \mathcal{L}^{\text{der}} + \mathcal{L}^{\text{em}} \tag{5}$$

with the individual parts defined as

(1) the *free Lagrangian*

$$\mathcal{L}^{\rm free} = \bar{\psi}(\mathrm{i}\gamma_\mu \partial^\mu - m)\psi \tag{6}$$

(2) the *four-fermion contribution*

$$\begin{aligned}\mathcal{L}^{\rm 4f} = &- \tfrac{1}{2}\,\alpha_{\rm S}(\bar{\psi}\psi)(\bar{\psi}\psi) - \tfrac{1}{2}\,\alpha_{\rm V}(\bar{\psi}\gamma_\mu\psi)(\bar{\psi}\gamma^\mu\psi) \\ &- \tfrac{1}{2}\,\alpha_{\rm TV}(\bar{\psi}\vec{\tau}\gamma_\mu\psi)\cdot(\bar{\psi}\vec{\tau}\gamma^\mu\psi),\end{aligned} \tag{7}$$

(3) the *higher-order terms*

$$\mathcal{L}^{\rm hot} = - \tfrac{1}{3}\,\beta_{\rm S}(\bar{\psi}\psi)^3 - \frac{1}{4}\,\gamma_{\rm S}(\bar{\psi}\psi)^4 - \tfrac{1}{4}\,\gamma_{\rm V}[(\bar{\psi}\gamma_\mu\psi)(\bar{\psi}\gamma^\mu\psi)]^2 \tag{8}$$

(4) the *derivative contributions*

$$\begin{aligned}\mathcal{L}^{\rm der} = &- \tfrac{1}{2}\,\delta_{\rm S}(\partial_\nu\bar{\psi}\psi)(\partial^\nu\bar{\psi}\psi) - \tfrac{1}{2}\,\delta_{\rm V}(\partial_\nu\bar{\psi}\gamma_\mu\psi)(\partial^\nu\bar{\psi}\gamma^\mu\psi) \\ &- \tfrac{1}{2}\,\delta_{\rm TV}(\partial_\nu\bar{\psi}\vec{\tau}\gamma_\mu\psi)\cdot(\partial^\nu\bar{\psi}\vec{\tau}\gamma^\mu\psi),\end{aligned} \tag{9}$$

(5) and finally the *photon part*

$$\mathcal{L}^{\rm em} = -eA_\mu\bar{\psi}[(1-\tau_3)/2]\gamma^\mu\psi - \tfrac{1}{4}\,F_{\mu\nu}F^{\mu\nu}, \tag{10}$$

from which it becomes apparent that we use the nuclear physics convention for the isospin, i. e., the neutron has isospin $\tau_3 = +1$ and the proton $\tau_3 = -1$.

As is stands this Lagrangian contains the ten coupling constants $\alpha_{\rm S}$, $\alpha_{\rm V}$, $\alpha_{\rm TV}$, $\beta_{\rm S}$, $\gamma_{\rm S}$, $\gamma_{\rm V}$, $\gamma_{\rm TV}$, $\delta_{\rm S}$, $\delta_{\rm V}$, and $\delta_{\rm TV}$. The subscripts indicate the symmetry of the coupling: "S" stands for scalar, "V" for vector, and "T" for isovector, while the symbols refer to the additional distinctions: α corresponds to four-fermion terms, δ to derivative couplings, and β and γ to third- and fourth order terms, respectively. Its best-fit force, called PC-F1, is discussed in the following section.

The crucial step from a field-theoretical Lagrangian to a workable nuclear-structure model invokes the *mean-field* and *no-sea* approximations. The field operators are expanded into single-particle states ϕ_k according to

$$\bar{\psi} = \sum_k \bar{\phi}_k b_k^\dagger, \qquad \psi = \sum_k \phi_k b_k \tag{11}$$

with $b_k^\dagger$ and b_k creation and annihilation operators, respectively. In the following the Lagrangian will also be treated in the *no-sea* approximation, which means that we consider only positive-energy states. In other words, the expansion of ψ in Eq. (11) contains only nucleon annihilation operators and no antinucleon creation operators. In modern terms, however, in the spirit of effective field theory vacuum physis, which is not resolved at the low energy scale which is relevant for the ground state of nuclei, is present and absorbed into the various terms and coupling constants of the model [8].

Using this expansion, the mean-field approximation to the many-body wave function is then sought by the variational principle, see Eq. (4)

$$\delta\langle\Phi|H|\Phi\rangle = 0. \tag{12}$$

where $|\Phi\rangle$ is restricted to a Slater determinant for A particles occupying the lowest available states:

$$|\Phi\rangle = \prod_{k=1}^{A} b_k^\dagger|0\rangle, \tag{13}$$

and H is the Hamiltonian derived from $\mathcal{L}$.

Assuming a stationary state with time-reversal invariance, a straightforward evaluation leads to Dirac equations for the single-particle states:

$$(\mathrm{i}\vec{\gamma}\cdot\vec{\partial} + m + V_\mathrm{S} + V_\mathrm{V}\gamma_0 + V_\mathrm{TV}\tau_3\gamma_0 + V_\mathrm{C}\tfrac{1-\tau_3}{2}\gamma_0)\phi_\alpha = \gamma_0\epsilon_\alpha\phi_\alpha,$$

where only the time-like components of the four-vectors appear.

The potentials appearing here result from the different types of coupling and are simple functions of the densities and their Laplacians:

(1) the isoscalar-scalar potential V_S, delivering the intermediate-range attraction, derived from the scalar density ρ_S

$$\begin{aligned} V_\mathrm{S} &= \alpha_\mathrm{S}\rho_\mathrm{S} + \beta_\mathrm{S}\rho_\mathrm{S}^2 + \gamma_\mathrm{S}\rho_\mathrm{S}^3 + \delta_\mathrm{S}\Delta\rho_\mathrm{S} \\ \rho_\mathrm{S}(\vec{r}) &= \sum_\alpha \bar{\phi}_\alpha(\vec{r})\phi_\alpha(\vec{r}), \end{aligned} \tag{14}$$

(2) the isoscalar-vector potential V_V, responsible for the short-range repulsion, with the vector density ρ_V

$$\begin{aligned} V_\mathrm{V} &= \alpha_\mathrm{V}\rho_\mathrm{V} + \gamma_\mathrm{V}\rho_\mathrm{V}^3 + \delta_\mathrm{V}\Delta\rho_\mathrm{V}, \\ \rho_\mathrm{V}(\vec{r}) &= \sum_\alpha \bar{\phi}_\alpha(\vec{r})\gamma_0\phi_\alpha(\vec{r}), \end{aligned} \tag{15}$$

(3) the isovector-vector potential V_TV, related to the asymmetry energy, and isovector-vector density ρ_TV

$$\begin{aligned} V_\mathrm{TV} &= \alpha_\mathrm{TV}\rho_\mathrm{TV} + \delta_\mathrm{TV}\Delta\rho_\mathrm{TV} \\ \rho_\mathrm{TV}(\vec{r}) &= \sum_\alpha \bar{\phi}_\alpha(\vec{r})\tau_3\gamma_0\phi_\alpha(\vec{r}). \end{aligned} \tag{16}$$

(4) Finally there is the Coulomb potential which, however, has to be calculated from the isoscalar and isovector densities using the Poisson equation:

$$V_C = eA_0, \quad \Delta A_0 = -4\pi\rho_C \tag{17}$$
$$\rho_C = \tfrac{1}{2}(\rho_V - \rho_{TV}) .$$

The attractive (negative) scalar and repulsive (positive) vector potentials as well as their sum are shown in Fig. 1. For the central potential, they almost cancel completely and add up to the well-known shell-model potential of $V = V_S + V_V \approx -50$ MeV. The isovector-vector and Coulomb contributions are small but important corrections tho this potential.

The spin-orbit term, however, which (in the nonrelativistic limit and spherical symmetry) is given by

$$V_{ls} \propto \frac{d}{dr}(V_S - V_V\gamma_0)\vec{l}\cdot\vec{s} \tag{18}$$

contains both potentials with the same sign, leading to the strong spin-orbit force required in nuclei. The spin-orbit potential is where relativity really shows up in these covariant models. The dynamics of the nucleons justifies very well a non-relativistic treatment.

In the relativistic framework, the spin-orbit term emerges without the need to introduce additional parameters from this structure of large scalar and vector fields. It has to be taken into account explicitely and with extra parameters in the Skyrme-Hartee-Fock approach.

Furthermore, the center-of-mass energy needs to be corrected, and pairing correlations are being added in the BCS framework. The pairing prescription introduces the pairing strength parameters V_P and V_N for protons and neutrons, respectively. These are not assumed *a priori* but fitted to experimental data simultaneously with the coupling constants appearing in the Lagrangian (see next section).

The coupled system of equations is iterated until convergence is achieved. As is apparent from the form of the potentials, the effective interaction in this model is density-dependent, which is a general feature of all mean-field approaches.

All properties of the nucleus, including its mass, its shape, and its single-particle structure, emerge from such a calculation. Thus, in contrast to other approaches such as the macroscopic-miscroscopic approach [1], no assumptions on the nuclear shape are prescribed.

As a first prediction of these models, it is worthwhile to look at properties of symmetric nuclear matter. In this model system, protons and neu-

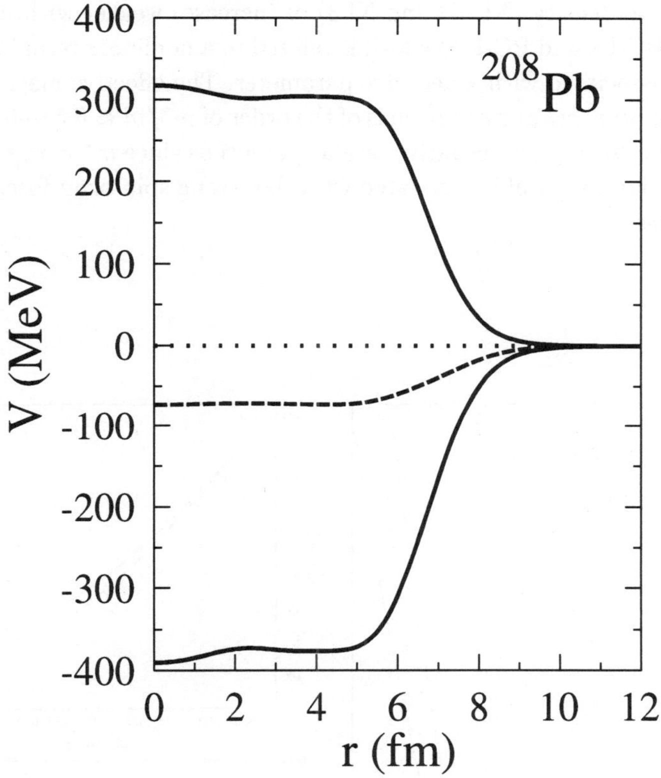

Fig. 1. Scalar (negative) and vector (positive) potentials are shown for the doubly-magic lead nucleus with solid lines. Their sum is plotted with the dashed line.

trons are assumed to be present in equal number and are described by plane waves. The isovector force is zero, and Coulomb effects, due to their long-range nature, are not included. Fig. 2 shows the predictions for two RMF forces with point couplings (PC-F1 and PC-LA) as well as two RMF forces employing effective meson fields (NL-Z2 and NL3). The various forces agree in the vicinity of the saturation point of nuclear matter, but disagree on the high-density behavior which is correlated with their compressibility. The behavior of PC-LA [7] is a peciuliarity which has been identified to be related to a special combination of the nonlinear parameters (see also the potentials) and should not be interpreted as a meaningful physical prediction [10]. In all four cases, the scalar potential reaches saturation for increasing density. The vector potential, however, increases either linearly with baryon density (NL-Z2 and NL3) or increases weaker with increasing density (PC-F1 and PC-LA) which is related to a nonlinear term in the vector field associated with a negative parameter. The effective mass for these relative approaches at saturation is of the order of $m^*/m \approx 0.58-0.60$ which reflects the strong (and negative) scalar potentials since $m^* = m + V_S$. This strong scalar potential is correlated with the strong spin-orbit force in these approaches.

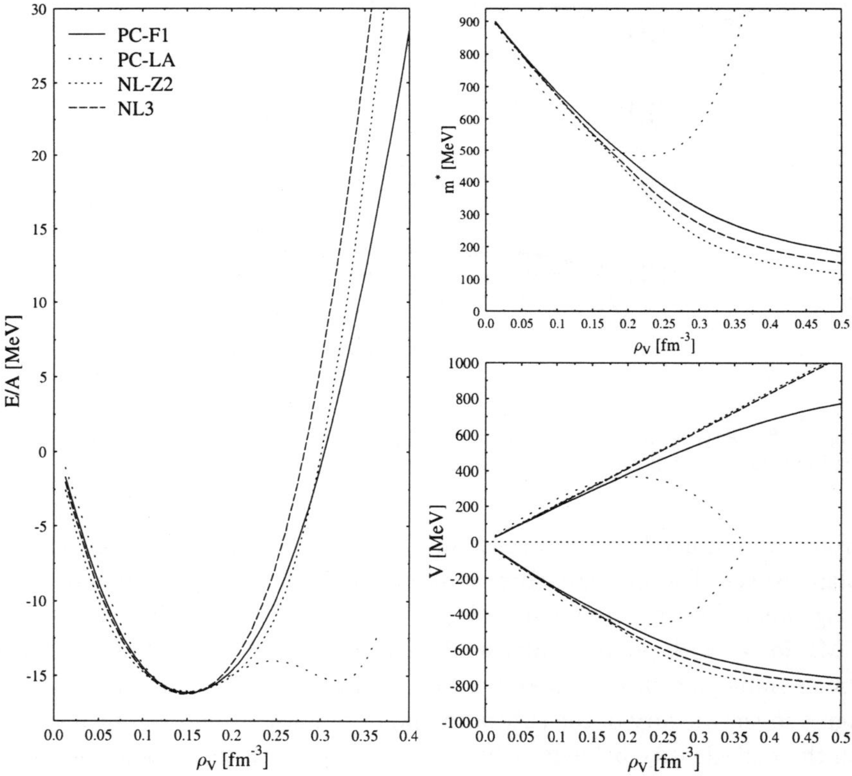

Fig. 2. Energy per particle of symmetric nuclear matter (left), effective mass (top right) and potentials (bottom right) for the forces as indicated.

3. Parameter adjustment and predictive power

The usual way of adjusting these mean-field models is a *least-squares minimization* of χ^2 defined as

$$\chi^2 = \sum_i \frac{(O_i^{th} - O_i^{expt})^2}{(\Delta O_i)^2}, \tag{19}$$

where O_i^{expt} are the experimental data and O_i^{th} denote the calculated values. [9] ΔO_i are the assumed errors of the observables which empirically express the demands on the accuracy of the model. They are in some cases (for example for binding energies) larger than the experimental errors. The observables chosen for the adjustment of the coupling constants for two RMF forces, namely NL-Z2 and PC-F1, are presented below [2]. The global minimum of χ^2 should correspond to the optimal set of coupling constants. Finding the global minimum is, however, a nontrivial and non-straightforward task. In practice, a combination of monte-carlo and downhill methods has been found to be quite successfull.

Table 1. Observables and chosen errors ΔO. E denotes the binding energy, $R_{\rm dms}$ the diffraction-radius, σ the surface thickness, r_{rms}^{ch} the rms charge radius, and $\Delta^{\rm p}$ and $\Delta^{\rm n}$ are the proton and neutron pairing gaps. A + indicates an observable contributing to the total χ^2.

observable	error	^{16}O	^{40}Ca	^{48}Ca	^{56}Ni	^{58}Ni	^{88}Sr	^{90}Zr	^{100}Sn	^{112}Sn	^{120}Sn	^{124}Sn	^{132}Sn	^{136}Xe	^{144}Sm	^{202}Pb	^{208}Pb	^{214}Pb
E	0.2 %	+	+	+	+	+	+	+	+	+	+	+	+	+	+	−	+	+
$R_{\rm dms}$	0.5 %	+	+	+	−	+	+	+	−	+	+	+	−	−	−	−	+	−
σ	1.5 %	+	+	+	−	−	−	+	−	−	−	−	−	−	−	−	+	−
$r_{\rm rms}^{\rm ch}$	0.5 %	−	+	+	+	+	+	+	−	+	−	+	−	−	−	+	+	+
Δ p	0.05 MeV	-	-	-	-	-	-	-	-	-	-	-	-	+	+	-	-	-
Δ n	0.05 MeV	-	-	-	-	-	-	-	-	+	+	+	-	-	-	-	-	-

As a first estimate of the quality of such a parameter adjustment, it is interesting to see how well the new force is able to describe the nuclei it was fitted to. In Fig. 3 we show the performance of the RMF-PC force PC-F1 and compare it to the RMF-FR force NL-Z2 which has been fitted with the same set of observables. We show comparisons for binding energy, diffraction radius, surface thickness and rms charge radius. One can see that the binding energy is described by both parametrizations with an accuracy of 4% or better for almost all nuclei. The density-related observables are reproduced within a few percent. Only for the diffraction radius both models remain within the desired 1.5% accuracy demanded in the adjustment

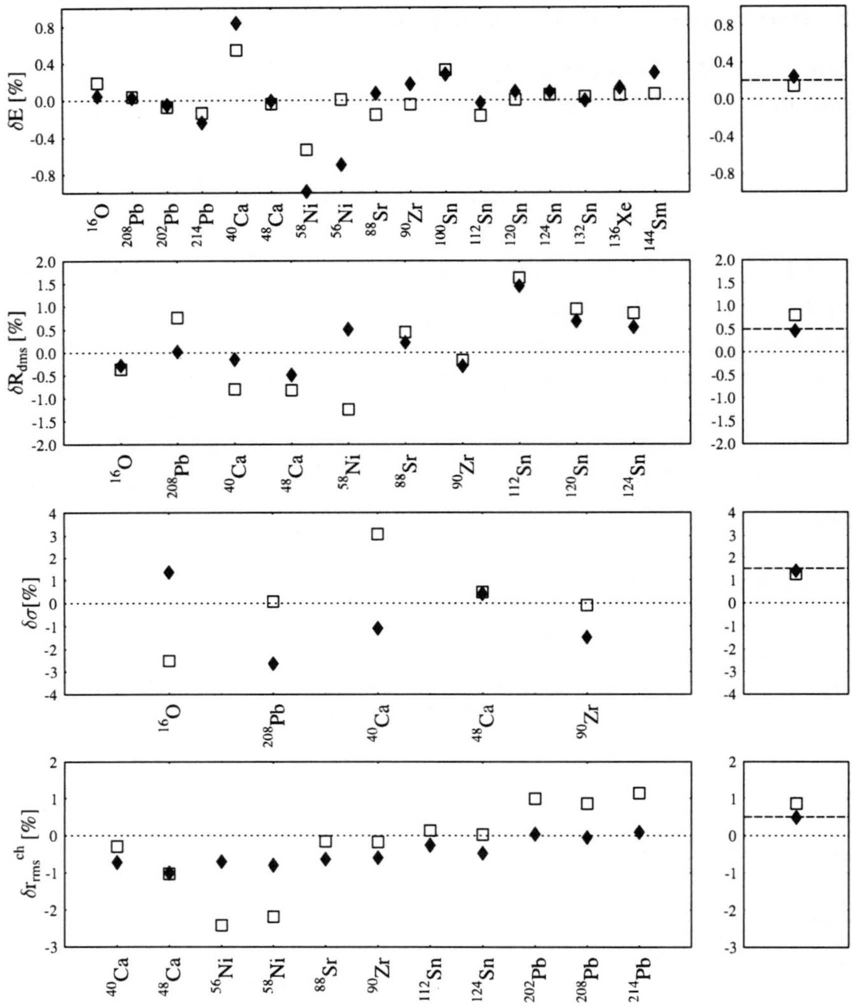

Fig. 3. Errors in percent for the observables binding energy, diffraction radius, surface thickness and rms charge radius for PC-F1 (filled diamonds) and NLZ2 (open squares) are seen on the left. The right panels show the absolute mean errors for the corresponding observables.

procedure. For the other observables the errors are larger than the values demanded in a few cases.

Taking a look at the mean errors in percent, the force NL-Z2 is clearly superior with respect to binding energies (it stays within the demanded acuracy in the fit) and with respect to surface thicknesses. It though cannot reach the demanded accuracy for the diffraction and mean square charge radii. Here PC-F1 performs better, staying within the tolerated errors. This appears to be a systematic difference in the forces which probably relates to the different parametrization of the density dependence.

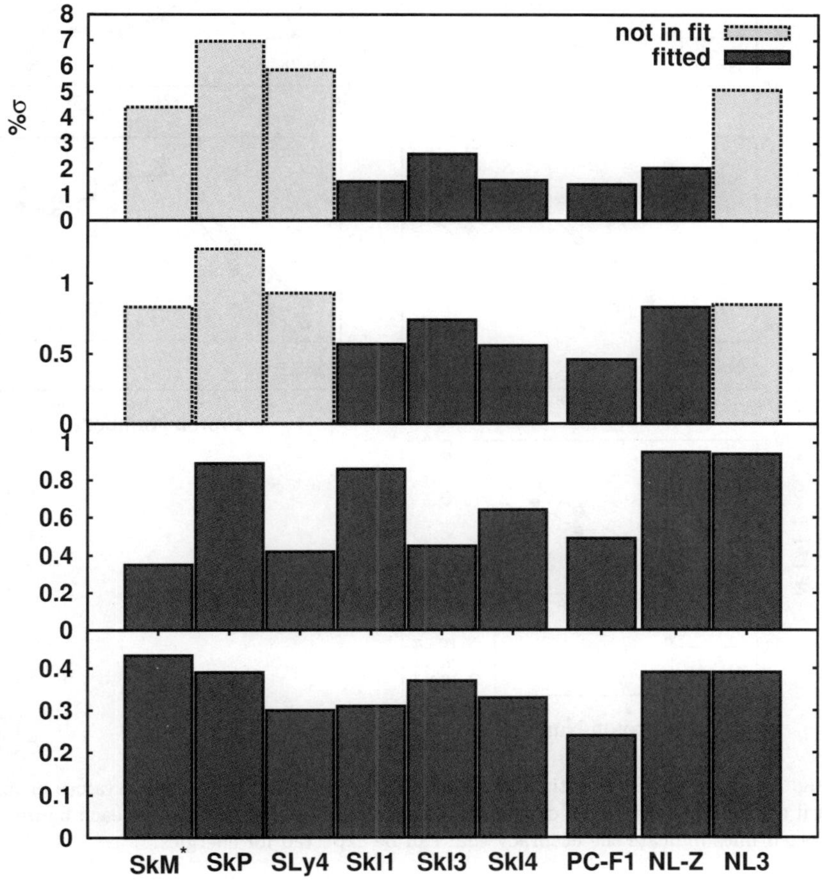

Fig. 4. Errors in percent for the mean-field forces as indicated and the observables (from top to bottom) surface thickness, diffraction radius, rms radius, and binding energy. Dark columns indicate that this observable has not been fitted, light colums indicate observables which have been part of the adjustment procedure.

It is further interesting to assess the quality of various forces with different adjustment procedures for various observables. Fig. 4 summarizes such a comparison for various Skyrme forces (SkM*, SkP, SLy4, SkI1, SkI3,

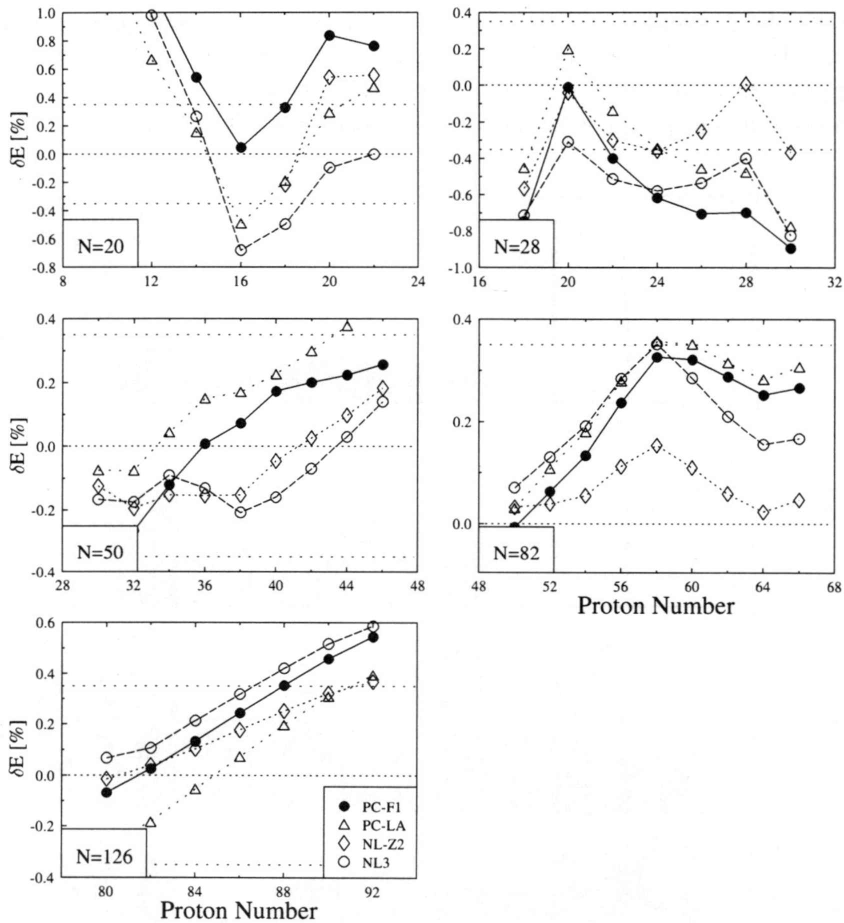

Fig. 5. Deviation in % of the calculated energies from the experimental values in spherical calculations of isotopic chains. Note that the scales are different for each figure. The dotted lines indicate the accuracy that can be expected for energies.

SkI4), a point-coupling RMF force (PC-F1) and two standard RMF forces (NL-Z and NL3). Shown is the error in percent for the observables (from top to bottom) surface thickness, diffraction radius, rms radius, and binding energy. These errors have been averaged over a selection of nuclei throughout the nuclear chart. Starting with the binding energy, this observable is always part of the adjustment procedure, and the quality of modern models can describe it with an accuracy below 0.5%. The rms radius as the prominent information about the nuclear density distribution is usually taken into account in the adjustment, and its mean error lies below 1 %. The other two form-factor-related observables, diffraction radius and surface thickness, are not taken into account in all adjustments. As can be inferred from the figure, its inclusion in the adjustment procedure can lead to a slightly better description. The biggest effect, however, is seen for the surface thickness. Firstly, the error scale is larger and ranges within several percent error. Secondly, an inclusion of this observable in the adjustment procedure can considerably decrease its error. Self-consistent mean-field models tend to underestimate the size of the surface thickness. This can be attributed to missing correlations and/or the missing long-range (pion) dynamics.

In Fig. 5 we show the results of spherical calculations of binding energies in isotopic chains. Shown are the deviations from the experimental values for the two RMF-PC forces and the RMF-FR forces disussed before. Note that the scales are different for each figure in order to emphasize the trends. The dotted lines at $\pm 0.35\%$ indicate the expected accuracy for nuclei that enter the fit.

To discuss the global trends, all forces have larger errors for the light systems, where the neglection of correlations in the mean-field approach has more serious consequences than for heavy nuclei. The heavier systems are described within an error of about 0.4%, with few exceptions. We also see that NL-Z2 performs best in most cases. Some slopes and bendings are also apparent in these plots for all forces that correspond to errors in the separation energies. They indicate yet unresolved isotopic and isotonic trends.

Another interesting observation can be made: the structure of the curves is, with differences in detail, similar for NLZ2 and PC-F1 in almost all cases (this is most striking for the Sn isotopes). It shows that the fitting strategy, and with it the choice of nuclei and observables, remains apparent in the binding energies, or, generally speaking, in the performance of the force.

Additionally one can observe that for some nuclei with proton or neutron numbers just beyond a magic number (see neutron number 84 for example in the Sn isotopes), errors get dramatically larger, indicating some deficiencies in the description of shell structure of these systems. Future improvements of these unresolved trends can be approached from at least three sides: On one hand, experimental data covering nuclei with a wide span of isospin can help to fix the isovector channel of our effective interaction by including them into the fit. Second, the isovector channel of the mean-field forces might not be flexibel enough at the present stage, so that additonal isovector terms together with an appropriate fitting strategy might cure the problem. Third, the inclusion of correlations into the calculations or, if possible, their absorption into the model ansatz might to some degree smoothen the large shell effects and the bends in the curves.

We now turn to superheavy nulei as an interesting application of these models.

4. Superheavy nuclei

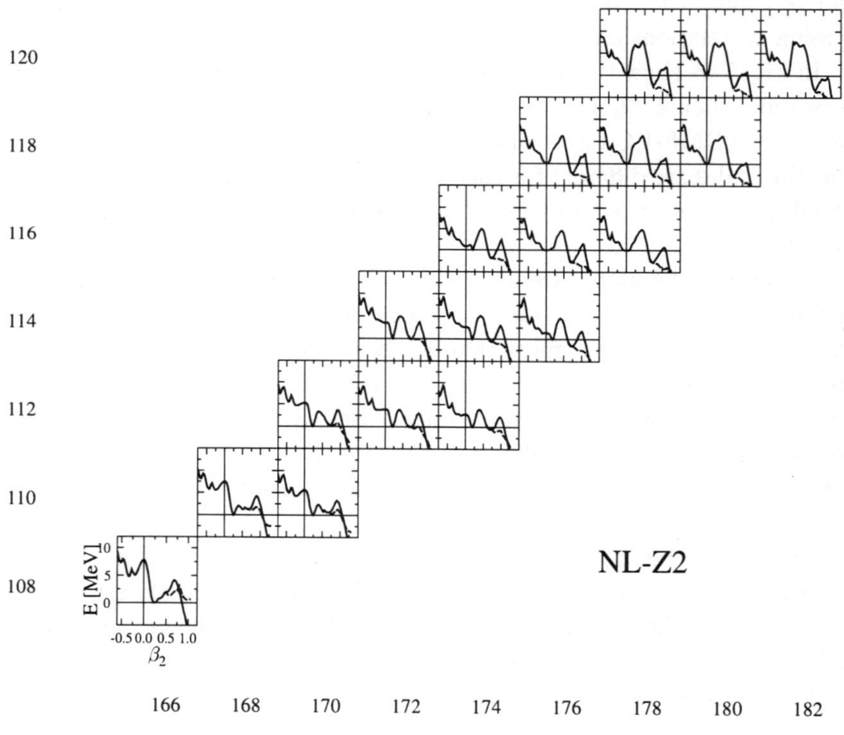

Fig. 6.

Superheavy nuclei are nuclei with proton numbers above $Z \approx 110$. These systems owe their stability only to quantum mechanical shell effects. While the nucleon-nucleon interaction is short-range and saturating, the Coulomb force, even though it is much weaker, has long-range nature. Thus, while a proton can only interact with its neighbors through the nuclear force, it feels all surrounding protons. Superheavy nuclei correspond to systems where the

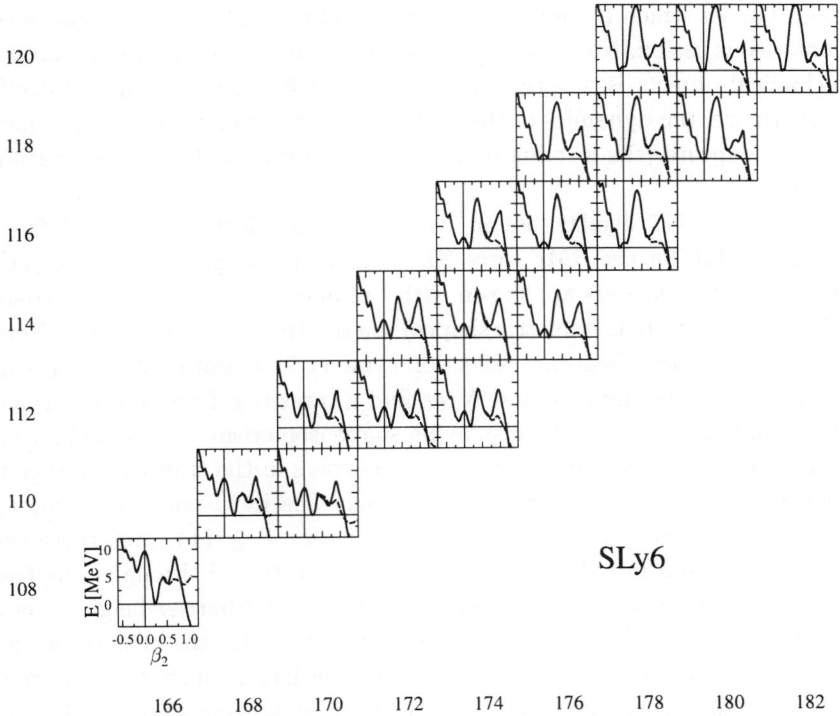

Fig. 7.

liquid-drop description predicts no stable ground state. Due to shell effects, however, the system can gain stability, and spontaneous fission proceeds through the fission barrier. For these systems, alpha decay and spontaneous fission become competing decay modes. While superheavy nuclei can on earth be only produced in experiments with heavy ions, they might come to existence in the universe in processes such as supernova explosions.

In order to gain insight into the stability towards spontaneous fission within the presented framework, once can compute fission barriers by employing a constraint on the total quadrupole moment β_2 of the nucleus, i.e., $\hat{H} \rightarrow \hat{H} - \lambda \hat{Q}_2$, where $\hat{H}$ denotes the nuclear Hamiltonian, and $\hat{Q}_2$ is the quadrupole operator. With increasing quadrupole deformation β_2, the nucleus deforms and assumes prolate (cigar-like) shapes. For very large quadrupole moments, one usually obtains two solutions: one corresponds to two separated fragments, the other solution corresponds to very elongated (but not yet separate) shapes. In the following we will focus on the (inner) fission barrier which the nucleus has to tunnel through in order to achieve spontaneous fission. Note that the fission dynamics of composite quantum-mechanical objects is a very fundamental and interesting topic by itself. Furthermore, the dynamics of the neck formation for elongated shapes and the neck rapture (the forming of two separated fragments) are fascinating topics.

Figs. 6 and 7 display fission barriers for a selection of nuclei from $Z = 108 - 120$ for the RMF force NL-Z2 and the Skyrme force SLy6 [11]. The symmetric barriers are shown with full lines, the asymmetric barriers (allowing reflection-asymmetric shapes) are shown with a dashed line. Note that the first barrier is always reflection-symmetric. In both cases the global trends agree: going up in Z and N we find a transition from well-deformed nuclei via transitional nuclei displaying shape isomerism to spherical nuclei at the upper right end of the chart. The barriers in the transitional region (especially for NL-Z2) become rather small. The inclusion of asymmetry makes the second (symmetric) barrier vanish for $Z \geq 114$, thus these superheavy systems only have one single barrier to tunnel through. The two forces, however, dramatically disagree on the actual barrier heights: they are up to a factor of two larger for the Skyrme force for the heaviest nuclei displayed – this would translate into several orders of magnitude discrepancy in predictions of life times. Investigations employing a larger selection of mean-field forces [11] revealed that these differences are not due to a specific force, but appear to be – at least for a similar adjustment protocol of the force – a model-dependent feature: RMF forces tend to smaller, SHF forces to quite high barriers.

Current research focusses on the various model ingredients and how they are related to these predictions, among them bulk properties such as asymmetry, and shell-structure-related model features.

Self-consistent models can yield predictions which are not possible with other models which prescribe the density or potential shape of a nucleus. A good example is the density distribution of the superheavy nucleus $^{292}120_{172}$ shown in Fig. 8, which is doubly magic in the relativistic approaches. The semi-bubble shape originates from the self-consistent procedure due to two effects: firstly, the nucleus is close to the proton dripline and thus contains a comparably small number of neutrons. Secondly, the last filled neutron single-particle states have large angular momenta and thus are localized at large radii, leaving a dip in the center of the nucleus. This explanation is supported by the fact that this semi-bubble structure fades away by adding more neutrons to the system. [12]

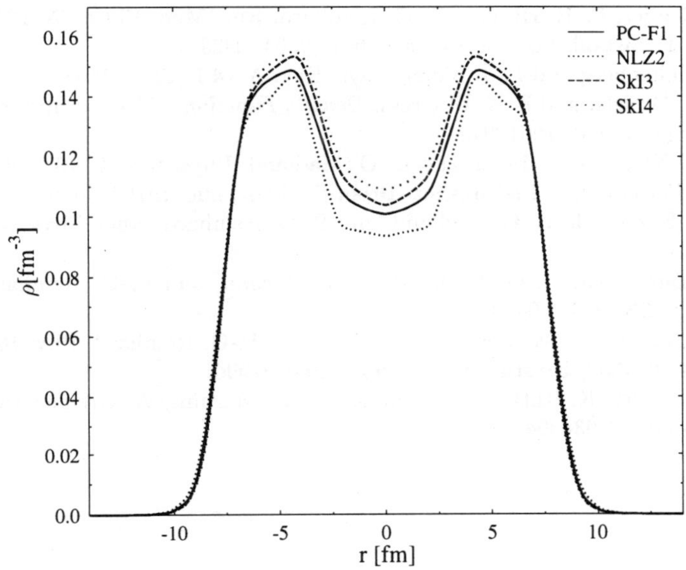

Fig. 8. Baryon density of the superheavy nucleus $^{292}120_{172}$ calculated with the two relativistic forces PC-F1 and NL-Z2 as well as the Skyrme forces SkI3 and SkI4.

5. Conclusions and Outlook

Self-consistent mean-field models have reached high predictive power and have become powerful and versatile tools to study nuclear ground states of medium, heavy, and super-heavy nuclei. Like any other nuclear model they possess, however, several short-comings that have to be addressed in future work. These models can be used as a basis for other schemes such as time-dependent Hartree-Fock, random phase approximation, generator coordinate method, which then extends their applicability to excited states, explicit treatment of correlations, and dynamical processes.

Acknowlegdement

The results reported go back to work that has been achieved together with my collaborators and mentors J. A. Maruhn, D. G. Madland, P.–G. Reinhard, and W. Greiner.

References

1. P. Ring, P. Schuck, *The Nuclear Many-Body Problem*, 2nd printing (Springer, 2000)

2. M. Bender, P.–H. Heenen, P.–G. Reinhard, Rev. Mod. Phys. **75**, 121 (2003)
3. R. J. Furnstahl, Lect.Notes Phys. 641 (2004) 1-29
4. W. Hohenberg and L. J. Sham, Phys. Rev. A **140** (1965) 1133
5. R. M. Dreizler and E. K. U. Gross, *Density Functional Theory* (Springer, 1990)
6. J. Engel, nucl-th/0610043
7. B. A. Nikolaus, T. Hoch, and D. G. Madland, Phys. Rev. C **46** (1992) 1757
8. R.J. Furnstahl, J. Piekarewicz, Brian D. Serot, nucl-th/0205048
9. T. J. Bürvenich, D. G. Madland, and P.–G. Reinhard, Nucl. Phys. **A 744**, 92 (2004)
10. T. Bürvenich, D. G. Madland, J. A. Maruhn, and P.–G. Reinhard, Phys. Rev. C **65** (2002) 044308
11. T. Buervenich, M. Bender, J. A. Maruhn, P.-G. Reinhard, Phys.Rev. C **69** (2004) 014307; Erratum-ibid. C **69** (2004) 029901
12. M. Bender, K. Rutz, P.-G. Reinhard, J. A. Maruhn, W. Greiner, Phys. Rev. C **60** (1999) 034304

II. NUCLEAR MULTIFRAGMENTATION AND EQUATION OF STATE

Isospin Transport in Heavy Ion Collisions and the Nuclear Equation of State

M.Di Toro

Laboratori Nazionali del Sud INFN, Physics-Astronomy Dept., Catania Univ.
Via S. Sofia 62, I-95123 Catania, Italy
E-mail: ditoro@lns.infn.it

Heavy Ion Collisions, HIC, represent a unique tool to probe the in-medium nuclear interaction in regions away from saturation and at high nucleon momenta. In this report we present a selection of reaction observables particularly sensitive to the isovector part of the interaction, i.e. to the symmetry term of the nuclear Equation of State, EoS. At low energies the behavior of the symmetry energy around saturation influences dissipation and fragment production mechanisms. A very good tracer appears to be the isospin transport during the reaction dynamics. Predictions are shown for deep-inelastic and fragmentation collisions induced by neutron rich projectiles. Differential flow measurements will also shed lights on the controversial neutron-proton effective mass splitting in asymmetric matter. The high density symmetry term can be derived from isospin effects on heavy ion reactions at higher energies, a few $AGeV$ range, that can even allow a *direct* study of the covariant structure of the isovector interaction in the hadron medium. We work within a relativistic transport frame, beyond a cascade picture, consistently derived from effective Lagrangians, where isospin effects are accounted for in the mean field and collision terms. Rather sensitive observables are proposed from collective flows and from pion-kaon production (π^-/π^+, K^0/K^+ yields). For the latter point relevant non-equilibrium effects are stressed. The possibility of the transition to a mixed hadron-quark phase, at high baryon and isospin density, is finally suggested. Some signatures could come from an expected *neutron trapping* effect.

1. Introduction

The symmetry energy E_{sym} appears in the energy density $\epsilon(\rho,\rho_3) \equiv \epsilon(\rho)+\rho E_{sym}(\rho_3/\rho)^2+O(\rho_3/\rho)^4+..$, expressed in terms of total ($\rho = \rho_p+\rho_n$) and isospin ($\rho_3 = \rho_p-\rho_n$) densities. The symmetry term gets a kinetic contribution directly from basic Pauli correlations and a potential part from the highly controversial isospin dependence of the effective interactions [1,2]. Both at sub-saturation and supra-saturation densities, predictions based of the existing many-body techniques diverge rather widely, see [3]. We remind that the knowledge of the *EoS* of asymmetric matter is very important at low densities (neutron skins, nuclear structure at the drip lines, neutron distillation in fragmentation, neutron star formation and crust..) as well as at high densities (transition to a deconfined phase, neutron star mass/radius, cooling, hybrid structure, formation of black holes...). We take advantage of new opportunities in theory (development of rather reliable microscopic transport codes for *HIC*) and in experiments (availability of very asymmetric radioactive beams, improved possibility of measuring event-by-event correlations) to present results that are severely constraining the existing effective interaction models. We will discuss dissipative collisions in a wide range of energies, from just above the Coulomb barrier up to a few $AGeV$. Low to Fermi energies will bring information on the symmetry term around (below) normal density, relativistic energies will probe high density regions, even testing the covariant structure of the isovector terms. The transport codes are based on mean field theories, with correlations included via hard nucleon-nucleon, elastic and inelastic, collisions and via stochastic forces, selfconsistently evaluated from the mean phase-space trajectory, see [2,4–6]. Stochasticity is essential in order to get distributions as well as to allow the growth of dynamical instabilities.

2. Isospin effects on Deep-Inelastic Collisions

Dissipative semi-peripheral collisions at low energies, including binary and three-body breakings, offer a good opportunity to study phenomena occurring in nuclear matter under extreme conditions with respect to shape, excitation energy, spin and N/Z ratio (isospin). At low energies the interaction times are rather long and therefore a large coupling among various mean-field modes is expected. In some cases, due to a combined Coulomb and angular momentum (deformation) effect, some instabilities can show up [7]. This can lead to 3-body breakings, where a light cluster is emitted from the neck region. Three body processes in collisions with exotic beams will allow to investigate how the development of surface (neck-like) instabilities, that would help ternary breakings, is sensitive to the structure of the symmetry term around (below) saturation.

In order to suggest proposals for the new *RIB* facility *Spiral* 2, [10] we have studied the reaction $^{132}Sn +^{64} Ni$ at $10AMeV$ in semicentral events, impact parameters $b = 6, 7, 8fm$, where one observes mostly binary exit channels, but still in presence of large dissipation [11]. Two different behaviors of the symmetry energy below saturation have been tested: one

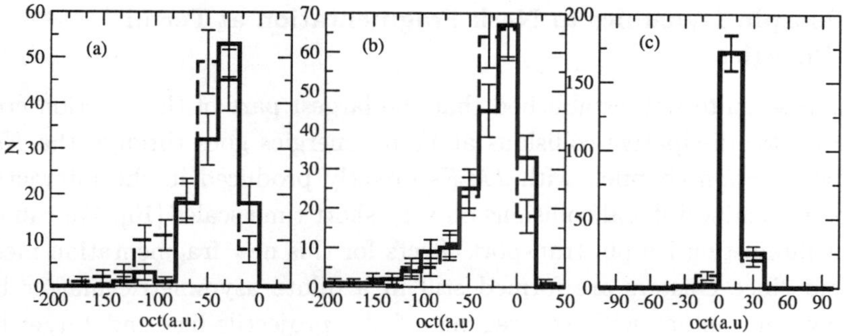

Fig. 1. Distribution of the octupole moment of primary fragments for the $^{132}Sn + ^{64}Ni$ reaction at 10 $AMeV$ (impact parameters (a):$b = 6fm$, (b):$7fm$, (c):$8fm$). Solid lines: asysoft. Dashed lines: asystiff

($asysoft$) where it is a smooth decreasing function towards low densities, and another one ($asystiff$) where we have a rapid decrease, [2]. The Wilczynski plots, kinetic energy loss vs. deflection angle, show slightly more dissipative events in the $asystiff$ case, consistent with the point that in the interaction at lower densities in very neutron-rich matter (the neck region) we have a less repulsive symmetry term [8]. In fact the neck dynamics is rather different in the two cases, as it can be well evidenced looking at the deformation of the PLF/TLF residues. The distribution of the octupole moment over the considered ensemble of events is shown in Fig.1 for the three considered impact parameters. Except for the most peripheral events, larger deformations, strongly suggesting a final 3-body outcome, are seen in the $asystiff$ case. In fact now, due to the lower value of the symmetry enrgy, the neutron-rich neck connecting the two systems survives a longer time leading to very deformed primary fragments, from which eventually small clusters will be dynamically emitted. Finally we expect to see effects of the different interaction times on the charge equilibration mechanism, probed starting from entrance channels with large N/Z asymmetries, like $^{132}Sn(N/Z = 1.64) +^{58} Ni(N/Z = 1.07)$. Moreover the equilibration mechanism is also directly driven by the strenght of the symmetry term. For more central collisions this can be studied via the direct measurement of the prompt Dynamical Dipole Emission, nucleus-nucleus collective bremsstrahlung radiation during the charge equilibration path, see [12–14].

3. Isospin Dynamics in Neck Fragmentation at Fermi Energies

It is now quite well established that the largest part of the reaction cross section for dissipative collisions at Fermi energies goes through the *Neck Fragmentation* channel, with IMFs directly produced in the interacting zone in semiperipheral collisions on very short time scales [15]. We can expect interesting isospin transport effects for this new fragmentation mechanism since clusters are formed still in a dilute asymmetric matter but always in contact with the regions of the projectile-like and target-like remnants almost at normal densities. Since the difference between local neutron-proton chemical potentials is given by $\mu_n - \mu_p = 4E_{sym}(\rho_3/\rho)$, we can expect a larger neutron flow to the neck clusters for a stiffer symmetry energy around saturation, [2,16]. The isospin dynamics can be directly extracted from correlations between N/Z, *alignement* and emission times of the IMFs. The alignment between $PLF - IMF$ and $PLF - TLF$ directions represents a very convincing evidence of the dynamical origin of the mid-rapidity fragments produced on short time scales [17]. The form of the Φ_{plane} distributions (centroid and width) can give a direct information on the fragmentation mechanism [18]. Recent calculations confirm that the light fragments are emitted first, a general feature expected for that rupture mechanism [19]. The same information can be derived from *direct* emission time measurements based on deviations from Viola systematics [21] observed in event-by-event velocity correlations between IMFs and the PLF/TLF residues [17,18,20]. We can figure out a continuous transition from fast produced fragments via neck instabilities to clusters formed in a dynamical fission of the projectile(target) residues up to the evaporated ones (statistical fission). Along this line it would be even possible to disentangle the effects of volume and shape instabilities. A neutron enrichment of the overlap ("neck") region is expected, due to the neutron migration from higher (spectator) to lower (neck) density regions. This effect is also nicely connected to the slope of the symmetry energy [19]. Neutron and/or light isobar measurements in different rapidity regions appear important.. A very nice new analysis has been performed on the $Sn + Ni$ data at 35 $AMeV$ by the Chimera Collab. [22], see Fig.2 left panel. The strong correlation between neutron enrichemnt and alignement (when the short emission time selection is enforced) can be well reproduced only with a stiff behavior of the symmetry energy [23], see the right panel. *This is the first clear evidence in favor of a relatively large slope (symmetry pressure) around saturation.*

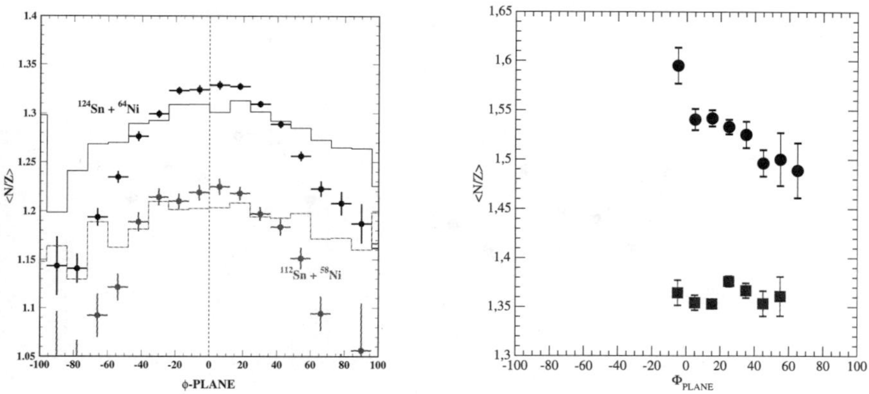

Fig. 2. Correlation between N/Z of IMF and *alignement* in ternary events of the $^{132}Sn + ^{64}Ni$ reaction at 35 $AMeV$. *Left panel.* Exp. results: points correspond to fast formed IMFs (Viola-violation selection); histogram for all IMFs at mid-rapidity (including statistical emissions). *Right Panel.* Simulation results: squares, asysoft; circles, asystiff

4. Effective Mass Splitting and Collective Flows

The problem of Momentum Dependence in the Isovector channel ($Iso-MD$) is still very controversial and it would be extremely important to get more definite experimental information, see the recent refs. [24–29]. Intermediate energies are important in order to have high momentum particles and to test regions of high baryon (isoscalar) and isospin (isovector) density during the reactions dynamics. Collective flows [30] are very good candidates since they are expected to be very sensitive to the momentum dependence of the mean field, see [2,31]. The transverse flow is given by $V_1(y, p_t) = \langle \frac{p_x}{p_t} \rangle$, where $p_t = \sqrt{p_x^2 + p_y^2}$ is the transverse momentum and y the rapidity along beam direction. It provides information on the anisotropy nucleon emission on the reaction plane. Very important for the reaction dynamics is the elliptic flow v_2 that can be written as $V_2(y, p_t) = \langle \frac{p_x^2 - p_y^2}{p_t^2} \rangle$ It measures the competition between in-plane and out-of-plane emissions. The sign of V_2 indicates the azimuthal anisotropy of emission: particles can be preferentially emitted either in the reaction plane ($V_2 > 0$) or out-of-plane ($squeeze - out$, $V_2 < 0$) [30,31].

We have then tested the $Iso - MD$ of the fields just evaluating the $Difference$ of neutron/proton transverse and elliptic flows $V_{1,2}^{(n-p)}(y, p_t) \equiv V_{1,2}^{n}(y, p_t) - V_{1,2}^{p}(y, p_t)$ at various rapidities and transverse momenta in semicentral ($b/b_{max} = 0.5$) $^{197}Au + ^{197}Au$ collisons at $250 AMeV$, where some proton data are existing from the $FOPI$ collaboration at GSI [32,33]. The

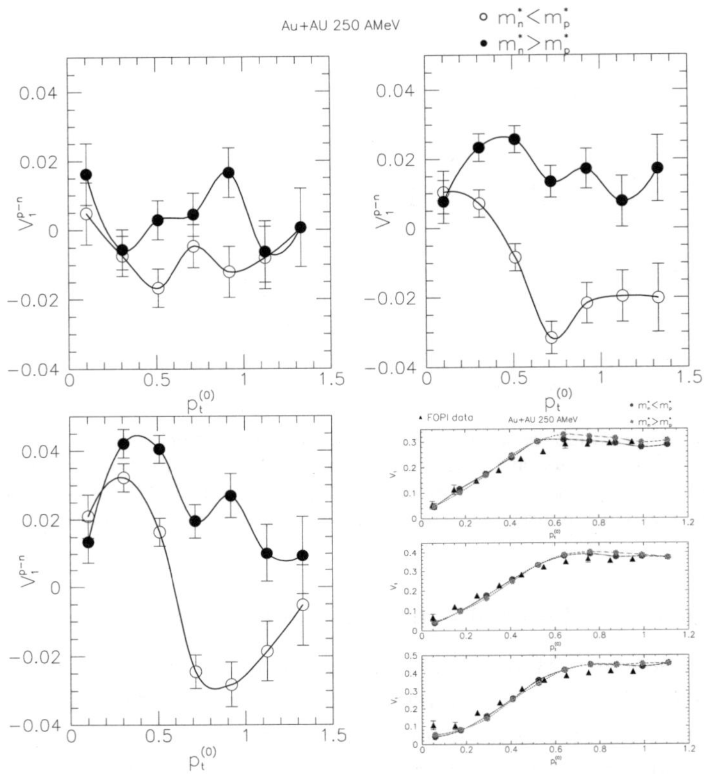

Fig. 3. Difference between proton and neutron V_1 flows in a semi-central reaction Au+Au at 250 AMeV for three rapidity ranges. Upper Left Panel: $|y^{(0)}| \leq 0.3$; Upper Right: $0.3 \leq |y^{(0)}| \leq 0.7$; Lower Left: $0.6 \leq |y^{(0)}| \leq 0.9$. Lower Right Panel: Comparison of the V_1 proton flow with FOPI data for three rapidity ranges. Top: $0.5 \leq |y^{(0)}| \leq 0.7$; center: $0.7 \leq |y^{(0)}| \leq 0.9$; bottom: $0.9 \leq |y^{(0)}| \leq 1.1$.

transport code has been implemented with a $BGBD - like$ [34,35] mean field with a different (n,p) momentum dependence, see [25–27], that allow to follow the dynamical effect of opposite n/p effective mass splitting while keeping the same density dependence of the symmetry energy.

For the difference of nucleon transverse flows, see Fig. 3, the mass splitting effect is evident at all rapidities, and nicely increasing at larger rapidities and transverse momenta, with more neutron flow when $m_n^* < m_p^*$. Just to show that our simulations give realistic results we compare in lower right panel of Fig. 3 with the proton data of the $FOPI$ collaboration, [32], for similar selections of impact parameters rapidities and transverse momenta. The agreement is rather good. We see a slightly reduced proton flow at high transverse momenta in the $m_n^* < m_p^*$ choice, but the effect is too small to be

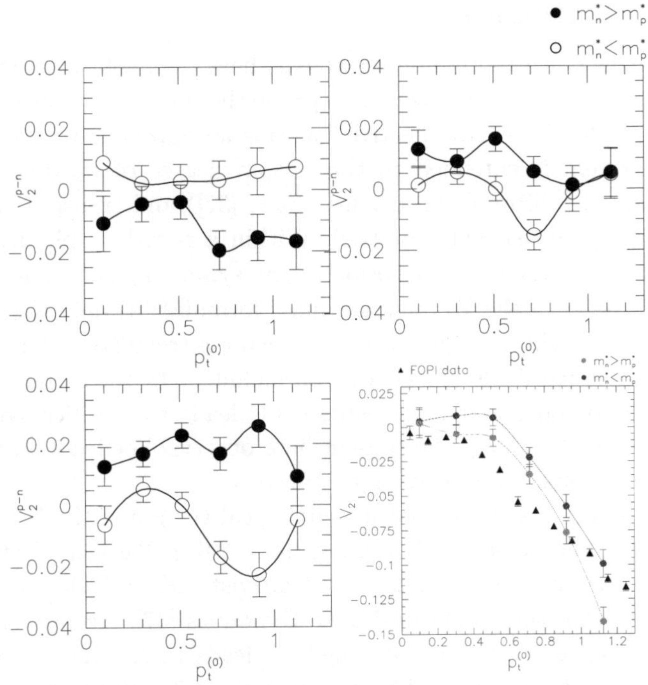

Fig. 4. Upper (left and right) and lower left panels: Difference between proton and neutron elliptic flows for the same reaction and rapidity ranges as in Fig. 3. Lower right panel: Comparison of the elliptic proton flow with FOPI data (Central bin, $|y^{(0)}| \leq 0.1$).

seen from the data. Our suggestion of measuring just the difference of n/p flows looks much more promising. The same analysis has been performed for the difference of elliptic flows, see Fig. 4. Again the mass splitting effects are more evident for higher rapidity and tranverse momentum selections. In particular the differential elliptic flow becomes negative when $m_n^* < m_p^*$, revealing a faster neutron emission and so more neutron squeeze out (more spectator shadowing). In the lower right panel we also show a comparison with recent proton data from the *FOPI* collaboration, [33] . The agreement is still satisfactory. As expected the proton flow is more negative (more proton squeeeze out) when $m_n^* > m_p^*$. It is however difficult to draw definite conclusions only from proton data. Again the measurement of a n/p flow difference appears essential. This could be in fact an experimental problem due to the difficulties in measuring neutrons. Our suggestion is to measure the difference between light isobar flows, like triton vs. ^3He and so on. We expect to clearly see the effective mass splitting effects, maybe even enhanced due to larger overall flows shown by clusters, see [2,36].

5. Relativistic Collisions

Finally we focus our attention on relativistic heavy ion collisions, that provide a unique terrestrial opportunity to probe the in-medium nuclear interaction at high densities. An effective Lagrangian approach to the hadron interacting system is extended to the isospin degree of freedom: within the same frame equilibrium properties (EoS, [37]) and transport dynamics [38,39] can be consistently derived. Within a covariant picture of the nuclear mean field, for the description of the symmetry energy at saturation (a_4 parameter of the Weizsäecker mass formula) (a) only the Lorentz vector ρ mesonic field, and (b) both, the vector ρ (repulsive) and scalar δ (attractive) effective fields [40,41] can be included. In the latter case the competition between scalar and vector fields leads to a stiffer symmetry term at high density [2,40]. We present here observable effects, in fact enhanced, in the dynamics of heavy ion collisions.
The starting point is a simple phenomenological version of the Non-Linear (with respect to the iso-scalar, Lorentz scalar σ field) Walecka effective theory which corresponds to the Hartree or Relativistic Mean Field (RMF) approximation within the Quantum-Hadro-Dynamics [37]. According to this model the presence of the hadronic medium leads to effective masses and momenta $M^* = M + \Sigma_s$, $k^{*\mu} = k^\mu - \Sigma^\mu$, with Σ_s, Σ^μ scalar and vector self-energies. For asymmetric matter the self-energies are different for protons and neutrons, depending on the isovector meson contributions. We will call the corresponding models as $NL\rho$ and $NL\rho\delta$, respectively, and just NL the case without isovector interactions. For the description of heavy ion collisions we solve the covariant transport equation of the Boltzmann type [38,39] within the Relativistic Landau Vlasov (RLV) method, using phase-space Gaussian test particles [42], and applying a Monte-Carlo procedure for the hard hadron collisions. The collision term includes elastic and inelastic processes involving the production/absorption of the $\Delta(1232MeV)$ and $N^*(1440MeV)$ resonances as well as their decays into pion channels, [43,44]. A larger repulsive vector contribution to the neutron energies is given by the ρ-coupling. This is rapidly increasing with density when the δ field is included [2,40]. As a consequence we expect a good sensitivity to the covariant structure of the isovector fields in nucleon emission and particle production data. Moreover the presence of a *Lorentz magnetic* term in the relativistic transport equation [2,38,39] will enhance the dynamical effects of vector fields [45].

Differential flows will be also directly affected. In Fig.5 transverse and elliptic differential flows are shown for the $^{132}Sn+^{124}Sn$ reaction at 1.5 $AGeV$

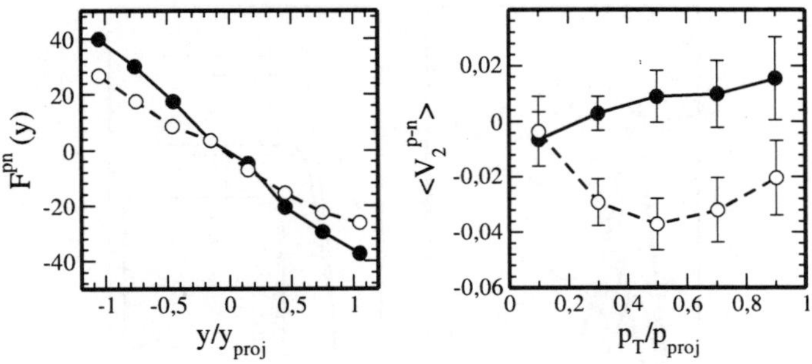

Fig. 5. Differential neutron-proton flows for the $^{132}Sn +^{124} Sn$ reaction at 1.5 $AGeV$ ($b = 6fm$) from the two different models for the isovector mean fields. Left: Transverse Flows. Right: Elliptic Flows. Full circles and solid line: $NL\rho\delta$. Open circles and dashed line: $NL\rho$.

($b = 6fm$), [45]. The effect of the different structure of the isovector channel is clear. Particularly evident is the splitting in the high p_t region of the elliptic flow. In the $(\rho + \delta)$ dynamics the high-p_t neutrons show a much larger *squeeze – out*. This is fully consistent with an early emission (more spectator shadowing) due to the larger ρ-field in the compression stage. We expect similar effects, even enhanced, from the measurements of differential flows for light isobars, like ^3H *vs.* ^3He.

6. Isospin effects on sub-threshold kaon production at intermediate energies

Kaon production has been proven to be a reliable observable for the high density EoS in the isoscalar sector [46,47] Here we show that the $K^{0,+}$ production (in particular the K^0/K^+ yield ratio) can be also used to probe the isovector part of the EoS. Using our RMF transport approach we analyze pion and kaon production in central $^{197}Au +^{197} Au$ collisions in the 0.8 – 1.8 $AGeV$ beam energy range, comparing models giving the same "soft" EoS for symmetric matter and with different effective field choices for E_{sym}. Here we also use a Lagrangian with density dependent couplings (DDF, see [41]), recently suggested for better nucleonic properties of neutron stars [48]. In the DDF model the ρ-coupling is exponentially decreasing with density, resulting in a rather "soft" symmetry term at high density. The hadron mean field propagation, which goes beyond the "collision cascade" picture, is essential for particle production yields: in particular the isospin dependence of the self-energies directly affects the energy balance

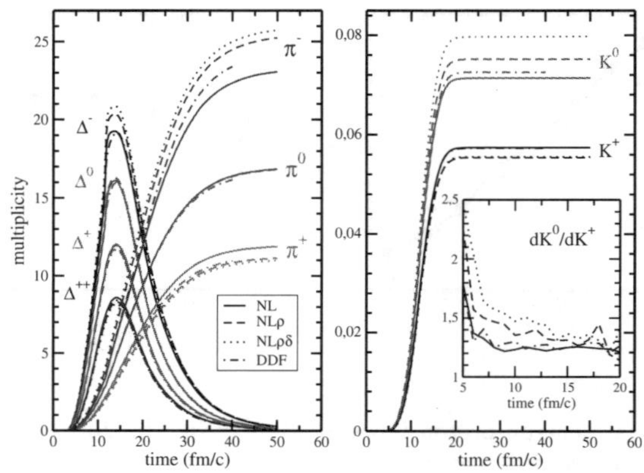

Fig. 6. Time evolution of the $\Delta^{\pm,0,++}$ resonances and pions $\pi^{\pm,0}$ (left), and kaons ($K^{+,0}$ (right) for a central ($b = 0$ fm impact parameter) Au+Au collision at 1 AGeV incident energy. Transport calculation using the $NL, NL\rho, NL\rho\delta$ and DDF models for the iso-vector part of the nuclear EoS are shown.

of the inelastic channels.

Fig. 6 reports the temporal evolution of $\Delta^{\pm,0,++}$ resonances, pions ($\pi^{\pm,0}$) and kaons ($K^{+,0}$) for central Au+Au collisions at $1AGeV$. It is clear that, while the pion yield freezes out at times of the order of $50fm/c$, i.e. at the final stage of the reaction (and at low densities), kaon production occur within the very early (compression) stage, and the yield saturates at around $20fm/c$. From Fig. 6 we see that the pion results are weakly dependent on the isospin part of the nuclear mean field. However, a slight increase (decrease) in the π^- (π^+) multiplicity is observed when going from the NL (or DDF) to the $NL\rho$ and then to the $NL\rho\delta$ model, i.e. increasing the vector contribution f_ρ in the isovector channel. This trend is more pronounced for kaons, see the right panel, due to the high density selection of the source and the proximity to the production threshold.

When isovector fields are included the symmetry potential energy in neutron-rich matter is repulsive for neutrons and attractive for protons. In a HIC this leads to a fast, pre-equilibrium, emission of neutrons. Such a *mean field* mechanism, often referred to as isospin fractionation [1,2], is responsible for a reduction of the neutron to proton ratio during the high density phase, with direct consequences on particle production in inelastic NN collisions. *Threshold* effects represent a more subtle point. The energy conservation in a hadron collision in general has to be formulated in terms of the canonical momenta, i.e. for a reaction $1 + 2 \rightarrow 3 + 4$ as $s_{in} = (k_1^\mu + k_2^\mu)^2 = (k_3^\mu + k_4^\mu)^2 = s_{out}$. Since hadrons are propagating with effective (kinetic) momenta and masses, an equivalent relation should be formulated starting from the effective in-medium quantities $k^{*\mu} = k^\mu - \Sigma^\mu$ and $m^* = m + \Sigma_s$, where Σ_s and Σ^μ are the scalar and vector self-energies. The self-energy contributions will influence the particle production at the level of thresholds as well as of the phase space available in the final channel. In neutron-rich colliding systems *Mean field* and *threshold* effects are acting in opposite directions on particle production. As an example, nn collisions excite $\Delta^{-,0}$ resonances which decay mainly to π^-. In a neutron-rich matter the mean field effect pushes out neutrons making the matter more symmetric and thus decreasing the π^- yield. The threshold effect on the other hand is increasing the rate of π^-'s due to the enhanced production of the Δ^- resonances: now the $nn \to p\Delta^-$ process is favored (with respect to $pp \to n\Delta^{++}$) since more effectively a neutron is converted into a proton. At lower energies the threshold effects (i.e. the self energy contributions) are dominant, as we see from our results. Such interplay between the two mechanisms cannot be fully included in a non-relativistic dynamics, in particular in calculations where the baryon symmetry potential is treated classically [49,50].

Finally the beam energy dependence of the π^-/π^+ (top) and K^0/K^+ (bottom) ratios is shown in Fig. 7. At each energy we see an increase of the yield ratios with the models $NL \to DDF \to NL\rho \to NL\rho\delta$. The effect is larger for the K^0/K^+ compared to the π^-/π^+ ratio. This is due to the subthreshold production and to the fact that the isospin effect enters twice in the two-step production of kaons, see [51]. Between the two extreme DDF and $NL\rho\delta$ isovector interaction models, the variations in the ratios are of the order of $14-16\%$ for kaons, while they reduce to about $8-10\%$ for pions. Interestingly the Iso-EoS effect for pions is increasing at lower energies, when approaching the production threshold.

We have to note that in a previous study of kaon production in excited nuclear matter the dependence of the K^0/K^+ yield ratio on the effective isovector interaction appears much larger (see Fig.8 of ref. [43]). The point is that in the non-equilibrium case of a heavy ion collision the asymmetry of the source where kaons are produced is in fact reduced by the $n \to p$ "transformation", due to the favored $nn \to p\Delta^-$ processes. This effect is almost absent at equilibrium due to the inverse transitions, see Fig.3 of ref. [43]. Moreover in infinite nuclear matter even the fast neutron emission is not present. This result clearly shows that chemical equilibrium models can lead to uncorrect results when used for transient states of an *open* system.

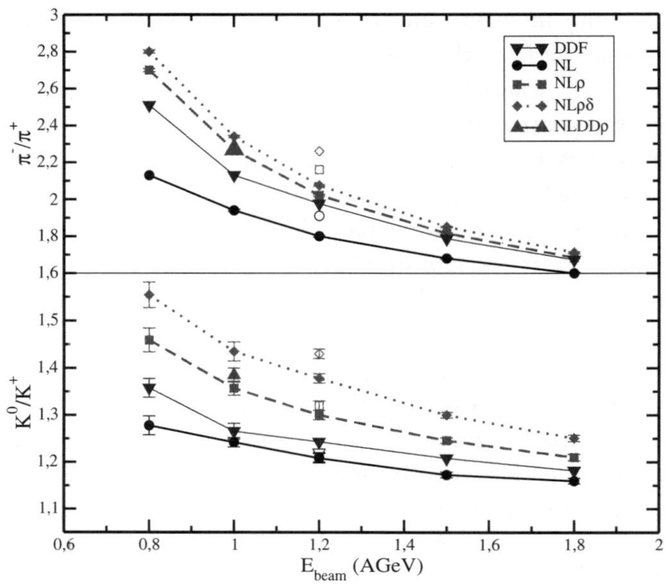

Fig. 7. π^-/π^+ (top) and K^+/K^0 (bottom) ratios as a function of the incident energy for the same reaction and models as in Fig. 6. In addition we present, for $E_{beam} = 1\ AGeV$, $NL\rho$ results with a density dependent ρ-coupling (triangles), see text. The *open* symbols at 1.2 $AGeV$ show the corresponding results for a $^{132}Sn +^{124} Sn$ collision, more neutron rich. Note the different scale for the π^-/π^+ ratios.

In order to further stress the distinction between effects of the stiffness of the symmetry energy and the detailed Lorentz structure of the isovector part of the effective Lagrangian, we also show the results for the K^0/K^+ with another parametrization of E_{sym}. This model, $NLDD\rho$, is a variant of $NL\rho$ with a density dependent ρ-coupling, built in such a way as to reproduce the same stiffer $E_{sym}(\rho_B)$ of the $NL\rho\delta$ model (see also ref. [45]). The results for the π^-/π^+ and K^0/K^+ ratios are shown in Fig. 7 for $E_{beam} = 1.0\ AGeV$ as triangles. We see that they are closer to the $NL\rho$ results (with a constant f_ρ) than to the ones of the $NL\rho\delta$ choice which has the same iso-stiffness. This nicely confirms that the differences observed going from the $NL\rho$ to the $NL\rho\delta$ parametrization are not due to the slightly increased stiffness of $E_{sym}(\rho_B)$, but more specifically to the competition between the attractive scalar δ-field and the repulsive vector ρ-field in the isovector channel, which leads to the increase of the vector coupling.. In the same Fig. 7 we also report results at 1.2 $AGeV$ for the $^{132}Sn +^{124} Sn$ reaction, induced by a radioactive beam, with an overall larger asymmetry (open symbols). The isospin effects are clearly enhanced. We note that the

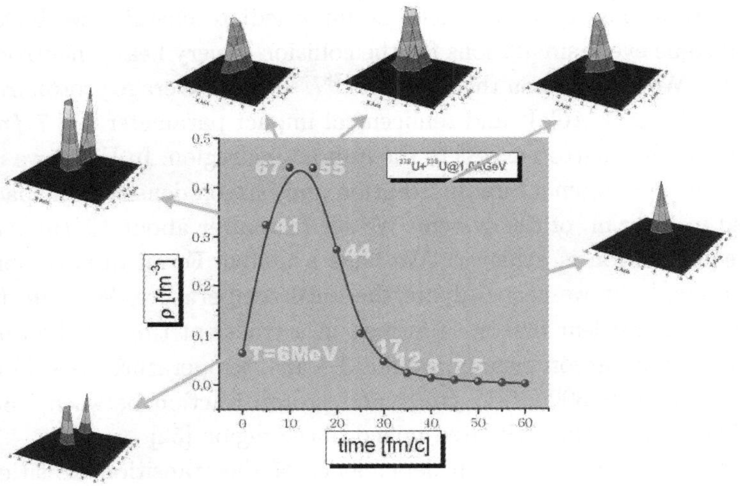

Fig. 8. $^{238}U +^{238} U$, 1 $AGeV$, semicentral. Correlation between density, temperature, momentum thermalization inside a cubic cell, 2.5 fm wide, in the center of mass of the system.

isospin effects on the kaon inclusive yield ratios at the freeze-out appear not too strong, although accessible. It seems important to select more exclusive kaon observables, in particular with a trigger related to an early time K-production (see the insert in Fig.6). A transverse momentum selection of pion yields, corresponding to a higher density source, should also be rather sensitive to isospin effects, in particular at lower energies, closer to the production threshold. A large asymmetry of the colliding matter is in any case of relevance. In this sense our work strongly supports the study of particle production at the new relativistic radioactive beam facilities.

7. Testing Deconfinement at High Isospin Density

The hadronic matter is expected to undergo a phase transition into a deconfined phase of quarks and gluons at large densities and/or high temperatures. On very general grounds, the transition critical densities are expected to depend on the isospin of the system, but no experimental tests of this dependence have been performed so far. Moreover, up to now, data on the phase transition have been extracted from ultrarelativistic collisions, when large temperatures but low baryon densities are reached. In order to check the possibility of observing some precursor signals of some new physics even in collisions of stable nuclei at intermediate energies we have performed some event simulations for the collision of very heavy, neutron-rich, elements. We have chosen the reaction $^{238}U + ^{238}U$ (average proton fraction $Z/A = 0.39$) at 1 $AGeV$ and semicentral impact parameter $b = 7\ fm$ just to increase the neutron excess in the interacting region. In Fig. 8 we report the evolution of momentum distribution and baryon density in a space cell located in the c.m. of the system. We see that after about 10 fm/c a nice local equilibration is achieved. We have a unique Fermi distribution and from a simple fit we can evaluate the local temperature. We note that a rather exotic nuclear matter is formed in a transient time of the order of 10 fm/c, with baryon density around $3 - 4\rho_0$, temperature $50 - 60\ MeV$, energy density $\approx 500\ MeV\ fm^{-3}$ and proton fraction between 0.35 and 0.40, likely inside the estimated mixed phase region [52] .

Here we study the isospin dependence of the transition densities [53] in a systematic way. Concerning the hadronic phase, we use the relativistic non-linear model of Glendenning-Moszkowski (in particular the "soft" $GM3$ choice) [54], where the isovector part is treated just with ρ meson coupling, and the iso-stiffer $NL\rho\delta$ interaction. For the quark phase we consider the MIT bag model with various bag pressure constants. In particular we are interested in those parameter sets which would allow the existence of quark stars [55], i.e. parameters sets for which the so-called Witten-Bodmer hypothesis is satisfied [56,57]. One of the aim of our work it to show that if quark stars are indeed possible, it is then very likely to find signals of the formation of a mixed quark-hadron phase in intermediate-energy heavy-ion experiments [52].

The structure of the mixed phase is obtained by imposing the Gibbs conditions [58] for chemical potentials and pressure and by requiring the conservation of the total baryon and isospin densities

$$\mu_B^{(H)} = \mu_B^{(Q)}, \quad \mu_3^{(H)} = \mu_3^{(Q)}, \quad P^{(H)}(T, \mu_{B,3}^{(H)}) = P^{(Q)}(T, \mu_{B,3}^{(Q)}),$$
$$\rho_B = (1-\chi)\rho_B^H + \chi\rho_B^Q, \quad \rho_3 = (1-\chi)\rho_3^H + \chi\rho_3^Q, \tag{1}$$

where χ is the fraction of quark matter in the mixed phase. In this way we get the *binodal* surface which gives the phase coexistence region in the (T, ρ_B, ρ_3) space [53,58]. For a fixed value of the conserved charge ρ_3 we will study the boundaries of the mixed phase region in the (T, ρ_B) plane. In the hadronic phase the charge chemical potential is given by $\mu_3 = 2E_{sym}(\rho_B)\frac{\rho_3}{\rho_B}$. Thus, we expect critical densities rather sensitive to the isovector channel in the hadronic EoS. In Fig. 9 we show the crossing density ρ_{cr} separating nuclear matter from the quark-nucleon mixed phase, as a function of the proton fraction Z/A. We can see the effect of the δ-coupling towards an *earlier* crossing due to the larger symmetry repulsion at high baryon densities. In the same figure we report the paths in the $(\rho, Z/A)$ plane followed in the c.m. region during the collision of the n-rich ^{132}Sn+^{132}Sn system, at different energies. At 300 $AMeV$ we are just reaching the border of the mixed phase, and we are well inside it at 1 $AGeV$. Statistical fluctuations could help in reducing the density at which drops of

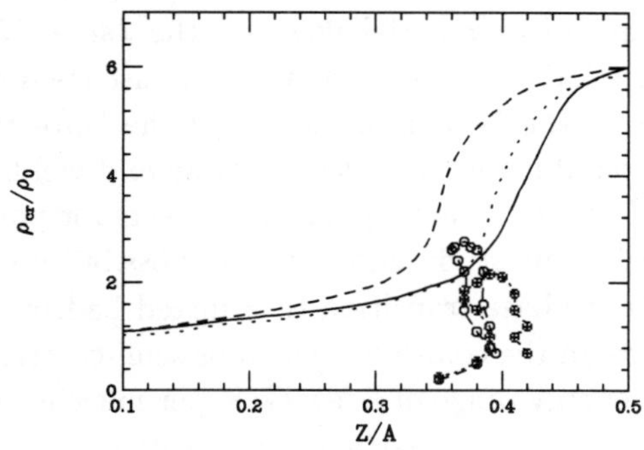

Fig. 9. Variation of the transition density with proton fraction for various hadronic EoS parameterizations. Dotted line: $GM3$ parametrization; dashed line: $NL\rho$ parametrization; solid line: $NL\rho\delta$ parametrization. For the quark EoS, the MIT bag model with $B^{1/4}$=150 MeV. The points represent the path followed in the interaction zone during a semi-central ^{132}Sn+^{132}Sn collision at 1 $AGeV$ (circles) and at 300 $AMeV$ (crosses).

quark matter form. The reason is that a small bubble can be energetically favored if it contains quarks whose Z/A ratio is *smaller* than the average value of the surrounding region. This corresponds to a *neutron trapping* effect, supported also by a symmetry energy difference in the two phases. In fact while in the hadron phase we have a large neutron potential repulsion (in particular in the $NL\rho\delta$ case), in the quark phase we only have the much smaller kinetic contribution. If in a pure hadronic phase neutrons are quickly emitted or "transformed" in protons by inelastic collisions, when the mixed phase starts forming, neutrons are kept in the interacting system up to the subsequent hadronization in the expansion stage [52]. Observables related to such neutron "trapping" could be an inversion in the trend of the formation of neutron rich fragments and/or of the π^-/π^+, K^0/K^+ yield ratios for reaction products coming from high density regions, i.e. with large transverse momenta. In general we would expect a modification of the rapidity distribution of the emitted "isospin", with an enhancement at mid-rapidity joint to large event by event fluctuations.

8. Perspectives

We have shown that *violent* collisions of n-rich heavy ions from low to relativistic energies can bring new information on the isovector part of the in-medium interaction, qualitatively different from equilibrium EoS properties. We have presented quantitative results in a wide range of beam energies. At low energies we have shown isospin effects on the dissipation in deep inelastic collisions, at Fermi energies the Iso-EoS sensitivity of the isospin transport in fragment reactions and finally at intermediate energies the dependence of differential flows on the $Iso - MD$ and effective mass splitting. In relativistic collisions we have discussed the possibility of a direct *measure* of the Lorentz structure of the isovector fields at high baryon density, from differential collective flows and yields of charged pion and kaon ratios. Important non-equilibrium effects for particle production are stressed. Finally our study supports the possibility of observing precursor signals of the phase transition to a mixed hadron-quark matter at high baryon density in the collision, central or semi-central, of neutron-rich heavy ions in the energy range of a few GeV per nucleon. As signatures we suggest to look at observables particularly sensitive to the expected different isospin content of the two phases, which leads to a neutron trapping in the quark clusters. The isospin structure of hadrons produced at high transverse momentum should be a good indicator of the effect. A large asymmetry of the colliding matter is in any case of relevance.
In conclusion the results presented here appear very promising for the possibility of exciting new results from dissipative collisions with radioactive beams.

Acknowledgements
This report is deeply related to ideas and results reached in very pleasant and fruitful collaborations with extremely nice people: V.Baran, M.Colonna, A.Drago, G.Fabbri, G.Ferini, Ch.Fuchs, Th.Gaitanos, V.Greco, G.Lalazissis, A.Lavagno, R.Lionti, B.Liu, S.Maccarone, F.Matera, T.Mikhailova, M.Zielinska-Pfabe', V.Prassa, J.Rizzo, E.Santini, L.Scalone, S.Yildirim and H.H.Wolter. I have learnt a lot from all of them in physics as well as in human relationships.

References

1. B.A. Li, W.U. Schroeder (Eds.), Isospin Physics in Heavy-Ion Collisions at Intermediate Energies, Nova Science, New York, 2001.
2. V.Baran, M.Colonna, V.Greco, M.Di Toro, *Phys. Rep.* **410** (2005) 335.
3. C.Fuchs, H.H.Wolter, *Modelization of the EoS,nucl-th/0511070*, *Eur. Phys.*

Jour. **A** 2006 in press
4. A.Guarnera, M.Colonna, P.Chomaz, *Phys. Lett.* **B373** (1996] 267.
5. M.Colonna et al., *Nucl. Phys.* **A642** (1998) 449.
6. P.Chomaz, M.Colonna, J.Randrup, *Phys. Rep.* **389** (2004) 263.
7. M.Colonna, M.Di Toro, A.Guarnera, *Nucl. Phys.* **A589** (1995) 160.
8. It is interesting to note that opposite effect of the symmetry stiffness on dissipation is expected at the Fermi energies, i.e. larger interaction times for the asysoft case, see [9]. This can be easily understood since at higher energies we are testing suprasaturation densities in the interacting region, with opposite trend of the symmetry repulsion.
9. M.Colonna, M.Di Toro, G.Fabbri, S.Maccarrone, *Phys. Rev.* **C57** (1998) 1410.
10. M.Lewitovicz, *Challenges of the SPIRAL 2 Project*, Proc. IX Nucleus-Nucleus Collisions, Rio de Janeiro 2006, *Nucl. Phys.* **A** in press.
11. Letter of Intent for Spiral2, *Dynamics and Thermodynamics of Exotic Nuclear Systems*, contact person G.Verde, October 2006.
12. C.Simenel, P.Chomaz, G.de France, *Phys. Rev. Lett.* **86** (2000) 2971.
13. V.Baran, D.M.Brink, M.Colonna, M.Di Toro, *Phys. Rev. Lett.* **87** (2001) 335.
14. D.Pierroutsakou et al., *Phys. Rev.* **C71** (2005) 054605.
15. M.Di Toro, A.Olmi, R.Roy, *Neck Dynamics,Eur. Phys. Jour.* **A** 2006 in press
16. V.Baran et al.,, *Phys. Rev.* **C72** (2005) 064620.
17. V.Baran, M.Colonna, M.Di Toro, *Nucl. Phys.* **A730** (2004) 329.
18. E. De Filippo et al. (Chimera Collab.), *Phys. Rev.* **C71** (2005) 044602; *Phys. Rev.* **C71** (2005) 064604.
19. R.Lionti, V.Baran, M.Colonna, M.Di Toro, *Phys. Lett.* **B625** (2005) 33.
20. J.Wilczynski et al. (Chimera Collab.), *Int. Jour. Mod. Phys.* **E14** (2005) 353.
21. It has been proposed to call such Viola-violation-correlation plot as *Wilczynski* − 2 *Plot*. Apart the origin from discussions with Janusz at the LNS/INFN Catania, in fact this correlation represents also a *chronometer* of the fragment formation mechanism. In this sense it is a nice Fermi energy complement of the famous *Wilczynski* − *Plot* which gives the time-scales in Deep-Inelastic Collisions.
22. E. De Filippo et al. (Chimera Collab.), Proc. IX Nucleus-Nucleus Collisions, Rio de Janeiro 2006, *Nucl. Phys.* **A** in press.
23. Analysis performed by V.Baran in August 06, using a Stochastic Mean Field Transport Code [2].
24. B.-A. Li, B.Das Champak, S.Das Gupta, C.Gale, *Nucl. Phys.* **A735** (2004) 563.
25. J.Rizzo, M.Colonna, M.Di Toro, V.Greco, *Nucl. Phys.* **732** (2004) 202.
26. M.Di Toro, M.Colonna, J.Rizzo, AIP Conf.Proc.**791** (2005) 70-83
27. J.Rizzo, M.Colonna, M.Di Toro *Phys. Rev.* **C72** (2005) 064609.
28. W.Zuo, L.G.Cao, B.-A.Li, U.Lombardo, C.W.Shen, *Phys. Rev.* **C72** (2005) 014005.
29. E. van Dalen, C.Fuchs, A.Fässler, *Phys. Rev. Lett.* **95** (2005) 022302.
30. J.Y. Ollitrault, *Phys. Rev.* **D46** (1992) 229.
31. P. Danielewicz, *Nucl. Phys.* **A673** (2000) 375.

32. A.Andronic et al., FOPI Collab., *Phys. Rev.* **C67** (2003) 034907.
33. A.Andronic et al., FOPI Collab., *Phys. Lett.* **B612** (2005) 173.
34. C.Gale, G.F.Bertsch, S.Das Gupta, *Phys.Rev.* **C41** (1990) 1545.
35. I.Bombaci et al., *Nucl.Phys.* **A583** (1995) 623.
36. L.Scalone, M.Colonna, M.Di Toro, *Phys. Lett.* **B461** (1999) 9.
37. B. D. Serot, J. D. Walecka, Advances in Nuclear Physics, **16**, 1, eds. J. W. Negele, E. Vogt, (Plenum, N.Y., 1986).
38. C. M. Ko, Q. Li, R. Wang, *Phys. Rev. Lett.* **59** (1987) 1084.
39. B. Blättel, V. Koch, U. Mosel, *Rep. Prog. Phys.* **56** (1993) 1.
40. B. Liu, V. Greco, V. Baran, M. Colonna, M. Di Toro, *Phys. Rev.* **C65** (2002) 045201.
41. T. Gaitanos, et al., *Nucl. Phys.* **A732** (2004) 24.
42. C. Fuchs. H.H. Wolter, *Nucl. Phys.* **A589** (1995) 732.
43. G. Ferini, M. Colonna, T. Gaitanos, M. Di Toro, *Nucl. Phys.* **A762** (2005) 147.
44. G. Ferini et al., *Isospin effects on sub-threshold kaon production..*, ArXiv:nucl-th/0607005.
45. V. Greco et al., *Phys. Lett.* **B562** (2003) 215.
46. C. Fuchs, *Prog.Part.Nucl.Phys.* **56** 1-103 (2006).
47. C.Hartnack, H.Oeschler, J.Aichelin, *Phys. Rev. Lett.* **96** (2006) 012302.
48. T.Klähn et al., *Phys. Rev.* **C74** 2006 035802.
49. B.A. Li, G.C. Yong, W. Zuo, *Phys. Rev.* **C71** 014608 (2005).
50. Q. Li et al., *Phys. Rev.* **C72** 034613 (2005).
51. In the energy range explored here, the main contribution to the kaon yield comes from the pionic channels, in particular from πN collisions, and from the $N\Delta$ channel, which together account for nearly 80% of the total yield, see [43].
52. M. Di Toro, A. Drago, T. Gaitanos, V. Greco, A. Lavagno, *Nucl. Phys.* **A775** (2006) 102-126..
53. H.Mueller, *Nucl. Phys.* **A618** (1997) 349.
54. N.K.Glendenning, S.A.Moszkowski, *Phys. Rev. Lett.* **67** (1991) 2414.
55. A.Drago, A.Lavagno, *Phys. Lett.* **B511** (2001) 229.
56. E.Witten, *Phys. Rev.* **D30** (1984) 272.
57. A.R.Bodmer, *Phys. Rev.* **D4** (1971) 1601.
58. N.K.Glendenning, *Phys. Rev.* **D46** (1992) 1274.

Statistical equilibrium in a dynamical multifragmentation path*

A. H. Raduta[1,2]

[1] *LNS-INFN, I-95123, Catania, Italy*
[2] *NIPNE, RO-76900 Bucharest, Romania*
E-mail: raduta@lns.infn.it

Various aspects of the nuclear multifragmentation phenomenon are discussed from the point of view of the Microcanonical Multifragmentation Model (MMM) model. This model provides results in very good agreement with experimental data and predicts a first order phase transition in nuclear matter. An analysis performed with MMM aiming to identify a statistically equilibrated stage in the dynamical path provided by a transport code (Stochastic Mean Field) is described. As a result, a distinct statistically equilibrated stage corresponding to the time of 140 fm/c was identified.

1. Introduction

1.1. *What is (nuclear) multifragmentation?*

Multifragmentation generally denotes the break-up of any physical system into many pieces. One may thus include in the above definition the break-up of various physical objects e.g. a porcelain plate, the sand grain formation or even a supernova explosion. From a thermodynamical view, the above phenomenas are interesting since they may be related to phase transitions.

Nuclear multifragmentation is obtainable in laboratories by violently colliding atomic nuclei. At its turn this phenomenon is interesting since it may be related to a liquid-gas phase transition in nuclear matter. Lots of theoretical and experimental studies have been performed trying to elucidate whether a liquid-gas phase transition is *really* taking place in nuclear matter

*This work was partly supported by the European Community under a Marie Curie fellowship, contract n. MEIF - CT - 2005 - 010403

1.2. *How can nuclear multifragmentation be measured?*

In laboratories one can obtain asymptotic fragments sizes and energy distributions. These observables are to be compared to the results of various dynamical and statistical models. Good agreements between the above theories and experiments lead to the conclusion that a fragmenting source is at the origin of the experimentally obtained yields.

1.3. *Theories*

One may basically divide the theories aiming to investigate this phenomenon into dynamical and statistical ones. From the dynamical models the most effective are the Landau Vlasov or the Quantum Molecular Dynamics ones. They mimic the evolution in time of the collision process.

The statistical theories are based on the assumption that a statistically equilibrated stage, occurs in the dynamical path. The common ingredient of statistical models is the equiprobability between all fragmentation probabilities.

These two categories of theories are complementary and represent a complete and unitary tool for analyzing the system's thermodynamics.

1.4. *Open questions*

From more than 20 years there have been *only* indirect evidences for the existence of a statistically equilibrated freeze-out: asymptotic fragment yields are similar with the predictions of various statistical models. The question then arises: Is there any stage in the dynamical evolution of the two heavy ion collision which is statistically equilibrated? If so, which is the time in the system's dynamical evolution corresponding to this equilibrium? How does this equilibrated freeze-out "looks like"? E.g. are fragments already completely formed and separated or does the statistical equilibrium correspond to a prefragment stage scenario? Which is the volume of the system corresponding to the equilibrium stage? Is there any phase transition "happening" in nuclear mater? How can one "measure" it? Which factors are influencing the phase transition / nuclear phase diagram? We will address some of these question in the present presentation.

2. Microcanonical Multifragmentation Model (MMM)

The MMM model concerns the disassembly of a statistically equilibrated nuclear source (A, Z, E, V) (i.e. the source is defined by the parameters:

mass number, atomic number, excitation energy and freeze-out volume respectively). Its basic hypothesis is equal probability between all configurations $C : \{A_i, Z_i, \epsilon_i, \mathbf{r}_i, \mathbf{p}_i, \quad i = 1, \ldots, N\}$ (the mass number, the atomic number, the excitation energy, the position and the momentum of each fragment i of the configuration C, composed of N fragments) which are subject to standard microcanonical constraints: $\sum_i A_i = A$, $\sum_i Z_i = Z$, $\sum_i \mathbf{p}_i = 0$, $\sum_i \mathbf{r}_i \times \mathbf{p}_i = 0$, E - constant. The fragment level density (entering the statistical weight of a configuration) is of Fermi-gas type adjusted with the cut-off factor $\exp(-\epsilon/\tau)$: $\rho(\epsilon) = \rho_0(\epsilon)\exp(-\epsilon/\tau)$ [7]. The above factor counts for the dramatic decrease of the lifetime of fragment excited states respective to the freeze-out specific time as the excitation energy increases (i.e. earlier freeze-outs should correspond to larger values of τ). The model can work within two freeze-out hypotheses: (1) fragments are treated as hard spheres placed into a spherical freeze-out recipient and are not allowed to overlap each-other or the recipient wall; (2) fragments may be deformed and a corresponding free-volume expression is approaching the integration over fragment positions [7]. Further a Metropolis-type simulation is employed for determining the average value of any system observable (see Refs. [5] for more details). The model includes secondary excited fragments deexcitation. This *secondary decay* stage is included through a Wiesskopf evaporation scheme.

3. General view on the MMM results

3.1. *Data description*

This model provides very good description of experimental data. An example in this direction is given in Figs. 1, 2 where various fragments yields, velocity correlations and kinematic experimental distributions corresponding to the Xe + Sn at 32 MeV/u and Gd + U at 36 MeV/u reactions are simultaneously reproduced [7].

3.2. *Liquid-gas phase transitions in nuclear matter*

MMM predicts a first order phase transition in nuclear matter [8]. This result is obtainable by analyzing its isobaric caloric curves or, equivalently, the probability distributions of isobaric canonical ensemble. A first order phase transition will be reflected in a backbanding in the first observable or, respectively, in a bimodal behavior of the second one. The first order phase transition signal can be nicely observed in Fig. 3 for the case of a A=200 nucleus without Coulomb interaction or in the case of a A=50

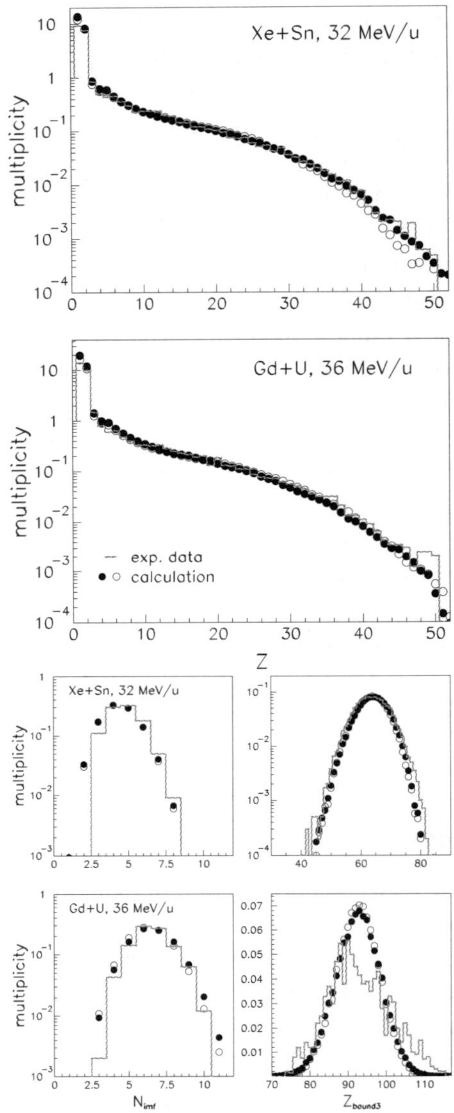

Fig. 1. Charge distribution (top); Intermediate mass fragments and *bound charge* multiplicities for reactions Xe+Sn at 32 MeV/u and Gd+U at 36 MeV/u. The MMM results are compared with the experimental results of the INDRA collaboration.

nucleus where Coulomb interaction is still present. Two conclusions may be drawn from here: 1) MMM model exhibits a 1st order phase transition. 2) Coulomb interaction acts towards lowering the system's critical point.

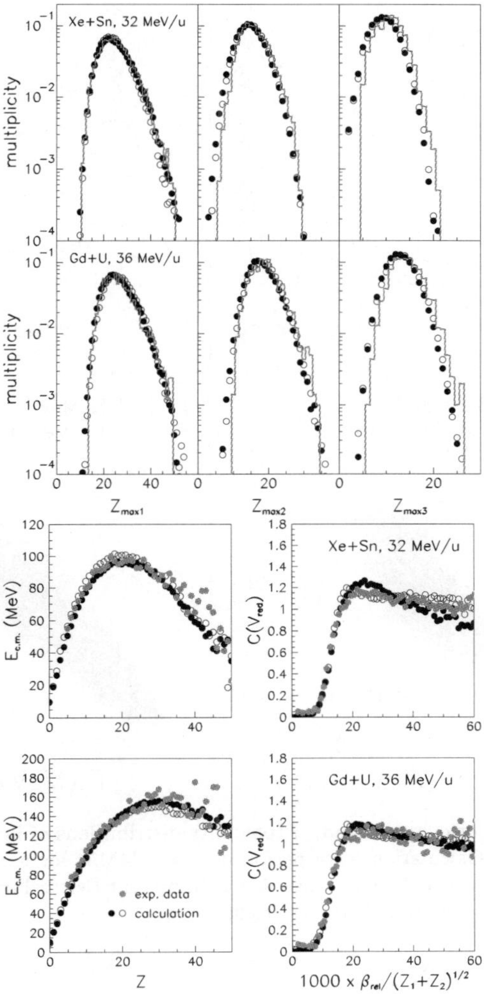

Fig. 2. Multiplicities of largest, second largest and third largest fragments (top); Kinetic energy distributions vs charge and velocity correlations (bottom) corresponding to the reactions Xe+Sn at 32 MeV/u and Gd+U at 36 MeV/u. The MMM results are compared with the experimental results of the INDRA collaboration.

In other words, Coulomb field may forbid the phase transition when it is too strong (e.g large nuclei like A=200 and Z=82). At smaller nuclei, where the Coulomb amount is smaller, the 1st order phase transition survives (see Fig. 4).

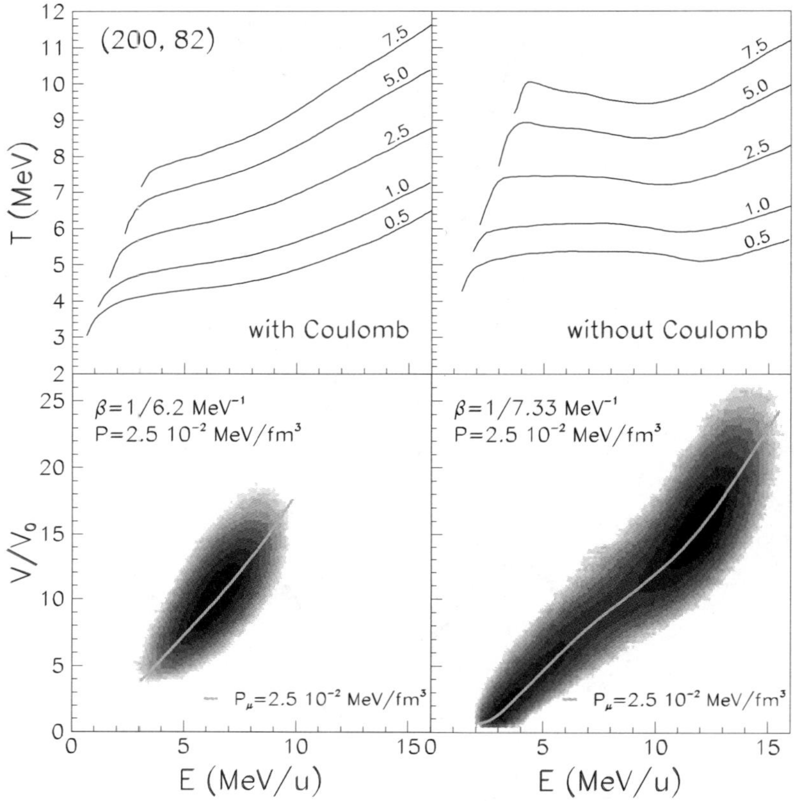

Fig. 3. Isobaric caloric curves (top); Probability distributions of isobaric canonical ensembles (bottom). Calculation were performed with MMM for a nuclear source with A=200 the case in which Coulomb interaction in present (left column) and in the case in which Coulomb interaction is absent (right column).

4. Statistical equilibrium in a dynamical multifragmentation path

4.1. *Preliminaries*

Good agreements between various observables related to the asymptotically resulted fragments and various models assuming statistical equilibrium [2–4] lead to the conclusion that a huge part of the available phase space is populated during the fragmentation process and thus a statistically equilibrated nuclear source could be at the origin of the experimentally observed nuclear multifragmentation. The source size, its excitation energy and its volume are thus quantities which can only be *indirectly* evaluated by

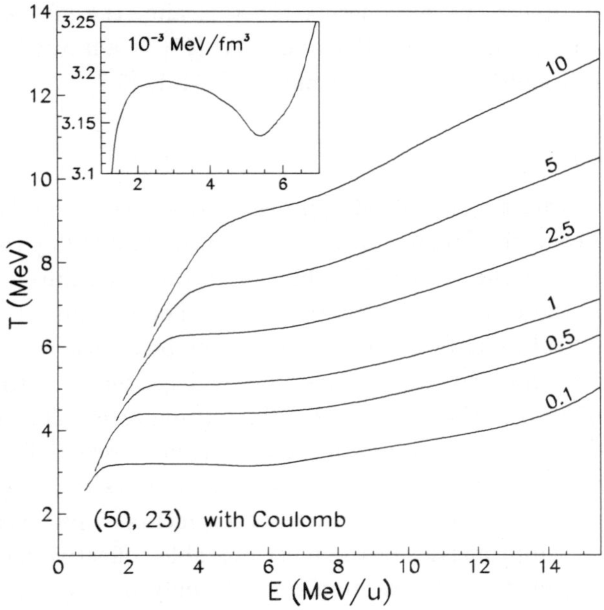

Fig. 4. Isobaric caloric curves for a nuclear source with A=50 calculated with MMM. Coulomb interaction is present.

comparisons between experimental data and statistical multifragmentation model predictions via a back-tracing procedure. The comparison process is complicated by the presence of several effects, such as pre-equilibrium particle emission, collective radial expansion, Coulomb propagation of the break-up primary fragments, secondary particle emissions. Moreover, there are intrinsic dynamical characteristics such as the freeze-out specific time, directly related to parameters such as the level density cut-off parameter, τ (see Ref. [7]) contributing to the weights of the system statistical ensemble. In the absence of any direct information about the freeze-out this last parameter has to be *employed as a fitting parameter* in a statistical model.

In Ref. [7], a very good agreement between statistical fragment distribution predictions and experimental data was obtained *assuming* $\tau = 9$ MeV. However, one has to have *direct* access to the freeze-out events in order to unambiguously decide on the value of such parameters and, if an equilibrated source exists, to find its location in space and time. This task can be achieved using "freeze-out" information from a dynamical model.

Herein we shall investigate whether statistical equilibration occurs in the dynamical path of two heavy ion collision and, if so, which are the corresponding freeze-out parameters. To this aim, the "freeze-out" data of a stochastic mean field (SMF) approach [11] is analyzed via a sharp microcanonical multifragmentation model (MMM) [5,7].

4.2. *Method*

We use the stochastic mean-field approach introduced in ref. [12]. According to this theory, the fragmentation process is dominated by the growth of volume (spinodal) and surface instabilities encountered during the expansion phase of the considered excited systems. The dynamical evolution of the system is still described in terms of the one-body distribution function (mean-field description), however this function experiences a stochastic evolution, in response to the action of the fluctuation term. The amplitude of the stochastic term incorporated in the treatment is essentially determined by the degree of thermal agitation present in the system. Hence fluctuations provide the seeds of the formation of fragments, whose characteristics are related to the properties of the most unstable collective modes [13]. In the model [12] fluctuations are implemented only in $\mathbf{r}$ space. Within the assumption of local thermal equilibrium, the kinetic fluctuations typical of a Fermi gas are projected on density fluctuations. Then fluctuations are propagated by the unstable mean-field, leading to the disassembly of the system.

Using MMM we investigate whether the "primary events" produced by the stochastic mean-field approach as a result of ^{129}Xe+^{119}Sn at 32 MeV/u reaction may correspond to the statistical equilibration of the compound system. According to the dynamical simulations

performed in Ref. [11], it is observed that, after the initial collisional shock, the system expands towards low densities entering the unstable region of the nuclear matter phase diagram (after about 100 fm/c from the beginning of the reaction). Then fragments are formed through spinodal decomposition. The freeze-out time is defined as the time when the fragment formation process is over. Hence average fragment multiplicities and distributions do not evolve anymore. For the reaction considered this time is 240 fm/c.

Our aim is to investigate whether fragment distributions are compatible with the MMM predictions. For washing-up pre-equilibrium effects which should appear in the dynamical simulation, only intermediate mass fragments (IMF) (i.e. fragments with $Z \geq 3$) are selected. Thus, all comparisons between MMM and stochastic mean-field results are to be restricted to IMF's. Due to the large Coulomb repulsion among primary fragments,

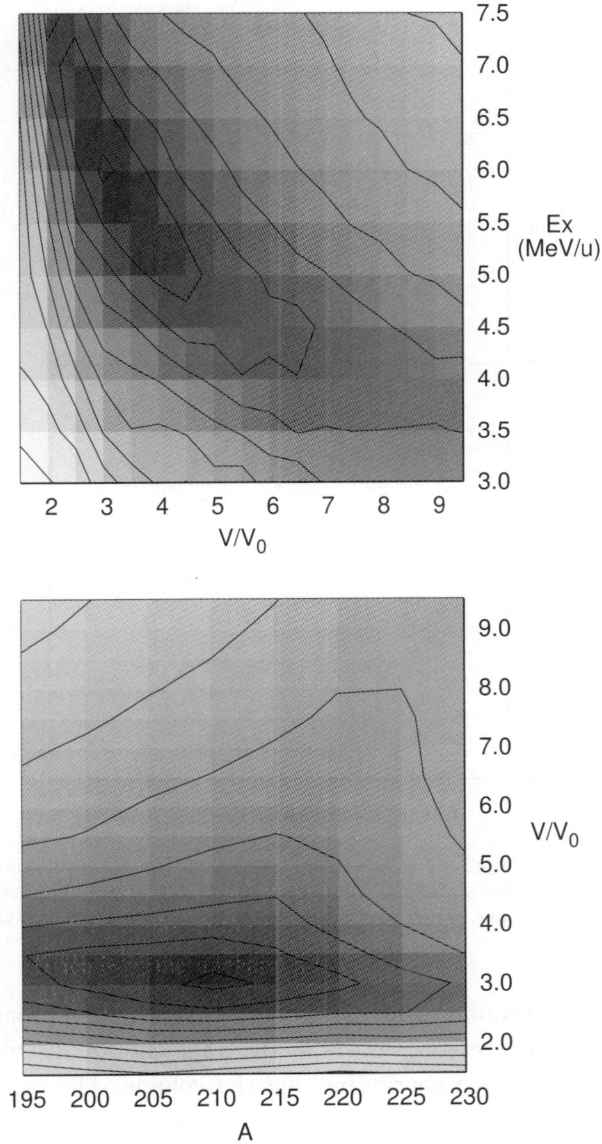

Fig. 5. Contour plots of the error function Err [see eq. (1)]: in the $(V/V_0, E)$ plane corresponding to $A = 210$ (upper panel); in the $(A, V/V_0)$ plane corresponding to $E = 5.7$ MeV/u (lower panel). Darker regions correspond to smaller Err; units are relative.

it is difficult to have a precise estimation of the freeze-out time (and consequently of freeze-out volume) so we assume that the largest uncertainty

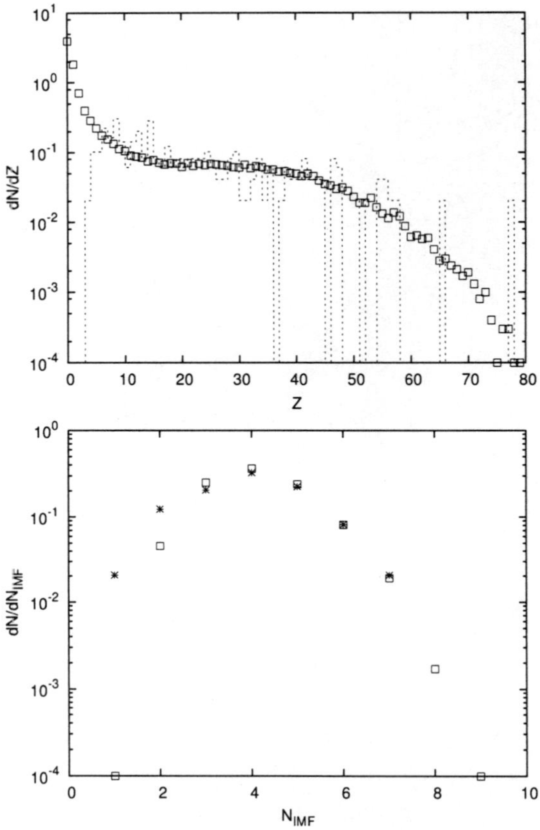

Fig. 6. Dynamical charge (upper panel) and number of IMF distributions (lower panel) in comparisons with the statistical ones. Statistical results are represented by open squares; histogram (upper panel) and stars (lower panel) corespond to the dynamical ones.

in the equilibrated source estimated from the dynamical approach concerns the volume. For this reason, we will fit the fragment size distributions and their internal excitation energy but *not* the volume. The best fit can be found by minimizing the following error function:

$$\begin{aligned} Err = \{ & 3\,[f(\langle A_{bound}\rangle) + f(\langle Z_{bound}\rangle)] \\ & + \left[\sum_{N_{IMF}} f[\langle \mathrm{d}N/\mathrm{d}N_{IMF}\rangle]/\sum_{N_{IMF}} 1\right] \\ & + f(\langle \epsilon_{IMF}\rangle) + \sum_{i=1}^{3} f(\langle Z_{\mathrm{max}i}\rangle)/3\}/9 \end{aligned} \tag{1}$$

where $\langle\cdot\rangle$ stands for average, A_{bound} and Z_{bound} are the bound mass and charge (sum of the mass number and, respectively, atomic number of all IMF's from a given event), N_{IMF} is the number of IMF's, ϵ_{IMF} is the excitation energy of one (IMF) fragment and $Z_{\max i}$ with $i = 1, 2, 3$ are the largest, second largest and third largest charge from one fragmentation event. Further, $f(x) = |2(x_s - x_d)/(x_s + x_d)|$, where the indexes s and d stand for "statistic" and "dynamic" and $|x|$ is the absolute value of x. MMM parameters A, E, V and τ are variated in wide ranges thus constructing a four-dimensional grid. The ranges are A : [195, 230], E : [3, 7.5] MeV/u, V/V_0 : [1.5, 9.5], $\tau = 12, 16, \infty$ MeV.

4.3. *Results*

An absolute minimum of the error function, Err, was found at $A = 210$, $Z = 87$, $V/V_0 = 3.4$, $E = 5.7$ MeV/u, $\tau = \infty$. Cuts in Err corresponding to $A = 210$ and $E = 5.7$ MeV/u are represented in the upper part and respectively the lower part of Fig. 7. The statistical source corresponding to the minimum value of Err yields the following results: $\langle A_{bound}\rangle = 199.03$, $\langle Z_{bound}\rangle = 83.74$, $\langle\epsilon_{IMF}\rangle = 4.21$ MeV/u, $\langle Z_{\max 1}\rangle = 42.34$, $\langle Z_{\max 2}\rangle = 24.35$, $\langle Z_{\max 3}\rangle = 11.4$. These are to be compared with the corresponding *dynamical* results: $\langle A_{bound}\rangle = 199$, $\langle Z_{bound}\rangle = 84$, $\langle\epsilon_{IMF}\rangle = 4.3$ MeV/u, $\langle Z_{\max 1}\rangle = 41.95$, $\langle Z_{\max 2}\rangle = 22.5$, $\langle Z_{\max 3}\rangle = 13.3$. Note the excellent agreement for all considered observables, proving the very good quality of the fit. This can be seen in Fig. 8, as well, where the *dynamical* Z and N_{IMF} distributions fit perfectly the MMM ones. At this stage the question still remains: *even if fragment size distributions and excitation energies of the fragments are very well reproduced, do the dynamically formed fragments come from an equilibrated source with freeze-out volume* $V = 3.4V_0$, *as predicted by MMM?*

4.4. *Consistency checking*

In the present work the freeze-out volume V is the volume of the smallest sphere which *totally* includes all fragments. We denote by $\widetilde{V}$ the volume of the smallest sphere which *totally* includes all fragments *and* has the center located in the center of mass of the system. Obviously, $\widetilde{V} > V$. The "dynamical" events have $\left\langle(\widetilde{V}/V_0)_{IMF}\right\rangle = 9.08$; the statistical ones have $\left\langle(\widetilde{V}/V_0)_{IMF}\right\rangle = 4.93$ (the IMF index indicates that we refer to the volume occupied by IMF fragments). This could mean that equilibration may

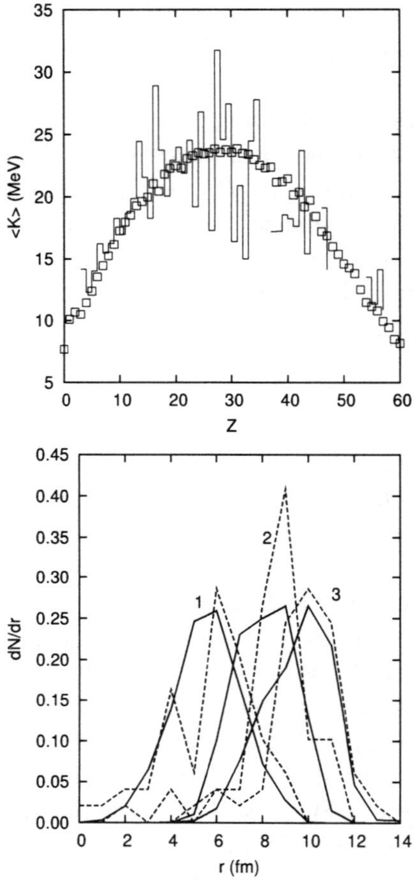

Fig. 7. Upper panel: "statistical" (open squares) and "dynamical" (histogram) fragment average kinetic energy versus charge. Lower panel: "statistical" (solid lines) and "dynamical" (dashed lines) radial distribution of fragments with largest (peak "1"), second largest (peak "2") and third largest (peak "3") charge in one fragmentation event. The plot corresponds to $\left\langle(\widetilde{V}/V_0)_{IMF}\right\rangle = 9.08$.

have occurred at a earlier time, i.e. , due to the uncertainties in the estimation of the freeze-out time fragments are already well defined at a earlier time and they are actually equilibrated inside a smaller volume. One can simply test this hypothesis: one just has to propagate the fragments in their mutual Coulomb field from the freeze-out positioning as generated by MMM up to $(\widetilde{V}/V_0)_{IMF} = 9.08$ (i.e. the value corresponding to the "dynamical" events) and then compare "dynamical" and "statistical" fragment kinetic energies and positions. However, in performing such a comparison,

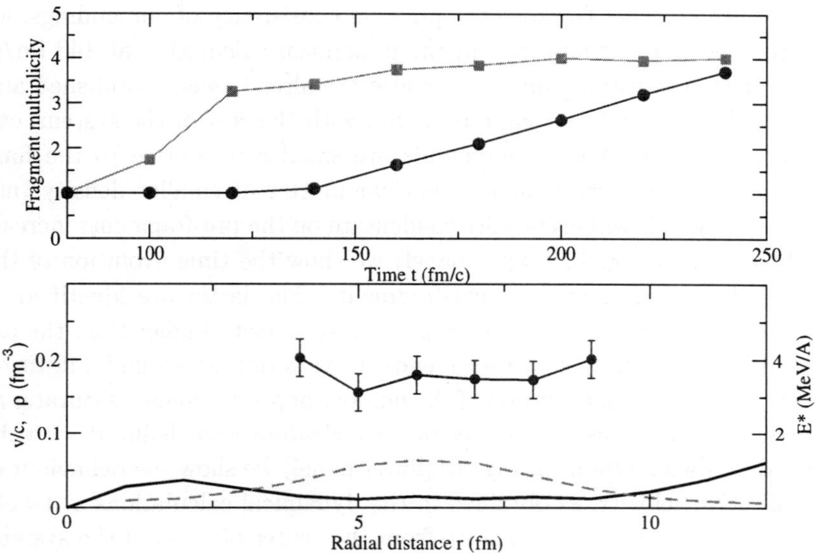

Fig. 8. Dynamical results. Upper panel: Evolution of IMF (circles) and pre-fragment (squares) multiplicities in time. Lower panel: Collective velocity (full line), nuclear density (dashed line) and excitation energy per nucleon (circles), as a function of the distance from the center of mass of the system, corresponding to 140 fm/c. The excitation energy is evaluated in the region occupied by pre-fragments.

one has to guess the initial fragment velocities, that could have a collective component. The best reproduction of the dynamical results is obtained for a radial collective flow equal to zero. The comparison is presented in Fig. 9: the fragment average kinetic energies versus charge is represented in Fig. 9, upper panel; the radial distribution of the fragment with largest, second largest and third largest charge is represented in Fig. 9, lower panel. The time of the Coulomb propagation is around 100 fm/c. A surprisingly good agreement is observed between "dynamical" and "statistical" data for both observables indicating the physical consistency of the obtained result. This means that the dynamical configurations could originate from an earlier *equilibrated* freeze-out ($t \approx 140$ fm/c), where fragments are located within a sphere (with the center in the center of mass of the system) of volume $\left\langle (\widetilde{V}/V_0)_{IMF} \right\rangle = 4.93$ and the collective flow is zero.

However, in order to prove the physical consistency of our findings, we have to check what "happens" in the dynamical calculation at 140 fm/c. At that time the leading unstable modes are already well established and some pre-fragments, in strong interaction with the rest of the system, can be recognized [14]. These compounds are smaller respective to the final fragments and are surrounded by nuclear matter at smaller density (nuclear "fog") which will eventually condensate on the pre-fragments increasing their size. In Fig. 10 (upper panel) we show the time evolution of the multiplicity of fragments and pre-fragments. The latter are identified as high density bumps, i.e. confined regions with density higher than the average. While the number of IMF's ends its variation at around 240 fm/c, remarquably the time evolution of the number of pre-fragments saturates at around 140 fm/c. This is precisely the *equilibration time* deduced with the above-described method. In Fig. 10 (lower panel) we show the behaviour of the collective velocity, as obtained in the dynamical calculations at t=140 fm/c, as a function of the distance from the center of mass of the system. One can observe that in the spatial region occupied by the prefragments (i.e. higher nuclear matter density, see the radial density profile on the same figure) the collective velocity is close to zero, which agrees with the MMM findings (see Fig.3). And finally, on the same figure one can observe that the excitation energy of the fragmenting source, evaluated in the region occupied by pre-fragments, is rather constant as a function of the position and moreover is quite close to the excitation energy of the final fragments, meaning that at $t = 140$ fm/c the system is in thermal equilibrium.

From the lower panel of Fig. 10 we also infer that the pre-fragments are formed near the turning point of the expansion. With increasing beam energy more radial flow is developed, the system expands in a hollow configuration and qualitatively different fragment partitions are dynamically populated [11].

4.5. *Remarks*

From a physical point of view one can conclude that (experimentally) freeze-out volumes are relatively small (for the present reaction $3.4V_0$) and primary fragment excitation energy is considerably large (4.3 MeV/u for this case). So, one deals with *hot* fragmentation at *small* freeze-out volumes. Also, it seems that there is no need for a cut-off factor in the fragments' level density, the minimum *Err* being obtained for $\tau = \infty$. (This was mainly dictated by the high excitation energy of the dynamically formed primary fragments.) And, finally, the formed compound system is completely equilibrated without any extra flow energy. The *full* physical consistency of the obtained result confirms the accuracy of the MMM model. To our knowledge, this is the first time when it is completely *demonstrated* that in the dynamical paths of violent heavy ion collisions there is a stage of complete statistical equilibration of the compound system. We can conclude that along the fragmentation path, as described by a dynamical model, a huge part of the available phase space is filled. It is remarkable that the equilibrated source obtained *only* by fitting fragment size distributions and fragment internal energies proves its physical consistency subject to the dynamical calculations. Indeed, after propagating the fragments from the freezeout volume specific to the identified equilibrated source to the one corresponding to the dynamical events, both kinematic and fragment position observables fit the corresponding dynamical data very well. Of course, some inherent uncertainties related to the MMM approaches for Coulomb and fragment deformation should be considered. Due to these reasons for example at the moment identified by MMM as freeze-out some flow may still be present. To our knowledge, this is the first time when in dynamical paths of violent heavy ion collisions a stage of statistical equilibrium of the compound system is identified.

5. Conclusions

Various aspects of the nuclear multifragmentation phenomenon have been discussed from the point of view of the MMM model. From this perspective, nuclear multifragmentation originates from a statistically equilibrated source and may reflect a liquid-gas phase transition. A distinct statistically equilibrated stage was identified by the MMM model in a dynamical multifragmentation path as provided by the SMF model. This result is particularly important as, while implicitly *assumed* from a long time in various nuclear multifragmentation studies, this is probably the first time when a statistically equilibrated stage stage is identified in a dynamical path.

References

1. W. Bauer and A. Botvina, *Phys. Rev.* **C52**, R1760 (1995).
2. D. H. E. Gross, *Rep. Progr. Phys.* **53**, 605 (1990).
3. J. P. Bondorf, A. S. Botvina, A. S. Iljinov, I. N. Mishustin and K. Sneppen, *Phys. Rep.* **257**, 133 (1995).
4. J. Randrup and S. Koonin, *Nucl. Phys.* **A471**, 355c (1987).
5. Al. H. Raduta and Ad. R. Raduta, *Phys. Rev.* **C55**, 1344 (1997).;Al. H. Raduta and Ad. R. Raduta, *Phys. Rev.* **C61**, 034611 (2000).

6. F. Gulminelli and Ph. Chomaz, *Nucl. Phys.* **A734**, 581 (2004).
7. Al. H. Raduta and Ad. R. Raduta, *Phys. Rev.* **C65**, 054610 (2002).
8. Al. H. Raduta and Ad. R. Raduta, *Phys. Rev. Lett.* **87**, 202701 (2001).
9. J.P.Bondorf et al., *Phys. Rev. Lett.* **73**, 628 (1994).
10. J.D. Frankland et al., *Nucl. Phys.* **A689**, 940 (2001); G. Tabacaru et al., *Eur. Phys. J.* **A18**, 103 (2003).
11. M. Colonna, G. Fabbri, M. Di Toro, F. Matera, H. H. Wolter, *Nucl. Phys.* **A742**, 337 (2004).
12. M. Colonna et al, *Nucl. Phys.* **A642**, 449 (1998).
13. Ph. Chomaz, M. Colonna, J. Randrup, *Phys. Rep.* **389**, 263 (2004).
14. A. Guarnera et al., *Phys. Lett.* **B403**, 191 (1997).
15. C.O. Dorso and J. Randrup, *Phys. Lett.* **B301**, 328 (1993).
16. M. Parlog et al. *Eur. Phys. J* **A25**, 223 (2005).

THE ROLE OF INSTABILITIES IN NUCLEAR FRAGMENTATION

V. BARAN*

University of Bucharest, Department of Theoretical Physics and Mathematics, Physics Faculty and National Institute for Physics and Nuclear Engineering-Horia Hulubei Bucharest/Romania
** E-mail: baran@lns.infn.it*

M. COLONNA, M. DI TORO

University of Catania and Istituto Nazional di Fisica Nucleare, Laboratori Nazionali del Sud,
Catania/Italy
E-mail: ditoro@lns.infn.it, colonna@lns.infn.it

The role of spinodal instabilities in nuclear fragmentation is investigated. A thermodynamical and dynamical analysis based on Landau theory of Fermi liquids is employed. It is shown that in the low density region of the phase diagram asymmetric nuclear matter can be characterized by a unique spinodal region, defined by the instability against isoscalarlike fluctuation, as in symmetric nuclear matter. Everywhere in this density region the system is stable against isovectorlike fluctuations related to the species separation tendency. Nevertheless, this instability in asymmetric nuclear matter induce isospin distillation leading to a more symmetric liquid phase and a more neutron rich gas phase.

Keywords: Nuclear Multifragmentation, Neck fragmentation, Fermi liquids theory, Spinodal decomposition

1. Introduction

An important dissipation mechanism in heavy ions reactions at intermediate energies is represented by the fragments production. A relevant phenomenon is the liquid-gas phase transition, very often invoked in discussing the nuclear multifragmentation. In such collisions the reaction times can be comparable to the fragment formation time and has relevance to discuss about the kinetic of the phase transition. The violent collision and fast expansion may quench the system inside the instability region of the phase diagram.

A binary system, as it is asymmetric nuclear matter (ANM) (see [1]), manifest a richer thermodynamical behaviour, since it has to accommodate one more conservation law.

In this lecture we will discuss first about the nature of the instabilities and of the related fluctuations in such systems. Then, we will describe the kinetics of phase transition in ANM both in the linear and nonlinear regime. Finally in the last section we will focus on the relevance of these results on nuclear multifragmentation and neck fragmentation in heavy ion collisions at intermediate energies.

2. Instabilities and fluctuations in ANM

2.1. *Thermodynamical approach*

One-component systems may become unstable against density fluctuations as the result of the mean attractive interaction between constituents. In symmetric binary systems, like symmetric nuclear matter (SNM), one may encounter two kinds of density fluctuations: i) isoscalar, when the densities of the two components oscillate in phase with equal amplitude, ii) isovector when the two densities fluctuate still with equal amplitude but out of phase. The mechanical instability is associated with instability against isoscalar fluctuations leading to cluster formation while chemical instability is related to instability against isovector fluctuations, leading to species separation.

In ANM, there is no longer a one to one correspondence between isoscalar (resp. isovector) fluctuations and mechanical (resp. chemical) instability. An appropriate framework for the study of instabilities is provided by the Fermi liquid theory [2], which has been applied, for instance, to symmetric binary systems as is SNM (the two components being protons and neutrons) [3], the liquid ^3He (spin-up and spin-down components), [4,5], or to proto-neutron stars to calculate neutrino propagation [6].

For the thermodynamical analysis, the starting point is an extension to the asymmetric case, [7], of the formalism introduced in [4]. The distribution functions for protons and neutrons are:

$$f_q(\epsilon_p^q) = \Theta(\mu_q - \epsilon_p^q) \ , \qquad q = n, p \tag{1}$$

where μ_q are the corresponding chemical potentials. The nucleons interaction is characterized by the Landau parameters:

$$F^{q_1 q_2} = N_{q_1} V^2 \frac{\delta^2 H}{\delta f_{q_1} \delta f_{q_2}} = N_{q_1} \frac{\delta^2 H}{\delta \rho_{q_1} \delta \rho_{q_2}} \tag{2}$$

$$N_q(T) = \int \frac{-2\, d\mathbf{p}}{(2\pi\hbar)^3} \frac{\partial f_q(T)}{\partial \epsilon_p^q} \tag{3}$$

where H is the energy density, V is the volume and N_q is the single-particle level density at the Fermi energy. At $T = 0$ this reduces to

$$N_q(0) = m p_{F,q}/(\pi^2 \hbar^3) = 3\rho_q/(2\epsilon_{F,q}),$$

where $p_{F,q}$ and $\epsilon_{F,q}$ are Fermi momentum and Fermi energy of the q-component. Thermodynamical stability for $T = 0$ requires the energy of the system to be an absolute minimum for the undistorted distribution functions, so that the relation:

$$\delta H - \mu_p \delta\rho_p - \mu_n \delta\rho_n > 0 \tag{4}$$

is satisfied when we deform proton and neutron Fermi seas. Only monopolar deformations will be taken into account, since we consider here momentum independent interactions, so that $F_{l=0}^{q_1 q_2}$ are the only non-zero Landau parameters.

Then, up to second order in the variations, the condition Eq.(4) becomes

$$\delta H - \mu_p \delta\rho_p - \mu_n \delta\rho_n = \frac{1}{2}(a\delta{\rho_p}^2 + b\delta{\rho_n}^2 + c\delta\rho_p\delta\rho_n) > 0 \tag{5}$$

where

$$a = N_p(0)(1 + F_0^{pp}) \;\; ; \;\; b = N_n(0)(1 + F_0^{nn}) \;\; ;$$
$$c = N_p(0)F_0^{pn} + N_n(0)F_0^{np} = 2N_p(0)F_0^{pn}. \tag{6}$$

The r.h.s. of Eq.(5) is diagonalized by the following transformation:

$$u = cos\beta\ \delta\rho_p + sin\beta\ \delta\rho_n,$$
$$v = -sin\beta\ \delta\rho_p + cos\beta\ \delta\rho_n, \tag{7}$$

where the *mixing* angle $0 \le \beta \le \pi/2$ is given by

$$tg\ 2\beta = \frac{c}{a-b} = \frac{N_p(0)F_0^{pn} + N_n(0)F_0^{np}}{N_p(0)(1+F_0^{pp}) - N_n(0)(1+F_0^{nn})}. \tag{8}$$

Then Eq.(5) takes the form

$$\delta H - \mu_p \delta\rho_p - \mu_n \delta\rho_n = Xu^2 + Yv^2 > 0 \tag{9}$$

with

$$X = \frac{1}{2}(\ a + b + sign(c)\sqrt{(a-b)^2 + c^2}\ )$$
$$\equiv \frac{(N_p(0) + N_n(0))}{2}(\ 1 + F_{0g}^s\ ) \tag{10}$$

and

$$Y = \frac{1}{2}(\ a + b - sign(c)\sqrt{(a-b)^2 + c^2}\ )$$
$$\equiv \frac{(N_p(0) + N_n(0))}{2}(\ 1 + F_{0g}^a\ ), \tag{11}$$

defining the new generalized Landau parameters $F_{0g}^{s,a}$.

It was possible to separate the total variation Eq.(4) into two independent contributions, called the "normal" modes, and characterized by the "mixing angle" β, which depends on the density of states and the details of the interaction. Thus the thermodynamical stability requires $X > 0$ *and* $Y > 0$. Equivalently, the following conditions have to be fulfilled:

$$1 + F_{0g}^s > 0 \quad and \quad 1 + F_{0g}^a > 0, \tag{12}$$

They represent Migdal-Pomeranchuk stability conditions extended to asymmetric binary systems.

The new stability conditions, Eq.(12), are equivalent to mechanical and chemical stability of a thermodynamical state, [8], i.e.

$$\left(\frac{\partial P}{\partial \rho}\right)_{T,y} > 0 \quad and \quad \left(\frac{\partial \mu_p}{\partial y}\right)_{T,P} > 0 \tag{13}$$

where P is the pressure and y the proton fraction. In fact, mechanical and chemical stability are very general conditions, deduced by requiring that the principal curvatures of thermodynamical potentials surfaces, such as the free energy (or the entropy) with respect to the extensive variables are positive (negative). In the following, we will show that spinodal instability and phase transition in ANM should be discussed in terms of isoscalar and isovector like instabilities. In the case discussed here, it can be proved that [7]:

$$XY = N_p(0)N_n(0)[(1 + F_0^{nn})(1 + F_0^{pp}) - F_0^{np}F_0^{pn}]$$

$$= \frac{[N_p(0)N_n(0)]^2}{(1-y)\rho^2}\left(\frac{\partial P}{\partial \rho}\right)_{T,y}\left(\frac{\partial \mu_p}{\partial y}\right)_{T,P} \tag{14}$$

and:

$$\left(\frac{\partial P}{\partial \rho}\right)_{T,y} = \frac{\rho y(1-y)}{N_p(0)N_n(0)}(\ t\ a + \frac{1}{t}\ b + c\ )$$
$$\propto X(\sqrt{t}cos\beta + \frac{1}{\sqrt{t}}sin\beta)^2 + Y(\sqrt{t}sin\beta - \frac{1}{\sqrt{t}}cos\beta)^2$$
$$with \quad t = \frac{y}{1-y}\frac{N_n(0)}{N_p(0)}. \tag{15}$$

Let us assume that in the density range we are considering the quantities a and b remain positive. In this way one can study the effect of the interaction between the two components, given by c, on the instabilities of the mixture. If $c < 0$, i.e. for an attractive interaction between the two components, from Eq.(11) one sees that the system is stable against isovector-like fluctuations. It becomes isoscalar unstable if $c < -2\sqrt{ab}$ (see Eq.(10)). However thermodynamically this instability against isoscalar-like fluctuations will show up as a chemical instability if $(-ta - b/t) < c < -2\sqrt{ab}$ or as a mechanical instability if $c < (-ta - b/t) < -2\sqrt{ab}$ (see Eq.(15)). This last observation is very interesting: it tells us that the nature of the thermodynamically instabilities can be related to the relative strength of the various interactions among the species. In other words, if is possible to determine experimentally for a binary systems the signs of $\left(\frac{\partial P}{\partial \rho}\right)_{T,y}$ and/or $\left(\frac{\partial \mu_p}{\partial y}\right)_{T,P}$ we can learn about the inequalities, at a given density, between species interactions.

On the other hand the distinction between the two kinds of instability (mechanical and chemical) is not really relevant regarding the nature of unstable fluctuations, it being essentially the same, i.e. isoscalar-like. The relevant instability region is defined in terms of instabilities against isoscalar fluctuations and we can speak therefore about a unique spinodal region. If $c > 0$, i.e. when the interaction between the components is repulsive, the thermodynamical state is always stable against isoscalar-like fluctuation, but can be isovector unstable if $c > 2\sqrt{ab}$. Since, with our choices the system is mechanically stable ($a, b, c > 0$, see Eq. (15)), the isovector instability is now always associated with chemical instability. Such situation will lead to a component separation of the liquid mixture.

2.2. *Asymmetric nuclear matter case*

We exemplify the previous general analysis to the case of asymmetric nuclear matter. Let us consider a potential energy density of Skyrme type:

$$H_{pot}(\rho_n, \rho_p) = \frac{A}{2}\frac{(\rho_n+\rho_p)^2}{\rho_0} + \frac{B}{\alpha+2}\frac{(\rho_n+\rho_p)^{\alpha+2}}{\rho_0^{\alpha+1}} + (C_1 - C_2(\frac{\rho}{\rho_0})^{\alpha})\frac{(\rho_n-\rho_p)^2}{\rho_0} \tag{16}$$

where $\rho_0 = 0.16$ fm^{-3} is the nuclear saturation density, [9,10]. The values of the parameters $A = -356.8$ MeV, $B = 303.9$ MeV, $\alpha = 1/6$, $C_1 = 125$ MeV, $C_2 = 93.5$ MeV are adjusted to reproduce the saturation properties of symmetric nuclear matter and the symmetry energy coefficient.

In Fig.1 we focus on the low density region, where phase transitions of the liquid-gas type are expected to happen, in agreement with the experimental evidences of multifragmentation in charge asymmetric systems [11,12]. Since $a, b > 0$ and $c < 0$, we deal only with instability against isoscalar-like fluctuations, as for symmetric nuclear matter.

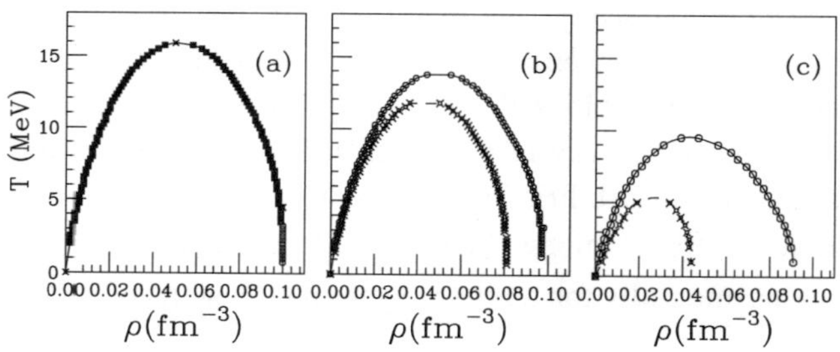

Fig. 1. Spinodal line corresponding to isoscalar-like instability of asymmetric nuclear matter (circles) and mechanical instability (crosses) for three proton fractions: $y = 0.5$ (a), $y = 0.25$ (b), $y = 0.1$ (c).

The circles represent the spinodal line corresponding to isoscalar-like instability, as defined above, for three values of the proton fraction. For asymmetric matter, $y < 0.5$, under this border one encounters either chemical instability, in the region between the two lines, or mechanical instability, under the inner line (crosses). The latter is defined by the set of values (ρ, T) for which $\left(\frac{\partial P}{\partial \rho}\right)_{T,y} = 0$.

Let us now focus on the direction of the instability in the ρ_p, ρ_n plane. If the eigenvector associated to the unstable mode is along $y = \rho_p/\rho$=const then the instability does not change the proton fraction. For symmetry reasons pure isoscalar and isovector modes appears only for SNM so it is interesting to introduce a generalization of isoscalar-like and isovector-like modes by considering if the protons and neutrons move in phase ($\delta\rho_n \delta\rho_p > 0$) or out of phase ($\delta\rho_n \delta\rho_p < 0$).

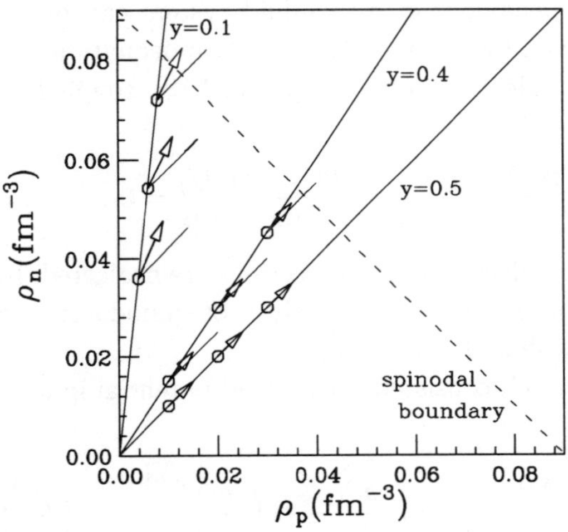

Fig. 2. The direction of unstable modes in the liquid phase.

Fig. 2 shows the direction of instabilities inside the spinodal border for the liquid phase and some iso-instability lines. The instability direction is between the y=const line and the ρ=const direction. This shows that the unstable direction is of isoscalar nature as expected from the attractive interaction between proton-neutron. Moreover, the unstable eigenvector drives the dense phase (i.e. the liquid) toward a more symmetric point in the density plane. In the gas phase the arrows will point toward a more charge asymmetric region leading to the fractionation phenomenon.

We want to stress that those qualitative conclusions are very robust and have been reached for all the Skyrme and Gogny forces we have tested (SGII, SkM*, RATP, D1, D1S, D1P) including the most recent one (SLy230a, D1P) as well as the original one (like SIII, D1) [13].

We eventually mention that also various relativistic mean field hadron models were also involved for the study of the phase transition from liquid to gas phases in ANM [14–16]. Was concluded that the largest differences between different parameterizations, regarding unstable behaviour in low density region occur at finite temperature and in the high isospin asymmetry region.

3. The kinetics of phase transition

3.1. *The linear response*

The dynamical behaviour of a two-fluid system can be described, at the semi-classical level, by considering two Vlasov equations, for neutrons and protons in the nuclear matter case [9,10,17,18], coupled through the self-consistent nuclear field :

$$\frac{\partial f_q(\mathbf{r},\mathbf{p},t)}{\partial t} + \frac{\mathbf{p}}{m}\frac{\partial f_q}{\partial \mathbf{r}} - \frac{\partial U_q(\mathbf{r},t)}{\partial \mathbf{r}}\frac{\partial f_q}{\partial \mathbf{p}} = 0\,, \qquad q = n,p. \tag{17}$$

For simplicity effective mass corrections are neglected. In fact in the low density region, of interest for our analysis of spinodal instabilities, effective mass corrections should not be large.

$U_q(\mathbf{r},t)$ is the self-consistent mean field potential in a Skyrme-like form [9,10] :

$$\begin{aligned} U_q = \frac{\delta H_{pot}}{\delta \rho_q} &= A\left(\frac{\rho}{\rho_0}\right) + B\left(\frac{\rho}{\rho_0}\right)^{\alpha+1} + C\left(\frac{\rho_3}{\rho_0}\right)\tau_q \\ &+ \frac{1}{2}\frac{dC(\rho)}{d\rho}\frac{\rho_3^2}{\rho_0} - D\Delta\rho + D_3\tau_q\Delta\rho_3\,, \end{aligned} \tag{18}$$

where

$$\begin{aligned} H_{pot}(\rho_n,\rho_p) &= \frac{A}{2}\frac{\rho^2}{\rho_0} + \frac{B}{\alpha+2}\frac{\rho^{\alpha+2}}{\rho_0^{\alpha+1}} \\ &+ \frac{C(\rho)}{2}\frac{\rho_3^2}{\rho_0} + \frac{D}{2}(\nabla\rho)^2 - \frac{D_3}{2}(\nabla\rho_3)^2 \end{aligned} \tag{19}$$

is the potential energy density (see Eq.16), where also surface terms are included; $\rho = \rho_n + \rho_p$ and $\rho_3 = \rho_n - \rho_p$ are respectively the total (isoscalar) and the relative (isovector) density; $\tau_q = +1$ $(q = n)$, -1 $(q = p)$.

The value of the parameter $D = 130$ MeV·fm^5 is adjusted to reproduce the surface energy coefficient in the Bethe-Weizsäcker mass formula $a_{surf} =$ 18.6 MeV. The value $D_3 = 40$ MeV·fm^5 $\sim D/3$ is chosen according to Ref. [19].

Let us now discuss the linear response analysis to the Vlasov Eqs. (17), corresponding to a semiclassical *RPA* approach [9,10].

The dispersion relation

$$(1 + F_0^{nn}\chi_n)(1 + F_0^{pp}\chi_p) - F_0^{np}F_0^{pn}\chi_n\chi_p = 0 \ , \tag{20}$$

where χ_n and χ_p are the corresponding long-wavelength limit of Linhard functions, is quadratic in ω and one finds two independent solutions (isoscalar-like and isovector-like solutions): ω_s^2 and ω_i^2.

The dispersion relation, Eq.(20), have been solved for various choices of the initial density, temperature and asymmetry of nuclear matter. Fig.3 reports the growth rate $\Gamma = \text{Im}\,\omega(k)$ as a function of the wave vector k, for three situations inside the spinodal region. Results are shown for symmetric ($I = 0$) and asymmetric ($I = 0.5$) nuclear matter.

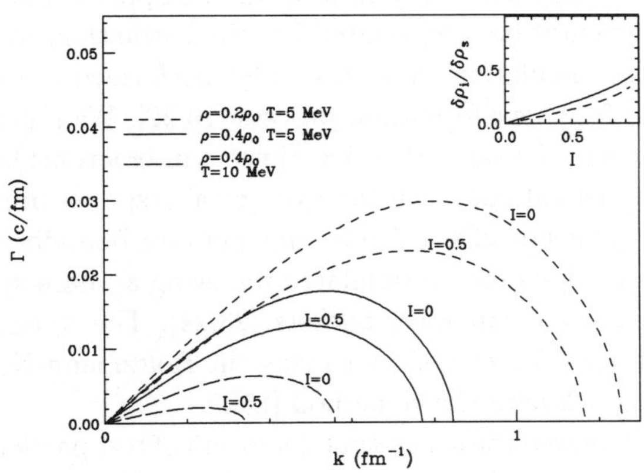

Fig. 3. Growth rates of instabilities as a function of the wave vector, as calculated from the dispersion relation Eq.(20), for three situations inside the spinodal region. Lines are labeled with the asymmetry value I. The insert shows the asymmetry of the perturbation $\delta\rho_I/\delta\rho_S$, as a function of the asymmetry I of the initially uniform system, for the most unstable mode, in the case $\rho = 0.4\rho_0$, $T = 5\ MeV$.

The growth rate has a maximum $\Gamma_0 = 0.01 \div 0.03$ c/fm corresponding to a wave-vector value around $k_0 = 0.5 \div 1\ \text{fm}^{-1}$ and becomes equal to zero at $k \simeq 1.5k_0$, due to k-dependence of the Landau parameters, as discussed above. One can see also that instabilities are reduced when increasing the temperature, an effect also present in the symmetric N = Z case [20]. At larger initial asymmetry decrease of the maximum of the growth rate indicates that the development of the spinodal instabilities is slower. One should expect also an increase of the size of the produced fragments in view of decrease of the wave number corresponding to the maximum growth rate. From the long dashed curves of Fig.3 we can predict the asymmetry effects to be more pronounced at higher temperature, when in fact the system is closer to the boundary of the spinodal region.

A full quantal investigation of spinodal instabilities and related phase diagram was applied to finite nuclear systems, corresponding to Ca and Sn isotopes [21]. So the first mode to become unstable is the low lying octupole vibration. Diluted systems are unstable against low multipole deformations of the surface. Was shown that also in this case the instabilities are mostly of isoscalar nature, with an isovector component leading to isospin distillation, in agreement with the previous predictions for nuclear matter case [7].

3.2. *Spinodal decomposition: Numerical simulations*

The previous analytical study in a linearized approach is restricted to the onset of fragmentation, and related isospin distillation, in nuclear matter.

Numerical calculations have been also performed in order to study all stages of the fragment formation process [10,22]. We report on the results of ref. [10] where the same effective Skyrme interactions have been used.

In the numerical approach the dynamical response of nuclear matter is studied in a cubic box of size L imposing periodic boundary conditions. The Landau-Vlasov dynamics is simulated following a phase-space test particle method, using gaussian wave packets [23,24]. The dynamics of nucleon-nucleon collisions is included by solving the Boltzmann-Nordheim collision integral using a Monte-Carlo method [25].

We have followed the space-time evolution of test particles in a cubic box with side $L = 24 fm$ for three values of the initial asymmetry $I = 0,\ 0.25$ and 0.5, at initial density $\rho^{(0)} = 0.06 fm^{-3} \simeq 0.4\rho_0$ and temperature $T = 5$ MeV. The initial density perturbation is created automatically due to the random choice of test particle positions.

The spinodal decomposition mechanism leads to a fast formation of the liquid (high density) and gaseous (low density) phases in the matter. Indeed this dynamical mechanism of clustering will roughly end when the variance saturates [26], i.e. around 250 fm/c in the asymmetric cases.

We can also discuss the "chemistry" of the liquid phase formation. In Fig.4 we report the time evolution of neutron (thick histogram in Fig. 4a) and proton (thin histogram in Fig.4a) abundances and of asymmetry (Fig.4b) in various density bins. The dashed lines respectively shows the initial uniform density value $\rho \simeq 0.4\rho_0$ (Fig.4a) and the initial asymmetry $I = 0.5$ (Fig.4b). The drive to higher density regions is clearly different for neutrons and protons: at the end of the dynamical clustering mechanism we have very different asymmetries in the liquid and gas phases (see the

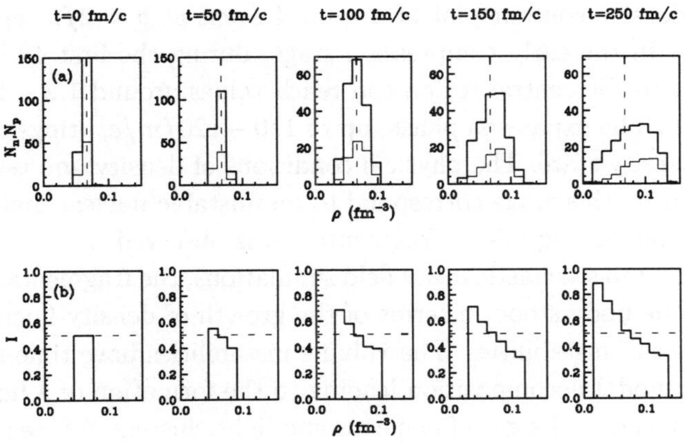

Fig. 4. Time evolution of neutron (*thick lines*) and proton (*thin lines*) abundances (a) and of asymmetry (b) in different density bins. The calculation refers to the case of $T = 5MeV$, with initial average density $\rho = 0.06fm^{-3}$ and asymmetry $I = 0.5$ (see the first panel of the (b) plots).

panel at $250fm/c$ in Fig.4b).

The conclusion is that the fast spinodal decomposition mechanism in neutron-rich matter will dynamically form more symmetric fragments surrounded by a less symmetric gas. Some recent experimental observations from fragmentation reactions with neutron rich nuclei at the Fermi energies seem to be in agreement with this result on the fragment isotopic content : nearly symmetric Intermediate Mass Fragments (IMF) have been detected in connection to very neutron-rich light ions [11], [12].

4. From bulk to neck fragmentation

4.1. *Multifragmentation*

Since dynamical instabilities are playing an important role in the reaction dynamics at Fermi energies it is essential to employ a stochastic transport theory. An approach has been adopted based on microscopic transport equations of Boltzmann-Nordheim-Vlasov (BNV) type where asymmetry effects are suitably accounted for and the dynamics of fluctuations is included, for more details see [27].

In particular we report on a study of the $50AMeV$ collisions of the systems $^{124}Sn +^{124} Sn$ $^{112}Sn +^{112} Sn$ and $^{124}Sn +^{112} Sn$, [28], where data are available from $NSCL - MSU$ exps. for fragment production. One can identify quite generally three main stages of the collision, as observed also from the density contour plot of a typical event at $b = 2fm$ displayed in Fig. 5: (1.) In the early compression stage, during the first $40 - 50fm/c$, the density in the central region can reach values around $1.2 - 1.3$ normal density; (2.) The expansion phase, up to $110 - 120fm/c$, brings the system to a low density state. The physical conditions of density and temperature reached during this stage correspond to an unstable nuclear matter phase; (3.) In the further expansion fragmentation is observed.

According to stochastic mean field simulations, the fragmentation mechanism can be understood in terms of the growth of density fluctuations in the presence of instabilities. The volume instabilities have time to develop through spinodal decomposition leading to the formation of a liquid phase in the fragments and a gas of nucleons and light clusters. As seen in the figure, the fragment formation process typically takes place up to a *freeze-out* time (around $260 - 280fm/c$). This time is well defined in the simulations since it is the time of saturation of the average number of excited primary fragments. The clusters are rather far apart with a negligible nuclear interaction left among them.

Guided by the density contour plots we can investigate the behaviour of some characteristic quantities which give information on the isospin dynamics in fragment formation. From the beginning of the fragment formation phase of the evolution, between 110 and $280fm/c$, was remarked the peculiar trends of the liquid and gas phase asymmetry. In the "central region" the liquid asymmetry decreases while an ***isospin burst*** of the gas phase is observed. This behaviour is consistent with the kinetic spinodal mechanism in dilute asymmetric nuclear matter leading to the ***Isospin Distillation*** between the liquid and the gas phase.

The charge distribution of primary fragments has rapidly decreasing trend, typical of a multifragmentation process. The effects of this process were clearly seen in the IMF isospin content, in both cases lower than at the beginning of the spinodal decomposition. Opposite trends for fragments with charge above and below $Z \approx 15$ can be observed. For heavier products the average asymmetry increases with the charge, a Coulomb related effect. However, the asymmetry rises again for lighter fragments. This can be a result of the differences in density and isospin between the regions in which the fragments grow, due to the fact that not all of them form simultaneously, as shown in the density contour plot.

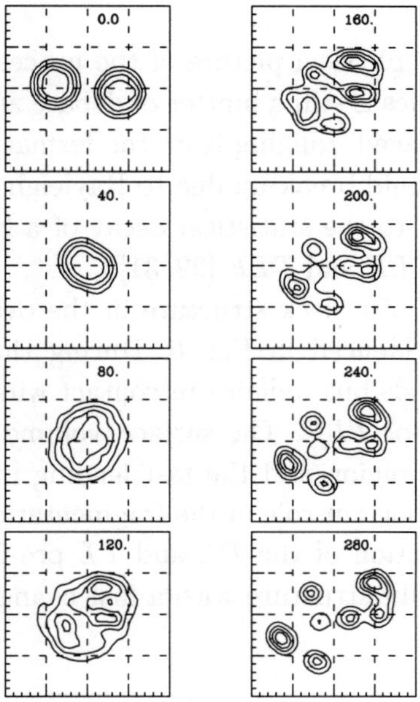

Fig. 5. Central $b = 2fm$ $^{124}Sn +^{124} Sn$ collision at $50AMeV$: time evolution of the nucleon density projected on the reaction plane: approaching, compression and expansion phases. The times are written on each figure. The iso-density lines are plotted every $0.02fm^{-3}$ starting from $0.02fm^{-3}$.

4.2. *Neck fragmentation*

Summarizing the main experimental observations, we enumerate the following features of a "dynamical" IMF production mechanism in semi-peripheral collisions:

- An enhanced emission is localized in the mid-rapidity region, intermediate between projectile-like fragment (PLF) and target-like fragments (TLF) sources, especially for IMF's with charge Z from 3 to 15 units.

- The IMF's relative velocity distributions with respect to PLF (or TLF) cannot be explained in terms of a pure Coulomb repulsion following a statistical decay. A high degree of decoupling from the PLF (TLF) is also invoked.

- Anisotropic IMF's angular distributions are indicating preferential emission directions and an alignment tendency.

- For charge asymmetric systems the light particles and IMF emissions keep track of a neutron enrichment process that takes place in the neck region.

A fully consistent physical picture of the processes that can reproduce observed characteristics is still a matter of debate and several physical phenomena can be envisaged, ranging from the formation of a transient neck-like structure that would break-up due to Rayleigh instabilities or through a fission-like process, to the statistical decay of a hot source, triggered by the proximity with PLF and TLF [29–31].

The development of a neck structure in the overlap region of the two colliding nuclei is evidenced in Fig. 6. During the interaction time this zone heats and expands but remains in contact with the denser and colder regions of PLF and/or TLF. The surface/volume instabilities of a cylindrically shaped neck region and the fast leading motion of the PLF and TLF will play an important role in the fragmentation dynamics. We notice the superimposed motion of the PL and TL pre-fragments linked to the formation of a neck-like structure with a fast changing geometry.

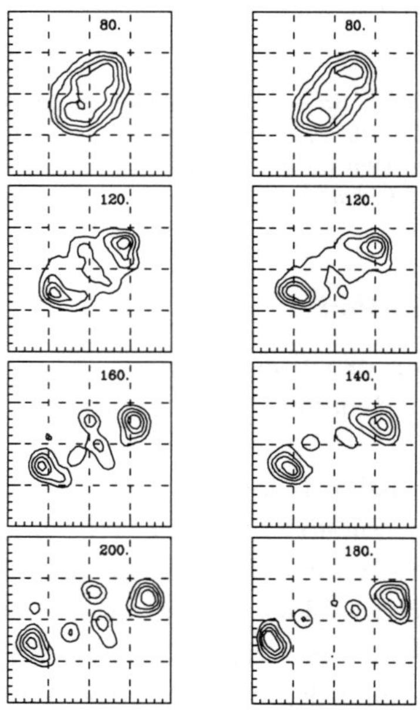

Fig. 6. $^{124}Sn +^{124} Sn$ collision at $50AMeV$: time evolution of the nucleon density projected on the reaction plane. Left column: $b = 4fm$. Right column: $b = 6fm$.

At the freeze-out time, with the neck rupture at about $140fm/c$, intermediate mass fragments are produced in the mid rapidity zone. In some events fragments form very early while, in others, they can remain for a longer time attached to the leading PLF's or TLF's. A transition behavior between multifragmentation and neck fragmentation we observed at $b = 4fm$.

From the simulations we can extract an interesting information on the time scale of the Neck-IMF production. Analysing the time distribution between the instant of the first separation of the dinuclear system and the moment when a Neck-IMF is identified (scission-to-scission time) we concluded that a large part of the $NIMF$s are formed in short time scales, within $50fm/c$. We would like to remark that the neck fragmentation shows a dependence on the nucleon-nucleon cross sections and the EOS-compressibility. The latter point is particularly interesting since it seems to indicate the relevance of volume instabilities even for the dynamics of neck. This appears consistent with the short time scales shown before, see also the discussion in ref. [32]. An interesting related effect we evidenced

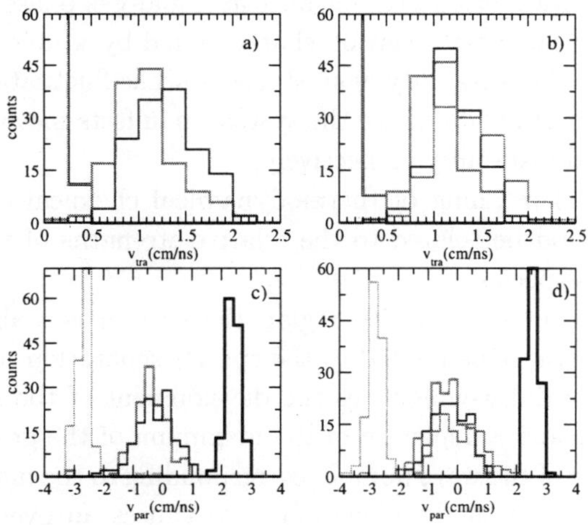

Fig. 7. Transversal velocity distributions for events with four fragments (a) and five fragments (c). Longitudinal velocity distributions for events with four fragments (b) and five fragments (d).

recently in neck fragmentation is a hierarchy in transversal velocities distribution. We considered an intrinsic axis of each event defined by the relative velocity of the two heaviest fragments, corresponding to PLF and TLF residues. Then we classified the events in terms of the total number of primary fragments. For each class of events, we constructed the transversal and longitudinal velocities distributions, in respect to the intrinsic axis, of the heaviest fragment, second heaviest and so on up to the lightest. We report in figure 7 (a),(c) these distributions for events with four fragments (two IMF) following the reaction $^{124}Sn + ^{124}Sn$ at $b = 4$fm. In 7 (b),(d) the same distributions are plotted for events with five fragments (three IMF). A clear correlation between the transversal velocity distribution and the rank of the fragment in the event is evidenced: lighter is the fragment more shifted toward higher values is its transversal velocity distribution. This may represent a signature of the neck fragmentation process. No hierarchy effect we can identify in the longitudinal velocity distributions.

5. Conclusions

In this work we investigated several properties of the asymmetric nuclear matter in low density region of phase diagram. The thermodynamical and dynamical analysis was based on Landau theory of Fermi liquid extended to binary systems. Was concluded that:

- at low densities, of interest for nuclear liquid-gas phase transition, the asymmetric nuclear matter can be characterized by a unique spinodal region, defined by the instability against isoscalarlike fluctuations; inside this we can identify the region where the system manifests mechanical instability and chemical instability respectively;

- the physical meaning of thermodynamical chemical and mechanical instabilities should be related to the relative strengths of the interactions among different species.

- everywhere in this density region the system is stable against the isovectorlike fluctuations related to the species separation tendency.

- at larger initial asymmetries the development of the spinodal instabilities is slower and a depletion of the maximum of the growth rate takes place.A decrease of the wave number corresponding to the maximum growth rate was deduced. Also the Coulomb force causes an overall decrease of growth rates. In this case the wave vector should exceed a threshold value in order to observe the instabilities.

- during the time development of the spinodal instabilities in ANM the fragment formation is accompanied by the isospin distillation leading to a more symmetric liquid phase and more neutron rich gas phase.

We have made a connection of these features with isospin transport properties in simulations of fragmentation reactions based on stochastic BNV transport models. The presence and the role of the instabilities along the reaction dynamics in bulk fragmentation and neck fragmentation were discussed.

The results discussed here refer to the formation processes of primary fragments. i.e. at the freeze-out time. We explored the possibility that IMF appear as a result of a mechanism that initially started as spinodal decomposition triggered by isoscalar-like instabilities. These fragments are excited and certainly the subsequent statistical decay modify the signal. Therefore it is important to search for various observables still keeping informations about the early stages of the fragments formation, for example those related to the kinematical properties (velocity distributions, angular distributions) and correlations between these observables and isospin content.

Moreover the neck dynamics and corresponding isospin transport shows distinctive features related to the interplay between volume and surface instabilities. These should be better clarified in the future since they can contribute to a proper understanding of intermediate mass fragments production at Fermi energies.

Acknowledgments

Virgil Baran acknowledges support of the Romanian Ministry for Education and Research for this work under the contract No. CEx-05-D10-02 and CEx-05-D11-03.

References

1. M. Barranco and J. R. Buchler, Phys. Rev. C22 *Phys. Rev.* **C 22** 1729 (1980).
2. L.D.Landau, *Soviet Physics JETP* **5** 101 (1957).
3. A.B.Migdal, *Theory of finite Fermi systems and applications to atomic nuclei*, (Wiley & Sons, N.Y. 1967).
4. G.Baym and C.J.Pethick in *The physics of Liquid and Solid Helium* edited by K.H.Bennemann and J.B.Ketterson, Vol 2, (Wiley, New-York, 1978), p.1.
5. C.J.Pethick, D.G.Ravenhall, *Ann.Phys. (N.Y.)* **183** 131 (1988).
6. N. Iwamoto, C.J. Pethick, *Phys. Rev.* **D 25** 313 (1982).
7. V.Baran, M.Colonna, M.Di Toro, V.Greco, *Phys.Rev.Lett.* **86** 4492 (2001).
8. L.D.Landau and E.M.Lifshitz, *Statistical Physics*, (Pergamonn Press, 1989).
9. M.Colonna, M.Di Toro, A.B.Larionov, *Phys.Lett.* **B 428** 1 (1998).
10. V.Baran, M.Colonna, M.Di Toro, A.B.Larionov, *Nucl.Phys.* **A 632** 287 (1998).
11. H.S.Xu et al., *Phys.Rev.Lett.* **85** 716 (2000).
12. S.Yennello et al., this Proceeding.
13. J. Margueron and Ph. Chomaz, *Phys. Rev.* **C 67** 041602(R) (2003).
14. H.Müller, B.D.Serot, *Phys.Rev.* **C 52** 2072 (1995).
15. B. Liu, V. Greco, V. Baran, M. Colonna, M. Di Toro *Phys.Rev.* **C 65** 045201 (2002).

16. S.S.Avancini, L. Brito, D.P. Menezes,C. Providencia *Phys.Rev.* **C 70** 015203 (2004).
17. P.Haensel, *Nucl.Phys.* **A 301** 53 (1978).
18. F.Matera, V.Yu.Denisov, *Phys.Rev.* **C 49** 2816 (1994).
19. G.Baym, H.A.Bethe, C.J.Pethick, *Nucl.Phys.* **A 175** 225 (1971).
20. Ph.Chomaz, M.Colonna, J.Randrup, *Phys.Rep.* **389** 263 (2004).
21. M.Colonna, Ph. Chomaz and S. Ayik, *Phys.Rev.Lett.* **88** 122701 (2002).
22. B.-A.Li, A.T.Sustich, M.Tilley, B.Zhang, *Nucl.Phys.* **A 699** 493 (2002).
23. Ch.Gregoire et al., Nucl.Phys. A465 (1987) 315.
24. V.Baran, A.Bonasera, M.Colonna, M.Di Toro, A.Guarnera, *Prog.Part.Nucl.Phys.* **38** 263 (1997).
25. A.Bonasera, F.Gulminelli, J.Mollitoris *Phys.Rep.* **243** 1 (1994).
26. M.Colonna, M.Di Toro, A.Guarnera, *Nucl.Phys.* **A 580** 312 (1994).
27. V.Baran, M.Colonna, V. Greco, M.Di Toro, *Physics Reports* **410** 335 (2005).
28. V.Baran, M.Colonna, V.Greco, M.Di Toro, M.Zielinska Pfabe, H.H. Wolter, Nucl.Phys. **A 703** 603 (2002).
29. U. Brosa, S. Grossman, A. Muller, *Phys. Rep.* **197** 167 (1990).
30. J. Lukasik et al., *Phys. Lett.* **B 566** 76 (2003).
31. A.S. Botvina et al., *Phys.Rev.* **C 59** 3444 (1999).
32. V.Baran, M.Colonna, M.Di Toro, *Nucl.Phys.* **A 730** 329 (2004).

THERMAL PROPERTIES OF NUCLEAR SYSTEMS: FROM NEUTRON STARS TO FINITE NUCLEI

FRANCESCA GULMINELLI†

LPC/Ensicaen, 6 Bd du Maréchal Juin
14050 Caen, France

Thermal and phase properties of nuclear systems are briefly reviewed within an information theory approach. Such theory allows treating on the same ground extended systems at the thermodynamic limit, as nuclear matter in the inner crust of neutron stars, and finite size, short-lived systems, as excited nuclei produced in heavy ion collisions. Different related issues including the pertinence of equilibrium in systems finite in size and time, ensemble inequivalence, and the effect of Coulomb interactions are discussed.

1. Phase properties of supernovae and neutron stars

Supernovae explosions, powered by the release of gravitational energy of a massive star which has exhausted its nuclear fuel, can lead to the formation of a most interesting dense stellar object: a neutron star[1].

Due to the lack of observational data, the composition and structure of a neutron star are still highly hypothetical[2]. In the outer part of the star, the stellar crust extending over about one kilometer, the matter density is expected to be comparable to normal nuclear matter density, and the star can be modelized as essentially composed of neutrons, protons, electrons and neutrinos in thermal and chemical equilibrium. In a few minutes after its birth, the proto-neutron star formed at a temperature of $\sim 10^{11}$ K becomes transparent to neutrinos, and cools via neutrino emission to temperatures which are small on the nuclear scale. The cooling process occurs via heat conduction and convection through the envelope to the surface on a time scale too short for the system to be in global thermal equilibrium; however local thermal equilibrium and beta-equilibrium should be well verified during the whole evolution of the proto-neutron star[3].

Crust matter is therefore very similar to nuclear matter, which is known to exhibit a complex phase diagram including first and second order phase transitions[4-6]. A recent calculation of the nuclear matter phase diagram using realistic Skyrme interactions[7] is shown in Figure 1.

---

† Member of the Institut Universitaire de France.

1.1. *Nuclear matter versus stellar matter*

The analogy between stellar matter and nuclear matter hides however an important difference. If nuclear matter is by definition neutral, only global charge neutrality is required by thermodynamic stability for stellar matter.

An important consequence of the charge neutrality constraint is that the canonical free energy density f is defined only for $\rho_c = \rho_p - \rho_e = 0$. Hence $f(T, \rho_n, \rho_p, \rho_e) = f(T, \rho_n, \rho)$ with $\rho = (\rho_p + \rho_e)/2$, and the chemical potential μ_c associated to ρ_c can not be defined, since the free energy is not differentiable in the total charge direction[2].

This constraint affects the thermodynamics directly, since it changes the number of degrees of freedom of the thermodynamic potentials[8].

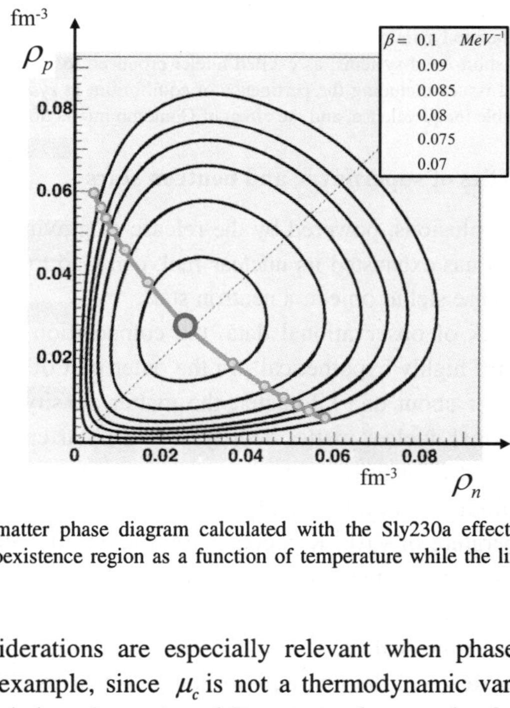

Figure 1. Nuclear matter phase diagram calculated with the Sly230a effective interaction. The contours give the coexistence region as a function of temperature while the line is a second order transition line[6].

These considerations are especially relevant when phase transitions are concerned. For example, since μ_c is not a thermodynamic variable, the Gibbs rule for two coexisting phases A and B $\mu_A = \mu_B$ does not imply that each of the chemical potentials μ_e and μ_p are identical in the two phases. The difference in chemical potentials of charged particles is counterbalanced by the Coulomb force: as some electrons move from one phase to the other driven by the chemical-potential difference, the Coulomb force reacts forbidding a macroscopic charge to appear.

To illustrate the consequence of charge neutrality for the phase diagram of stellar matter, we present in the following section mean field calculations[6,9] using the Sly230a effective interaction, which has been optimized to describe exotic nuclei and pure neutron matter[7].

1.2. *Instabilities in stellar matter*

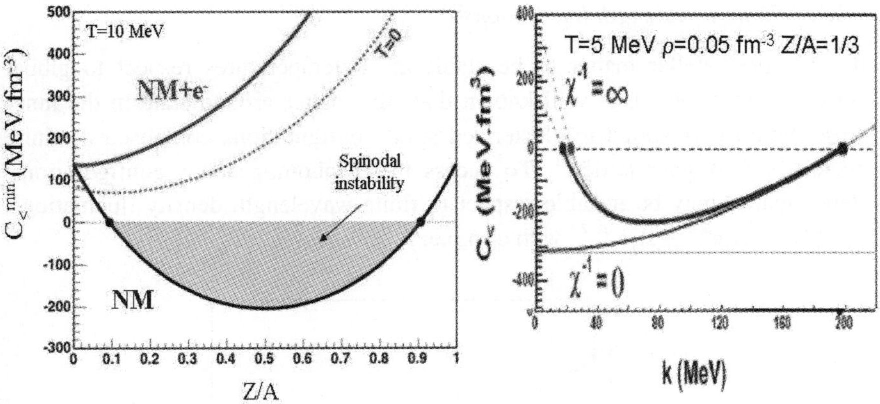

Figure 2. Left part: minimal curvature of the matter mean field free energy at T=10 MeV as a function of the isospin. Lower curve: normal nuclear matter; upper curve: electrons included; dashed line: same as the upper curve, but for T=0. Right part: minimal curvature of the matter mean field free energy in the presence of a plane wave density fluctuation as a function of the wave number. The limiting cases of a completely incompressible, and fully compressible electron background are also given. All calculations are performed with the Sly230a force[9].

The matter instability to density fluctuations can be spotted looking at the curvature matrix[10]

$$C = \begin{vmatrix} \frac{\partial^2 f}{\partial\rho_n\partial\rho_n} & \frac{\partial^2 f}{\partial\rho_n\partial\rho} \\ \frac{\partial^2 f}{\partial\rho\partial\rho_n} & \frac{\partial^2 f}{\partial\rho\partial\rho} \end{vmatrix} = \begin{vmatrix} \frac{\partial\mu_n}{\partial\rho_n} & \frac{\partial\mu_n}{\partial\rho_p} \\ \frac{\partial\mu_p}{\partial\rho_n} & \frac{\partial\mu_p}{\partial\rho_p} + \frac{\partial\mu_e}{\partial\rho_e} \end{vmatrix}, \tag{1.1}$$

where the free energy density is $f(\rho_n,\rho) = f_N(\rho_n,\rho_p=\rho) + f_e(\rho_e=\rho)$. The additional term $1/\chi = \partial\mu_e/\partial\rho_e$ in the matrix modifies the stability conditions with respect to the nuclear matter part, i.e. to the curvature of f_N. Due to the small electron mass, i.e. high Fermi energy, the electron fluid is highly incompressible, leading to a quenching of the instability: the instability conditions detC<0 or TrC<0 are more difficult to fulfill.

The result for the Sly230a interaction is shown in Figure 2: the lowest eigenvalue of the curvature matrix is always positive, independent of the temperature and proton fraction, meaning that the spinodal zone is suppressed in stellar matter[9].

1.2.1. *Finite wavelength fluctuations*

If we expect stellar matter to be stable at all temperatures respect to global density fluctuations, it is well known that the matter ground state in the inner crust should correspond to clusterized solid configurations composed of finite nuclei on a Wigner lattice[11]. To access these inhomogeneous configurations, stellar matter may be instable respect to finite wavelength density fluctuations, $\delta\rho_q(kr) = A_q e^{i\vec{k}\cdot\vec{r}} + A_q^* e^{-i\vec{k}\cdot\vec{r}}$, with q=n,p,e.

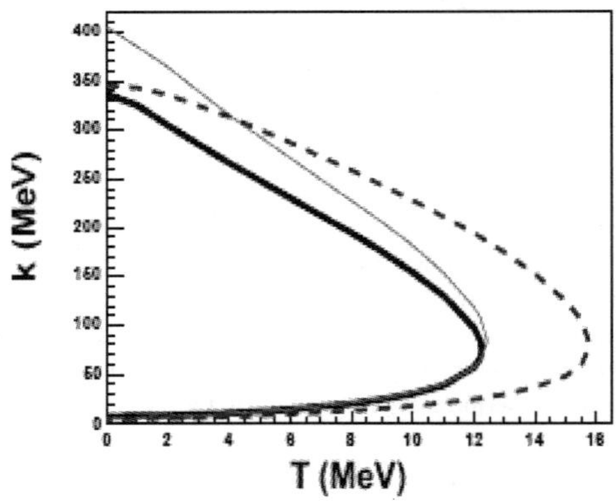

Figure 3. Instability interval of stellar matter respect to finite wavelength fluctuations in the temperature versus wave-number plane, with different Skyrme forces. Thick line: Sly230a; thin line: SGII; dashed line: SIII.

The mean field response to such a fluctuation induces two extra terms in the curvature matrix eq.(1.1)[10]. The gradient term of the Skyrme functional[7] produces a term $\propto k^2$ coupling proton and neutron densities, which tends to suppress high k fluctuations; the direct Coulomb interaction produces a term $\propto k^{-2}$, coupling proton and electron densities, which tends to suppress low k fluctuations. The resulting effect is the appearance of a finite k interval where one eigenvalue turns negative, i.e. matter becomes unstable. An example is shown in the right part of Figure 2[9].

We can see that an important instability region exists even at high temperatures, and almost independent of the effective interaction employed, suggesting that clusterized configurations may be important also in the proto-neutron star evolution. The finite wavelength instability region associated to different Skyrme interactions is shown in Figure 3.

1.2.2. *Beta-equilibrium and phase structure*

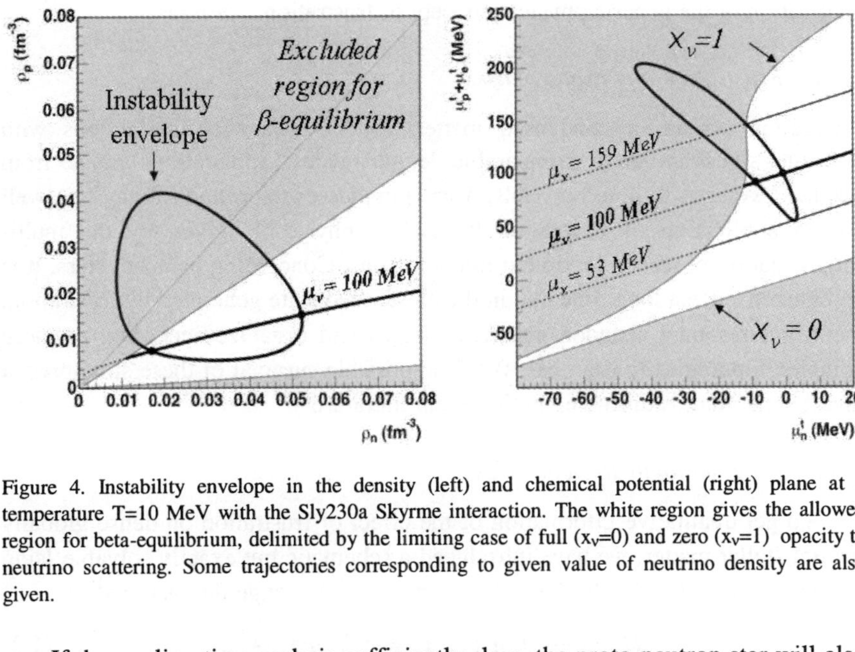

Figure 4. Instability envelope in the density (left) and chemical potential (right) plane at a temperature T=10 MeV with the Sly230a Skyrme interaction. The white region gives the allowed region for beta-equilibrium, delimited by the limiting case of full (x_ν=0) and zero (x_ν=1) opacity to neutrino scattering. Some trajectories corresponding to given value of neutrino density are also given.

If the cooling time scale is sufficiently slow, the proto-neutron star will also be subject to the constraint of β-equilibrium, $\mu_p^t + \mu_e^t = \mu_n^t + \mu_\nu$, with $\mu_q^t = \mu_q + m_q$. This is certainly the case in the final step of the star evolution, when the temperature is low enough for the matter to be completely transparent to neutrino emission, $\mu_\nu = 0$, which becomes then the most effective cooling process of the star[1].

Chemical equilibrium is generally assumed in the thermodynamic studies of stellar matter[3]. The chemical equilibrium constraint defines an accessible region in the phase diagram shown for a given temperature in Figure 4[9]. The lower limit is given by the equation $\mu_\nu = 0$, while lepton number conservation defines an upper limit. We can see in Figure 4 that, depending on μ_ν, i.e. on the neutrino opacity, the instability region can be crossed under the constraint of chemical equilibrium.

1.3. *The effect of charge density fluctuations*

In the previous section we have seen that hot stellar matter can show an instability respect to finite wavelength density fluctuations, eventually leading to the formation of clusterized configurations. This instability is due to the interplay between the surface and Coulomb interaction terms. This effect is a specific application of the generic physical concept of frustration.

1.3.1. *Frustration and exotic phases*

Frustration occurs in condensed matter physics whenever interactions with opposite signs act on a comparable length scales; applications range from magnetic systems to liquid crystals, from spin glasses to protein folding[12]. A well known application of frustration in nuclear physics is given by the multi-fragmentation observed in violent ion collisions. Concerning compact stars, it is recognized since a long time that in the clusterized state generated by frustration, the structures may abandon spherical shapes and organize according to more complex topologies (pasta phases)[13]. The possible survival of these structures at high temperature is discussed in the recent literature[14,15].

1.3.2. *Ising analogue to stellar matter*

To get qualitative information of the effect of frustration on dense globally neutral stellar matter, we have introduced a schematic but exactly solvable Ising model with long Coulomb-like and short nuclear-like range interactions[16]

$$H_N = -\frac{\varepsilon}{2}\sum_{\langle ij\rangle} n_i n_j \; ; \; H_C = \frac{\chi}{2}\sum_{ij} q_i q_j \frac{1}{r_{ij}} \quad , \tag{1.2}$$

where each site of a three dimensional lattice of $N = L^3$ sites is characterized by an occupation number $n_i = 0,1$ and an effective charge $q_i = n_i - \sum_{k=1}^{N} n_k / N$. The effective charge represents the proton distribution screened by a uniform electron background[17,a].

A statistical tool to deal with frustrated systems is given by the multi-canonical ensemble[18]. The two energy components E_N,E_C are treated as two

---

[a] To accelerate thermodynamic convergence, the finite lattice is repeated in all three directions of space a large number of times.

independent observables associated to two Lagrange multipliers β_N, β_C. A generalized grand potential is defined by

$$Z_{\beta_N \beta_C \vec{\alpha}} = \int W(E_N, V_C, \vec{N}) e^{-\beta_N E_N - \beta_C V_C - \sum_q \alpha_q N_q} \tag{1.3}$$

where $e^{\alpha q}$ is the fugacity of particle type q.

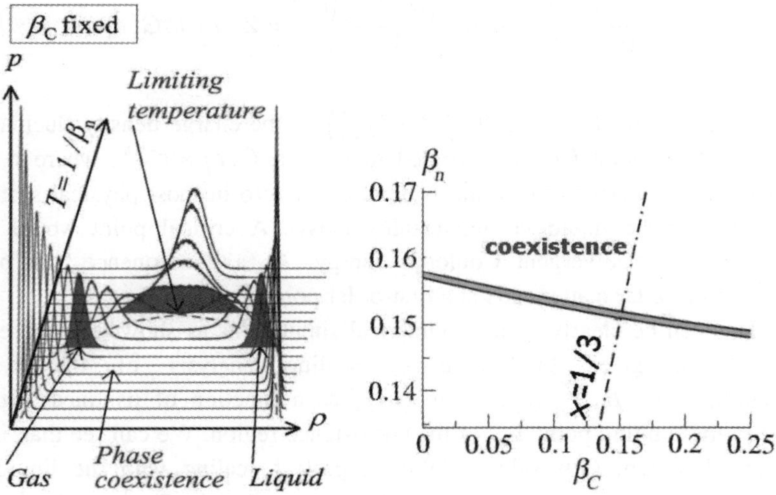

Figure 5. Phase analysis of the charged Ising system. Left side: representation in the temperature-density plane. Full lines: density distribution at different temperatures; dashed line: coexistence zone. Right side: phase diagram in the temperature versus effective charge plane. The dashed-dotted line gives the β_c value corresponding to a proton fraction x=1/3.

If E_C represents the Coulomb energy and E_N the nuclear term, the choice $\beta_C = \beta_N$ gives the usual (grand)-canonical thermodynamics for charged systems, $\beta_C = 0$ leads to the uncharged thermodynamics, while all intermediate values $0 \le \beta_C \le \beta_N$ correspond to interpolating ensembles, or equivalently to physical systems with an effective charge $\left(q_{eff} / q_0\right)^2 = \beta_C / \beta_N$. The multi-(grand)-canonical ensemble allows to construct a single unified phase diagram for neutral and charged matter, and is therefore an ideal statistical tool to make some connections between nuclear and stellar matter.

The resulting phase diagram is shown in Figure 5. We can see that, at variance with the case of multi-fragmentation of finite charged nuclei, the effect of the Coulomb interaction is an expansion of the coexistence zone. This result can be understood[16] as an electron screening effect minimizing the Coulomb energy of the dense phase, and thus making it accessible even at high temperature at variance with the finite nucleus case. It implies that pasta phases may be relevant in a wide range of thermodynamic conditions.

Another important effect of the Coulomb interaction in stellar matter is to suppress the critical point of neutral nuclear matter and the related phenomenon of critical opalescence. The Coulomb energy density can be expressed as

$$\frac{\langle V_{coul}\rangle}{V}=\frac{q^2}{2V}\int d\vec{r}d\vec{r}\,'\frac{\langle \rho_c(\vec{r})\rho_c(\vec{r}\,')\rangle}{|\vec{r}-\vec{r}\,'|}=2\pi\alpha\int drrG(r) \quad , \tag{1.4}$$

where $G(r)=\langle(\rho_p(\vec{r})-\langle\rho_e\rangle)(\rho_p(\vec{r})-\langle\rho_e\rangle)\rangle$ is the charge density fluctuation. At the critical point G(r) is expected to scale as $G(r)\propto r^{-1-\eta}$, where η is a critical exponent which turns out to be close to zero in most physical systems. (η=0.017 for the liquid-gas universality class). A critical point would then correspond to a divergent Coulomb energy. As a consequence, the phase transition in stellar matter ends at a first-order point.

This can be clearly seen in numerical simulations as shown in Figure 6[16]. This figure gives the finite-size scaling analysis of the system susceptibility $\chi=\beta_N\sum_{ij}^N\langle n_i n_j\rangle-\langle n_i\rangle\langle n_j\rangle$, as a function of its linear size L, close to the limiting point $T_{\lim}$ of the coexistence region. We can see that, if the uncharged system (left side) exhibits a perfect scaling with the liquid-gas universality class critical exponents, in the presence of the Coulomb interaction the situation is completely different. The transition rounding effect on $T_{\lim}$ is compatible with ν=0, which means that the divergence of the correlation length (which should scale as $\xi\propto t^{-\nu}$) is quenched. More important, scaling is violated.

Such an effect has been already observed in Ising models with long-range frustrating interactions, where the coexistence region was seen to end at a first-order point[12]. We have here shown that the quenching of criticality is a generic effect of Coulombic systems, as long as the charge density has a non zero correlation with the order parameter of the transition[8].

Since the Fourier transform of the correlation function gives the enhancement of the in medium scattering cross section respect to its free value, the important physical implication of this result is that we expect hot stellar matter to show a small opacity to neutrino scattering[15], which may have important consequences on the cooling dynamics[19].

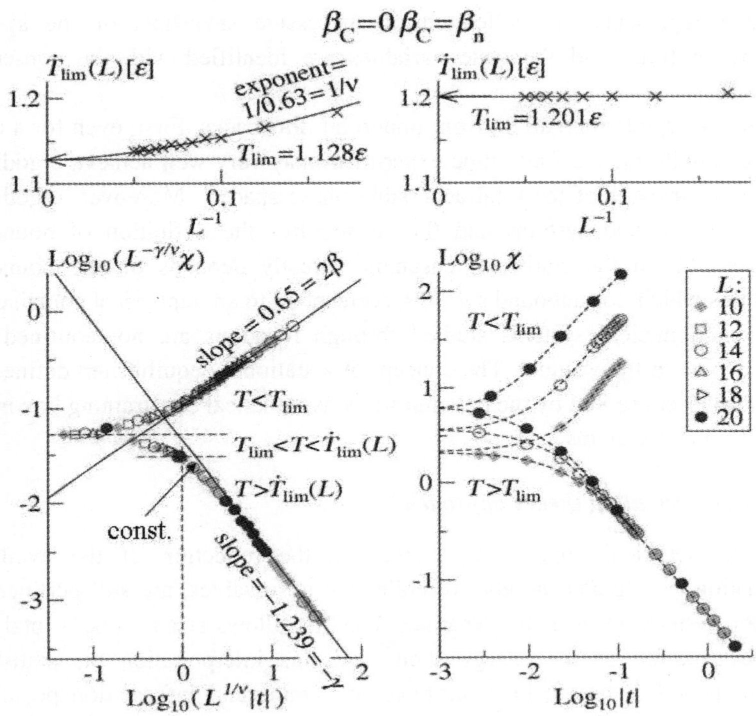

Figure 6. Finite size scaling analysis of the charged Iisng model (see text). Left part: uncharged system. Right: charged system.

2. Thermodynamics of finite nuclei

The microscopic foundations of thermodynamics are well established using the Gibbs hypothesis of statistical ensembles maximizing the Shannon entropy[20].

At the thermodynamic limit, the various Gibbs ensembles converge to a unique thermodynamic equilibrium. Conversely, in finite systems as atomic nuclei, the various Gibbs ensembles are not equivalent[21], and the physical meaning and relevance of these different equilibria has to be investigated.

A common interpretation of a statistical ensemble for a finite system is given by the Boltzmann ergodic assumption. In this interpretation the statistical ensemble represents the collection of successive snapshots of the system evolving in time, and the state variables are identified with the conserved observables.

This interpretation suffers from important drawbacks. First, even for a truly ergodic Hamiltonian, a finite time experiment may very well achieve ergodicity only on a subspace of the total accessible phase space[22]. Moreover, ergodicity applies to confined systems and thus it requires the definition of boundary conditions. Then the statistical ensemble directly depends on the boundary conditions which, for unbound systems, correspond to an unphysical container.

Excited nuclear systems studied through reactions are not confined but freely evolve in the vacuum. The concept of a stationary equilibrium defined by the variables conserved by the dynamics in a hypothetical constraining box is not useful for these systems.

2.1. *An information theory approach*

However statistical approaches, expressing the reduction of the available information to a limited number of collective observables, are still pertinent to complex systems even if the dynamics does not allows at any time a total and even exploration of the energy shell[23]. In this interpretation the statistical MaxEnt postulate has to be interpreted as a minimum information postulate, which finds its justification in the complexity of the dynamics independent of any time scale[20,23].

When the system is characterized by L observables known in average, $\langle \hat{A}_l \rangle = Tr\hat{D}\hat{A}_l$, statistical equilibrium corresponds to the maximization of the constrained entropy

$$S_c = -\mathrm{Tr}\hat{D}\log\hat{D} - \sum_l \lambda_l \langle \hat{A}_l \rangle \ , \qquad (2.1)$$

where $\hat{D} = \sum_n |\psi^{(n)}\rangle p^{(n)} \langle \psi^{(n)}|$ is the density matrix, and $\vec{\lambda} = \{\lambda_l\}$ are Lagrange multipliers. Gibbs equilibrium is then given by

$$\hat{D}_{\vec{\lambda}} = \frac{1}{Z_{\vec{\lambda}}} \exp -\vec{\lambda} \cdot \vec{\hat{A}} \ , \qquad (2.2)$$

where Z is the associated partition sum. It should be noticed that micro-canonical thermodynamics[24] can also be obtained from the variation of the Shannon entropy, in the special case of a fixed energy subspace. In this case the maximum of the Shannon entropy can be identified with the Boltzmann entropy max (S) = logW (E), where W is the total state density with the energy E. In the following we shall confine ourselves to the Gibbs formulation, which is the most general.

This information theory approach is a very powerful extension of the classical Gibbs equilibrium: any arbitrary observable can act as a state variable, and all statistical quantities can be unambiguously defined for any number of particles[25].

2.2. *Ensemble inequivalence*

Each choice of the L independent observables A_l leads to the definition of a different statistical ensemble. A fundamental theorem in statistical mechanics, the Van Hove theorem, guarantees the equivalence between ensembles at the thermodynamic limit. However the theorem does not apply in finite systems. The non-equivalence of statistical ensembles[21] has important conceptual consequences. It implies that the value of thermodynamic variables for the very same system depends on the type of experiment which is performed (i.e. on the ensemble of constraints which are put on the system), contrary to the standard thermodynamic viewpoint that water heated in a kettle is the same as water put in an oven at the same temperature.

Let us consider energy as the extensive observable and inverse temperature as the conjugated intensive one. The definition of the canonical partition sum and entropy is

$$Z_\beta = \sum_n \exp\left(-\beta E^{(n)}\right) \quad ; \quad S_{can}\left(\langle E\rangle\right) = \log Z_\beta + \beta\langle E\rangle \, , \qquad (2.3)$$

where the sum runs over the available eigenstates n of the Hamiltonian. Here, we assume that the partition sum converges; this is not always the case[a]. If energy can be treated as a continuous variable, eq.(2.3) can be written as:

$$Z_\beta = \int_0^\infty dE W(E) \exp\left(-\beta E\right), \qquad (2.4)$$

where energies are evaluated from the ground state. Eq.(2.4) is a Laplace transform between the canonical partition sum and the micro-canonical density of states linked to the entropy by S(E) = logW(E). If the integrand is a strongly peaked function, the integral can be replaced by its maximum, times a gaussian integral (saddle point approximation):

$$Z_\beta \approx W(\bar{E}) \exp\left(-\beta \bar{E}\right) \sqrt{2\pi\sigma_E^2} \quad ; \quad \log Z_\beta \approx S(\bar{E}) - \beta \bar{E} \, . \qquad (2.5)$$

Eq.(2.5) has the structure of an approximate Legendre transform, similar to the exact expression (2.3). This shows that in the lowest order saddle point approximation, ensembles differing at the level of constraints acting on a specific observable (here energy), lead to the same entropy, i.e. they are equivalent. We

---

[a] The possible divergence of the thermodynamic potential of the intensive ensemble is already a known case of ensemble inequivalence[26].

will see in the next section that however the saddle point approximation eq.(2.5) can be highly incorrect close to a phase transition, for the simple reason that the integrand is bimodal making a unique saddle point approximation inadequate. In this case eq.(2.4) is the only correct transformation between the different ensembles, and ensemble inequivalence naturally arises.

2.3. *Continuum states and boundary conditions*

The statistical physics formalism is valid for any system size. However, as soon as one A_l contains differential operators such as a kinetic energy, eq.(2.2) is not defined, unless boundary conditions are specified. Only at the thermodynamic limit boundary conditions are irrelevant, as only in this limit surface effects are negligible. Dealing with finite nuclei, a fictitious container is generally introduced[24,27,28]. The volume and shape of this box has no influence for self-bound systems, but in the presence of continuum states the situation is different. Let us consider the standard case of the annulation of the wave-function on the surface S of a containing box V: Introducing the projector, $\hat{P}_S$, over S and its exterior, the boundary conditions reads $\hat{P}_S\left|\psi^{(n)}\right\rangle = 0$ for all microstates (n), or $Tr\hat{D}\hat{P}_S = 0$ [b]. The density matrix then reads:

$$\hat{D}_{\lambda S} = \frac{1}{Z_{\lambda S}} \exp\left(-\vec{\lambda}\vec{\hat{A}} - b_S \hat{P}_S\right) , \tag{2.6}$$

which shows that the thermodynamics of the system depends on the whole surface S. To specify the density matrix, the projector P_S has to be exactly known, and this is in fact impossible. The nature of P_S is intrinsically different from the usual global observables A_l. Not only it is a many-body operator, but it requires the exact knowledge of each point of the boundary surface, while no or few parameters are sufficient to define the A_l.

This infinity of points corresponds to an infinite amount of information to be known to define the density matrix (2.6). This requirement is in contradiction with the statistical mechanics principle of minimum information. Thus eq.(2.6) is unphysical, and the same is true for the standard (N,T,V) or (N,E,V) ensembles, when dealing with finite unbound unconfined systems.

One way to get around the difficulties encountered to take into account our incomplete knowledge on the boundaries is to introduce a hierarchy of observables, describing the size and shape of the matter distribution. For example, if only the average system size $\langle R^2\rangle$ is known, the minimum information principle implies

$$\hat{D}_{\lambda\omega} = \frac{1}{Z_{\lambda\omega}} \exp\left(-\vec{\lambda}\vec{\hat{A}} - \omega\hat{R}^2\right). \tag{2.7}$$

---

[b] We have used the projector property $\hat{P}_S^2 = \hat{P}_S$.

A typical application of this concept is the so called freeze-out hypothesis in nuclear collisions: at a given time t_0 the main evolution (i.e. the main entropy creation) is assumed to stop, and partitions are supposed to be essentially frozen. Typically thermal and chemical equilibrium is assumed, meaning that the information at t_0 on the energetics and particle numbers is limited to the observables $\langle H \rangle$ and $\langle N_f \rangle$ for the different species f [27,28].

Freeze-out occurs when the system has expanded to a finite size. Then at least one measure of the system's compactness should be included. The limited knowledge of the system extension leads to a minimum biased density matrix given by eq.(2.7)[25,29].

2.4. *Time dependent statistical ensembles*

As soon as one of the constraining observables A_l is not a constant of the motion, the statistical ensemble (2.2) is not stationary. A single time description may still looks appropriate in the freeze-out configuration discussed in the last section. Indeed in many physical cases one can clearly identify a specific time at which the information concentrated in a given observable is frozen (i.e. the observable expectation value ceases to evolve).

However this freeze-out time is in general fluctuating, and different for different observables. For example for the ultra-relativistic heavy ion reactions two freeze-out times are discussed[28], one for the chemistry, and one for the thermal agitation. We need therefore to define a statistical ensemble constrained by information coming from different times.

Let us now suppose that the different information on the system $\langle A_l \rangle$ is known at different times t_l. A generalization of the Gibbs idea is that at a time t the least biased state of the system is the maximum of the Shannon entropy, where observables are constrained at former times t_l:

$$-Tr\left(\hat{D}(t)\ln\hat{D}(t)\right)-\sum_{l=0}^{L}\lambda_l\left\langle\hat{A}_l\right\rangle_{t_l}=\max\,, \qquad (2.8)$$

where the λ_l are the Lagrange parameters associated with all the constraints. This maximization will lead to a density matrix which can be considered as a generalization to time dependent processes of Gibbs ensembles (2.2).

Let us consider the case of a deterministic evolution $\partial\hat{D}/\partial t=-i\left[\hat{H},\hat{D}\right]$. The minimum biased density matrix is given by[29]

$$\hat{D}_{\vec{\lambda}}(t)=\frac{1}{Z_{\vec{\lambda}}(t)}\exp\text{-}\vec{\lambda}\cdot\vec{\hat{A}}' \qquad (2.9)$$

with

$$\hat{A}_l^{'} = e^{-i(t-t_l)\hat{H}}\,\hat{A}_l e^{i(t-t_l)\hat{H}} = \hat{A}_l + \sum_{p=1}^{\infty}\frac{(t-t_l)^p}{p!}\hat{B}_l^{(p)} \,, \qquad (2.10)$$

where the additional constraints associated with the time evolution of the system are recursively defined by

$$\hat{B}_l^{(p)} = -i\left[\hat{H}, \hat{B}_l^{(p-1)}\right] \;\; ; \;\; \hat{B}_l^{0} = \hat{A}_l \,. \qquad (2.11)$$

Eqs.(2.9-2.11) give in principle an exact solution of the complete many body evolution problem with a minimum information hypothesis on the final time t, having made few observations $<A_l>$ at previous times t_l.

However, in the general case, an infinite amount of information, i.e. an infinite number of Lagrange multipliers are needed. Different interesting physical situations exist though, for which the series can be analytically summed up. In this case, the knowledge of a small number of average observables will be sufficient to describe the whole density matrix at any time, under the unique hypothesis that the information was finite at a given time.

2.5. *The dynamics of the expansion*

Let us now apply the above formalism to transient unconfined systems. We consider a scenario often encountered experimentally: a finite system of loosely interacting particles with a finite extension in an open space. We shall assume that at a given freeze-out time t_0 the system can be modelized as a non-interacting ensemble of n=1,…,N particles or fragments, and a definite value for the mean square radius $<R^2>$ characterizes the ensemble of states according to eq.(2.7). Since $\omega<R^2>$ is not an external confining potential, but only a finite size constraint, the minimum biased distribution (2.7) is not stationary. To take into account the time evolution, we must introduce additional constraining observables

$$\hat{B}^{(1)} = \frac{1}{m}\left(\hat{\vec{R}}\cdot\vec{P} + \hat{\vec{P}}\cdot\vec{R}\right) \;\; ; \;\; \hat{B}^{(2)} = \frac{2}{m}\hat{\vec{P}}^2 \,. \qquad (2.12)$$

Since $[H,B^{(2)}]=0$, all the other $B^{(p)}$ with p>2 are zero. The canonical density matrix is given by

$$\hat{D}_{\beta\lambda h}(t) \propto \exp\left(-\beta(t)\frac{\sum_i(\hat{\vec{p}}_i - h(t)\hat{\vec{r}}_i)^2}{2m} - \lambda(t)\hat{R}^2\right), \qquad (2.13)$$

where the time dependent temperature and hubblian factor and are given by:

$$\beta(t) = \beta_0 + \frac{2\omega_0(t-t_0)^2}{m} \;\; ; \;\; h(t) = \frac{2\omega_0(t-t_0)}{\beta m + 2\omega_0(t-t_0)^2} \,, \qquad (2.14)$$

while the radius constraint varies in time according to $\omega(t) = \omega_0 - m\beta h^2/2$.

In this simple case of an ideal non-interacting gas, the infinite information which is a priori needed to follow the time evolution of the density matrix according to eq.(2.9), reduces to the three observables r^2, r.p, p^2. Indeed these operators form a closed Lie algebra, and the exact evolution of (2.9) preserves it algebraic structure.

The description of the time evolution when describing unconfined finite systems has introduced a new phenomenon: the expansion. The important consequence is that radial flow is a necessary ingredient of any statistical description of unconfined finite systems: the static (canonical or micro-canonical) Gibbs ansatz in a confining box which is often employed[27], misses this crucial point. On the other hand, if a radial flow is observed in the experimental data, the formalism we have developed allows associating this flow observation to a distribution at a former time when flow was absent. This initial distribution corresponds to a standard static Gibbs equilibrium in a confining harmonic potential, i.e. to an isobar ensemble.

The validity of the ideal gas approximation eq.(2.13) for the expansion dynamics is tested in Figure 7[30] in the framework of classical molecular dynamics[31]. A Lennard-Jones system is initially confined in a small volume and successively freely expanding in the vacuum.

We can see that after a first phase of the order of ~10 Lennard-Jones time units, where inter-particle interactions cannot be neglected, the time evolution predicted by eq.(2.9) is remarkably fulfilled for all total energies. This result is due to the fact that the system's size and dynamics are dominated by the free particles, while deviations from a self similar flow can be seen if the analysis is restricted to bound particles[30]. We expect eq.(2.9) to describe the system evolution even better if the degrees of freedom are changed from particles to clusters, as suggested by the Fisher model of condensation[32].

Eq.(2.13) shows that, in the hypothesis of negligible interaction between the system's constituents, the expansion is self-similar. This in turn implies that the situation is equivalent to a standard Gibbs equilibrium in the local rest frame. In the expanding canonical ensemble the total average kinetic energy per particle is simply the sum of the thermal energy $\langle e_{th} \rangle = 3/(2\beta)$, and the radial flow $< e_{flow} >= h^2 < r^2 > /(2m)$.

This scenario is often invoked in the literature[27,28] to justify the treatment of flow as a collective radial velocity superimposed on thermal motion. However it is interesting to note that the decoupling between intrinsic and collective motion is true only in the canonical ensemble. In the case of isolated system like expanding nuclei, the two motions are coupled by the total energy conservation constraint:

$$\hat{D}_{\lambda h}(E) \propto \exp\left(-\beta \frac{\sum_i (\hat{\vec{p}}_i - h\hat{\vec{r}}_i)^2}{2m} - \beta \hat{V} - \lambda \hat{R}^2 \right) \delta\left(E - \hat{H}\right) \qquad (2.15)$$

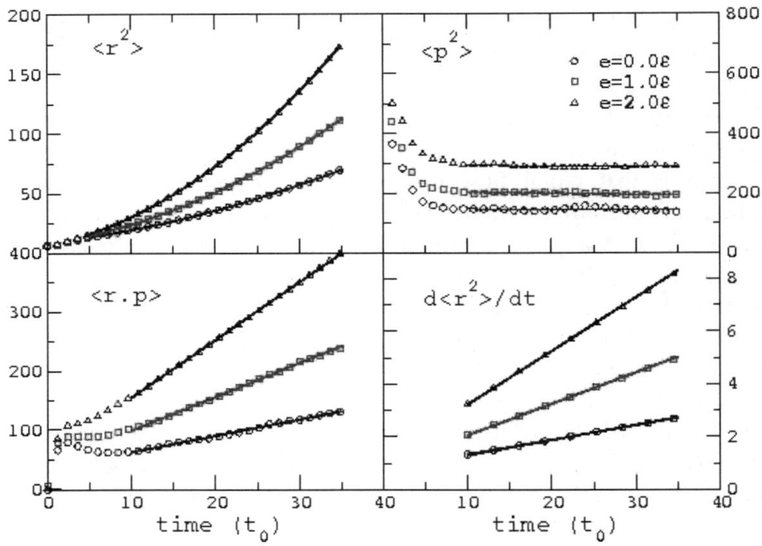

Figure 7. Time evolution of a 147 atoms system, initially compressed and freely expanding in the vacuum. The average mean suare radius, square momentum, radial flow and radius variation are plotted as a function of time. Points: molecular dynamics simulation. Lines: time dependent Gibbs ensemble estimate from eq.(2.15).

Eq.(2.15) is an exact expression for a non-interacting system, and a minimum bias ansatz in the other cases. To explore the effect of this coupling, Figure 8 shows the potential energy and largest fragment size distribution numerically evaluated for a Lennard-Jones system[30] with the micro-canonical ensemble under flow eq.(2.15). Two different energies in the liquid phase and close to the liquid-gas transition point are chosen, while the values of h and λ are typical values extracted from dynamical simulations at the freeze out time.

From Figure 8 we can see that, if the effect of flow is negligible at the lower energy, the same is not true close to the transition region. There, the collective energy acts as a heat bath allowing the system to explore the two different phases within the same total energy.

This example shows that, even in the self-similar approximation, collective flow can have dramatic effects on the system's partitions and has to be consistently accounted for in the statistical treatment of an open fragmenting nucleus.

3. Phase transitions in finite systems and applications in nuclear physics

Phase transitions are universal properties of matter in interaction. In macroscopic physics, they are singularities (i.e. non-analytical behaviors) in the system equation of state (EoS), and hence classified according to the degree of non-analyticity of the EoS at the transition point. Then, a phase transition is an intrinsic property of the system, and not of the statistical ensemble used to describe the equilibrium.

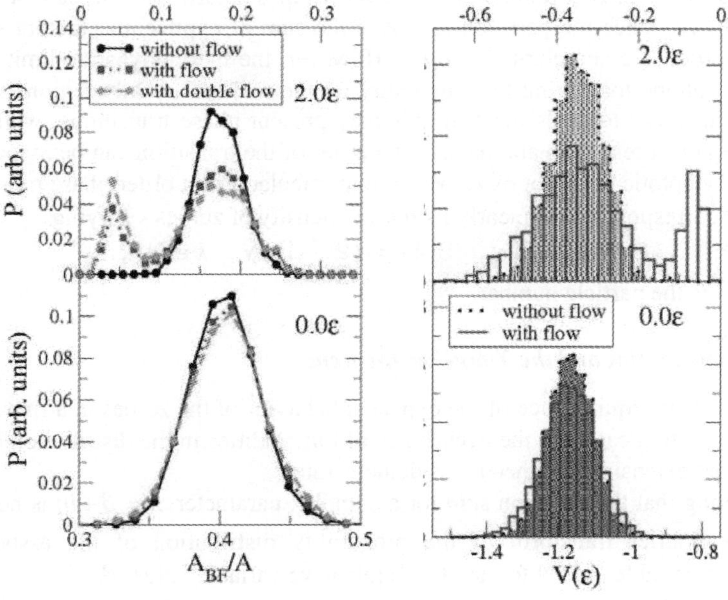

Figure 8. Potential energy (right) and size of the largest cluster (left) distributions for a Lennard Jones 147 atoms system at two different total energies (in Lennard-Jones units) in the microcanonical ensemble with and without radial flow. The amount of flow is estimated at each energy from the free evolution in the vacuum.

Indeed, at the thermodynamic limit all the possible statistical ensembles converge towards the same EoS, and the various thermodynamic potentials are related by simple Legendre transformations, leading to a unique thermodynamics. On the other side for finite systems, two ensembles which put different constraints on the fluctuations of the order parameter lead to qualitatively different EoS close to a first order phase transition[21,25]. Thermodynamic observables like heat capacities can therefore be completely different depending on the experimental conditions of the measurement.

3.1. *Phase transitions in finite systems*

When we consider a finite physical system, the analysis of the singularities of the thermodynamic potential has no meaning, since it is an analytical function. The standard statistical physics textbooks thus conclude that rigorously speaking there is no phase transitions in finite systems. For intensive ensembles, since the pioneering work of Yang and Lee[33], another definition was proposed considering the zeroes of the partition sum in the complex intensive parameter plane $\lambda = \beta + i\eta$ [34]. The idea is simple: the zeroes of Z are the singularities of log Z and so phase transitions, which are singularities, must come from the zeroes of the partition sum. In a finite system the zeroes of the partition sum cannot be on the real axis, since the partition sum Z is the sum of exponential factors which cannot produce a singularity of log Z. However, the thermodynamic limit of an infinite volume may bring the singularity on the real axis. Only regions where zeroes converge towards the real axis may present phase transitions, while the other regions present no anomalies. The order of the transition can be associated to the asymptotic behavior of zeroes[34]. In particular a first order phase transition at $\beta=\beta_0$ corresponds to a linearly increasing density of zeroes verifying

$$\Re(\lambda) = \beta_0 \quad ; \quad \Im(\lambda) \propto i(2k+1)/N \quad k \in \mathbb{N} \ , \tag{3.1}$$

where N is the particle number.

3.2. *Bimodalities and the Yang-Lee theorem*

In refs.[35,36] the equivalence of the expected behavior of the zeroes in a first order phase transition case and the occurrence of bimodalities in the distribution of the associated extensive parameter was demonstrated.

Noting that the partition sum for a complex parameter $\lambda = \beta + i\eta$ is nothing but the Fourier transform of the probability distribution of the associated extensive variable e=E/N for any (real) intensive variable value β_0:

$$Z_\lambda = \int de Z_{\beta_0} P_{\beta_0}(e) e^{-i\eta e} , \tag{3.2}$$

a connection can be established between the partition sum's zeroes and the convexity properties of the associated distribution. In particular if the distribution is normal (asymptotically Gaussian), the partition sum has no zeroes, while a bimodal distribution corresponds to zeroes $\lambda_k = \beta + i\eta_k$ defined by

$$\eta_k = i(2k+1)/\Delta E , \tag{3.3}$$

where ΔE is the distance between the two maxima[35]. This expression recovers the Yang Lee theorem if $\Delta E \propto N$. As shown in ref.[36] the reciprocal is also true: a first order phase transition is univocally defined by a bimodal distribution of the order parameter in the corresponding finite system, where "bimodality" means that the extensive variable distribution can be splitted at the transition point into two (arbitrary) distributions of equal height, with the distance between the two maxima scaling like the system size.

3.3. *Phase transition versus channel opening*

The bimodality of the order parameter distribution at a first order transition point implies that the underlying density of states presents a convex region. Indeed taking E as the extensive observable and β as the conjugated Lagrange field, the distribution reads

$$P_\beta(E) = Z_\beta^{-1} W(E) e^{-\beta E} = Z_\beta^{-1} e^{S(E)-\beta E} : \tag{3.4}$$

a convex E distribution at the transition temperature necessarily implies a convex region in the underlying Boltzmann entropy S=logW[24]. Such a situation naturally occurs each time that a new channel opens at a threshold value of E, if this channel is associated to a high degeneracy, i.e. a rapid increase of the associated entropy with increasing energy. It has therefore sometimes proposed in the literature[24] that any channel opening can be associated to a phase transition. It is however important to stress that the equivalence with the Yang-Lee theorem, and consequently the connection with the thermodynamic definition of phase transitions, is preserved only if the distance between the two peaks, i.e. the extension of the convex entropy region, linearly scales with the system size (see eq.(3.3)).

A typical example in molecular and cluster physics is given by isomerization, that may lead to accidents in the concavity of the entropy without being connected to a phase change in the bulk. Let us consider the simplest possible case: the state change from one dimer to two monomers[25]. This is a system with only two particles in a box of volume V, which can exist in two states: (i) a bound system of mass 2m bounded by an energy -ε, and (ii) an unbound state with two free monomers, corresponding to the state densities W_{bound} and W_{free} respectively

$$W_{bound}(e) = \frac{\pi^{3/2} V}{(3/2)!\, h^3} \left(4m(e+\varepsilon)\right)^{3/2} \theta(e+\varepsilon)$$

$$W_{free}(e) = \frac{1}{3!} \left(\frac{\pi^{3/2}}{h^3} V (2me)^{3/2} \right)^2 \theta(e) \tag{3.5}$$

The total density of states is $W_{tot}=W_{bound}+W_{free}$. As it can be seen from Figure 9, at sufficiently low density (i.e. large volume) the micro-canonical entropy presents a convex intruder. If we however consider a larger and eventually infinite number of constituents is straightforward (and still only two body bounds are allowed), the anomaly in the entropy disappears going towards the thermodynamic limit, similarly to the ionization phenomenon which is known to present a smooth cross-over from a dimer gas to a plasma. This behavior is due to the fact that the distance between the successive thresholds (measured in energy per particle) is a decreasing function of the number of constituents.

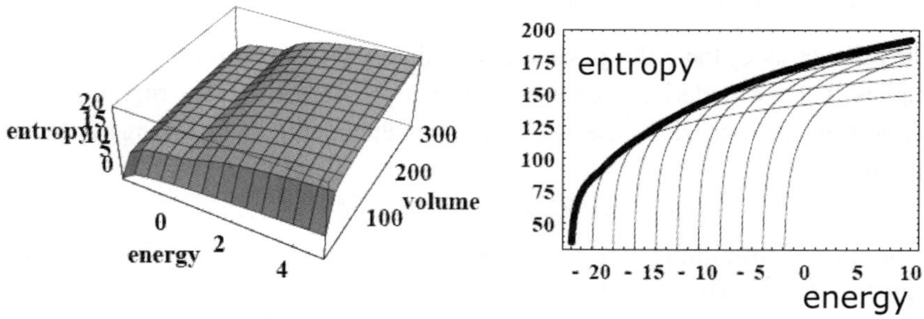

Figure 9. Entropy surface for the transition from one dimer to two monomers. Left side: two particles. Right side: thermodynamic limit. The different curves give the entropy associated to each pair breaking.

From this simple example we can get an intuitive understanding of the microscopic origin of phase transitions. A channel opening in a finite system corresponds to a phase transition in the bulk if

(i) it is sudden enough to lead to a convex intruder in the entropy,

(ii) the corresponding order parameter is an observable collective enough to scale linearly with the number of constituents.

3.4. *Bimodalities in fragmentation distributions*

Bimodalities have been recently observed in the fragmentation pattern of atomic nuclei. Figure 10 shows the distribution of the largest A_{max} and second largest A_{second} fragment produced in the fragmentation of a Au nucleus upon collisions with a Au target at 80 MeV/u detected by the Indra@GSI collaboration[37]. Data are sorted in bins of transverse energy of light charged particles emitted on the quasi-target side, which is a measure of the centrality of the collisions. In the most peripheral collisions (upper left), the partitions are constituted of a large biggest fragment, and much smaller other fragments, which is typical of the evaporation mechanism. In the most central ones (lower right) the largest fragment is small and close in size to the second largest one, which corresponds to multi- fragmentation.

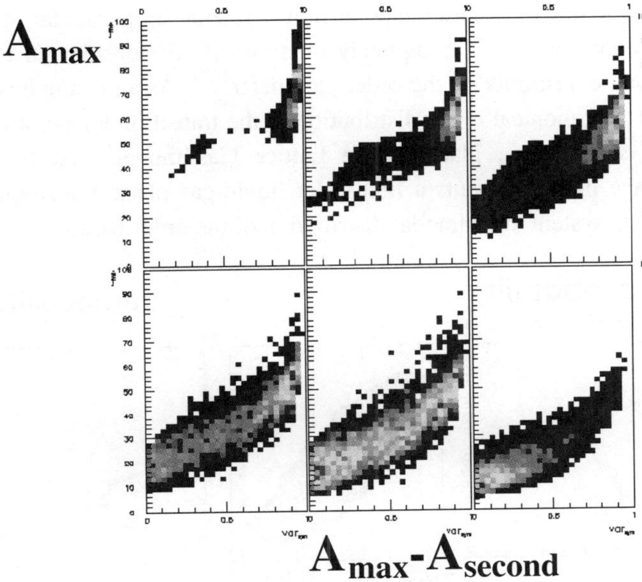

Figure 10. Event distribution for Au quasi-projectile fragmentation measured by Indra. A_{max} (A_{second}) are the size of the largest (second largest) fragment respectively. Data are sorted in bins of transverse energy of particles emitted by the associated quasi-target.

Between these extreme configurations a region exists (lower middle) where both solutions are present and intermediate configurations are suppressed, i.e. the distribution is bimodal. This bimodal behavior is relatively clear in the asymmetry A_{max}-A_{second} direction, but does not appear once the distribution is projected over the A_{max} axis[38]. This is surprising, considering that the largest fragment size should be the order parameter of the transition if this latter belongs to the liquid-gas[25] or to the fragmentation-aggregation[39] universality class. The scaling properties of the experimental A_{max} distribution have been analyzed in detail. Based on these studies, if A_{max} can be considered as an order parameter, the transition should be second order or more probably a cross over[38]. The nature and order of the multi-fragmentation transition is therefore still a subject of debate.

3.5. *The effect of constraints*

The apparent incoherence between the information extracted from the A_{max} and the A_{max}-A_{second} distribution can be explained, if one considers the irreducible inequivalence of statistical ensembles in finite systems. In particular, the phase transition phenomenology is completely modified if we consider an ensemble that puts strong constraints on the order parameter[21,25]. As an example we show in Figure 11 the canonical event distribution at the transition temperature in the total energy versus A_{max} plane in the Lattice Gas model[25]. At the chosen temperature the model presents a first order liquid-gas phase transition, giving rise in the finite system to a bimodal distribution of the order parameter.

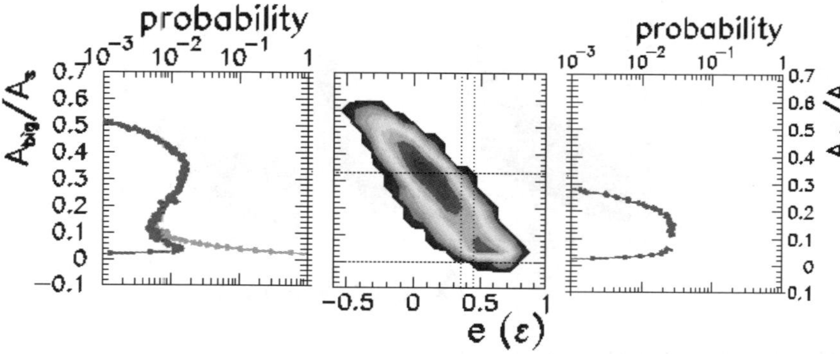

Figure 11. Event distribution in the total energy versus plane in the isobar Lattice Gas Model. Central panel: canonical distribution at the transition temperature. Left panel: projection over the largest fragment size direction. Right panel: as the left panel, but after the energy cut given by the central panel dashed lines.

Since the transition has a non-zero latent heat, energy is an order parameter, and the energy distribution is bimodal. The largest cluster size being strongly correlated to total energy, the distribution is bimodal in the A_{max} direction also (left side of Figure 11). Because of this correlation between energy and largest fragment size, in the micro-canonical ensemble, where the energy is constrained, the A_{max} distribution loses its bimodality. The same is true for any intermediate ensemble (right side of Figure 11) allowing energy fluctuations less wide than the energy distance separating the two coexisting phases.

This simple example is the prototype of a generic situation. As we have recalled in section 3.2, in a first order phase transition the order parameter distribution is bimodal only in the "intensive" ensemble where the constraint acts on the conjugated Lagrange, namely the canonical ensemble for any transition with non-zero latent heat. The ensemble explored in the experiment is certainly not canonical, and the associated energy distribution has a shape determined by the entrance channel dynamics and data selection criteria. The centrality sorting leads to an ensemble where the energy deposit is relatively well defined, which may prevent the observation of bimodality.

3.6. *Getting rid of the effect of constraints*

The experimental excitation energy distribution $p_{exp}(E)$ associated to the collision of the Au system with the Au target is shown in the upper left part of Figure 12[40].

Because of the entrance channel dynamics essentially determined by the impact parameter geometry, such distribution is rapidly decreasing and dominated by peripheral events. Considering the two correlated variables E and A_{max} we can formally write

$$p_{exp}\left(E,A_{max}\right)=p_{\beta}\left(E,A_{max}\right)\frac{p_{exp}(E)}{p_{\beta}(E)} \tag{3.6}$$

where p_β is the canonical distribution. It is clear that no information on the convexity properties of the density of states $S(E,A_{max})$ in the E direction can be extracted from this function because of the arbitrary shape of $p_{exp}(E)$, and the same my be true in the A_{max} direction, if the two variables are correlated.

To get rid of the entrance channel dependence we can define a re-weighted distribution that gives an even weight to all excitation energy bins:

$$p_{W}\left(E,A_{max}\right)=\frac{p_{exp}\left(E,A_{max}\right)}{p_{exp}(E)}=\frac{W\left(E,A_{max}\right)}{W(E)} \tag{3.7}$$

In the case of a first order transition with non-zero latent heat, W(E) is convex. If E and A_{max} are strongly correlated, the normalization by W(E) will suppress the bimodality also in the A_{max} direction. If however the correlation is sufficiently loose, a bimodality signal may still survive after renormalization.

The effect of the renormalization on the experimental distribution is shown in the upper right part of Figure 12. The resulting largest fragment distribution (lower left) after this procedure is clearly bimodal. The same procedure is applied to Lattice Gas event in Figure 13. If each mono-energetic distribution is clearly mono-modal, the sum of all distributions with an even weight recovers the bimodal shape characteristic of the canonical distribution at (or close to) the transition temperature.

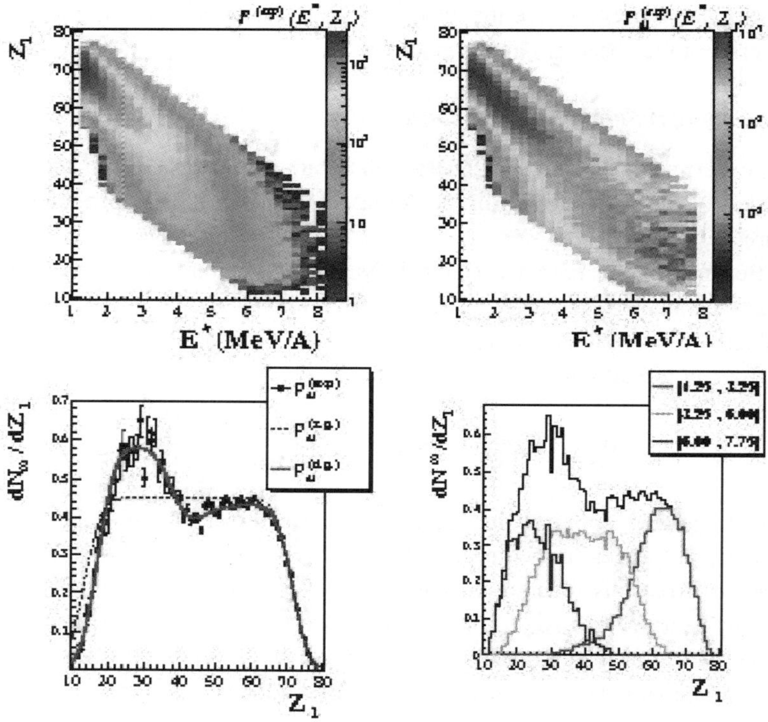

Figure 12. Event distribution of fragmenting Au quasi-projectiles measured by Indra. Upper left: total distribution of projectile events in the largest fragment Z_1 versus excitation energy E^* plane. Upper right: same after an event renormalization to have a flat energy distribution. Lower left: projection of the reweighted distribution over the Z_1 axis and fit with a concave (dashed line) and convex (full line) entropy ansatz . Lower right: Reweighted distribution in three different excitation energy regions.

Of course such re-weighted distribution is not equivalent to the canonical one. In particular the simple link eq.(3.4) between the distribution and the underlying state density is replaced by the more involved expression (3.7). Contrary to the canonical case, the micro-canonical thermodynamics cannot then be directly inferred from such distribution.

The presence of a convexity region in the underlying entropy can however be directly deduced from the convexity of the renormalized distribution. The best fit of the experimental renormalized distribution with a convex and concave ansatz for the density of states is also shown in Figure 12[40]. It is clear that only a convex entropy can explain the hull in the distribution.

4. Conclusions

In these lectures we have presented a non exhaustive review of different phenomenologies associated to nuclear thermodynamics

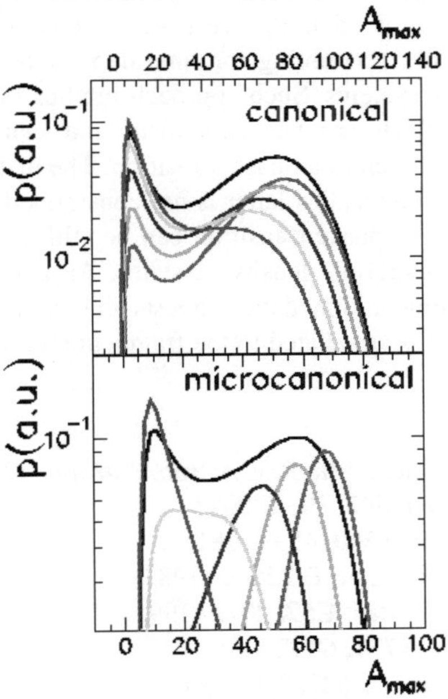

Figure 13. Distributions of the largest fragment size in the isobar Lattice gas model for a system of 108 particles. Upper part: canonical calculation at different temperatures close to the transition temperature (black line). Lower part: microcanonical calculations at different energies inside the coexistence zone. Black line: the different distributions are summed up with even weights.

The phase diagram of nuclear matter has been studied in detail in the literature and is known to present a wide first order coexistence region and a critical line of second order phase transitions belonging to the liquid-gas universality class. This idealized model has to be modified in many respects when dealing with physical nuclear objects.

Compact objects in the universe as neutron stars crusts and supernovae cores are extended physical pieces of nuclear matter but, because of the presence of protons and electrons, are subject to the Coulomb force. The presence of a long range non-saturating interaction modifies the thermodynamic properties of the matter in analogy with other physical systems subject to frustration: the critical behavior is quenched; the first order phase transition is suppressed, and replaced by a transition to clusterized matter.

Going to the thermodynamic properties of finite nuclei, an extra complication arises from the absence of a thermodynamic limit and the importance of boundary conditions, which implies that the concept of equilibrium itself has to be dealt with care. An information theory approach leads to a coherent treatment of such systems, which recovers the correct thermodynamic limit and allows accounting for time dependent phenomena as collective motions in the finite systems. Such approach implies the definition of many different ensembles, which are not equivalent and can lead to very different phenomena at the approach of a phase transition. The distribution of the order parameter is bimodal as long as this latter is not constrained; otherwise the bimodality is suppressed, but the phase transition can be still inferred from the convexity properties of the underlying density of states. An application of this approach to recent multi-fragmentation data suggests that a first order phase transition with finite latent heat is associated to the fragmentation phenomenon.

References

1. D.G.Yakovlev and C.J.Pethick, *Ann. Rev. Astron. Astroph.* **42,** 169 (2004).
2. N.K.Glendenning, *Phys.Rep.* **342,** 393 (2001).
3. D.Q.Lamb et al., *Nucl.Phys.* **A360,** 459 (1981).
4. G.Bertsch, P.J.Siemens, *Phys. Lett.***B 126,** 9 (1983).
5. H.Muller and B.Serot, *Phys. Rev.* **C52,** 2072 (1995).
6. C.Ducoin et al.,*Nucl.Phys.* **A771,** 68 (2006).
7. E.Chabanat et al., *Nucl. Phys.* **A 627,** 710 (1997).
8. Ph. Chomaz et al., *arXiv:astro-ph/*0507633.
9. C.Ducoin et al., *arXiv :nucl-th/*0606034.
10. J.M.Lattimer et al., *Nucl. Phys.* **A432,** 646 (1985).
11. J.W.Negele and D.Vautherin, *Nucl.Phys.* **A207,** 298 (1973).
12. M. Grousson et al., *Phys. Rev.* **E64,** 036109 (2001).
13. C.J. Pethick and D.G. Ravenhall, *Ann.Rev.Nucl.Part.Sci.* **45,** 429 (1995).
14. G. Watanabe et al., *Phys. Rev. Lett.* **94,** 031101 (2005).
15. C.J. Horowitz, and J. Piekarewicz, *Phys. Rev.* **C72,** 035801 (2005).
16. P.Napolitani et al, *arXiv :nucl-th/*0610007.
17. P. Magierski and P.H. Heenen, *Phys. Rev.* **C65,** 045804 (2001).
18. F. Gulminelli et al., *Phys. Rev. Lett.* **91,** 202701 (2003).
19. J. Margueron, J. Navarro and P. Blottiau, *Phys. Rev.***C70,** 28801 (2004).
20. E.T.Jaynes, *Statistical Physics, Brandeis Lectures*, **vol.3**, 160 (1963).
21. F. Bouchet, J. Barré, *Journ. Stat. Phys.* **118**, 1073 (2005).
22. W. Thirring, H. Narnhofer, H. A. Posch, *Phys.Rev.Lett.* **91,** 130601 (2003).
23. R. Balian, *'From microphysics to macrophysics'*, Springer Verlag (1982).
24. D. H. E. Gross, *Lecture Notes in Physics* **vol.66**, Springer (2001).
25. F. Gulminelli, *Ann.Phys.Fr.* **29**, 6 (2004).
26. P.H. Chavanis and M. Rieutord, *Astron.Astrophys.* **412**, 1 (2003).

27. J. P. Bondorf et al., *Phys. Rep.* **257**, 133 (1995).
28. F.Becattini et al., *Phys.Rev.* **C69**, 024905 (2004).
29. Ph. Chomaz, F. Gulminelli, O. Juillet, *Ann. Phys.* **320**, 135 (2005).
30. M.Ison et al., in preparation.
31. A. Chernomoretz et al., *Nucl.Phys.* **A723**, 229 (2003).
32. M. E. Fisher, *Physics* **vol. 3**, No.5, 255 (1967).
33. C.N.Yang and T.D.Lee, *Phys.Rev.* **87**, 404 (1952).
34. P.Borrmann et al., *Phys.Rev.Lett.* **84**, 3511 (2000).
35. K.C. Lee, *Phys. Rev.* **E 53**, 6558 (1996).
36. Ph. Chomaz, F. Gulminelli, *Physica* **A 330**, 451 (2003).
37. M.Pichon et al., *ArXiv:nucl-ex/0602003*
38. J.Frankland et al., *Phys.Rev.* **C71,** 034607 (2005).
39. R.Botet and M.Ploszajczak, *Phys.Rev.* **E62**, 1825 (2000).
40. E.Bonnet, *PhD Thesis,* University of Paris XI (2006).

Multifragmentation, Phase Transitions and the Nuclear Equation of State

S.J. Yennello[*]

Cyclotron Institute, Texas A&M University College Station, TX 77845-3366, USA E-mail:yennello@comp.tamu.edu

The nuclear equation-of-state is fundamental to both understanding systems as diverse as nuclei and neutron stars. Light-ion-induced reactions have served well to elucidate the behavior of excited nuclear material near the valley-of-stability. Heavy-ion reactions with systems of varying neutron-to-proton ratio (N/Z) are currently being used to gain a greater understanding of the equation-of-state away from the valley-of-stability.

1. Light-ion induced reactions

One of the fundamental goals of the studies of nuclear collisions in the nucleonic regime has been to attain an understanding of the nuclear phase diagram. An extensive program was carried out with light-ion beams to study the multifragmentation of heavy nuclei in an effort to better understand the nuclear liquid-gas phase transition.

Light-ion induced reactions are quite useful in this regard as they discriminate toward thermally – rather than dynamically – driven fragmentation. These reactions tend to produce predominately one equilibrium source, after limited preequilibrium emission. This single source does not have a large velocity in the laboratory frame so there is not a significant transformation that must occur to the center of mass frame. The angular momentum and the radial flow in these reactions are much smaller than from heavy-ion collisions.

[*] Work partially supported by the Department of Energy and the Robert A Welch Foundation through grant A-1266

1.1. *ISiS Experimental program*

The ISiS experimental program [1] was a series of experiments of pion, antiproton and ^{3}He induced reactions that were performed at the Brookhaven AGS and LNS Saturn. Figure 1 shows the ISiS detector. The ISiS detector consists of 162 particle-identification telescopes arranged in a spherical geometry covering from 14°- 86.4° and 93.6° – 166°. Each telescope contains a gas-ionization chamber followed by a 300um silicon detector and a CsI(Tl) crystal read out by a photodiode.

Fig. 1 Photograph of ISiS detector.

Table 1.1 Beam and target combinations

Beam	Target	Energy / Momentum
π^-	^{197}Au	5.0 GeV/c, 8.0 GeV/c, 8.2 Gev/c, 9.2 GeV/c
p	^{197}Au	6.2 GeV/c, 9.2 GeV/c, 12.8 GeV/c, 14.6 GeV/c
p bar	^{197}Au	8.0 GeV/c
^{3}He	^{nat}Ag	1.8 GeV, 3.6GeV, 4.8GeV
^{3}He	^{197}Au	1.8 GeV, 4.8GeV

Table 1 shows the beam and target combinations for the various experiments. The multifragmentation data of ISiS show many indications of a phase transition.

1.2. *Charge Distributions*

Early work by the Perdue group [2] studied the fragment formation from p + Xe reactions and fit the inclusive charge distributions with a power law distribution in the spirit of the Fischer droplet model. A minimum in the tau parameter extracted from the data was to be expected for a phase transition. Many experiments since then related the tau parameter to a change in reaction mechanism. Most recently the Fischer droplet model has been used by the Berkeley group to study exclusive charge distributions including the ISiS data [3]. One of the experimental signatures of a change in reaction mechanism that would be consistent with a phase transition is the decrease in the tau parameter from high values at low excitation energies to a minimum of approximately 2 near the phase transition.

In figure 2 the parameter τ from power-law fits to the charge distributions are plotted as a function of *E*/A*. Values of τ decrease steadily as the system is heated i.e. the probability for forming larger fragments increases. A minimum is reached at $\tau \sim 2$ near *E*/A* ~ 6 MeV, followed by a slight increase. This signifies that maximum cluster sizes are obtained around 5-7 MeV/nucleon. Thereafter, additional excitation appears to produce a hotter environment, leading to an increased yield of lighter particles and clusters.

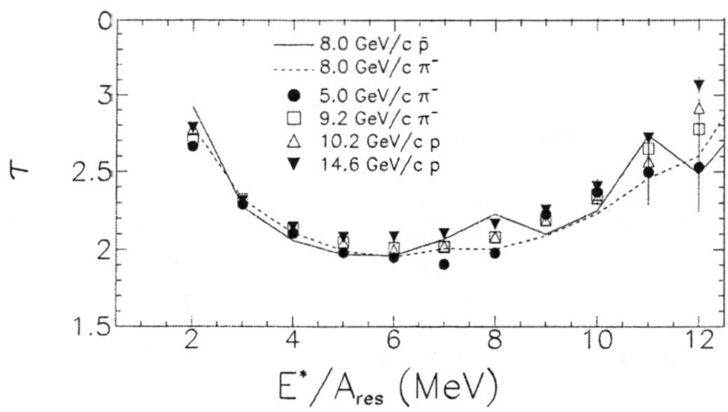

Fig. 2. The power law parameter tau from the charge distributions versus the excitation energy per nucleon of the excited system.

Caloric Curve

The Aladin group first showed the correlation of temperature with excitation energy and the flattening of this "caloric curve" at about 5 MeV/nucleon[4]. The isotope-ratio temperatures *T*app corresponding to a given excitation energy per nucleon, were calculated with the double-isotope-ratio thermometer [5]. According to Albergo *et al.*, the temperature for a system in chemical and thermal equilibrium can be extracted from a double-isotope ratio

$$T_{app} = \frac{B}{\ln(aR)} \tag{1}$$

where *B* is the binding-energy parameter, *a* is a factor that depends on statistical weights of the ground state nuclear spins, and *R* is the ground state population ratio at freeze-out. Measured yield ratios differ from the primary yield ratios due to sequential decay of the excited fragments. Tsang *et al.*[6] have proposed an empirical method to account for these effects by defining a correction factor κ for each isotope ratio

$$\frac{1}{T_{app}} = \frac{1}{T_o} + \frac{\ln(\kappa)}{B} \tag{2}$$

where To is the temperature from the primary fragments.

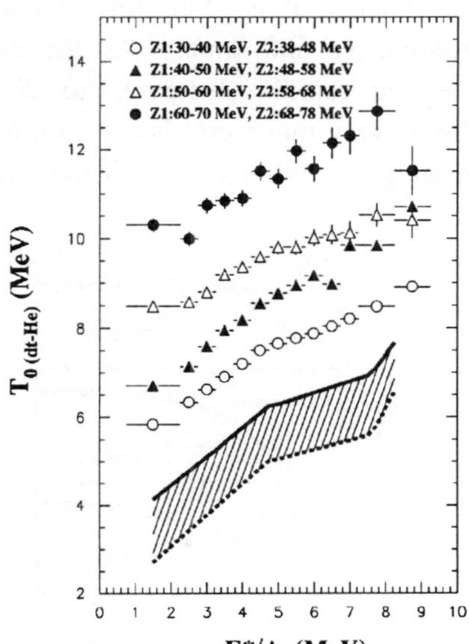

Fig. 3. Caloric curve from p + Au reaction.

Figure 3 shows the temperature versus excitation energy for the 8.0GeV π^- + Au data. A flattening of the caloric curve is evidenced at around 5 MeV/nucleon. Similar behavior is observed with the antiproton and ^{3}He induced reactions. In addition to the open circles which is the temperature for the lowest energy fragments we detected we have measured the temperature for fragments of increasing kinetic energy so we could map out the "cooling curve" for this reaction. One can see that as we gate on higher and higher kinetic energies of the fragments the apparent temperature extracted gets larger. The hatched region represents the temperature that would be correlated to the coulomb peaks of the energy spectra. The ISiS data allows one to look at the caloric curve all within a single data set.

1.3. *Timescale of fragment emission*

Fragment-fragment correlations can be used to extract the relative time between emission of fragments from a deexciting system [7]. In a surface emitting - or sequentially decaying system – the time between collisions may be long. In a system that is undergoing bulk multifragmentation the time between emission of fragments must decrease. In the ISiS data we have used fragment-fragment correlations to measure the timescale of fragment emission. In figure 4 the correlation function is plotted against the reduced velocity for three different excitation energy bins. For low excitation energies the timescale is approximately 500 fm/c. As the excitation energy is increased the timescale rapidly decreases to 20-30 fm/c above 5 MeV/nucleon. This is very much consistent with a change in reaction mechanism transitioning to a bulk multifragmentation.

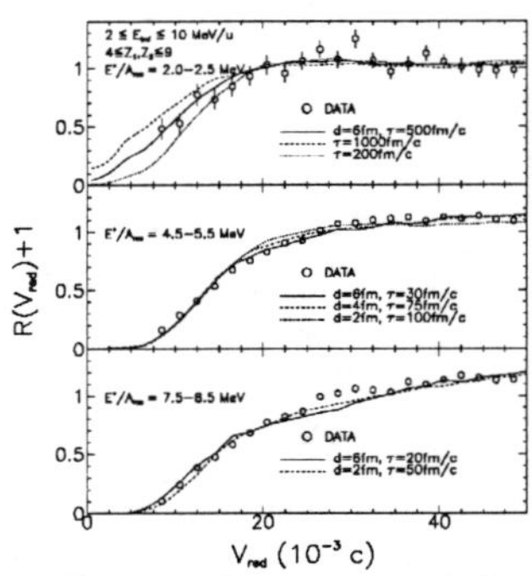

Fig. 4. Fragment fragment correlations.

1.4. *Indications of a Phase Transition*

The ISiS data examined many different signatures for a phase transition. Figure 5 gives a composite look at the changes in the charge distribution, the caloric curve and the timescale for fragment emission versus the measured excitation energy. Taken together these many observables are all consistent with a change in reaction mechanism or a transition from surface emission at low excitation energy to bulk emission at energies above approximately 5 MeV/nucleon. The is perhaps the most convincing evidence of a liquid-gas phase transitions within a single set of data where there multiple observations have occurred.

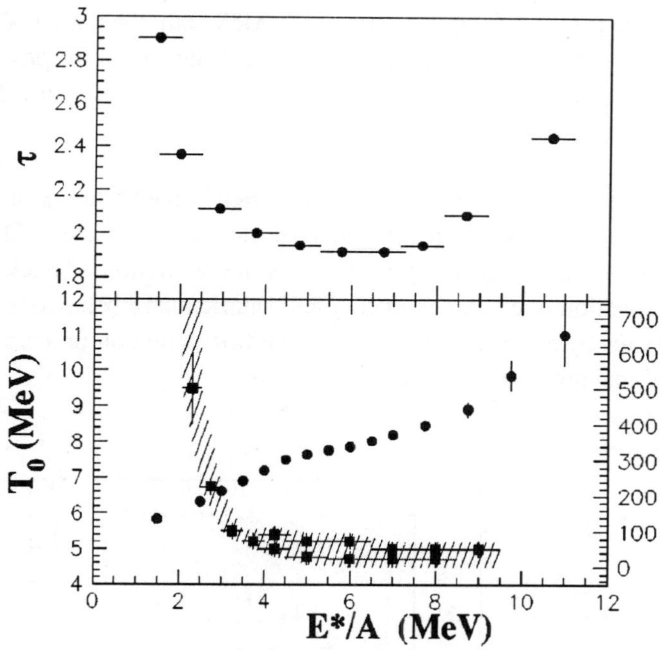

Fig. 5. The powerlaw parameter(top), the temperature and emission time (bottom) versus excitation energy.

1.5. *Breakup density*

Breakup densities have been deduced from the systematic trends in the Coulomb observables for IMF spectra produced in light-ion-induced reactions on ^{197}Au. Because ISiS has a gas ionization chamber as the first detector the energy threshold for detection of charged particles is low. This has enabled the measurement of the Coulomb peaks in the energy spectra. The ISiS data has been coupled with some lower energy data that also has low energy thresholds to extract out the density at the time of fragmentation. The relative Coulomb parameter as a function of excitation energy is shown on the top left in figure 6. This Coulomb parameter has been transformed into a relative density which is shown in the bottom left of figure 6. The extracted average densities are consistent with ρ/ρ_o ~1.0 up to E^*/A ~ 2 MeV but then gradually decrease. Above E^*/A ~ 5 MeV, the obtained constant value of ρ/ρ_o ~ 0.3 is consistent with the breakup density assumed in the multifragmentation models.

Using a Fermi gas relation this density can be transformed into a plot of temperature versus excitation energy and a caloric curve extracted. This curve is shown on the right in figure 6. Thus, the evolution of nuclear density as a function of excitation energy and caloric-curve behavior can be accounted for by a mechanism in which the fragmentation process is driven by thermal pressure and Coulomb effects.

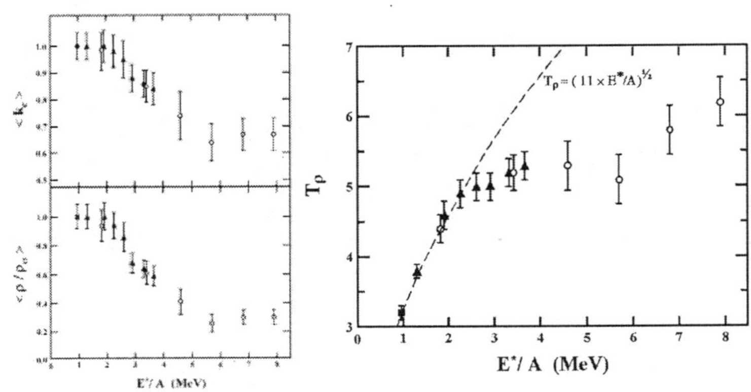

Fig. 6. Coulomb barrier (top) and density (bottom) versus excitation energy (left). Temperature extracted from the density versus excitation energy (right).

Projectile Fragmentation

Another avenue to produce mainly thermal disintegration is projectile fragmentation. In a study of the projectile fragmentation of ^{28}Si [8] we have used the fragment charge distributions to investigate the dependence of the onset of the phase transition on the N/Z of the system.

This experiment was done with a beam of ^{28}Si impinging on 1mg/cm2 of ^{112,124}Sn self-supporting targets. The beam was delivered at 30, 40, and 50 MeV/nucleon by the K500 superconducting cyclotron at the Cyclotron Institute of Texas A & M University. The detector setup for this experiment was FAUST [9]. It is composed of an arrangement of 68 silicon-CsI(Tl) Telescopes covering angles from 1.64° to 33.6° in the laboratory. Each element is composed of a 300- mm silicon detector followed by a 3-cm CsI(Tl) crystal. The detectors are arranged in five concentric rings. The geometrical efficiency is more than 90% for each ring. These detectors allow for isotopic identification of light charged particles and intermediate-mass fragments up to a charge of Z=5. Charged fragments from the quasiprojectile are detected and the quasiprojectile is reconstructed in both Z and A. Figure 7 shows the power law parameter tau as a function of excitation energy for reconstructed quasiprojectiles of various N/Z. As you can see the location of the minimum in the tau parameter is sensitive to the N/Z of the fragmenting system. The value of tau at the minimum also varies as a function of the N/Z of the reconstructed quasiprojectile.

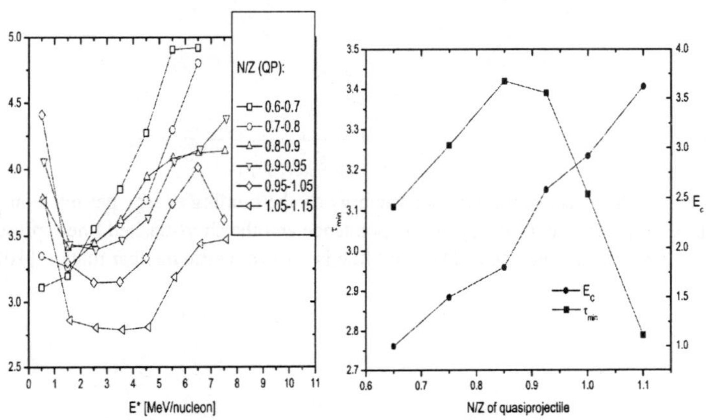

Fig. 7. The powerlaw parameter from the charge distributions as a function of the excitation energy for various values of N/Z of the quasiprojectile (left). The position and value of the minimum of the tau distribution as a function otf the N/Z of the quasiprojectile (right).

2. Symmetry Energy of the Nuclear Equation-of-State.

The study of the nuclear symmetry energy is currently a topic of intense theoretical and experimental work. It is well established that the symmetry energy plays a central role in a variety of astrophysical phenomena, including the structure and evolution of neutron stars and the dynamics of supernova explosions. In addition, the symmetry energy determines the nuclear structure of neutron-rich or neutron deficient rare isotopes.

The symmetry energy at normal nuclear density is reasonably well understood. However, its values at densities below or above the normal nuclear density are not adequately constrained. Indeed, the experimental determination of the symmetry energy and its density dependence is a challenging scientific endeavor. Information on the symmetry energy can be gleaned from heavy-ion collisions. A great deal of effort is currently devoted to identifying observables sensitive to the nuclear symmetry energy and its density dependence. A schematic plot showing the effect of the symmetry energy on the total binding energy is shown in firgure8.

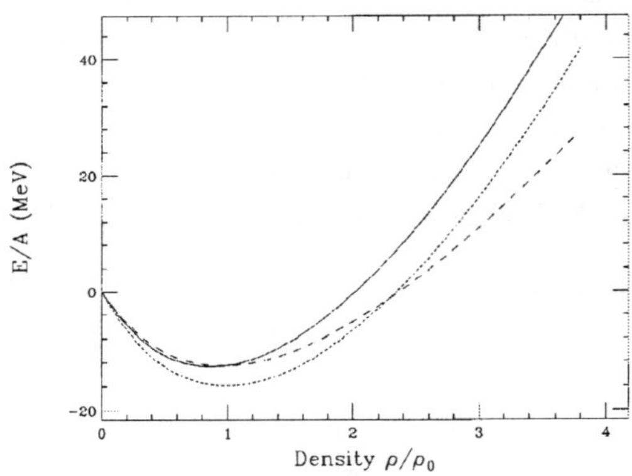

Fig. 8. The effect of the symmetry energy on the binding energy per nucleon. The dashed line represents an asy-soft dependence and the short-dashed line represents an asy-stiff dependence. The solid line is for symmetric nuclear matter. [10]

While we were able to see a shift in the position of the phase transition in the quasiprojectile fragmentation we have also been able to investigate the symmetry energy term in the nuclear equation of state using heavy-ion collisions.

2.1. *Isoscaling*

One important observable in heavy-ion collisions is the fragment isotopic composition. Investigations with the recently developed isoscaling approach [11] attempt to isolate the effects of the nuclear symmetry energy in the fragment yields, thus allowing a direct study of the role of this term of the nuclear binding energy in the formation of hot fragments. Isoscaling refers to a general exponential relation between the yields of a given fragment from two reactions that differ only in their isospin asymmetry (N/Z). In particular, if two reactions, 1 and 2, lead to primary fragments having approximately the same temperature but different isospin asymmetry, the ratio R21(N,Z) of the yields of a given fragment (N,Z) from these primary fragments exhibits an exponential dependence on the neutron number N and the atomic number Z of the following form:

$$R_{21}(N,Z) = C \exp(\alpha N + \beta Z), \quad (3)$$

where α and β are the scaling parameters and C is a normalization constant.

Since the fragments that are emitted during the deexcitation of an excited nuclear system often carry enough excitation energy to themselves deexcite measuring the final fragments is affected by the secondary decay of the primary fragments. Isoscaling was a technique developed to try to find an observable that was insensitive to secondary decay.

2.2. *Experimental Details of NiFe data*

Experiments examining the fragment formation from a set of isobaric reactions of ^{58}Fe, ^{58}Ni + ^{58}Fe + ^{58}Ni were performed at the Cyclotron Institute in Texas A&M University [12]. The targets were placed in the center of a scattering chamber that was housed inside the TAMU 4π neutron ball detector. Fragments from the reaction were measured in six discrete particle telescopes placed inside the scattering chamber and centered at laboratory angles of 10°, 44°, 72°, 100°, 128°, and 148°. Each telescope consisted of a gas ionization chamber followed by a pair of silicon detectors and a CsI scintillator detector thus providing three distinct detector pairs IC-Si, Si-Si, and Si-CsI for fragment identification. Good *Z* identification was achieved for fragments that punched through the IC detector but were stopped in the first silicon detector. Fragments were measured in the Si-Si detector pair with very good isotopic separation. The Si-CsI detector provided good isotopic separation of light charged particles up to ^{4}He. The following is an analysis of only the isotopically identified fragments [13].

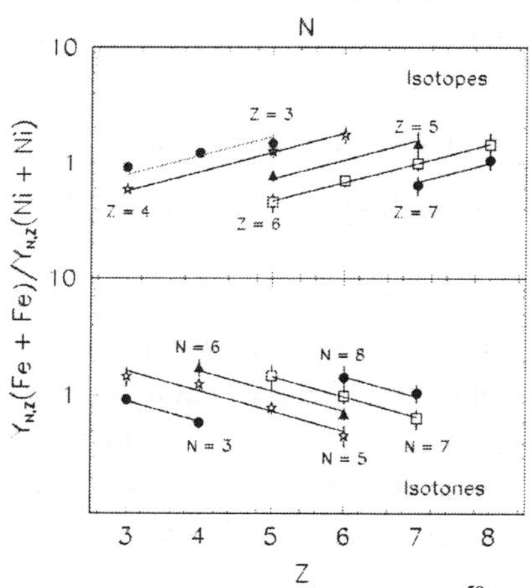

Fig. 9. Double ratio of fragment yields from the reactions ^{58}Fe+^{58}Fe and ^{58}Ni+^{58}Ni at 30 Mev/nucleon.

2.3. *Isoscaling of Intermediate Mass Fragments*

In figure 9 we show the double ratio of the yield of fragments from the reaction of ^{58}Fe + ^{58}Fe relative to ^{58}Ni + ^{58}Ni reaction. The lines connect the yield ratios of the various isotope and isotones. Linear scaling is observed. From this data the reduced densities of the free neutrons and free protons can be extracted.

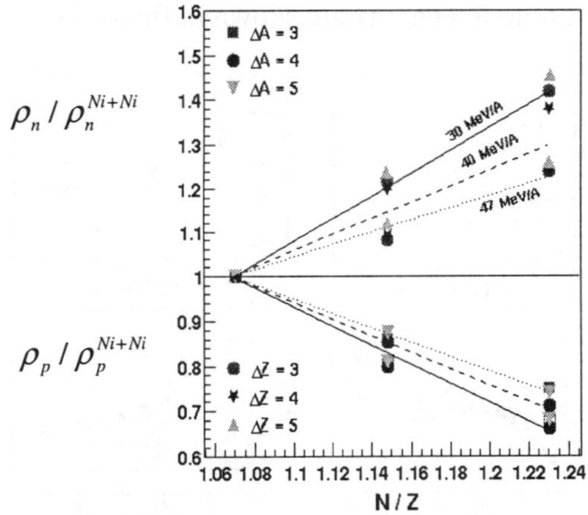

Fig. 10. Relative free neutron (top) and proton(bottom) densities.

Figure 10 shows the relative free nucleon densities obtained and plotted as a function of N/Z of the composite systems for ^{58}Ni + ^{58}Ni, ^{58}Fe + ^{58}Ni, and ^{58}Fe + ^{58}Fe reactions at 30, 40, and 47 MeV/nucleon. The top (bottom) part of the figure shows mean relative free neutron (proton) density obtained from the measured yields of He, Li, Be, B, and N fragments and averaged over various isotopes. The densities shown in the figure are all relative to the ^{58}Ni + ^{58}Ni (N/Z=1.07) reaction.

From the isoscaling we can extract the isoscaling parameter alpha. This can be related to the symmetry energy. Using Botvina's Statistical Multifragmentation Model (SMM) we can then study what is happening in the fragmenting system that would be consistent with the observed data.[14] The symmetry energy in the SMM calculation was varied until the isoscaling parameter alpha was reproduced. The comparison of primary and secondary alphas as well as the symmetry energy as a function of the excitation energy are shown in figure 11.

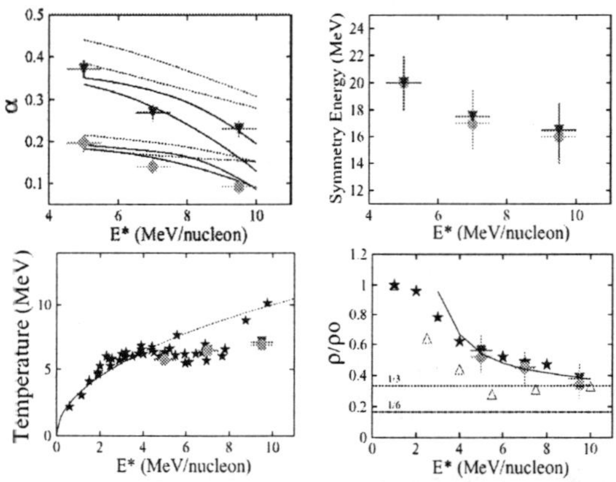

Fig. 11. The isoscaling parameter alpha, the symmetry energy, temperature and density from SMM calculations that describe the data

Also shown in figure 11 is the temperature as a function of excitation energy obtained from the SMM calculation that uses the excitation energy dependence of the break-up density to explain the observed isoscaling parameters. These are shown by the solid circles and inverted triangle symbols. These data are compared with the experimentally measured caloric curve data compiled by Natowitz from various measurements for A~100. The data from these measurements are shown collectively by solid star symbols. Finally we attempt to extract the density of the fragmenting system as a function of excitation energy. By assuming that the decrease in the breakup density can be approximated by the expanding Fermi gas model, one can extract the density as a function of excitation energy using the relation

$$T = \sqrt{K_o \left(\frac{\rho}{\rho_o} \right)^{\frac{2}{3}} E^*} \quad (4)$$

The extracted density is shown in the lower right panel of figure 11. It is evident that the decrease in the experimental isoscaling parameter, symmetry energy, break-up density, and the flattening of the temperature with increasing excitation energy are all correlated. One can thus conclude that the expansion of the system during the multifragmentation process leads to a decrease in the isoscaling parameter, decrease in the symmetry energy and density, and the flattening of the temperature with excitation energy. From the above correlation between the symmetry energy as a function of excitation energy, and the density as a function of excitation energy, we obtain the symmetry energy as a function of density. This is shown in Fig. 12. for both pairs of systems. The symmetry energy as a function of density extracted from this data is consistent with the the density dependence of the symmetry energy extracted by comparison to dynamical AMD calculations [15]

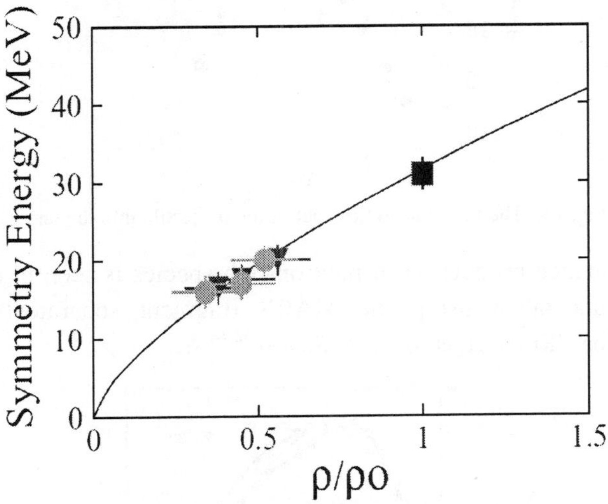

Fig. 12 the symmetry energy as a function of the density. The solid line is corresponds to $(\rho/\rho_o)^{.69}$.

2.4. *Production of Neutron-rich Nuclei by DIC Collisions*

The driving forces present during the interactions stage of deep inelastic collisions produce more neutron-rich quasiprojectiles than would be expected. This enhancement is has been seen in both the reconstructed quasiprojectiles that have been detected with FAUST following multifragmentation as well as the residues measured with a fragment separator. This neutron enhancement is sensitive to the transport and hence should be a way to elucidate the symmetry energy of the nuclear equation-of-state.

In figure 13 we show the reconstructed quasiprojectile from the reactions of ^{20}Na + Au. The data is compared to hybrid calculations of DIT plus a fragmentation code. The theoretical data has all been filtered through a complete replica of the experimental setup.

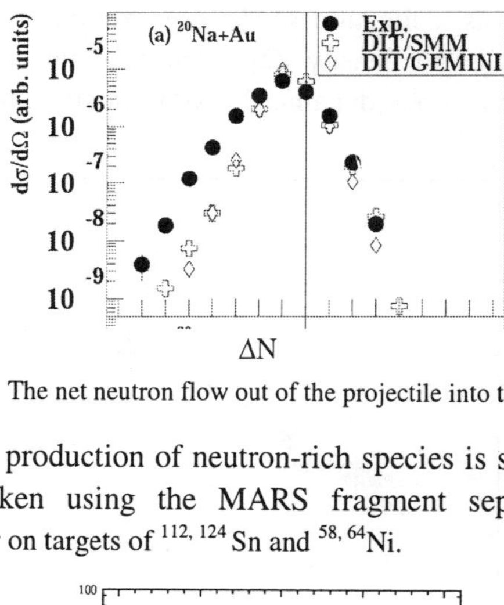

Fig. 13. The net neutron flow out of the projectile into the target.

Similar enhance production of neutron-rich species is seen in our heavy residue data taken using the MARS fragment separator using 25 MeV/nucleon ^{36}Kr on targets of $^{112,\,124}$Sn and $^{58,\,64}$Ni.

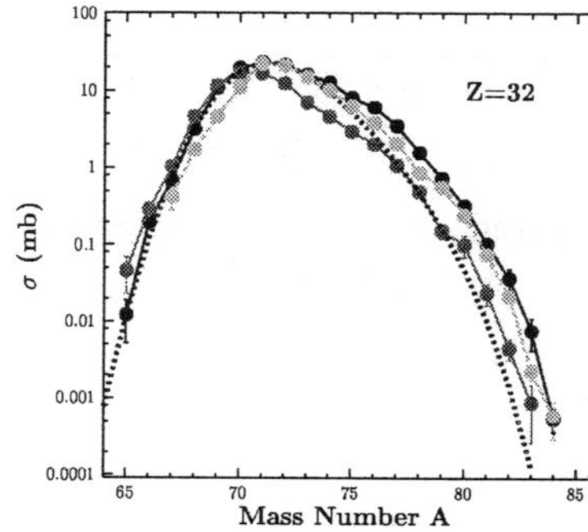

Fig. 14. Cross section for production of isotopes of Z=32 residues.

Fragments were accepted in the angular range 2.7° –5.4°. This angular range lies inside the grazing angle of 6.5° of the Kr + Sn reactions and mostly outside the grazing angle of 3.5° of the Kr + Ni reactions at 25 MeV/nucleon The effect of this difference is noticeable in the isoscaling of the residue data. For the Kr + Ni data there is a fairly constant value of alpha that is extracted from the isoscaling. In contrast the Kr + Sn data shows an evolution in the R_{12} being somewhat flat – or having little dependence on the difference in target to a steeper, and constant value, for R_{12} for fragments further away from the projectile in Z. From this trend we can map the isotopic equilibration in this process.

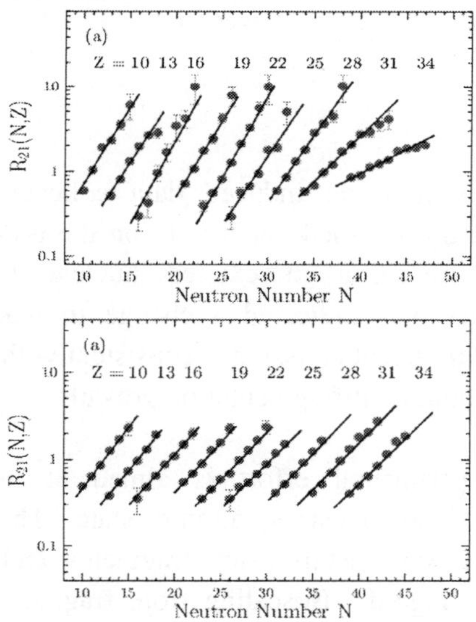

Fig. 15. Isoscaling parameter alpha extracted from Kr fragments and residues.

The values of the isoscaling parameter alpha that were extracted from the residue data are plotted as a function of the fragment size in figure 15. The lower data is for Kr + Ni and shows no appreciable deviation from the average. The higher data set is for Kr + Sn and you can clearly see the evolution going from a system where the N/Z of the fragments is largely controlled by the projectile to one in which the composite system is the controlling variable.

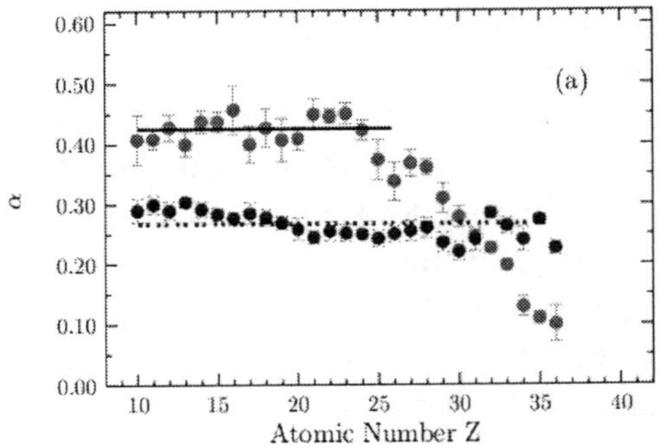

Fig. 16. The isoscaling parameter alpha from residues detected by the fragment separator.

3. Summary

The high statistics light-ion induced data collected from the ISiS experimental program has shed much light on the issues of a possible liquid-gas phase transition in excited nuclear material. Many experimental observables indicated a change in reaction mechanism from one in which fragment emission is consistent with surface emission to a regime where bulk multifragmentation prevails.

The currently experimental effort is aimed at understanding the symmetry energy of the nuclear equation of state. The DIC mechanism seems to produced very neutron-rich fragments and should be very informative in this regard. Isoscaling from fragments produced from heavy-ion collisions has already enabled us to narrow down the choice of the density dependence of the symmetry energy.

4. Acknowledgements:

The author would like to thank all of the members of the ISiS collaboration and the Texas A&M nuclear chemistry group that have worked on this data in particular V.E. Viola, K. Kwiatkowski, L. Beaulieu, D.S. Bracken, H. Breuer, J. Brzychczk, R.T. de Souza, D.S. Ginger, W-c. Hsi, A.L. Keksis, R.G. Korteling, T.Lefort, W.G. Lynch, K.B. Morley, R. Legrain, L. Pienkowski, E.C. Pollacco,

E. Ramakrishnan, E. Renshaw, A. Ruangma, D. Shetty, G. Souliotis, M.B. Tsang, C. Volant, G.Wang. In addition thanks are extended to colleagues who have allowed the use of their theoretical codes including L. TasenGot, R. Charity and A. Botvina. This work was supported in part by the Department of Energy and the Robert A. Welch Foundation through grant A-1266.

References

[1] V.E.Viola *et al.*, Physics Reports 434, 1 – 46 (2006).
[2] J.E. Finn, S. Agarwal, A. Bujak, J. Chuang, L.J. Gutay, A.S. Hirsch, R.W. Minich, N.T. Porile, R.P. Scharenberg, and B.C. Stringfellow, Phys. Rev. Lett. 49, 1321 (1982).
[3] J.B. Elliott, *et al.*, Phys. Rev. Lett. 88, 042701 (2002).
[4] J. Pochodzalla *et al.*, Phys. Rev. Lett. 75, 1040 (1995).
[5] S. Albergo, S. Costa, E. Costanzo, and A. Rubbino, Nuovo Cimento 89, 1 (1985).
[6] M.B. Tsang, W.G. Lynch, H. Xi, and W.A. Friedman, Phys. Rev. Lett. 78, 3836 (1997).
[7] T. Glasmacher, *et al.,* Phys. Rev. C 50, 952 (1994).
[8] M. Jandel *et al.*,, phys rev C in press (2006).
[9] F. Gimeno Nogues *et al.*, Nucl. Instrum. Methods Phys. Res. A **399**, 94 (1997).
[10] M. Colonna *et al.*, PRC 57, 1410 (1998).
[11] M.B. Tsang *et al.*, Phys. Rev. Lett. 86, 5023 (2001).
[12] E. Ramakrishnan *et al.*, Phys Rev C 57, 1803 (1998).
[13] D.V. Shetty *et al.*, Phys Rev C 68, 021602(R) (2003).
[14]D.V. Shetty *et al.*, nucl-ex/0606032 (2006).
[15] D.V. Shetty *et al.*, Phys. Rev. C 70, 011601 (2004).
[16] D.J. Rowland *et al.*, Phys Rev C 67, 064602 (2003).
[17]G.A. Suliotis *et al.*, Phys Rev C 68, 024605 (2003).

Transport Description of Heavy Ion Collisions and Dynamic Fragmentation

Hermann H. Wolter
Department für Physik, University of Munich,
85748 Garching, Germany
E-mail: Hermann.Wolter@lmu.de

In this lecture we review the theoretical investigation of heavy ion collisions in order to obtain information on the nuclear equation-of-state (EOS). We discuss the present knowledge of the EOS, and stress, in particular, the large uncertainty about the density dependence of the symmetry energy. We develop the treatment of heavy ion collisions with transport theory and non-equilibrium effects. We then discuss investigations both of the high density EOS with intermediate energy collisions and of the low density EOS in the Fermi energy regime. At the high density we make connections with neutron stars. At low density we discuss the fragmentation process and, in particular, the role and treatment of fluctuations and the dynamical fragment formation.

Keywords: Heavy ion collisions, transport theory, fragmentation, fluctuations, nuclear equation-of-state, symmetry energy, neutron stars

1. Introduction

A primary motivation for the study of heavy ion collisions has been the investigation of the phase diagram of strongly interacting matter. Here we expect a phase transition of hadronic matter to deconfined quark-gluon matter at high temperatures and/or densities, which is, however, not a point of discussion in this talk. Rather here we are interested in the hadronic sector of the phase diagram, where we study the equation-of-state (EOS) of nuclear matter as a function of density and temperature, but nowadays also of asymmetry or proton/neutron ratio. There are several features of the nuclear EOS, that make its study interesting. Since the NN-interaction behaves very much like a van-der-Waals type interaction at densities below saturation density and lower temperatures one expects a phase transition of the liquid-gas type. On the other hand, the asymmetry dependence of the EOS is of great interest today, in connection with radioactive beam physics, and because of the astrophysical implications. Indeed, the nuclear EOS, and in particular, its asymmetry dependence, is of large interest in the physics of supernova explosions and the structure of neutron stars. The question, whether neutron stars may be exotic objects in their interior, depends largely on the symmetry energy at high density. Recently the point of view has also changed considerably, as to whether relativistic and field theoretical concepts are more appropriate to formulate the EOS.

The nuclear phase diagram can be explored in the laboratory by heavy ion collisions. Here one has the advantage of being able to choose different experimental conditions, like the incident energy, the size and the asymmetry of the colliding nuclei, to explore different features. In particular, it will be of large interest to fully exploit the asymmetry degree of freedom in collisions of very exotic nuclei. A heavy ion collision can be thought of as exploring a path in the phase diagram depending on the incident energy and the colliding nuclei. It proceeds from initial normal nuclei at saturation density at more or less asymmetry, to high density and heated nuclear matter at the maximum compression to dilute nuclear matter in the expansion stage with the possible observation of the phase transition. Different observables are appropriate to study these different phases. Flow and particle production are best suited for the high density phase, while fragmentation observables carry information about the phase transition.

On the other hand, a heavy ion collision is a transient and very non-equilibrated process. One usually does not "see" directly the nuclear EOS - a static concept - in a heavy ion collisions, but rather has to infer back to it via theoretical tools, in particular transport theory. While semiclassical transport theory has been used very successfully in the last decades to correlate many heavy ion data, there are still open questions about it foundation. The usual transport approach is essentially classical, and does not take into account quantum phenomena, such as the finite life time of particle, i.e. the finite spectral functions. Also transport theory is essentially a theory for the one-body distribution function. Especially in the production of complex particles and generally in fragmentation phenomena, however, higher order correlations become important. On the lowest level these have be treated as a fluctuation term in the transport description.

In this lectures, being given at a school, I will try to discuss some of these questions on a more pedagogical level, to outline the difficulties and the chances of heavy ion physics. On the other hand, space and time is limited. However, there are also other lectures, which which there exists a strong overlap, as the ones by Christian Fuchs on the EOS, Massimo

Di Toro on the symmetry energy, Virgil Baran on phase transitions, and George Lalazissis on relativistic mean field theory. In fact, my lecture is based on material which was produced in close collaboration with these and other colleagues, in particular T. Gaitanos, M. Colonna, D. Blaschke, S. Typel, and I want to take the opportunity right now, to thank them all.

As an outline, in a first chapter I will discuss the modelization of the EOS, in particular, of its density and asymmetry dependence. In the second chapter I will introduce transport theory and discuss its extensions and problems. I will then present some investigations of the high density behavior of the EOS, in particular of flow variables and kaon production, as well as the consistency with neutron star observables. In the following chapter I will address the low density behavior, and fragmentation reactions. Finally I will make a summary and try to give an outlook.

2. The Nuclear Equation-of-State

Nuclear Matter is an idealized concept. Approximately, it exists in the interior of heavy nuclei, and its properties at saturation are extracted from the mass formula and its charge density from electron scattering. Very asymmetric nuclear matter can most likely be found in the outer core of neutron stars, however, their inner structure is far from certain. Thus theory has to try to give a guidance at the behavior of nuclear matter for high densities, temperatures, and asymmetries.

In recent years a relativistic formulation for nuclear matter has become increasingly popular. In a non-relativistic treatment one starts with a Hamiltonian $H = \sum_i T_i + \sum_{ij} V_{ij} + \sum_{ijk} V_{ijk}$ with two-body and eventually three-body NN-forces. The forces, depending on their construction, contain explicitly the underlying interaction mechanism, e.g. via the exchange of mesons. In a relativistic, field theoretical approach one starts with a hadronic Lagrangian density, which is taken of the minimal coupling form.

$$\mathcal{L}(\psi;\sigma,\omega,\rho,\delta,\ldots) = \bar{\psi}[\gamma_\mu D^\mu - (M - \Gamma_\sigma\sigma - \Gamma_\delta\vec{\tau}\cdot\vec{\delta}\,)]\psi + \mathcal{L}^0_{mes} \tag{1}$$

$$D^\mu = (i\partial^\mu - \Gamma_\omega\omega^\mu - \Gamma_\rho\vec{\rho}^\mu\cdot\vec{\tau}) \tag{2}$$

$$\mathcal{L}^0_{mes} = \frac{1}{2}(\partial_\mu\sigma\partial^\mu\sigma - m_s^2\sigma^2) - \frac{a}{3}\sigma^3 - \frac{b}{4}\sigma^4 \tag{3}$$

$$- \frac{1}{4}W_{\mu\nu}W^{\mu\nu} + \frac{1}{2}m_v^2\omega_\nu\omega^\nu$$

$$+ \frac{1}{2}(\partial_\mu\vec{\delta}\cdot\partial^\mu\vec{\delta} - m_\delta^2\delta^2) - \frac{1}{4}\vec{R}_{\mu\nu}\vec{R}^{\mu\nu} + \frac{1}{2}m_\rho^2\vec{\rho}_\nu\vec{\rho}^\nu \tag{4}$$

Here, ψ is the nucleon field, and σ, ω, ρ, and δ are the various meson fields, with their corresponding field tensors. This approach is also called Quantumhadrodynamics (QHD), and was pioneered by Walecka [1]. The coupling constants Γ_i, resp. the two-body forces, are fixed in such a way, as to explain as well as possible the two-body data. This is meant, when one talks of realistic forces. When these forces are treated by advanced many-body techniques, e.g. in Brueckner theory, then one predicts the properties of nuclear matter. Results of such calculations are shown in Fig. 1, where the saturation density and the binding energy per particle are given for different forces and approaches. The empirical saturation point with its uncertainties is shown by the square. The open symbol represent the results of non-relativistic Brueckner calculations with different realistic two-body forces. It is seen that these calculations miss the saturation point, and tend to lie on a line, which is called the Coester line, which was already described in the 70's. Thus calculations give the correct binding, but to high density, or the right density but too low binding. The hatched symbols are non-relativistic calculations including 3-body forces, which shift the results much closer to the saturation point. It was also shown [2] that calculations of light nuclei get very close to experimental data, when 3-body forces are included.

The full symbols in Fig. 1 give the results of relativistic Brueckner calculations (called Dirac-Brueckner Hartree-Fock, DBHF or short DB). These

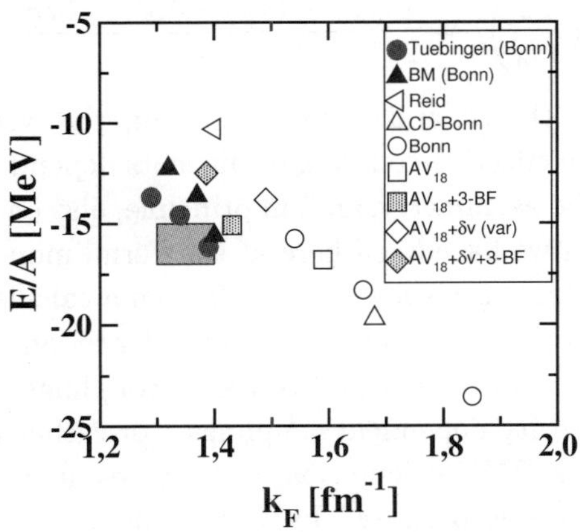

Fig. 1. Binding energy - Fermi momentum plane of nuclear matter. The hatched square represents the empirical saturation point. The symbols are the results of various theoretical calculations of the saturation point of symmetric nuclear matter, which are discussed in the text

are now considerably closer to the empirical point, using only two-body vertices. This desirable behavior is interpreted as signifying, that the explicit relativistic treatment includes effective 3-body effects, which have to parametrized in an non-relativistic treatment. In this lecture I will primarily use the relativistic formulation, even though for some of the calculations at energies below 100 MeV per nucleon a non-relativistic program has been used.

Brueckner calculations can be performed for nuclear matter, but they are difficult for finite nuclei or – *a fortiori* – for heavy ion collisions. In finite nuclei calculations are done in the Hartree– (usually called Relativistic Mean Field, RMF), or sometimes in the Hartree–Fock–approximation. Then the coupling coefficients are not the "realistic" ones from the two-body data, but they are fixed in this approximation to the properties of nuclear matter, and/or to a number of finite nuclei. However, a connection can be established between the many-body DB calculation and the RMF approach. The DB self energies can be decomposed into Lorentz invariants, which in nuclear matter at rest, is $\Sigma(p) = \Sigma^s(p) - \gamma^0\ \Sigma^0(p) + \vec{\gamma}\cdot\vec{p}\ \Sigma^v(p)$, in asymmetric systems for protons and neutrons separately. From these invariants one may then extract effective coupling constants used in a Hartree treatment, as

$$\left(\frac{\Gamma_\sigma}{m_\sigma}\right)^2 = \frac{1}{2}\frac{\Sigma_n^{s(DB)}(k_F) + \Sigma_p^{s(DB)}(k_F)}{\rho_n^s + \rho_p^s} \tag{5}$$

$$\left(\frac{\Gamma_\delta}{m_\delta}\right)^2 = \frac{1}{2}\frac{\Sigma_n^{s(DB)}(k_F) - \Sigma_p^{s(DB)}(k_F)}{\rho_n^s - \rho_p^s}\ , \tag{6}$$

and similarly for the ω and ρ mesons using the vector self energies $\Sigma_{p,n}^{0(DB)}(k_F)$. These effective coupling coefficients depend on density (Fermi momentum), on the asymmetry, and, in principle, also on the momentum of the particle, which we have fixed here at the Fermi momentum. The effective coupling constants are shown in Fig. 2 from a calculation of Hofmann, et al. [3]. It is seen that the asymmetry dependence, once the coupling has been decomposed into isoscalar and isovector couplings, is very weak.

When these density-dependent couplings – or rather vertex functions – are to be used in a RMF calculation, one arrives at a density-dependent RMF approach, as was proposed by Fuchs and Lenske [4]. The density dependence of the vertices has to be operator valued, in order to maintain the covariance of the theory, e.g. as the vector density $\Gamma_i(\hat{\rho}); \hat{\rho} = \sqrt{j^\mu j_\mu}$, where j^μ is the current operator. Then additional terms appear in the field equations due to the variation of the vertex functions. The functional

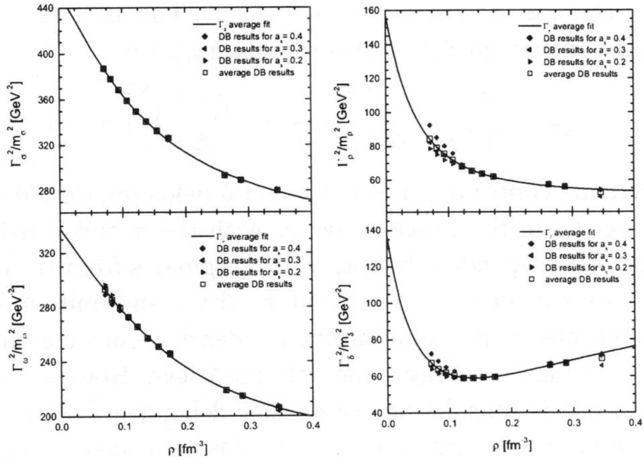

Fig. 2. Density dependence of the σ (upper left), ω (lower left), ρ (upper right), and δ (lower right) effective meson-nucleon vertices. Shown are results extracted from DB selfenergies from ref. [3] calculated for asymmetry ratios $a_s 0.2, 0.3, 0.4$. The solid line is an asymmetry independent fit.

dependence can be taken directly from DB calculations, as was done e.g. in ref. [3], or they can be adjusted by a fit to data of nuclear matter and some spherical nuclei, as was done e.g.. by Typel and Wolter [5] and also more recently by Lalazissis et al. [6]. Parametrizations of this type provide highly precise descriptions of nuclear ground states, but also of collective features of nuclei.

This approach can be thought of as a density functional approach. Indeed, the density dependence of the vertex functions effectively contains correlations beyond the mean field approach. If it is taken from DB calculations, it is seen explicitly that it contains Pauli and two-body correlations from the ladder summation of the G-matrix approach. It is thus a concrete example, of how the density dependence can be fixed in a density functional approach.

In the formulation of a Lagrangian in Eq. (4) we have included also a δ-meson exchange, or rather a δ-like field. This is an isovector–scalar field, the analogon of the σ field in the isovector sector. It should not necessarily be thought of as the exchange of a physical δ meson, which, in fact, rather massive. Rather it takes care of the full isospin nature of the interaction. It is seen from Fig. 2 that it is naturally predicted in a microscopic DB approach, even without introducing explicitly a δ-meson there. Such a field has usually not been used in empirical RMF parametrization, such as in ref. [5], since there is not real evidence in for it. This is understood, if one looks at the symmetry energy in this parametrization

$$E_{sym} = \frac{1}{6}\frac{k_F^2}{E_F} + \frac{1}{2}\left[\frac{\Gamma_\rho}{m_\rho} - \frac{\Gamma_\delta}{m_\delta}\left(\frac{m^*}{E_F}\right)^2\right]\rho_B \quad . \tag{7}$$

It is seen that the contribution of the ρ and δ fields interfere destructively just as this occurs in the isoscalar sector with the σ and ω fields. The δ-field has a density–dependent factor, which originates from the coupling to the scalar (isovector) density. At a fixed density – and finite nuclei receive mainly contibutions from around saturation density – one may parametrize the empirical symmetry energy using only the ρ field. However, a difference appears when one is considering the density dependence of the symmetry energy, because the contribution of the δ field is weakened at higher density since it couples to the scalar isoscalar density. Thus, a $\rho\delta$ parametrization of the isovector part of the EOS predicts a stiffer density dependence.

A compilation and assessment of theoretical predictions of the nuclear EOS has been attempted in an article for the so-called WCI group by Fuchs and Wolter [7]. From this paper we reproduce the Fig. 3, which give the nuclear matter (symmetric) and the neutron EOS from various theoretical calculations, mentioned in the caption. It is seen that large differences exist between different theoretical models. The uncertainty in the symmetric EOS today has been considerably reduced due to heavy ion experiments, some of which will be referred to later on. The difference between the neutron and nuclear EOS is precisely the symmetry energy, which is shown in Fig. 4 for the models given in Fig. 3 and also some Skyrme forces. While different theories agree fairly well at saturation energy – actually somewhat below saturation density – the density dependence is largely unconstrained both for densities below and above saturation. It is one of the aims of heavy ion physics today to determine this behavior. We will come back to this in later sections.

3. Transport Descriptions of Heavy Ion Collisions

Heavy ion collisions represent a strongly non-equilibrium process, which is extremely difficult to describe fully in a quantum–mechanical way. For many purposes such a description is also not needed, since, because of the complicated final state, ensemble averages are appropriate to describe the data. For certain questions it may be justified to use thermal or hydrodynamical models. However, if one would like to follow the complete evolution of the process from the initial to the final state, which would be desirable in

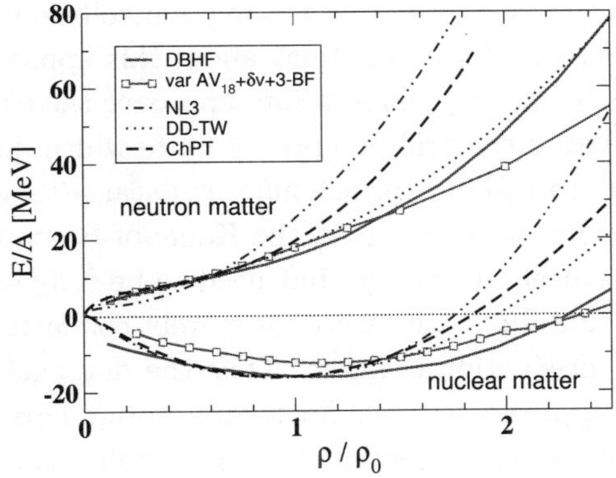

Fig. 3. EOS in nuclear matter and neutron matter from different theoretical calculations: DBHF (Brueckner-HF [8]), variational with 3-body forces (var AV18 [9]), RMF (NL3 [10]), density-dep. RMF (DD-TW [5]), chiral perturbation (ChPT [11])

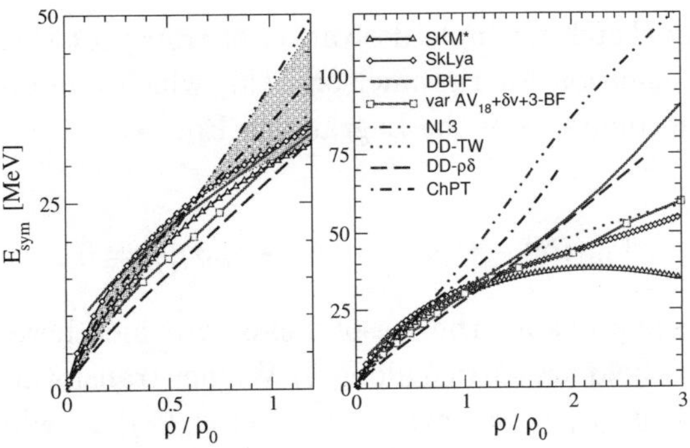

Fig. 4. (right panel) Symmetry energy as a function of density as predicted by different models, discussed partially in Fig. 3. The left panel shows the low density region won a larger scale.

order not to depend on a-prior assumptions, one has to resort to transport approaches. On the lowest level this describes the evolution of the one-body phase space distribution. This is a semiclassical approximation to the time-dependent HF method (Vlasov equation). However, the HF approach neglects the dissipation due to two-body collisions, and thus the Vlasov equation is supplemented by a collision term of the Boltzmann type, with a modification due to the Pauli principle. This approach is known today as the Boltzmann-Uehling-Uhlenbeck (BUU) or Boltzmann-Nordheim-Vlasov (BNV) approach [12]. It has been used extensively and alltogether successfully in the last decades to interpret heavy ion collisions.

However, there are basic questions about this approach, which are still not fully answered. In a non-equilibrium approach the particles are actually quasiparticles with a spectral function of finite width, i.e. with a finite life time, either due to collision broadening or because of decay processes. A quantum transport theory exists in the Kadanof-Baym theory [13]. Implementations have been attempted, but need to be fully explored [14].

Dissipation and fluctuations are intimately connected, as expressed by the fluctuation–dissipation theorem. Thus the dissipative transport equation has to be supplemented by a fluctuating term. This is often referred to as the Boltzmann-Langevin equation [15]. The fluctuation term is usually not very important, since it leads to small fluctuation about the average phase space density. However, in the investigation of phase transitions (of the liquid-gas type) the systems enters regions of thermodynamical instability and fluctuations become dominant in deciding the evolution of the system. Thus in this region the inclusion of fluctuations is of vital importance.

I will briefly sketch a simple derivation of transport theory. One starts from the field equation for the nucleon field, which one obtains from the Euler-Lagrange equations of the Lagrangian Eq. (4).

$$[\gamma_\mu(i\partial^\mu - \Gamma_\omega \omega^\mu) - (M - \gamma_\sigma \sigma)]\Psi = 0 \ , \tag{8}$$

together with equations for the meson fields, of which here, for simplicity, only the σ and ω field are carried along. A Wigner transform is performed of the one-body density $\rho_{\beta\alpha}(x_1, x_2) = < \bar{\psi}_\beta(x_1)\ \psi_\alpha(x_2) >$, which is a Fourier transformation with respect to the coordinate difference $r = (x_1 - x_2)$

$$F_{\beta\alpha}(x,p) = \frac{1}{(2\pi)^4} \int dr\ e^{-ip_\mu r^\mu} < \bar{\psi}_\beta(x + \frac{r}{2})\ \psi_\alpha(x - \frac{r}{2}) >, \tag{9}$$

where $x = \frac{1}{2}(x_1 + x_2)$. The equations of motion for the Wigner transform are separated in real and imaginary parts, and into its different Lorentz invariants. One equation leads to $((p^*)^2 - (m^*)^2)F(x,p) = 0$, which is a mass shell constraint, meaning that energy and momentum are connected by the relativistic relation. The other equation is the transport equation for the scalar part of the Wigner transform $F_s \equiv m^* f(x, p^*)$

$$\left[p^{*\mu}\partial_\mu^x + (p_\nu^* F^{\mu\nu} + m^*\partial_x^\mu m^*)\,\partial_\mu^{p^*}\right] f(x,p^*) = 0, \qquad , \tag{10}$$

where $p^*_\mu = p_\mu - \Sigma_\mu$ is the kinetic momentum, $m^* = m - \Sigma_s$ the effective (Dirac) mass, and $F^{\mu\nu} = \partial^\mu \Sigma^\nu - \partial^\nu \Sigma^\mu$ is the field strength tensor, equivalent to the electric and magnetic field tensor in electrodynamics. As it stands this is the Vlasov equation of the evolution of the phase space distribution in a mean field, namely the scalar and vector self energies Σ_s, Σ^μ. To describe also dissipation due to two-body collisions on the average, a collision term $\mathcal{I}_C$ is added on the rhs of Eq. (10)

$$\begin{aligned}\mathcal{I}_C = \frac{1}{2}\int &\frac{d^4p_2}{E^*_{p_2}(2\pi)^3}\frac{d^4p_3}{E^*_{p_3}(2\pi)^3}\frac{d^4p_4}{E^*_{p_4}(2\pi)^3}W(pp_2|p_3p_4)\\ \times &\Big[f(x,\mathbf{p}_3)f(x,\mathbf{p}_4)(1-f(x,\mathbf{p}))\ (1-f(x,\mathbf{p}_2))\\ - &\ f(x,\mathbf{p})f(x,\mathbf{p}_2)(1-f(x,\mathbf{p}_3))\ (1-f(x,\mathbf{p}_4))\Big] \quad . \end{aligned} \tag{11}$$

Here $W(pp_2|p_3p_4) = (p^* + p_2^*)^2 \frac{d\sigma}{d\Omega}_{inmed} \delta^4(p + p_2 - p_3 - p_4)$ is the in–medium energy-momentum conserving transition probability, and the last terms are the occupation and blocking factors, where the latter represent the action of the Pauli principle. $\frac{d\sigma}{d\Omega}_{inmed}$ is the in-medium cross section.

This transport equation is a non-linear partial integro-differential equation, which has not been solved directly as such. Rather, so so-called test particle method is used to simulate it [12]. The phase space distribution is represented as

$$f(x,p^*) = \frac{1}{N}\sum_{i=1}^{A\cdot N}\int_{-\infty}^{\infty} d\tau g(x - x_i(\tau))(\tilde{g}(p^* - p_i^*(\tau)), \tag{12}$$

i.e. as a collection of test particles centered at $\{x_i(\tau), p_i^*(\tau)\}$. The shape g of the test particles is often given as *delta*-functions, but can also be of finite width, as in our formulation [16]. N is the number of test particles per nucleon, which is chosen sufficiently large to give a good representation of the phase space. It is then shown, that the test particles, under the action of the lhs of Eq. (10) obey Hamiltonian equations of motion.

$$\begin{aligned}\frac{d}{d\tau}x_i^\mu &= u_i^\mu(\tau)\\ \frac{d}{d\tau}u_i^\mu &= \frac{1}{m_i^*}\left(u_{i\nu}F^{\mu\nu} + \partial^\mu m_i^* - (\partial^\nu m_i^*)u_{i\nu}u_i^\mu\right) \quad . \end{aligned} \tag{13}$$

τ is the eigen time of the test particles. The collision term is usually evaluated stochastically. Roughly speaking a criterion is set, that two test-particles perform an elastic collision, if they approach within a distance depending on the in-medium cross section $d = \sqrt{\sigma_{inmed}/N\pi}$. Their direction after the collision is chosen according to the angular distribution of the cross section.

At higher energies new particles can be produced through inelastic NN collisions. Above about 300 MeV cm energy Δ resonances are produced, which decay into nucleons and pions, which in turn interact strongly with nucleons. Then one has to set up coupled transport equations for nucleons, Δ's and pions, which are coupled through inelastic cross sections in the collision term. At higher energies also strangeness is produced in form of hyperons and K mesons. Again many different channels have to be considered. A great many inelastic cross sections are needed to describe these processes which only partially can be taken from experiment, but are otherwise obtained from theoretical considerations. It would be too much to go into details here, which can be found in the literature. Here I just want to give an impression what is involved in describing a heavy ion collision in all its detail in a transport description.

As mentioned above, the derivation of a transport equation sketched above, is only the lowest order. In particular the collision term has been added "by hand" from empirical arguments. There is a deeper approach to quantum transport theory by Kadanof and Baym [13], which explicitly deals with the irreversibility effects. It would be too far to go into details here. One result is that in addition to the distribution function one also has a spectral function for the test particles which describes the off-shellness. This has been further discussed in ref. [14]. It seems that these effects are only important for extreme subthreshold production of particles, but this remains to be investigated in more detail. Such an approach also consistently leads to a collision term, which links the mean field terms in the Vlasov part to the in-medium cross section in the collision term. In a Brueckner approach to the self energy they are both connected through the G-Matrix. This has been employed in several works [17].

The G-Matrix depends on the density, or more generally on the phase space distribution, since the intermediate propagator depends on this. Thus one would have to calculate the G-Matrix for the phase space configuration of the heavy ion collision, i.e. for a non-equilibrium configuration, which is clearly not possible. Usually the non–equilibrium effect are ignored, but this bears the danger that the EOS extracted from heavy ion collisions effectively includes such effects. There have been attempts in our group to take this into account in an approximate way [18].

4. The EOS at High Density

The high density EOS can be investgated with relativistic collision energies, where up to three times the saturation density is reached. In particular, the high density symmetry energy can be studied. This question touches intimately with the question of the structure of neutron stars. In this section I will briefly study these topics.

The primary observable sensitive to the pressure at the maximum density is the nucleon (or particle) flow, i.e. the momentum distributions of the hadrons in the final state. The momentum distribution is usually expanded in terms of a Fourier series for the azimuthal distribution of the yield

$$N(\Theta, y, p_t; b) = N_0(1 + v_1(y, p_t)\cos(\Theta) + v_2(y, p_t)\cos(2\Theta) + \ldots) \quad (14)$$

as a function of the rapidity y (longitudinal) and transverse momentum p_t for a given impact parameter b. The first coefficient v_1 is called the sideward flow and describes the flow in the reaction plane, while the second coefficient, elliptic flow, describes the flow out of plane.

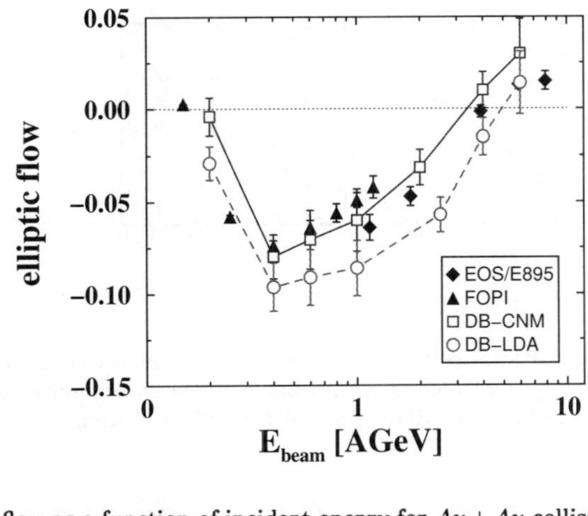

Fig. 5. Elliptic flow as a function of incident energy for $Au + Au$ collisions at intermediate impact parameter. Data are from the FOPI and EOS collaborations. The calculation are done with DB mean fields (DBF [8]), in different approximations: LDA (local density) and CNM (colliding nuclear matter).

In Fig. 5 the elliptic flow has been represented as a function of incident energy from the SIS to the AGS energy regime, i.e. from about 50 MeV to several GeV per nucleon. The data are not quite consistent between the two regimes. The calculation used fields from DB calculations [8]. Without non-equilibrium effects, i.e. in the local density approximation, LDA, the elliptic flow is overpredicted. However, when these are taken into account (CNM), the flow is quantitatively described in the whole energy regime. It should be mentioned that the corresponding directed flow is also described well, but only below about 1 GeV. This is understood, because the mean fields become too repulsive for higher energies. Similar, systematic investigations have been performed by Danielewicz et al. in a non-relativistic momentum-dependent parametrization of the EOS [19]. From this one may obtain limits of the EOS, which are compatible with the data.There it is seen that several equations discussed in the literature fall outside this region. In this reference one may also see that the neutron EOS is much less constrained, which points to the need to obtain more information experimentally.

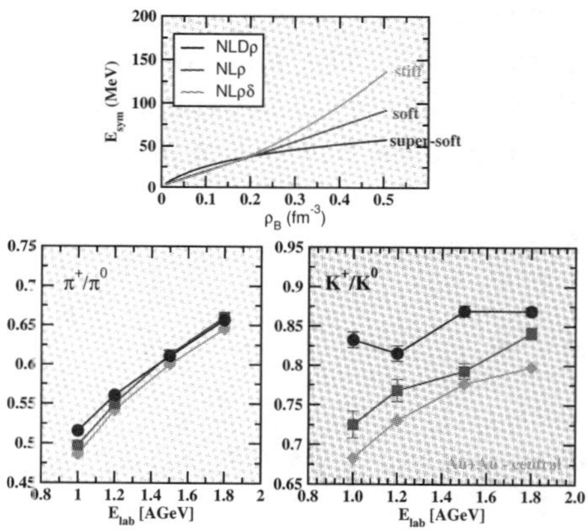

Fig. 6. (lower part) Ratios of π^+/π^0 and K^+/K^0 yields in central $Au + Au$ collisions as a function of energy for different assumptions on the symmetry energy shown in the upper part: $NL\rho\delta$ (stiff, green diamonds), $NL\rho$ (soft, red squares), $NLD\rho$ (supersoft, blue disks)

The other, probably most stringent determination of the EOS has been obtained from kaon production near threshold energies. As discussed above, other particles are produced via inelastic NN-collisions. The first and most important process is the excitation of the Δ resonance with a threshold of about 300 MeV, which in turn decays with producing pions. Thus pions and Δs are the most copious particles in intermediate energy collisions. For strangeness production at energies below the kaon threshold the primary source are secondary reactions, of which the most important ones are associated strangeness production $N\Delta \rightarrow N\Lambda K$ and the strangeness exchange

$N\pi \to \Lambda K$. In ref. [20] (see also lecture by Chr. Fuchs) the ratio of kaon production in a heavy system ($Au + Au$) with a high compression is compared to the production in a light system ($C+C$). It is seen that this clearly favors a soft EOS.

The sensitivity seen in the investigation in ref. [20] arises from two effects: from the closeness of the collision energy to the production threshold and to the fact, that K^+ mesons interact weakly with nuclear matter, and thus carry information from the dense phase of the collision. There are, however, different species of kaons with different charges, which have slightly different masses and also different mean field potentials, and thus different thresholds Therefore the ratio of differently charged kaons, i.e. the ration K^+/K^0, should be an interesting observable to investigate the high density symmetry energy. This is demonstrated in Fig. 6 from ref. [21], where the ratios of π^+/π^0 and K^+/K^0 are calculated for isovector EOS's of different stiffness (exemplified here by the RMF EOS discussed above with and without a δ-meson). Pions interact strongly with nuclear matter and thus carry information about the whole evolution of the collision and not only the high density phase, and thus their sensitivity is low. In contrast, the sensitivity of the kaon ratio is appreciable, and it should be a suitable variable to investigate the symmetry energy at high density. This is discussed more in detail in the lecture of Prof. Di Toro.

The high density isovector EOS is also of direct relevance to the structure of neutron stars (NS). The solution of the Tolman-Oppenheimer-Volkov equation yields the NS mass for a given EOS for a given starting central density. On the other hand, the proton fraction $x = Z/N$ from β–equilibrium and charge neutrality is a direct function of the symmetry energy. In Fig. 7 we show this dependence for the NS mass and the proton fraction for a number of different EOS's. These are not complete, but represent a fair selection of recent theoretical results. The details on these EOS's are given in ref. [22], where also the following results are taken from. The masses obtained have to be compared to observed NS masses, which are typically in the range of 1 to 1.5 solar masses. All EOS's obtain such masses. Recently a very heavy neutron star of 2.1 solar masses was discovered. Depending on the error assigned to this observation, some of the EOS's begin to have problems, explaining such a heavy NS.

In addition one has to consider the cooling behavior of a NS. Above a proton fraction of about 11% the direct URCA process, i.e. the β-decay of the proton, can take place, which leads to emission of neutrinos and thus to a rapid cooling of the NS, such that it is not observable. Thus one has

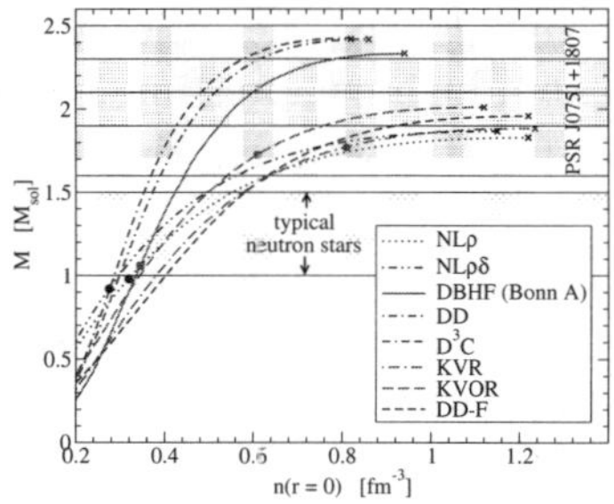

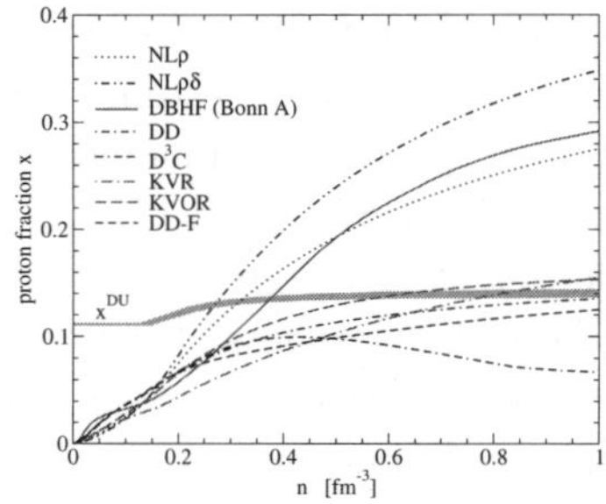

Fig. 7. (left panel) Neutron star masses as a function of the central density for different EOS (see text). Shown are limits for typical neutron star masses and those for the newly discovered heavy NS with one and two σ errors. (right panel) Proton fraction $x = Z/A$ for the same EOS's as a function of central density. Indicated is also the region of the onset of the direct URCA process. This point is indicated on the curves of the left hand panel as a dot.

to require, that the URCA limit is not reached for stars with masses of the heavy NS, and much less for typical NS masses. The point, where the URCA limit is reached in the right hand panel is marked as a dot in the curves on the left hand panel. It is seen that several of the EOS's reach this limit even in the range of typical NS masses, and are thus not compatible with the cooling curves. These considerations are somewhat qualitative, because additional considerations for the heat conduction through the crust (see lecture of N. Sandulescu) are important, but this does not change the qualitative result.

Other checks have been performed in ref. [22] against further observables of NS, such as mass-radius relations, gravitational vs. baryonic mass, etc. On the other hand, also the results for heavy ion collisions have been taken into account, such as to whether the EOS falls into acceptable region of ref. [19], and for kaon production. It was found that none of all the EOS's considered satisfies all checks. This may be interpreted in several ways: it shows the usefulness of such a simultaneous comparison to NS and heavy ion data, and stresses the need to find even better EOS's. However, if one would not succeed to find a satisfactory hadronic EOS, then this could be an indication that other phenomena, such as strangeness condensation, quark cores or hybrid stars have to be considered more seriously in NS.

5. Low Density EOS and Dynamical Fragmentation

As discussed above the low density behavior of the EOS can be investigated in heavy ion collisions in the final expansion stage of a central collision or in the decay of the spectator in a peripheral collision. In particular, one may enter the region of thermodynamical instability, i.e. the region of the liquid-gas type phase transition. Since gravity does not play a role in nuclear systems, the liquid-gas coexistence should correspond to fragmentation processes, i.e. to a separation of the dilute system into heavier fragments ("liquid") embedded in free nucleons or light nuclei ("gas"). Fragmentation is in fact the dominant final state in heavy ion reactions in this regime. This has been amply demonstrated for central collisions by the INDRA collaboration, and for spectator decay by the ALADIN collaboration. An indication of a phase transition in a finite system is a bimodality in special order variables. A striking example of a bimodality in the correlation between the largest and the second largest fragment has recently been demonstated by B. Tamain [23]. An extensive review of the work of the Catania group in this domain by V. Baran et al. is found in ref. [24].

The description of heavy ion reactions by transport theories of the BUU type treats the evolution of the one-body density under the influence of a mean field and the average effect of two-body collisions. Fragmentation, on the other hand, is intrinsically a many-body correlation, which is not contained in a one-body description. While a full quantum many-body transport theory is our of reach, the viewpoint has been taken, that fragment formation in a thermodynamically instable region is triggered by fluctuations of the one-body density. These fluctuation are then the seeds of fragment formation and are exponentially enhanced by mean field dynamics in the instable region. It is then of primary importance to control the physically correct amount of fluctuations in a transport theory.

Transport theory with fluctuations is described by the Boltzmann-Langevin (BL) equation, which in addition to the collision term also includes a fluctuation term. The BL equation has been studied numerically in model systems [15] but is too complicated for realistic applications. Thus, various approximation schemes to such a full treatment have been proposed: In the BOB (Brownian One-Body dynamics, [25]) approach the fluctuations have been gauged to the most unstable modes of the system. In the Stochastic Mean Field (SMF) approach, fluctuations are injected into the phase space density under the assumption of a local statistical equilibrium [26]. Molecular Dynamics approaches, like AMD, QMD, or FMD, include fluctuations by describing many body dynamics on the classical or quantum level. Finally, the most often employed method is to use the numerical fluctuations, which are due to the solution of the transport equation using the test particle method, as seeds of fluctuations. Here the number of test particles

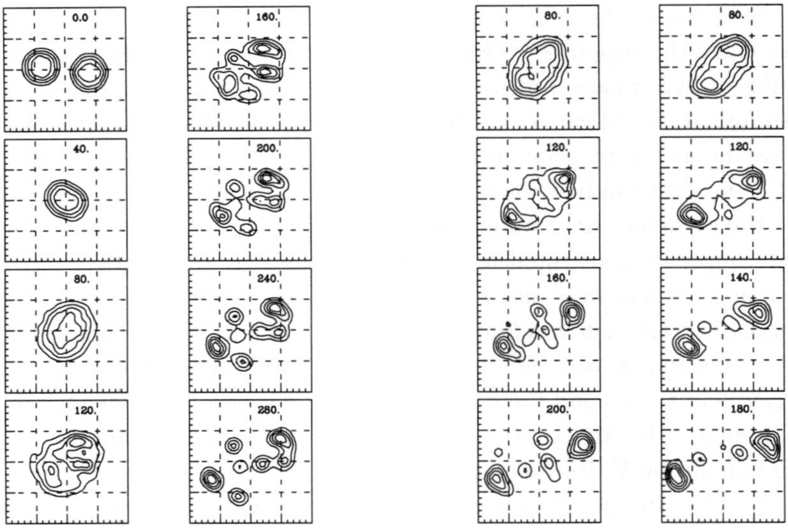

Fig. 8. Density contour plots of the time evolution of $Sn + Sn$ collisions at 50 AMeV for impact parameter $b = 2fm$ (left, central) and $b = 6fm$ (right, semi-peripheral). The examples correspond to bulk and neck fragmentation (see text).

controls the amount of fluctuations.

Schematically the BL equation has the following form

$$\frac{df}{dt} = I_{coll}(f) + \delta I_{fluc} \ , \tag{15}$$

where the left hand side is the Vlasov term and there is a fluctuation term in addition to the usual collision term. It has zero average but non-zero correlation functions $< \delta I_{fluc}(t) \delta I_{fluc}(t') >= D(t)\delta(t - t')$. Due to the fluctuation the distribution function can be split as $f(r, p, t) = \bar{f}(r, p, t) + \delta f$, i.e. into a mean field part $\bar{f}$, which is described by the dissipative BUU equation, and a fluctuation part δf. One may argue [28] that it is a good approximation to assume that the fluctuations are given locally in space and time by the statistical fluctuations of an equilibrated system of the corresponding density and temperature. Then the equilibrium variance of the distribution function is given as

$$\sigma^2_{equil} =< (f - \bar{f})^2 >= \bar{f}(1 - \bar{f}). \tag{16}$$

In the Stochastic Mean Field model we have implemented such a scheme. The fluctuations of Eq. (16) are projected on coordinate space, and the resulting density fluctuations are inserted into the phase space distribution "by hand" at appropriate times during the instability phases of the evolution. With a fragment recognition (coalescence) algorithm the fragments are, finally, identified. It was shown that this procedure is consistent in schematic models with the BL equation [27].

An example of such calculations in this model is given in Fig. 8, where the time evolution of central (left) and peripheral (right) $Au+Au$ collisions at 50 AMeV is shown. In the central collisions one observes bulk or spinodal fragmentation, i.e. the dilute system decomposes into intermediate mass fragments (IMF) and a gas of nucleons or light fragments (not seen in this representation). In the peripheral collision, on the other hand, one observes the formation of a neck between the two leading spectator fragments, which finally ruptures into IMF's. This process has been called neck fragmentation [29], and involves also surface instabilities. One should note, however, that the IMF produced in such transport calculations are still excited and will further decay be secondary evaporation.

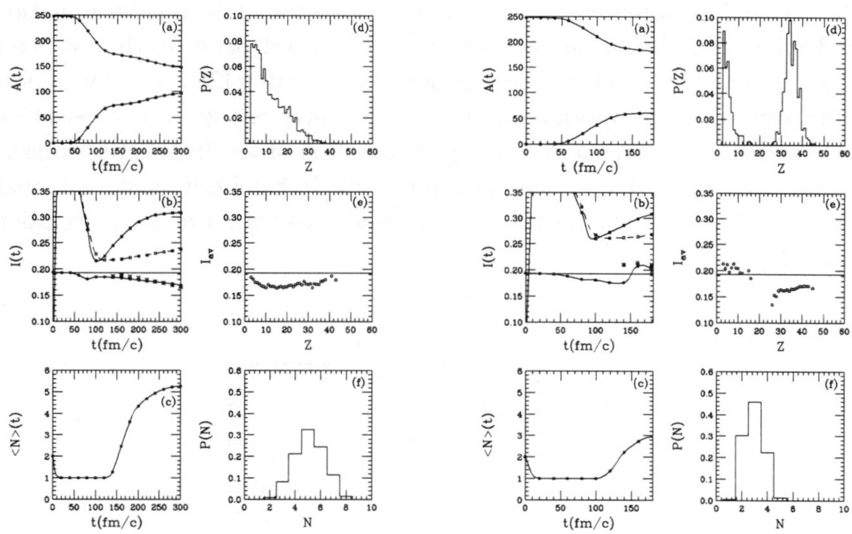

Fig. 9. Fragmentation observables in $^{124}Sn+^{124}Sn$ reactions for central (left, $b=2fm$) and semiperipheral (right, $b=6fm$) collisions as in Fig. 8. In each case the following quantities are shown: as a function of time (a) mass of liquid and gas, (b) asymmetry of gas and liquid, (c) number of fragments ($Z \geq 3$); at freeze–out (d) charge distribution of fragments, (e) mean asymmetry $I=(N-Z)/A$ of fragments as a function of fragment charge, (f) fragment multiplicity distribution. In the text we only discuss panel (e).

These fragmentation processes may also be used to investigate the isovector EOS. The initial asymmetry of the system (here $Au + Au, I = (N - Z)/A = .19$) can be compared with the asymmetry of the produced IMF's. This is shown in Fig. 9, where (among other quantities) the mean asymmetry of the IMF is shown as a function of the charge of the fragment for central ()left) and peripheral (right) collisions. The initial asymmetry is indicated. It is seen that in the central collision the asymmetry of the fragments is lower than the initial asymmetry. Thus a charge fractionation has taken place, in that the dilute system, seen in Fig. 8 splits into more symmetric fragments ("liquid") and into neutrons and neutron-rich light particles ("gas"). On the other hand, the fragment asymmetry is different for the neck fragmentation, also shown in the figure. The spectator-like fragments behave similarly as in the other case, but the neck IMF are actually more asymmetric that the original nuclei. This is due to a neutron flow into the dilute neck region, which has been called isospin migration. Both these phenomena can be well interpreted starting from the behavior of the chemical potential for protons and neutrons as a function of density. Since these chemical potentials depend on the isovector EOS, the fractionation and migration phenomena described above, also depend on the isovector EOS. Indeed, this is what is seen in actual calculation [28]. It is thus a possible way to obtain information on the symmetry energy at low densities. In practice, however, the secondary evaporation of the fragments tends to reduce the sensitivity to the isovector EOS. It has been suggested, that fragment-fragment correlations may be a way to obtain more information here [29].

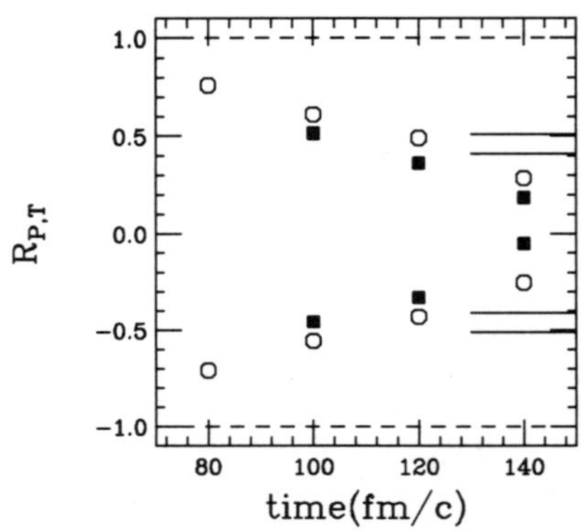

Fig. 10. Isospin transport coefficient (see text) for symmetric and asymmetric $Sn + Sn$ collisions with mass 112 and 124 as a function of interaction time, i.e. impact parameter.

Recently an observable has been suggested, which does not suffer from this difficulty. One considers peripheral collisions between nuclei of different asymmetry, as e.g. ^{124}Sn and ^{112}Sn. Then the isospin should be transported or diffused through the neck during the time of contact. Following the FOPI collaboration an isospin transport ratio has been defined as

$$R_P = \frac{2I_P^M - I_P^H - I_P^L}{I_P^H + I_P^L} \ . \tag{17}$$

The quantity I here is the asymmetry of the projectile-like fragment (PLF), but could also be any other isospin sensitive quantity. The upper indices M, H, L signify the mixed, heavy-heavy, and light-light reaction system, respectively. A similar ratio R_T could also be defined for the TLF. The above ratio R_P is equal to one when the asymmetry of the PLF in the mixed reaction is equal to that of the projectile, i.e. when no isospin has been transported. It is zero, when the the asymmetry is the mean of the partners in the mixed reaction, i.e. when the isospin has been fully equilibrated. This ratio has been calculated for the above reaction and is represented in Fig. 10 as a function of the interaction time, i.e. the impact parameter for different isovector EOS's. It is seen that it reduces from the value one at large impact parameters as expected toward smaller values at close collisions. For the measured system with $b \approx 6fm$ the value is about 0.5, corresponding to not complete equilibration. The effect depends on the isovector EOS, in that the softer EOS allows more equilibration. The data seem to favor a more stiff EOS. In ref. [30] the roles of density and concentration gradients in this result have been analyzed in detail. In ref. [31] the influence of a momentum dependent isovector EOs was investigated with the result that this leads to a softer isovector EOS. While thus the final conclusion is not quite settled, these examples show, how fragmentation processes can be used to probe the low density symmetry energy.

6. Summary

In this lecture we have reviewed how heavy ion collisions can be used to investigate various aspects of the nuclear equation-of-state, such as the high and low density behavior, the isospin part of the EOS, and phase transitions. Transport theory is used to describe the very non-equilibrated collision process. It foundations and numerical simulations are discussed. Microscopic input can be used in such transport descriptions, making a link to nuclear many-body theory, as well as to phenomenological approaches in the vein of density functional theory. Many observational variables of heavy ion collisions have been correlated. The density dependence of the symmetry energy is discussed in particular, as well as connections to neutron star physics.

Fragmentation is a process that appears at low density nuclear matter and which can be seen as a consequence of a liquid-gas phase transition. The role of fluctuations is essential to describe fragmentation, and it is discussed how these can be treated in a transport approach.

Even though we thus have a good grasp of heavy ion collisions there are many aspects that still need to be understood better. These concern theoretical aspects, like non-equilibrium transport theory and off-shell effects, and also the availability of exclusive heavy ion observable, in particular of very asymmetric, exotic beams, in order to obtain more constraints on the symmetry energy.

References

1. B.D.Serot, J.D.Walecka in *Advances in Nuclear Physics* Vol. 16, Eds. J.M.Negele and E.Vogt, Plenum, New York, 1986.
2. S.C. Pieper, R.B. Wiringa, J. Carlson, Phys.Rev. C70 (2004) 054325.
3. F. Hofmann, C. M. Keil, H. Lenske, Phys.Rev. C64 (2001) 034314.
4. C. Fuchs, H. Lenske, H. H. Wolter, Phys. Rev. C 52, 3043 (1995); H. Lenske and C. Fuchs, Phys. Lett. B 345, 355 (1995).
5. S. Typel, H.H. Wolter, Nucl. Phys. A 656 (1999) 331.
6. G. Lalazissis, T. Niksic, D. Vretenar, P. Ring, Phys. Rev. C71 (2005) 024312.
7. C. Fuchs, H.H. Wolter, in *Dynamics and Thermodynamics with Nuclear Degrees of Freedom*, ed. F. Gulminelli, et al., Eur. Phys. J. A 30 (2006) 5.
8. T. Gross-Boelting, C. Fuchs, and A. Faessler, Nucl. Phys. A 648 (1999) 105; E. van Dalen, C. Fuchs, A. Faessler, Nucl. Phys. A 744 (2004) 227.
9. V.R. Pandharipande and R.B. Wiringa, Rev. Mod. Phys. 51 (1979) 821.
10. G.A. Lalazissis, J. König, P. Ring, Phys. Rev. C 55 (1997) 540.
11. P. Finelli, N. Kaiser, D. Vretenar, W. Weise, Eur. Phys. J. A 17 (2003) 573; Nucl. Phys. A 735 (2004) 449.
12. G.F.Bertsch, S.Das Gupta, Phys.Rep. 160 (1988) 189.
13. L. P. Kadanof and G. Baym, *Quantum Statistics Mechanics* (Benjamin, New York, 1962).
14. M. Effenberger, U. Mosel, Phys.Rev. C60 (1999) 051901; W. Cassing, S. Juchem, Nucl.Phys. A665 (2000) 377.
15. P.G. Reinhard, E. Suraud, S. Ayik, Ann. Phys. 213 (1992) 2004.
16. C. Fuchs, H.H. Wolter, Nucl. Phys. A589 (1995) 732.
17. T. Gaitanos, C. Fuchs, H.H. Wolter, Phys.Lett. B609 (2005) 241.
18. T. Gaitanos, C. Fuchs, H. H. Wolter, Nucl.Phys. A650 (1999) 97.
19. P. Danielewicz, R. Lacey, W.G. Lynch, Science 298 (2002) 1592.
20. C. Fuchs, Amand Faessler, E. Zabrodin, Y.M. Zheng, Phys. Rev. Lett. 86 (2001) 1974.

21. G.Ferini, M.Colonna, T.Gaitanos, M.Di Toro, Nucl.Phys. A, in print; nucl-th/0504032.
22. T. Klaehn, D. Blaschke, S. Typel, E.N.E. van Dalen, A. Faessler, C. Fuchs, T. Gaitanos, H. Grigorian, A. Ho, E.E. Kolomeitsev, M.C. Miller, G. Ropke, J. Truemper, D.N. Voskresensky, F. Weber, H.H. Wolter, Phys.Rev. C74 (2006) 035802
23. M. Pichon, B. Tamain, R. Bougault, F. Gulminelli, O. Lopez, Nucl.Phys. A779 (2006) 267.
24. V.Baran, M.Colonna, V.Greco, M.Di Toro, Phys. Rep. 410 (2005) 335.
25. A. Guarnera, et al., Phys. Lett. B 403 (1997) 191.
26. M. Colonna, M. Di Toro, A. Guarnera, S. Maccarone, M. Zielinska-Pfabe, H.H. Wolter, Nucl. Phys. A 642 (1998) 449
27. M. Colonna, G. Fabbri, M. Di Toro, F. Matera, H.H. Wolter, Nucl. Phys. 742 (2004) 337.
28. V. Baran, M. Colonna, M.Di Toro, V. Greco, M. Zielinska-Pfabe, H.H. Wolter, Nucl.Phys. A703 (2002) 603.
29. V. Baran, M. Colonna, M. Di Toro, Nucl.Phys. A730 (2004) 329.
30. V.Baran, M.Colonna, M.Di Toro, M.Zielinska-Pfabe, H.H.Wolter, Phys.Rev. C72 (2005) 064620
31. Lie-Wen Chen, Che Ming Ko, Bao-An Li, Phys.Rev.Lett. 94 (2005) 032701.

The Nuclear Equation of State at high densities

Christian Fuchs

Institut für Theoretische Physik,
Universität Tübingen,
D-72076 Tübingen, Germany
E-mail: christian.fuchs@uni-tuebingen.de

Ab inito calculations for the nuclear many-body problem make predictions for the density and isospin dependence of the nuclear equation-of-state (EOS) far away from the saturation point of nuclear matter. I compare predictions from microscopic and phenomenological approaches. Constraints on the EOS derived from heavy ion reactions, in particular from subthreshold kaon production, as well as constraints from neutron stars are discussed.

1. Introduction

Heavy ion reactions provide the only possibility to reach nuclear matter densities beyond saturation density $\rho_0 \simeq 0.16 \text{ fm}^{-3}$. Transport calculations indicate that in the low and intermediate energy range $E_{\text{lab}} \sim 0.1 \div 1$ AGeV nuclear densities between $2 \div 3\rho_0$ are accessible while the highest baryon densities ($\sim 8\rho_0$) will probably be reached in the energy range of the future GSI facility FAIR between $20 \div 30$ AGeV. At even higher incident energies transparency sets in and the matter becomes less baryon rich due to the dominance of meson production. The isospin dependence of the nuclear forces which is at present only little constrained by data will be explored by the forthcoming radioactive beam facilities at FAIR/GSI [1], SPIRAL2/GANIL and RIA [2]. Since the knowledge of the nuclear equation-of-state (EOS) at supra-normal densities and extreme isospin is essential for our understanding of the nuclear forces as well as for astrophysical purposes, the determination of the EOS was already one of the primary goals when first relativistic heavy ion beams started to operate in the beginning of the 80ties. In the following I briefly discuss the knowledge about the nuclear EOS at *moderate densities and temperatures*. For more details see e.g. Ref. [3].

Models which make predictions on the nuclear EOS can roughly be divided into three classes: ***phenomenological density functionals*** such as Gogny or Skyrme forces [4–6] and relativistic mean field (RMF) models [7], ***effective field theory*** (EFT) and ***ab initio approaches.*** In EFT a systematic expansion of the EOS in powers of density, respectively the Fermi momentum k_F is performed. EFT can be based on density functional theory [8,9] or e.g. on chiral perturbation theory [10–12]. Ab initio approaches are ased on high precision free space nucleon-nucleon interactions and the nuclear many-body problem is treated microscopically. Predictions for the nuclear EOS are essentially parameter free. Examples are variational calculations [13], Brueckner-Hartree-Fock (BHF) [14,15] or relativistic Dirac-Brueckner-Hartree-Fock (DBHF) [16–18] and Greens functions Monte-Carlo approaches [19,20]. In the follwoing I will mainly concentrate on the DBHF approach.

2. The EOS from ab inito calculations

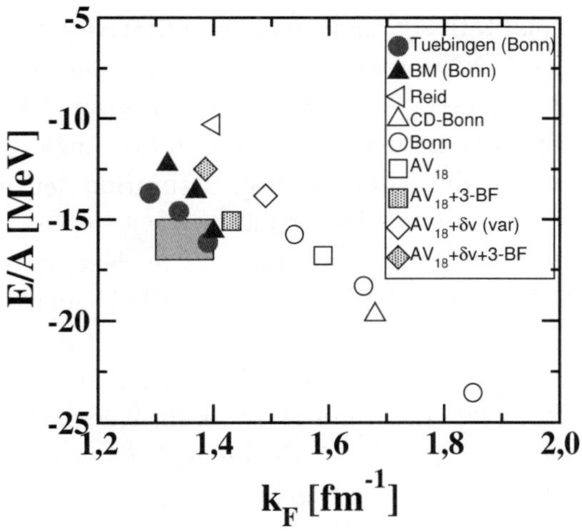

Fig. 1. Nuclear matter saturation points from relativistic (full symbols) and non-relativistic (open symbols) Brueckner-Hartree-Fock calculations based on different nucleon-nucleon forces. The diamonds show results from variational calculations. Shaded symbols denote calculations which include 3-body forces. The shaded area is the empirical region of saturation. Figure is taken from Ref. [21].

In *ab initio* calculations based on many-body techniques one derives the energy functional from first principles, i.e. treating short-range and many-body correlations explicitely. A typical example for a successful many-body approach is Brueckner theory [22]. In the relativistic Brueckner approach the nucleon inside the medium is dressed by the self-energy Σ. The in-medium T-matrix which is obtained from the relativistic Bethe-Salpeter (BS) equation plays the role of an effective two-body interaction which contains all short-range and many-body correlations of the ladder approximation. Solving the BS-equation the Pauli principle is respected and intermediate scattering states are projected out of the Fermi sea. The summation of the T-matrix over the occupied states inside the Fermi sea yields finally the self-energy in Hartree-Fock approximation. This coupled set of equations states a self-consistency problem which has to be solved by iteration.

In contrast to relativistic DBHF calculations which came up in the late 80ties non-relativistic BHF theory has already almost half a century's history. The first numerical calculations for nuclear matter were carried out by Brueckner and Gammel in 1958 [22]. Despite strong efforts invested in the development of improved solution techniques for the Bethe-Goldstone (BG) equation, the non-relativistic counterpart of the BS equation, it turned out that, although such calculations were able to describe the nuclear saturation mechanism qualitatively, they failed quantitatively. Systematic studies for a large number of NN interactions were always allocated on a so-called *Coester-line* in the $E/A - \rho$ plane which does not meet the empirical region of saturation. In particular modern one-boson-exchange (OBE) potentials lead to strong over-binding and too large saturation densities where relativistic calculations do a much better job.

Fig. 1 compares the saturation points of nuclear matter obtained by relativistic Dirac-Brueckner-Hartree-Fock (DBHF) calculations using the Bonn potentials [23] as bare NN interactions to non-relativistic Brueckner-Hartree-Fock calculations for various NN interactions. The DBHF results are taken from Ref. [24] (BM) and more recent calculations based on improved techniques are from Ref. [17] (Tübingen). Several reasons have been discussed in the literature in order to explain the success of the relativistic treatment (see e.g. discussion in Ref. [25]). Three-body forces (3-BFs) have extensively been studied within non-relativistic BHF [14] and variational calculations [13]. Both approaches shwon in Fig. 2 are based on the latest AV_{18} version of the Argonne potential. The variational results shown contain boost corrections (δv) which account for relativistic kinematics and lead to additional repulsion [13].

The contributions from 3-BFs are in total repulsive which makes the EOS harder and non-relativistic calculations come close to their relativistic counterparts. The same effect is observed in variational calculations [13]

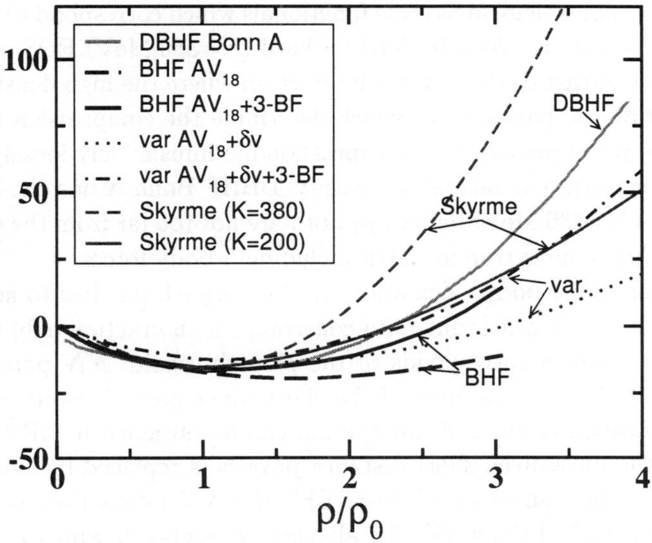

Fig. 2. Predictions for the EOS of symmetric nuclear matter from microscopic ab initio calculations, i.e. relativistic DBHF [17], non-relativistic BHF [14] and variational [13] calculations. For comparison also soft and hard Skyrme forces are shown. Figure is taken from Ref. [21].

shown in Fig. 2. It is often argued that in non-relativistic treatments 3-BFs play in some sense an equivalent role as the dressing of the two-body interaction by in-medium spinors in Dirac phenomenology. Both mechanisms lead indeed to an effective density dependent two-body interaction V which is, however, of different origin. One class of 3-BFs involves virtual excitations of nucleon-antinucleon pairs. Such Z-graphs are in net repulsive and can be considered as a renormalization of the meson vertices and propagators. A second class of 3-BFs is related to the inclusion of explicit resonance degrees of freedom. The most important resonance is the $\Delta(1232)$ isobar which provides at low and intermediate energies large part of the intermediate range attraction.

Fig. 2 compares the equations of state from the different approaches: DBHF from Ref. [17] based the Bonn A interaction[a] [23], BHF [14] and variational calculations [13]. The latter ones are based on the Argonne AV_{18} potential and include 3-body forces. All the approaches use modern high precision NN interactions and represent state of the art calculations.

[a]The high density behavior of the EOS obtained with different interaction, e.g. Bonn B or C is very similar. [17]

Two phenomenological Skyrme functionals which correspond to the limiting cases of a soft (K=200 MeV) and a hard (K=380 MeV) EOS are shown as well. In contrast to the Skyrme interaction where the high density behavior is fixed by the parameteres which determine the compression modulus, in microscopic approaches the compression modulus is only loosely connected to the curvature at saturation density. DBHF Bonn A has e.g. a compressibility of K=230 MeV. Below $3\rho_0$ both are not too far from the soft Skyrme EOS. The same is true for BHF including 3-body forces.

When many-body calculations are performed, one has to keep in mind that elastic NN scattering data constrain the interaction only up to about 400 MeV, which corresponds to the pion threshold. NN potentials differ essentially in the treatment of the short-range part. A model independent representation of the NN interaction can be obtained in EFT approaches where the unresolved short distance physics is replaced by simple contact terms. In the framework of chiral EFT the NN interaction has been computed up to N^3LO [26,27]. An alternative approach which leads to similar results is based on renormalization group (RG) methods [28]. In the $V_{\mathrm{low\ k}}$ approach a low-momentum potential is derived from a given realistic NN potential by integrating out the high-momentum modes using RG methods. When applied to the nuclear many-body problem low momentum interactions do not require a full resummation of the Brueckner ladder diagrams but can already be treated within second-order perturbation theory [29]. However, without repulsive three-body-forces isospin saturated nuclear matter was found to collapse. Including 3-BFs first promising results have been obtained with $V_{\mathrm{low\ k}}$ [29], however, nuclear saturation is not yet described quantitativley.

2.1. *EOS in symmetric and asymmetric nuclear matter*

Fig. 3 compares now the predictions for nuclear and neutron matter from microscopic many-body calculations – DBHF [18] and the 'best' variational calculation with 3-BFs and boost corrections [13] – to phenomenological approaches and to EFT. As typical examples for relativistic functionals we take NL3 [30] as one of the best RMF fits to the nuclear chart and a phenomenological density dependent RMF functional DD-TW from Ref. [31]. ChPT+corr. is based on chiral pion-nucleon dynamics [11] including condensate fields and fine tuning to finite nuclei. As expected the phenomenological functionals agree well at and below saturation density where they are constrained by finite nuclei, but start to deviate substantially at supranormal densities. In neutron matter the situation is even worse since the

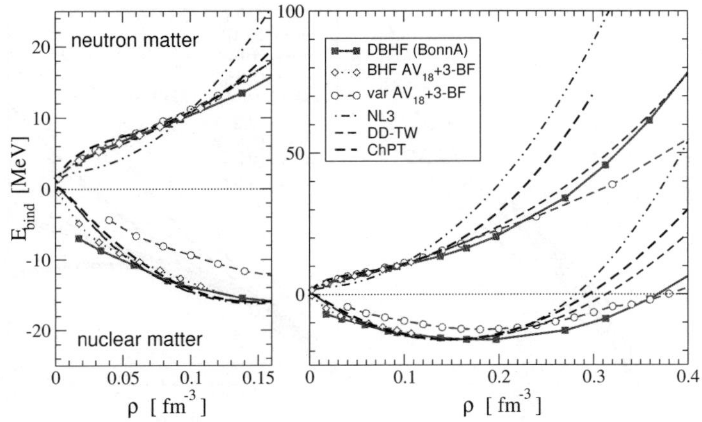

Fig. 3. EOS in nuclear matter and neutron matter. BHF/DBHF and variational calculations are compared to phenomenological density functionals NL3 and DD-TW and ChPT+corr.. The left panel zooms the low density range. The Figure is taken from Ref. [3].

isospin dependence of the phenomenological functionals is less constrained. The predictive power of such density functionals at supra-normal densities is restricted. *Ab initio* calculations predict throughout a soft EOS in the density range relevant for heavy ion reactions at intermediate and low energies, i.e. up to about three times ρ_0. There seems to be no way to obtain an EOS as stiff as the hard Skyrme force shown in Fig. 2 or NL3. Since the nn scattering lenght is large, neutron matter at subnuclear densities is less model dependent. The microscopic calculations (BHF/DBHF, variational) agree well and results are consistent with 'exact' Quantum-Monte-Carlo calculations [20].

In isospin asymmetric matter the binding energy is a functional of the proton and neutron densities, characterized by the asymmetry parameter $\beta = Y_n - Y_p$ which is the difference of the neutron and proton fraction $Y_i = \rho_i/\rho$, $i = n, p$. The isospin dependence of the energy functional can be expanded in terms of β which leads to a parabolic dependence on the asymmetry parameter

$$E(\rho,\beta) = E(\rho) + E_{\rm sym}(\rho)\beta^2 + \mathcal{O}(\beta^4) + \cdots$$
$$E_{\rm sym}(\rho) = \frac{1}{2}\frac{\partial^2 E(\rho,\beta)}{\partial\beta^2}|_{\beta=0} = a_4 + \frac{p_0}{\rho_0^2}(\rho-\rho_0) + \cdots \tag{1}$$

Fig. 4 compares the symmetry energy predicted from the DBHF and variational calculations to that of the empirical density functionals already

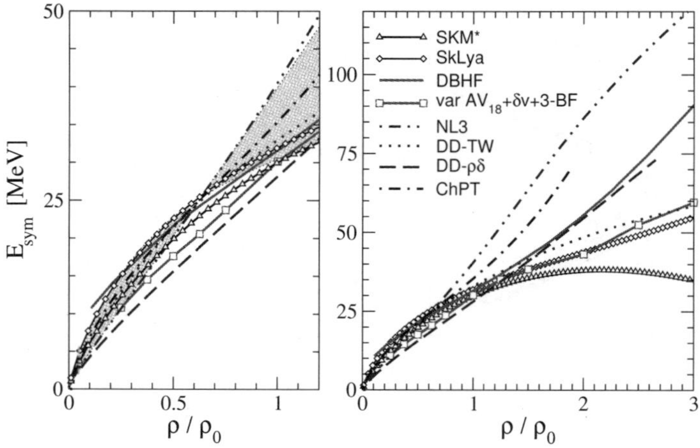

Fig. 4. Symmetry energy as a function of density as predicted by different models. The left panel shows the low density region while the right panel displays the high density range. The Figure is taken from Ref. [3].

shown in Fig. 3 In addition the relativistic DD-$\rho\delta$ RMF functional [32] is included. Two Skyrme functionals, SkM* and the more recent Skyrme-Lyon force SkLya represent non-relativistic models. The left panel zooms the low density region while the right panel shows the high density behavior of $E_{\rm sym}$. Remarkable is that most empirical models coincide around $\rho \simeq 0.6\rho_0$ where $E_{\rm sym} \simeq 24$ MeV. This demonstrates that constraints from finite nuclei are active for an average density slightly above half saturation density. However, the extrapolations to supra-normal densities diverge dramatically. This is crucial since the high density behavior of $E_{\rm sym}$ is essential for the structure and the stability of neutron stars (see also the discussion in Sec. V.5). The microscopic models show a density dependence which can still be considered as *asy-stiff*. DBHF [18] is thereby stiffer than the variational results of Ref. [13]. The density dependence is generally more complex than in RMF theory, in particular at high densities where $E_{\rm sym}$ shows a non-linear and more pronounced increase. Fig. 4 clearly demonstrates the necessity to constrain the symmetry energy at supra-normal densities with the help of heavy ion reactions.

The hatched area in Fig. 4 displays the range of $E_{\rm sym}$ which has been obtained by constructing a density dependent RMF functional varying thereby the linear asymmetry parameter a_4 from 30 to 38 MeV [33]. In Ref. [33] it was concluded that charge radii, in particular the skin thickness $r_n - r_p$ in heavy nuclei constrains the allowed range of a_4 to $32 \div 36$ MeV for relativistic functionals.

2.1.1. *Effective nucleon masses*

The introduction of an effective mass is a common concept to characterize the quasi-particle properties of a particle inside a strongly interacting medium. In nuclear physics exist different definitions of the effective nucleon mass which are often compared and sometimes even mixed up: the non-relativistic effective mass m^*_{NR} and the relativistic Dirac mass m^*_D. These two definitions are based on different physical concepts. The non-relativistic mass parameterizes the momentum dependence of the single-particle potential. The relativistic Dirac mass is defined through the scalar part of the nucleon self-energy in the Dirac field equation which is absorbed into the effective mass $m^*_D = M + \Sigma_S(k, k_F)$. The Dirac mass is a smooth function of the momentum. In contrast, the nonrelativistic effective mass - as a model independent result - shows a narrow enhancement near the Fermi surface due to an enhanced level density [34]. For a recent review on this subject and experimental constraints on m^*_{NR} see Ref. [35].

While the Dirac mass is a genuine relativistic quantity the effective mass m^*_{NR} is determined by the single-particle energy

$$m^*_{NR} = k[dE/dk]^{-1} = \left[\frac{1}{M} + \frac{1}{k}\frac{d}{dk}U\right]^{-1} \quad . \tag{2}$$

m^*_{NR} is a measure of the non-locality of the single-particle potential U (real part) which can be due to non-localities in space, resulting in a momentum dependence, or in time, resulting in an energy dependence. In order to clearly separate both effects, one has to distinguish further between the so-called k-mass and the E-mass [37]. The spatial non-localities of U are mainly generated by exchange Fock terms and the resulting k-mass is a smooth function of the momentum. Non-localities in time are generated by Brueckner ladder correlations due to the scattering to intermediate states which are off-shell. These are mainly short-range correlations which generate a strong momentum dependence with a characteristic enhancement of the E-mass slightly above the Fermi surface [34,37,38]. The effective mass defined by Eq. (2) contains both, non-localities in space and time and is given by the product of k-mass and E-mass [37]. In Fig. 5 the nonrelativistic effective mass and the Dirac mass, both determined from DBHF calculations [36], are shown as a function of momentum k at different Fermi momenta of $k_F = 1.07$, 1.35, 1.7 fm^{-1}. m^*_{NR} shows the typical peak structure as a function of momentum around k_F which is also seen in BHF

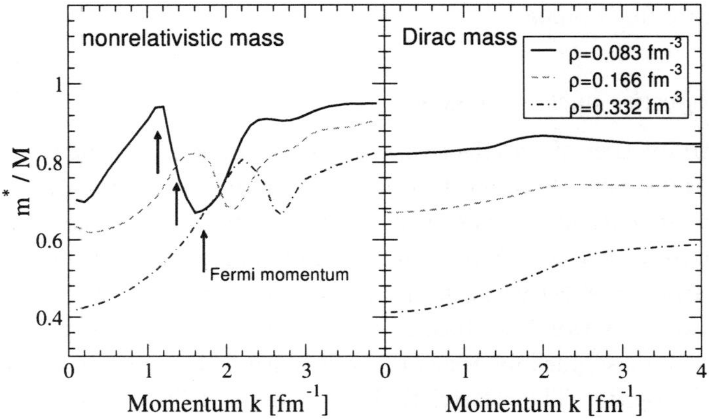

Fig. 5. The effective mass in isospin symmetric nuclear matter as a function of the momentum k at different densities determined from relativistic Brueckner calculations. Figure is taken from Ref. [36].

calculations [38]. The peak reflects the increase of the level density due to the vanishing imaginary part of the optical potential at k_F which is also seen, e.g., in shell model calculations [34,37]. One has, however, to account for correlations beyond mean field or Hartree-Fock in order to reproduce this behavior. Fig. 6 compares the density dependence of the two effective

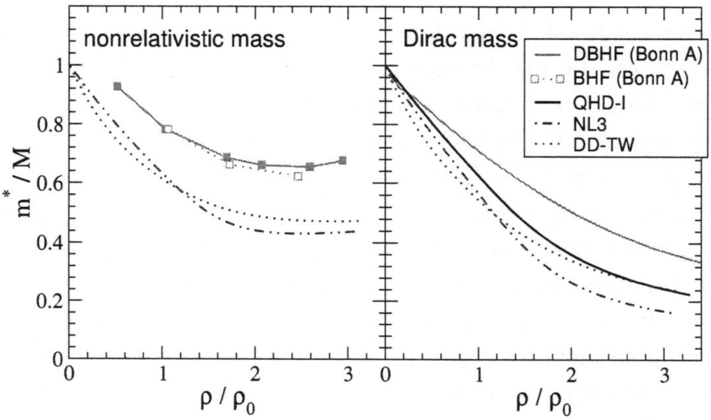

Fig. 6. Nonrelativistic and Dirac effective mass in isospin symmetric nuclear matter as a function of the density for various models.

masses determined at k_F. Both masses decrease with increasing density, the Dirac mass continously, while m^*_{NR} starts to rise again at higher densities.

Phenomenological density functionals (QHD-I, NL3, DD-TW) yield systematically smaller values of m^*_{NR} than the microscopic approaches. This reflects the lack of nonlocal contributions from short-range and many-body correlations in the mean field approaches.

2.1.2. *Proton-neutron mass splitting*

A heavily discussed topic is in the moment the proton-neutron mass splitting in isospin asymmetric nuclear matter. This question is of importance for the forthcoming new generation of radioactive beam facilities which are devoted to the investigation of the isospin dependence of the nuclear forces at its extremes. However, presently the predictions for the isospin dependences differ substantially. BHF calculations [14,38] predict a proton-neutron mass splitting of $m^*_{NR,n} > m^*_{NR,p}$. This stands in contrast to relativistic mean-field (RMF) theory. When only a vector isovector ρ-meson is included Dirac phenomenology predicts equal masses $m^*_{D,n} = m^*_{D,p}$ while the inclusion of the scalar isovector δ-meson, i.e. $\rho+\delta$, leads to $m^*_{D,n} < m^*_{D,p}$ [32]. When the effective mass is derived from RMF theory, it shows the same behavior as the corresponding Dirac mass, namely $m^*_{NR,n} < m^*_{NR,p}$ [32]. Conventional Skyrme forces, e.g. SkM*, lead to $m^*_{NR,n} < m^*_{NR,p}$ [39] while the more recent Skyrme-Lyon interactions (SkLya) predict the same mass splitting as RMF theory. The predictions from relativistic DBHF calculations are in the literature still controversial. They depend strongly on approximation schemes and techniques used to determine the Lorentz and the isovector structure of the nucleon self-energy. Projection techniques are involved but more accurate and yield the same mass splitting as found in RMF theory when the δ -meson is included, i.e. $m^*_{D,n} < m^*_{D,p}$ [18,40]. Recently also the non-relativistic effective mass has been determined with the DBHF approach and here a reversed proton-neutron mass splitting was found, i.e. $m^*_{NR,n} > m^*_{NR,p}$ [36]. Thus DBHF is in agreement with the results from nonrelativistic BHF calculations.

2.1.3. *Optical potentials*

The second important quantity related to the momentum dependence of the mean field is the optical nucleon-nucleus potential. At subnormal densities the optical potential $U_{\rm opt}$ is constraint by proton-nucleus scattering data [41] and at supra-normal densities constraints can be derived from heavy ion reactions, see Refs. [42–44]. In a relativistic framework the optical Schroedinger-equivalent nucleon potential (real part) is defined as

$$U_{\rm opt} = -\Sigma_S + \frac{E}{M}\Sigma_V + \frac{\Sigma_S^2 - \Sigma_V^2}{2M} \ . \tag{3}$$

One should thereby note that in the literature sometimes also an optical potential, given by the difference of the single-particle energies in medium and free space $U = E - \sqrt{M^2 + \mathbf{k}^2}$ is used [42] which should be not mixed up with (3). In a relativistic framework momentum independent fields $\Sigma_{S,V}$ (as e.g. in RMF theory) lead always to a linear energy dependence of $U_{\rm opt}$. As seen from Fig. 7 DBHF reproduces the empirical optical potential [41] extracted from proton-nucleus scattering for nuclear matter at ρ_0 reasonably well up to a laboratory energy of about 0.6-0.8 GeV. However, the saturating behavior at large momenta cannot be reproduced by this calculations because of missing inelasticities, i.e. the excitation of isobar resonances above the pion threshold. When such continuum excitations are accounted for optical model caculations are able to describe nucleon-nucleus scattering data also at higher energies [45]. In heavy ion reactions at incident energies above 1 AGeV such a saturating behavior is required in order to reproduce transverse flow observables [44]. One has then to rely on phenomenological approaches where the strength of the vector potential is artificially suppressed, e.g. by the introduction of additional form factors [44] or by energy dependent terms in the QHD Lagrangian [46] (D^3C model in Fig. 7).

The isospin dependence, expressed by the isovector optical potential $U_{\rm iso} = (U_{\rm opt,n} - U_{\rm opt,p})/(2\beta)$ is much less constrained by data. The knowledge of this quantity is, however, of high importance for the forthcoming radioactive beam experiments. The right panel of Fig. 7 compares the predictions from DBHF [18] and BHF [47] to the phenomenological Gogny and Skyrme (SkM* and SkLya) forces and a relativistic $T-\rho$ approximation [49] based on empirical NN scattering amplitudes [50]. At large momenta DBHF agrees with the tree-level results of Ref. [49]. While the dependence of $U_{\rm iso}$ on the asymmetry parameter β is found to be rather weak [18, 47], the predicted energy and density dependences are quite different, in particular between the microscopic and the phenomenological approaches. The energy dependence of $U_{\rm iso}$ is very little constrained by data. The old analysis of optical potentials of scattering on charge asymmetric targets by Lane [51] is consistent with a decreasing potential as predicted by DBHF/BHF, while more recent analyses based on Dirac phenomenology [52] come to the opposite conclusions. RMF models show a linearly increasing energy dependence of $U_{\rm iso}$ (i.e. quadratic in k) like SkLya, however generally with a smaller

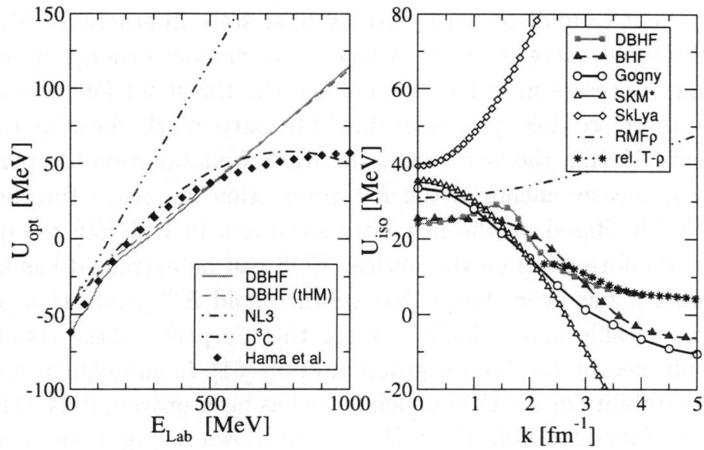

Fig. 7. Nucleon optical potential in nuclear matter at ρ_0. On the left side DBHF calculations for symmetric nuclear matter from [16] and [17] are compared to the phenomenological models NL3 and D^3C [46] and to the p-A scattering analysis of [41]. The right panel compares the iso-vector optical potential from DBHF [18] and BHF [47] to phenomenological RMF [48] , Gogny and Skyrme forces and to a relativistic $T-\rho$ approximation [49].

slope (see discussion in Ref. [32]). To clarify this question certainly more experimental efforts are necessary.

2.2. *Probing the EOS by kaon production in heavy ion reactions*

With the start of the first relativistic heavy ion programs the hope was that particle production would provide a direct experimental access to the nuclear EOS [53]. It was expected that the compressional energy should be released into the creation of new particles, primarily pions, when the matter expands [53]. However, pions have large absorption cross sections and they turned out not to be suitable messengers of the compression phase. They undergo several absorption cycles through nucleon resonances and freeze out at final stages of the reaction and at low densities. Hence pions loose most of their knowledge on the compression phase and are not very sensitive probes for stiffness of the EOS.

After pions turned out to fail as suitable messengers, K^+ mesons were suggested as promising tools to probe the nuclear EOS [54]. At subthreshold energies K^+ mesons are produced in the high density phase and due to the absence of absorption reactions they have a long mean free path and leave the matter undistorted by strong final state interactions. Moreover, at subthreshold energies nucleons have to accumulate energy by multiple scattering processes in order to overcome the threshold for kaon production and therefore these processes should be particularly sensitive to collective effects. Within the last decade the KaoS Collaboration has performed systematic measurements of the K^+ production far below threshold, see Refs. [55–59]. Based on the new data situation, in Ref. [60] the question if valuable information on the nuclear EOS can be extracted has been revisited and it has been shown that subthreshold K^+ production provides indeed a suitable and reliable tool for this purpose. These results have been confirmed by the Nantes group later on [61]. In subsequent investigations the stability of the EOS dependence has been proven, Refs. [21,62,63]. Excitation functions from KaoS [57,59] are shown in Fig. 8 and compared

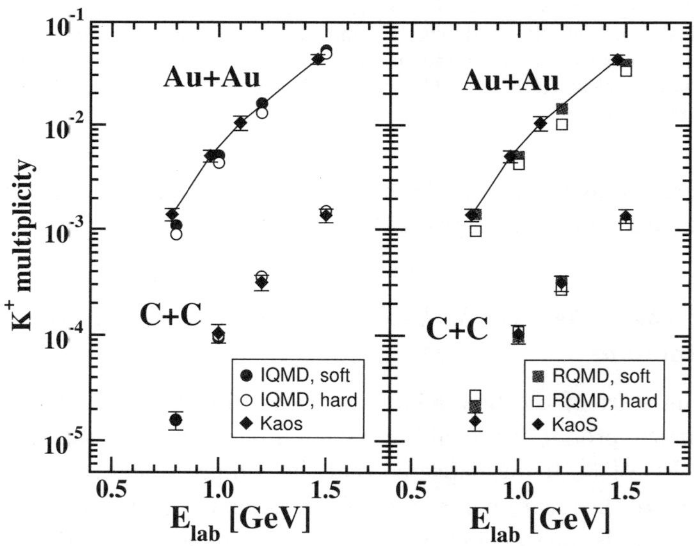

Fig. 8. Excitation function of the K^+ multiplicities in $Au + Au$ and $C + C$ reactions. RQMD [60] and IQMD [63] with in-medium kaon potential and using a hard/soft nuclear EOS are compared to data from the KaoS Collaboration [59].

to RQMD [21,60] and IQMD [63] calculations. As expected the EOS dependence is pronounced in the Au+Au system while the light C+C system serves as a calibration. The effects become even more evident when the ratio R of the kaon multiplicities obtained in Au+Au over C+C reactions (normalised to the corresponding mass numbers) is built [59,60]. Such a ratio has the advantage that possible uncertainties which might still exist in the theoretical calculations should cancel out to large extent. This ratio is shown in Fig. 9. Both, soft and hard EOS, show an increase of R with decreasing energy down to 1.0 AGeV. However, this increase is much less pronounced when the stiff EOS is employed. The comparison to the experimental data from KaoS [59], where the increase of R is even more pronounced, strongly favours a soft equation of state. Fig. 9 demonstrates also

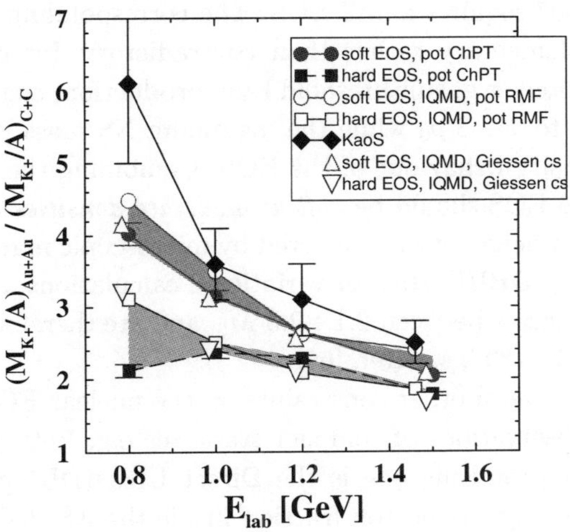

Fig. 9. Excitation function of the ratio R of K^+ multiplicities obtained in inclusive Au+Au over C+C reactions. RQMD [60] and IQMD [63] calculations are compared to KaoS data [59]. Figure is taken from [21].

the robustness of this observable. Exploring the range of uncertainty in the corresponding transport calculations the stability of the conclusions drawn from this observable has been demonstrated in Ref. [63]. This concerns elementary input, in particular the elementary production cross sections $N\Delta; \Delta\Delta \mapsto NYK^+$ which are not constrained by data.

2.3. *Constraints from neutron stars*

Measurements of "extreme" values, like large masses or radii, huge luminosities etc. as provided by compact stars offer good opportunities to gain deeper insight into the physics of matter under extreme conditions. There has been substantial progress in recent time from the astrophysical side.

The most spectacular observation was probably the recent measurement [64] on PSR J0751+1807, a millisecond pulsar in a binary system with a helium white dwarf secondary, which implies a pulsar mass of $2.1 \pm 0.2 \, (^{+0.4}_{-0.5})$ $M_\odot$ with 1σ (2σ) confidence. Therefore, a reliable EOS has to describe neutron star (NS) masses of at least 1.9 $M_\odot$ (1σ) in a strong, or 1.6 $M_\odot$ (2σ) in a weak interpretation. This condition limits the softness of EOS in NS matter. One might therefore be worried about an apparent contradiction between the constraints derived from neutron stars and those from heavy ion reactions. While heavy ion reactions favor a soft EOS, PSR J0751+1807 requires a stiff EOS. The corresponding constraints are, however, complementary rather than contradictory. Intermediate energy heavy-ion reactions, e.g. subthreshold kaon production, constrains the EOS at densities up to $2 \div 3$ ρ_0 while the maximum NS mass is more sensitive to the high density behaviour of the EOS. Combining the two constraints implies that the EOS should be *soft at moderate densities and stiff at high densities.* Such a behaviour is predicted by microscopic many-body calculations (see Fig. 2). DBHF, BHF or variational calculations, typically, lead to maximum NS masses between $2.1 \div 2.3$ $M_\odot$ and are therefore in accordance with PSR J0751+1807, see Ref. [65].

There exist several other constraints on the nuclear EOS which can be derived from observations of compact stars, see e.g. Refs. [65–67]. Among these, the most promising one is the Direct Urca (DU) process which is essentially driven by the proton fraction inside the NS [68]. DU processes, e.g. the neutron β-decay $n \to p + e^- + \bar{\nu}_e$, are very efficient regarding their neutrino production, even in superfluid NM [69,70], and cool NSs too fast to be in accordance with data from thermally observable NSs. Therefore, one can suppose that no DU processes should occur below the upper mass limit for "typical" NSs, i.e. $M_{DU} \geq 1.5$ $M_\odot$ (1.35 $M_\odot$ in a weak interpretation). These limits come from a population synthesis of young, nearby NSs [71] and masses of NS binaries [64].

3. Summary

The status of theoretical models which make predictions for the EOS can roughly be summarized as follows: phenomenological density functionals such as Skyrme, Gogny or relativistic mean field models provide high precision fits to the nuclear chart but extrapolations to supra-normal densities or the limits of stability are highly uncertain. A more controlled way provide effective field theory approaches which became quite popular in recent time. Effective chiral field theory allows e.g. a systematic generation of two- and many-body nuclear forces. However, these approaches are low momentum expansions and when applied to the nuclear many-body problem, low density expansions. Ab initio calculations for the nuclear many-body problem such as variational or Brueckner calculations have reached a high degree of sophistication and can serve as guidelines for the extrapolation to the regimes of high density and/or large isospin asymmetry. Possible future devellopments are to base such calculations on modern EFT potentials and to achieve a more consistent treatment of two- and three-body forces.

If one intends to constrain these models by nuclear reactions one has to account for the reaction dynamics by semi-classical transport models of a Boltzmann or molecular dynamics type. Suitable observables which have been found to be sensitive on the nuclear EOS are directed and elliptic collective flow pattern and particle production, in particular kaon production, at higher energies. Heavy ion data suggest that the EOS of symmetric nuclear matter shows a soft behavior in the density regime between one to about three times nuclear saturation density, which is consistent with the predictions from many-body calculations. Conclusions on the EOS are, however, complicated by the interplay between the density and the momentum dependence of the nuclear mean field. Data which constrain the isospin dependence of the mean field are still scare. Promising observables are isospin diffusion, iso-scaling of intermediate mass fragments and particle ratios (π^+/π^- and eventually K^+/K^0 [72]). Here the situation will certainly improve when the forthcoming radioactive beam facilities will be operating.

References

1. GSI Conceptual Design Report, http://www.gsi.de/GSI-Future
2. RIA hoempage, http://www.orau.org/ria
3. C. Fuchs, H.H. Wolter, [nucl-th/0511070] Euro. Phys. J. A in press.
4. M. Kleban, B. Nerlo-Pomorska, J. F. Berger, J. Decharge, M. Girod, S. Hilaire, Phys. Rev. C **65**, 024309 (2002).
5. B. Cochet, K. Bennaceur, J. Meyer, P. Bonche, T. Duguet, Int. J. Mod. Phys. **E13**, 187 (2004).
6. P.-G. Reinhard, M. Bender, Lect. Notes Phys. **641**, 249 (2004).
7. P. Ring, Prog. Part. Nucl. Phys. **73**, 193 (1996); Lect. Notes Phys. **641**, 175 (2004).
8. B.D. Serot, J.D. Walecka, Int. J. Mod. Phys. **E6**, 515 (1997).
9. R.J. Furnstahl, Lect. Notes Phys. **641**, 1 (2004).
10. M. Lutz, B. Friman, Ch. Appel, Phys. Lett. **B474**, 7 (2000).
11. P. Finelli, N. Kaiser, D. Vretenar, W. Weise, Eur. Phys. J. **A17**, 573 (2003); Nucl. Phys. **A735**, 449 (2004).

12. D. Vretenar, W. Weise, Lect. Notes Phys. **641**, 65 (2004).
13. A. Akmal, V.R. Pandharipande, D.G. Ravenhall, Phys. Rev. C **58**, 1804 (1998).
14. W. Zuo, A. Lejeune, U. Lombardo, J.F. Mathiot, Nucl. Phys. **A706**, 418 (2002).
15. X.R. Zhou, G.F. Burgio, U. Lombardo, H.-J. Schulze, W. Zuo, Phys. Rev. C **69**, 018801 (2004).
16. B. ter Haar, R. Malfliet, Phys. Rep. **149**, 207 (1987).
17. T. Gross-Boelting, C. Fuchs, A. Faessler, Nucl. Phys. **A648**, 105 (1999).
18. E. van Dalen, C. Fuchs, A. Faessler, Nucl. Phys. **A744**, 227 (2004); Phys. Rev. C **72**, 065803 (2005).
19. W.H. Dickhoff, C. Barbieri, Prog. Part. Nucl. Phys. **52**, 377 (2004).
20. J. Carlson, J. Morales, V.R. Pandharipande, D.G. Ravenhall, Phys. Rev. C **68**, 025802 (2003).
21. C. Fuchs, Prog. Part. Nucl. Phys. **56**, 1 (2006).
22. K.A. Brueckner, J.L. Gammel, Phys. Rev. **107**, 1023 (1958)
23. R. Machleidt, K. Holinde, Ch. Elster, Phys. Rep. **149**, 1 (1987)
24. R. Brockmann, R. Machleidt, Phys. Rev. C **42**, 1965 (1990).
25. C. Fuchs, Lect. Notes Phys. **641**, 119 (2004).
26. D.R. Entem, R. Machleidt, Phys. Rev. C **68**, 041001 (2003).
27. E. Epelbaum, W. Glöckle, U.-G. Meissner, Nucl. Phys. **A747**, 362 (2005).
28. S.K. Bogner, T.T.S. Kuo, A. Schwenk, Phys. Rep. **386**, 1 (2003).
29. S.K. Bogner, A. Schwenk, R.J. Furnstahl, A. Nogga, Nucl. Phys. **A763**, 59 (2005).
30. G.A. Lalazissis, J. König, P. Ring, Phys. Rev. C **55**, 540 (1997).
31. S. Typel, H.H. Wolter, Nucl. Phys. **A656**, 331 (1999).
32. V. Baran, M. Colonna, V. Greco, M. Di Toro, Phys. Rep. **410**, 335 (2005).
33. T. Nikšić, D. Vretenar, P. Ring, Phys. Rev. C **66**, 064302 (2002).
34. C. Mahaux, P.F. Bortignon, R.A. Broglia, C.H. Dasso, Phys. Rep. **120**, 1 (1985).
35. D. Lunney, J.M. Pearson, C. Thibault, Rev. Mod. Phys. **75**, 1021 (2003).
36. E. van Dalen, C. Fuchs, A. Faessler, Phys. Rev. Lett. **95**, 022302 (2005).
37. M. Jaminon, C. Mahaux, Phys. Rev. C **40**, 354 (1989).
38. T. Frick, Kh. Gad, H. Müther, P. Czerski, Phys. Rev. C **65**, 034321 (2002).
39. J.M. Pearson, S, Goriely, Phys. Rev. C **64**, 027301 (2001).
40. F. de Jong, H. Lenske, Phys. Rev. C **58**, 890 (1998).
41. S. Hama *et al.*, Phys, Rev. C **41**, 2737 (1990); E.D. Cooper *et al.*, Phys. Rev. C **47**, 297 (1993).
42. P. Danielewicz, Nucl. Phys. **A673**, 275 (2000).
43. T. Gaitanos, C. Fuchs, H.H. Wolter, A. Faessler, Eur. Phys. J. **A12**, 421 (2001).
44. P.K. Sahu, W. Cassing, U. Mosel, A. Ohnishi, Nucl. Phys. **A672**, 376 (2000).
45. H.F. Arellano, H.V. von Geramb, Phys. Rev. C **66**, 024602 (2002).
46. S. Typel, Phys. Rev. C **71**, 064301 (2005).
47. W. Zuo, L.G. Cao, B.A. Li, U. Lombardo, C.W. Shen, Phys. Rev. C **72**, 014005 (2005).

48. T. Gaitanos, M. Di Toro, S. Typel, V. Baran, C. Fuchs, V. Greco, H.H. Wolter Nucl. Phys. **A732**, 24 (2004).
49. L.-W. Chen, C.M. Ko, B.-A. Li, Phys. Rev. C **72**, 064606 (2005).
50. J.A. McNeil, J.R. Shepard, S.J. Wallace, Phys. Rev. C **27**, 2123 (1983).
51. A.M. Lane, Nucl. Phys. **35**, 676 (1962).
52. R. Kozack, D.G. Madland, Phys. Rev. C **39**, 1461 (1989); Nucl. Phys. **A509**, 664 (1990).
53. R. Stock, Phys. Rep. **135**, 259 (1986).
54. J. Aichelin and C.M. Ko, Phys. Rev. Lett. **55**, 2661 (1985).
55. D. Miskowiec et al. [KaoS Collaboration], Phys. Rev. Lett. **72**, 3650 (1994).
56. R. Barth et al. [KaoS Collaboration], Phys. Rev. Lett. **78**, 4007 (1997).
57. F. Laue et al. [KaoS Collaboration], Phys. Rev. Lett. **82**, 1640 (1999).
58. F. Laue et al. [KaoS Collaboration], Eur. Phys. J. **A9**, 397 (2000).
59. C. Sturm et al. [KaoS Collaboration], Phys. Rev. Lett. **86**, 39 (2001).
60. C. Fuchs, Amand Faessler, E. Zabrodin, Y.M. Zheng, Phys. Rev. Lett. **86**, 1974 (2001).
61. Ch. Hartnack, J. Aichelin, J. Phys. G **28**, 1649 (2002).
62. C. Fuchs, A. Faessler, S. El-Basaouny, E. Zabrodin, J. Phys. G **28**, 1615 (2002).
63. Ch. Hartnack, H. Oeschler, J. Aichelin, Phys. Rev. Lett. **96**, 012302 (2006).
64. D.J. Nice, E.M. Splaver, I.H. Stairs, O. Löhmer, A. Jessner, M. Kramer, and J.M. Cordes, Astrophys. J. **634**, 1242 (2005).
65. T. Kähn et al., Phys. Rev. C **74**, 035802 (2006).
66. A.W. Steiner, M. Prakash, J.M. Lattimer, P.J. Ellis, Phys. Rep. **411**, 325 (2005).
67. B.-A. Li, A.W. Steiner, [nucl-th/0511064].
68. J.M. Lattimer, C.J. Pethick, M. Prakash, and P. Haensel, Phys. Rev. Lett. **66**, 2701 (1991).
69. D. Blaschke, H. Grigorian, and D. Voskresensky, Astron. Astrophys. **424**, 979 (2004).
70. E.E. Kolomeitsev, and D.N. Voskresensky, Nucl.Phys. A **759**, 373 (2005).
71. S. Popov, H. Grigorian, R. Turolla and D. Blaschke, Astron. Astrophys. **448**, 327 (2006).
72. G. Ferini, T. Gaitanos, M. Colonna, M. Di Toro, H.H. Wolter, [nucl-th/0607005].

III. ALPHA DECAY, NUCLEAR REACTIONS, COLD FISSION AND NUCLEAR FUSION

NUCLEAR MOLECULAR STRUCTURE

G. G. ADAMIAN[1], N. V. ANTONENKO[1,2], Z. GAGYI-PALFFY[2], S. P. IVANOVA[1], R. V. JOLOS[1], YU. V. PALCHIKOV[1], W. SCHEID[2], T. M. SHNEIDMAN[1], A. S. ZUBOV[1]

[1]*Joint Institute for Nuclear Research, 141980 Dubna (Moscow Region), Russia*
[2]*Institut für Theoretische Physik der Universität, 35392 Giessen, Germany*
werner.scheid@theo.physik.uni-giessen.de

The concept of a nuclear molecule or a dinuclear system assumes two touching nuclei which carry out motion in the internuclear distance and exchange nucleons by transfer. The dinuclear model can be applied to nuclear structure, to fusion reactions leading to superheavy nuclei and to multi-nucleon transfer reactions.

Keywords: Dinuclear model, mass asymmetry motion, parity splitting, rotation modes, super- and hyperdeformed nuclei, superheavy nuclei, fusion, quasifission, nucleon transfer, master equations.

1. Introduction

Nuclear molecular structures were for the first time observed by Bromley, Kuehner and Almqvist [1] in the scattering of ^{12}C on ^{12}C and then seen up to the system Ni + Ni [2]. A nuclear molecule or a dinuclear system (DNS) as named by V. V. Volkov [3] is a configuration of two touching nuclei (clusters) which keep their individuality. Such a system has two main degrees of freedom which govern its dynamics: (i) the relative motion between the nuclei describing molecular resonances in the internuclear potential and the decay of the dinuclear system which is called quasifission and (ii) the transfer of nucleons between the nuclei leading to a dependence of the dynamics on the mass and charge asymmetries in fusion and fission reactions. The latter processes are described by the mass and charge asymmetry coordinates

$$\eta = \frac{A_1 - A_2}{A_1 + A_2} \quad \text{and} \quad \eta_Z = \frac{Z_1 - Z_2}{Z_1 + Z_2}. \tag{1}$$

These coordinates can be assumed as continuous or discrete quantities. For $\eta = \eta_Z = 0$ we have a symmetric clusterization with two equal nuclei, and if η approaches the values ± 1 or if A_1 or A_2 is equal to zero, a fused system has been formed. The importance of these coordinates for nuclear reactions was pointed out by V. V. Volkov [3] in his dinuclear system concept.

In this article we consider some aspects of dinuclear configurations playing a role in nuclear structure and reactions. We give a short and concise review on several applications of the DNS model, namely we discuss normal- and superdeformed bands, hyperdeformed states, the fusion dynamics in producing superheavy nuclei, multi-nucleon transfer between nuclei, and the decay of the dinuclear system, i. e. the quasifission where no compound nucleus is formed.

2. The dinuclear system model

The dinuclear configuration describes quadrupole- and octupole-like deformations related with normal, super- and hyperdeformed states. To demonstrate the deformation of the dinuclear configuration, we calculated [4] the mass and charge multipole moments of a nucleus described by a dinuclear configuration with a (mass and charge) density $\varrho(\mathbf{r}) = \varrho_1(\mathbf{r}_1)+\varrho_2(\mathbf{r}_2)$, where ϱ_i $(i = 1, 2)$ is the density of the individual nucleus i. These moments are compared with those of an axially deformed nucleus by use of a shape expansion with multipole deformation parameters $\beta_\lambda = \beta_0, \beta_1, \beta_2, \beta_3 ...$ Then one obtains these parameters β_λ as functions of η or η_Z. For spherical clusters they are nearly independent of A. Realistic clusters yield a specific dependence on the surface thickness, the radius parameters and their deformations [4].

The dinuclear system model can be applied in the range of $\eta = 0 - 0.3$ to hyperdeformed (HD) states (nuclei with large quadrupole deformations), in the range of $\eta = 0.6 - 0.9$ to superdeformed (SD) states (similar quadrupole and octupole deformations) and around $\eta \approx 1$ to the parity splitting of bands (linear increase of deformations). As example let us discuss the ^{152}Dy system [4]. The potential energy of the DNS as a function of η shows significant minima for $\eta = 0.34$ (^{50}Ti + ^{102}Ru), $\eta = 0.66$ (^{26}Mg + ^{126}Xe) and $\eta = 0.71$ (^{22}Ne + ^{130}Ba). The DNS ^{50}Ti + ^{102}Ru is compatible with HD properties, the dinuclear systems ^{26}Mg + ^{126}Xe and ^{22}Ne + ^{130}Ba have SD properties. For ^{26}Mg + ^{126}Xe we calculated a moment of inertia of J=104 $\hbar^2$/MeV and an electric quadrupole moment of Q_2=24 eb in comparison with the experimental values of SD states with J=(85±3) $\hbar^2$/MeV and Q_2=(18±3) eb.

3. Normal- and superdeformed bands

The DNS model can be used to describe the normaldeformed (ND) and superdeformed (SD) bands of various nuclei. We applied this model to the structure of ^{60}Zn [5] and of ^{190,192,194}Hg and ^{192,194,196}Pb [6].

The ^{60}Zn nucleus has a threshold of 2.7 MeV above the ground state for its decay into ^{56}Ni + α. Therefore, we can assume that the ground state band contains an α-component. Further thresholds are positioned at 10.8 and 11.2 MeV above the ground state for the decays into ^{52}Fe + ^{8}Be and ^{48}Cr + ^{12}C, respectively. The extrapolated band head of the SD band has an energy of 7.5 MeV and a moment of inertia of (692-795)M fm^2 in comparison with the moment of inertia of the ^{52}Fe + ^{8}Be system of 750 M fm^2. Hence, the SD band of ^{60}Zn contains a prominent ^{8}Be component.

The observed strong collective dipole transitions between the excited SD band and the lowest-energy SD band in ^{150}Gd, ^{152}Dy, ^{190,194}Hg, ^{196,198}Pb and between the SD and ND bands in ^{194}Hg and ^{194}Pb indicate a decay out of pronounced octupole deformed states. The measured properties of the excited SD bands in ^{152}Dy and ^{190,192,194}Hg have been interpreted in terms of rotational bands built on collective octupole vibrations [7].

Configurations with large quadrupole and octupole deformation parameters and low-lying collective negative parity states are strongly related to a clustering describable with heavy and light clusters within the DNS model. So the above mentioned ND and SD bands can be consistently treated by assuming a collective dynamics in the mass or charge asymmetry coordinate η or η_Z, respectively. To achieve this aim, we formulate a conventional collective Schrödinger equation in η_Z (or η):

$$\left(-\frac{\hbar^2}{2}\frac{d}{d\eta_Z}\frac{1}{B_{\eta_Z}}\frac{d}{d\eta_Z}+U(\eta_Z,I)\right)\psi_n(\eta_Z,I)=E_n(I)\psi_n(\eta_Z,I). \qquad (2)$$

In Fig. 1 we show the calculated potential (histogram) U of ^{194}Hg as a function of the charge number Z_2 of the lighter cluster for two nuclear spins $I = 0$ and 10. The potentials have minima for α-type clusterizations, namely for $Z_2 = 2, 4, 6, 8...$ In addition Fig. 1 presents the probability $|\psi_n(\eta_Z, I)|^2$ expressed with the intrinsic wave functions of the ND and SD states. This probability is peaked around the minima of the potential indicating a corresponding cluster structure of the states. In Fig. 2 we show as example the calculated level spectra of ^{194}Hg and ^{194}Pb in comparison with the experimental data. We note that the shift of the negative parity states is reproducible with the dynamics in η_Z and is related to the properties of the octupole degree of freedom. Also electromagnetic transition probabilities can be evaluated [5,6] with the intrinsic wave functions which agree well with the experimental ones.

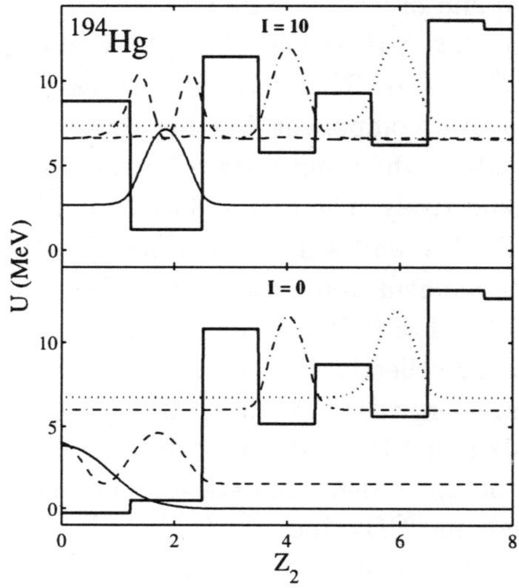

Fig. 1. Potential energy (histogram) U of ^{194}Hg for the spins $I = 0$ and 10. The curves are the absolute squares of the wave functions of the ground (solid) and first excited (dashed) ND bands and ground (dashed-dotted) and first excited (dotted) SD bands.

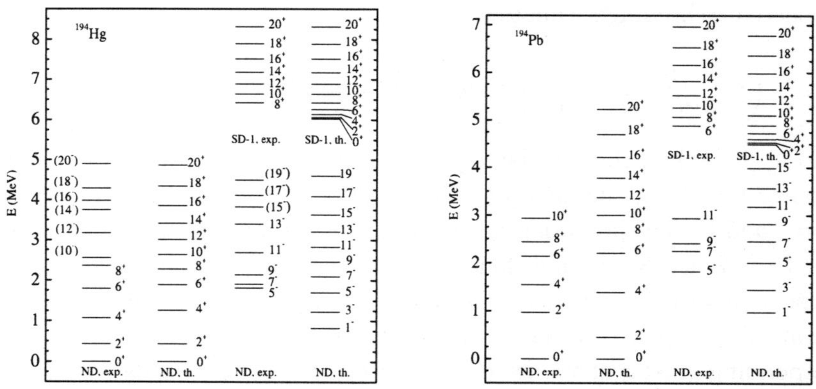

Fig. 2. Calculated and experimental levels of the ground state and superdeformed bands of ^{194}Hg (l.h.s.) and ^{194}Pb (r.h.s).

4. Hyperdeformed states in heavy ion collisions

The question arises whether heavier nuclei have excited states with the properties of molecular (or cluster) states. Such states could be the hyperdeformed (HD) states which are usually explained by nuclear shapes with a ratio of axes of 1 : 3 caused by a third minimum in the potential energy surfaces (PES) of the corresponding nuclei. An interesting observation in shell model calculations was made [8] that the third minimum of the PES of actinide nuclei belongs to a molecular configuration of two touching nuclei (clusters) which is a dinuclear configuration. We showed that dinuclear systems have quadrupole moments and moments of inertia as those measured for superdeformed states and estimated for HD states [4].

Under the assumption that hyperdeformed states can be considered as quasimolecular states, it should be possible to excite them by forming a hyperdeformed configuration in the scattering of heavy ions. In the following we discuss the systems ^{48}Ca + ^{140}Ce and ^{90}Zr + ^{90}Zr as possible candidates for exploring the properties of hyperdeformed states [9]. First, we calculated the potentials $V(R, \eta, L)$ as a function of the relative distance for various angular momenta. These potentials are shown in Fig. 3. They have a minimum around 11 fm at a distance $R_m \approx R_1 + R_2 + 0.5$ fm where R_1 and R_2 are the radii of the nuclei. The depth of this molecular minimum decreases with growing angular momentum and vanishes for $L > 100$ in the considered systems.

The potential pocket has virtual and quasibound molecular resonance states situated above and below the barrier, respectively. The nuclei stay in the potential minimum without changing the mass and charge asymmetries if they are spherical and stiff (magic and double magic nuclei). Approximating the potential in the neighborhood of the minimum by a harmonic oscillator potential, we can estimate the positions of one to three quasibound states with an energy spacing of $\hbar\omega \approx 2.2$ MeV for $L > 40$. For example, in the ^{90}Zr + ^{90}Zr system we find the lowest quasibound state for $L = 50$ lying 1.1 MeV above the potential minimum.

The charge quadrupole moments of (40-50)$\cdot 10^2$ e fm^2 and the moments of inertia of (160-190) $\hbar^2$/MeV of the quasibound dinuclear configurations ^{48}Ca + ^{140}Ce and ^{90}Zr + ^{90}Zr are close to those estimated for hyperdeformed states. Therefore, we can assume that the quasibound states are HD states and propose to produce these states in heavy ion reactions of ^{48}Ca on ^{140}Ce and ^{90}Zr on ^{90}Zr.

The cross section for penetrating the barrier and populating quasibound

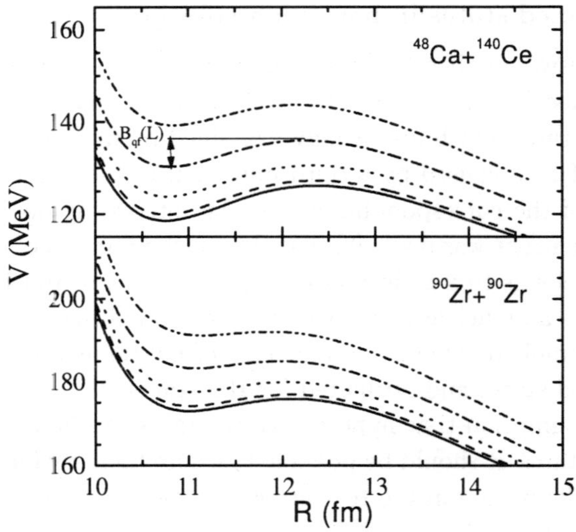

Fig. 3. The potential $V(R, L)$ for the systems ^{48}Ca + ^{140}Ce (upper part) and ^{90}Zr + ^{90}Zr (lower part) as a function of R for $L = 0, 20, 40, 60, 80$ presented by solid, dashed, dotted, dashed-dotted and dashed-dotted-dotted curves, respectively.

states can be written as

$$\sigma(E_{c.m.}) = \frac{\pi\hbar^2}{2\mu E_{c.m.}} \sum_{L=L_{min}}^{L_{max}} (2L+1)T_L(E_{c.m.}). \tag{3}$$

Here, $E_{c.m.}$ is the incident energy in the center of mass system and $T_L(E_{c.m.})$ the transmission probability through the entrance barrier at R_b which is approximated by a parabola. The angular momentum quantum numbers L_{min} and L_{max} in Eq. (3) fix the interval of angular momenta contributing to the excitation of HD states. The range of partial waves leading to the excitation of quasibound states constitutes the so called molecular window known in the theory of nuclear molecules with light heavy ions.

In the reaction ^{48}Ca on ^{140}Ce, cold and long living DNS states can be formed at an incident energy $E_{c.m} = 147$ MeV and $90 < L < 100$, and in the reaction ^{90}Zr on ^{90}Zr at $E_{c.m} = 180$ MeV and $40 < L < 50$. For both reactions we estimate a cross section (3) of about 1 μb. Also other reactions, namely ^{48}Ca + ^{144}Sm ($E_{c.m} = 149$ MeV, $80 < L < 90$), ^{48}Ca + ^{142}Nd ($E_{c.m} = 148$ MeV, $80 < L < 90$), and ^{38}Ar + ^{140}Ce, ^{142}Nd, ^{144}Sm ($E_{c.m} = 137, 141$ and 145 MeV, respectively, $80 < L < 90$) can be thought to be applied for a possible observation of cluster-type HD states.

The spectroscopic investigation of the HD structures is difficult because of the small formation cross section and the high background produced by fusion-fission, quasifission and other reactions. However, the latter processes have characteristic times much shorter than the life-time of the HD states which is of the order of 10^{-16} s. Therefore, the HD states should show up as sharp resonance lines as a function of the incident energy.

5. Rotation modes in the dinuclear model

The clusters of the dinuclear system model are assumed to be deformed and to rotate. In this case we have to discriminate three different coordinate systems: (i) the space-fixed system with its origin at the center of mass, (ii) the molecular system where the z'-axis is defined by the direction of the internuclear distance $\boldsymbol{R}$, (iii) body-fixed systems of clusters with axes x_1'', y_1'', z_1'' and x_2'', y_2'', z_2'' which are the principal axes for the tensors of the moments of inertia of clusters 1 and 2. If we include the mass asymmetry motion and β- and γ-vibrations, we can write the general Hamiltonian as follows:

$$
\begin{aligned}
H = & \frac{p_R^2}{2\mu_{RR}} + \frac{\boldsymbol{L}^2 - L_{z'}^2}{2\mu_{RR}R^2} + \frac{p_\eta^2}{2B_{\eta\eta}} + \sum_{i=1}^{3} \frac{I_i^{(1)2}}{2\Im_i^{(1)}} + \sum_{i=1}^{3} \frac{I_i^{(2)2}}{2\Im_i^{(2)}} \\
& + T_{vib}(\beta_1, \gamma_1) + T_{vib}(\beta_2, \gamma_2) + \text{kinetic coupling terms} \\
& + U(R, \eta, \text{angles}, \beta_1, \gamma_1, \beta_2, \gamma_2). \qquad (4)
\end{aligned}
$$

Here, $\boldsymbol{L}$ is the angular momentum of the molecular system and $I_{i=1,2,3}^{(1)\text{ or }(2)}$ are the angular momentum components for the rotation of the clusters 1 or 2, respectively, with respect to the axes of the body-fixed systems of the clusters. The angular momentum $\boldsymbol{L}$ of the molecular system is connected with the reduced mass μ_{RR} of the internuclear motion. The total angular momentum of the system is given as

$$
\boldsymbol{J} = \boldsymbol{L} + \boldsymbol{I}^{(1)} + \boldsymbol{I}^{(2)}. \qquad (5)
$$

The rotation energy can be transformed to the molecular coordinate system

$$
\begin{aligned}
T_{rot} = & \frac{1}{2\mu_{RR}R^2} \left((J_{x'} - I_{x'}^{(1)} - I_{x'}^{(2)})^2 + (J_{y'} - I_{y'}^{(1)} - I_{y'}^{(2)})^2 \right) \\
& + \sum_{i=1}^{3} \frac{I_i^{(1)2}}{2\Im_i^{(1)}} + \sum_{i=1}^{3} \frac{I_i^{(2)2}}{2\Im_i^{(2)}}, \qquad (6)
\end{aligned}
$$

where $I_{x',y',z'}^{(1)\text{ or }(2)}$ are the components of the angular momenta $\boldsymbol{I}^{(1)}$ and $\boldsymbol{I}^{(2)}$ of the clusters and are related to $I_{i=1,2,3}^{(1)\text{ or }(2)}$ by an orthogonal transformation, depending on the angles of the body-fixed systems with respect to the molecular system:

$$I^{(j)}_{x'(y',z')} = \sum_{i=1}^{3} R^{(j)}_{x'(y',z'),i} I^{(j)}_i \quad \text{with} \quad j = 1 \text{ or } 2. \tag{7}$$

The terms containing the products of $J_{x',y'}$ and $I^{(1) \text{ or } (2)}_{x',y'}$ in T_{rot} constitute the so called Coriolis interaction which is strongly contributing in molecular systems and can be partly approximated by introducing the total moment of inertia of the system.

The Hamiltonian (4) without the mass asymmetry motion and the γ-vibrations was applied to resonances observed in the scattering of ^{24}Mg on ^{24}Mg [10,11]. These resonances have widths of about 200 keV and angular momenta of 36-42 $\hbar$ at incident energies $E_{c.m.} = 42 - 56$ MeV [12]. They can be explained by molecular states in pole-to-pole-like configurations of the prolately deformed ^{24}Mg nuclei.

6. Rotation modes and the structure of ^{238}U

In this Section we apply the Hamiltonian of the dinuclear model to the case of large mass asymmetries and describe low-lying bands in ^{238}U. We assume that the clusterization can vary and consists of a heavy cluster with an axially symmetric quadrupole deformation β and of a spherical light cluster, e.g. an alpha-particle. The internuclear distance coordinate $R = R_m$ is chosen as fixed at the touching configuration which is determined by the minimum of the potential of the DNS in the internuclear coordinate R. Then the degrees of freedom are: (i) rotation of the heavy cluster about an axis perpendicular to its symmetry axis, where the latter axis is fixed by the angles ϑ_1 and φ_1 in the space-fixed system, (ii) rotation of the molecular system, defined by the direction of $\boldsymbol{R}$ with the angles ϑ_2 and φ_2 in the space-fixed system, (iii) mass asymmetry motion described by a new mass asymmetry coordinate with positive values only:

$$\xi = 2A_2/A = 1 - \eta. \tag{8}$$

The total Hamiltonian is assumed as

$$H = \frac{J_1^2}{2\Im_h} + \frac{J_2^2}{2\mu_{RR}(\xi)R_m^2} - \frac{\hbar^2}{2B}\frac{1}{\xi}\frac{\partial}{\partial\xi}\xi\frac{\partial}{\partial\xi} + U(\xi,\varepsilon), \tag{9}$$

where

$$J_i^2 = -\hbar^2\left(\frac{1}{\sin\vartheta_i}\frac{\partial}{\partial\vartheta_i}\sin\vartheta_i\frac{\partial}{\partial\vartheta_i} + \frac{1}{\sin^2\vartheta_i}\frac{\partial^2}{\partial\varphi_i^2}\right) \quad \text{with} \quad i = 1,2. \tag{10}$$

Here, $\Im_h$ is the moment of inertia of the heavy cluster fixed from a comparison with the energy of the lowest experimental and calculated 2^+ state. The spherical light cluster has $\Im_\ell = 0$. Further, J_1^2 and J_2^2 are the squares of the angular momentum operators of the rotation of the heavy cluster and of the internuclear distance $\boldsymbol{R}$ (rotation in the relative motion of light cluster), respectively, described in the space-fixed coordinate system. The potential energy is a sum of two terms: a power series expansion in the mass asymmetry coordinate ξ and an interaction energy depending on the difference angle ε between (ϑ_1, φ_1) and (ϑ_2, φ_2) which is the angle between the internuclear distance $\boldsymbol{R}$ and the symmetry axis of the heavy fragment:

$$U = \sum_{n=0}^{3} a_n \xi^{2n} + \frac{C_0}{2} \xi \sin^2(\varepsilon) \tag{11}$$

with

$$\sin^2(\varepsilon) = \frac{2}{3}\left(1 - \frac{4\pi}{\sqrt{5}}[Y_2(\vartheta_1, \varphi_1) \otimes Y_2(\vartheta_2, \varphi_2)]_{(0,0)}\right). \tag{12}$$

If the parameter C_0 is small, the two rotation degrees of freedom are approximately independent. For large C_0, the symmetry axis of the heavy deformed cluster is essentially directed towards the light cluster and bending oscillations of the heavy cluster around the molecular axis occur. Spectra resulting from smaller and larger C_0 values will be discussed below.

The Hamiltonian (9) can be diagonalized. Then the wave function results in the form

$$\Psi_{JM} = \sum_{\kappa, J_1, J_2} c_{\kappa, J_1, J_2, J} \phi_\kappa(\xi)[Y_{J_1}(\vartheta_1, \varphi_1) \otimes Y_{J_2}(\vartheta_2, \varphi_2)]_{(J,M)}, \tag{13}$$

where the functions $\phi_\kappa(\xi)$ form a basis set for the bound mass asymmetry motion. Since the heavy cluster can have only even nuclear spin values, the parity of Ψ_{JM} is determined by the wave function of the molecular motion consisting of a rotation in the relative motion of the light cluster: $P = (-1)^{J_2}$.

Let us consider the level spectrum for $C_0 = 0$ with a fixed value of $\xi = \xi_0$. In this case the ground state band with states of positive parity originates from the rotation of the heavy cluster only with $J_1 = 0, 2, 4, ...$, and the rotation of the molecular axis is zero ($J_2 = 0$). We note that the relation of moments of inertia is $\Im_h(\xi) > \mu_{RR}(\xi) R_m^2$. Therefore, the first excited band with states of negative parity is built on a rotation of the molecular axis with $J_2 = 1$. These states are degenerated with spins

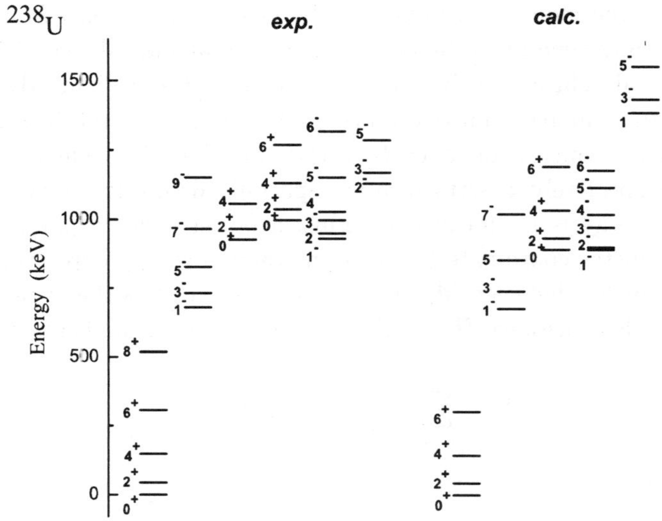

Fig. 4. The experimental and calculated level spectra of ^{238}U. In the diagonalization of H (Eq. (9)) a smaller parameter C_0 is used. The mass asymmetry motion is included. Experimental data are taken from http://www.nndc.bnl.gov/nndc/ensdf/.

$1^-, (1^-, 2^-, 3^-), (3^-, 4^-, 5^-), \ldots$ If C_0 starts to increase, the considered negative parity states lose their property of degeneracy and are shifted. This effect is recognizable in Fig. 4 where we compare the experimental energy spectrum of ^{238}U with the spectrum calculated within this model. Here, the mass asymmetry motion is included and the parameter C_0 has a smaller value. The first excited 0^+ state results from the first excited state in the mass asymmetry motion on which the lower spectrum is approximately repeated again (ground state band and 1^- band starting at 1386 keV).

For larger C_0 values, the first negative parity band is shifted downwards and we obtain bending oscillations of the heavy cluster with a small angle ε. The rotational part of the Hamiltonian can be transformed for fixed mass asymmetry ($C = C_0\xi_0$) as follows:

$$\tilde{H} = H_{rot} + H_{bend} + V_{int}, \tag{14}$$

where

$$H_{rot} = \frac{1}{2\mu_{RR}R_m^2}\left(\boldsymbol{J}^2 - 2J_{z'}^2\right), \tag{15}$$

$$H_{bend} = -\frac{\hbar^2}{2\Im_b}\frac{1}{\varepsilon}\frac{\partial}{\partial\varepsilon}\varepsilon\frac{\partial}{\partial\varepsilon} + \frac{1}{2\Im_b\varepsilon^2}J_{z'}^2 + \frac{C}{2}\varepsilon^2, \tag{16}$$

$$V_{int} = -\frac{1}{2\mu_{RR}R_m^2}\left(J_{x'}I_{x'}^{(1)} + I_{x'}^{(1)}J_{x'} + J_{y'}I_{y'}^{(1)} + I_{y'}^{(1)}J_{y'}\right). \tag{17}$$

The moment of inertia of the bending motion is

$$\Im_b = \Im_h\mu_{RR}R_m^2/(\Im_h + \mu_{RR}R_m^2). \tag{18}$$

Approximate eigenenergies can be written as

$$E_{J,M,K,n} = \frac{\hbar^2}{2(\Im_h + \mu_{RR}R_m^2)}(J(J+1) - K^2) + \hbar\omega_b(2n + |K| + 1) \tag{19}$$

with the oscillator energy of the bending mode

$$\hbar\omega_b = \hbar\sqrt{C/\Im_b}\ . \tag{20}$$

In order to prove (19), we calculated the energy spectrum by diagonalizing the Hamiltonian (9) disregarding the mass asymmetry motion by assuming an alpha-clusterization of ^{238}U as example. The resulting spectrum is well approximated by (19). The ground state band is an unperturbed alternating parity band as expected in this limit because we have a stable reflection-asymmetric shape. This band does not describe the experimental ground state band of ^{238}U which contains only states with even spin and positive parity. The next bands have $K = 1$ and $n = 1$.

The mass asymmetry coordinate used in this Section assumes positive values, and the light cluster is transferred to the other side of the heavy cluster by rotating it by an angle $\varepsilon = \pi(-\pi)$. This makes it possible to study the dynamics of axially asymmetric dinuclear shapes in greater detail. Depending on the stiffness parameter C_0, the system has small angular vibrations of the clusters around their equilibrium position for larger values of C_0 and a rotation of the light cluster around the heavy one over a potential barrier at $\varepsilon = \pi/2$ for smaller values of C_0. The latter case gives a good description of the spectrum of ^{238}U.

7. Dynamics of fusion in the dinuclear system model

Heavy and superheavy nuclei can be produced by fusion reactions with heavy ions. We discriminate Pb or Bi based or cold fusion reactions, e. g. ^{70}Zn + ^{208}Pb $\rightarrow$ ^{278}112 $\rightarrow$ ^{277}112 + n with an evaporation residue cross section of $\sigma = 1$ pb and an excitation energy of the $^{278}112$ compound nucleus of about 11 MeV, and actinide based or hot fusion reactions, e. g. $^{48}\text{Ca} + {}^{244}\text{Pu} \rightarrow {}^{288}114 + 4\text{n}$, with the emission of more neutrons. The cross sections are small because of a strong competition between complete fusion and quasifission and small survival probabilities of the excited compound nucleus.

7.1. *Models with adiabatic and diabatic potentials*

The models for the production of superheavy nuclei can be discriminated by the dynamics in the most important collective degrees of freedom of the system, i. e. the relative and mass asymmetry motions, and depend sensitively whether adiabatic or diabatic potentials in the internuclear coordinate R are assumed.

a) *Models using adiabatic potentials:* These models minimize the potential energy. In that case the nuclei first change their mass asymmetry in the direction to more symmetric clusters and then they fuse together by crossing a smaller fusion barrier in the relative coordinate around $\eta = 0$. The models tend to give larger probabilities for fusion if similar target and projectile nuclei are taken which contradicts the exponential fall-off of the evaporation residue cross section with increasing projectile nuclei in Pb based reactions.

b) *Dinuclear system concept:* The fusion proceeds by a transfer of nucleons between the nuclei in a touching configuration, i. e. in the dinuclear configuration. Here, mainly a dynamics in the mass asymmetry degree of freedom occurs. The potential is of diabatic type with a minimum in the touching range and a repulsive part towards smaller internuclear distances prohibiting the dinuclear system to amalgamate to the compound nucleus in the relative coordinate. Such a potential, achieved with a diabatic two-center shell model [13], has a survival time of the order of the reaction time of 10^{-20} s. It can also be justified with structure calculations based on group theoretical methods [14].

The distinction between the models is based on a different dynamics forced by the potentials. If more collective coordinates like orientation angles and vibration coordinates of the fragments and the neck coordinate are included in the dynamical treatment, the differences could be diminished and the above two approaches may converge in an unique description of fusion.

Similar important like the difference between adiabatic and diabatic potentials is the coordinate-dependence of the various masses. As explored by

Fink *et al.* [15] for the case of $^{12}C + ^{12}C$ scattering, a coordinate-dependence of the mass of relative motion can be transformed into a constant mass and an energy-dependent repulsive potential which screens the outer range from the inner one and, therefore, has similar properties as the diabatic potential.

7.2. *Evaporation residue cross section*

The cross section for the production of superheavy nuclei can be written

$$\sigma_{ER}(E_{c.m.}) = \sum_{J=0}^{J_{max}} \sigma_{cap}(E_{c.m.}, J) P_{CN}(E_{c.m.}, J) W_{sur}(E_{c.m.}, J). \quad (21)$$

The three factors are the capture cross section, the probability for complete fusion and the survival probability. The maximal contributing angular momentum J_{max} is of the order of 15 - 20. The capture cross section σ_{cap} describes the formation of the dinuclear system at the initial stage of the reaction when the kinetic energy of the relative motion is transferred into potential and excitation energies. The DNS can decay by crossing the quasifission barrier B_{qf} which is of the order of 0.5 - 5 MeV.

After its formation the DNS evolves in the mass asymmetry coordinate. The center of the mass distribution moves towards more symmetric fragmentations and its width is broadened by diffusion processes. The part of the distribution, which crosses the inner fusion barrier B^*_{fus} of the driving potential $U(\eta)$, yields the probability P_{CN} for complete fusion. The DNS can also decay by quasifission during its evolution. Therefore, the fusion probability P_{CN} and the mass and charge distributions of the quasifission have to be treated simultaneously. The fusion probability can be quantitatively estimated with the Kramers formula and results as $P_{CN} \sim \exp(-(B^*_{fus} - \min[B_{qf}, B_{sym}])/T)$ where the temperature T is related to the excitation energy of the DNS, and B_{sym} is the barrier in η to more symmetric configurations. B_{sym} is 4-5 MeV ($> B_{qf}$) in cold reactions and 0.5-1.5 MeV ($< B_{qf}$) in hot reactions. Since the inner fusion barrier increases with decreasing mass asymmetry, we find an exponential depression of the fusion probability towards symmetric projectile and target combinations in lead based reactions. In hot fusion reactions with ^{48}Ca projectiles, P_{CN} drops down with increasing mass and charge of the target nucleus. These systems run easier towards symmetric fragmentations and undergo quasifission there.

The excited compound nucleus decays by fission and emits neutrons besides negligible emissions of other particles and photons. The probability to reach the ground state of the superheavy nucleus by neutron emission is denoted as survival probability W_{sur}. In the case of the one-neutron emission in Pb-based reactions the survival probability is roughly the ratio Γ_n/Γ_f of the widths for neutron emission and for fission because of $\Gamma_f \gg \Gamma_n$. The survival probability depends sensitively on the nuclear structure properties of the superheavy elements like level density, fission barriers and deformations.

With the DNS concept we reproduced the measured evaporation residue cross sections of the Pb- and actinide-based reactions with a precision of a factor of two. This concept also yields the excitation energies of the superheavy compound nuclei at the optimal bombarding energies, where the production cross sections are largest, in agreement with the experimental data. As an example we give results of calculations for the reactions ^{86}Kr+^{134,138}Ba in Table 1 and compare them with recently measured data.

Table 1. The calculated evaporation residue cross sections in the indicated most probable channels of the reactions ^{86}Kr+^{134,138}Ba are compared with experimental data of Satou *et al.*[16].

Reaction	$E_{c.m.}$ (MeV)	Channel	σ_{ER}^{th}	σ_{ER}^{exp}
^{86}Kr+^{138}Ba	218.6	αn	18 nb	8^{+10}_{-6} nb
	225.3	$2\alpha n$	94 nb	50^{+42}_{-32} nb
	225.3	$\alpha 2n$	59 nb	19^{+38}_{-19} nb
	232.3	$\alpha 2n$	64 nb	140^{+120}_{-90} nb
	237.4	$3n + \alpha 3n$	156 nb	180^{+130}_{-100} nb
^{86}Kr+^{134}Ba	220	$2n + \alpha 2n$	1 nb	2^{+5}_{-2} nb
	220.9	$np + \alpha np$	0.7 nb	6^{+12}_{-6} nb
	227	$np + \alpha np$	1.7 nb	6^{+13}_{-6} nb
	229	$np + \alpha np$	3 nb	

7.3. *Isotopic dependence of production cross section*

Whether the production cross section of isotopic superheavy nuclei is increasing or decreasing with the neutron number, depends on the fusion and survival probabilities. For example, the evaporation residue cross section of Ds ($Z = 110$) increases with the neutron number. The reactions ^{62}Ni + ^{208}Pb $\rightarrow$ ^{269}Ds + n and ^{64}Ni + ^{208}Pb $\rightarrow$ ^{271}Ds + n have cross sections of 3.5 and 15 pb, respectively. Let us discuss the isotopic dependence of the fusion and survival probabilities. When the neutron number of the projectile is increasing, the dinuclear fragmentation gets more symmetrically and the fusion probability decreases if the more symmetric DNS does not consists of more stable nuclei. Also the survival probability is of importance. For compound nuclei with closed neutron shells the survival probability is larger. Hence, the product of P_{CN} and W_{sur} determines whether the production cross section increases or decreases with increasing neutron number. Fig. 5 shows examples for cold and hot fusion reactions [17]. These calculations are very valuable and support an adequate choice of projectile and target nuclei in experiment.

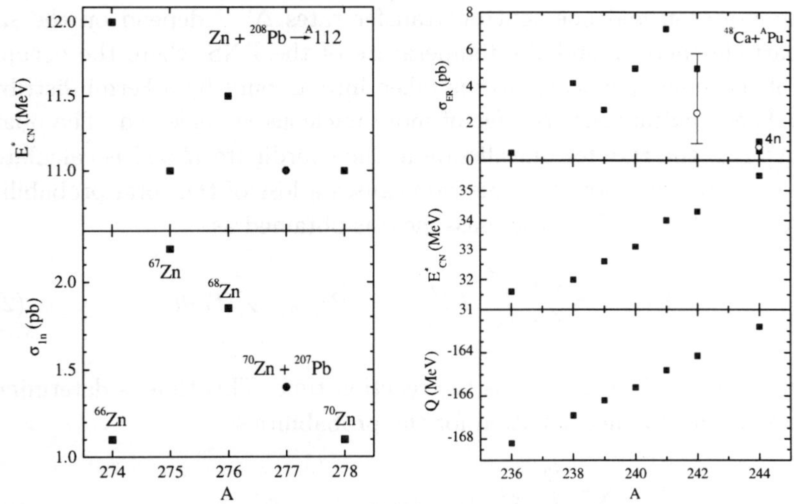

Fig. 5. Excitation energy E^*_{CN}, evaporation residue cross section σ_{1n}, $\sigma_{3n,4n}$ and Q-value for Zn + ^{208}Pb → ^{A}112 (l.h.s.) and ^{48}Ca + ^{A}Pu (r.h.s.). The experimental points are from Oganessian *et al.*[18].

8. Quasifission as signature for mass transfer

The process of quasifission is the decay of the DNS. Since quasifission leads to a large quantity of observable data like mass and charge distributions, distributions of total kinetic energies (TKE), variances of total kinetic energies and neutron multiplicities, a comparison of the theoretical description with experimental data provides sensitive information about the applicability and correctness of the used model. For this reason we studied the dynamics of mass and charge transfer and the succeeding quasifission with master equations [19]. At the starting point we consider the shell model

Hamiltonian of all dinuclear fragmentations of the nucleons. This Hamiltonian can be used to derive master equations for the probability $P_{Z,N}(t)$ to find the dinuclear system in a fragmentation with $Z_1 = Z$, $N_1 = N$ and $Z_2 = Z_{tot} - Z$, $N_2 = N_{tot} - N$. The master equations are

$$\begin{aligned} \frac{d}{dt}P_{Z,N}(t) &= \Delta^{(-,0)}_{Z+1,N}P_{Z+1,N}(t) + \Delta^{(+,0)}_{Z-1,N}P_{Z-1,N}(t) \\ &\quad +\Delta^{(0,-)}_{Z,N+1}P_{Z,N+1}(t) + \Delta^{(0,+)}_{Z,N-1}P_{Z,N-1}(t) \\ -\left(\Delta^{(-,0)}_{Z,N} + \Delta^{(+,0)}_{Z,N} + \Delta^{(0,-)}_{Z,N} + \Delta^{(0,+)}_{Z,N}\right) &P_{Z,N}(t) - \Lambda^{qf}_{Z,N}P_{Z,N}(t). \end{aligned} \tag{22}$$

The one-proton and one-neutron transfer rates $\Delta^{(\cdot,\cdot)}$ depend on the single particle energies and the temperature of the DNS where the occupation of the single particle states is taken into account by a Fermi distribution. The simultaneous transfer of more nucleons is neglected. The quantity $\Lambda^{qf}_{Z,N}$ is the rate for quasifission in the coordinate R and is calculated with the Kramers formula. This rate causes a loss of the total probability $\sum_{Z,N} P_{Z,N}(t) \leq 1$. Then the mass yield is obtained as

$$Y(A_1) = \sum_{Z_1} \int_0^{t_0} \Lambda^{qf}_{Z_1,A_1-Z_1} P_{Z_1,A_1-Z_1}(t)\, dt, \tag{23}$$

where $t_0 \approx (3-5) \times 10^{-20}$ s is the reaction time. This time is determined by solving the balance equation for the probabilities:

$$\sum_{Z,N} \int_0^{t_0} \Lambda^{qf}_{Z,N} P_{Z,N}(t) dt = 1 - P_{CN}. \tag{24}$$

Here, $P_{CN} = \sum_{Z<Z_{BG},N<N_{BG}} P_{Z,N}(t_0)$ is the probability for fusion, defined by the fraction of probability existing for $Z < Z_{BG}$ and $N < N_{BG}$ at time t_0, where Z_{BG} and N_{BG} determine the inner fusion barrier in the charge and neutron asymmetry coordinates. The DNS with $Z < Z_{BG}$ and $N < N_{BG}$ evolves to the compound nucleus in a time of 10^{-21}s which is short compared with the decay time of the compound nucleus. The DNS dynamics was also studied by Li *et al.* [20] with similar master equations.

a) *Results for quasifission* [19]: We calculated quasifission distributions, TKEs, variances of TKE and neutron multiplicities for cold and hot fusion reactions and found satisfying agreement with the experimental data of Itkis *et al.* [21]. For heavier systems, e. g. ^{48}Ca + ^{248}Cm, the contribution of fusion-fission products to the mass distribution can be neglected since the probability for forming a compound nucleus is very small.

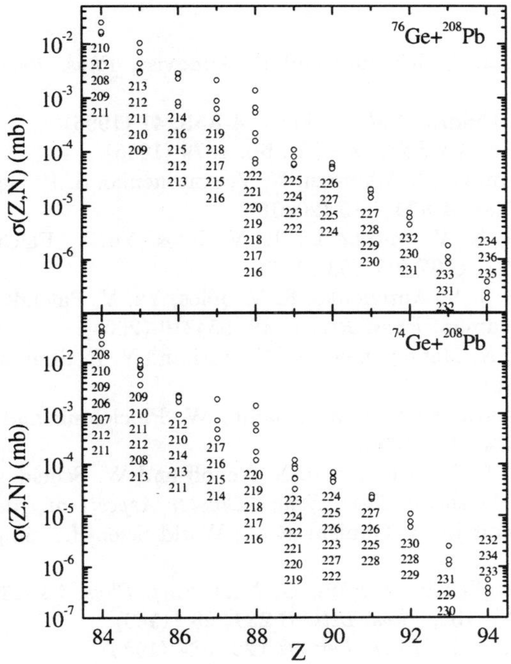

Fig. 6. Production cross sections in the reaction $^{76}Ge + ^{208}Pb$ and $^{74}Ge + ^{208}Pb$ as a function of Z and A of the heavier fragment.

b) *Production cross sections for asymmetric systems:* The master equations give also probabilities for more asymmetric systems than the initial one. In Fig. 6 we present production (transfer) cross sections for asymmetric fragmentations in the reactions $^{74,76}Ge + ^{208}Pb$ [22]. The measurement of these observable cross sections would be a proof for the fusion dynamics in the dinuclear system concept. In the asymmetric fragmentation in the reactions $^{48}Ca+^{238}U$, ^{243}Am and $^{244,246,248}Cm$ one can produce new isotopes of superheavy nuclei with $Z = 104 - 108$ which fill the gap between the isotopes of heaviest nuclei obtained in cold and hot complete fusion reactions [23].

Acknowledgements

We thank DFG (Bonn), VW-Stiftung (Hannover) and RFBR (Moscow) for supporting this work. We thank Prof. Junqing Li (Lanzhou), Prof. Enguang Zhao (Beijing) and Prof. Nikolay Minkov (Sofia) for valuable discussions.

References

1. D. A. Bromley, J. A. Kuehner and E. Almqvist, *Phys. Rev. Lett.* **4**, 365 (1960).
2. L. Vannucci, N. Cindro *et al.*, *Z. Phys. A* **355**, 41 (1996).
3. V. V. Volkov, *Izv. AN SSSR ser. fiz.* **50**, 1879 (1986).
4. T. M. Shneidman, G. G. Adamian, N. V. Antonenko, S. P. Ivanova and W. Scheid, *Nucl. Phys. A* **671**, 119 (2000).
5. G. G. Adamian, N. V. Antonenko, R. V. Jolos, Yu. V. Palchikov and W. Scheid, *Phys. Rev. C* **67**, 054303 (2003).
6. G. G. Adamian, N. V. Antonenko, R. V. Jolos, Yu. V. Palchikov, W. Scheid and T. M. Shneidman, *Phys. Rev. C* **69**, 054310 (2004).
7. T. Nakatsukasa, K. Matsuyanagi, S. Mizutori and Y. R. Shimizu, *Phys. Rev. C* **53**, 2213 (1996).
8. S. Cwiok, W. Nazarewicz, J. X. Saladin, W. Plociennik and A. Johnson, *Phys. Lett. B* **322**, 304 (1994).
9. G. G. Adamian, N. V. Antonenko, N. Nenoff and W. Scheid, *Phys. Rev. C* **64**, 014306 (2001) and in *Proc. Symp. Cluster Aspects of Quantum Many-Body Systems*, edit by A. Ohnishi *et al.*, World Scientific, Singapore, 2002, p. 215.
10. R. Maass and W. Scheid, *J. Phys. G: Nucl. Part. Phys.* **16**, 1359 (1990).
11. E. Uegaki and Y. Abe, *Phys. Lett. B* **231**, 28 (1989).
12. R. W. Zurmühle *et al.*, *Phys. Lett. B* **129**, 384 (1983).
13. A. Diaz-Torres, G. G. Adamian, N. V. Antonenko and W. Scheid, *Phys. Lett. B* **481**, 228 (2000).
14. G. G. Adamian, N. V. Antonenko and Yu. M. Tchuvil'sky, *Phys. Lett. B* **451**, 289 (1999).
15. H. J. Fink, W. Scheid and W. Greiner, *J. Phys. G (Nucl. Phys.)* **1**, 685 (1975).
16. K. Satou *et al.*, *Phys. Rev. C* **73**, 034609 (2006).
17. G. G. Adamian, N. V. Antonenko and W. Scheid, *Phys. Rev. C* **69**, 011601, 014607 (2004).
18. Yu. Ts. Oganessian *et al.*, *Eur. Phys. J. A* **13**, 135 (2002); *A* **15**, 201 (2002).
19. G. G. Adamian, N. V. Antonenko and W. Scheid, *Phys. Rev. C* **68**, 034601 (2003).
20. W. Li, N. Wang, J. F. Li, H. Xu, W. Zuo, E. Zhao, J. Q. Li and W. Scheid, *Europhys. Lett.* **64**, 750 (2003).
21. M. G. Itkis *et al.*, *Nucl. Phys. A* **734**, 136 (2004).
22. G. G. Adamian and N. V. Antonenko, *Phys. Rev. C* **72**, 064617 (2005).
23. G. G. Adamian, N. V. Antonenko and A. S. Zubov, *Phys. Rev. C* **71**, 034603 (2005).

NEW SPECTROSCOPY WITH COLD FISSION

D. S. DELION
National Institute of Physics and Nuclear Engineering,
Bucharest-Măgurele, POB MG-6, Romania,
e-mail: delion@theory.nipne.ro

A. SĂNDULESCU
Center for Advanced Studies in Physics, Romanian Academy
Calea Victoriei 125, Bucharest, Romania

We investigate the spectroscopic information which can be extracted from low-lying rotational yields in the cold fission of ^{252}Cf. A fissioning state is considered as a resonance in the potential well between the emitted fragments. As fissioning states we select those resonances which are oriented close to the pole-to-pole configuration in the overlapping region. We predict a strong dependence of decay yields upon the quadrupole and hexadecapole deformation parameters. Predictions of rotational yields for ten possible cold splittings of ^{252}Cf are given.

1. Introduction

In the last years an intense experimental activity was performed in order to investigate cold (neutron-less) binary and ternary fission process of ^{252}Cf [1,2,3,4,5,6,7,8,9]. It involved modern facilities, as the Gammasphere and Eurogam, which were able to identify this rare process using the triple γ-rays coincidence technique.

The measurements confirmed the idea that this process is a natural extension of the cluster radioactivity [10,11]. A very convincing theoretical evidence that it has a sub-barrier character was the WKB penetration calculation, using a double folding potential with M3Y plus Coulomb nucleon-nucleon forces. This simple estimate was able to reproduce the gross features of the binary cold fragmentation isotopic yields of ^{252}Cf [12].

The yields of rotational states were extracted from the intensities of γ-rays emitted in coincidence during the deexcitation of fragments for ^{104}Mo-^{148}Ba and ^{106}Mo-^{146}Ba [2]. It was shown that the cold fission population is centered around the low-lying 2^+ and 4^+ states and the states higher than

6^+ are practically not populated. This proves the assumption concerning the cold rearrangement of nucleons during the cold fission.

The role of Coulomb excitations under various assumptions of the initial spin distribution at the scission point was studied in Ref. [13] within a semi-classical time dependent formalism. Here only the quadrupole-quadrupole Coulomb term was considered.

In two recent papers [14,15] we analyzed the double fine structure of emitted fragments within the stationary scattering formalism. The aim of this lecture is to simplify this approach by using a common rotational basis, as in our recent Ref. [16]. This will allow us to take into account all degrees of freedom, generalizing our previous approach.

2. Theoretical background

Fission process is described by the stationary Schrödinger equation

$$H\Psi(\mathbf{R},\Omega_1,\Omega_2) = E\Psi(\mathbf{R},\Omega_1,\Omega_2) \ , \tag{1}$$

where $\mathbf{R} = (R, \Omega)$ denotes the distance between the centers of two deformed nuclei. The orientation of their major axes in the laboratory system is given by Euler angles $\Omega_k = (\varphi_k, \theta_k, 0)$, $k = 1, 2$. The most favourable fissioning configuration is the pole-to-pole $(p-p)$ one, with $\Omega_1 = \Omega_2 = \Omega$, where the Coulomb barrier has the lowest possible value.

The Hamiltonian can be written as follows

$$H = -\frac{\hbar^2}{2\mu}\nabla_R^2 + H_1(\Omega_1) + H_2(\Omega_2) + V(\mathbf{R},\Omega_1,\Omega_2) \ , \tag{2}$$

where μ is the reduced mass of the dinuclear system and H_k are the Hamiltonians describing the rotation of the fragments. We estimate the interaction between nuclei in terms of the double folding between the nuclear densities [17] by using the M3Y nucleon-nucleon [18] plus Coulomb force. This potential can be divided into a spherical and a deformed component

$$V(\mathbf{R},\Omega_1,\Omega_2) = V_0(R) + V_D(\mathbf{R},\Omega_1,\Omega_2) \ . \tag{3}$$

By expanding the nuclear densities in multipoles one obtains the deformed part of the interaction as follows

$$V_D(\mathbf{R},\Omega_1,\Omega_2) = \sum_{\lambda_0\lambda_1\lambda_2} V_{\lambda_0\lambda_1\lambda_2}(R)\mathbf{Y}_{\lambda_0\lambda_1\lambda_2}(\Omega,\Omega_1,\Omega_2) \ , \tag{4}$$

where the term $(\lambda_0\lambda_1\lambda_2) = (000)$ is excluded from summation. Here the angular part of the wave function has the following ansatz

$$\mathbf{Y}_{\lambda_0\lambda_1\lambda_2}(\Omega,\Omega_1,\Omega_2) = \left\{Y_{\lambda_0}(\Omega)\otimes\left[Y_{\lambda_1}(\Omega_1)\otimes Y_{\lambda_2}(\Omega_2)\right]_{\lambda_0}\right\}_0 \ . \tag{5}$$

If the rotational states of fragments belong to the ground band the wave function is given by a similar superposition, i.e.

$$\Psi(\mathbf{R},\Omega_1,\Omega_2,) = \frac{1}{R}\sum_{lI_1I_2} f_{lI_1I_2}(R)\mathbf{Y}_{lI_1I_2}(\Omega,\Omega_1,\Omega_2) , \tag{6}$$

In the "molecular" intrinsic system of coordinates, defined by the Euler angle Ω as follows

$$\Omega'_k = \Omega^{-1}\Omega_k, \quad k = 1,2 , \tag{7}$$

one obtains

$$\mathbf{Y}_{lI_1I_2}(\Omega,\Omega'_1,\Omega'_2) = \frac{(-)^l}{\sqrt{4\pi}}[Y_{I_1}(\Omega'_1)\otimes Y_{I_2}(\Omega'_2)]_{l0} . \tag{8}$$

By using the orthonormality of angular functions entering superposition (6) one obtains in a standard way the coupled system of differential equations for radial components

$$\frac{d^2 f_{lI_1I_2}(R)}{d\rho_{12}^2} = \sum_{l'I'_1I'_2} A_{lI_1I_2;l'I'_1I'_2}(R) f_{l'I'_1I'_2}(R) , \tag{9}$$

where the coupling matrix is given by

$$A_{lI_1I_2;l'I'_1I'_2}(R) = \left[\frac{l(l+1)}{\rho_{12}^2} + \frac{V_0(R)}{E_{12}} - 1\right]\delta_{ll'}\delta_{I_1I'_1}\delta_{I_2I'_2} + \frac{1}{E_{12}}\langle\mathbf{Y}_{lI_1I_2}|V_D(R)|\mathbf{Y}_{l'I'_1I'_2}\rangle . \tag{10}$$

Here we introduced the following short-hand notations

$$\rho_{12} = \kappa_{12}R , \quad \kappa_{12} = \sqrt{\frac{2\mu E_{12}}{\hbar^2}} , \quad E_{12} = E - E_{I_1} - E_{I_2} , \tag{11}$$

where E_{I_1}, E_{I_2} are the ground band energies of the emitted fragments. The matrix element entering Eq. (10) is given by the Appendix of the Ref. [15]. The procedure to integrate numerically this system of equations and to find resonant states is also given here.

The total decay width is a sum over partial channel widths. It can be derived from the continuity equation in a straightforward way as follows

$$\Gamma = \sum_{lI_1I_2}\Gamma_{lI_1I_2} = \sum_l \hbar v_{lI_1I_2}\lim_{R\to\infty}|f_{lI_1I_2}(R)|^2 , \tag{12}$$

where $v_{lI_1I_2}$ is the center of mass velocity at infinity in the channel (l, I_1, I_2), i.e.

$$v_{lI_1I_2} = \frac{\hbar\kappa_{12}}{\mu} . \tag{13}$$

3. Numerical application

We selected among the binary splittings of ^{252}Cf the most intense even-even events, given in Table 1 with deformation parameters taken from [19].

Table 1. The charge and mass numbers for ten different splittings of ^{252}Cf. The quadrupole and hexadecapole deformations of the two fragments are given in columns 6-9 and the Q-value (in MeV) in the 10-th column. In the last two columns are given the logarithms of the total decay widths (in MeV) by using one and 15 states in the basis, respectively. We considered the first resonant states. By stars are given the decay widths of second resonant states.

No.	Z_1	A_1	Z_2	A_2	$b_2^{(1)}$	$b_4^{(1)}$	$b_2^{(2)}$	$b_4^{(2)}$	Q	$lg\Gamma_1$	$lg\Gamma_{15}$
1	38	98	60	154	0.357	0.056	0.270	0.114	208.90	-1.094	-1.175
2	38	100	60	152	0.368	0.033	0.262	0.128	206.41	-2.679	-0.246
2*											-1.692
3	40	100	58	152	0.358	0.039	0.261	0.111	212.42	-1.803	-1.515
4	40	102	58	150	0.369	0.017	0.252	0.126	212.76	-1.497	-0.369
4*											-0.780
5	40	104	58	148	0.381	0.005	0.216	0.109	213.35	-2.322	-2.542
6	42	104	56	148	0.349	0.030	0.236	0.131	214.67	-2.778	-1.020
7	42	106	56	146	0.361	0.002	0.199	0.100	217.38	-2.282	-1.164
8	42	108	56	144	0.333	0.027	0.164	0.078	219.15	-2.403	-2.384
9	44	110	54	142	0.250	0.044	0.145	0.075	220.94	-5.735	-4.852
9*											-3.987
10	44	112	54	140	0.258	0.064	0.116	0.070	224.37	-3.606	-2.481

In expanding the double-folding interaction (4) and the wave function (6) we used the same rotational basis (l, I_1, I_2), given in Table 2.

Table 2. The rotational basis (l, I_1, I_2).

No.	l	I_1	I_2
1	0	0	0
2	0	2	2
3	0	4	4
4	2	2	0
5	2	0	2
6	2	2	2
7	2	4	4
8	2	4	2
9	2	2	4
10	4	4	0
11	4	0	4
12	4	2	2
13	4	4	4
14	4	4	2
15	4	2	4

In Figure 1 we plotted by a solid line the pole-to-pole radial component of the potential, defined as

$$V_{p-p}(R) = V(R, \Omega, \Omega, \Omega), \tag{14}$$

for the splitting 6 in Table 1, i.e. ^{104}Mo+^{148}Ba. From Eq. (8) it becomes clear that indeed this interaction does not depend upon Euler angles.

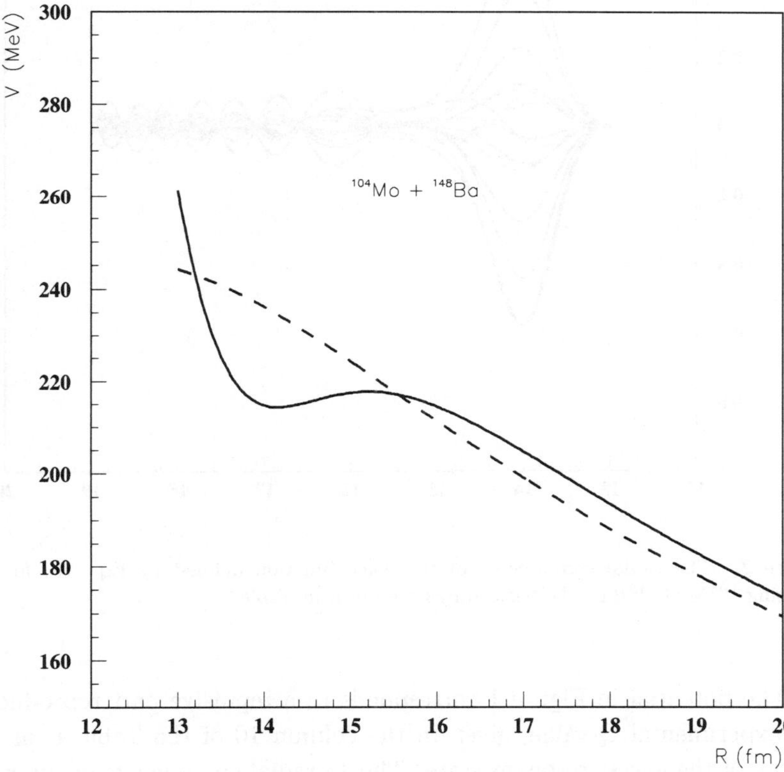

Figure 1. The double folding potential $V_{p-p}(R)$ for the pole-to-pole configuration defined by Eq. (14) (solid line) and the spherical component $V_0(R)$ in Eq. (3) (dot-dashed line) for the splitting ^{104}Mo+^{148}Ba.

This potential gives a much lower Coulomb barrier in comparison with the pure spherical component $V_0(R)$, plotted by a dashed line. Therefore in our calculation we considered as a spherical component the pole-to-pole interaction $V_{p-p}(R)$ until the intersection with the true spherical part $V_0(R)$. Beyond this point $V_0(R)$ becomes energetically more favourable and the two fragments start to rotate separately.

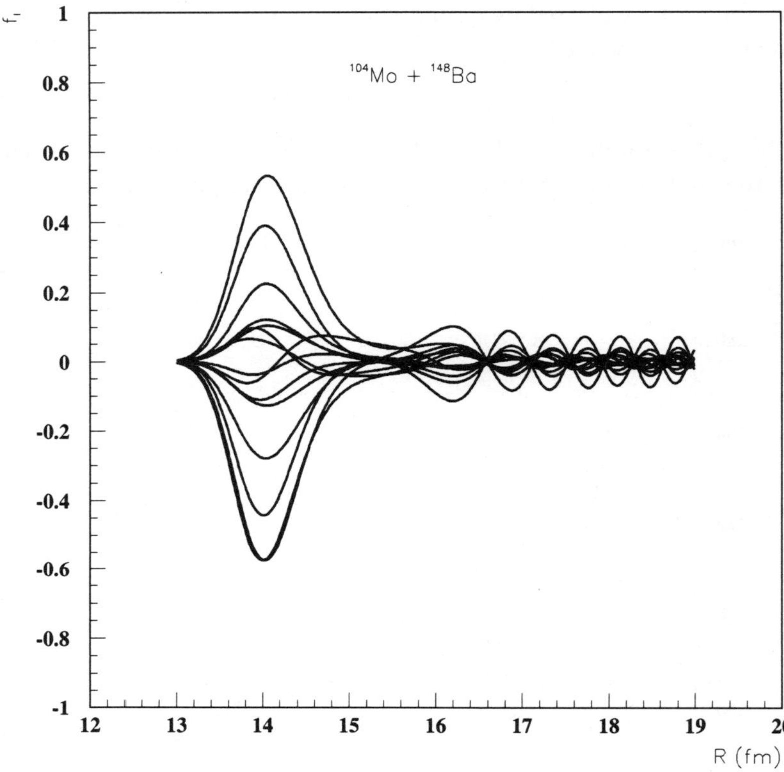

Figure 2. The radial components of the wave function defined by Eq. (6) for the splitting ^{104}Mo+^{148}Ba. The basis states are given in Table 2.

The potential in Figure 1 corresponds to a repulsive part reproducing the experimental Q-value, given in the column 10 of the Table 1, as the energy of the lowest resonant state. The 15 radial components of the wave function $f_l(R)$, $l \to (l, I_1, I_2)$ are plotted in Figure 2.

First of all we estimated total decay widths, using Eq. (12). In Ref. [12] they are called total yields. The results are given by the last column of the Table 1. We compared these values with the results obtained using only the first component of the Table 2, i.e. the penetration through a spherical barrier. These values are given in the column 11. As a rule they are lower up to one order of magnitude in comparison with the full-basis calculation. Exceptions are the lines 2 and 4.

In order to understand this feature we computed the angular distribution of the wave function (6) for a radius corresponding to maximal values in Figure 2, i.e. in to the minimum of the pocket-like potential. We plotted in Figure 3 by dashed lines the angular probability depending upon the angle θ_1 for various values of $\theta_2 = 0^o, 30^o, 60^o, 90^0$ and $\varphi_1 = \varphi_2 = 0$. Here we considered all angles in the "molecular" intrinsic system of coordinates defined by Eq. (7), i.e.

$$\Omega'_k = (\varphi_k, \theta_k, 0)\ , \quad k = 1, 2\ . \tag{15}$$

For the sake of simplicity we dropped the prime upper index for angles.

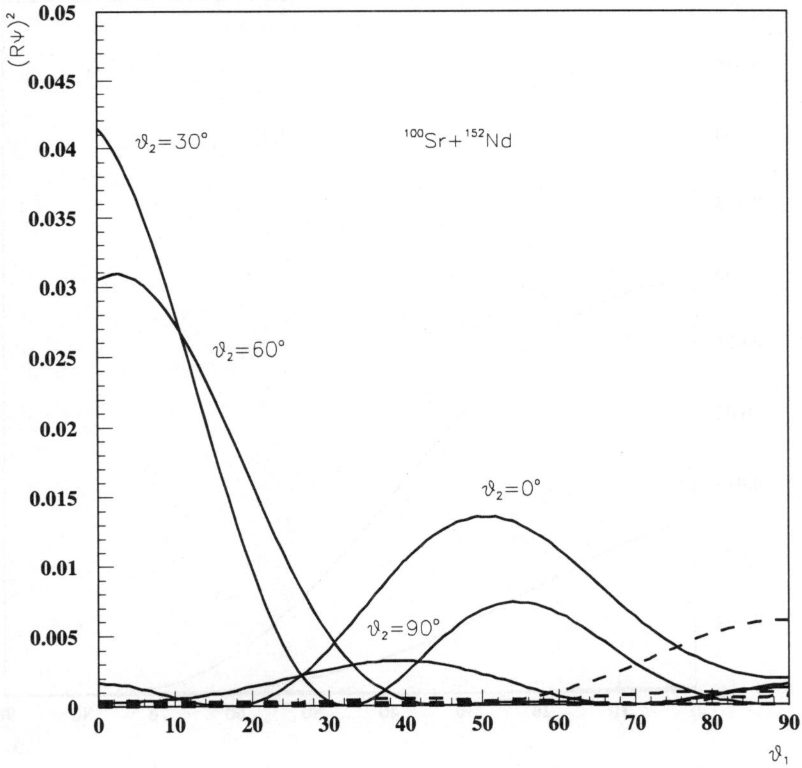

Figure 3. The angular part of the wave function versus the angle θ_1 for different angles θ_2. By solid lines are given the values corresponding to the second resonant state and by dashes to the first resonance of the system ^{100}Sr+^{152}Nd.

All these curves are centered around 90°, i.e. the fissioning system is not concentrated around the pole-to-pole configuration. On the contrary, for the second resonant state, giving the value on the line 2*, the angular probability, plotted in Figure 3 by solid lines, is mainly concentrated in a region of $\pm 30^{\circ}$ around the pole-to-pole configuration.

We found out that for the cases 4 and 9 the situation is similar, i.e. the angular probability of the second resonant state belongs mainly to the forward region, in contrast with the first resonance. Moreover, for the splittings 7-10, where the Coulomb barrier is larger, the angular probabilities are more concentrated around the pole-to-pole configuration. This is illustrated in Figure 4 for the splitting 7 (^{106}Mo+^{146}Ba).

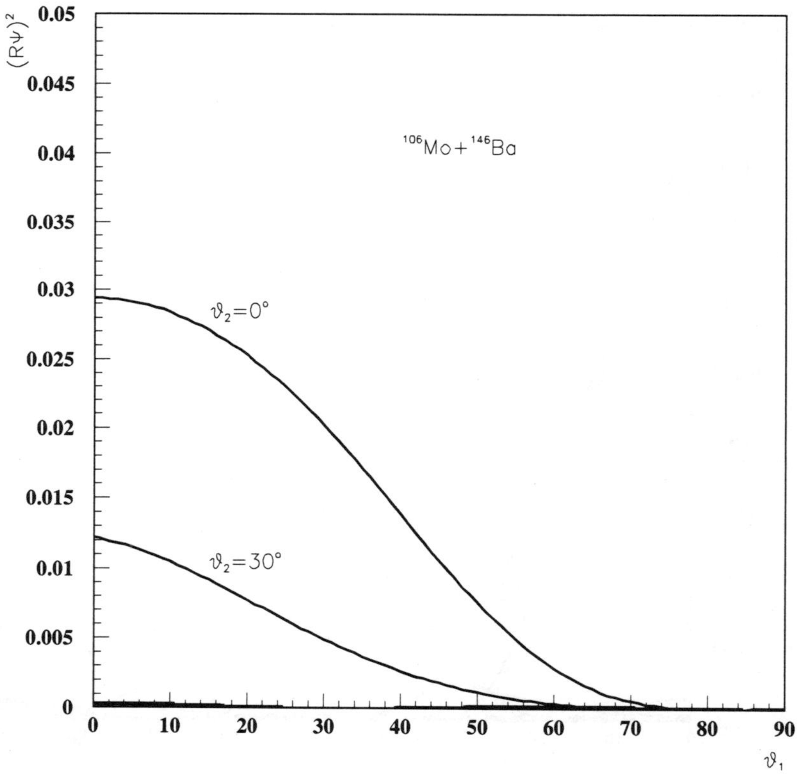

Figure 4. The angular part of the wave function versus the angle θ_1 for different angles θ_2. The values correspond to to the first resonance of the system ^{106}Mo+^{146}Ba.

Based on this analysis we selected as fissioning states the lowest possible resonant states which are concentrated around the pole-to-pole configurations. This corresponds to the first resonance, except the cases 2, 4, 9, where we considered the second resonant state (2*, 4*, 9*).

In our previous Refs. [14,15] we used in the internal overlapping region a vibrational basis, depending only on "planar" coordinates θ_k. This is connected with the strong increasing of the potential along θ_k-coordinates, starting from the pole-to-pole configuration. As a result of the integration procedure we obtained in the internal region one predominant vibrational component. In terms of the rotational basis our analysis showed that a vibrational state corresponds to several rotational components with comparable amplitudes. This is indeed our case. In Table 3 we give the mean squared amplitudes, computed according to the following relation

$$\overline{f}_{lI_1I_2} = \left[\int_{R_0}^{R_2} |f_{lI_1I_2}(R)|^2 dR \right]^{1/2} . \tag{16}$$

One can see that most of amplitudes have comparable values, in agreement with our previous calculation within the internal vibrational basis.

Table 3. The averaged amplitudes $\overline{f}_{lI_1I_2}$ on internal region, defined by Eq. (16) for the basis components given in Table 2. The labels in the first column correspond to the splittings given in Table 1.

No.	1	2	3	4	5	6	7	8	9	10	11	12	13	14	15
1	.03	.04	.04	.27	.04	.09	.04	.10	.27	.64	.47	.29	.04	.19	.27
2	.47	.13	.07	.50	.48	.28	.07	.08	.11	.17	.25	.18	.08	.13	.17
2*	.12	.05	.07	.32	.15	.11	.07	.09	.34	.54	.39	.38	.06	.18	.28
3	.07	.05	.05	.36	.11	.09	.05	.12	.31	.55	.45	.33	.04	.21	.26
4	.05	.55	.06	.05	.07	.62	.08	.20	.26	.04	.04	.14	.09	.28	.29
4	.05	.09	.08	.33	.05	.17	.09	.09	.30	.54	.43	.37	.05	.15	.32
5	.07	.02	.06	.45	.10	.05	.06	.16	.30	.55	.42	.22	.05	.22	.25
6	.08	.03	.10	.47	.06	.07	.10	.08	.32	.46	.45	.34	.09	.18	.23
7	.25	.27	.14	.29	.36	.30	.14	.21	.23	.22	.30	.42	.15	.19	.22
8	.22	.17	.27	.36	.14	.12	.29	.29	.45	.18	.19	.28	.25	.15	.28
9	.15	.04	.01	.22	.24	.15	.01	.49	.35	.21	.17	.52	.01	.23	.28
9*	.27	.26	.12	.35	.43	.30	.12	.20	.22	.19	.28	.38	.11	.17	.19
10	.26	.26	.14	.41	.35	.30	.15	.23	.24	.22	.22	.37	.13	.19	.20

In spite of the fact the amplitudes have different signs, as can be seen from the Figure 2, we notice that the dependence upon the angular variable θ_1, θ_2 is mainly concentrated in the forward direction. This feature we can understand from the Table 3. The largest components correspond to the basis elements $(l, l, 0)$ or $(l, 0, l)$. They contribute coherently in the wave function (6), giving a diffraction-like pattern around the "planar" pole-to-pole direction for the selected resonant states.

Our method includes also the angles φ_k. We mention here that the angular distribution upon $\varphi_1 - \varphi_2$ for $\theta_1 = \theta_2 = 0$ is centered around 0°

or 180°. Thus, most of the "torsional" angular distribution is also concentrated around the pole-to-pole configuration.

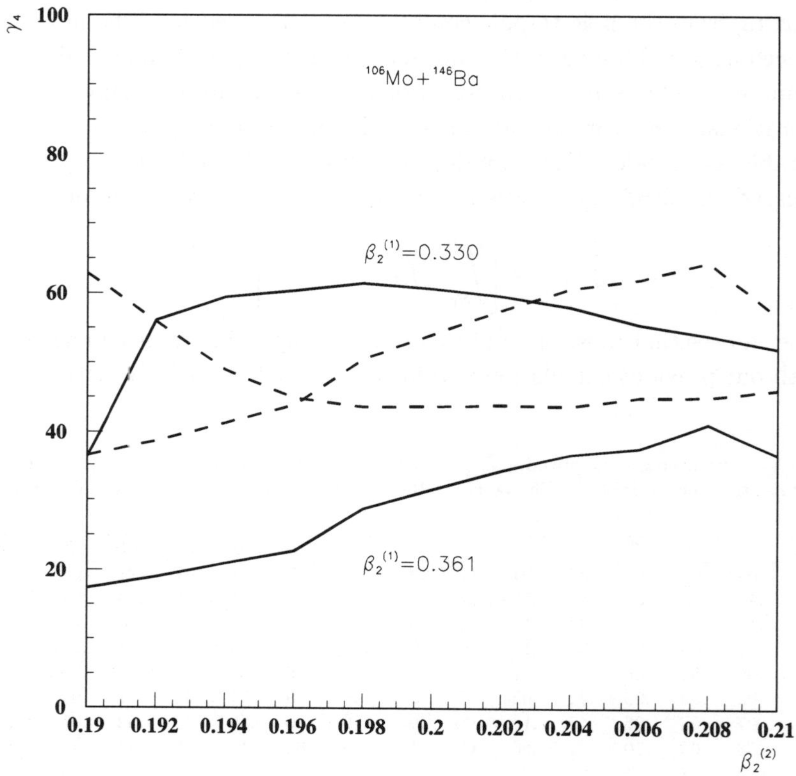

Figure 5. The relative hexadecapole yields γ_4 defined by Eq. (18) versus the quadrupole parameter of the second fragment, for two different quadrupole deformations of the first fragment. The splitting is $^{106}Mo+^{146}Ba$. By solid lines are given the yields of the ^{106}Mo and by dashes those of ^{146}Ba.

Let us now analyse the most "controversial" part of our method, namely the internal repulsive core. The procedure to fix the decay energy using an internal repulsive potential was widely used as a phenomenological approach to describe α and heavy cluster decays. The external barrier, as in our approach, can be estimated on microscopic grounds, using the double folding procedure. From this point of view our analysis is an extension of this approach and the internal repulsive core has only the role in adjusting the energy of some resonant state.

A similar pocket-like interaction was used to describe the so-called quasimolecular resonances in the α and heavy-ion collisions. In this case the existence of such states is connected with rather large relative angular momenta. Thus, the centrifugal force together with the Pauli principle keeps away the two fragments from the strong nuclear attraction. A real "giant molecule" can exist in our case only for very large relative angular momenta. Our belief is that this possibility really occurs if the initial ^{252}Cf is in a strongly superdeformed state with a high angular momentum.

Relative small decay widths in our case correspond to rather large half-lives, but they cannot be interpreted in general as the time the fragments live in a molecular configuration. A small width, as in the usual cluster decay, has the meaning of a small probability for a fission event. This is consistent with the latest experimental findings concerning the gamma-emission from ternary emitted particles in flight [20].

It is already known the important effect of the nuclear deformation on the total decay width in cluster decay processes. Its role increases by increasing the product between the charges of emitted fragments [21]. As it was shown in Ref. [12] in the case of cold fission, where the charges of emitted fragments are large, the effect of the hexadecapole deformation on the total widths becomes also important.

Therefore we expect an important role of the quadrupole and hexadecapole deformation parameters on the partial decay widths $\Gamma_{lI_1I_2}$ in Eq. (12), describing the double fine structure. To this purpose we investigated the partial yields for each fragment, defined as follows

$$\Gamma_I^{(1)} = \sum_l (\Gamma_{lI2} + \Gamma_{lI4}) \ , \quad I = 2,4, \tag{17}$$

$$\Gamma_I^{(2)} = \sum_l (\Gamma_{l2I} + \Gamma_{l4I}) \ ,$$

i.e. by excluding the contribution coming from the transition to the gs, as determined in experiments. Here the upper index (1) denotes the first fragment of the splitting in Table 1, while (2) the second one.

In Figure 5 we studied the role of the quadrupole deformation on the hexadecapole relative yields, defined as

$$\gamma_4^{(k)} = 100 \frac{\Gamma_4^{(k)}}{\Gamma_2^{(k)} + \Gamma_4^{(k)}} \ , \quad k = 1,2. \tag{18}$$

We analyzed here the splitting 7 (^{106}Mo + ^{146}Ba) of the Table 1. We considered a fixed quadrupole deformation for the first fragment and changed the deformation of the second fragment around the value in Table 1. One can see first of all that $\gamma_4^{(1)}$ (solid line) is smaller than $\gamma_4^{(2)}$

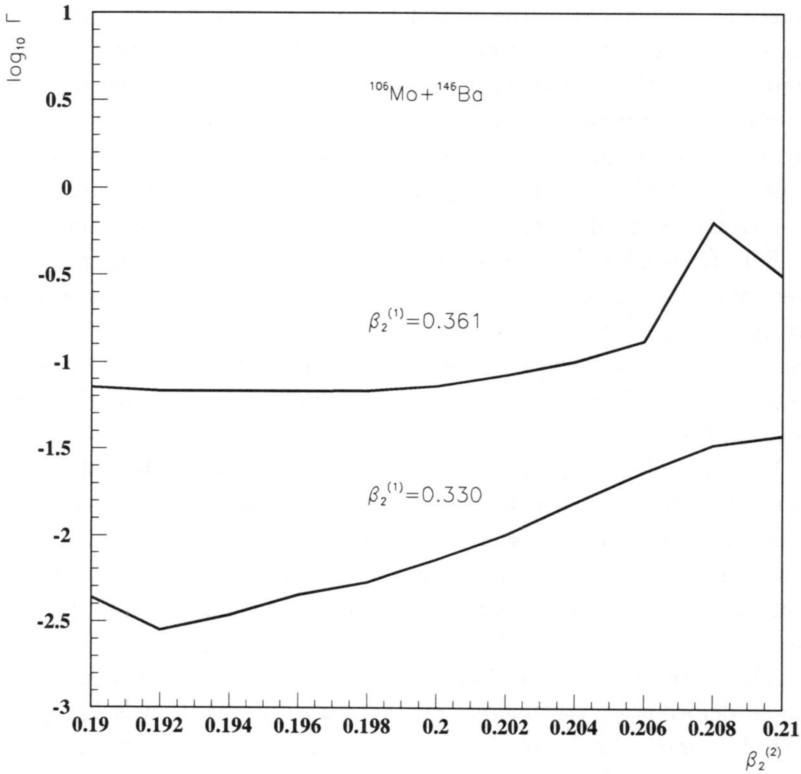

Figure 6. The logarithm of the total width versus the quadrupole parameter of the second fragment, for two different quadrupole deformations of the first fragment. The splitting is $^{106}Mo+^{146}Ba$.

(dashed line). Thus, the experimental situation is reproduced qualitatively. The experimental yields are $\gamma_4^{(1)} \leq 23, \gamma_4^{(2)} = 90[+10,-20]$. On the other hand both yields have a significant variation over a relative small interval $\beta_2^{(2)} \in [0.19, 0.21]$. Moreover, by changing the deformation of the first fragment from its value $\beta_2^{(1)} = 0.361$ up to $\beta_2^{(1)} = 0.330$ the ordering of the two yields becomes inverse. Therefore the relative yields are very sensitive to the quadrupole deformation.

This is also illustrated in Figure 6, where we plotted the logarithm of total widths in the interval $\beta_2^{(2)} \in [0.19, 0.21]$ for the same two values of $\beta^{(1)}$. An increasing of the quadrupole deformation by 0.03 increases the total yield by 1.5 orders of magnitude. We mention that this effect is by one order of magnitude more pronounced than for the α-decay [21].

We investigated in a similar way the role of the hexadecapole deforma-

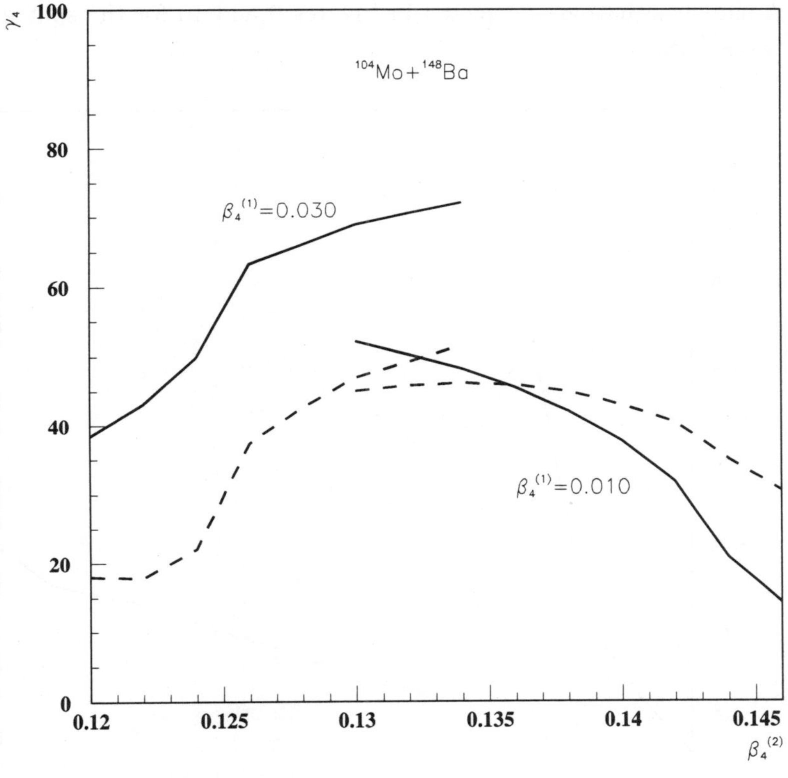

Figure 7. The relative hexadecapole yields γ_4 versus the hexadecapole parameter of the second fragment, for two different hexadecapole deformations of the first fragment. The splitting is ^{104}Mo+^{148}Ba. By solid lines are given the yields of the ^{104}Mo and by dashes those of ^{148}Ba. The quadrupole deformations are $\beta_2^{(1)} = 0.349$, $\beta_2^{(2)} = 0.236$.

tion. In Figure 7 we plotted the yields $\gamma_4^{(1)}$ (solid line) and $\gamma_4^{(2)}$ (dashed line) versus $\beta_4^{(2)}$, in the interval where we found a resonant state. We considered two different values of $\beta_4^{(1)}$ for the splitting 6 (^{104}Mo+^{148}Ba) in Table 1. First of all we obtained an inverse ratio of the yields for the two fragments in comparison with the splitting 7 (^{106}Mo+^{146}Ba). In this case the experimental values are $\gamma_4^{(1)} = 80 \pm 20, \gamma_4^{(2)} \leq 15$. This is again in a qualitative agreement with the experimental situation.

Then we can see a rather strong dependence upon this deformation parameter. Moreover, from Figure 8 one sees that an increasing of the hexadecapole deformation by 0.02 increases the total decay width by almost two orders of magnitude.

A similar situation is depicted in Figures 9 and 10 for the splitting 7 (^{106}Mo+^{146}Ba).

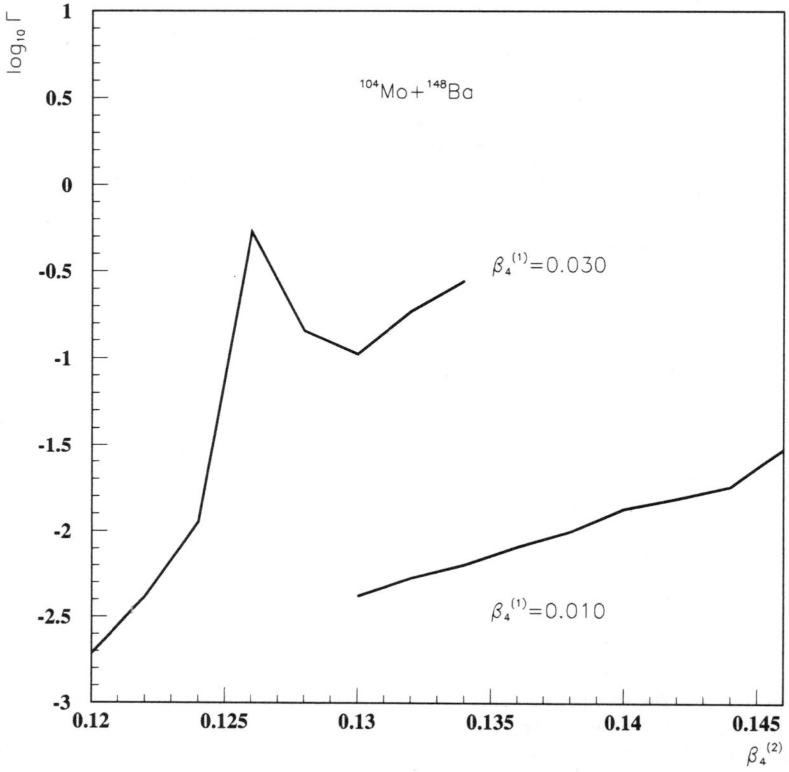

Figure 8. The logarithm of the total width versus the hexadecapole parameter of the second fragment, for two different hexadecapole deformations of the first fragment. The splitting is ^{104}Mo+^{148}Ba.

By taking into account relative large experimental errors we conclude that the experimental situation can be satisfactorily described within our formalism. The differences with respect to experimental values can be also explained by larger than gs deformations of fragments gained in the overlapping region.

Based on this fact we performed some predictions concerning all splittings given in Table 1. Namelly, we estimated in Table 4 the relative yields for all possible pairs of angular momenta (I_1, I_2), computed as follows

$$\gamma_{I_1 I_2} = 100 \frac{\sum_l \Gamma_{l I_1 I_2}}{\Gamma} . \qquad (19)$$

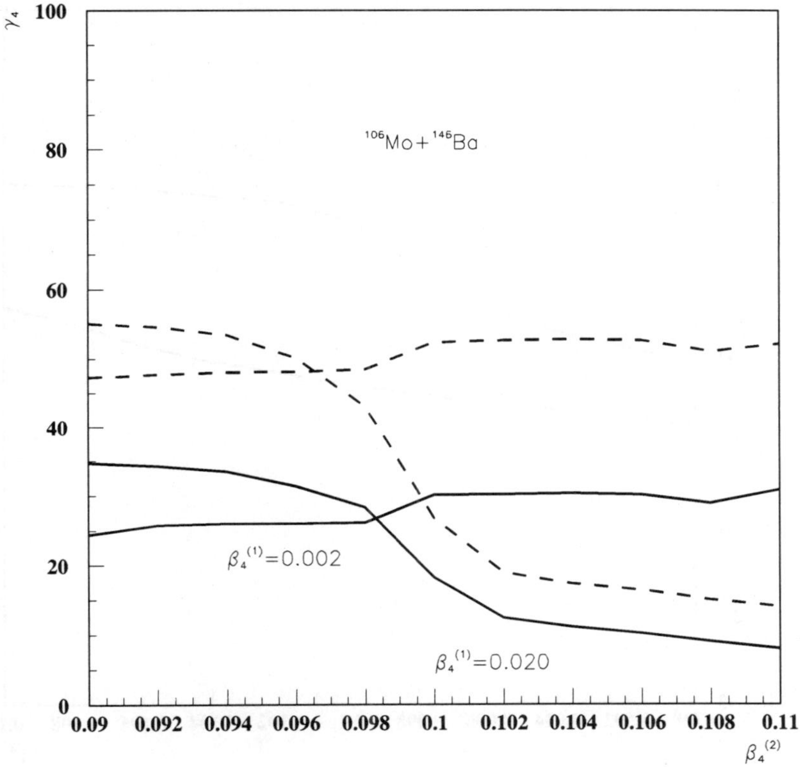

Figure 9. The same as in Figure 7, but for the splitting ^{106}Mo+^{146}Ba. The quadrupole deformations are $\beta_2^{(1)} = 0.361$, $\beta_2^{(2)} = 0.199$.

We mention first of all that the situation is very different from case to case. On the second hand we can see that in the cases 1, 2, 2*, 3, 5, 7 the gs-gs combination is larger than 10%. On the third hand the relative yields in the pair cases (2,2*); (4,4*) and (9,9*) are significantly different. Therefore a proper experimental setup, measuring the above relative yields, would be some kind of "microscope" able to "see" into the relative wave-function components inside the potential barrier. The situation is similar for proton or α-decay fine structure, revealing the details of the internal structure. Here the correlation with the nuclear structure details is enhanced due to the following three factors:

1) the double fine structure of the measured widths,
2) large fragment charges,
3) a relative low Coulomb barrier.

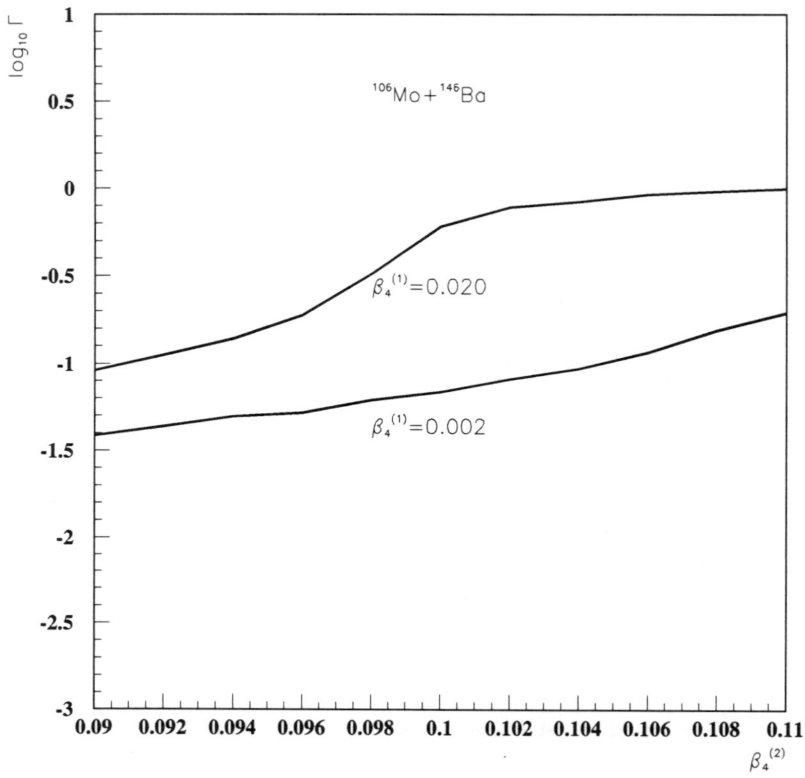

Figure 10. The same as in Figure 8, but for the splitting ^{106}Mo+^{146}Ba.

4. Conclusions

In this lecture we improved and simplified our previous formalism to compute the decay probabilities to final rotational states from the fissioning nucleus ^{252}Cf. To this purpose we used a common rotational basis in order to describe resonant states within the stationary scattering theory.

We estimated the mutual interaction of the two fragments by using the double folding procedure with M3Y two-body plus Coulomb forces. We reproduced the experimental position of the resonant state by an internal repulsive delta-force core. By changing the repulsive strength it is possible to obtain different resonant states inside the resulting pocket-like potential.

We showed that some resonant states have an angular distribution of the probability which is mainly concentrated around the pole-to-pole configuration both in the "planar" and "torsional" degrees of freedom. We choosed them as the best candidates describing cold fission process.

Table 4. The relative yields $\gamma_{I_1 I_2}$, defined by Eq. (19), for the pairs of fragment angular momenta. The labels in the first column correspond to the splittings given in Table 1.

No.	(0,0)	(0,2)	(0,4)	(2,0)	(2,2)	(2,4)	(4,0)	(4,2)	(4,4)
1	13.19	13.13	10.80	30.68	4.56	6.50	12.13	6.71	2.30
2	35.33	22.45	4.08	17.90	13.05	2.80	1.02	2.40	0.97
2*	42.80	27.21	5.36	8.14	8.32	2.90	0.67	2.87	1.72
3	11.13	15.66	15.84	0.45	11.90	0.66	20.55	3.73	20.07
4	1.80	1.32	0.12	0.07	75.73	8.28	0.04	10.03	2.60
4*	4.39	6.98	0.07	29.80	9.25	11.09	2.43	30.39	5.60
5	16.72	6.92	1.17	39.18	9.65	0.97	4.74	5.43	15.22
6	3.25	2.14	14.04	2.32	21.61	0.24	39.52	4.43	12.44
7	6.41	10.36	7.89	21.28	15.17	16.20	4.01	7.03	11.65
8	37.45	12.17	6.86	5.38	27.20	5.59	5.05	0.06	0.24
9	2.21	2.78	0.37	23.42	10.26	3.35	7.87	41.37	8.37
9*	5.47	8.19	1.09	1.87	53.52	3.75	2.93	21.32	1.85
10	8.64	11.87	0.19	1.56	36.31	6.48	2.37	5.13	27.46

We analyzed the role played by quadrupole and hexadecapole deformation parameters. We showed that the relative yields are very sensitive to these parameters. The total decay width increases by one order of magnitude if one increases one of the deformation parameters by 0.02.

We were able to describe qualitatively the experimental situation for two measured splittings, assuming the deformation parameters given by independent calculations. The differences beyond experimental errors can be explained by larger deformations of fragments in the overlapping region. We also predicted relative yields/pairs of fragment spins for ten most intense splittings of ^{252}Cf.

Thus, the cold fission process involving transitions to low-lying states is an useful tool to investigate nuclear structure details. We hope that the conclusions of this lecture will encourage experimentalists to perform an extensive analysis concerning the double fine structure of fission products.

References

1. G.M. Ter-Akopian, et al., *Phys. Rev. Lett.* **73** 1477 (1994).
2. A. Săndulescu, A. Florescu, F. Carstoiu, W. Greiner, J.H. Hamilton, A.V. Ramayya, and B.R.S. Babu, *Phys. Rev.* **C54** 258 (1996).
3. A.V. Ramaya, J.H. Hamilton, J.K. Hwang, and G.M. Ter-Akopian, "Heavy Elements and Related New Phenomena", Vol I, Eds. R.K. Gupta and W. Greiner (World Scientific, Singapore, 1999) p. 477.
4. A.V. Ramayya et al., in Proceedings of the Second International Conference "Fission and Properties of Neutron-Rich Nuclei", Eds. J.H. Hamilton, W.R. Philips, and H.K. Carter, St. Andrews, Scottland, June 28-July 3, 1999, (World Scientific, Singapore, 1999) p. 246.
5. J.H. Hamilton, et al., in Proceedings of the Symposium on "Fundamental Issues in Elementary Matter" Ed. W. Greiner, Bad Honnef, Germany, September 25-29, 2000 (Ep Systema, Debrecen, Hungary, 2000) p. 151.
6. S. -C. Wu, et. al., *Phys. Rev.* **C62** 041601 (2000).
7. A.V. Ramayya, et. al., *Progr. Part. Nucl. Phys.* **46** 221 (2001).
8. M. Jandel, et. al., *J. Phys. (London)* **G28** 2893 (2002).
9. J.H. Hamilton, et. al., *Yad. Fiz.* **65** 677 (2002); *Phys. At. Nucl.* **65** 695 (2002).
10. A. Săndulescu and W. Greiner, *J. Physics* (London) **G3** L189 (1977).
11. A. Săndulescu and W. Greiner, *Rep. Prog. Phys.* **55** 1423 (1992).
12. A. Săndulescu, Ş. Mişicu, F. Carstoiu, A. Florescu, and W. Greiner, *Phys. Rev.* **C58** 2321 (1998).
13. J.O. Rasmussen, et al., in Proceedings of the third International Conference "Dynamical Aspects of Nuclear Fission", Eds. J. Kliman and B.I. Pustylnik Častá-Papiernička, Slovak Republic, August 30-Sept 4, 1996, (JINR, Dubna, 1996) p. 289.
14. D.S. Delion, A. Sandulescu, S. Misicu, F. Carstoiu, and W. Greiner, *Phys. Rev.* **C64** 041303 (2001).
15. D.S. Delion, A. Sandulescu, S. Misicu, F. Carstoiu, and W. Greiner, *J. Phys. (London)* **G28** 289 (2002).
16. D.S. Delion, A. Sandulescu, and W. Greiner, *Phys. Rev.* **C68** 041303(R) (2003).
17. F. Carstoiu and R.J. Lombard, *Ann. Phys. (N.Y.)* **217** 279 (1992).
18. G. Bertsch, J. Borysowicz, H. McManus, and W.G. Love, *Nucl. Phys.* **A284** 399 (1977).
19. P. Möller, J.R. Nix, W.D. Myers, and W.J. Swyatecki, *At. Data Nucl. Data Tables* **59** 185 (1995).
20. J.K. Hwang, et. al., Proceedings of the "Symposium on Nuclear Clusters", Eds. R. Jolos and W. Scheid, Rauischholzhausen, Germany, August 5-9, 2002, (Ep Systema, Debrecen, Hungary, 2003) p. 257.
21. D.S. Delion, A. Insolia, and R.J. Liotta, *Phys. Rev. Lett.* **78** 4549 (1997).

HINDRANCE IN DEEP SUB-BARRIER FUSION REACTIONS

Ş. MIŞICU

National Institute for Nuclear Physics-HH,
Bucharest-Magurele, P.O.Box MG6, Romania
** E-mail:misicu@theor1.theory.nipe.ro*
http://theor1.theory.nipne.ro/ misicu/

H. ESBENSEN

Physics Division, Argonne National Laboratory,
Argonne, Illinois 60439, USA
E-mail: esbensen@phy.anl.gov

The steep falloff in the fusion cross section at energies far below the Coulomb barrier is discussed in the frame of the Coupled Channels metod for various symmetric and asymmetric projectile-target combinations where this phenomenon was very recently discovered. We incorporate a repulsive core in the nuclear potential which accounts for the saturation of nuclear matter and therefore provides the correct value of the incompressibility. The result of the incorporation of the nuclear Equation of State is that the internal part of the sudden heavy-ion potential becomes more shallow and consequently the fusion cross sections decreases at bombarding energies far below the barrier. The importance of the inclusion of the low-lying 2^+ and 3^- states in both target and projectile as well as mutual and two-phonon excitations of these states is highlighted. In overall we obtain a very good fit to the data for the cases ^{58}Ni+^{58}Ni, ^{64}Ni+^{64}Ni and ^{64}Ni+^{74}Ge and a satisfactory fit for the more asymmetric case ^{64}Ni+^{100}Mo. We also present the analysis of the data using diagnostic tools well suited for deep sub-barrier energies, such as the astrophysical factor and the logarithmic derivative ans spin distribution. We predict, in particular, a distinct double peaking in the S-factor for the far subbarrier fusion of ^{58}Ni+^{58}Ni which should be tested experimentally. The relation between anti-resonances and the maximum in the average angular momentum is discussed.

Keywords: heavy-ion fusion; coupled channels; collective states; astrophysical factor; nuclear equation of state.

1. Introduction

Fusion is classicaly forbidden below the Coulomb barrier, but due to quantum tunneling it can nevertheless occur. Sub-barrier fusion depends on energy and projectile-target combination and cannot be explained by standard one-dimensional barrier penetration calculations [1]. The nuclear matter distributions of projectile and target are crucial in determining the heavy-ion potential inside the Coulomb barrier and therefore also the capture rates. In the late '70 it was established that internal degrees of the reacting nuclei play an important role in explaining the enhancement of sub-barrier fusion cross sections. For example, in a sub-barrier fusion reaction of a spherical projectile and a rotational target, the systematic decrease in the slopes of the excitation function were attributed to the effect of increasing ground-state deformations [2]. Apart of this it was later on proved that also the inclusion of zero-point motion of collective surface vibrations is essential in getting an excellent agreement with measured fusion cross sections for energies below the Coulomb barrier [3]. A better understanding of the sub-barrier fusion mechanism is important in at least different areas of nuclear reactions at low-energy, e.g.. the synthesis and the spontaneous cold decay (cluster radioactivity) of heavy and superheavy nuclei on one hand, and the problem of extrapolating the near-barrier data at lower energies for the reaction cross sections of light nuclei, relevant for astrophysical applications.

Very recent investigations performed on several medium-heavy nuclei provided some indications of a steep decrease of the excitation functions at the lowest bombarding energies, This unexpected trend was reported by C. L. Jiang et al. [4–8] . The reported cross sections are measured down to 10 nb : ^{60}Ni+ ^{89}Y [4] ($\sigma_f \geq 100$ nb), ^{64}Ni+ ^{64}Ni [6] ($\sigma_f \geq 10$ nb), ^{64}Ni+^{100}Mo [7] ($\sigma_f > 10$ nb). The suppression of the measured low-energy fusion cross sections with respect to model calculations was later on characterized by a reference energy E_s where the S-factor for fusion develops a maximum at low energy [5] . At first some authors were suspecting a possible failure in the deep sub-barrier regime of the standard theoretical approach to treat capture reactions,

In order to check if there is a break-down of the C. C formalism we proposed at the beginning of 2006 an explanation of the hindrance observed in the sub-barrier fusion of ^{64}Ni+^{64}Ni [9] , based on the same standard C. C. formalism as before [5] but with amendments that concern the potential. Essential in getting a good description of the data was to take into account the saturation of nuclear matter and to use realistic neutron and proton distributions of the reacting nuclei. These two ingredients are naturally incorporated in a potential calculated via the double-folding method with tested effective nucleon-nucleon forces and with realistic charge and nuclear densities, a fact which is often overlooked or only indirectly included in the Woods-Saxon parametrization. In subsequent publications we confirmed this scenario for other combinations: ^{58}Ni+^{58}Ni and ^{64}Ni+^{100}Mo [10] and ^{28}Si+^{64}Ni [11].

We review in this talk the theoretical framework and the results reported in the above mentioned references.

2. Coupled-Channels Approach

We used the same approach as in previous publications (see [12] and references therein), i.e. coupled-channels calculations performed in the so-called iso-centrifugal or rotating-frame approximation, where it is assumed that the orbital angular momentum L for the relative motion of the dinuclear system is conserved. The rotating frame approximation (RFA) allows a drastic reduction of the number of channels used in the calculations. If for example we consider a phonon structure for one of the participating nuclei with account of up to N=3 phonons, then we are facing 33 channels whereas after applying RFA we end-up with only 10 channels (see Fig.1).

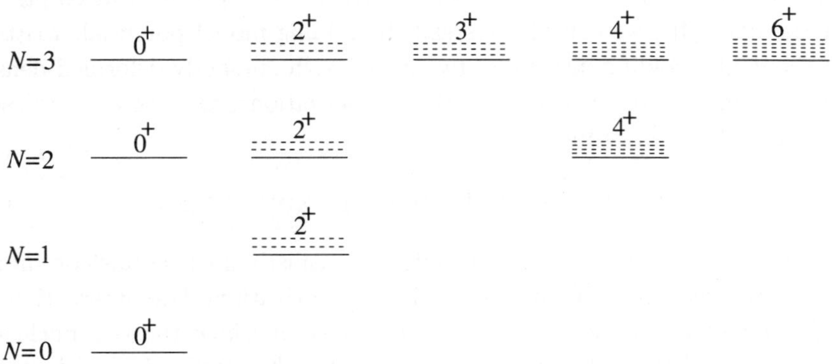

Fig. 1. Three-phonon excitation scheme in a vibrational nucleus. RFA selects only the states denoted with solid lines.

The set of coupled channels reads:

$$\left(\frac{\hbar^2}{2M_0}\left[-\frac{d^2}{dr^2}+\frac{L(L+1)}{r^2}\right]+\frac{Z_1Z_2e^2}{r}+V(r)+\sum_{n_1,n_2}\varepsilon_{n_1,n_2}-E\right)u_{n_1n_2}(r)$$

$$= - \sum_{m_1 m_2} \langle n_1 n_2 \mid \delta V_C + \delta V_N \mid m_1 m_2 \rangle u_{m_1 m_2}(r), \tag{1}$$

where E is the relative energy in the center of mass frame, L is the conserved orbital angular momentum, and M_0 is the reduced mass of the dinuclear system. The C. C. equations (1) are written for two coupled vibrators of eigenenergy ε_{n_1,n_2} and consequently the radial wave function $u(r)$ is labeled by the quantum numbers n_1 and n_2.

As for the spherical part of the potential, $V(r)$, the "proximity" approximation allows us to express it as a function of the shortest distance between the nuclear surfaces of the reacting nuclei:

$$s = r - R_1 - R_2 - \delta R \ , \tag{2}$$

where

$$\delta R = R_1 \sum_{\lambda\mu} \alpha^{(1)}_{\lambda\mu} Y^*_{\lambda\mu}(\hat{r}) + R_2 \sum_{\lambda\mu} \alpha^{(2)}_{\lambda\mu} Y^*_{\lambda\mu}(-\hat{r}), \tag{3}$$

Above $\hat{r}$ specifies the spatial orientation of the projectile-target system in the laboratory frame and $\alpha^{(i)}_{\lambda\mu}$ are the deformation parameters. In the RFA the direction of $\boldsymbol{r}$ defines the z-axis. The only vibrational excitations that can take place are therefore given by $\mu = 0$ components of the distortion amplitudes, since $Y_{\lambda\mu}(\hat{z}) \propto \delta_{\mu,0}$. For the nuclear part of the ion-ion potential, used in a Woods-Saxon parametrization in previous attempts to fit the data [4] , we consider instead the folding model potential, written as the double folding integral of two nuclei with one-body deformed densities ρ_1 and ρ_2, subjected to vibrational fluctuations, and center of masses separated by the distance $\boldsymbol{r}$,

$$V_N(\boldsymbol{r}) = \int d\boldsymbol{r}_1 \int d\boldsymbol{r}_2 \ \rho_1(\boldsymbol{r}_1) \ \rho_2(\boldsymbol{r}_2) \ v(\boldsymbol{r}_{12}), \tag{4}$$

where $\boldsymbol{r}_{12} = \boldsymbol{r} + \boldsymbol{r_2} - \boldsymbol{r}_1$. For the effective realistic nucleon-nucleon interaction we take the M3Y in the Reid parametrization. This interaction is independent of the density of nuclear matter in which the two nucleons are embedded [13] . The nuclear potential for the elastic channel is most conveniently calculated by the Fourier transform [14] ,

$$V(r) = \frac{1}{2\pi^2} \int dq \ q^2 \ \rho_1(q) \ \rho_2(q) \ v(q) \ j_0(qr). \tag{5}$$

Here $\rho_i(q)$ is the Fourier transform of the spherical ground state density of ion i, $v(q)$ the Fourier transform of the effective $N-N$ interactions, and $j_0(qr)$ stands for the spherical Bessel function. The ion-ion potential based on M3Y effective $N-N$ forces is providing a good starting point to evaluate fusion barriers but its interior part is unphysically deep. Because of the lack of the density dependence in the $N-N$ force the nuclear matter is collapsing instead of saturating! In order to account for the effect of the density of the surrounding medium we add to the above double folding integral a second one which has the role to simulate the weakening of the nuclear attraction as the two density start to overlap. The actual form of the repulsive core and its strength depend strongly on the extent to which the collision is adiabatic or sudden. We assumed in our investigations a pure sudden potential. A further uncertainty with regard to the core parameters is the influence of individual characteristics of the considered nuclei, including binding energies, shapes, and the nucleon distributions. The models of ion-ion potentials which provide a repulsive core lead to quantitatively different estimates of the height, radius and diffuseness of the core potential. Details on the effective $N-N$ M3Y interaction and on how one incorporate the repulsive core can be found in ref. [10] .

The non-spherical part of the nuclear potential results from the difference between the total interaction and the potential in the elastic channel. Since linear and quadratic interactions are necessary and often sufficient to fit the data at least in the intermediate energy region (see [12] and references therein) $\delta V(r)$ is expanded up to second-order in the surface distortion (3),

$$\delta V_N(r) = -\frac{\partial V}{\partial r}\delta R + \frac{1}{2}\frac{\partial^2 V}{\partial r^2}\ [(\delta R)^2 - \langle 0|(\delta R)^2|0\rangle]. \tag{6}$$

It is seen that the ground state expectation of this interaction, $\langle 0|\delta V_N|0\rangle$, is zero, but the second order term will give a non-zero contribution to the diagonal matrix element in an excited state, this prescription being exact for a harmonic oscillator up to second order in the deformation amplitudes. We include a similar expansion of the Coulomb field, δV_C, but only to first order in the deformation amplitudes [12] . Expressions for the matrix elements of δV in the double-oscillator basis are given in [15] . These expressions are inserted into the C. C. formalism in the RFA which singles out only axially symmetric distortions ($\alpha_{\lambda\mu=0}$).

The C. C. equations, as written down in eq.(1), are solved with the usual boundary conditions at large distances whereas on the left side of the barrier the so-called in-going-wave boundary condition (IWBC) is imposed, more precisely, at the radial separation $r = r_b$ where the potential pocket attains its minimum. At this point the wave function in channel $n \equiv (n_1, n_2)$ is

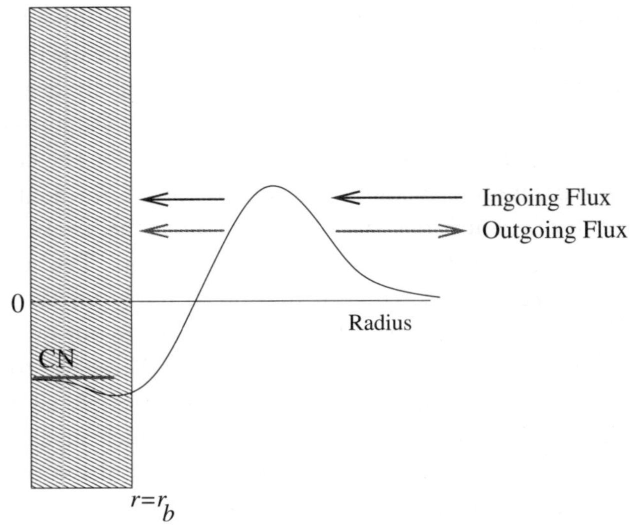

Fig. 2. In-coming wave boundary conditions. Inside the barrier there are only in-going waves in all the channels, whereas outside the barrier there are both in-coming and out-going waves in the entrance channel and only out-going waves in all the channels.

taken in the following WKB form

$$u_n(r) \sim \frac{1}{\sqrt{k_n(r)}} \exp\left[-i\int_{r_b}^{r} k_n(r')dr'\right] \tag{7}$$

to insure that there will be no reflected wave inside the barrier. In the same time on the outer flank of the barrier the quantum flux has ingoing as well as reflected components (see Fig.2).

In the asymptotic region usual scattering conditions are set [15]

$$u_n^{LM}(r) \longrightarrow \delta_{n0}F_L(k_n r) + \mathcal{T}_n{}^L(G_L(k_n r) + iF_L(k_n r)) \tag{8}$$

where the regular, F_L, and irregular, G_L, Coulomb functions contain in the argument the wave-number in channel n,

$$k_n(r) = k_n(\infty)\sqrt{1 - \frac{V(r) + \frac{\hbar^2}{2M_0}\frac{L(L+1)}{r^2}}{E - \varepsilon_n}} \tag{9}$$

and $k_n(\infty) = (2\mu(E - \varepsilon_n)/\hbar^2)^{\frac{1}{2}}$. Once we get the solution of the C. C. equations, together with the boundary conditions (7) and (8), the reaction matrix $\mathcal{T}_n$ is determined. This coefficient can be related to the S-matrix through the relation

$$\mathcal{T}_n{}^L = \frac{1}{2\pi}(\delta_{n0} - S_{n0}{}^L) \tag{10}$$

Eventually the transmission probability is obtained through the formula

$$T_L = 1 - \sum_n | S_{n0}{}^L |^2 \tag{11}$$

3. Diagnostic Tools for the Sub-barrier Fusion

To investigate the complete fusion at extreme sub-barrier there are several tools available. We recall that at energies near or above the barrier the barrier distribution [19] is a usefull tool. However this quantity vanishes deep bellow the barrier.

Before presenting each tool in detail we note that the calculations include one-phonon excitations of the lowest 2^+ and 3^- states in target and projectile, and all two-phonon and mutual excitations of these states up to a 7.2 MeV excitation energy. This energy cutoff was chosen so that all of the two-phonon states were included in the calculations for ^{64}Ni+^{64}Ni, ^{64}Ni+^{74}Ge and ^{64}Ni+^{100}Mo, whereas the two-phonon octupole states were excluded in the calculations for ^{58}Ni+^{58}Ni. The necessary structure input for ^{64}Ni, ^{74}Ge and ^{100}Mo is given in Ref [16]. The input for ^{58}Ni is from Ref. [15] .

3.1. *Cross-Sections*

Once the transmission factors (11) are determined, the fusion cross section is given by the total in-going flux

$$\sigma_F = \frac{\pi}{k_0^2} \sum_L (2L+1) T_L \tag{12}$$

In Fig.3 the calculated as well the experimental fusion excitation functions are displayed for the systems ^{58}Ni+^{58}Ni (experimental data from [20]), ^{64}Ni+^{64}Ni (experimental data from [6]) ^{64}Ni+^{74}Ge (experimental data from [21]) and ^{64}Ni+^{100}Mo (experimental data from [7]) . For the sake of comparison with previous C. C. calculations we present also the case when one uses the Akyüz-Winther potential and the no-coupling case (NOC). We see that when we use the potential with shallower depth the low-energy points are well and even very well reproduced.

The agreement with data, when using the M3Y+repulsion potential, is sensitively better than the one provided by the Akyüz-Winther starting at 90 MeV, for ^{64}Ni+^{64}Ni, and not only for the 4 lowest experimental data points. The excitation function obtained with the M3Y+repulsion potential has the right slope, not only because the potential attains a higher-lying

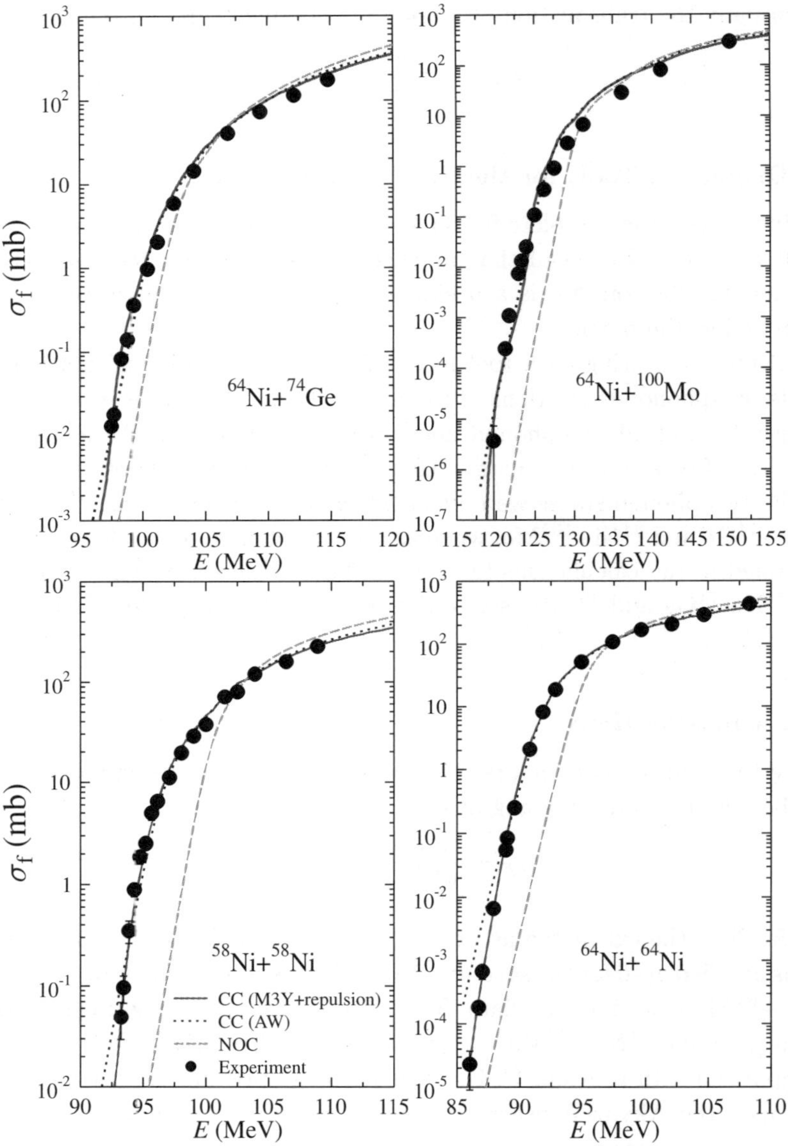

Fig. 3. Experimental fusion excitation functions for the systems ^{58}Ni+^{58}Ni, ^{64}Ni+^{64}Ni, ^{64}Ni+^{64}Ge and ^{64}Ni+^{100}Mo are compared to various C. C. calculations described in the text, and to the no-coupling (NOC) limit for the AW potential.

pocket but also because the curvature of the barrier is different, with a thicker barrier in the overlapping region. For the reaction ^{64}Ni+^{64}Ni, for example, the best χ^2 per point is only $\chi^2/N = 0.86$. This value is obtained by applying the energy shift $\Delta E = 0.16$ MeV to the calculated excitation function. The best fit obtained with AW potential, on the other hand, gives a $\chi^2/N = 10$ and requires an energy shift of $\Delta E = 0.9$ MeV.

However, for the case ^{64}Ni+^{100}Mo the agreement with the data seen in the upper right panel of Fig. 3 is clearly not as good as in the other three cases shown in the same figure, although in order to improve the fit to the data we included up to three phonon excitations of the quadrupole mode in ^{100}Mo using the structure parameters given in Ref. [16] . The reason is that the C. C. effects are very strong for this heavy-ion system and the calculations have not fully converged with respect to multiphonon excitations, as discussed in Ref. [16] . Another problem is that the nuclear structure properties of multiphonon states are often poorly known, so we did not try to improve the fit to the data here.

3.2. *S-factors*

Since the cross section is decreasing by several orders of magnitude in the energy range bellow the barrier (7 orders of magnitude for ^{64}Ni+ ^{64}Ni), two representations are used instead of the cross section in order to remove partly the strong energy dependence caused by barrier penetration. The first one is the "astrophysical" or "nuclear cross-section" factor, $S(E)$ defined by [17]

$$S = E\sigma(E)\ \exp(2\pi\eta), \tag{13}$$

where E is the center-of-mass energy, η the Sommerfeld parameter. The factor $\exp(2\pi\eta)$ is essentially the ratio of quantum probability densities in the asymptotic region and at the minimum of the pocket inside the barrier. It accounts for that part of the quantum tunneling which is independent of nuclear properties such as deformations, nuclear radii, diffuseness. The effect of the ansatz (13) on $\sigma(E)$ is to make it nearly constant at the lowest energy.

At the same time, the S-factor displays typical molecular resonant structures in the excitation function of systems like ^{12}C+^{12}C [18] . The series of narrow and prominent resonances was associated with quasi-bound, long-lived states of the ^{24}Mg nucleus. Thus, the S-factor is a quantity that magnifies structures in the excitation function at energies below the barrier, and it is also an instrument for exploring the inner part of the barrier in low-energy, heavy-ion fusion reactions.

The experimental value of S increases with decreasing bombarding energy and has the tendency to develop a maximum for the systems of interest (see Fig.4). The account of the density dependence in the nuclear interaction is obviously able to explain the maximum in the experimental S-factor.

For the case ^{64}Ni+ ^{100}Mo we reproduce roughly the trend of the S-factor to develope a maximum, when using the potential with repulsive core.

3.3. *Logarithmic Derivative*

Another diagnostic tool proposed in Ref. [5] is the logarithmic derivative,

$$L(E) = \frac{d[\ln(E\sigma)]}{dE} = \frac{1}{E\sigma}\frac{d}{d\,E}(E\sigma). \tag{14}$$

The fact that earlier attempts to reproduce the low-energy data points of the measured fusion cross sections, employing the Akyüz-Winther potential, were unsuccesfull is most clearly seen from the inspection of the logarithmic derivative $L(E)$ (see Fig.5). The inclusion of the repulsive core in the dynamic model explain the divergency in $L(E)$ for all cases (solid curve in Fig.5).

For the case ^{64}Ni+ ^{100}Mo we reproduce the tendency of the logarithmic derivative to develop a divergent behavior, when using the potential with repulsive core. The appearance of a local maximum at 124 MeV, on the other hand, is most likely caused by a poor convergence with respect to multiphonon excitations.

3.4. *Average spin for fusion*

Another tool that we propose to investigate the sub-barrier fusion is the spin distribution whose first moment (average angular momentum) is given by

$$\langle L \rangle = \frac{1}{\sigma_f}\sum_L L\sigma_f(L) \tag{15}$$

In the past it has always been believed that $\langle L \rangle$ for fusion would approach a constant at low energy. However if we use the potential with shallower depth we obtain a narrowing of the spin distribution for fusion as the center-of-mass energy decreases and approaches the pocket energy. An example is shown in Fig. 6, where the measurements of the γ-ray multiplicity from the compound nucleus formed in the fusion of ^{64}Ni+^{64}Ni have been converted into an average angular momentum for fusion [22]. The thin dashed curve

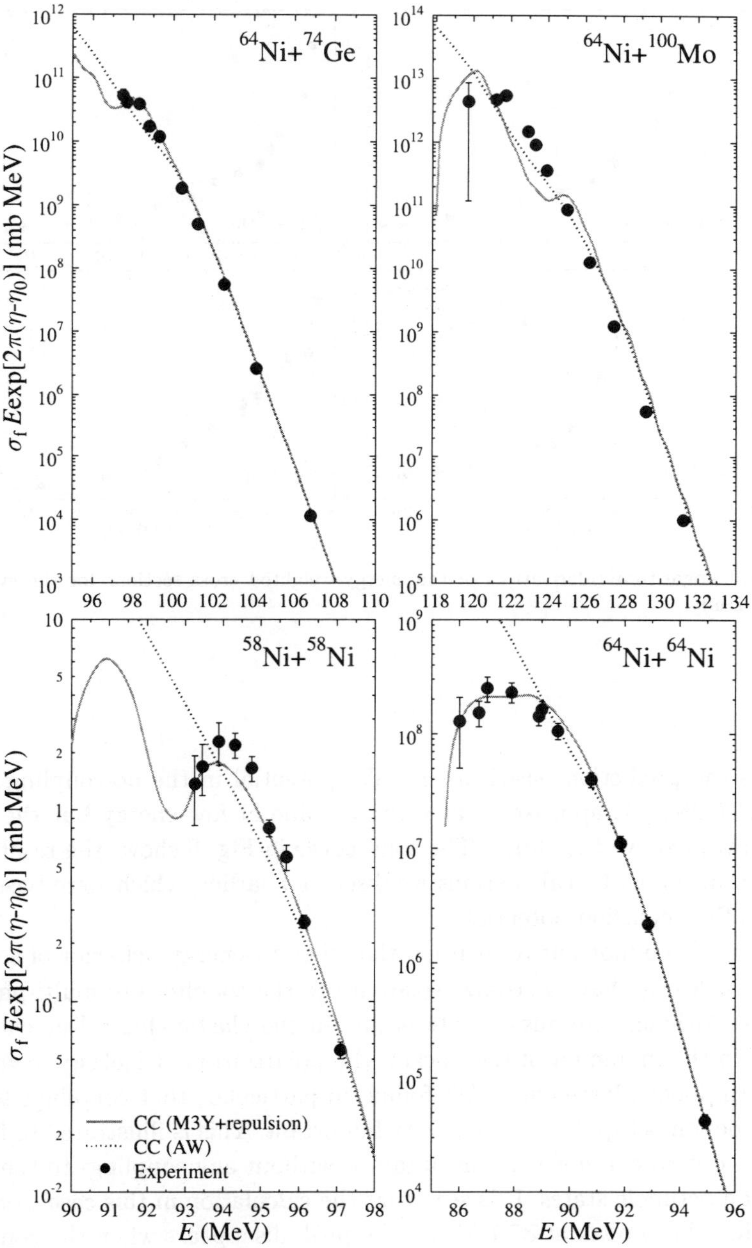

Fig. 4. Experimental S-factors for the systems under study are indicated by solid circles. They are compared to the coupled-channels calculations performed with the M3Y+repulsion (solid curve) and Akyüz-Winther (dashed curve) potentials.

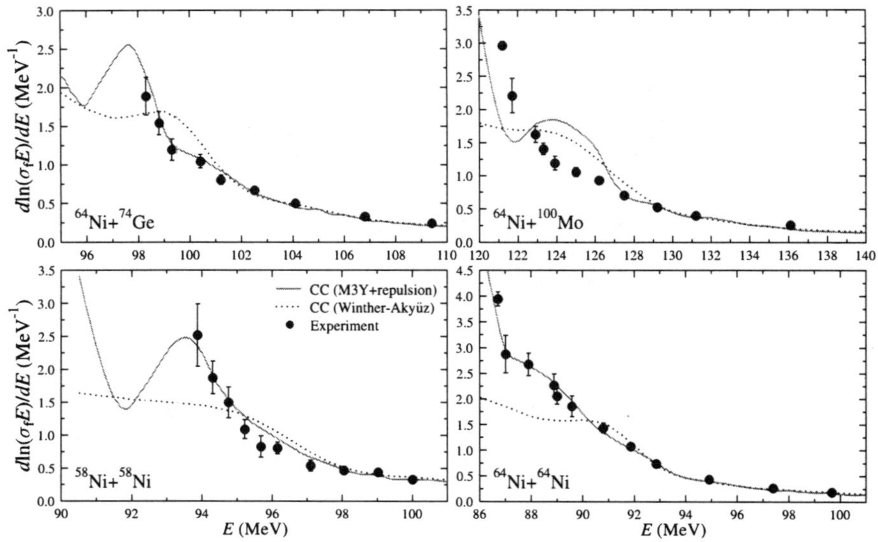

Fig. 5. Logarithmic derivatives of the energy-weighted cross sections for the systems under study.

shows the prediction based on the AW potential in the no-coupling limit (NOC (AW)). It approaches a constant value at low energy but the data are always above that limit. The solid curve in Fig. 6 shows the results we obtain in the C. C. calculations we discussed earlier, which were based on the M3Y+repulsion potential.

It is also important to mention that the low-energy behavior of several observables can have a strong sensitivity to the couplings to multi-phonon states. Although the fusion only occurs in the elastic channel at energies close to the minimum of the pocket, the polarization of inelastic channels can still have a large effect. We found, in particular, that couplings to the two-phonon octupole states are very important. This is illustrated in Fig. 6 by the dotted curve which was obtained without any couplings to the two-phonon octupole states. It is seen that the calculation in this case develops a rather sharp peak at 87.7 MeV. The peak disappears when the coupling to the two-phonon octupole states is included, as illustrated by the solid curve. Unfortunately, the data cannot tell us which of these two calculations is the most realistic.

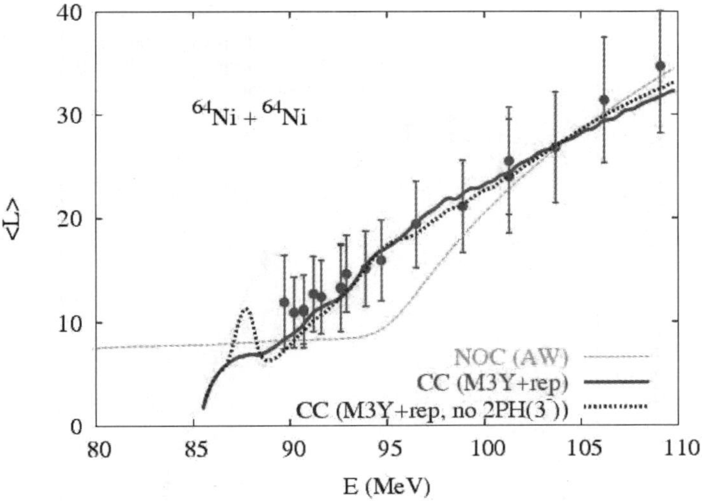

Fig. 6. Average angular momentum for the fusion of ^{64}Ni+^{64}Ni obtained in C. C. calculations based on the M3Y+repulsion potential with (solid curve) and without (thick dotted curve) the effect of couplings to the two-phonon octupole states. The thin dashed curve was obtained in the no-coupling limit (NOC) using the AW potential. The data are from [22] .

4. Other Features of Sub-Barrier Fusion

4.1. *Anti-resonance*

The local maximum that appears in the dotted curve in Fig. 6 is not a resonance. It is the result of an anti-resonance for which the wave function in the elastic channels vanishes at the radial separation where the IWBC are imposed. When this condition is fulfilled at low energy, where fusion is restricted to the elastic channel, the fusion probability will vanish. When it occurs for a range of low angular momenta, it results in a large average angular momentum for fusion and this explains the appearance of the peak in Fig. 6.

The suppression of the fusion probability, which is caused by the anti-resonance conditions, produces a local minimum in the S-factor. This is illustrated by the dotted curve in Fig. 7. The solid curve, which includes the effect of couplings to the two-phonon octupole states, shows a single, broad maximum. In this particular case there is a clear preference for the solid curve which makes an excellent fit to the data.

In the calculations we presented for ^{58}Ni+^{58}Ni we did not include the

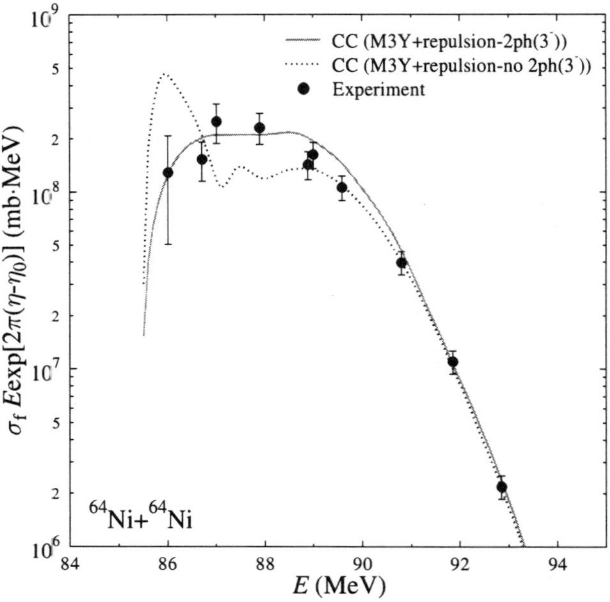

Fig. 7. S factors for the fusion of ^{64}Ni+^{64}Ni obtained in the C. C. calculations based on the M3Y+repulsion potential with (solid curve) and without (dotted curve) the effect of couplings to the two-phonon octupole states. The data are from Ref. [6] .

two-phonon octupole states in the C. C. calculations because the excitation energy is very large, almost 9 MeV, so we did not expect these states would be important. In view of the above discussion it is now understandable why the calculated S-factor for ^{58}Ni+^{58}Ni shown in the left panel of Fig. 4 develops a double-peaked structure at low energy. We have therefore repeated the calculations and included the two-phonon octupole states. The resulting S-factor exhibits a single, broad peak, just as we saw in the fusion of ^{64}Ni+^{64}Ni.

It is not clear why the couplings to the two-phonon octupole states play such an important role as discussed above. However, the analysis of the ^{64}Ni+^{64}Ni data shows a perfect agreement with a single, broad S factor peak, whereas the analysis of the existing fusion data for ^{58}Ni+^{58}Ni fusion has a strong preference for the double-peaked S-factor curve. In view of these findings it is of interest to continue the fusion measurements for ^{58}Ni+^{58}Ni to even lower energies because the existing data [20] shown in Fig. 4 do not verify explicitly the double-peaked structure.

4.2. *Correlation between the pocket energy and E_s*

The results obtained using a potential with a shallower depth motivated us to search for a correlation between the minimum of the potential pocket in the spherical channel and the reference energy that signalize the onset of hindrance in fusion. This correlation can be first guessed it we plot the E_s on the graphic displaying the radial dependence of the total potential. From the inspection of Fig.8, we see that in all four cases E_s is a few MeV above the minimum of the pocket

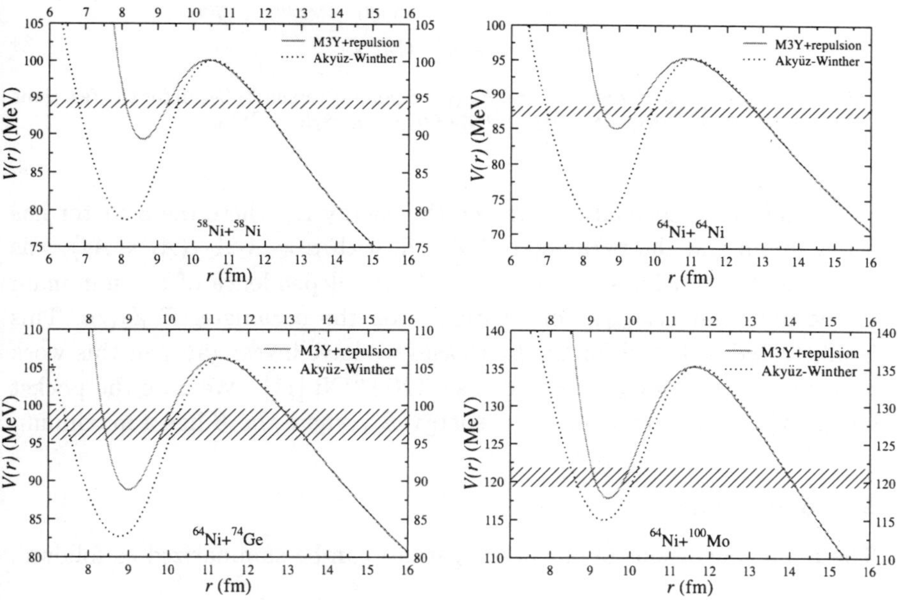

Fig. 8. Ion-ion potentials for ^{58}Ni+^{58}Ni, ^{64}Ni+^{64}Ni, ^{64}Ni+^{74}Ge, ^{64}Ni+^{100}Mo. The solid curves are the potentials employed in the present work. The short-dashed curves are based on the Akyüz-Winther potential, which was used in Ref. [4] . The dashed strips show for each system the experimental boundaries of the threshold energy E_s [8] .

The solid curves in Fig. 8 are based on the M3Y+repulsion potentials we use in this work, whereas the short-dashed curves are based on the AW potential. It is seen that the pockets predicted for the M3Y+repulsion potentials are much shallower than predicted by the AW potential, in particular for the lighter systems. The dashed region in Fig. 8 shows the energy E_s where the experimental S-factor develops a maximum, and the width of it illustrates the uncertainty in the extracted value of E_s [8] .

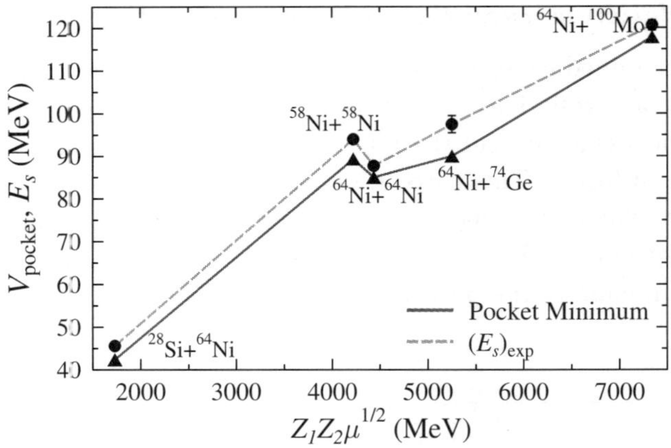

Fig. 9. Plot of $V_{\rm pocket}$ (triangles) and E_s (solid points) versus $Z_1Z_2\sqrt{\mu}$ for the reactions : ^{28}Ni+^{64}Ni, ^{58}Ni+^{58}Ni, ^{64}Ni+^{64}Ni, ^{64}Ni+^{74}Ge and ^{64}Ni+^{100}Mo.

In Ref. [5] a correlation between the energy E_s where the S-factor has a maximum and the parameter $Z_1Z_2\sqrt{\mu}$, with $\mu = A_1A_2/(A_1 + A_2)$, was reported. We considered useful to study the dependence of the minimum of the potential $V_{\rm pocket}$, along with E_s, on the parameter $Z_1Z_2\sqrt{\mu}$. This is illustrated in Fig. 9 for the four cases we have investigated in this work togther with the newly reported case ^{28}Ni+^{64}Ni [11] . We note the pocket energy $V_{\rm pocket}$ follows closely the energy E_s, where the S has a maximum.

5. Conclusions

The main conclusions of our investigations can be summarized as follows :

(1) In order to explain the experimental fusion data at extreme sub-barrier energies we used the double-folding potential based on the Reid parametrization of the M3Y interaction, and realistic parameters of the proton and neutron distributions of both target and projectile, supplemented with a repulsive potential that takes into account the incompressibility of the nuclear matter.

(2) It is necessary to define the fusion in terms of IWBC. When these consitions are imposed at the minimum of the potential pocket, it is possible to reproduce the steep energy dependence of the fusion data at extreme subbarrier energies. Simulating the fusion by the absorption in an imaginary potential, on the other hand, does not allow us to

reproduce the steep falloff of data at the lowest energies.

(3) The recent data from ATLAS for the S-factor exhibits a single maximum for the cases ^{64}Ni+^{64}Ni and ^{64}Ni+^{100}Mo. Our calculations show a single broad maximum in the S-factor at the lowest measured energies in agreement with the measurements. In contrast, our calculations for ^{58}Ni+^{58}Ni show a double-peaked structure of the S-factor. We find that the predicted shape of the low-energy S-factor is very sensitive to the couplings to the two-phonon octupole states. Thus if we ignore these couplings, the S-factor develops a double-peaked structure at low energy, whereas a strong coupling to the these states (as in a harmonic vibration) tends to produce a single, broad low-energy peak. It is therefore important to test the predicted S-factor experimentally at lower energies, in particular in the case of ^{58}Ni+^{58}Ni.

(4) We noticed for the four fusion reactions studied in this paper, together with the case ^{28}Si+^{64}Ni reported in [11] , a correlation between the minimum of the potential pocket $V_{\rm min}$ and the experimentally extracted reference energy E_s, where the S-factor reaches a maximum. A systematic study of fusion reactions over a wider range of values of the parameter $Z_1 Z_2 \sqrt{\mu}$ is necessary to confirm this apparent correlation, and the conjecture of a repulsive core for overlapping configurations.

(5) We also mentioned another feature typical for sub-barrier fusion, which is the narrowing of the spin distribution at energies below E_s. This conjecture could be tested by measurements of the γ-ray multiplicity emitted from the compound nucleus.

Acknowledgements

One of the authors (Ş.M.) is grateful to the Fulbright Commission for financial support and for the hospitality of the Physics Division at Argonne National Laboratory. H.E. acknowledge the support of the U.S. Department of Energy, Office of Nuclear Physics, under Contract No. W-31-109-ENG-38. We are also grateful to C. L. Jiang, B. B. Back, R. V. F. Janssens and K. E. Rehm for usefull discussions.

References

1. H. Esbensen, J. Q. Wu and G. F. Bertsch, Nucl. Phys. **A411**, 275 (1983).
2. R. G. Stokstad et al., Phys. Rev. Lett. **41**, 465 (1978).
3. H. Esbensen, Nucl. Phys. **A352**, 147 (1981).
4. C. L. Jiang *et al.*, Phys. Rev. Lett. **89**, 052701 (2002).
5. C. L. Jiang, H. Esbensen, B. B. Back, R. V. F. Janssens, and K. E. Rehm, Phys. Rev. C **69**, 014604 (2004).
6. C. L. Jiang *et al.*, Phys. Rev. Lett. **93**, 012701 (2004).
7. C. L. Jiang *et al.*, Phys. Rev. C **71**, 044613 (2005).
8. C. L. Jiang, B. B. Back, H. Esbensen, R. V. F. Janssens, and K. E. Rehm, Phys. Rev. C **73**, 014613 (2006).
9. Ş. Mişicu and H. Esbensen, Phys. Rev. Lett. **96**, 112701 (2006).
10. Ş. Mişicu and H. Esbensen, Preprint ANL, PHY-11523-TH-2006.
11. C. L. Jiang *et al.*,Phys. Lett. **B640**, 18 (2006).
12. H. Esbensen, Prog. Theor. Phys. (Kyoto), Suppl. **154**, 11 (2004).
13. M. E. Brandan and G. R. Satchler, Phys. Rep. **285** (1997) 143.
14. Ş. Mişicu and W. Greiner, Phys. Rev. C **66**, 044606 (2002).
15. H. Esbensen and S. Landowne, Phys. Rev. C **35**, 2090 (1987).
16. H. Esbensen, Phys. Rev. C **72**, 054607 (2005).
17. E. M. Burbridge, E. Margaret, G. R. Burbridge, W. A. Fowler, and F. Hoyle, Rev. Mod. Phys. **29**, 547 (1957).
18. K. A. Erb and D. A. Bromley, in *Treatise in Heavy-Ion Science*, Vol. 3, ed. D. A. Bromley (Plenum, 1985).
19. N. Rowley, G. R. Satchler and P. H. Stelson, Phys. Lett. **B254**, 25 (1991).
20. M. Beckerman, J. Ball, H. Enge, M. Salomaa, A. Sperduto, S. Gazes, A. Di Rienzo and J. D. Molitoris, Phys. Rev. C **23**, 1581 (1982).
21. M. Beckerman, M. Salomaa, A. Sperduto, J. D. Molitoris and A. Di Rienzo, Phys. Rev. C **25**, 837 (1982).
22. D. Ackerman *et al.*, Nucl. Phys. **A609**, 91 (1996).

QUESTIONS OF THE MICROSCOPICAL OPTICAL POTENTIAL FOR ALPHA-PARTICLES AT LOW ENERGIES

M. Avrigeanu*

"Horia Hulubei" National Institute for Physics and Nuclear Engineering,
P.O. Box MG-6, 76900 Bucharest, Romania
**E-mail: mavrig@ifin.nipne.ro*
www.nipne.ro, tandem.nipne.ro/~mavrig/

The main questions which are still open concerning the optical model potential for α-particles at low energies are discussed with respect to the α-particle elastic scattering as well as α-induced reactions and α-particle emission. In order to understand the differences between the optical potential parameters used to describe the α-particle emission from excited compound residual nuclei, as compared to those determined by analysis of α-particle elastic scattering, the double-folding model is involved within semi-microscopic analysis of the α-particle elastic scattering on medium-mass nuclei, at energies below 30 MeV, and next involved within calculations of (n, α) reaction cross sections. The corresponding phenomenological optical potentials which match each other in the outer limit of the nuclear surface are shown to be described by a temperature-dependent shape of the nuclear density, which produces decreased central values and a larger diffuseness of the DF real potential.

Keywords: nuclear matter density, optical potential, double-folding model, statistical compound-nucleus reactions, (α, α_0) elastic scattering, (n, α) reactions.

1. Introduction

The α-nucleus interaction is essential for the understanding of nuclear structure and nuclear reactions. The concept of the α-particle mean field has been widely used to unify the bound and scattering α-particle states in a similar way to use of the nuclear mean field to calculate the properties of bound single-particle states and also the scattering of unbound nucleons by nuclei. Moreover, at positive energies the α-particle mean field is simply the familiar α-particle optical model potential (OMP), which describes properly the differential cross sections for the elastic scattering of α-particles by nuclei. Unlike the nucleon case, there are however no global optical potentials for α-particles that fit to good accuracy the scattering from many nuclei over a wide range of energies, except a global parameter set [1] for α-particles with energies above 80 MeV which has been moreover extended to lower energies and proved appropriate to describe the (n, α) reactions [2].

Actually, McMahan and Alexander reemphasized earlier [3] that not only fusion reactions are the inverse of evaporative decay, but also the general expectation for elastic-scattering sensitivity to the real potential at distances well beyond the s-wave barrier distance while fusion cross sections at low energy are considered to be especially sensitive to the height of the s-wave barrier and its penetrability. From comparison of calculated to measured α-particle evaporation spectra they found a significant effective barrier reduction between ground-states and excited states of nuclei. The need for new physics in potentials to describe nuclear deexcitation within the statistical model calculations was next pointed out by La Rana *et al.* [4]. A tail in the nuclear density distribution was introduced with an effective increased heavy-ion mean evaporation radius, in a manner unattainable by axially symmetric deformations alone. It was thus suggested that particle evaporation occurs from a transient nuclear stratosphere of the emitter nucleus, with a density that differs from cold nuclei and which has not yet relaxed to the density profile expected for complete equilibration [5]. This additional degree of freedom was proposed to be accounted for by effective temperature-dependent potentials for particle emission channels as, *e.g.*, a contracted Woods-Saxon shape plus a tail at the surface region [6]. Furthermore, the formation of a diffuse nuclear surface in hot nuclei has been confirmed by self-consistent Hartree-Fock calculations [7]. These results as well as the need of lowering the barrier in the statistical model [8,9] supports the conclusion that the temperature dependence of the nuclear density distribution function could be the missing degree of freedom that needs to be included in conventional statistical model calculations.

On the other hand, results from the analysis of the low-energy elastic scattering data suffer from discrete and continuous ambiguities in the OMP parameters, whose uncertainties vary for various target nuclei and for different incident energies due to the precision of the data analysed. In order to avoid too much phenomenology in the description of these data, numerous attempts have been made to replace the phenomenological real potential of Woods-Saxon (WS) type by a more microscopic α-nucleus potential using an effective interaction. Thus, the double-folding (DF) model [10] has become widely used with an effective nucleon-nucleon (NN) interaction folded with the mass distributions of both the target nucleus and the projectile. While the M3Y Reid [11] and Paris [12] are the most familiar interactions [10,13,14] a density dependence has been also incorporated especially for the description of refractive nuclear scattering at higher energies. The scattering data proved to be in this case sensitive to a wide radial domain and various specific forms have been adopted in the M3Y-interactions [15–17] for the corresponding density-dependent factor so as to reproduce consistently the same saturation properties of nuclear matter in Hartree-Fock calculations. Nevertheless, the M3Y interaction can be used only to obtain the real potential, and the imaginary term must be parameterized independently (*e.g.*, Ref. [14]) or simply taken from a phenomenological [10] OMP. However, the former approach may reduce the number of the optical potential parameters and corresponding uncertainties, with the success proved in the description [14] of the elastic scattering of many systems. The DF method has actually been widely used during the past two decades in order to generate the real part of the potential for nucleons, α-particles and heavy-ion optical potentials (*e.g.*, Refs. [10,13,17]). Nevertheless, the situation is considered less clear for the α-nucleus OMP at low energies, where the imaginary-potential is strongly energy dependent and nuclear structure effects should be taken into account [18] while the data are mainly sensitive to the potential at the nuclear surface.

The suitable constraints for OMP parameters at α-particle energies around the Coulomb barrier were concerned in addition to the results of Khoa [17] by means of a semi-microscopic OMP with a DF real part and no adjustable parameter or normalization constant [19]. An additional check for use of more suitable interaction and density distribution for the α-particle was also performed through an α-α elastic-scattering analysis, as shown in Sec. 2. Semi-microscopic analysis of the experimental α-particle elastic scattering on medium-mass nuclei at energies below 32 MeV have next been performed [19–21] and reviewed in Sec. 3, in order to adopt a proper energy-dependent phenomenological imaginary part. These studies have been completed by a full phenomenological analysis of the same data, making possible a comparison of the corresponding real potential part with the DF potential. While the available α-induced or (n, α) reaction data were not taken into account within these analyses, in order to avoid the question arising due to the remaining parameters of the statistical model, a further step has concerned just this different kind of data [21,22] in Sec. 4. Finally, effects due to changes of the nuclear density at a finite temperature and considered within DF formalism [22] have been proved as an important aspect to be included in statistical-model calculations even for temperatures $<$2 MeV, as shown in Sec. 5, while conclusions are summarised in Sec. 6.

2. The semi-microscopic DF approach

2.1. *Double-Folding real potential*

The DF formalism for the α-nucleus optical potential has more recently been revised at α-particle energies above 80 MeV, in order to study the exchange effects and density dependence of the effective NN-interaction [17]. On the other hand, a review of the elastic α-nucleus scattering and three improved global α-particle OMPs at low energies have recently been published [23] with the aim to determine a global potential over the whole mass region that is able to reproduce all the existing experimental data for α-particle elastic scattering as well as α-induced and (n, α) reactions. Three global semi-microscopic OMPs finally derived within this comprehensive study have the real part obtained by using the DDM3Y interaction within the DF procedure, and WS imaginary parts with either a purely volume imaginary term (I), or a volume plus surface imaginary potential (II), as well as a damped surface potential together with the dispersive contribution to the real DF potential (III). It was however found, in spite of the quite distinct assumptions, that overall these three OMPs lead to cross sections which do not exhibit any substantial differences apart for some cases at backward angles. Thus, it has been concluded [23] that they can be regarded as providing at the uncertainty of a factor 10 up to which it is possible today to predict globally α-induced reaction cross sections. Recent measurements [24,25] of the α-particle elastic scattering with almost the same high accuracy over five orders of magnitude, pointed also to ambiguities which do not allow the unique determination of the shape of OMP imaginary part unless a wider energy range will be involved.

The basic formulae for calculations of the real part of the optical potential based on the DF model (DFM) are given in the following only in close connection with the changes involved within our approach [19,22] while the rest of the model assumptions were discussed previously [26,27]. The direct $U^D_{0(1)}$ and exchange $U^{EX}_{0(1)}$ real parts of the microscopic optical potential $U(\mathbf{R}, E)$ are obtained in terms of the projectile and target nuclear densities $\rho_1(\mathbf{r}_1)$ and $\rho_2(\mathbf{r}_2)$, respectively, folded with the density- and energy-dependent effective NN-interaction $v^X_{ij}(\rho, E, \mathbf{s}{=}\mathbf{R}{+}\mathbf{r}_2{-}\mathbf{r}_1)$ [10,15], where $\mathbf{s}$ is the distance between nucleons, and $\mathbf{R}$ is the distance vector between the two densities. Thus we have

$$U^D_0(E, \mathbf{R}) = \int d\mathbf{r}_1 \int d\mathbf{r}_2\, \rho_1(\mathbf{r}_1)\, \rho_2(\mathbf{r}_2)\, v^D_{00}(\rho, E, \mathbf{s}), \tag{1.a}$$

$$U_1^D(E,\mathbf{R}) = \int d\mathbf{r}_1 \int d\mathbf{r}_2 \, (\rho_{n1}(\mathbf{r}_1) - \rho_{p1}(\mathbf{r}_1))\,(\rho_{n2}(\mathbf{r}_2) - \rho_{p2}(\mathbf{r}_2))$$
$$v_{01}^D(\rho, E, \mathbf{s}), \tag{1.b}$$

and for the exchange terms:

$$U_0^{EX}(E,\mathbf{R}) = \int d\mathbf{r}_1 \int d\mathbf{r}_2 \, \rho_1(\mathbf{r}_1, \mathbf{r}_1 + \mathbf{s})\rho_2(\mathbf{r}_2, \mathbf{r}_2 - \mathbf{s})$$
$$v_{00}^{EX}(\rho, E, \mathbf{s}) \exp\left[\frac{i\mathbf{s}\mathbf{k}(\mathbf{R})}{M}\right], \tag{1.c}$$

$$U_1^{EX}(E,\mathbf{R}) = \int d\mathbf{r}_1 \int d\mathbf{r}_2 \, (\rho_{n1}(\mathbf{r}_1, \mathbf{r}_1 + \mathbf{s}) - \rho_{p1}(\mathbf{r}_1, \mathbf{r}_1 + \mathbf{s}))$$
$$\times(\rho_{n2}(\mathbf{r}_2, \mathbf{r}_2 - \mathbf{s}) - \rho_{p2}(\mathbf{r}_2, \mathbf{r}_2 - \mathbf{s}))$$
$$\times v_{01}^{EX}(\rho, E, \mathbf{s}) \exp\left[\frac{i\mathbf{s}\mathbf{k}(\mathbf{R})}{M}\right], \tag{1.d}$$

where $M = A_1 A_2/(A_1 + A_2)$, while $v_{00(01)}^D(\rho, E, \mathbf{s})$ and $v_{00(01)}^{EX}(\rho, E, \mathbf{s})$ are the direct and exchange isoscalar and isovector components of the effective NN-interaction, respectively, $\rho_{n(p)1(2)}(\mathbf{r}_{1(2)})$ are the neutron- and respectively the proton-density distributions of the projectile (1) and target (2) nuclei, and $\rho_{1(2)}(\mathbf{r}_i, \mathbf{r}_j)$ in Eqs. (1.c) and (1.d) are the corresponding density matrices. The knock-on exchange term of the folded potential has been calculated by using the approximation of Campi and Bouyssy [28], which preserves the first term of the expansion given by Negele-Vautherin [29] for the realistic density-matrix:

$$\rho(\mathbf{R}, \mathbf{R} + \mathbf{s}) = \rho(\mathbf{R} + \frac{\mathbf{s}}{2})\,\hat{j}_1\left(k_{av}(\mathbf{R} + \frac{\mathbf{s}}{2})\,s\right), \tag{2}$$

where $\hat{j}_1(x) = 3\,(\sin x - x \cos x)/x^3$. Here k_{av} defines the average relative momentum as a function of $\mathbf{R}$, the density distribution $\rho(r)$, and of the kinetic-energy density $\tau(r)$ for each participant in the interaction [28].

Moreover, $\mathbf{k}(\mathbf{R})$ is the relative momentum of the nucleons, while the average relative momenta of nucleons in the α-particle [26,30] and target [27] have been considered within the modified Thomas-Fermi approximation of Krivine and Treiner [31] in terms of the kinetic-energy density τ and the local Fermi momentum [19,22] $k_F(r)$.

Basic ingredients of the DFM calculations as the nuclear-density distributions of the interacting nuclei and the effective NN-interaction, have been discussed previously [19]. Thus, based on the analysis of the $\alpha - \alpha$ elastic-scattering angular distributions measured at incident energies between 9 and 30 MeV, the use of the density distribution of the α-particle of Baye *et al.* [32] has been introduced,

$$\rho_\alpha(R) = 4\left(\frac{3\pi\, b^2}{4}\right)^{-3/2} exp(-4r^2/3b^2)\ , \tag{3}$$

with the parameter b=1.28. A better agreement with the data has been thus obtained as compared to the Satchler-Love [10] and Tanihata *et al.* [33] forms. Next, we found that the data are better described by using the M3Y-Reid [34] than by the M3Y-Paris [35] effective NN-interaction, both folded with the Baye *et al.* α-particle density distribution (Fig. 1). A density dependence of the M3Y effective NN-interaction which accounts for the reduction of the interaction strength with increasing density was also chosen. From the analysis of the angular distributions of the elastically scattered α-particles on ^{90}Zr at energies between 21 and 25 MeV, the best description of the experimental data was found for the density-dependence of the effective NN-interaction in the linear BDM3Y form. On the other hand, an energy-dependent factor [15] has been considered as a linear function of the incident energy per nucleon E, of the form $g(E)$=1-0.002E. Finally, the nuclear density distribution of the target nuclei which has been described by means of a two-parameter Fermi-type function with the parameters chosen to reproduce the electron scattering data [36,37] and the shell model calculations [38].

A basic point of the present approach is the fact that no adjustable parameter or normalization constant have been involved in order to obtain the microscopic DF real potential. However, in the following section it will be shown that a temperature-dependent nuclear density distribution for the excited nuclei is needed. We have been guided in this respect by the Fermi-type local density distribution $\rho(r,T)$ determined within the extended Thomas-Fermi (ETF) method by Antonov *et al.* [42]:

$$\rho_{ETF}(r,T) = \rho_0(T)\ /\ [1 + exp[(r - R(T))/\alpha(T)]]^{\gamma(T)}\ . \tag{4}$$

Since they calculated and listed the values of the four temperature-dependent parameters $(\rho_0, R, \alpha, \gamma)$ entering this equation, for the two kinds of nucleons in the nucleus ^{208}Pb and for temperatures up to T=4 MeV, one

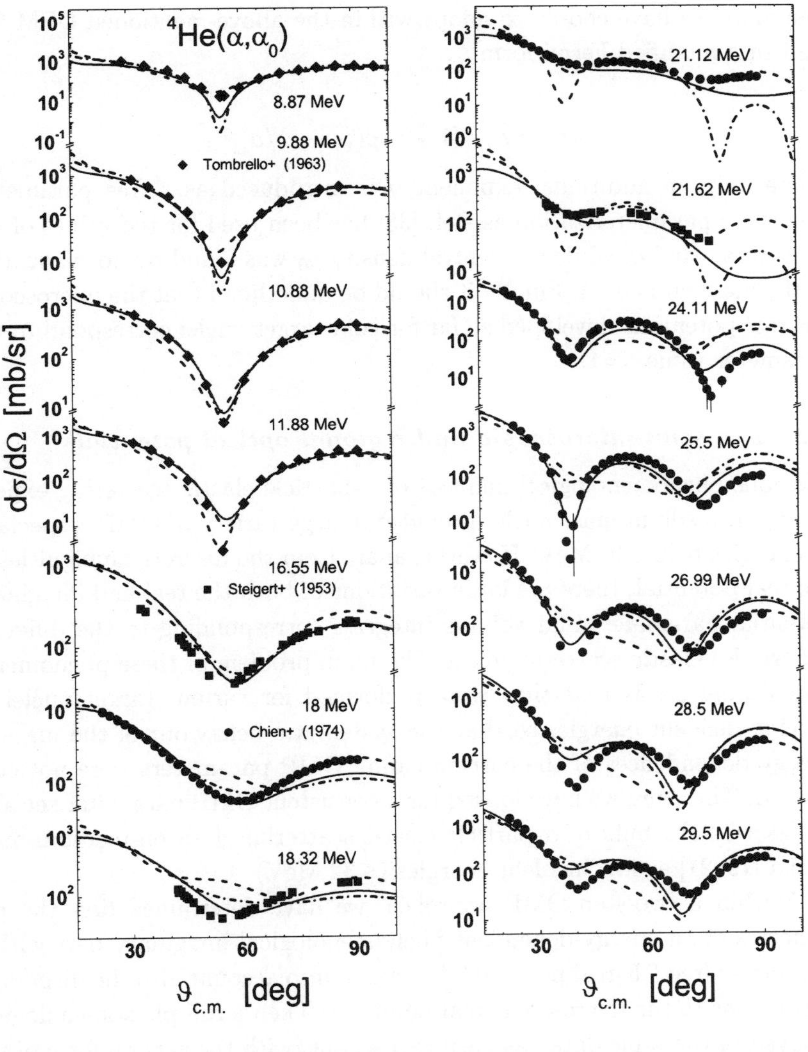

Fig. 1. Comparison of measured [39–41] and calculated elastic-scattering differential cross sections for the $\alpha+\alpha$ system by using the M3Y-Reid (solid curves) and M3Y-Paris (dashed curves) effective NN interactions at the same time with the α-particle density distribution of Baye *et al.* [32], as well as the M3Y-Reid effective interaction and the density distribution of Satchler and Love [10] (dot-dashed curves).

may see that variation of the $\rho_0(T)$, $R(T)$ and $\alpha(T)$ parameters within this temperature range has been generally below 5%, while the exponent $\gamma(T)$ decreases by ~30-50% with increasing temperature in the same range.

Therefore we have chosen to adopt within the above-mentioned DFM formalism a modified Fermi form:

$$\rho(r) = \rho_0 \,/\, [1 + exp((r-c)/a)]^{\gamma} \,, \tag{5}$$

where only an additional exponent γ is introduced as a free parameter. The same parameterization as Ref. [38] has been used for the values of the constants c and a, while the central density ρ_0 was found by normalization to the mass number A. Finally it should be underlined that the microscopic DF real potentials developed so far for cold target nuclei correspond to the parameter value γ=1.

2.2. *The semi-microscopic and regional optical potentials*

Previous phenomenological analysis of α-particle elastic scattering experiments on medium-mass nuclei provided a large variety of OMPs especially at energies below 40 MeV. However, apart from the discrete ambiguities in the real potential, there are large variations of both the real and imaginary potential parameters and volume integrals corresponding to the different α-particles elastic scattering data. The main problem of these phenomenological analyses is that they were performed for various target nuclei at specific incident energies so that the systematic behaviour of the mass or energy dependences of the corresponding OMP parameters were not considered. Therefore we have looked for a consistent OMP parameter set able to describe the bulk of α-particle elastic scattering data on medium-mass nuclei [19–21] at low incident energies ($\leq$32 MeV).

Within a two-step OMP approach, we have determined first the parameters of an energy-dependent phenomenological imaginary part with a parameter free DF-real potential, by taking into account also the dispersive corrections to the microscopic real potential. Then a full phenomenological analysis of the same data has been carried out, with the general form of the optical potential with Woods-Saxon form factor also for the real part, and the imaginary part unchanged from the former semi-microscopic analysis:

$$\begin{aligned} U(r) = V_C(r) + V_R\, f(r, R_R, a_R) \; + \; iW_V\, f(r, R_V, a_V) \\ + iW_D\, g(r, R_D, a_D) \,, \end{aligned} \tag{6}$$

where $f(r,R_i,a_i)$=(1+exp[(r-R_i)/a_i])$^{-1}$, $g(r,R_i,a_i)$=-4a_id/dr[$f(r,R_i,a_i)$], and R_i=r_i $A^{1/3}$, A being the target-nucleus mass number. $V_C(r)$ is the Coulomb potential of a uniformly charged sphere of radius R_C while r_C=

1.30 fm. The advantage of having well settled already at least half of the usual OMP parameters increases obviously the accuracy of the local fit of data. The corresponding average mass-, charge-, and energy-dependent OMP parameters, similar to those introduced by Nolte *et al.* [1] above 80 MeV, have been obtained in the end of this latter step providing a regional optical potential (ROP).

3. (α,α_0) semi-microscopic and phenomenological analyses

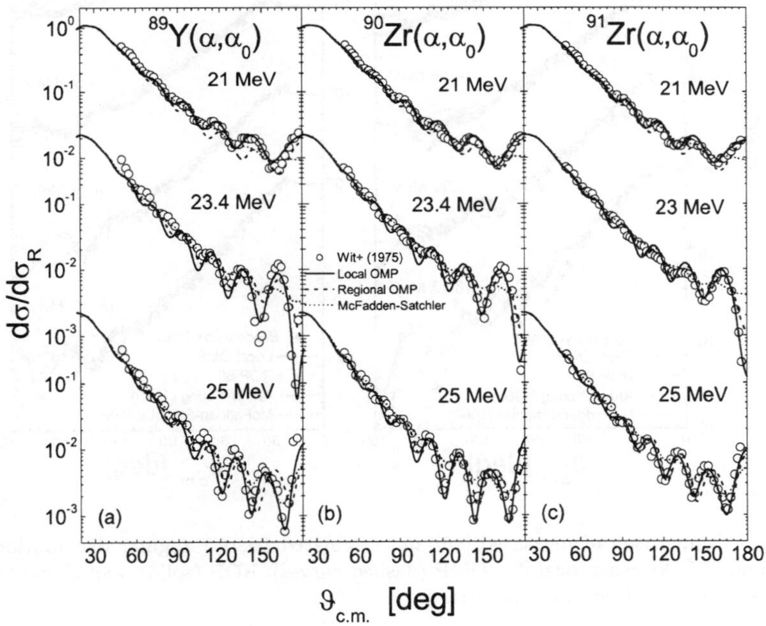

Fig. 2. Comparison of measured [44] and calculated (α, α_0) angular distributions on ^{89}Y, ^{90}Zr, and ^{91}Zr at 21, 23.4 and 25 MeV, using the local phenomenological OMP based on the semi-microscopic analysis (solid curves), the corresponding ROP parameter set [19] (dashed curves) and the global OMP [45] (dotted curves).

Semi-microscopic and phenomenological optical potential analyses of the α-particle elastic scattering at energies below 32 MeV have been completed by using the above-mentioned potentials for A$\sim$100 nuclei [19] and A$\sim$60 nuclei [21]. The scattering cross sections were calculated by using the computer code SCAT2 [43], modified to include the semi-microscopic DF potential as an option for the OMP real part. As can be seen in Fig. 2 for reference data sets [44], a much better description is obtained in comparison with the well-known global parameter set of McFadden and Satchler [45]. This is true for both the local parameters sets obtained by fit of each elastic-scattering angular distribution, and the regional parameter set provided by the mass-, charge-, and energy-dependent average of the local values.

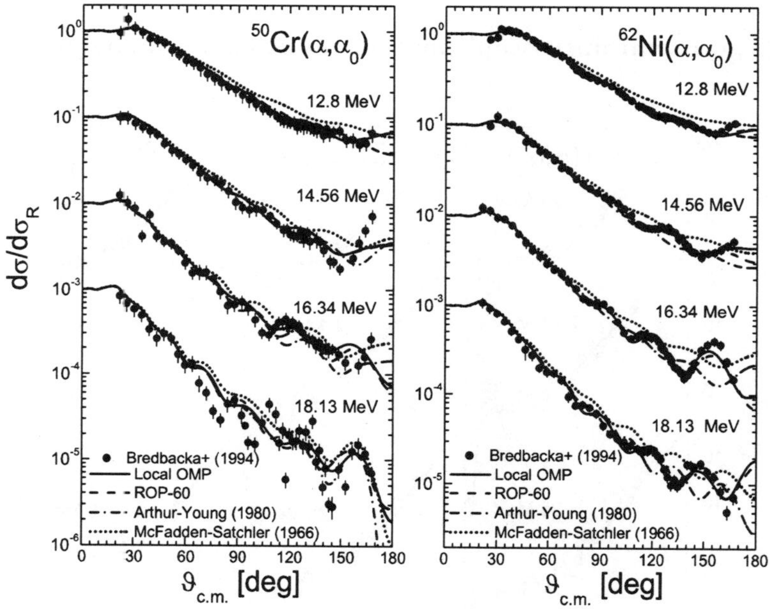

Fig. 3. Comparison of measured [46] and calculated (α, α_0) angular distributions on ^{50}Cr and $^{58,62}Ni$ using local [21] OMP (dashed curves), ROP (solid), and global OMPs of Refs. [45,47] (dotted, dash-dotted).

One can see that the regional potentials [19,21,22] is able to describe appropriately even nuclear absorption for the particular case of Wit *et al.* data [44] at 23 MeV. A similar behaviour is proved for A∼60 nuclei and α-particle energies below 20 MeV (Fig. 3), of first astrophysical interest, in comparison with the predictions of the also well-known global potential of Arthur and Young established [47] within an analysis of fast neutron-induced reactions on $^{54,56}Fe$. Therefore we find that a rather suitable description of the analyzed data is provided by OMPs with real-potential diffuseness a_R notably lower than the a_R-values which are needed in order to describe the α-particle emission from excited compound nuclei [2].

4. Total α-reaction and (α,n) cross sections

A basic point of the ROP for A∼60 based on the semi-microscopic analysis [21] consisted in the analysis of the particularly accurate (α,n) and total α-reaction cross sections for ^{48}Ti and ^{51}V measured and respectively established by Vonach *et al.* [48]. These data are described [48] within ∼10% by the global OMPs [45,47] but overestimated by the OMP for α-particle emission [2].

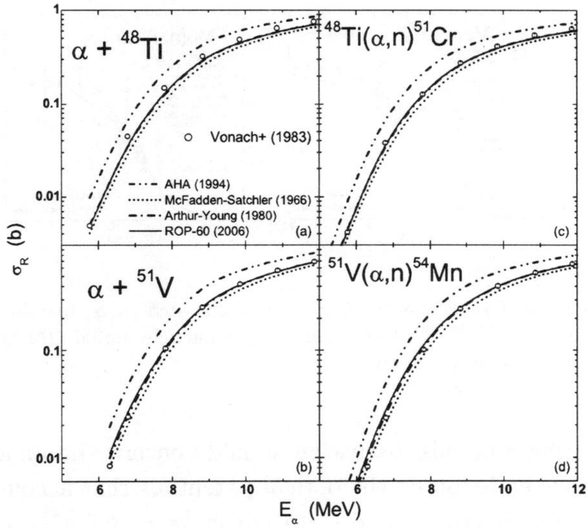

Fig. 4. Comparison of measured [48] and calculated (α,n) and σ_R cross sections for ^{48}Ti and ^{51}V using present ROP (solid) and global OMPs [2,45,47] (dotted, dash-dotted and dot-dot-dashed curves, respectively).

The Hauser-Feshbach statistical model calculations of these (α,n) reaction cross sections have been carried out similarly to the analysis [22] of (n,α) reaction cross section for A∼90 (see next section) except eventually different potentials used in the incident and emergent α-channels. The consistent local parameter set for the mass range A∼60, adopted recently [49] through an analysis of various independent data, was used in this respect. The reaction cross sections (Fig. 4) calculated with the same OMP parameters involved in discussion of the (α,α_0) angular distributions show that both scattering and α-induced reaction data are well described by above-mentioned ROP.

5. α-particle emission cross section analysis

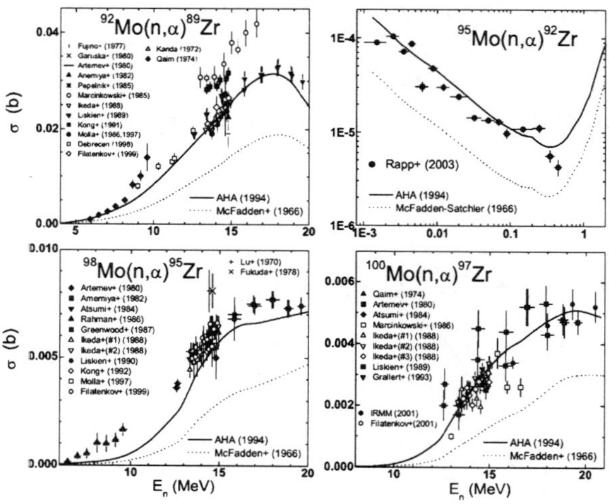

Fig. 5. Comparison of experimental [52,53] and calculated (n, α) reaction cross sections for the target nuclei ^{92,95,98,100}Mo by using the α-particle global OMPs [2,45] (dotted curves and solid curves, respectively).

The next phase of this discussion should concern the underestimation of the α-particle emission by the optical potentials that account for elastic scattering on the (cold) ground-state nuclei (*e.g.* Ref. [2]) and the need for new physics in potentials to describe nuclear de-excitation within the statistical model calculations [4]. Therefore, the ROP for A$\sim$90 based on the semi-microscopic analysis [19] has been used to replace the global potential of Ref. [2] used in a recent analysis of inelastic scattering, pre-equilibrium and Hauser-Feshbach statistical-model calculations of the available fast-neutron reaction cross sections for Mo isotopes in Ref. [50]. A consistent local parameter set was adopted in this respect throughout the analysis of independent experimental data for various quantities, and next used within an updated version of the computer code STAPRE-H95 [51] for the description of the whole body of available experimental neutron activation data for all stable isotopes of molybdenum (*e.g.*, Fig. 5). The compensation effects of various parameter inaccuracies or bad model assumptions are thus well decreased.

However, in spite of the previous careful choice of the ROP of Ref. [19], the cross sections calculated using this potential do not describe, with one

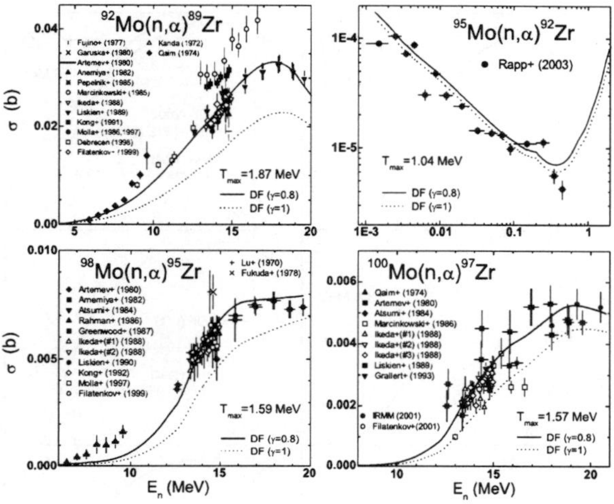

Fig. 6. As for Fig. 5, but for the DF-equivalent potentials corresponding to the parameter γ–values 1 and 0.8 (dotted curves and solid curves, respectively).

exception, the (n, α) reaction experimental data available for the target nuclei ^{92,95,98,100}Mo (Fig. 6). The results range between those obtained by using the global OMP parameter set of McFadden and Satchler [45], which generally underestimates the reaction cross sections, and the global OMP established especially for emitted α-particles [2]. The exception of a good description concerns the recent measurement [52] of the ^{95}Mo$(n,\alpha)^{92}$Zr reaction cross sections in the low energy range from 1 to 500 keV. The available experimental data at higher neutron energies for the other three isotopes ^{92,98,100}Mo are all underestimated. The case of the lightest isotope ^{92}Mo could be considered especially illustrative due to the well increased PE contribution for the heaviest isotopes ^{98,100}Mo, including the effect of the OMP imaginary part which enters in the calculation of the corresponding intranuclear transition rates. The largest underestimation of the data is found just for this ^{92}Mo case when the present regional OMP is used. Thus, there remains the open question regarding the appropriate potential in these calculations for lowest energy α-particles.

A further progress has been made possible by the comparison between the radial dependence of the semi-microscopic DF potential, including the dispersive component, and ROP of Ref. [19] in Fig. 7(a-c). It was underlined that these two potentials are almost identical only in the tail region. In order to emphasize this point, they are shown in Fig. 7(d) as well as the

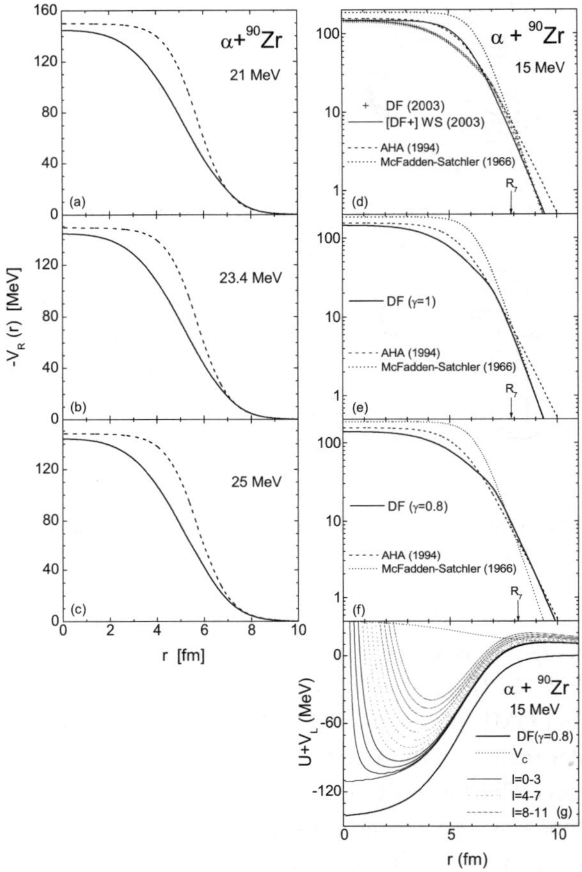

Fig. 7. Comparison between radial dependence of the DF potential (dashed curves) and real part of ROP (solid curves) for scattering of α-particles on ^{90}Zr at 21, 23.4 and 25 MeV (a-c), as well as the global OMPs [2,45] (dotted and dashed curves, respectively) and the DF (crosses) and DF-equivalent potentials (solid curves) corresponding to the γ-values 1 (d,e) and 0.8 (f). (g) The sum of microscopic DF real potential corresponding to γ=0.8 (solid thick curve), of the Coulomb potential (dotted curve) and of centrifugal potential for various partial waves. The arrows indicate the radius R_7 for the DF real potentials where $V_R(R_7)$=-7 MeV.

radius R_7 for the DF real potential. The last quantity has been introduced by Fernandez and Blair [54] through the requirement that $V_R(R_7)$=-7 MeV, a potential value corresponding to the region near the top of the barrier of the total potential (Coulomb, centrifugal and real nuclear potentials) which was found to be highly important for scattering analysis. Nevertheless, neither the global [45] nor the regional [19] phenomenological optical potentials established by the analysis of the elastic-scattering data describe well the potentials at the nuclear surface. Moreover, the semi-microscopic DF potential shows that the real WS potential has not a suitable form over the whole radial domain, despite of its usefulness proved at higher energies [1]. Actually, the inadequacy of the WS exponential tail was demonstrated early by Mailandt *et al.* [55] in folding model calculations, while Goldberg [56] has shown that with the squared Woods-Saxon potential (WS2) functional dependence, the surface and tail regions can be optimized simultaneously. Also, Kobos *et al.* [57] found that the WS2 shape is very close over a wide radial range to the DFM results.

The need of an increased diffuseness for the OMP which are able to describe the (n,α) reaction cross sections, is supported by the previously mentioned suggestions [3–7] that the temperature dependence of the nuclear density distribution function could be the missing degree of freedom that have to be included in conventional statistical model calculations. Therefore we have looked, in completion to the previous semi-microscopic analysis [19], for the effect of the temperature dependence of the nuclear density used for calculation of the DF real potential. According to Eq. (5) the parameter γ is introduced, with the value γ=1 corresponding to the cold target nuclei within α-particle elastic scattering. The microscopic DF potentials obtained by using various γ-values decreasing from unity have been appropriately fitted by WS and WS2 shapes. In this respect, the volume integral of the corrected DF real potential has also been considered fixed. Then, these real DF-equivalent potentials have been used together with the imaginary part of the ROP [19] for the calculation of the (n,α) reaction cross sections. The general trend is an increase of the diffuseness parameter with the decrease of the γ-value, and consequently an increase of the α-particle transmission coefficients for the (n,α) reaction cross sections. Finally, we have found for γ=0.8, that the DF potential as well as DF-equivalent WS and WS2 potentials shown in Fig. 7(f) are in close agreement with the previous global potential [2] for the α-particle emission. The corresponding (n,α) reaction cross sections also agree with the experimental data for the Mo isotopes (Fig. 6). It is important to note that a further decrease of the parameter γ to a value of 0.75 is followed by an additional increase of the (n,α) reaction cross-section of $\sim$12%. Thus a large sensitivity results for the present analysis to the temperature dependence of the nuclear matter density. However, the low nuclear temperature (<2 MeV) involved in these reactions (see the temperatures corresponding to the highest energies in the excitation functions in Fig. 6) allows one to find only a global value of this main temperature-dependent parameter of the nuclear matter density involved within the calculation of the DF real potential.

6. Conclusions

This work provides additional evidence of the recent consideration that (n, α) cross section studies offer a good opportunity for enabling improvements in the alpha-nucleus optical potential for astrophysics applications [58–61]. The application of the detailed-balanced theorem in order to obtain the reverse reaction cross sections without taking into account the different OMP behavior of α-particles in the incident and emergent channels respectively, may also prove significant. Finally, we note that DF-equivalent WS and WS2 forms were used in this work in the both stages of underprediction and suitable description of the (n, α) reaction cross sections. This method has been useful for a corresponding discussion at the level of the present knowledge in the field. Nevertheless, next it will be more appropriate to use the microscopic DF real potential directly in such calculations. Further work will be devoted to semi-microscopic description of α-particle emission for A$\sim$60.

Acknowledgments

The work reported here was carried out mainly together with Professor W. von Oertzen and Dr. V. Avrigeanu, and supported in part by the Contract No. ERB-5005-CT-990101 of Association EURATOM–MEdC and the Research Contract No. 12422 of the International Atomic Energy Agency.

References

1. M. Nolte, H. Machner and J. Bojowald, *Phys. Rev. C* **36**, 1312 (1987).
2. V. Avrigeanu, P.E. Hodgson and M. Avrigeanu, *Phys. Rev. C* **49**, 2136 (1994).
3. M.A. McMahan and J.M. Alexander, *Phys. Rev. C* **21**, 1261 (1980).
4. G. La Rana, D.J. Moses, W.E. Parker, M. Kaplan, D. Logan, R. Lacey, J.M. Alexander and R.J. Welberry, *Phys. Rev. C* **35**, 373 (1987); G. D.J. Moses, M. Kaplan, G. La Rana, W.E. Parker, R. Lacey and J.M. Alexander, *Phys. Rev. C* **36**, 422 (1987).
5. R. Lacey *et al.*, *Phys. Lett. B* **191**, 253 (1987).
6. G. Batko and O. Civitarese, *Phys. Rev. C* **37**, 2647 (1988).
7. G. Bozzolo, O. Civitarese and J.P. Vary, *Phys. Lett. B* **219**, 161 (1989).
8. J.M. Alexander, M.T. Magda and S. Landowne, *Phys. Rev. C* **42**, 1092 (1990).

9. M. Kildir, G. La Rana, *et al.*, *Phys. Rev. C* **46**, 2264 (1992).
10. G.R. Satchler and W.G. Love, *Phys. Rep.* **55**, 183 (1979).
11. G. Bertsch, J. Borysowicz, H. McManus and W.G. Love, *Nucl. Phys. A* **284**, 399 (1977).
12. N. Anantaraman, H. Toki and G. Bertsch, *Nucl. Phys. A* **398**, 279 (1983).
13. M.E. Brandan and G.R. Satchler, *Phys. Rep.* **285**, 143 (1997).
14. Dao T. Khoa and G.R. Satchler, *Nucl. Phys. A* **668**, 3 (2000).
15. Dao T. Khoa and W. von Oertzen, *Phys. Lett. B* **304**, 8 (1993); **342**, 6 (1995); Dao T. Khoa, W. von Oertzen and H. G. Bohlen, *Phys. Rev. C* **49**, 1652 (1994); Dao T. Khoa, W. von Oertzen and A. A. Ogloblin, *Nucl. Phys. A* **602**, 98 (1996).
16. D.T. Khoa, G.R. Satchler and W. von Oertzen, *Phys. Rev. C* **56**, 954 (1997).
17. Dao T. Khoa, *Phys. Rev. C* **63**, 034007 (2001).
18. S. Goriely, in *Proc. Int. Conf. on Nuclear Data for Science and Technology*, edited by R.C. Haight, M.B. Chadwick, T. Kawano and P. Talou, AIP Conf. Proc. No. 769 (American Institute of Physics, New York, 2005), p. 1263.
19. M. Avrigeanu, W. von Oertzen, A.J.M. Plompen and V. Avrigeanu, *Nucl. Phys. A* **723**, 104 (2003).
20. M. Avrigeanu and V. Avrigeanu, *Phys. Rev. C* **73**, 038801 (2006).
21. M. Avrigeanu, W. von Oertzen, F.L. Roman and V. Avrigeanu, in *Proc. Int. Symp. on Nuclear Astrophysics Nuclei in the Cosmos IX*, June 25-30, 2006, CERN, Geneva, PoS(NIC-IX).
22. M. Avrigeanu, W. von Oertzen and V. Avrigeanu, *Nucl. Phys. A* **764**, 246 (2006).
23. P. Demetriou, C. Grama and S. Goriely, *Nucl. Phys. A* **707**, 253 (2002).
24. Zs. Fülöp, Gy. Gyürky, Z. Máté, E. Somorjai, L. Zolnai, D. Galaviz, M. Babilon, P. Mohr, A. Zilges, T. Rauscher, H. Oberhummer and G. Staudt, *Phys. Rev. C* **64**, 065805 (2001) and Refs. therein.
25. D. Galaviz, Zs. Fülöp, Gy. Gyürky, Z. Máté, P. Mohr, T. Rauscher, E. Somorjai and A. Zilges, *Phys. Rev. C* **71**, 065802 (2005).
26. M. Avrigeanu, G.S. Anagnostatos, A.N. Antonov and J. Giapitzakis, *Phys. Rev. C* **62**, 017001 (2000).
27. M. Avrigeanu, A.N. Antonov, H. Lenske and I. Stetcu, *Nucl. Phys. A* **693**, 616 (2001).
28. X. Campi and A. Bouyssy, *Phys. Lett. B* **73**, 263 (1978).
29. W. Negele and D. Vautherin, *Phys. Rev. C* **5**, 1472 (1972).
30. I.V. Krouglov, M. Avrigeanu and W. von Oertzen, *Eur. Phys. J. A* **12**, 399 (2001).
31. H. Krivine and J. Treiner, *Phys. Lett. B* **88**, 212 (1979).
32. D. Baye, L. Desorgher, D. Guillain and D. Herschkowitz, *Phys. Rev. C* **54**, 2563 (1996).
33. I. Tanihata, D. Hirata, T. Kobayashi, S. Shimoura, K. Sugimoto and H. Toki, Preprint RIKEN-AF-NP-123, 1992; *Phys. Lett. B* **289**, 261 (1992).
34. G. Bertsch, J. Borysowicz, H. McManus and W.G. Love, *Nucl. Phys. A* **284**, 399 (1977).
35. N. Anantaraman, H. Toki and G. Bertsch, *Nucl. Phys. A* **398**, 279 (1983).

36. J.W. Negele, *Phys. Rev. C* **1**, 1260 (1970).
37. J.W. Lightbody, Jr., S. Penner, S.P. Fivozinsky, P.L. Hallowell and H. Crannell, *Phys. Rev. C* **14**, 952 (1976).
38. M. El-Azab Farid and G.R. Satchler, *Nucl. Phys. A* **438**, 525 (1985).
39. F.E. Steigert and M.B. Samson, *Phys. Rev.* **92**, 660 (1953).
40. T.A. Tombrello and L.S. Senhouse, *Phys. Rev.* **129**, 2252 (1963).
41. W.S. Chien and R.E. Brown, *Phys. Rev. C* **10**, 1767 (1974).
42. A.N. Antonov, J. Kanev, I.Zh. Petkov and M.V. Stoitsov, *Il Nuovo Cimento A* **101**, 525 (1989); A.N. Antonov, P.E. Hodgson and I.Zh. Petkov, *Nucleon Correlations in Nuclei*, Springer-Verlag, Berlin Heidelberg, 1993.
43. O. Bersillon, Centre d'Etudes de Bruyeres-le-Chatel, Note CEA-N-2227, 1981.
44. M. Wit *et al.*, *Phys. Rev. C* **12**, 1447 (1975).
45. L. McFadden and G.R. Satchler, *Nucl. Phys. A* **84**, 177 (1966).
46. L.R. Gasques *et al.*, *Phys. Rev. C* **67** 024602 (2003); A. Bredbacka *et al.*, *Nucl. Phys. A* **574**, 397 (1994).
47. E. Arthur and P.G. Young, LANL Report LA-8626-MS (ENDF-304), 1980.
48. H. Vonach, R.C. Haight and G. Winkler, *Phys. Rev. C* **28**, 2278 (1994).
49. V. Semkova, V. Avrigeanu, T. Glodariu, A.J. Koning, A.J.M. Plompen, D.L. Smith and S. Sudar, *Nucl. Phys. A* **730**, 255 (2004).
50. M. Avrigeanu, V. Avrigeanu and A.J.M. Plompen, *J. Nucl. Sci. Technol.* Suppl. **2**, 803 (2002); P. Reimer *et al.*, *Phys. Rev. C* **71**, 044617 (2005).
51. M. Avrigeanu and V. Avrigeanu, Report NP-86-1995, Bucharest, IPNE, 1995; *News NEA Data Bank* **17**, 22 (1995).
52. W. Rapp, P.E. Koehler, F. Kappeler and S. Raman, *Phys. Rev. C* **68**, 015802 (2003).
53. EXFOR Nuclear Reaction Data, http://www-nds.iaea.or.at/exfor .
54. B. Fernandez and J.S. Blair, *Phys. Rev. C* **1**, 523 (1970).
55. P. Mailandt, J.S. Lilley and G.W. Greenlees, *Phys. Rev. C* **8**, 2189 (1973).
56. D.A. Goldberg, *Phys. Lett.* **55B**, 59 (1975).
57. A.M. Kobos, B.A. Brown, P.E. Hodgson, G.R. Satchler, and A. Budzanowski, *Nucl. Phys. A* **384**, 65 (1982).
58. Yu.M. Gledenov, P.E. Koehler, J. Andrzejewski, K.H. Guber and T. Rauscher, *Phys. Rev. C* **62**, 042801(R) (2000).
59. P.E. Koehler, Yu.M. Gledenov, J. Andrzejewski, K.H. Guber, S. Raman and T. Rauscher, *Nucl. Phys. A* **688**, 86c (2001).
60. T. Rauscher, *Nucl. Phys. A* **719**, 73c (2003); *ibid.* **725**, 295(E) (2003); T. Rauscher, C. Fröhlich and K.H. Guber, nucl-th/0302046, 13 Jun 2003.
61. P.E. Koehler, Yu.M. Gledenov, T. Rauscher and C. Fröhlich, *Phys. Rev. C* **69**, 015803 (2004).

NUCLEAR-SURFACE EFFECTS IN PRE-EQUILIBRIUM PROCESSES

V. Avrigeanu*

"Horia Hulubei" National Institute for Physics and Nuclear Engineering, P.O. Box MG-6, 76900 Bucharest, Romania

** E-mail: vavrig@ifin.nipne.ro*

www.nipne.ro, tandem.nipne.ro/~vavrig/

The final nuclear states in pre-equilibrium (PE) reactions, which link the extreme mechanisms of compound nucleus and direct reactions, lie usually in the continuum of the nuclear excitation spectrum. Since both semi-classical models and quantum-statistical theories describe the PE processes as passing through a series of particle-hole excitations caused by two-body interactions, the initial target-projectile interactions within the diffuse nuclear surface limit the energy of the possible hole excitation due to the shallower nuclear potential in this region. The radial dependences of the nucleon's mean free path and the probability for the first PE nucleon-nucleon collision are pointing out the surface character of this interaction even at low energies, while improved cross-section calculations are made possible by use of partial level densities including surface effects as well as energy-dependent single-particle level densities, in the framework of the Geometry-Dependent Hybrid PE model.

Keywords: pre-equilibrium and statistical emission, model calculations, local density approximation, partial nuclear level density.

1. Introduction

The energy equilibration in nuclear reactions at low energies is described within semi-classical models as well as quantum-statistical theories by a sequence of nucleon-nucleon (NN) interactions leading to particle-hole excitations (*e.g.*, Ref. [1]). The complexity of nuclear multistep processes has inhibited however the use of microscopic methods which are already standard for the description of reactions leading to discrete final states and giant resonances. The inherent complications on how medium effects, induced interactions and the whole set of many-body effects contribute to statistical reactions leading to highly excited nuclei are still open questions. This is the main reason why schematic interaction models are still used even within the multistep-direct (MSD) and multistep-compound (MSC) reaction theory of Feshbach, Kerman, and Koonin [2] (FKK) which supplied finally an unified description of nuclear reaction mechanisms, including the so-called pre-equilibrium emission (PE) phase. Thus, the simplest 1 fm range Yukawa interaction was chosen in the frame of the FKK theory in order to describe the effective NN interaction leading to particle-hole excitations. Moreover, its strength V_0 has been the only MSD free parameter adjusted to reproduce the experimental data. There exist however quite large discrepancies in the systematics of the phenomenological V_0 values although a consistent model-parameter set as well as several other effects has been considered [3,4]. A particular meaning in this respect has the eventual scaling of V_0 in order to compensate for some effects which have been neglected and should be added to the theory [5,6]. Here we should mention that the microscopic calculations [7] of V_0 lead to unsatisfactory results for the real proton-nucleus scattering potentials. Meanwhile, it has been found that the so-called M3Y interaction may perhaps not be as good as assumed [6] while the actual aim of the FKK analyses [8,9] is to include the most sophisticated ingredients from other independent nuclear structure and reaction studies, which may lead to a better test of the underlying quantum-statistical assumptions. On the other hand, the often low predictive power of the PE and statistical nuclear reaction models [10] was underlined more recently by an outstanding study [11] of both absolute cross sections and shape of excitation functions of fast neutron reactions on medium mass nuclei.

A further step forward has concerned the possibility to take into account some effects which have formerly been neglected [5] especially in relation with the particle-hole state densities. Thus, it has been the case of the less exact but global semi-classical state densities, assumed not to be so crucial in the past while errors in them could be compensated by rescaling of the effective NN-interaction strength [6]. Actually, combinatorial calculations of the partial state density performed in the space of realistic shell model single-particle levels (s.p.l.) have already been used in MSC calculations by Herman *et al.* [12] or developed in this respect [13,14]. Partial state densities calculated by a shell model are involved in MSD studies as well [12]. However, the strong dependence of the microscopic state densities on the basic set of single-particle levels is the main among the several shortcomings inherent in the method (*e.g.* [15,16]), which may explain the general use of the semi-classical Williams-type [17] formulas even at present within a quantum formalism.

At the same time, it is well known that the direct reaction amplitude is dominated by contributions from the nuclear surface region due to the radial localization of the important partial waves, and also by considering (for short wavelengths) the flux patterns associated with the ingoing and outgoing distorted waves [18]. Thus, Watanabe *et al.* [3] estimated the incident local energies at the radius R=$1.25A^{1/3}$, where A is the mass number, in the analysis of the V_0 values extracted from both (n,n') and (p,p') scattering at 12 to 26 MeV. Since the first step of the PE processes is essentially an extension of direct reactions in the continuum, a limited energy of the possible hole excitation due to the shallower nuclear potential within the nuclear surface was assumed for this step even in semi-classical models [19–22]. The dependence of the effective NN-interaction strength on the realistic density of finite nuclei was involved [23] also in order to obtain consistent MSD and MSC results. More recently, it has been shown that the radial dependences of the nucleon's mean free path and the probability for the first NN collision are pointing out the surface character of the first NN interaction in multistep reactions even at low energies [24]. Next, improved cross-section calculations have been proved possible on the basis of partial level density (PLD) including energy-dependent single-particle level densities [21,25] as well as surface effects [19–21,24] within a reviewed formalism [26,27]. Fortunately the surface effects have been increasingly taken into account within various modern approaches [28–35] so that it becomes possiblw to review hereafter their contribution to an increased accuracy of the nuclear-reaction model predictions.

2. Nuclear surface localization of the first NN collision

Within the semi-classical distorted wave (SCDW) model [36–38] it has been shown that the NN collision in the nucleus may be considered as localized when the final nuclear states lie in the continuum of the excitation spectrum as is usually the case of PE reactions. However, since the use of a constant reduced potential depth [21] or the local incident energy [3] calculated just at the nuclear radius R could be difficult to justify, the surface localization of the first NN collision in low-energy PE reactions was analyzed [24] by using the semi-classical method to follow the incoming particle's path within the nuclear target. The probability of the first interaction between the projectile and one of the target nucleons to occur in the diffuse nuclear surface is often considered by using the mean free path (MFP) λ of the incident particle in nuclear matter. In this case the phenomenological nucleon λ-values [39] which are large compared with the nuclear size better support the assumption that the reaction takes place in the nuclear volume. The microscopic models have provided larger values of the MFP too (especially at energies lower than, *e.g.*, 50 MeV) when the final-states Pauli blocking [40] and the nonlocality of the nuclear optical potential are taken into account [41–43]. One should note however, that first, even the phenomenological central $\lambda(r=0)$ is ambiguous and depends on the parametrization chosen for the radial shape of the optical model potential (OMP), smaller values being yet possible [43,44]. Second, it has been shown [45] that only a local MFP $\lambda(r)$ should be considered for finite nuclei and its values can differ enormously from the value in the central region of the nucleus. Alternative quantities like the absorption probability have been considered to be more meaningful.

It is known that the semi-classical approximation is applicable then the wavelength of the incident particle fulfills the condition $\lambda \ll R$ and when the incident energy E_i is much larger than the depth of the respective potential. However, the results obtained by using this high-energy approximation are qualitatively correct [46,47] even below, *e.g.*, E_i=100 MeV. The SCDW model has been applied successfully for 62 MeV proton-induced reactions [37,38]. This result seems to be due to the correctness of the assumption that the distorting potential and the density of the nucleus are slowly varying radial functions in comparison with the rapid oscillation of the distorted waves at intermediate and even lower energies. Moreover, we should note the early comparison [48] of the quantal and classical flux of neutrons in a complex optical potential for incident energies from 5 to 30 MeV which showed that the behaviour of both the local average flux and its divergence (giving the probability of a collision to occur at a point) can be described quite well at the *higher* limit of this energy range by the geometrical-optics approximation.

The probability of the incident particle to reach the point $\mathbf{r}$ is obtained by means of the strict eikonal approximation [49]. Hence the high-energy approximation to the distorted wave in the incident channel with the wave number $k_i = (2\mu E_i)^{1/2}/\hbar$

$$\chi_i^{(+)}(\mathbf{r}) \sim \exp\left\{ i \left[\mathbf{k}_i \cdot \mathbf{r} + \int^{\mathbf{r}} d\mathbf{s} \cdot \mathbf{k}_i(\mathbf{s}) \right] \right\} \tag{1}$$

involves the path of integration along the classical-particle curved trajectory that passes through a given point with the direction of the local wave number vector $\mathbf{k}_i(\mathbf{s})$ given by the tangent to the path. The complex incident momentum $\hbar\mathbf{k}_i(\mathbf{r})$ inside the target nucleus and the local kinetic energy $E_i(\mathbf{r}) = \hbar^2 k_i^2(\mathbf{r})/2\mu$ are connected within the local energy approximation

$$E_i(\mathbf{r}) + U_{i,eff}(\mathbf{r}) = E_i \tag{2}$$

with the effective potential energy [36] evaluated for the complex distorting potential $V(r) + iW(r)$:

$$U_{i,eff}(r) = \frac{1}{2}\left\{\frac{\hbar^2 k_i^2}{2\mu} + V(r) - \left[\left(\frac{\hbar^2 k_i^2}{2\mu} - V(r)\right)^2 + W^2(r)\right]^{1/2}\right\} \tag{3}$$

where μ is the reduced mass. Finally, after the averaging over the partial waves too, it results the probability of the first NN collision to occur within a small ds interval along the curved trajectory of the form [24]

$$P(r) = C\sum_l (2l+1)\frac{k_i(r)}{k_i}\exp\left[-\int^{\mathbf{r}} ds/\lambda(s)\right]\left\{1 - \frac{k_i(r')}{k_i(r)}\exp\left[-\int_{\mathbf{r}}^{\mathbf{r}'} ds/\lambda(s)\right]\right\} \tag{4}$$

where the normalization constant C is determined by the condition

$$\int_0^\infty drP(r) = 1 \tag{5}$$

which is related to our interest in the distribution of this probability along the nuclear radius. Eq. (4) clearly shows the dependence of the first NN-collision probability on the MFP $\lambda(r)$ which is determined by the imaginary and the real part of the distorting potential. The general trend of the results given by this formulation has formerly [24] been illustrated for neutrons incident on ^{93}Nb at various energies from 10 to 50 MeV, while in Fig. 1 is shown the case of protons on ^{93}Nb between the energies of 25 and 71 MeV involved elsewhere [50]. It follows the energy- and radial dependences of the phenomenological imaginary potential which are responsible for the corresponding behaviour of the local MFP. The surface peaking of the imaginary potential at lower energies results in a minimum of $\lambda(r)$ in the region of the nuclear surface. The central imaginary potential depth is small at the lowest energies due to the Pauli-blocking effect and thus $\lambda(0)$ is higher. On the other hand, $\lambda(r)$ is increasing quickly while the nuclear density and consequently the potential well are vanishing. The surface part of the imaginary OMP decreases with the increase of the energy, so that the radial dependence of λ diminishes gradually. At medium energies where the imaginary potential has only a volume component, the MFP values become constant and equal to $\lambda(0)$. The microscopic calculations of the nucleon MFP in finite nuclei using the Thomas-Fermi theory for multiparticle-multihole configurations [51] describe well these trends. It has been thus shown that the interplay between the smaller imaginary potential depth and the smaller local momenta at the nuclear surface determines the MFP at low energies, i.e. there are smaller λ-values than in the central nuclear region.

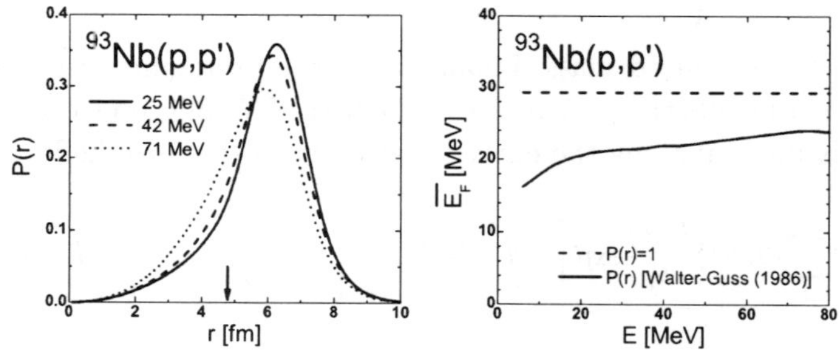

Fig. 1. Radial dependence (left) of the first NN-collision probability for incident protons on ^{93}Nb with energies of 25 (full curves), 42 (dashed), and 71 MeV (dotted), and energy dependence (right) of the average local Fermi energy calculated for the first two-body collision in the same case (solid curve) as well as for the assumption of an equiprobable site of interaction, i.e. to the use of the constant unity value for the first NN-collision probability $P(r)$ (dashed curve). The OMP parameter set of Walter-Guss [52] is used while the arrow indicates the half-density radius [53] of the nuclear matter density distribution for ^{93}Nb.

The maximum of the $P(r)$ distribution within the nuclear surface, which becomes broader and moves slightly to smaller radii as the energy increases, is firstly due to the maximum of the imaginary potential and the minimum of the MFP, respectively. Second, the averaging over the partial waves yields a small additional change of the maximum of $P(r)$ to larger radii and a significant decrease of its width especially for medium energies. This stronger surface character in comparison with the central collisions [54] is just related to the contributions of the peripheral trajectories. The correctness of this method to take into account the surface character of the first NN collision by means of the radial dependence $P(r)$ can now be better estimated. The use of the quantal distorted waves makes it possible indeed to consider contributions from the regions which are inaccessible by the classical trajectories too. However, even the cross sections in the forbidden regions are still less significant [37] while the analysis of $P(r)$ suggests that such regions have lowest contribution to this quantity. This is also the meaning of the agreement found between the classical and quantal average flux and its divergence [48] even for the neutron incident energies between 10 and 30 MeV where a gradual transition to the pattern described by geometrical optics was noted (with the strongest change from 5 to 10 MeV).

Actually, DeVries and DiGiacomo [45] showed that, due to the large variations of the local MFP, one must carefully identify the region in which the reaction takes place. Moreover, they found that it is necessary to look for a quantity which should include effects of the changing nuclear density and interactions along the path of the projectile. Actually, l-dependent average Fermi energies have been involved in the framework of the Geometry-Dependent Hybrid (GDH) semi-classical model [19] to calculate the particle-hole state density in the first stage of the pre-equilibrium emission. While the Fermi energy at the saturation nuclear density ρ_0 is E_F = 40 MeV, a value around 30 MeV was found for l=0 and for higher partial waves it decreases as a function of the incident energy [22]. Kalbach also took into account the surface effects for the first NN interaction by using a mean effective Fermi energy within the exciton model and found empirical values firstly [21] between 11 and 25 MeV in a broad target mass and incident energy range.

Finally, the local Fermi energy [40,55] (see also microscopical studies of the optical potential [51,56–59]) can be expressed within the local density approximation (LDA, *e.g.*, Ref. [60]) in terms of the Fermi momentum and the nuclear density $\rho(r)$ as

$$E_F(r) = \hbar^2 k_F^2(\frac{\mathbf{r}}{)}2m = \frac{\hbar^2}{2m}\left[\frac{3\pi^2}{2}\rho(\mathbf{r})\right]^{2/3} \tag{6}$$

Consequently, the Fermi energy averaged along the trajectory of the incident nucleon with respect to both the nuclear density and the first NN-collision probability becomes

$$\overline{E}_F = \frac{\int d\mathbf{r}\,\rho(\mathbf{r})P(r)E_F(r)}{\int d\mathbf{r}\,\rho(\mathbf{r})P(r)} = \frac{\int_0^{R_s} dr\, r^2\rho(r)P(r)E_F(r)}{\int_0^{R_s} dr\, r^2\rho(r)P(r)} \tag{7}$$

The calculated average Fermi energy is also shown in Fig. 1 for protons incident on ^{93}Nb. The parametrization of Negele [53] has been used to describe the realistic nuclear matter distribution. As an independent test a separate calculation has been carried out by omitting the first NN-collision probability $P(r)$ in Eq. (7). This assumption of an equiprobable site of the interaction led to a constant average Fermi energy of ~30 MeV (Fig. 1). It reproduces the result of Blann for the complete s-wave penetration in the nucleus. At the same time, the comparison of the two average values of the Fermi energy illustrates the surface-localization effect for the first NN interaction in pre-equilibrium reactions. It is rather constant above the nucleon incident energy of 50 MeV (actually very slowly decreasing with energy) and stronger for lower energies. In fact, the physics comprised within the imaginary part of the optical potential is rejoined with respect to a quantity usually related to nuclear structure, in general agreement with the unified description of the mean field [51,57,58]. By using the first NN-collision probability $P(r)$ it has also been deduced the average strength of the effective NN interaction $\overline{V}_0$ along the trajectory of the incident nucleon [24], in good agreement with the phenomenological V_0 values which have been underestimated at low energies when the trend of the optical-model potential has only been followed.

3. The average energy-dependent partial-level density

Once the nuclear surface localization of the first NN collision within PE processes is proved, the consideration of the surface effects within the PE models by means of the PLD formulas can be discussed. It has followed the account of the finite depth of the real nuclear potential well [61] in the frame of the s.p.l. equidistant spacing model [62] (ESM) for p excited particles above the Fermi level and h holes below it ($n = p + h$), at the total excitation energy E. Thus, Kalbach [21] proposed the extension in this respect of the finite well depth correction by replacing the central well depth V_0=38 MeV with an average effective well depth $\overline{V}_1$ for the surface region, for the particle-hole states populated through the initial target-projectile interaction (with h_0=1). At the same time it was shown that the PE surface effects are also tied to the single-particle energy dependence of the s.p.l. density $g(\varepsilon)$ due to the interdependence of the respective assumptions. Actually, the valid use of energy-dependent s.p.l. densities within the ESM particle-hole state density formula, even when corrected for the finite depth of the real nuclear potential well [61], has recently [26] been proved by means of recursive relations particularly using the Fermi-gas model (FGM) s.p.l. density. The latter form has also been compared with and thus supported the results of the ESM formula [26,27]

$$\omega(p,h,E) = \frac{[g_p(p,h)]^p[g_h(p,h)]^h E^{n-1}}{p!h!(n-1)!} f_K(p,h,E,F)\,, \tag{8}$$

modified by using s.p.l. densities different for excited particles and holes with respect to the Fermi energy (whose previous notation E_F is replaced by the shorter one F), respectively

$$g_p(p,h) = g(F + \overline{u}_p) \tag{9}$$

$$g_h(p,h) = g(F - \overline{u}_h)\,, \tag{10}$$

obtained from the FGM at the respective average-excitation energies

$$\overline{u}_p = \frac{1}{p}\int_0^E \frac{u \cdot g_p(F+u)\,\omega(p-1,h,E-u,F)}{\omega(p,h,E,F)}\,du\,, \tag{11}$$

$$\overline{u}_h = \frac{E - p\overline{u}_p}{h}\,, \tag{12}$$

where the $f_K(p,h,E,F)$ functions including the finite-depth, Pauli-blocking and pairing corrections, as well as eventually the bound-state condition, are given elsewhere [26,27]. Moreover, it has been shown at the same time that quantum-mechanical s.p.l. densities and the *continuum effect* can also match a corresponding FGM formula, suitable for use within the average energy-dependent partial state density in multistep reaction models.

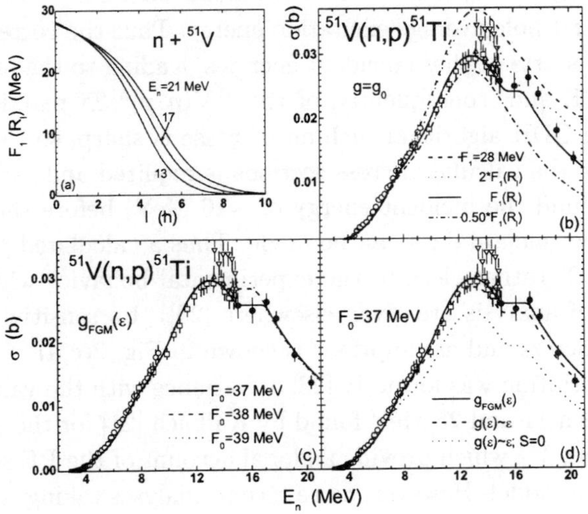

Fig. 2. (a) Average local-density Fermi energies $\bar{F}_1(R_l)$ versus partial wave of neutrons on ^{51}V, for F_0=37 MeV, at the incident energies of 13, 17, and 21 MeV, corresponding to the central well depth value F_0=37 MeV, and (b-d) comparison of experimental [66] and calculated cross sections of the reaction $^{51}V(n,p)^{51}Ti$ (see text).

This PLD formalism has been fully included within the GDH model [19] which has adopted a sum of the contributions of various partial waves involved in PE processes. While this sum accounts for the influence of the nucleus density distribution on first NN collision, the relevant PE model parameters are averaged over the nuclear densities corresponding to the particle trajectories instead of the entire nucleus. Therefore, within the GDH version in the computer code [63] STAPRE-H including the average energy-dependent PLD formula [26,27], the surface effects are determined by the GDH reduced local-density Fermi energies $\bar{F}_1(R_l)$ staying for finite-depth correction in PLD for the hole number h_0=1. These quantities have been determined as a function of orbital angular momentum by a trajectory average in the local density approximation using an average imaginary optical potential. The averages $\bar{F}_1(R_l)$ for l^{th} partial-wave neutrons on ^{51}V, arbitrarily evaluated from the point at which the nuclear density is $\sim$1/150 of its saturation value to the radius $R_l = l\lambda/2\pi$ corresponding to the lth partial wave [64], are shown in Fig. 2 (see also Fig. 5 of Ref. [22]). The curves shown in this figure correspond to the incident energies from maximum of the ^{51}V$(n,p)^{51}$Ti reaction excitation function to the upper end of 21 MeV. Since the average excitation energy of the holes within the exciton configurations (2p,1h) and (1p,1h) varies between 7 and 9 MeV in this incident energy range, it is the l=5 $\hbar$ partial wave for which $\bar{F}_1(R_l)$ just crosses the first hole average excitation energy. Thus the corresponding PE contribution starts at these incident energies, leading to the increase of the calculated PE and, consequently, of the ^{51}V$(n,p)^{51}$Ti reaction cross sections. In the GDH algorithm such an increase is sharp, so that a physical smoothing of the calculated cross sections is required and carried out over $\sim$2 MeV around the incident energy of $\sim$16 MeV, before the comparison with the experimental data can be made. Thus a calculated plateau is obtained (Fig. 2) rather close to the experimental behavior while references and details of analysis are given elsewhere [29]). Its sensitivity to various model parameters and assumptions is shown in Fig. 2(c,d).

This description was formerly [22] at variance with the values empirical values between 11 and 25 MeV found by Kalbach [21] for the average effective well depth $\overline{V}_1$ which provide a global account of the PE surface effects in the exciton model. However, more recent analyses taking into account a larger data basis [28] proved lower values $\overline{V}_1$ $\sim$7 MeV for neutron-induced reactions at least up to an incident energy of 30 MeV. Thus one may consider now the two approaches rather consistent.

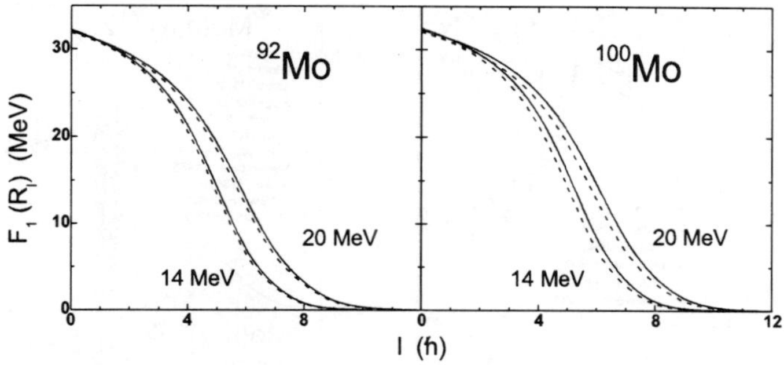

Fig. 3. As for Fig. 2, but for neutrons (solid curves) as well as protons (dashed curves), and the target nuclei ^{92,100}Mo, at the incident energies of 14 (lower curves) and 20 MeV (upper curves).

4. Systematic analysis of PE surface effects

Since the semi-classical calculations can be applied rather easily for all reaction excitation functions, including reactions with the same residual nuclei in different channels, a proper description of a large body of data without free parameters may validate the adopted nuclear model assumptions. The use of a consistent set of model parameters based on different types of independent experimental data is a first basic demand in this respect. A second point of model consistency lies in the use of the same model parameters for the description of the various processes in the framework of the direct interaction, PE and compound-nucleus statistical models. Therefore, similar analyses of this kind have been performed recently for fast-neutron induced reactions on all stable isotopes of V [29], Co and Ni [31], Mo [33], Hf and Ta [65], and W isotopes [35] by using the GDH model within the STAPRE-H code.

Some characteristic results are shown in the following, in order to provide a realistic evaluation of the PE surface effects description. Thus, in Fig. 3 it is shown the change of the angular-momentum dependence of local-density Fermi energies with projectiles, incident energy, and target-nucleus mass. The above-mentioned values of the l=5 $\hbar$ partial-wave PE *thresholds*, leeding to a suitable description of the (n,p) reaction excitation functions (Figs. 4-5), correspond to energies from ~16 to ~13 MeV for ^{92,100}Mo isotopes. Their different positions with respect to the maxima of (n,p) reaction excitation functions, moving from above (^{92}Mo) to below them (^{98}Mo), are obviously related to the excitation-function shape in addition to the corresponding Q-value effects.

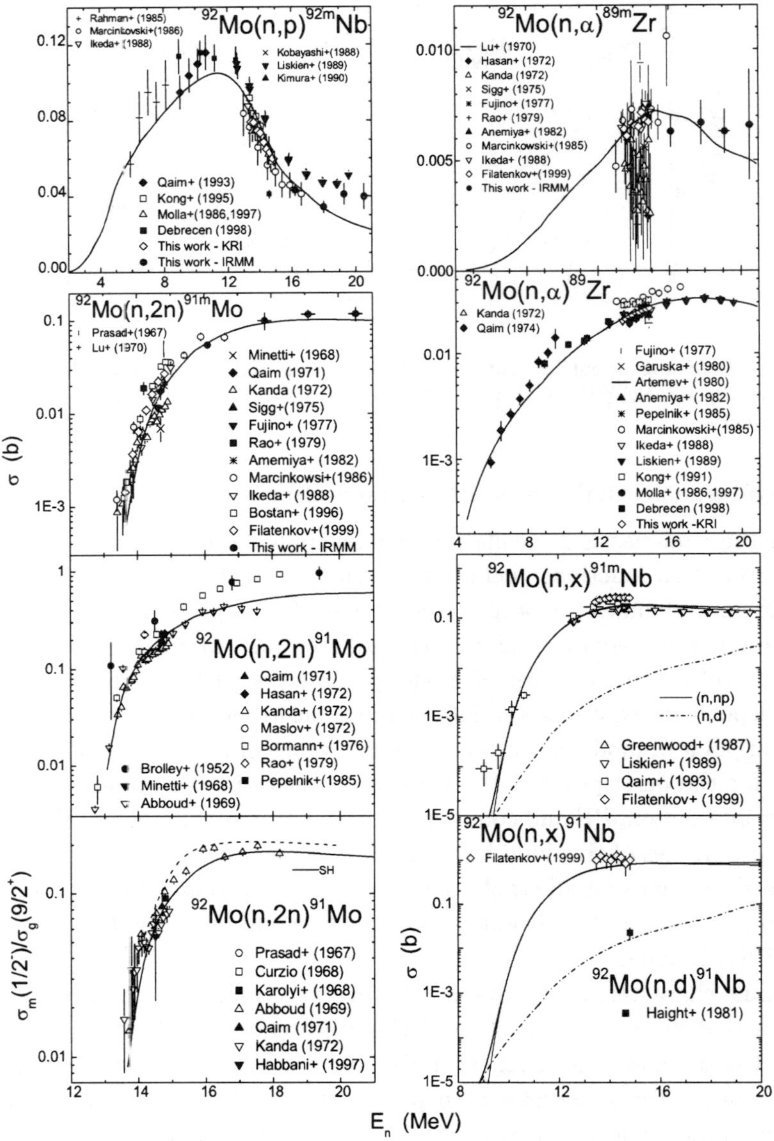

Fig. 4. Comparison of measured [66] and calculated neutron-activation cross sections for the target nucleus ^{92}Mo.

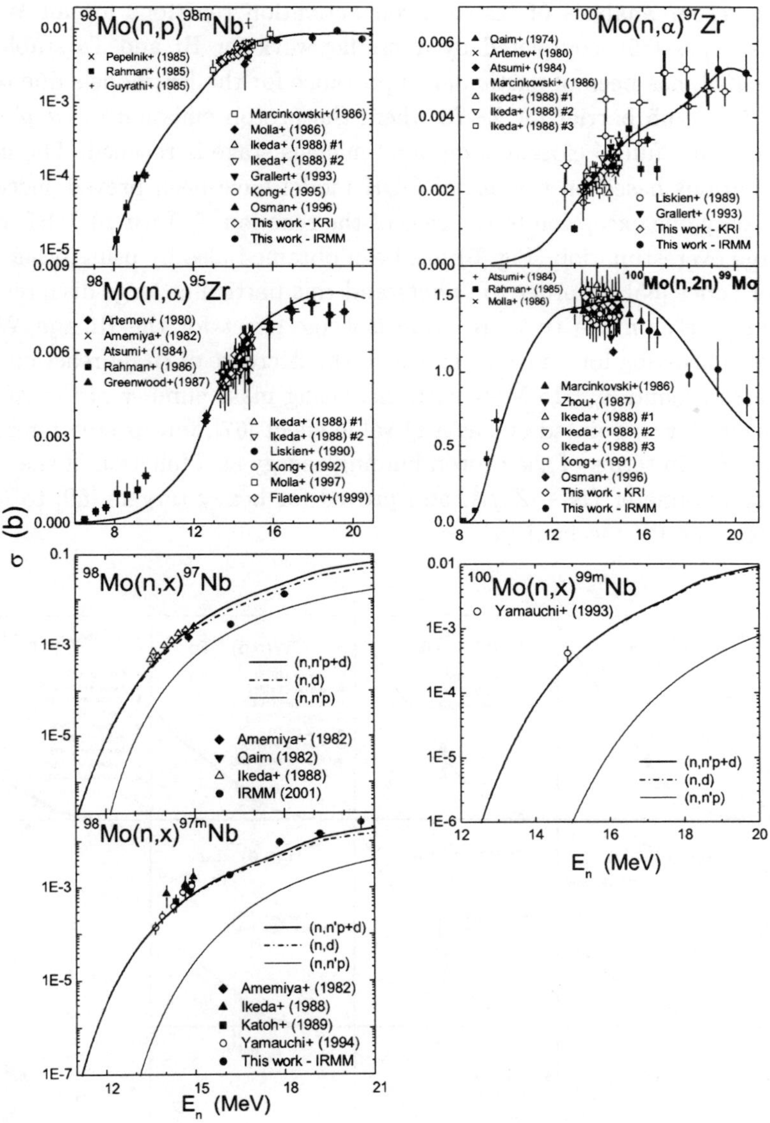

Fig. 5. As for Fig. 4, but for the target nuclei ^{98,100}Mo.

A recent analysis of fast-neutron activation reactions for all W stable isotopes [35], completed by a similar work for Hf and Ta stable isotopes [65], has been of particular importance for the PE models due to the high Coulomb barrier for Z∼73 where the proton emission in (n,p) reactions occurs fully before a compound-nucleus stage is reached. The model calculations based on the use of GDH model have been proved successful with only one exception in the case of the reaction $^{181}Ta(n,p)^{181}Hf$, where a large overestimation (Fig. 6) has been obtained also by using other computer codes [65]. In order to understand this particular large disagreement we may rely on the (n,p) reaction isotopic effect for the Hf and W isotopes, consisting for a given element in the decrease of (n,p) reaction cross sections around E_n=15 MeV with increasing mass number of the isotope. This effect was pointed out as a Q-value effect [67], interpreted for lighter nuclei [68] in terms of the proton binding energy as a function of the asymmetry parameter $(N-Z)/A$, and proved for heavy targets [69] to follow entirely the PE mechanism.

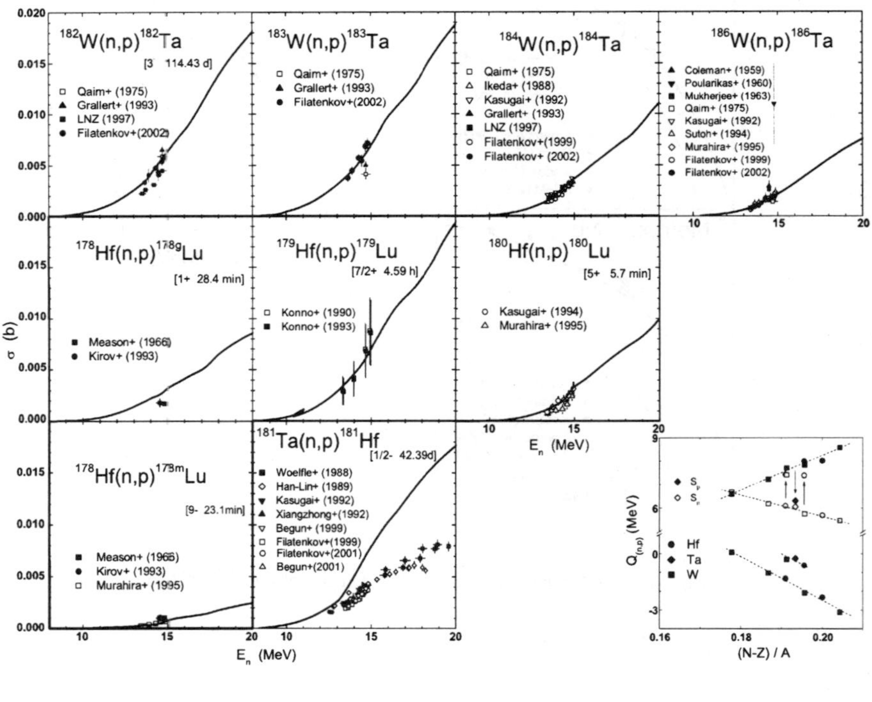

Fig. 6. Comparison of experimental [66] and calculated (n,p) reaction cross sections of Hf, Ta, and W stable isotopes, and the related Q-values versus the asymmetry $(N-Z)/A$.

The corresponding (n,p) reaction data for the Hf and W isotopes are shown in Fig. 6 as well as the data for the ^{181}Ta target nucleus, and the corresponding proton and neutron separation energies (top in the bottom-right corner, with lines connecting them only for eye guiding) as a function of the asymmetry parameter $(N-Z)/A$. The two excitation functions for the ground and isomeric states of ^{178}Hf should be considered together, in the row of Hf isotopes. First, it can be seen that data corresponding to Hf and W odd isotopes do not follow exactly the isotopic effect, having larger values but well-accounted for by the GDH model due to the improved pairing correction of the PLD formula [26,27]. Second, the quite similar Q-values for the ^{179}Hf, ^{181}Ta, and ^{183}W target nuclei, corresponding also to close $(N-Z)/A$-values, suggest comparable (n,p) reaction cross sections which are predicted at variance with the lower measured cross sections for ^{181}Ta. This can be resolved by the assumption of a longer-lived isomer of the ^{181}Hf nucleus, not yet observed, but also in agreement with the Hf isotopes systematics. Further experiments to investigate this possibility would therefore be useful. Nevertheless the present results have proved that, since the (n,p) reaction occurs fully at the PE stage, this reaction analysis is a powerful tool for PE-model validation in the A∼180 range provided that accurate calculations of reactions cross sections used no re-normalization or free parameters but involved (i) the unitary use of common model parameters for different mechanisms, (ii) the use of consistent sets of input parameters which are determined by analyses of various independent experimental data, and (iii) the unitary account of a whole body of related experimental data for isotope chains and neighboring elements.

5. Conclusions

Since the first step of PE reactions are essentially an extension of direct reactions in the continuum, while the direct reaction amplitude is dominated by contributions from the nuclear surface region due to the radial localization of the important partial waves, the nuclear-surface effects of a corresponding shallower potential were normally included in PE models. However only later it has been shown that the radial dependences of the nucleon's mean free path and the probability for the first NN collision are pointing out the surface character of the first NN interaction in multistep reactions even at low energies [24]. Next, improved cross-section calculations have been proved possible on the basis of partial level density including energy-dependent single-particle level densities [21,25] as well as surface effects [19–21,24] within a reviewed formalism [26,27]. Within the

GDH version in STAPRE-H code [63] the surface effects are determined by the GDH reduced local-density Fermi energies $\bar{F}_1(R_l)$ staying for finite-depth correction in PLD for the hole number h_0=1. These quantities have been determined as a function of orbital angular momentum by a trajectory average in the local density approximation using an average imaginary optical potential. Finally, a proper description of a large body of data without free parameters has been involved in order to validate the adopted nuclear model assumptions.

Acknowledgments

The work reported here has been carried out mainly together with M. Avrigeanu. We gratefully acknowledge valuable discussions with M. Blann as well as within the EAF international working group on fast-neutron activation. Work was supported in part by the Contract No. ERB-5005-CT-990101 of Association EURATOM-MEdC.

References

1. E. Gadioli and P.E. Hodgson, *Pre-Equilibrium Nuclear Reactions* (Oxford University Press, Oxford, 1992).
2. H. Feshbach, A. Kerman, and S. Koonin, *Ann. Phys. (N.Y.)* **125**, 429 (1980).
3. Y. Watanabe *et al.*, *Phys. Rev. C* **51**, 1891 (1995).
4. Y. Watanabe, *Acta Phys. Slovaca* **45**, 749 (1995).
5. P.E. Hodgson, *Acta Phys. Slovaca* **45**, 673 (1995).
6. R. Lindsay, *Acta Phys. Slovaca* **45**, 717 (1995).
7. G. Bartnitzky, H. Clement, P. Czerski, H. Muther, F. Nuoffer and J. Siegler, *Phys. Lett. B* **386**, 7 (1996).
8. A.J. Koning and M.B. Chadwick, *Phys. Rev. C* **56**, 970 (1997).
9. M.B. Chadwick *at al.*, *Acta Phys. Slovaca* **49**, 365 (1999).
10. P. Oblozinsky, *Development of Reference Input Parameter Library for Nuclear Model Calculations of Nuclear Data*, Report IAEA-TECDOC-1034, IAEA, Vienna, 1998; http://www-nds.iaea.or.at/ripl/ .
11. A. Fessler, E. Wattecamps, D.L. Smith and S.M. Qaim, *Phys. Rev. C* **58**, 996 (1998).
12. M. Herman, G. Reffo, H. Lenske and H. Wolter, in *Proceedings of the International Conference on Nuclear Data for Science and Technology*, Gatlinburg, Tennessee, 1994, J.K. Dickens (ed.) (American Nuclear Society, 1994), p. 466.
13. K. Sato, Y. Takahashi and S. Yoshida, *Z. Phys. A* **339**, 129 (1991).
14. S. Yoshida, M. Abe and K. Sato, *Acta Phys. Slovaca* **45**, 757 (1995).
15. M. Blann and G. Reffo, in International Atomic Energy Agency Report No. INDC(NDS)-214/LJ, Vienna, 1989, p. 75.
16. G. Reffo and M. Herman, as Ref. [12], p. 473.
17. F.C. Williams, *Nucl. Phys. A* **166**, 231 (1971).

18. N. Austern, *Direct Nuclear Reaction Theories* (Wiley-Interscience, New York, 1970), p. 122.
19. M. Blann, *Phys. Rev. Lett.* **28**, 757 (1972); *Nucl. Phys. A* **213**, 570 (1973).
20. M. Blann and H.K. Vonach, *Phys. Rev. C* **28**, 1475 (1983).
21. C. Kalbach, *Phys. Rev. C* **32**, 1157 (1985).
22. M. Avrigeanu and V. Avrigeanu, *J. Phys. G* **20**, 613 (1994).
23. R. Bonetti and L. Colombo, *Phys. Rev. C* **28**, 980 (1983).
24. M. Avrigeanu, A. Harangozo, V. Avrigeanu and A. N. Antonov, *Phys. Rev. C* **54**, 2538 (1996); *ibid.* **56**, 1633 (1997).
25. M. Herman, G. Reffo, M. Rosetti, G. Giardina and A. Italiano, *Phys. Rev. C* **40**, 2870 (1989).
26. A. Harangozo, I. Stetcu, M. Avrigeanu, V. Avrigeanu, *Phys. Rev. C* **58**, 295 (1998).
27. M. Avrigeanu and V. Avrigeanu, *Comp. Phys. Comm.* **112**, 191 (1998); Computer code PLD, Catalogue Id. ADIK (1998), CPC Program Library, Queen's University of Belfast.
28. C. Kalbach, *Phys. Rev. C* **62**, 44608 (2000); *ibid.* **69**, 014605 (2004).
29. P. Reimer, V. Avrigeanu, A.J.M. Plompen and S.M. Qaim, *Phys. Rev. C* **65**, 014604 (2001).
30. P. Reimer, M. Hult, A.J.M. Plompen, P.N. Johnston, S.M. Qaim and V. Avrigeanu, *Nucl. Phys. A* **705**, 265 (2002).
31. V. Semkova, V. Avrigeanu, T. Glodariu, A.J. Koning, A.J.M. Plompen, D.L. Smith and S. Sudar, *Nucl. Phys. A* **730**, 255 (2004).
32. A.J. Koning and M.C. Duijvestijn, *Nucl. Phys. A* **744**, 15 (2004).
33. P. Reimer, V. Avrigeanu, S. Chuvaev, A.A. Filatenkov, T. Glodariu, A.J. Koning, A.J.M. Plompen, S.M. Qaim, D.L. Smith and H. Weigmann, *Phys. Rev. C* **71**, 044617 (2005).
34. M. Avrigeanu, W. von Oertzen, and V. Avrigeanu, *Nucl. Phys. A* **764**, 246 (2006).
35. V. Avrigeanu, S.V. Chuvaev, R. Eichin, A.A. Filatenkov, R.A. Forrest, H. Freiesleben, M. Herman, A.J. Koning and K. Seidel, *Nucl. Phys. A* **765**, 1 (2006).
36. M. Kawai, *Prog. Theor. Phys.* **27**, 155 (1962).
37. Y.L. Luo and M. Kawai, *Phys. Rev. C* **43**, 2367 (1991).
38. M. Kawai and H.A. Weidenmüller, *Phys. Rev. C* **45**, 1856 (1992); Y. Watanabe and M. Kawai, *Nucl. Phys. A* **560**, 43 (1993); M. Kawai, Y. Watanabe and H. Shinohara, *Acta Phys. Slovaca* **45**, 693 (1995).
39. A. Nadasen *et al.*, *Phys. Rev. C* **23**, 1023 (1981).
40. K. Kikuchi and M. Kawai, *Nuclear Matter and Nuclear Reactions* (North-Holland, Amsterdam, 1968).
41. J.W. Negele and K. Yazaki, *Phys. Rev. Lett.* **47**, 71 (1981).
42. V.R. Pandharipande and S.C. Pieper, *Phys. Rev. C* **45**, 791 (1992).
43. R. Crespo, R.C. Johnson, J.A. Tostevin, R.S. Mackintosh and S.G. Cooper, *Phys. Rev. C* **49**, 1091 (1994).
44. H.O. Meyer and P. Schwandt, *Phys. Lett. B* **107**, 353 (1981).
45. R.M. DeVries and N.J. DiGiacomo, *J. Phys. G* **7**, L51 (1981).

46. H.C. Chiang and J. Hüfner, *Nucl. Phys. A* **349**, 466 (1980).
47. H. Feshbach, *Theoretical Nuclear Physics* (John Wiley & Sons, Inc., New York, Chichester, Brisbane, Toronto, Singapore, 1992), p. 409.
48. I.E. McCarthy, *Nucl. Phys.* **10**, 583 (1959); *ibid.* **11**, 574 (1959).
49. R.O. Newton, *Scattering Theory of Waves and Particles* (McGraw-Hill, New York, 1966), p. 583.
50. M. Avrigeanu, A.N. Antonov, H. Lenske and I. Stetcu, *Nucl. Phys. A* **693**, 616 (2001).
51. R.W. Hasse and P. Schuck, *Nucl. Phys. A* **445** 205 (1985).
52. R.L. Walter and P.P. Guss, *Rad. Effects* **95**, 73 (1986).
53. J.W. Negele, *Phys. Rev. C* **1**, 1260 (1970).
54. M. Avrigeanu, A. Harangozo, V. Avrigeanu and A. N. Antonov, *Acta Phys. Slovaca* **45**, 617 (1995).
55. K. Chen, Z. Fraenkel, G. Friedlander, J.R. Grover, J.M. Miller and Y. Shimamoto, *Phys. Rev.* **166**, 949 (1968).
56. M. Abe, S. Yoshida and K. Sato, *Phys. Rev. C* **52**, 837 (1995).
57. R.W. Hasse and P. Schuck, *Nucl. Phys. A* **438**, 157 (1985).
58. C. H. Johnson, D. J. Horen and C. Mahaux, *Phys. Rev. C* **36**, 2252 (1987).
59. K. Sato and S. Yoshida, *Phys. Rev. C* **49**, 1099 (1994).
60. A.N. Antonov, P.E. Hodgson and I.Zh. Petkov, *Nucleon Momentum and Density Distributions in Nuclei* (Clarendon Press, Oxford, 1988), p. 22.
61. E. Běták and J. Dobeš, *Z. Phys. A* **279**, 319 (1976).
62. T. Ericson, *Adv. in Phys.* **9**, 425 (1960).
63. M. Avrigeanu and V. Avrigeanu, Report NP-86-1995, Bucharest, IPNE, 1995; *News NEA Data Bank* **17**, 22 (1995).
64. M. Avrigeanu, V. Avrigeanu and A.J.M. Plompen, *J. Nucl. Sci. Technol.* Suppl. **2**, 803 (2002).
65. M. Avrigeanu, R.A. Forrest, F.L. Roman and V. Avrigeanu, in *Proc. of PHYSOR-2006 Topical Meeting on Advances in Nuclear Analysis and Simulation*, Vancouver, Sept. 10-14, 2006 (in press).
66. EXFOR Nuclear Reaction Data, http://www-nds.iaea.or.at/exfor .
67. D.G. Gardner, *Nucl. Phys.* **29**, 373 (1962).
68. N.I. Molla and S.M. Qaim, *Nucl. Phys. A* **283**, 269 (1977).
69. R. Caplar, Lj. Udovicic, E. Holub, D. Pocanic and N. Cindro, *Z. Phys. A* **313**, 227 (1983).

ALPHA HALF-TIME ESTIMATES FOR THE SUPERHEAVY ELEMENTS

I. SILIŞTEANU, A. ŞANDRU, A.O. SILIŞTEANU
B. POPOVICI, A. NEACŞU
Horia Hulubei National Institute of Physics and Nuclear Engineering, RO-077125, Bucharest-Magurele, Romania, E-mail:silist@theory.nipne.ro

B.I. CIOBANU
Gh. Asachi Technical University, Iassy, Romania

We present a coupled channel method for calculating the emission rates based on selfconsistent models for nuclear structure and low-energy dynamics. Phenomenological adjustment of model parameters is discussed in detail. The α-decay properties of some new superheavy elements under current experimental research are estimated using the shell model preformation amplitude and resonant reaction amplitudes for open decay channels. Some extensions and applications of the method to resonant particle spectroscopy technique in studies of α-clustering and fine structure with position-sensitive charge particle detectors are discussed.

1. Introduction

Progress in obtaining the most complete information decay properties of superheavy elements (SHEs) is being made on several fronts: 1) Improving the structure models in order to describe essential features and obtain spectroscopic information on a number of nuclei that can then tested against data extracted from decay and in-beam studies; 2) Extending the range of applicability of reaction models by using accurate reaction channels methods (e.g. via additional channels including, deformation, exchange effects, antysimmetrisation, etc; 3) Including microscopic structure information in coupled channel reaction models and treating more carefully of "intermediate" systems that are more or less bound or have mixed composition. Current research topics in field includes production of new nuclei in bound and less unbound states, decay and in-beam studies, probes of nuclear matter at limits, macroscopic static and dynamical nuclear properties and microscopic simulations. In this work we present the basic formulas for calculating the

α- emission width in the single channel and many channel cases. We stress some of the resonance features of the single channel solution that we shall require for the extension to the resonance solution in many channel case. It is our aim to discuss the results obtained in microscopic shell-model approaches for the α-decaying spherical and deformed nuclei together with experimental facts and previous predictions.The objective of this work was to discuss various challenges in theoretical nuclear structure, especially in the context of SHE physics.

2. Description of the formalism

2.1. *Single channel. Spherical system*

In the simplest case of α-decay of a single resonance state k into a single decay channel n, the decay width is[1]:

$$\Gamma_n^k = 2\pi \left| \frac{\int_{r_{\min}}^{r_{\max}} I_n^k(r) u_n^0(r) dr}{\int_{r_{\min}}^{r_{\max}} I_n^k(r) u_n^k(r) dr} \right|^2 \tag{1}$$

where $I_n^\nu(r)$ is the particle (cluster) formation amplitude (FA) and $u_n^k(r)$ and $u_n^0(r)$ are the solutions of the system of differential equations

$$\left[\frac{\hbar^2}{2m}\left(\frac{d^2}{dr^2} - \frac{l(l+1)}{r^2}\right) - V_n(r) + Q_n\right] u_n^0(r) = 0 \tag{2}$$

$$\left[\frac{\hbar^2}{2m}\left(\frac{d^2}{dr^2} - \frac{l(l+1)}{r^2}\right) - V_n(r) + Q_n\right] u_n^k(r) = I_n^k\,(r) \tag{3}$$

The radial functions in Eqs.(2) and (3) describe the radial motion of the fragments at large and small separations, respectively by using the reduced mass m, the kinetic energy of emitted particle $Q_n = E - E_D - E_p$, the FA $I_n^k\,(r)$, and the interaction potential $V(r)$. The FA is the antisymmetrized projection of the parent wave function (WF) $\mid \Psi_k >$ on the channel WF $\mid n\rangle =\mid [\Phi_D(\eta_1)\Phi_p(\eta_2)Y_{lm}(\hat{r})]_n\rangle$:

$$I_n^k(r) = r\langle \Psi_k \mid \mathcal{A}\left\{[\Phi_D(\eta_1)\Phi_p(\eta_2)Y_{lm}(\hat{r})]_n\right\}\rangle \tag{4}$$

where $\Phi_1(\eta_1)$ and $\Phi_2(\eta_2)$ are the internal (space-spin) wave functions of the fragments, $Y_{lm}(\hat{r})$ is the wave function of the angular motion, $\mathcal{A}$ is the inter-fragment antisymmetrizer, r connects the centers of mass of the fragments, and the symbol $<|>$ means integration over the internal coordinates and angular coordinates of relative motion. The potential continues the nuclear, Coulomb, and spin-orbit parts:

$$V_n(r) =< n \mid \left[V^{nucl.}(r) + V^{Coul.}(r) + V^{so}(r)\right] \mid n\rangle \tag{5}$$

The equations (2,3) are solved with usual boundary conditions for the decay problem:

$$u_n^0\,(r \longrightarrow 0) = 0; \qquad u_n^0(r \to \infty) \to i/2(k_n/\pi Q_n)^{1/2}\{u_n^{(-)}(r) + S_n u_n^{(+)}\} \tag{6}$$

$$u_n^{\nu}(r \longrightarrow 0) = 0; \quad u_n^{\nu}(r \to \infty) = 0 \tag{7}$$

where $\hbar k_n = (2mQ_n)^{1/2}$, S_n is the scattering amplitude, $u_n^{(\pm)}(r) = G_n(r) \pm iF_n(r)$ and F_n and G_n are the regular and irregular Coulomb functions. The lower limit in integrals (1) is an arbitrary small radius $r_{\min} > 0$, while the upper limit $r_{\max}$ is close to the first exterior node of $u_n^0(r)$. To avoid the usual ambiguities encountered in formulating the potential for the resonance tunneling of the spherical barrier we iterate directly the nuclear potential in equations of motion[2].

The "one-body" (o.b.) resonance width in the single channel problem can be expressed only with the eigenvalues and eigenfunctions of the system:

$$\Gamma_n^{o.b.} = 2\pi \left| \frac{\int_{r_{\min}}^{r_{\max}} G_n(r) u_n^0(r) dr}{\int_{r_{\min}}^{r_{\max}} G_n(r) u_n^{o.b.}(r) dr} \right|^2 . \tag{8}$$

where $u_l^{o.b.}(r)$ is a solution of Eq.(3) in which $I_n^k(r)$ is merely replaced by $G_n(r)$.

2.2. *Coupled channels. Deformed system*

The relative motion of the fragments can be strongly influenced by couplings of the relative motion of the fragments to several nuclear intrinsic motions. The usual way to address the effects of coupling between the intrinsic degrees of freedom and relative motion is to numerically solve the coupled channel equations, including all the relevant channels. The total decay width for the multichannel decay of the state k into a set of $\{n\}$ different channels is[3]:

$$\Gamma^k = \sum_n \Gamma_n^k, \qquad where: \tag{9}$$

$$\Gamma_n^k = 2\pi \left| \frac{\int_{r_{\min}}^{r_{\max}} I_n^k(r) u_n^0(r) dr}{\int_{r_{\min}}^{r_{\max}} I_n^k(r) u_n^k(r) dr} \right|^2 . \tag{10}$$

In Eq. (9) $I_n^\nu(r)$ is the particle (cluster) formation amplitude (FA) and $u_n^k(r)$ and $u_n^0(r)$ are the solutions of the systems of differential equations

$$\left[\frac{\hbar^2}{2m}\left(\frac{d^2}{dr^2} - \frac{l(l+1)}{r^2}\right) - V_{nn}(r) + Q_n\right] u_n^0(r) + \sum_{m \neq n} V_{nm}(r) u_m^0(r) = 0 \tag{11}$$

$$\left[\frac{\hbar^2}{2m}\left(\frac{d^2}{dr^2} - \frac{l(l+1)}{r^2}\right) - V_{nn}(r) + Q_n\right] u_n^\nu(r) + \sum_{m \neq n} V_{nm}(r) u_m^\nu(r) = I_n^\nu(r) \tag{12}$$

Now, the matrix elements $V_{nm}(r)$ of the interaction potential depends on the radial distance between the fragments and also on nuclear deformations of involved nuclei . The matrix elements of the interaction potential are given by:

$$V_{nm}(r) = < n \mid V^{nucl.}(r) + V^{Coul.}(r) + V^{so}(r) \mid m > \tag{13}$$

The potential we use is taken as . The nuclear Hamiltonian is generated by changing the radius of the daughter nucleus R_0 to a dynamical operator[4]:

$$R_0 \longrightarrow R_0 + \hat{O} = R_0 + \beta_2 R_D Y_{20} + \beta_4 R_D Y_{40} \tag{14}$$

where β_2 and β_4 being the quadrupole and hexadecapole deformation parameters. The resulting nuclear coupling matrix elements between states$\mid n >=\mid I0 >$ and $\mid m >=\mid I^{'}0 >$ are

$$V_{nm}^{nucl.}(r) = < I0 \mid V^{nucl.}(r, \hat{O}) \mid I^{'}0 > - \delta_{nm} V_0^{nucl.}(r)$$

$$= < \alpha \mid I0 \mid \alpha > < \alpha \mid I^{'}0 > V^{nucl.}(r, \lambda_\alpha) - \delta_{nm} V_0^{nucl.}(r); \tag{15}$$

$$\hat{O}_{II'} = \left[\frac{5(2I+1)(2I^{'}+1)}{4\pi}\right]^{1/2} \beta_2 R_D \left(C_{000}^{I2I^{'}}\right)^2$$

$$+ \left[\frac{9(2I+1)(2I^{'}+1)}{4\pi}\right]^{1/2} \beta_4 R_D \left(C_{000}^{I4I^{'}}\right)^2 . \tag{16}$$

Similarly, the Coulomb matrix elements are then given by

$$V_{nm}^{Coul.}(r) = \frac{3Z_DZ_pR_D^2}{5r^3}\left[\frac{5(2I+1)(2I^{'}+1)}{4\pi}\right]^{1/2}\left(\beta_2 + \frac{2}{7}(5/\pi)^{1/2}\beta_2^2\right)\left(C_{000}^{I2I^{'}}\right)^2$$

$$+\frac{3Z_DZ_pR_D^2}{9r^5}\left[\frac{9(2I+1)(2I^{'}+1)}{4\pi}\right]^{1/2}\left(\beta_4 + \frac{9}{7}\beta_2^2\right)\left(C_{000}^{I4I^{'}}\right)^2. \qquad (17)$$

In the case in which all exit channels are open the boundary conditions should be:

$$u_n^0\,(r\to 0) = 0; \quad u_n^0(r\to\infty) \to i/2(k_n/\pi Q_n)^{1/2}\{\delta_{nm}\exp\left[-i(k_n - l\pi/2)\right]$$

$$-S_{nm}\exp\left[i(k_nr - l\pi/2)\right]\} \qquad (18)$$

$$u_n^{\nu}(r\to 0) = 0; \qquad u_n^{\nu}(r\to\infty) = 0 \qquad (19)$$

where S_{nm} is the scattering matrix. The solutions $u_n^0(r\ )$may be matched to the boundary conditions at two values of r large enough so the terms $V_{nm}$are negligible. A special type of eigenvalue solution will be considered here for which the behavior of solution in each separate channel is similar to that of $G_n$in the one channel problem.

3. Model estimates for the α- cluster formation amplitude

For the α- formation amplitude we employ the microscopic shell model wave functions. This method has been initiated[5,6], for the harmonic oscillator s.p. wave functions and extended to Woods-Saxon wave functions[7,8,9]. Following [8] we use two kinds of FAs: one-body resonance amplitude that results in Breit procedure and the shell model FAs given on the basis of shell model s.p. wave functions[10,11]. In the first case we obtain the asymptotic formation amplitude $I_n^{o.b.}(r) \cong G_n(r)$ and from Eq.(8) $\Gamma_n^{o.b.}$. In the second case using in Eq.(4) shell model w.f.$|\ \Psi_k >= \det \parallel \psi_{nlj}(r_i) \parallel$where i=1,A and $|\ \Phi_D >= \det \parallel \psi_{nlj}(r_i) \parallel$where i=1,A-4. one obtains a shell model estimation of formation amplitude $I_n^k(r)$ and from Eq.(1) a shell model width Γ_n^k. The CFA is related to the amplitude of reduced width[8]

$$\gamma_n^k(r) = (\hbar^2/2mr)^{1/2}I_n^k(r). \qquad (20)$$

The spectroscopic factor is simply defined as

$$S_n^k = \Gamma_n^k / \Gamma_n^{o.b.} = T_n^{o.b.} / T_n^k \tag{21}$$

is a measure of the contribution of shell effects, finite sizes of nucleons and α-particle which are neglected in the one body approximation where α is a pointlike particle. The inverse of decay reactions and transfer cluster reactions provide similar information[12,13].

4. Decay spectroscopy

As nuclei move further from β stability on the proton rich side, their binding energy rapidly decreases, do to increasing Coulomb repulsion and reaction Q-values, which leads to major difficulties in their production and also in study of decay properties. The relatively large Q-values cause high excitations in nuclear systems involved and open up many competing decay channels favoring the nuclei closer to stability. Damping these excitations can be very crucial for nuclei produced near the limit of proton stability. Very weak reaction channels can be studied by using the resonant particle spectroscopy (RPS) method[14], and the so called recoil decay tagging (RDT) method [15]. Using these methods in studies of long α-chains of isotopes far of stability, make possible to access the basic nuclear-ground state properties: their masses, lifetimes, energy levels, spins, moments and sizes.The classical fission barriers of the heaviest elements with Z >100 approaches zero because of the large Coulomb energy. However, a series of measurements has established that the elements with Z up to 118 are sufficiently bound against fission to preferentially decay by α-emission. A large shell correction energy leads to additional binding and, hence, create sizable fission barrier of up to 8 MeV [16]. The $\alpha - SF$ competing channels have been observed[17,18,19,20] in SHEs.

5. The Element 118 and its α-descendants

Fig.1 shows[21] the three α-chains originated from the even-even isotope $^{294}118$ (E_α=11.65±0.06 MeV, $T_\alpha = 0.89^{+1.70}_{-0.31}$ ms) produced in the 3n-evaporation channel of the $^{249}Cf +^{48} Ca$ reaction with a maximum cross section of $0.5^{+1.6}_{-0.3}$ pb. For calculating T_α and S_α we use the s.p. shell model states[10,11] (protons: $1i_{13/2}$, $2f_{7/2}$, $3p_{3/2}$, neutrons: $2g_{7/2}$, $3d_{5/2}$, $3d_{5/2}$), the

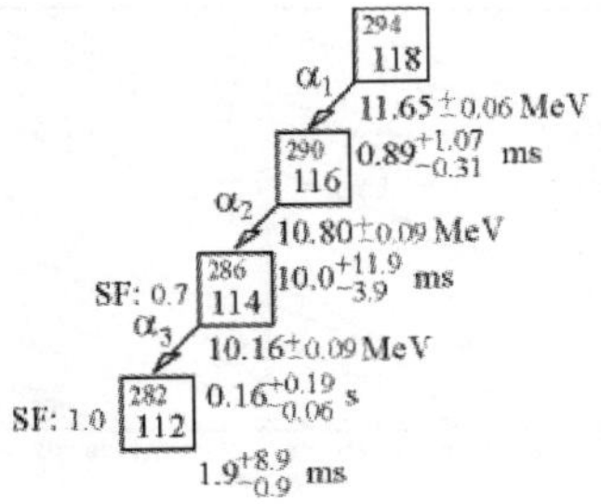

Figure 1. Time sequences in the decay chain of $^{294}118$ observed in the $^{249}Cf + ^{48}Ca$ reaction[21]. The average measured -particle energies, half-lives, and SF branching ratios of the observed nuclei are shown.

Table 1. Decay properties of some heaviest nuclei produced at JINR-Dubna[21].

Nuclide	Decay mode	E_α (MeV) Exp.	T_n^k Emp. est.	T_n^k Theor.	$S_n^k = T_n^{o.b.}/T_n^k$
$^{294}118$	α	11.65 ± 0.06	$0.89^{+1.07}_{-0.31}$ ms	$0.72^{+1.98}_{-0.27}$ms	$0.262 * 10^{-2}$
$^{290}116$	α	10.80 ± 0.09	$10^{+11.9}_{-3.9}$ ms	21^{+16}_{-8}ms	$0.412 * 10^{-2}$
$^{286}114$	α/SF	10.16 ± 0.09	$0.16^{+0.19}_{-0.06}$ s	$0.28^{+0.17}_{-0.13}$ s	$1.484 * 10^{-1}$
$^{282}112$	α	$10.25 \pm 0.25^*$		$40.8^{+154.2}_{-31.92}$ms	$1.000 * 10^{-1}$
$^{282}112$	SF	209	$1.9^{+8.9}_{-0.9}$ ms		

Note: $^*E_\alpha \pm \delta E_\alpha$ are the limits for the HFB estimations[22,23,24,25,26].

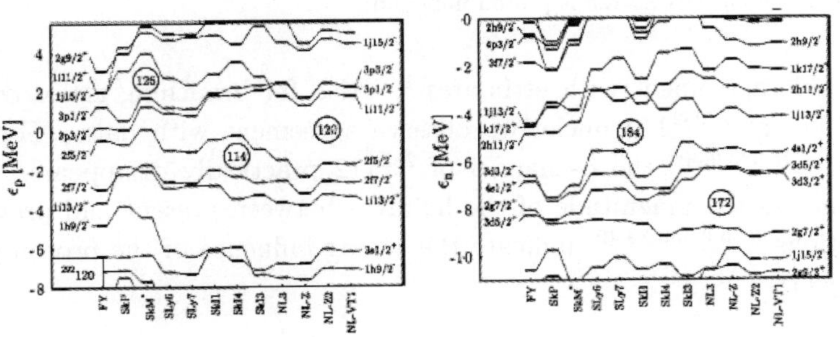

Figure 2. Single particle Fermi levels[10] used in the overlap integral $I_n^k(r)$ for $^{294}118$ and $^{290}116$ nuclei.

deformation parameters[34], the measured[21] $E_\alpha^{\exp}$ for $^{294}118$, $^{290}116$, $^{286}114$ and calculated ones[22,23,24,25,26] for $^{282}112$. Our results are shown in Table 1 and Fig.5. For $^{294}118$, $^{290}116$ and $^{286}114$ we can observe a

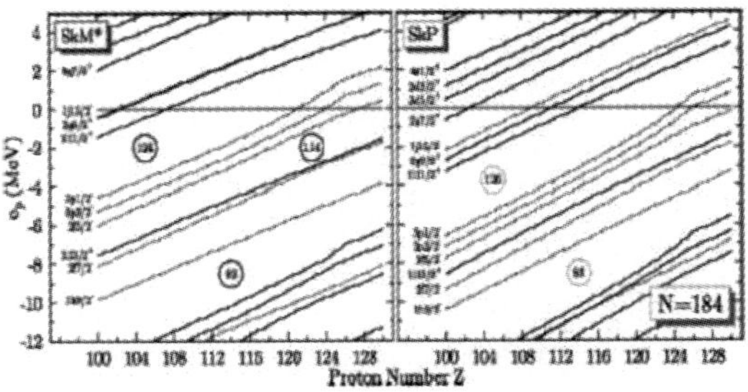

Figure 3. Single-proton Fermi levels used in the overlap integral $I_n^k(r)$ for 114 and 112 isotopes. Single-proton levels isotopes are calculted[11] in the RMF approach for 160 $\leq$N$\leq$190 with NL3 (left) and NL-Z2 no linear parameterizations.

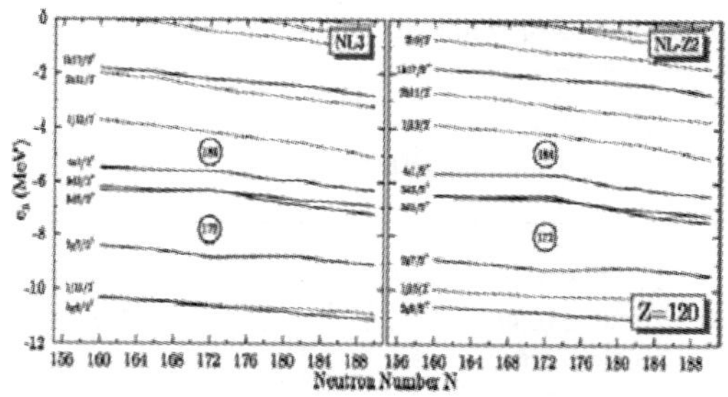

Figure 4. Single-neutron Fermi levels[11] used in the overlap integral $I_n^k(r)$ for 114 and 112 isotopes. The levels for N=184 isotones with 110=Z=130 , are obtained with Skyrme-Hartree-Fock model with SKN(left) and SKP effective interactions. Positive (negative) parity levels are indicated by solid (dashed) lines and by their spherical labels (nlj) In both cases the nucleus Z=126 is proton unbound.

reasonable agreement with estimates[21]. Also for branching ratios α/SF for $^{286}114$ and $^{282}112$ nuclei we observe agreement with data. For Q_α estimates[27,28,29,30], the α-channel of $^{282}112$ practically disappears. The difference in the magnitude of the halflives between present and previous predictions[35,36,37,38,39,40], indicate the strong influence of the proton shell at $Z \geq 118$.

6. Concluding remarks

Experimentally, there is evidence for α-clustering and fine structure (deformation) in SHEs. Aiming at confirmation of a part of recent results[21] for the decay chain of $^{294}118$, we estimate the α-decay rates using the shell model formation and resonance reaction amplitudes given by selfconsistent

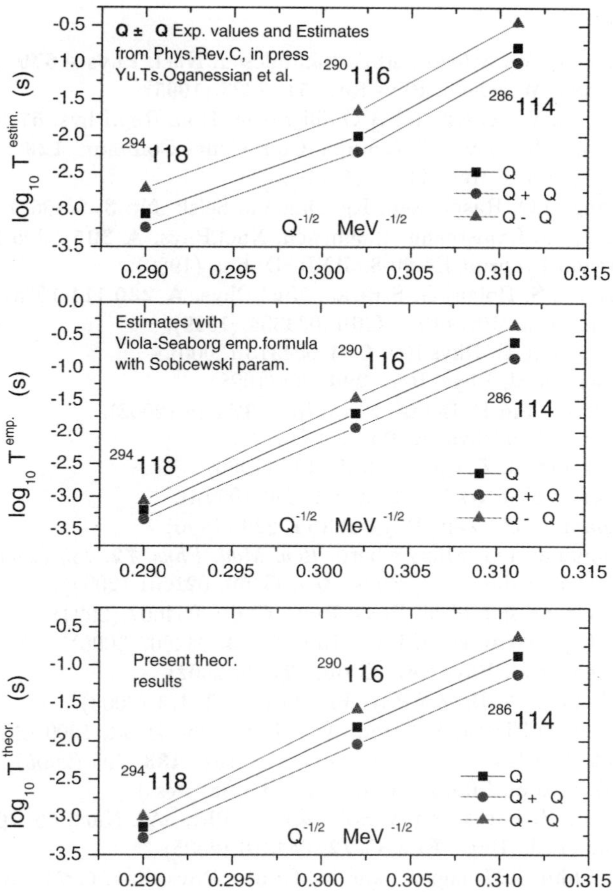

Figure 5. The Gamow plot for calculated α-halflives for the element $^{294}118$ and its descendants. Q_α-values are taken from [21]. Estimates[21] are obtained using the formula [31]with parameters[32,33] and empirical estimations are given using the formula[34].

models for nuclear structure and low-energy dynamics. In the most cases, the decay rates predicted agree well with the measured data. Systematic calculations of spectroscopic properties of SHEs has become feasible, as has been shown on the example of ground states of $^{294}118$ and its α-descendants.

We thank to Profs. S. Hofmann, G. Münzenberg, Yu.Ts. Oganessian, V.K. Utynkov, W. Scheid, M. Rizea, A. Sandulescu for many stimulating discussions. This work was supported from Contracts: CERES-3/40,41, 4/218 and CEEX1-D08-10.

References

1. I. Silişteanu , W. Scheid and A. Sandulescu, Nucl. Phys **A679**, 317 (2001).
2. I. Silişteanu, W.Scheid, Phys.Rev. **51**, 2023 (1995).
3. I. Silişteanu, W.Scheid and A.O.Silişteanu, Rom.Rep.Phys. **57**, 4 (2005).
4. K.Hagino, N.Rowley, A.T.Kruppa, Comp.Phys.Commun. **123** , 143(1999).
5. H.J. Mang, Phys. Rev. **119**, 1069(1960).
6. H.J. Mang, J.O. Rasmussen, Kgl.Dan.Vid.Selsk. **No.3**, 2(1962).
7. A.Sandulescu, I.Silişteanu, R.Wuench, Nucl.Phys. **A 305** , 205 (1978).
8. I. Silişteanu Preprint E4-80-8, JINR, Dubna (1980).
9. V.I.Furman, S. Holan, G. Stratan, Nucl.Phys. **A 239**,114(1975).
10. M. Bender et al. Phys.Rev. **C60**, 034304 (1999).
11. A.T.Kruppa et al. Phys.Rev.**C61** 034313 (2000).
12. R.G. Lovas et al, Phys. Rep. **294**, 265(1998).
13. P.E.Hodgson and E. Betak, Phys. Rep. **374**,89 (2003).
14. D. Robson, Nucl.Phys. **A204**, 204 (1973).
15. R.S. Simon et al, Z. Phys. **A325**, 197 (1986).
16. P. Reiter et al, Phys.Rev.Lett. **82**, 509(1999).
17. S. Hofmann et al., Zeit. Phys. **A354**, 229 (1996).
18. S. Hofmann and G. Münzenberg, *Rev. Mod. Phys.* **72**, 733 (2000).
19. Yu. Ts. Oganessian et al, Phys. Rev. **C 69**, 021601 (2004).
20. Yu. Ts. Oganessian et al, Phys. Rev. **C 69**, 054607 (2004).
21. Yu. Ts. Oganessian et al, Phys. Rev. **C 74**, 044602(2006).
22. S. Gorieely et al, Phys.Rev. **C 66**, 024326(2002).
23. S.Typel and B.A. Brown Phys.Rev **C 67**, 034313 (2003).
24. J.F.Berger, D.Hirata, M.Girod, Acta Phys. Pol. **B 34**, 1909 (2003).
25. S. Cwiok, P.H.Heenen,W. Nazarewicz, Nature **433**, 705 (2005).
26. M.Warda and al, Intern.J.Mod.Phys.E**15**,(in press).
27. I. Muntian, Z. Patyk and A. Sobiczewski, Phys. At. Nucl. 66, 1015 (2003).
28. A. Baran et al., Phys. Rev. **C 72**, 044310 (2005).
29. Y.K.Gambhir, A. Bhagwat, and M. Gupta, Phys. Rev. **C 71**, 037301 (2005).
30. M. Bender et al, Phys. Rev. **C 61**, 031302 (2000).
31. V. E. Viola and G. T. Seaborg, J. Inorg. Nucl. Chem. **28**, 741 (1966).
32. A. Sobiczewsky, Z. Patyk, and S Cwiok, Phys. Lett. **B 224**, 1(1989).
33. Z. Patyk and A. Sobiczewski, Nucl. Phys. **A354**, 229 (1996).
34. P. Möller et al, At. Data and Nucl. Data Tab. **66**, 131 (1997).
35. V.Yu.Denisov and H.Ikezoe, Phys. Rev. **C 72**, 064613 (2005).
36. P.Mohr, Phys. Rev **C 73**, 031301 (2006).
37. P.Roy.Chowdhury, C.Samanta and D.N.Basu Phys.Rev. **C 73**, 014612(2006).
38. Chang Xu and Zhongzhou Ren, Phys. Rev. **C 74**, 014304 (2006).
39. Takatoshi Ichikawa et al., Phys. Rev. **C 71**, 044608 (2005).
40. I. Silisteanu et al, Proc. CSSP-2005 Exotic nuclei and Nuclear/ Particle Astrophysics (World Scientific,2006, Eds. Stoica, Trache, Tribble) pag.423.

IV. DARK MATTER, DOUBLE BETA DECAY, AND POSSIBLE MECHANISMS FOR DILEPTON PRODUCTION

DARK MATTER IN THE COSMOS-EXPLOITING THE SIGNATURES OF ITS INTERACTION WITH NUCLEI

J.D. VERGADOS

Physics Department, University of Ioannina, Gr 451 10, Ioannina, Greece
** E-mail:vergados@cc.uoi.gr*

We review various issues related to the direct detection of constituents of dark matter, which are assumed to be Weakly Interacting Massive Particles (WIMPs). We specifically consider heavy WIMPs such as: 1) The lightest supersymmetric particle LSP or neutralino. 2) The lightest Kaluza-Klein particles in theories of extra dimensions and 3) other extensions of the standard model. In order to get the event rates one needs information about the structure of the nucleon as well as as the structure of the nucleus and the WIMP velocity distribution. These are also examined Since the expected event rates for detecting the recoiling nucleus are extremely low and the signal does not have a characteristic signature to discriminate against background we consider some additional aspects of the WIMP nucleus interaction, such as the periodic behavior of the rates due to the motion of Earth (modulation effect). Since, unfortunately, this is characterized by a small amplitude we consider other options such as directional experiments, which measure not only the energy of the recoiling nuclei but their direction as well. In these, albeit hard, experiments one can exploit two very characteristic signatures: a)large asymmetries and b) interesting modulation patterns. Furthermore we extended our study to include evaluation of the rates for other than recoil searches such as: i) Transitions to excited states, ii) Detection of recoiling electrons produced during the neutralino-nucleus interaction and iii) Observation of hard X-rays following the de-excitation of the ionized atom.

Keywords: WIMP; Dark Matter; CDM; Neutralino; Supersymmetry; LSP; Kaluza-Klein WIMPs; Modulation; Directional event rate.

1. Introduction

The combined MAXIMA-1 [1] , BOOMERANG [2] , DASI [3] , COBE/DMR Cosmic Microwave Background (CMB) observations [4] , the recent WMAP data [5] and SDSS [6] imply that the Universe is flat [7] and that most of the matter in the Universe is dark, i.e. exotic.These results have been confirmed and improved by the recent WMAP data [8]. The deduced cosmological expansion is consistent with the luminosity distance as a function of redshift of distant supernovae [9–11]. According to the scenario favored by the observations there are various contributions to the energy content of our Universe. The most accessible energy component is baryonic matter, which accounts for $\sim 5\%$ of the total energy density. A component that has not been directly observed is cold dark matter (CDM)): a pressureless fluid that is responsible for the growth of cosmological perturbations through gravitational instability. Its contribution to the total energy density is estimated at $\sim 25\%$. The dark matter is expected to become more abundant in extensive halos, that stretch up to 100–200 kpc from the center of galaxies. The component with the biggest contribution to the energy density has an equation of state similar to that of a cosmological constant and is characterized as dark energy. The ratio $w = p/\rho$ is negative and close to -1. This component is responsible for $\sim 70\%$ of the total energy density and induces the observed acceleration of the Universe [9]$^-$ [11] . The total energy density of our Universe is believed to take the critical value consistent with spatial flatness. Additional indirect information about the existence of dark matter comes from the rotational curves [12] . The rotational velocity of an object increases so long is surrounded by matter. Once outside matter the velocity of rotation drops as the square root of the distance. Such observations are not possible in our own galaxy. The observations of other galaxies, similar to our own, indicate that the rotational velocities of objects outside the luminous matter do not drop. So there must be a halo of dark matter out there.

Since the non exotic component cannot exceed 40% of the CDM [13] , there is room for exotic WIMP's (Weakly Interacting Massive Particles). In fact the DAMA experiment [14] has claimed the observation of one signal in direct detection of a WIMP, which with better statistics has subsequently been interpreted as a modulation signal [15] . These data, however, if they are due to the coherent process, are not consistent with other recent experiments, see e.g. EDELWEISS and CDMS [16] . It could still be interpreted as due to the spin cross section, but with a new interpretation of the extracted nucleon cross section.

Since the WIMP is expected to be very massive, $m_\chi \geq 30 GeV$, and extremely non relativistic with average kinetic energy $T \leq 100 KeV$, it can be directly detected mainly via the recoiling of a nucleus in the WIMP-nucleus elastic scattering.

The above developments are in line with particle physics considerations.

(1) Dark matter in supersymmetric theories

The lightest supersymmetric particle (LSP) or neutralino is the most natural WIMP candidate. In the most favored scenarios the LSP can be simply described as a Majorana fermion, a linear combination of the neutral components of the gauginos and Higgsinos [12]$^{-}$ [17] .
In order to compute the event rate one needs an effective Lagrangian at the elementary particle (quark) level obtained in the framework of supersymmetry [12,17,18] . One starts with representative input in the restricted SUSY parameter space as described in the literature, e.g. Ellis *et al* [19] , Bottino *et al* , Kane *et al* , Castano *et al* and Arnowitt *et al* [18] as well as elsewhere [20]$^{-}$ [21] . We will not, however, elaborate on how one gets the needed parameters from supersymmetry.
Even though the SUSY WIMPs have been well studied, for tor the reader's convenience we will give a description in sec. 2 of the basic SUSY ingredients needed to calculate LSP-nucleus scattering cross section.

(2) Kaluza-Klein (K-K) WIMPs.
These arise in extensions of the standard model with compact extra dimensions. In such models a tower of massive particles appear as Kaluza-Klein excitations. In this scheme the ordinary particles are associated with the zero modes and are assigned K-K parity +1. In models with Universal Extra Dimensions one can have cosmologically stable particles in the excited modes because of a discreet symmetry yielding K-K parity −1 (see previous work [22–24] as well as the recent review by Servant [25]).
The kinematics involved is similar to that of the neutralino, leading to cross sections which are proportional μ_r^2, μ_r being the WIMP-nucleus reduced mass. Furthermore the nuclear physics input is independent of the WIMP mass, since for heavy WIMP $mu_r \simeq Am_p$. There are appear two differences compared to the neutralino, though, both related to its larger mass.
i) First the density (number of particles per unit volume) of a WIMP falls inversely proportional to its mass. Thus, if the WIMP's considered are much heavier than the nuclear targets, the corresponding event rate takes the form:

$$R(m_{WIMP}) = R(A)\frac{A \text{ GeV}}{m_{WIMP}} \tag{1}$$

where $R(A)$ are the rates extracted from experiment up to WIMP masses of the order of the mass of the target.
ii) Second the average WIMP energy is now higher. In fact one finds

that $\langle T_{WIMP}\rangle = \frac{3}{4}m_{WIMP}v_0^2 \simeq 40\,(m_{WIMP}/(100\text{ GeV}))\text{keV}$ ($v_0 \simeq 2.2\times 10^5$km/s). Thus for a K-K WIMP with mass 1 TeV, the average WIMP energy is 0.4 MeV. Hence, due to the high velocity tail of the velocity distribution, one expects **an energy transfer to the nucleus in the MeV region. Thus many nuclear targets can now be excited by the WIMP-nucleus interaction and the de-excitation photons can be detected.**

In addition to the particle model one needs the following ingredients:

- A procedure in going from the quark to the nucleon level, i.e. a quark model for the nucleon. The results depend crucially on the content of the nucleon in quarks other than u and d. This is particularly true for the scalar couplings as well as the isoscalar axial coupling [26]$^-$ [27] . Such topics will be discussed in sec. 3.
- computation of the relevant nuclear matrix elements [28]$^-$ [29] using as reliable as possible many body nuclear wave functions. By putting as accurate nuclear physics input as possible, one will be able to constrain the SUSY parameters as much as possible. The situation is a bit simpler in the case of the scalar coupling, in which case one only needs the nuclear form factor.
- Convolution with the LSP velocity Distribution. To this end we will consider here Maxwell-Boltzmann [12] (MB) velocity distributions, with an upper velocity cut off put in by hand. The characteristic velocity of the M-B distribution can be increased by a factor n ($v_0 \to nv_0$, $n \geq 1$)by considering the interaction of dark matter and dark energy [30].
  Other distributions are possible, such as non symmetric ones, like those of Drukier [31] and Green [32] , or non isothermal ones, e.g. those arising from late in-fall of dark matter into our galaxy, like Sikivie's caustic rings [33] . In any event in a proper treatment the velocity distribution ought to be consistent with the dark matter density as, e.g., in the context of the Eddington theory [34] .

Since the expected rates are extremely low or even undetectable with present techniques, one would like to exploit the characteristic signatures provided by the reaction. Such are:

(1) The modulation effect, i.e the dependence of the event rate on the velocity of the Earth
(2) The directional event rate, which depends on the velocity of the sun

around the galaxy as well as the the velocity of the Earth. has recently begun to appear feasible by the planned experiments [35,36] .

(3) Detection of signals other than nuclear recoils, such as
   - Detection of γ rays following nuclear de-excitation, whenever possible [37,38] . This seems to become feasible for heavy WIMPs especially in connection with modified M-B distributions due to the coupling of dark matter and dark energy ($\langle T_{WIMP} \rangle \simeq n^2 40\,(m_{WIMP}/(100\text{ GeV}))\,,\ n \geq 1\text{keV}$)
   - Detection of ionization electrons produced directly in the LSP-nucleus collisions [39,40] .
   - Observations of hard X-rays produced [41] , when the inner shell electron holes produced as above are filled.

In all calculations we will, of course, include an appropriate nuclear form factor and take into account the influence on the rates of the detector energy cut off. We will present our results a function of the LSP mass, m_χ, in a way which can be easily understood by the experimentalists.

2. The Feynman Diagrams Entering the Direct Detection of WIMPS

2.1. *The Feynman Diagrams involving the neutralino*

The neutralino is perhaps the most viablle WIMP candidate and has been extensively studied (see, e.g., our recent review [42]). Here we will give a very brief summary of the most important aspects entering the direct neutralino searches. In currently favorable supergravity models the LSP is a linear combination [12] of the neutral four fermions $\tilde{B}, \tilde{W}_3, \tilde{H}_1$ and $\tilde{H}_2$ which are the supersymmetric partners of the gauge bosons B_μ and W^3_μ and the Higgs scalars H_1 and H_2. Admixtures of s-neutrinos are expected to be negligible. The relevant Feynman diagrams involve Z-exchange, s-quark exchange and Higgs exchange.

2.1.1. *The Z-exchange contribution*

The relevant Feynman diagram is shown in Fig. 1. It does not lead to coherence, since $\bar{\Psi}\gamma_\lambda\Psi = 0$ for a majorana fermion like the neutralino (the majorana fermions do not have electromagnetic properties). The coupling $\bar{\Psi}\gamma_\lambda\gamma_5\Psi$ yields negligible contribution for a non relativistic particle in the case of the spin independent cross section [43]. It may be important in the case of the spin contribution, which arises from the axial current).

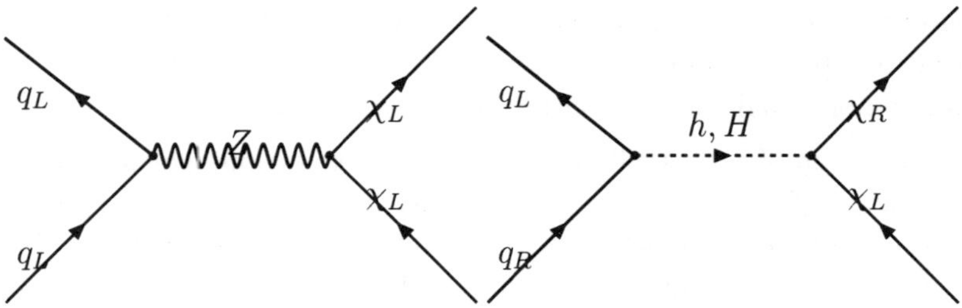

Fig. 1. The LSP-quark interaction mediated by Z and Higgs exchange.

2.1.2. *The s-quark Mediated Interaction*

The other interesting possibility arises from the other two components of χ_1, namely $\tilde{B}$ and $\tilde{W}_3$. Their corresponding couplings to s-quarks (see Fig. 2) can be read from the appendix C4 of Ref. [17] and our earlier review [42]. Normally this contribution yields vector like contribution, i.e it does not

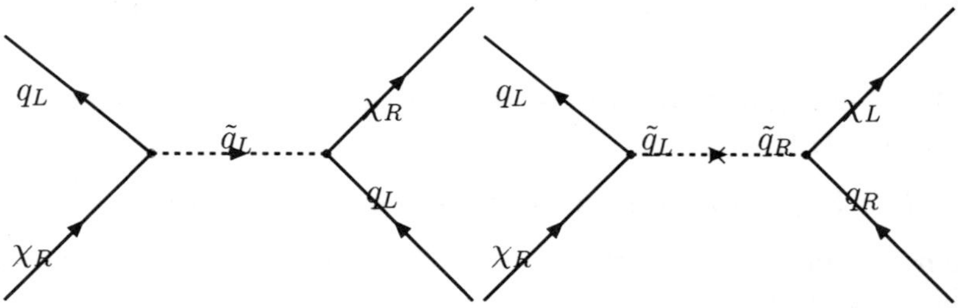

Fig. 2. The LSP-quark interaction mediated by s-quark exchange. Normally it yields V-A interaction which does not lead to coherence at the nuclear level. If, however, the isodoublet s-quark is admixed with isosinglet one to yield a scalar interaction at the quark level.

lead to coherence. If, however, there exists mixing between the s-quarks with isospin 1/2 ($\tilde{q}_L$) and the isospin 0 ($\tilde{q}_R$), the s-quark exchange may lead to a scalar interaction at the quark level and hence to coherence over all nucleons at the nuclear level [42].

2.1.3. *The Intermediate Higgs Contribution*

The most imporatant contribution to coherent scattering can be achieved via the intermediate Higgs particles which survive as physical particles. In supersymmetry there exist two such physical Higgs paricles, one light h with a mass $m_h \leq$120 GeV and one heavy H with mass m_H, which is much larger. The relevant interaction can arise out of the Higgs-Higgsino-gaugino interaction [42] leading to a Feynman diagram shown in Fig. 1.

In the case of the scalar interaction the resulting amplitude is proportional to the quark mass.

2.2. *The Feynman Diagrams involving the K-K WIMPs*

2.2.1. *The Kaluza-Klein Boson as a dark matter candidate*

We will assume that the lightest exotic particle, which can serve as a dark matter candidate, is a gauge boson B^1 having the same quantum numbers and couplings with the standard model gauge boson B, except that it has K-K parity -1. Thus its couplings must involve another negative K-K parity particle. In this work we will assume that such a particle can be one of the K-K quarks, partners of the ordinary quarks, but much heavier [22–24] .

- Itermediate K-K quarks.
  this case the relevant Feynman diagrams are shown in fig. 3.

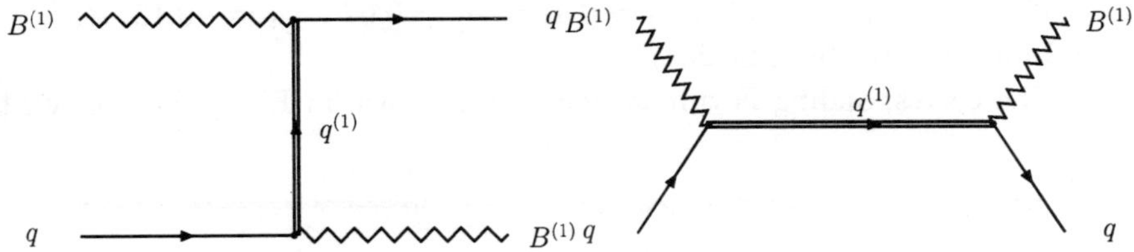

Fig. 3. K-K quarks mediating the interaction of K-K gauge boson B^1 with quarks at tree level.

The amplitude at the nucleon level can be written as:

$$\mathcal{M}_{coh} = \Lambda(\boldsymbol{\epsilon}^{*'}.\boldsymbol{\epsilon})N\left[\frac{11+12\tau_3}{54}\frac{m_p m_W}{(m_{B^{(1)}})^2}f_1(\Delta) + \frac{1+\tau_3}{3}\frac{m_W}{m_{B^{(1)}}}f_2(\Delta)\right]N \tag{2}$$

$$\Lambda = i4\sqrt{2}G_F m_W \tan^2\theta_W, f_1(\Delta) = \frac{1+\Delta+\Delta^2/2}{\Delta^2(1+\Delta/2)^2},$$

$$f_2(\Delta) = \frac{1+\Delta}{\Delta(1+\Delta/2)}\ , \ \Delta = \frac{m_{q^{(1)}}}{m_{B^{(1)}}} - 1$$

We see that the amplitude is very sensitive to the parameter Δ ("resonance effect").

In going from the quark to the nucleon level the best procedure is to replace the quark energy by the constituent quark mass $\simeq 1/3m_p$, as opposed to adopting [22–24] a procedure related to the current mass encountered in the neutralino case [42].
In the case of the spin contribution we find at the nucleon level that:

$$\mathcal{M}_{spin} = -i4\sqrt{2}G_F m_W \tan^2\theta_W \frac{1}{3}\frac{m_p m_W}{(m_{B^{(1)}})^2} f_1(\Delta) i(\boldsymbol{\epsilon}^{*\prime} \times \boldsymbol{\epsilon}).$$
$$[N\boldsymbol{\sigma}(g_0 + g_1\tau_3)N] \tag{3}$$

$$g_0 = \frac{17}{18}\Delta u + \frac{5}{18}\Delta d + \frac{5}{18}\Delta s\ ,\ g_1 = \frac{17}{18}\Delta u - \frac{5}{18}\Delta d$$

for the isoscalar and isovector quantities [42]. The quantities Δ_q are given by [42]

$$\Delta u = 0.78 \pm 0.02\ ,\ \Delta d = -0.48 \pm 0.02\ ,\ \Delta s = -0.15 \pm 0.02$$

We thus find $g_0 = 0.26\ ,\ g_1 = 0.41 \Rightarrow a_p = 0.67\ ,\ a_n = -0.15$.

- Intermediate Higgs Scalars.
The corresponding Feynman diagram is shown in Fig. 4 The relevant

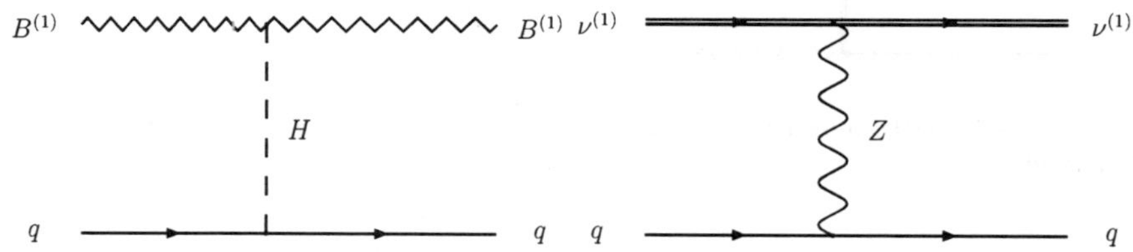

Fig. 4. The Higgs H mediating interaction of K-K gauge boson B^1 with quarks at tree level (on the left). The Z-boson mediating the interaction of K-K neutrino $\nu^{(1)}$ with quarks at tree level (on the right).

amplitude is given by:

$$\mathcal{M}_N(h) = -i\ 4\sqrt{2}G_F m_W^2 \tan^2\theta_W \left[\frac{1}{4}\frac{m_p}{m_h^2}\left(-\boldsymbol{\epsilon}^{*\prime}.\boldsymbol{\epsilon}\right) \prec N|N \succ \sum_q f_q\right] \tag{4}$$

In going from the quark to the nucleon level we follow a procedure analogous to that of the of the neutralino, i.e. $\prec N|m_q q\bar{q}|N \succ \Rightarrow f_q m_p$

2.2.2. *K-K neutrinos as dark matter candidates*

The other possibility is the dark matter candidate to be a heavy K-K neutrino. We will distinguish the following cases:

- Process mediated by Z-exchange.
  The amplitude associated with the diagram of Fig. 4 becomes:

$$\mathcal{M}_{\nu^{(1)}} = -\frac{1}{2\sqrt{2}} G_F J^{\lambda}(\nu^{(1)}) J_{\lambda}(NNZ) \tag{5}$$

  with $J_{\lambda}(NNZ)$ the standard nucleon neutral current and

$$J_{\lambda}(\nu^{(1)}) = \bar{\nu}^{(1)} \gamma_{\lambda} \gamma_5 \nu^{(1)} \ , \ J_{\lambda}(\nu^{(1)}) = \bar{\nu}^{(1)} \gamma_{\lambda} (1 - \gamma_5) \nu^{(1)}$$

  for Majorana and Dirac neutrinos respectively.
- Process mediated by right handed currents via Z'-boson exchange.
  The process is similar to that exhibited by Fig. 4, except that instead of Z we encounter Z', which is much heavier. Assuming that the couplings of the Z' are similar to those of Z, the above results apply except that now the amplitudes are retarded by the multiplicative factor $\kappa = m_Z^2/m_{Z'}^2$
- Process mediated by Higgs exchange.
  In this case in Fig 4 the Z is replaced by the Higgs particle. Proceeding as above we find that the amplitude at the nucleon level is:

$$\mathcal{M}_{\nu^{(1)}}(h) = -2\sqrt{2} G_F \frac{m_p m_{\nu^{(1)}}}{m_h^2} \bar{\nu}^{(1)} \ \nu^{(1)} \prec N|N \succ \sum_q f_q \tag{6}$$

  In the evaluation of the parameters f_q one encounters both theoretical and experimental errors.

3. Going from the Quark to the Nucleon Level

In going from the quark to the nucleon level one has to be a bit more careful in handling the quarks other than u and d. This is especially true in the case of the scalar interaction, since in this case the coupling of the WIMP to the quarks is proportional to their mass [42] . Thus one has to consider in the nucleon not only sea quarks ($u\bar{u}, d\bar{d}$ and $s\bar{s}$) but the heavier quarks as well due to QCD effects [44] . This way one obtains the scalar Higgs-nucleon coupling by using effective parameters f_q defined as follows:

$$\left\langle N|m_q\bar{q}q|N\right\rangle = f_q m_N \tag{7}$$

where m_N is the nucleon mass. The parameters f_q, $q = u, d, s$ can be obtained by chiral symmetry breaking terms in relation to phase shift and dispersion analysis (for a recent review see [42]). We like to emphasize here that since the current masses of the u and d quarks are small, the heavier quarks tend to dominate even though the probability of finding them in the nucleus is quite small. In fact the s quark contribution may become dominant, e.g. allowed by the above analysis is the choice:

$$f_d = 0.046, f_u = 0.025, f_s = 0.400, f_c = 0.050, f_b = 0.055, f_t = 0.095$$

The isoscalar and the isovector axial current in the case of K-K theories has already been discussed above. In the case of the neutralino these couplings at the nucleon level, f_A^0, f_A^1, are obtained from the corresponding ones given by the SUSY models at the quark level, $f_A^0(q)$, $f_A^1(q)$, via renormalization coefficients g_A^0, g_A^1, i.e. $f_A^0 = g_A^0 f_A^0(q), f_A^1 = g_A^1 f_A^1(q)$. The renormalization coefficients are given terms of Δq defined above [45], via the relations

$$g_A^0 = \Delta u + \Delta d + \Delta s = 0.77 - 0.49 - 0.15 = 0.13 \ , \ g_A^1 = \Delta u - \Delta d = 1.26$$

We see that, barring very unusual circumstances at the quark level, the isoscalar contribution is negligible. It is for this reason that one might prefer to work in the isospin basis.

4. The allowed SUSY Parameter Space

It is clear from the above discussion that the LSP-nucleon cross section depends, among other things, on the parameters of supersymmetry. One starts with a set of parameters at the GUT scale and predicts the low energy observables via the renormalization group equations (RGE). Conversely starting from the low energy phenomenology one can constrain the input parameters at the GUT scale. The parameter space is the most crucial. In SUSY models derived from minimal SUGRA the allowed parameter space is characterized at the GUT scale by five parameters:

- two universal mass parameters, one for the scalars, m_0, and one for the fermions, $m_{1/2}$.
- $tan\beta$, i.e the ratio of the Higgs expectation values, $\langle H_2 \rangle / \langle H_1 \rangle$.
- The trilinear coupling A_0 (or m_t^{pole}) and
- The sign of μ in the Higgs self-coupling $\mu H_1 H_2$.

The experimental constraints [42] restrict the values of the above parameters yielding the **allowed SUSY parameter space**.

5. Event rates

The differential non directional rate can be written as

$$dR_{undir} = \frac{\rho(0)}{m_\chi}\frac{m}{Am_N}d\sigma(u,\upsilon)|\upsilon| \tag{8}$$

where A is the nuclear mass number, $\rho(0) \approx 0.3GeV/cm^3$ is the WIMP density in our vicinity, m is the detector mass, m_χ is the WIMP mass and $d\sigma(u,\upsilon)$ is the differential cross section.

The directional differential rate, i.e. that obtained, if nuclei recoiling in the direction $\hat{e}$ are observed, is given by [42,46] :

$$dR_{dir} = \frac{\rho(0)}{m_\chi}\frac{m}{Am_N}|\upsilon|\hat{\upsilon}.\hat{e}\ \Theta(\hat{\upsilon}.\hat{e})\ \frac{1}{2\pi}\ d\sigma(u,\upsilon\ \delta(\frac{\sqrt{u}}{\mu_r\upsilon\sqrt{2}} - \hat{\upsilon}.\hat{e})$$

where $\Theta(x)$ is the Heaviside function.

The differential cross section is given by:

$$d\sigma(u,\upsilon) == \frac{du}{2(\mu_r b\upsilon)^2}[(\bar{\Sigma}_S F(u)^2 + \bar{\Sigma}_{spin}F_{11}(u)] \tag{9}$$

where u the energy transfer Q in dimensionless units given by

$$u = \frac{Q}{Q_0}\ ,\quad Q_0 = [m_p Ab]^{-2} = 40A^{-4/3}\ MeV \tag{10}$$

with b is the nuclear (harmonic oscillator) size parameter. $F(u)$ is the nuclear form factor and $F_{11}(u)$ is the spin response function associated with the isovector channel.

The scalar contribution is given by:

$$\bar{\Sigma}_S = (\frac{\mu_r}{\mu_r(p)})^2\sigma^S_{p,\chi^0}A^2\left[\frac{1+\frac{f^1_S}{f^0_S}\frac{2Z-A}{A}}{1+\frac{f^1_S}{f^0_S}}\right]^2 \approx \sigma^S_{N,\chi^0}(\frac{\mu_r}{\mu_r(p)})^2A^2 \tag{11}$$

(since the heavy quarks dominate the isovector contribution is negligible). σ^S_{N,χ^0} is the LSP-nucleon scalar cross section.

The spin contribution is given by:

$$\bar{\Sigma}_{spin} = (\frac{\mu_r}{\mu_r(p)})^2\sigma^{spin}_{p,\chi^0}\ \zeta_{spin},\zeta_{spin} = \frac{1}{3(1+\frac{f^0_A}{f^1_A})^2}S(u) \tag{12}$$

$$S(u) \approx S(0) = [(\frac{f^0_A}{f^1_A}\Omega_0(0))^2 + 2\frac{f^0_A}{f^1_A}\Omega_0(0)\Omega_1(0) + \Omega_1(0))^2\,] \tag{13}$$

The couplings f^1_A (f^0_A) and the nuclear matrix elements $\Omega_1(0)$ ($\Omega_0(0)$) associated with the isovector (isoscalar) components are normalized so that, in the case of the proton at $u = 0$, they yield $\zeta_{spin} = 1$.

With these definitions in the proton neutron representation we get:

$$\zeta_{spin} = \frac{1}{3} S^{'}(0) \ , \ S^{'}(0) = \left[(\frac{a_n}{a_p} \Omega_n(0))^2 + 2 \frac{a_n}{a_p} \Omega_n(0) \Omega_p(0) + \Omega_p^2(0) \right] \quad (14)$$

where $\Omega_p(0)$ and $\Omega_n(0)$ are the proton and neutron components of the static spin nuclear matrix elements. In extracting limits on the nucleon cross sections from the data we will find it convenient to write:

$$\sigma_{p,\chi^0}^{spin} \ \zeta_{spin} = \frac{\Omega_p^2(0)}{3} |\sqrt{\sigma_p} + \frac{\Omega_n}{\Omega_p} \sqrt{\sigma_n} e^{i\delta}|^2 \quad (15)$$

In Eq. (15) δ the relative phase between the two amplitudes a_p and a_n, which in most models is 0 or π, i.e. one expects them to be relatively real. The static spin matrix elements are obtained in the context of a given nuclear model. Some such matrix elements of interest to the planned experiments can be found in [42].

The spin ME are defined as follows:

$$\Omega_p(0) = \sqrt{\frac{J+1}{J}} \prec J\ J|\sigma_z(p)|J\ J \succ \ , \ \Omega_n(0) = \sqrt{\frac{J+1}{J}} \prec J\ J|\sigma_z(n)|J\ J \succ \quad (16)$$

where J is the total angular momentum of the nucleus and $\sigma_z = 2S_z$. The spin operator is defined by $S_z(p) = \sum_{i=1}^{Z} S_z(i)$, i.e. a sum over all protons in the nucleus, and $S_z(n) = \sum_{i=1}^{N} S_z(i)$, i.e. a sum over all neutrons. Furthermore $\Omega_0(0) = \Omega_p(0) + \Omega_n(0) \ , \ \Omega_1(0) = \Omega_p(0) - \Omega_n(0)$

6. The WIMP velocity distribution

To obtain the total rates one must fold with WIMP velocity distribution and integrate the above expressions over the energy transfer from Q_{min} determined by the detector energy cutoff to Q_{max} determined by the maximum LSP velocity (escape velocity, put in by hand in the Maxwellian distribution), i.e. $\upsilon_{esc} = 2.84\ \upsilon_0$ with υ_0 the velocity of the sun around the center of the galaxy(229 Km/s).

For a given velocity distribution f($\boldsymbol{\upsilon}'$), with respect to the center of the galaxy, one can find the velocity distribution in the Lab f($\boldsymbol{\upsilon}$,$\boldsymbol{\upsilon}_E$) by writing $\boldsymbol{\upsilon}' = \boldsymbol{\upsilon} + \ \boldsymbol{\upsilon}_E$, $\boldsymbol{\upsilon}_E$=$\boldsymbol{\upsilon}_0$+ $\boldsymbol{\upsilon}_1$, with $\boldsymbol{\upsilon}_1$ the Earth's velocity around the sun.

It is convenient to choose a coordinate system so that $\hat{x}$ is radially out in the plane of the galaxy, $\hat{z}$ in the sun's direction of motion and $\hat{y} = \hat{z} \times \hat{x}$.

Since the axis of the ecliptic lies very close to the x, y plane ($\omega = 186.3^0$) only the angle $\gamma = 29.8^0$ becomes relevant. Thus the velocity of the earth around the sun is given by

$$\boldsymbol{\upsilon}_E = \boldsymbol{\upsilon}_0 \hat{z} + \boldsymbol{\upsilon}_1 (\ sin\alpha\,\hat{\mathbf{x}} - cos\alpha\, cos\gamma\,\hat{\mathbf{y}} + cos\alpha\, sin\gamma\,\hat{\mathbf{z}}\,) \quad (17)$$

where α is phase of the earth's orbital motion.

The WIMP velocity distribution f($\boldsymbol{v}'$) is not known. Many velocity distributions have been used. The most common one is the M-B distribution with characteristic velocity v_0 with an upper bound $v_{esc} = 2.84 v_0$.

$$f(\boldsymbol{v}') = \frac{1}{(\sqrt{\pi} \boldsymbol{v}_0)^3} e^{-(\boldsymbol{v}'/\boldsymbol{v}_0)^2} \tag{18}$$

Modifications of this velocity distribution have also been considered such as: i) Axially symmetric M-B distribution [31,47]. and ii) modifications of the characteristic parameters of the M-B distribution by considering a coupling between dark matter and dark energy [30] ($v_0 \rightarrow n v_0, v_{esc} \rightarrow n v_{esc}$). Other possibilities are adiabatic velocity distribution following the Eddington approach [48]$^-$ [49] , caustic rings [50]$^-$ [51] and Sagittarius dark matter [32] .

For a given energy transfer the velocity v is constrained to be

$$v \geq v_{min} \ , \ v_{min} = \sqrt{\frac{Q A m_p}{2}} \frac{1}{\mu_r}. \tag{19}$$

7. The Direct detection rate

The event rate for the coherent WIMP-nucleus elastic scattering is given by [42,46,52,53]:

$$R = \frac{\rho(0)}{m_{\chi^0}} \frac{m}{m_p} \sqrt{\langle v^2 \rangle} \left[f_{coh}(A, \mu_r(A)) \sigma^S_{p,\chi^0} + f_{spin}(A, \mu_r(A)) \sigma^{spin}_{p,\chi^0} \ \zeta_{spin} \right] \tag{20}$$

with

$$f_{coh}(A, \mu_r(A)) = \frac{100\text{GeV}}{m_{\chi^0}} \left[\frac{\mu_r(A)}{\mu_r(p)} \right]^2 A \ t_{coh} \left(1 + h_{coh} cos\alpha\right) \tag{21}$$

$$f_{spin}(A, \mu_r(A)) = \left[\frac{\mu_r(A)}{\mu_r(p)} \right]^2 \frac{t_{spin}(A)}{A} t_{spin} \left(1 + h_{spin} cos\alpha\right) \tag{22}$$

with σ^S_{p,χ^0} and σ^{spin}_{p,χ^0} the scalar and spin proton cross sections ζ_{spin} the nuclear spin ME. In the above expressions h is the modulation amplitude.

The number of events in time t due to the scalar interaction, which leads to coherence, is:

$$R \simeq 1.60 \ 10^{-3} \frac{t}{1\text{y}} \frac{\rho(0)}{0.3 GeV cm^{-3}} \frac{m}{1\text{Kg}} \frac{\sqrt{\langle v^2 \rangle}}{280\text{k}ms^{-1}} \frac{\sigma^S_{p,\chi^0}}{10^{-6} \text{ pb}} f_{coh}(A, \mu_r(A)) \tag{23}$$

In the above expression m is the target mass, A is the number of nucleons in the nucleus and $\langle v^2 \rangle$ is the average value of the square of the WIMP velocity.

In the case of the spin interaction we write:

$$R \simeq 16 \frac{t}{1\text{y}} \frac{\rho(0)}{0.3GeV cm^{-3}} \frac{m}{1\text{Kg}} \frac{\sqrt{\langle v^2 \rangle}}{280 \text{k}ms^{-1}} \frac{\sigma^S_{p,\chi^0}}{10^{-2}\text{ pb}} f_{spin}(A, \mu_r(A)) \quad (24)$$

Note the different scale for the proton spin cross section. The parameters $f_{coh}(A, \mu_r(A))$, $f_{spin}(A, \mu_r(A))$, which give the relative merit for the coherent and the spin contributions in the case of a nuclear target compared to those of the proton, have already been tabulated [42] for energy cutoff $Q_{min} = 0$, 10 keV. It is clear that for large A the coherent process is expected to dominate unless for some reason the scalar proton cross section is very suppressed.

In the case of directional experiments the event rate is given by Eqs (23) and (24) except that now:

$$f_{coh}(A, \mu_r(A)) = \frac{100\text{GeV}}{m_{\chi^0}} \left[\frac{\mu_r(A)}{\mu_r(p)} \right]^2 A \frac{\kappa}{2\pi} t_{coh} \left(1 + h_m(coh) cos(\alpha + \alpha_m \pi)\right) \quad (25)$$

$$f_{spin}(A, \mu_r(A)) = \frac{100\text{GeV}}{m_{\chi^0}} \left[\frac{\mu_r(A)}{\mu_r(p)} \right]^2 \frac{\kappa}{2\pi} \frac{t_{spin}}{A} \left(1 + h_m(spin) cos(\alpha + \alpha_m \pi)\right) \quad (26)$$

In the above expressions h_m is the modulation amplitude and α_m the shift in the phase of the modulation (in units of π) relative to the phase of the Earth. $\kappa/(2\pi)$, $\kappa \leq 1$, is the supression factor entering due to the restriction of the phase space. κ, h_m and α_m depend on the direction of observation. It is precisely this dependence as well as the large values of h_m, which can be exploited to reject backround [42], that makes the directional experiments quite attractive in spite of the spression factor relative to the standard experiments.

8. Bounds on the scalar proton cross section

Using the above formalism one can obtain the quantities of interest t and h both for the standard as well as the directional experiments. Due to lack of space we are not going to present the obtained results here. The interested reader can find some of these results elsewhere [42,46] . Here we are simply going to show how one can employ such results to extract the nucleon cross section from the data. Due to space considerations we are not going to discuss the limits extracted from the data on the spin cross sections, since in this case one has to deal with two amplitudes (one for the proton and one for the neutron). We will only extract some limits imposed on the scalar nucleon cross section (the proton and neutron cross section are essentially the same). In what follows we will employ for all targets [54]$^-$ [55] the limit of CDMS II for the Ge target [56] , i.e. < 2.3 events for an exposure of 52.5 Kg-d with a threshold of 10 keV. This event rate is similar to that for other systems [57]. The thus obtained limits are exhibited in Fig. 5. For larger WIMP masses one can extrapolate these curves, assuming an increase as $\sqrt{m_\chi}$.

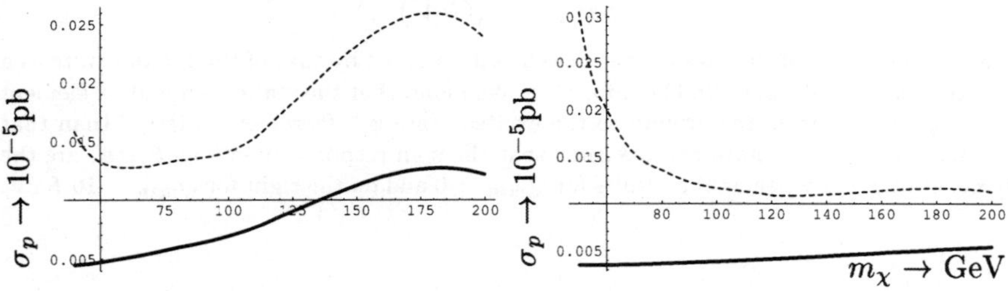

Fig. 5. The limits on the scalar proton cross section for A= 127 on the left and A= 73 on the right as functions of m_χ. The continuous (dashed) curves correspond to Q_{min} = 0 (10) keV respectively. Note that the advantage of the larger nuclear mass number of the A= 127 system is counterbalanced by the favorable form factor dependence of the A= 73 system.

9. Transitions to excited states

The above formalism can easily be extended to cover transitions to excited states. Only the kinematics and the nuclear physics is different. In other words one now needs:

- The inelastic scalar form factor.
  The transition amplitude is non zero due to the momentum transfer involved. The relevant multipolarities are determined by the spin and parity of the final state.
- Spin induced transitions.
  In this case one can even have a Gamow-Teller like transition, if the final state is judiciously chosen.

In the case of ^{127}I the static spin matrix element involving the first excited state around 50 keV is twice as large compared to that of the ground state [38]. The spin response function was assumed to be the same with that of the ground state. The results obtained [38] are shown in Fig. 6. These results are very encouraging, since, as we have mentioned, for

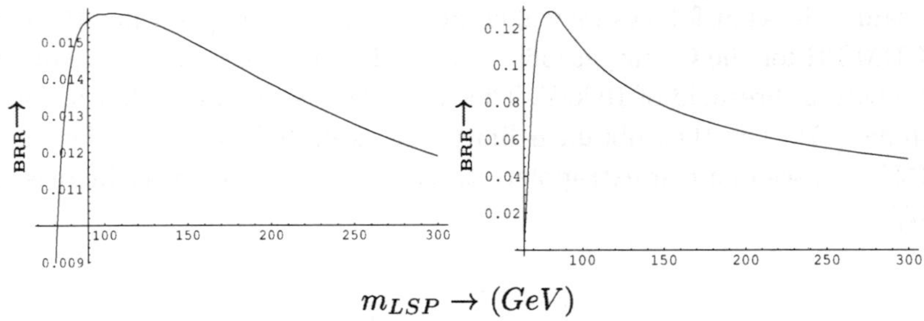

Fig. 6. The ratio of the rate to the excited state divided by that of the ground state as a function of the LSP mass (in GeV) for ^{127}I. We found that the static spin matrix element of the transition from the ground to the excited state is a factor of 1.9 larger than that involving the ground state and assumed that the spin response functions $F_{11}(u)$ are the same. On the left we show the results for $Q_{min} = 0$ and on the right for $Q_{min} = 10\ KeV$.

heavier WIMPs like those involved in K-K theories, the branching ratios are expected to be much larger. Thus one may consider such transitions, since the detection of de-excitation γ rays is much easier than the detection of recoiling nuclei.

10. Other non recoil experiments

As we have already mentioned the nucleon recoil experiments are very hard. It is therefore necessary to consider other possibilities. One such possibility is to detect the electrons produced during the WIMP-nucleus collisions [39,40] employing detectors with low energy threshold with a high Z target. Better yet one may attempt to detect the very hard X-rays generated when the inner shell electron holes are filled [41]. The relative X-ray to nucleon recoil probabilities $[Z\sigma_K/\sigma_r]_i$, for $i = L(m_\chi \leq 100\text{GeV})$, $M(100\text{ GeV} \leq m_\chi \leq 200\text{ GeV})$ and $H(m_\chi \simeq 200\text{ GeV})$ are shown in table 1. For even heavier WIMPs, like those expected in K-K theories, the relative probability is expected to be even larger.

The K_α and K_β lines can be separated experimentally by using good energy-resolution detectors, but the sum of all K lines can be measured in modest energy-resolution experiments.

K X-ray	$E_K(K_{ij})$ keV	$[\frac{Z\sigma_K(K_{ij})}{\sigma_r}]_L$	$[\frac{Z\sigma_K(K_{ij})}{\sigma_r}]_M$	$[\frac{Z\sigma_K(K_{ij})}{\sigma_r}]_H$
$K_{\alpha 2}$	29.5	0.0086	0.0560	0.0645
$K_{\alpha 1}$	29.8	0.0160	0.1036	0.1196
$K_{\beta 1}$	33.6	0.0047	0.0303	0.0350
$K_{\beta 2}$	34.4	0.0010	0.0067	0.0077

11. Conclusions

We examined the various signatures expected in the direct detection of WIMPs via their interaction with nuclei. We specially considered WIMPs predicted in supersymmetric models (LSP or neutralino) as well as theories with extra dimensions. We presented the formalism for the modulation amplitude for non directional as well as directional experiments. We discussed the role played by nuclear physics on the extraction of the nucleon cross sections from the data. We also considered non recoil experiments, such as measuring the γ rays following the de-excitation of the nucleus and/or the hard X-rays after the de-excitation of the inner shell electron holes produced during the WIMP nucleus interaction. These are favored by very heavy MIMPs in the TeV region and velocity distributions expected in models allowing interaction of dark matter and dark energy.

Acknowledgements: This work was supported in part by the European Union contract MRTN-CT-2004-503369. Special thanks to Professor Raduta for support and hospitality during the Predeal Summer School.

References

1. S. Hanary *et al*: *Astrophys. J.* **545**, L5 (2000);
   J.H.P Wu *et al*: *Phys. Rev. Lett.* **87**, 251303 (2001);
   M.G. Santos *et al*: *Phys. Rev. Lett.* **88**, 241302 (2002).
2. P. D. Mauskopf *et al*: *Astrophys. J.* **536**, L59 (2002);
   S. Mosi *et al*: *Prog. Nuc.Part. Phys.* **48**, 243 (2002);
   S. B. Ruhl *al*, astro-ph/0212229 and references therein.
3. N. W. Halverson *et al*: Astrophys. J. **568**, 38 (2002)
   L. S. Sievers *et al*: astro-ph/0205287 and references therein.
4. G. F. Smoot et al (COBE Collaboration): Astrophys. J. **396**, L1 (1992).
5. D. N. Spergel et al: Astrophys. J. Suppl. **148**, 175 (2003).
6. M. Tegmark et al: Phys.Rev. D **69**, 103501 (2004).
7. A. H. Jaffe et al: Phys. Rev. Lett. **86**, 3475 (2001).
8. D.N. Spergel *et al*, Three-Year WMAP Results: Implications for Cosmology, astro-ph/0603449;

L. Page *et al*, Three-Year WMAP Results: Polarization Analysis, astro-ph/0603450;
G. Hinsaw *et al*, Three-Year WMAP Observations: Implications Temperature Analysis, astro-ph/0603451;
N Jarosik *et al*, Three-Year WMAP Observations: Beam Profiles, Data Processing, Radiometer Characterization and Systematic Error Limits, astro-ph/0603452.

9. A. G. Riess *et al.* [Supernova Search Team Collaboration], Astron. J. **116** (1998) 1009 [arXiv:astro-ph/9805201].
10. S. Perlmutter *et al.* [Supernova Cosmology Project Collaboration], Astrophys. J. **517** (1999) 565 [arXiv:astro-ph/9812133].
11. W. J. Percival *et al.* [The 2dFGRS Collaboration], Mon. Not. Roy. Astron. Soc. **327** (2001) 1297 [arXiv:astro-ph/0105252].
12. G. Jungman, M. Kamionkowski, and K. Griest: Phys. Rep. **267**, 195 (1996).
13. D. P. Bennett et al.: Phys. Rev. Lett. **74**, 2867 (1995).
14. R. Bernabei et al: Phys. Lett. B **389**, 757 (1996).
15. R. Bernabei et al: Phys. Lett. B **424**, 195 (1998).
16. A. Benoit *et al*, [EDELWEISS collaboration]: Phys. Lett. B **545**, 43 (2002);
V. Sanglar,[EDELWEISS collaboration] arXiv:astro-ph/0306233;
D. S. Akerib *et al*,[CDMS Collaboration]: Phys. Rev D **68**, 082002 (2003); arXiv:astro-ph/0405033.
17. G. L. Kane et al: Phys. Rev. D **49**, 6173 (1994).
18. A. Bottino *et al.*, *Phys. Lett B* **402**, 113 (1997).
R. Arnowitt. and P. Nath, *Phys. Rev. Lett.* **74**, 4592 (1995); *Phys. Rev. D* **54**, 2374 (1996); hep-ph/9902237;
V. A. Bednyakov, H.V. Klapdor-Kleingrothaus and S.G. Kovalenko, *Phys. Lett. B* **329**, 5 (1994).
19. J. Ellis, K. A. Olive, Y. Santoso, and V. C. Spanos: Phys.Rev. D **70**, 055005 (2004).
20. M. W. Goodman and E. Witten: Phys. Rev. D **31**, 3059 (1985).
21. P. Ullio and M. Kamioknowski: JHEP **0103**, 049 (2001).
22. G. Servant and T. M. P. Tait: Nuc. Phys. B **650**, 391 (2003).
23. G. Servant and T. M. P. Tait: New Jour. Phys. **4**, 99 (2002).
24. H. C. Cheng, J. L. Feng, and K. T. Matchev: Phys. Rev. Lett. **89**, 211301 (2002).
25. See, e.g., G. Servant, in Les Houches :Physics at TeV Colliders 2005" Beyond the Standard Model working group: summary report, B.C. Allanach (ed.), C. Grojean (ed.), P. Skands (ed.), *al*, section 25, p.164; hep-ph/0602198.
26. M. Drees and M. M. Nojiri, *Phys. Rev. D* **48**, 3843 (1993); *Phys. Rev. D* **47**, 4226 (1993).
27. T. P. Cheng, *Phys. Rev. D* **38**, 2869 (1988); H-Y. Cheng, *Phys. Lett. B* **219**, 347 (1989).
28. M. T. Ressell *et al.*, *Phys. Rev. D* **48**, 5519 (1993); M.T. Ressell and D. J. Dean, Phys. Rev. C **56**, 535 (1997).
29. E. Homlund and M. Kortelainen and T. S. Kosmas and J. Suhonen and J. Toivanen, Phys. Lett B, **584**,31 (2004); Phys. Atom. Nucl. **67**, 1198 (2004).

30. N. Tetradis and J.D. Vergados , Detection Rates of Dark Matter Coupled to Dark Energy, to be published; hep-ph/0609078.
31. A. K. Drukier, K. Freeze and D. N. Spergel, *Phys. Rev.* D, **33**, 3495 (1986); J.I. Collar et al., *Phys. Lett* B **275**, 181 (1992).
32. A. M. Green: Phys. Rev. D **66**, 083003 (2002).
33. P. Sikivie, I. Tkachev and Y. Wang, *Phys. Rev. Let.* **75**, 2911 (1995); *Phys. Rev.* **D 56**, 1863 (1997)
P. Sikivie, *Phys. Let.* **b 432**, 139 (1998); astro-ph/9810286.
34. D. Owen and J. D. Vergados: Astrophys. J. **589**, 17 (2003); astro-ph/0203923.
35. K. N. Buckland, M. J. Lehner and G. E. Masek, in Proc. *3nd Int. Conf. on Dark Matter in Astro- and part. Phys.* (Dark2000), Ed. H.V. Klapdor-Kleingrothaus, Springer Verlag (2000).
36. The NAIAD experiment B. Ahmed *et al*, Astropart. Phys. **19** (2003) 691; hep-ex/0301039
B. Morgan, A. M. Green and N. J. C. Spooner, Phys. Rev. D **71** (2005) 103507; astro-ph/0408047.
37. H. Ejiri, K. Fushimi, and H. Ohsumi: Phys. Lett. B **317**, 14 (1993).
38. J. D. Vergados, P. Quentin, and D. Strottman: IJMPE **14**, 751 (2005); hep-ph/0310365.
39. J. D. Vergados and H. Ejiri: Phys. Lett. B **606**, 305 (2005); hep-ph/0401151.
40. C. C. Moustakidis, J. D. Vergados, and H. Ejiri: Nucl. Phys. B **727**, 406 (2005); hep-ph/0507123.
41. H. Ejiri and Ch. C. Moustakidis and J. D. Vergados, Dark matter search by exclusive studies of X-rays following WIMPs nuclear interactions, (to appear in Phys. Lett.); hep-ph/0507123.
42. J. D. Vergados, On The Direct Detection of Dark Matter- Exploring all the signatures of the neutralino-nucleus interaction, hep-ph/0601064.
43. J. D. Vergados: J. of Phys. G **22**, 253 (1996).
44. A. Djouadi and M. K. Drees, *Phys. Lett. B* **484**, 183 (2000); S. Dawson, *Nucl. Phys. B* **359**, 283 (1991); M. Spira it et al, *Nucl. Phys.* **B453**, 17 (1995).
45. The Strange Spin of the Nucleon, J. Ellis and M. Karliner, hep-ph/9501280.
46. J. D. Vergados: J.Phys. G **30**, 1127 (2004); 0406134.
47. J. D. Vergados: Phys. Rev. D **62**, 023519 (2000).
48. A. S. Eddington: NRAS **76**, 572 (1916).
49. J.D. Vergados and D. Owen, Dark Matter Velocity Distribution in the Context of Eddington's Theory, astro-ph/0603704.
50. P. Sikivie: Phys. Rev. D **60**, 063501 (1999).
51. G. Gelmini and P. Gondolo: Phys. Rev. D **64**, 123504 (2001).
52. J. D. Vergados: Phys. Rev. D **63**, 06351 (2001).
53. J. D. Vergados: Phys. Rev. D **57**, 103003 (2003); hep-ph/0303231.
54. P. Belli, R. Cerulli, N. Fornego, and S. Scopel: Phys. Rev. D **66**, 043503 (2002); hep-ph/0203242.
55. M. M. Pavan, I. I. Strakovsky, R. L. Workman, and R. A. Arndt: PiN Newslett. **16**, 110 (2002); arXiv:hep-ph/0111066.
56. D. S. Akerib et al (CDMS Collaboration): Phys.Rev.Lett. **93**, 211301 (2004).
57. C. Savage, P. Gondolo, and K. Freese: Phys. Rev. D **70**, 123513 (2004).

Neutrinos, Dark Matter and Nuclear Structure

JOUNI SUHONEN

Department of Physics, University of Jyväskylä
P.O.Box 35, FIN-40014, Jyväskylä, Finland
E-mail: jouni.suhonen@phys.jyu.fi

In the first part of this review I concentrate on the nuclear-structure problems related to nuclear double beta decay. The present status of the nuclear matrix element calculations of the neutrinoless double beta decay is presented. Independent probes of the involved virtual transitions are discussed. In the second part I concentrate on nuclear-structure calculations related to detection of the cold dark matter of the Universe. In particular, I discuss the scattering of the SUSY predicted stable neutralino off nuclei. The results can be contrasted with the claimed WIMP detection of the DAMA experiment.

1. Introduction

The existence of non-zero neutrino mass has been established by the neutrino-oscillation experiments Super-Kamiokande, SNO, KamLAND, and CHOOZ [1]. These experiments probe the differences of the squares of the masses, instead of the absolute mass scale of the neutrino. The most feasible way to get information on the mass scale of neutrinos is the neutrinoless double beta ($0\nu\beta\beta$) decay. This information can be gained through the effective neutrino mass, $\langle m_\nu \rangle$, extracted from the results of the underground double-beta-decay experiments. To determine the neutrino masses we need reliable nuclear matrix elements [2] together with information on neutrino mixing [3] and the associated CP phases [4]. Last but not least, Majorana nature of the neutrino would be confirmed if the $0\nu\beta\beta$ decay were detected.

Reliable nuclear matrix elements (NME) are the key to obtain quantitative information about the important neutrino properties potentially probed by the $0\nu\beta\beta$ decay. Lack of accuracy in the values of these matrix elements is reflected as inaccuracies in the information on neutrino properties derived from the $0\nu\beta\beta$-decay experiments. Since the 1980's many different approaches have been used to compute values of the NME's [2,5–7].

Beside this, the calculation of half-lives and the corresponding NME's of the two-neutrino double beta ($2\nu\beta\beta$) decay has served as a first test for theories aimed at description of the $0\nu\beta\beta$ decay process [2]. The $2\nu\beta\beta$ decay is a second-order perturbative process within the standard model and the associated virtual transitions are of the Gamow–Teller type going through the 1^+ intermediate states.

Vast amounts of astrophysical observations support the idea that the mass of the universe mostly consists of matter which cannot be seen by detection of electromagnetic radiation. The contribution of this dark matter is roughly 90 percent of the total mass and its true nature is still unknown. Part of the dark matter can be explained by postulating the existence of the so-caled massive compact halo objects (MACHO), like brown and white dwarfs, Jupiter-like planets, neutron stars and black-hole remnants, which belong to the baryonic component of the dark matter. Some contribution can also come from the relativistic light neutrinos, this component being called the hot dark matter. Besides these, the dominant component stems from the non-baryonic cold dark matter, CDM, the constituents of which are particles that are heavy, non-relativistic and interact weakly with the ordinary baryonic matter. An interesting subgroup are the weakly interacting massive particles, WIMP's, and, in particular, the lightest supersymmetric particle, LSP. The LSP results from the inclusion of the supersymmetry into the standard model. LSP's are stable particles in GeV-TeV scale that interact only via the weak interaction. The possible LSP candidates include sneutrinos, neutralinos, Higgsino-like and mixed gaugino-Higgsino LSPs.

The non-relativistic LSP's form a halo enclosing the Milky Way and they should be detectable, for instance, via a neutral-current elastic LSP-nucleus scattering. Because of the Earth's motion around the Sun, the rate of the recoiling nuclei is assumed to reflect the seasonal changes in the relative velocity of the detector and the halo. The measured signal can be ionisation, phonons or light produced in the target by the recoiled nucleus. The present experiments aiming to detect this process apply targets such as NaI (DAMA [8]) and ^{73}Ge (HDMS [9], EDELWEISS [10]). Theoretical description of the target nuclei is needed for the interpretation of the LSP-nucleus scattering results. This is where the nuclear-structure calculations are setting in.

2. Nuclear Matrix Elements for Double Beta Decay

Both for the $2\nu\beta\beta$ and $0\nu\beta\beta$ decay the initial and final nuclei are even-even ones, two charge units apart from each other. The $2\nu\beta\beta$ decay can be thought of as two successive Gamow–Teller beta-decay transitions through the intermediate 1^+ states [2,11], whereas for the $0\nu\beta\beta$ decay intermediate states of all possible multipolarities appear if the involved two-body transition densities are expanded as one-body densities [12]. In this case allowed and forbidden beta transitions coexist.

The double beta decays can proceed to the final ground state and excited states [13–18]. The most studied decay mode is the double β^- decay, including the most well known example, namely the decay of ^{76}Ge. Some work has also been done on the positron side of the decay [14,15,18–20]. It is also interesting that for the case of ^{48}Ca decay the double beta decay has to compete with highly forbidden beta decays [21]. Such highly forbidden beta decays can be found in other systems as well [22]. Here I consentrate on decays to the 0^+ ground state of the final nucleus.

For the neutrino-mass mode of the $0\nu\beta\beta$ decay the half-life can be written as [2,23]

$$\left[t_{1/2}^{(0\nu)}(0_i^+ \to 0_f^+)\right]^{-1} = G^{(0\nu)} \left|M_{\mathrm{FGT}}^{(0\nu)}\right|^2 \frac{\langle m_\nu \rangle^2}{m_{\mathrm{e}}^2} , \tag{1}$$

where m_{e} is the electron rest mass and $\langle m_\nu \rangle$ is the effective neutrino mass

$$\langle m_\nu \rangle = \sum_{i=1}^{3} m_i \lambda_i |U_{ei}|^2 , \tag{2}$$

with the CP phases λ_i, mass eigenstates m_i, and the mixing matrix elements U_{ei} for the electron neutrino. The quantity $M_{\mathrm{FGT}}^{(0\nu)}$ is the nuclear matrix element

$$M_{\mathrm{FGT}}^{(0\nu)} = M_{\mathrm{GT}}^{(0\nu)} - \left(\frac{g_{\mathrm{V}}}{g_{\mathrm{A}}}\right)^2 M_{\mathrm{F}}^{(0\nu)} , \tag{3}$$

containing the double Gamow–Teller NME, $M_{\mathrm{GT}}^{(0\nu)}$, and the double Fermi NME, $M_{\mathrm{F}}^{(0\nu)}$. The factor $G^{(0\nu)}$ is a leptonic phase-space factor [2]. The constants g_{V} and g_{A} are the vector and axial-vector coupling constants of the effective nucleon current within a nucleus.

The most frequently used nuclear models in the evaluation of the NME of Eq. (3) are the nuclear shell model (SM) and the proton-neutron quasiparticle random-phase approximation (pnQRPA), along with many of its higher-RPA extensions. The pnQRPA calculations are characterized by an adjustable particle-particle part of the proton-neutron two-body interaction [2,11,24–27]. Determination of the value of the corresponding strength parameter, g_{pp}, has been a key issue since the mid 80's. The NME of the $2\nu\beta\beta$ decay is very sensitive to the value of this parameter, leading to the so-called $g_{\rm pp}$ problem of the pnQRPA. An example is shown in Fig. 1 for the decay of ^{76}Ge. The value $g_{\rm pp} = 1$ corresponds to the bare proton-neutron two-body interaction and the value $g_{\rm pp} = 0$ to the disappearance of this interaction. As seen from the figure, the increasing strength of the proton-neutron interactions drives the $2\nu\beta\beta$ NME through zero to the negative side. Hereby it reaches the values allowed by the measured $2\nu\beta\beta$ decay half-life. These experimentally allowed regions of the NME, extracted from several different measurements with their ranges of error included, are shown in the figure as two shaded horizontal stripes. A heated debate on the proper way of determining the value of $g_{\rm pp}$ has started [11,28,29].

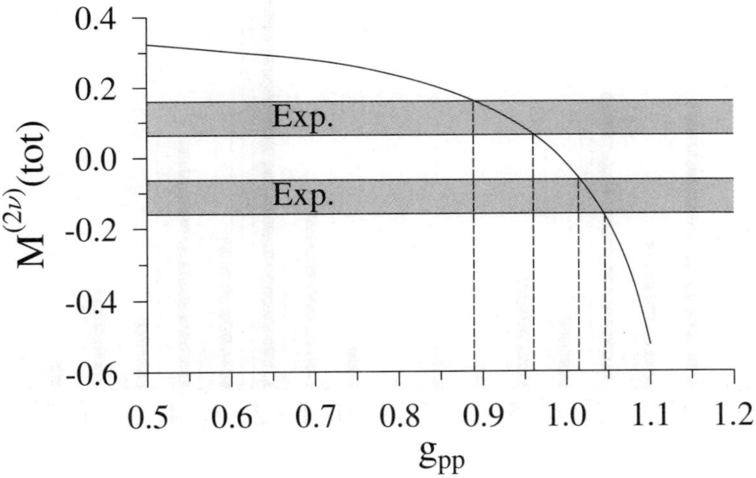

Fig. 1. The NME for the $2\nu\beta\beta$ decay of ^{76}Ge as a function of the interaction strength parameter $g_{\rm pp}$. The two horizontal stripes define the experimentally allowed values.

Contrary to the $2\nu\beta\beta$ decay, the NME of the $0\nu\beta\beta$ decay is much less dependent on the value of $g_{\rm pp}$ within the framework of the pnQRPA. This has been shown in Fig. 2 for the decay of ^{76}Ge. Here the four different values of $g_{\rm pp}$ correspond to the borders of the experimentally allowed regions of the $2\nu\beta\beta$ NME in Fig. 1. The decomposition to different multipoles has been done according to

$$M_{\rm GT}^{(0\nu)} = \sum_{J^\pi} M_{\rm GT}^{(0\nu)}(J^\pi) , \tag{4}$$

where the different multipoles stem from the Fourier-Bessel expansion of the neutrino-exchange potential corresponding to the propagator of the Majorana neutrino between the two decaying neutrons [23].

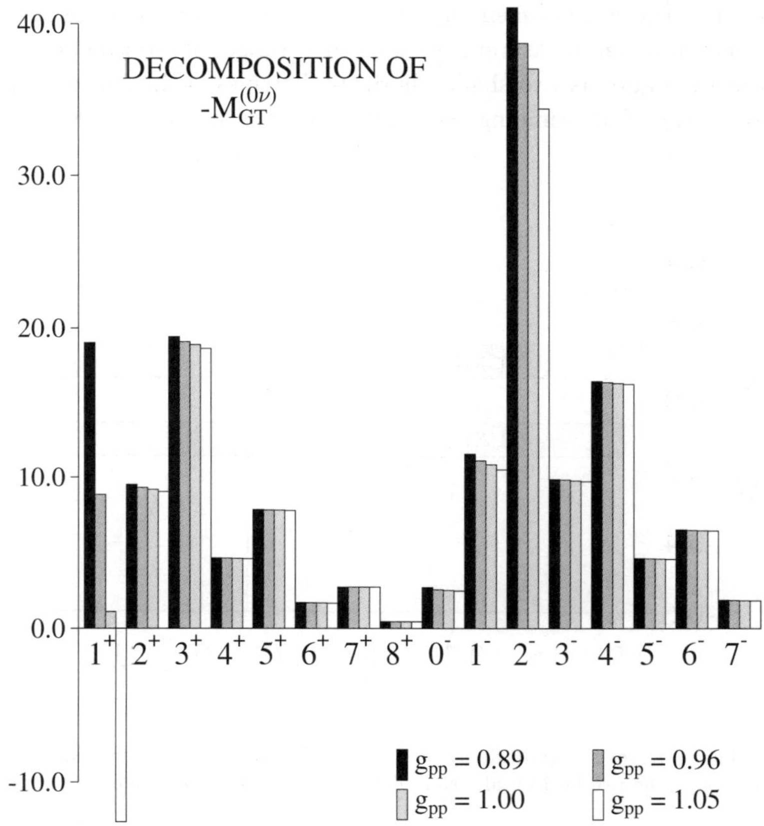

Fig. 2. Multipole decomposition of the negative of the $0\nu\beta\beta$ NME $M_{\rm GT}^{(0\nu)}$ of Eq. (4) for the decay of ^{76}Ge within the pnQRPA description. Four different values of $g_{\rm pp}$, indicated in the figure, have been included in the analysis.

From Fig. 2 we notice that the 1^+ component of the NME (4) is strongly dependent on $g_{\rm pp}$, as was the case for the $2\nu\beta\beta$ NME. This is the main uncertainty of the $0\nu\beta\beta$ NME with respect to the value of $g_{\rm pp}$. On the other hand, the multipoles other than $J^\pi = 1^+$ depend only weakly on $g_{\rm pp}$ reducing the overall dependence of $M_{\rm GT}^{(0\nu)}$ on $g_{\rm pp}$. In addition, we notice that the multipoles 2^-, 3^+ and possibly 4^- give a substantial contribution to the total $M_{\rm GT}^{(0\nu)}$. In fact, studying several other $0\nu\beta\beta$ NME's of medium-mass nuclei it seems that the 2^- contribution is by far the most important and describes a kind of universal behaviour of the $0\nu\beta\beta$ NME's [30]. In any case, the problem remains: how to probe independently the involved virtual transitions and to obtain useful information to constrain the parameter spaces of the used nuclear models. This problem is addressed in the following two sections.

3. Beta Decay as a Probe of Virtual Transitions

It is important to find independent probes of the virtual transitions involved in the NME's of double beta decays. The probe which most likely first enters one's mind is the measurement of single beta decays from the intermediate odd-odd nucleus of the double beta decay to the adjacent even-even nuclei. Extensive theoretical work on this probe has been done along the years [14,31–35].

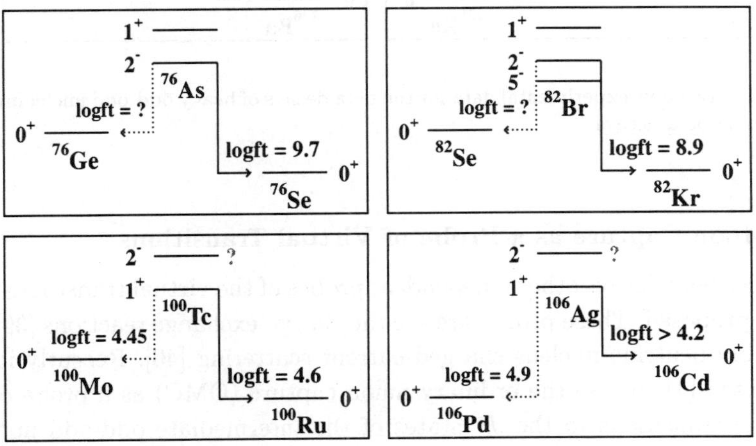

Fig. 3. Available experimental data for the beta decays of medium-mass odd-odd nuclei involved in double beta decays.

Typical examples of the available data are shown in Figs. 3 and 4 for double-beta chains involving medium-mass and heavy-mass nuclei, respectively. As seen from the figures the available data on relevant beta decays are rather scarse and more beta-decay measurements, along with other independent probes, are called for to access higher intermediate multipoles (see Fig. 2) and higher intermediate energies. However, in case of a possible single-state dominance [36], i.e. when the lowest state of the intermediate odd-odd nucleus is a 1^+ state and it exhausts practically all of the $2\nu\beta\beta$-decay amplitude, beta decays are a powerful tool to experimentally determine the $2\nu\beta\beta$-decay rate [37,38].

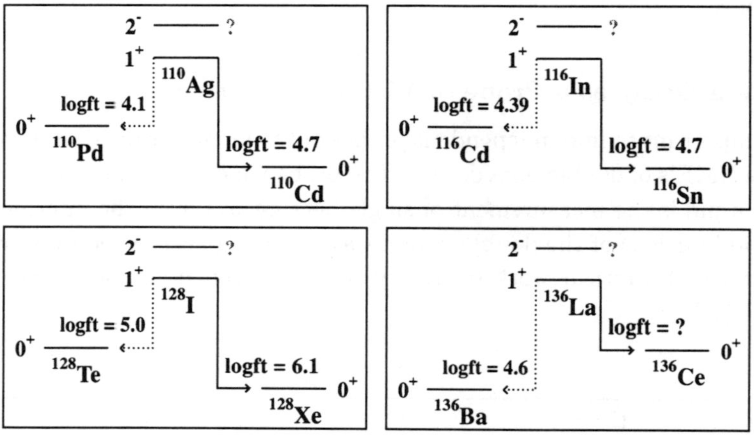

Fig. 4. Available experimental data for the beta decays of heavy odd-odd nuclei involved in double beta decays.

4. Muon Capture as a Probe of Virtual Transitions

Besides beta decays other independent probes of the virtual transitions have been proposed. These probes are are the charge-exchange reactions [39] and possibly neutrino-nucleus charged-current scattering [40]. Recently it was proposed [41] to use the ordinary muon capture (OMC) as a probe of the virtual transitions to the J^π states of the intermediate odd-odd nucleus. Due to the large mass of the captured muon (roughly 100 MeV) the OMC can lead to highly excitated states of the final nucleus. However, the large muon mass also leads to a more complicated theoretical treatment of this process as compared to the beta decay. In particular, the induced terms of the nucleonic weak current are activated [42]. Of the various terms of the induced current, access to the magnitude of the induced pseudoscalar contribution has been problematic [43]. In contrast to beta decay, the large mass of the muon allows high transition rates also for the so-called forbidden transitions. This can be used to analyze the structure of other than 1^+ states in the intermediate nuclei, like the 2^- states which are very relevant in the $0\nu\beta\beta$ decay [30].

In particular in light nuclei the OMC seems to be a powerful probing tool [44]. In the calculations of [44] different well-established two-body interactions were used in the framework of the nuclear shell model. The discussed nuclei were double-beta-decaying nuclei for which the valence nucleons occupied either the sd or pf shells. Conclusions about the power of the OMC are based on the computational evidence that the leading contributions to the $2\nu\beta\beta$-decay matrix elements stem from intermediate 1^+ states which are strongly fed also in the OMC. In this way the OMC is able to probe wave functions of those states which are the most relevant for double beta decay.

An example of the OMC is given in Fig. 5 where the capture is on ^{48}Ti. The corresponding process is $^{48}\mathrm{Ti} + \mu^- \rightarrow {}^{48}\mathrm{Sc} + \nu_\mu$ which probes virtual transitions to the states of the intermediate nucleus ^{48}Sc of the ^{48}Ca double beta decay. The presented relative capture rates (relative to the capture rate to the first 2^- state) can be compared with the experimental ones in the near future when the analyses of the experiments performed recently at the PSI are finished (more information can be found in [45]).

From Fig. 5 we can make few interesting observations, namely

- the first-forbidden OMC to the 2^- states is not suppressed relative to the allowed OMC to the 1^+ states, as would be the case for the ordinary electron-capture (EC) transitions. In fact, the OMC to the 2_1^- state is the strongest of all transitions. Similar conclusions are also valid for the higher-forbidden OMC transitions: they are not as forbidden as the EC would suggest.
- The OMC can probe also intermediate states of high excitation energy due to the involved large Q value, which stems from the large mass of the captured muon. In particular, the OMC can access, in principle, the energy region of the Gamow–Teller giant resonance, which possibly plays a role in the double beta decays.

The study [44] on the relation of the OMC with the $2\nu\beta\beta$ decay can be extended to comparison of the OMC with the $0\nu\beta\beta$ decay. For the $0\nu\beta\beta$ decay of ^{48}Ca in Fig. 6 are shown the most relevant multipole contributions to the sum (4) as functions of the excitation energy of the intermediate state. The calculations have been done in the pf shell by using the FPBP

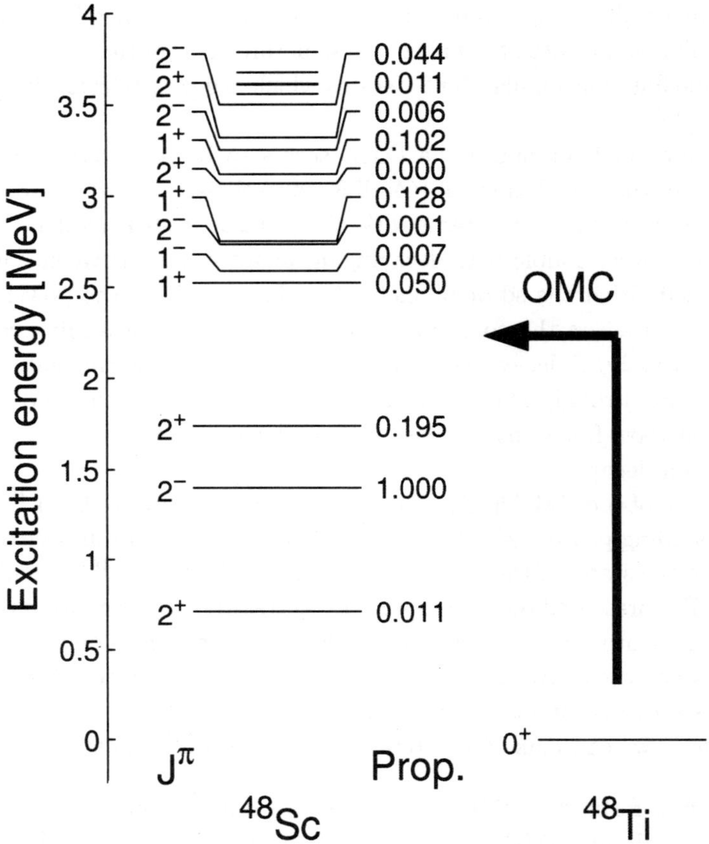

Fig. 5. Calculated rates for the OMC on the nucleus ^{48}Ti.

two-body interaction [46]. The limitation of the calculation is that only positive-parity states can be produced since no negative parity orbitals are included in the single-particle space. In addition, at least four nucleons are required to occupy the $0f_{7/2}$ orbital. Adding all the multipole contributions, including the even J's, produces the value $M_{\rm GT}^{(0\nu)} = -1.72$ for the double GT NME.

The relevant contributions to the cumulative sum of Fig. 6 can be contrasted against strong muon-capture transitions $^{48}\text{Ti} \rightarrow {}^{48}\text{Sc}(J^{\pi})$. The corresponding comparison of the OMC rates and the contributions to $M_{\rm GT}^{(0\nu)}$ are displayed in Fig. 7 for the indicated four multipolarities $J^{\pi} = 1^+, 2^+, 3^+, 4^+$ of states in ^{48}Sc. From the figure one can conclude that the strong OMC rates feed mainly the same J_n^{π} states which are relevant in building up the

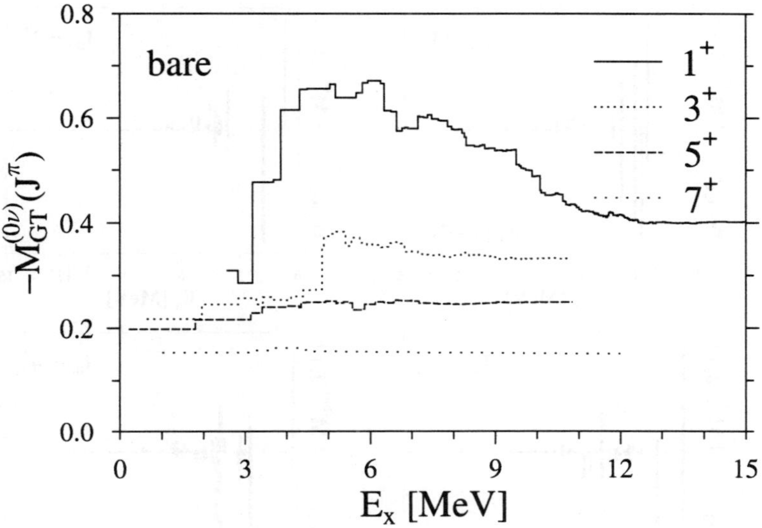

Fig. 6. The most relevant multipole components in (4) for the $0\nu\beta\beta$ decay of ^{48}Ca. Cumulative sums of the contributions are shown as functions of the excitation energy of the intermediate nucleus ^{48}Sc. The multipole J^π is indicated in the panel.

double Gamow–Teller matrix element $M_{\rm GT}^{(0\nu)}$. It is then obvious that, like in the $2\nu\beta\beta$ case, the OMC is a powerful probe of the intermediate states of the $0\nu\beta\beta$ decays, as well.

5. Nuclear Matrix Elements for Dark-Matter Detection

The starting point for the LSP detection rate calculation is the differential cross section $d\sigma(q,v)/dq^2$ of the LSP-nucleus scattering. Instead of the momentum transfer q, it is convenient to use the dimensionless variable $u = q^2b^2/2$, where b is the nuclear harmonic-oscillator size parameter. In this way the cross section can be written in the laboratory frame as follows [47–49]:

$$\frac{d\sigma(u,v)}{du} = \frac{1}{2}\sigma_0 \left(\frac{1}{m_{\rm p}b}\right)^2 \frac{c^2}{v^2}\frac{d\sigma_{\rm AS}(u,v)}{du} \tag{5}$$

with

$$\frac{d\sigma_{\rm AS}(u,v)}{du} = \left(f_{\rm A}^0\Omega_0\right)^2 F_{00}(u) + 2f_{\rm A}^0 f_{\rm A}^1 \Omega_0\Omega_1 F_{01}(u) + \left(f_{\rm A}^1\Omega_1\right)^2 F_{11}(u) + A^2\left(f_{\rm S}^0 - f_{\rm S}^1\frac{A-2Z}{A}\right)^2 |F(u)|^2 . \tag{6}$$

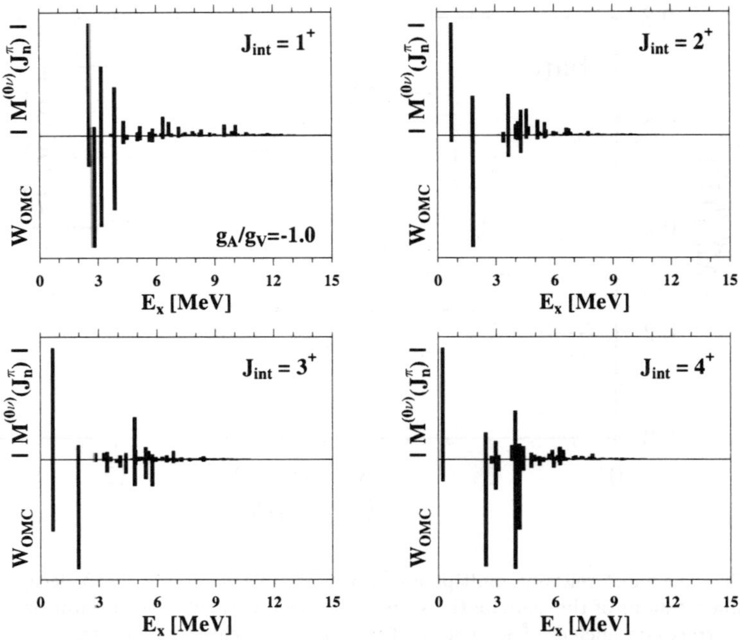

Fig. 7. Contributions to the matrix element $M_{\mathrm{GT}}^{(0\nu)}(J^{\pi})$ of (4) (upper part of each panel, in arbitrary units) for the decay of ^{48}Ca, and OMC rates to J^{π} states in ^{48}Sc (lower part of each panel, in arbitrary units). The multipole J^{π} is indicated in the panel. The horizontal axes gives the excitation energy in ^{48}Sc.

The values of the nucleonic-current parameters f_{A}^{ρ} and f_{S}^{ρ} depend on the specific SUSY model employed [50].

In order to calculate detection rates of the earthbound LSP detector one has to take account of the LSP velocity distribution in the galactic halo and the Earth's velocity with respect to the galactic center. By assuming a Maxwellian velocity distribution and folding it with the cross section one ends up with following expression for the event rate:

$$\langle R\rangle = \Big[\left(f_{\mathrm{A}}^{0}\right)^{2} D_1 + 2 f_{\mathrm{A}}^{0} f_{\mathrm{A}}^{1} D_2 + \left(f_{\mathrm{A}}^{1}\right)^{2} D_3 + A^2 \left(f_{\mathrm{S}}^{0} - f_{\mathrm{S}}^{1} \frac{A-2Z}{A} \right)^{2} D_4 \Big] m_{\mathrm{det}}[\mathrm{kg}] \,. \qquad (7)$$

Above $m_{\mathrm{det}}[\mathrm{kg}]$ is the detector mass in units of kg and the coefficients D_n contain all the information about the nuclear structure and halo profile. The coefficients D_1, D_2 and D_3 of Eq. (7) represent the spin-dependant incoherent channel of the LSP-nucleus scattering process whereas the coefficient D_4 represents the spin-independent coherent channel. For further details see [49].

The nuclear-structure calculations were performed by the shell-model code EICODE [51]. For ^{23}Na the USD [52] interaction was used. The pf-shell nuclei ^{71}Ga and ^{73}Ge were calculated in a model space consisting of the orbitals $1p_{1/2}$, $1p_{3/2}$, $0f_{5/2}$ and $0g_{9/2}$. For the two-body interaction an effective G-matrix [53] was used. For ^{127}I the used model space included orbitals $2s_{1/2}$, $1d_{3/2}$, $1d_{5/2}$, $0g_{7/2}$ and $0h_{11/2}$. In this case, too, the interaction was derived from the CD-Bonn potential [53]. In the shell-model calculations of heavier nuclei the model space was truncated either by setting an upper limit for the average energy of the proton and neutron configurations that were taken into account (^{71}Ga and ^{73}Ge) or by limiting the maximum amount of neutrons in the $0h_{11/2}$ orbital to be six (^{127}I).

The calculated values of the static spin matrix elements Ω_ρ are listed in Table 1. In this table the present results are compared with the ones calculated by using the microscopic quasiparticle-phonon model (MQPM) [47]. As can be seen the results are quite similar for ^{71}Ga and ^{73}Ge, but far apart for ^{127}I. The reason is that the MQPM predicts a rather pure one-quasiparticle character for the ground state of ^{127}I, leading to too large a magnetic moment, close to its single-particle estimate.

Table 1. Calculated results for the static spin matrix elements.

	^{71}Ga		^{73}Ge		^{127}I		^{23}Na	
	Ω_0	Ω_1	Ω_0	Ω_1	Ω_0	Ω_1	Ω_0	Ω_1
SM	0.905	0.830	0.912	−0.891	0.871	0.690	0.691	0.588
MQPM	0.919	0.925	0.978	−1.070	1.220	1.230	-	-

The calculated event rates depend on the threshold energy of the LSP detector. This is demonstrated in Fig. 8, where the annual averaged event rates have been plotted as functions of the detector threshold energy $Q_{\rm thr}$ for two SUSY models which are simply called A and B. These models represent different parts of the large available SUSY parameter space, and the parameters were chosen in such a way that model A emphasizes the role of the spin-dependant channel, whereas in model B the spin-independent coherent channel is the dominant one. This can be seen in the different scales of the plots.

If the LSP-nucleus scattering cross section is dominated by the spin-dependent channel, the heaviest nuclei are not automatically the best candidates for an LSP detector. In [54] the authors examine the possibility to explain the DAMA results [55] by dominance of the spin-dependant chan-

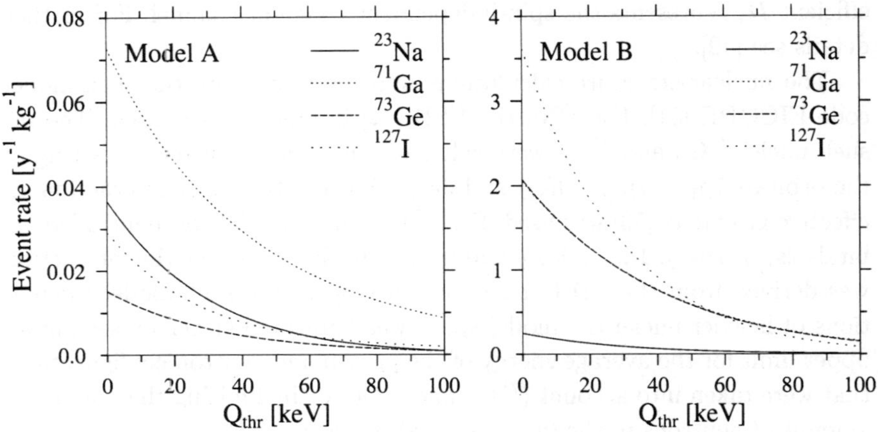

Fig. 8. Calculated event rates as functions of the detector energy threshold Q for SUSY models A and B.

nel. The DAMA claims to see the annual modulation signal coming from the WIMP-nucleus scatterings. This observaton contradicts the null results of the other dark-matter experiments. In the present calculations the example models A and B demonstrate that it is indeed possible to find SUSY models where the spin-dependent channel dominates.

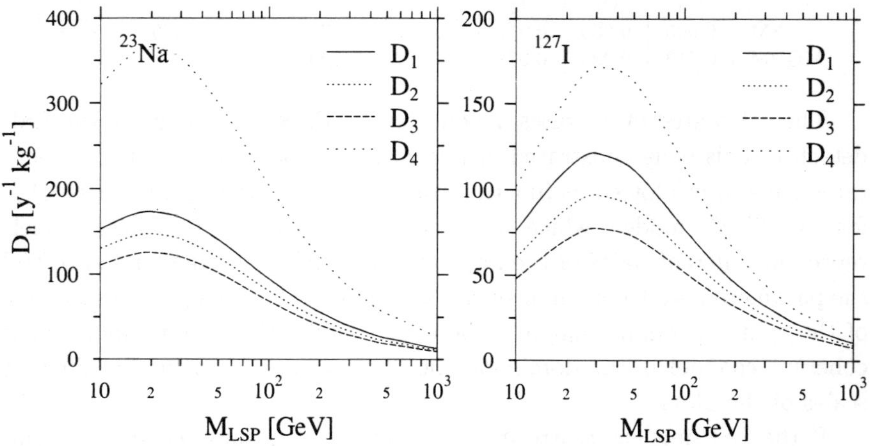

Fig. 9. Calculated coefficients D_n of Eq. (7) for ^{23}Na and ^{127}I as functions of the mass of the LSP. The detector threshold energy has been set to $Q_{thr} = 0$.

As mentioned earlier the coefficients D_n of Eq. (7) contain all the information about the nuclear structure and halo profile in the LSP-nucleus scattering. Values of these coefficients depend on the mass of the LSP. To illustrate this, in Fig. 9 the annual average values of the coefficients D_n are plotted as functions of the LSP mass, $M_{\rm LSP}$, for ^{23}Na and ^{127}I. In these plots the detector threshold has been set to $Q_{\rm thr} = 0$.

As can be seen from the plots all the D_n peak at a certain value of $M_{\rm LSP}$, and this value depends on the nucleus. For heavy LSP masses the magnitudes of the coefficients decrease. This is due to the fact that in the calculations it is assumed that the local LSP density is fixed at $\rho_0 = 0.3\,{\rm GeV\,cm^{-3}}$. Thus, the heavier the LSP, the more thinly populated is the dark-matter halo.

In addition to their dependence on $M_{\rm LSP}$ and $Q_{\rm thr}$ the coefficients D_n (and thus the event rates) depend also on time. Due to the Earth's rotation around the Sun the WIMP wind velocity toward an earthbound detector has an annual variation. This leads to annual modulation of the event rates. In [49] it was demonstrated that this is a few-per-cent effect.

6. Summary and conclusions

In the present article I have made a short review of the status of the double beta decay from the point of view of independent probes for the related nuclear matrix elements. The thus far used probes include charge-exchange reactions and beta decays. Currently a lot of effort is invested in probing double beta decays by the ordinary muon capture. What are needed urgently are experiments measuring these processes so that the nuclear matrix elements involved in double beta decays can be probed as thoroughly as possible before using them in predictions of the neutrino mass, neutrino hierarchy, CP phases, etc.

Second topic in this work was the description of nuclear structure in the context of detection of the cold dark matter of the Universe. Most recently the nuclear shell-model has been applied to calculate event rates for various dark-matter detectors. The event rate of the LSP-nucleus scattering has been presented in a form where the nuclear-structure aspects have been separated from the particle-physics aspects. This allows the study of different SUSY scenarios in a very efficient way.

Different SUSY models weigh differently the coherent spin-independent part and the incoherent spin-dependent part in the total event rate. The calculations of the present work indicate that the values of the SUSY parameters strongly affect the choice of the best detector candidate nuclei.

Also, it was noted that the coefficients D_n, and therefore the calculated event rates, depend on the mass of the LSP. It was found that the optimum mass for detection varies from nucleus to nucleus.

Acknowledgments

This work has been supported by the Academy of Finland under the Finnish Centre of Excellence Programme 2006-2011 (Nuclear and Accelerator Based Programme at JYFL).

References

1. Super-Kamiokande Collaboration, S. Fukuda et al., *Phys. Rev. Lett.* **86**, 5651 (2001) ; SNO Collaboration, Q. R. Ahmad et al., *Phys. Rev. Lett.* **89**, 011302 (2002) ; KamLAND Collaboration, K. Eguchi et al., *Phys. Rev. Lett.* **90**, 021802 (2003) ; M. Appollonio et al., *Phys. Lett.* **B466**, 415 (1999).
2. J. Suhonen and O. Civitarese, *Phys. Rep.* **300**, 123 (1998).
3. O. Civitarese and J. Suhonen, *Nucl. Phys.* **A729**, 867 (2003).
4. S. Pascoli, S. T. Petcov and W. Rodejohann, *Phys. Lett.* **B549**, 177 (2002).
5. J. Suhonen, *Phys. Atom. Nucl.* **61**, 1286 (1998) ; ibid **65**, 2176 (2002).
6. J. Suhonen, *Nucl. Phys.* **A752**, 53c (2005).
7. J. Suhonen, *Nucl. Phys.* **B143** (Proceedings Supplements), 240 (2005).
8. R. Bernabei *et al.*, *Nucl. Phys.* **B110** (Proceedings Supplements), 61 (2002).
9. L. Baudis, A. Dietz, B. Majorovits, F. Schwamm, H. Strecker and H.V. Klapdor-Kleingrothaus, *Phys. Rev.* **D63**, 022001 (2000).
10. The EDELWEISS Collaboration, *Phys. Lett.* **B513**, 15 (2001).
11. J. Suhonen, *Phys. Lett.* **B607**, 87 (2005).
12. T. Tomoda, *Rep. Prog. Phys.* **54**, 53 (1991).
13. J. Suhonen and O. Civitarese, *Phys. Lett.* **B308**, 212 (1993).
14. M. Aunola and J. Suhonen, *Nucl. Phys.* **A602**, 133 (1996).
15. M. Aunola and J. Suhonen, *Nucl. Phys.* **A643**, 207 (1998).
16. J. Suhonen, *Phys. Lett.* **B477**, 99 (2000).
17. J. Suhonen, *Phys. Rev.* **C62**, 042501(R) (2000).
18. J. Suhonen and M. Aunola, *Nucl. Phys.* **A723**, 271 (2003).
19. J. Suhonen, *Phys. Rev.* **C48**, 574 (1993).
20. J. Suhonen and O. Civitarese, *Phys. Lett.* **B497**, 221 (2001).
21. M. Aunola, J. Suhonen and T. Siiskonen, *Europhys. Lett.* **46**, 577 (1999).
22. M. T. Mustonen, M. Aunola and J. Suhonen, *Phys. Rev.* **C73**, 054301 (2006).
23. M. Doi, T. Kotani and E. Takasugi, *Prog. Theor. Phys. Suppl.* **83**, 1 (1985).
24. P. Vogel and M. R. Zirnbauer, *Phys. Rev. Lett.* **57**, 3148 (1986).
25. O. Civitarese, A. Faessler and T. Tomoda, *Phys. Lett.* **B194**, 11 (1987).
26. J. Toivanen and J. Suhonen, *Phys. Rev. Lett.* **75**, 410 (1995).
27. J. Toivanen and J. Suhonen, *Phys. Rev.* **C55**, 2314 (1997).
28. O. Civitarese and J. Suhonen, *Nucl. Phys.* **A761**, 313 (2005).
29. V. A. Rodin, A. Faessler, F. Šimkovic and P. Vogel, *Nucl. Phys.* **A766**, 107 (2006).

30. O. Civitarese and J. Suhonen, *Phys. Lett.* **B626**, 80 (2005).
31. A. Griffiths and P. Vogel, *Phys. Rev.* **C46**, 181 (1992).
32. J. Suhonen, *Nucl. Phys.* **A563**, 205 (1993).
33. J. Suhonen and O. Civitarese, *Phys. Rev.* **C49**, 3055 (1994).
34. J. Suhonen, *Nucl. Phys.* **A700**, 649 (2002).
35. J. Suhonen and O. Civitarese, *Nucl. Phys.* **A584**, 449 (1995).
36. O. Civitarese and J. Suhonen, *Phys. Rev.* **C58**, 1535 (1998) ; *Nucl. Phys.* **A653**, 321 (1999).
37. A. García et al., *Phys. Rev.* **C47**, 2910 (1993).
38. M. Bhattacharya et al., *Phys. Rev.* **C58**, 1247 (1998).
39. H. Ejiri, *Phys. Rep.* **338**, 265 (2000) ; S. Rakers et al., *Phys. Rev.* **C71**, 054313 (2005).
40. C. Volpe, *J. Phys.* **G31**, 903 (2005).
41. M. Kortelainen and J. Suhonen, *Europhys. Lett.* **58**, 666 (2002) ; ibid., *Nucl. Phys.* **A713**, 501 (2003).
42. T. Siiskonen, J. Suhonen, V.A. Kuz'min and T.V. Tetereva, *Nucl. Phys.* **A635**, 446 (1998) ; ibid. *Nucl. Phys.* **A651**, 435 (1999).
43. T. Siiskonen, J. Suhonen and M. Hjorth-Jensen, *Phys. Rev.* **C59**, R1839 (1998) ; *J. Phys.* **G25**, L55 (1999).
44. M. Kortelainen and J. Suhonen, *J. Phys.* **G30**, 2003 (2004).
45. H. O. U. Fynbo et al., *Nucl. Phys.* **A724**, 493 (2004).
46. W. A. Richter, M. G. Van Der Merwe, R. E. Julies and B. A. Brown, *Nucl. Phys.* **A523**, 325 (1991).
47. E. Holmlund, M. Kortelainen, T. S. Kosmas, J. Suhonen and J. Toivanen, *Phys. Lett.* **B584**, 31 (2004).
48. P. C. Divari, T. S. Kosmas, J. D. Vergados and L. D. Skouras, *Phys. Rev.* **C61**, 054612 (2000).
49. M. Kortelainen, T. S. Kosmas, J. Suhonen and J. Toivanen, *Phys. Lett.* **B632**, 226 (2006).
50. J. D. Vergados, *J. Phys.* **G22**, 253 (1996).
51. J. Toivanen, computer code EICODE, JYFL, Finland (2004).
52. B. H. Wildenthal, *Prog. Part. Nucl. Phys.* **11**, 5 (1984).
53. M. Hjorth-Jensen, T. T. S. Kuo and E. Osnes, *Phys. Rep.* **260**, 125 (1995) ; M. Hjorth-Jensen, private communication.
54. C. Savage, P. Gondolo and K. Freese, *Phys. Rev.* **D70**, 123513 (2004).
55. R. Bernabei et al., astro-ph/0405282.

PROBING IN-MEDIUM VECTOR MESONS BY DILEPTONS AT HEAVY-ION COLLIDERS

M. I. KRIVORUCHENKO

Institute for Theoretical and Experimental Physics, B. Cheremushkinskaya 25
117259 Moscow, Russia
E-mail: mikhail.krivoruchenko@itep.ru

An introduction to physics of in-medium hadrons with special emphasis towards modification of vector meson properties in dense nuclear matter is given. We start from remarkable analogy between the in-medium behavior of atoms in gases and hadrons in nuclear matter. Modifications of vector meson widths and masses can be registered experimentally in heavy-ion collisions by detecting dilepton spectra from decays of nucleon resonances and light unflavored mesons including ρ- and ω-mesons. Theoretical schemes for description of the in-medium hadrons are reviewed and recent experimental results of the NA60 and HADES collaborations on the dilepton production are discussed.

Keywords: Heavy-ion collisions; vector mesons; dileptons.

1. Introduction

Elementary particles as ground-state excitations are characterized differently in the vacuum and in the dense matter. This effect is known from the solid state physics where elementary excitations are traditionally called *quasi-particles* to indicate medium-induced modifications. The medium effects manifest themselves in a wide range of physical phenomena ranging from shifts and broadening of energy levels of pionic atoms to matter-affected neutrino oscillations.

In-medium modifications of hadron properties are of special interest in order to test QCD at finite densities and temperatures and, in particular, for better understanding of the chiral and quark-hadron phase transitions and QCD confinement problem. Heavy-ion collisions provide a unique possibility to create nuclear matter under extreme conditions and to study hadron physics around phase transitions points. One of the best direct probes to measure in-medium masses and widths of vector mesons are dileptons e^+e^- or $\mu^+\mu^-$ which, being produced, leave the reaction zone essentially undistorted by final-state interactions.

The key problem is the dependence of elementary excitations on density and temperature. The knowledge of such dependence allows to use elementary particles as probes to measure thermodynamical properties of highly compressed nuclear mater, improving thereby our knowledge of QCD, nuclear physics and the phenomenology of strong interactions.

In this lecture we discuss main effects leading to modifications of in-medium spectral functions of hadrons. In the next Sect., a remarkable analogy between atoms in gases and hadrons in nuclear matter is discussed. In Sect. 3, we show how the concept of collision broadening of particles, originated from the atomic physics, applies to behavior of nucleon resonances in nuclear matter. Sects. 4 and 5 provide a summary, respectively, of theoretical and experimental works on in-medium modifications of vector mesons and dilepton production. Sect. 6 deals with elementary sources of dileptons. The extended vector meson dominance model (eVMD) is formulated in Sect. 7 and its gauge invariance is proved in Sect. 8. Sect. 9 discusses the recent experimental results (summer 2006) from the NA60 and HADES collaborations.

2. Atoms in gases

The concept of collision broadening is discussed in atomic physics since last century. The physics behind the modifications of atomic spectral lines in gases represents the obvious theoretical interest, because it has significant features in common with physics of behavior of hadrons in nuclear matter.

2.1. *Radiation damping*

It is known that radiative damping results to finite widths of atomic spectral lines. The energy deposited during transition to a lower energy state can never be monochromatic, but distributed according to the Lorentz formula

$$dI(\omega) = \frac{I}{2\pi} \frac{\Gamma d\omega}{(\omega - \omega_0)^2 + (\Gamma/2)^2} \tag{1}$$

where

$$\Gamma = \Gamma_a + \Gamma_b, \tag{2}$$

with Γ_a and Γ_b being the vacuum widths of levels a and b, respectively, and $\omega_0 = E_a - E_b$ is the transition energy.

Recall its quantum-mechanical derivation: The wave function of state a equals $\Psi \sim \exp(-iE_a t - \frac{1}{2}\Gamma_a t)$, similarly for b. Transition amplitude $a \to b$

+ photon of energy ω is given by $A_{ba} \sim (\exp(-i\omega t)\Psi_b)^*\Psi_a$. Evaluation of the integral $\int_{-\infty}^{-\infty} |A_{ba}|^2 dt$ gives spectral density (1).

Radiation damping is inherent to any radiating system. It results to broadening of spectral lines of isolated atoms. Thermal motion of atoms and collisions of particles affect the profile of lines also.

2.2. *Doppler effect*

Doppler effect results to an additional broadening of spectral lines due to motion of atoms in gases. It has a purely kinematic origin.

2.3. *Collision broadening*

Lorentz attributed collision broadening of spectral lines to a decoherence effect accompanied scattering of a probing atom with surrounding atoms and electrons (see e.g. [1], Chap. 10). Consider an excited atom which experiences at random times t_i collisions with surrounding particles. Its emission intensity is summed up coherently between two collisions for $\tau \in [t_i, t_{i+1}]$ and decoherently with respect to time intervals $[t_i, t_{i+1}]$. One has

$$dI(\omega) \sim \frac{d\omega}{2\pi N} \sum_{i=1}^{N} \left| \int_{t_i}^{t_{i+1}} dt \exp(-i(\omega - \omega_0)t - \frac{\Gamma}{2}t) \right|^2 . \tag{3}$$

One has to assume further that waiting times for sequential collisions $\tau_i = t_{i+1} - t_i$ are distributed exponentially

$$dW(\tau) = \frac{d\tau}{\tau_0} \exp(-\tau/\tau_0). \tag{4}$$

Such a hypothesis is equivalent to the requirement of the Poisson distribution of surrounding uncorrelated particles, τ_0 has the meaning of collision time. Replacing the average over collisions $\frac{1}{N}\sum_{i=1}^{N} \ldots$ with the integral $\int dW(\tau)\ldots$, one arrives at the Lorentz formula (1), with Γ replaced by $\Gamma^{total} = \Gamma + 2/\tau_0$.

According to Eq.(2), the radiation width Γ represents the sum of radiation widths of levels a and b, so one has to split $2/\tau_0$ among two levels:

$$\Gamma_a^{total} = \Gamma_a + 1/\tau_0 \tag{5}$$

and similarly for b. The value

$$\Gamma_{coll} = 1/\tau_0 \tag{6}$$

is called *collision width.*

We wish to bring attention to two circumstances inherent to the above approach:

- Collision broadening is physically attributed to decoherence.
- The coefficient at $1/\tau_0$ in Eq.(6).

The modern schemes use field-theoretic methods which do not refer to decoherence, although predict additional broadening inverse proportional to τ_0 in the exact agreement with the Lorentz theory. The interplay between the decoherence assumption and field-theoretic methods is interesting. The collision width is determined as an imaginary part of (obviously coherent) sum of forward scattering amplitudes with surrounding particles. Due to the quantum-mechanical optical theorem, it is equal to noncoherent sum of cross sections with surrounding particles. The coefficient at $1/\tau_0$ appears to be exact, although Eq.(6) is derived using assumptions of the *instantaneous* interactions and the *decoherence* which are not immediately evident. In the first order to the density, however, there are no corrections to Eq.(6) provided cross sections are calculated quantum-mechanically. The results of the Lorentz theory are therefore recovered using the field-theoretic methods.

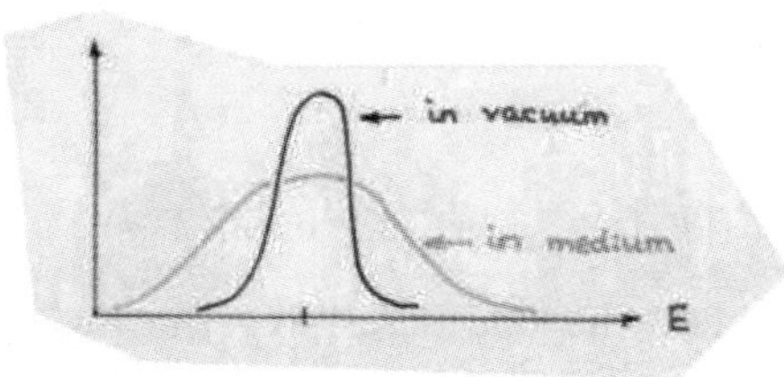

Fig. 1. Schematic representation of broadening of atomic energy levels in gases.

It is instructive to follow an elementary approach of Ref. [2].

The free path length ℓ_f of a hydrogen atom (or any other particle) in a gas with density n is determined by its cross section σ for scattering with other particles (atoms, molecules, ions, and electrons). There should be one particle inside of a cylinder of height ℓ_f and base σ. This condition gives

$$\ell_f = \frac{1}{n\sigma}. \tag{7}$$

The disappearance of atoms from atomic beam as a function of distance ℓ is described by an exponential law. Let a hydrogen atom in a state a moves with velocity v, then $\ell = vt$. One finds

$$N_a(\ell) = N_a(0)\exp(-\ell/\ell_f) \;\;\&\;\; \ell = vt \;\;\Rightarrow\;\; N_a(t) = N_a(0)\exp(-n\sigma vt).$$

In quantum mechanics, decays of a quasistationary state a are governed

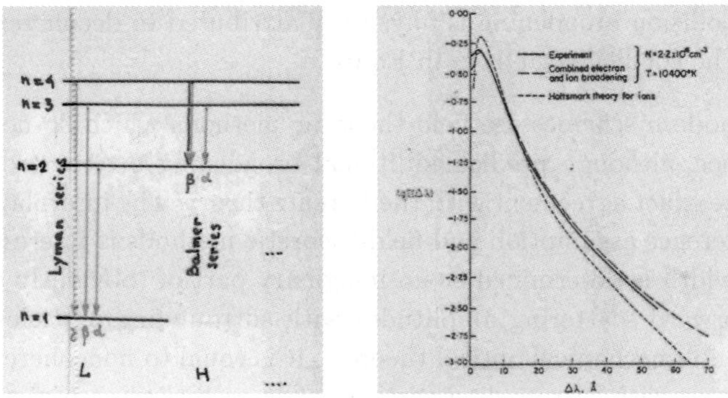

Fig. 2. **Left panel**: Schematic representation of hydrogen atom energy levels and the associated Lyman (L) and Balmer (H) series $\alpha, \beta, \gamma, \ldots$. **Right panel**: Logarithm of the spectral density of the H_β line of hydrogen-like atom HeI at density of $n = 2.2 \times 10^{16}$ cm^{-3} and temperature of $T = 10400$ K as a function of the wavelength shift $\Delta\lambda = \lambda - \lambda_0$ where $\lambda_0 = 1/\omega_0$. The solid line is the experiment. The dashed lines show theoretical predictions. The spectral density is broadened by a few $\AA$ and has two peaks at $\Delta\lambda \sim$ $\pm$ a few $\AA$ (from Ref. [3]).

by equation

$$|\Psi_a(t)|^2 = |\Psi_a(0)|^2 \exp(-\Gamma_a t). \tag{8}$$

This equation looks like the one describing the disappearance of particles from beam due to collisions. Since two distinct mechanisms exists for the disappearance and $N_a(t) \sim |\Psi_a(t)|^2$, the right idea is to combine two widths:

$$\Gamma_a^{total} = \Gamma_a + \Gamma_{coll} \tag{9}$$

where $\Gamma_{coll} = n\sigma v = 1/\tau_0$ is the collision width.

If resonance is placed in a medium, it acquires an additional broadening. Its spectral function modifies accordingly, as shown schematically on Fig. 1.

Collision broadening modifies spectral functions of resonances.

These ideas are in agreement with the uncertainty relation:

$$\Delta E \geq \frac{1}{\tau_0} = \frac{v}{\ell_f} = n\sigma v = \Gamma_{coll}. \tag{10}$$

Experimentally it is not possible to resolve transition energy between two atomic levels with accuracy ΔE better than the inverse collision time τ_0. If we would ask an experimentalist to construct an experimental distribution of photon energies (wavelengths) e.g. in the Balmer β-line, he could present something resembling the curves on Fig. 1.

2.4. *Experimental observations of modified profiles of atomic spectral lines*

Shifts of energy levels and collision broadenings of atoms are observed experimentally in gases. Let us provide a couple of illustrations. Fig. 2 (left) recalls the structure of the hydrogen atom energy levels and the associated transitions and shows (right) the *modified* spectral density of the H_β line of a hydrogen-like atom in a gas.

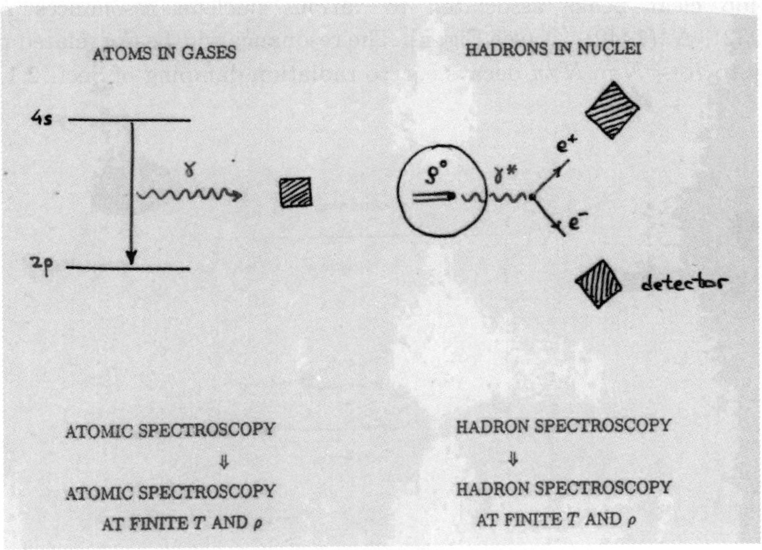

Fig. 3. Remarkable analogy between atomic spectroscopy and hadron spectroscopy.

3. Nucleon resonances in nuclei

After clarifying principal issues of spectroscopy of isolated atoms and molecules (as a result of which quantum mechanics appearred), atomic physics evolved towards studying the in-medium modifications of atoms at finite temperature and density. Physics of intermediate energies $\sim$ 1 GeV follows in its evolution the general trend of atomic physics with a time lag $\sim$ 70 years. Last 15 years, new possibilities occurred to accomplish experimentally similar program for hadron spectroscopy. This analogy is illustrated in Fig. 3.

In nuclear physics, an example of broadening of resonance profiles is delivered by experiments on photoabsorption on nuclei [4]:

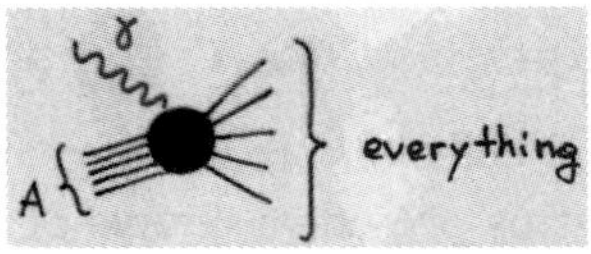

The photoabsorption cross sections on free nucleons ($A = 1$) develop clear peaks associated to various nucleon resonances $R = \Delta(1232)$, $N^*(1440), \dots$ (see Fig. 4). The resonance widths are related physically to $R \to N\pi$, $N\pi\pi$ decays e.g. to radiation damping of Sect. 2.1.

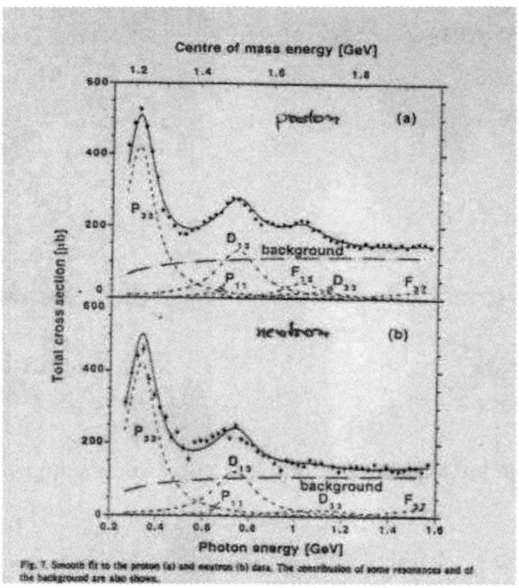

Fig. 4. Total photoabsorption cross section on proton (a) and neutron (b) vs photon energy. Contributions of various nucleon resonances and the background are shown (from Ref. [2]).

In nuclear environment, the Doppler effect (Sect. 2.2) related to the Fermi motion of nucleons and collision broadening (Sect. 2.3) come into play. The result is presented on Fig. 5.

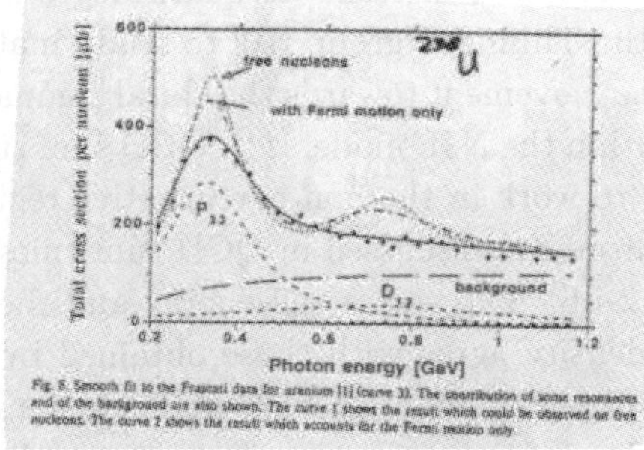

Fig. 5. Total photoabsorption cross section on ^{238}U versus photon energy in the laboratory system. Contributions of various nucleon resonances $R = \Delta(1232), N^*(1440), \ldots$ and the background (long-dashed curve) are shown. The additional broadening is related to the Doppler effect (Fermi motion) and collision broadening. The Pauli blocking effect is included. The modification of the in-medium spectral density of resonances is clearly seen. Resonances heavier than $\Delta(1232)$ are broadened and masked by the background (from Ref. [2]).

4. In-medium properties of vector mesons: Theoretical models

The change of the nucleon mass in nuclear matter was discussed first in the pioneering works by Walecka and Chin [5,6] already in the 1970's. The Mean Field and Relativistic Mean Field (RMF) approximations were developed to treat the dense matter self-consistently to all orders in the density. The effective nucleon mass was found to decrease significantly in nuclei at saturation density. This effect predicted by phenomenological models is confirmed by the developments of QCD.

During the last decade, the problem of the description of hadrons in dense and hot nuclear matter received new attention due to the possibility to test theoretical predictions with heavy-ion experiments. This problem is related to chiral symmetry breaking and QCD confinement. Being intrinsically non-perturbative, the number of available theoretical schemes is limited. The knowledge of the non-perturbative QCD comes in particular from lattice simulations. However, lattice results suffer still from uncertainties due to finite lattice size effects and the proper inclusion of fermions. An important role is played by field theoretical models which incorporate fundamental features of QCD.

The model proposed by Walecka [5] was further analyzed in Refs. [6–8] to study the in-medium vector mesons. The Nambu-Jona-Lasinio (NJL) model was introduced in the sixties as a theory of interacting nucleons [9] and later it was reformulated in terms of quark degrees of freedom. This model was used by the Tübingen group [10] to study matter under the extreme conditions. The movement towards the chiral symmetry restoration is reliably described within the NJL model [11]. QCD sum rules provide a successful analytic tool to work in the non-perturbative regime. The medium modifications of hadrons are discussed in QCD sum rules at finite temperature and density [12–16]. The vector self-energy and the effective nucleon mass at saturation density agree with those obtained by the conventional methods of nuclear physics. The values of QCD condensates and the nucleon expectation values of the various quark operators including the chiral order parameter can be determined within this approach [17,18].

Dispersion theory combined with the optical potential method has been used by Weise et al. [16,19,20] to calculate mass shifts, broadening and spectral densities for vector mesons in the presence of nucleons and pions at finite density and temperature. The experimental data on the total cross sections are used to saturate the scattering amplitude at low energies with resonances. However, the obtained results are at variance with [21]. Unitarized thermal chiral perturbation theory has been used in [22,23]. There one matches unitarized amplitudes smoothly with the ChPT loop expansion in the low-energy region. The estimates show, in particular, mass shifts and a clear increase of the thermal widths of unflavored mesons which have relevance within the context of ultra-relativistic heavy-ion collisions. Other attractive phenomenological models for the in-medium unflavored and charmed mesons have also been proposed in [21,24–27].

The current status of the Brown-Rho scaling [21] has recently been discussed by Brown and Rho [11].

5. In-medium properties of vector mesons: Experiments on dilepton production

The available experiments on the dilepton production are summarized in Table 1.

The dilepton spectra measured by the CERES [29] and HELIOS-3 [30] Collaborations at CERN SPS found a significant enhancement of the low-energy dilepton yield below the ρ and ω peaks [29] in heavy systems ($Pb+Au$) as compared to light systems ($S+W$) and proton induced reactions ($p+Be$). Theoretically, this enhancement can be explained assuming a dropping mass scenario for the ρ meson and the inclusion of in-medium spectral functions for the vector mesons [35,36]. The enhanced low energetic dilepton yield originates to most extent from an enhanced contribution of

Table 1. The enhanced production of dileptons in low and intermediate mass continuum. The fifth column shows intervals of invariant masses of dileptons. O/E is the ratio between integral numbers of observed and expected dileptons.

Collaboration	Ref.	E/A [GeV]	Dileptons	M [GeV]	O/E
NA38 at CERN	[28]	200 - 450	$\mu^+\mu^-$	0.6 - 6	~ 1.5
CERES	[29]	200 - 450	e^+e^-	0.05 - 1.4	5 ± 3
HELIOS-3	[30]	200 - 450	$\mu^+\mu^-$	0.3 - 4	~ 2.5
DLS at BEVALAC	[31]	1 - 5	e^+e^-	0.05 - 1	$2 - 3$
KEK	[32]	12	e^+e^-	0.05 - 2	~ 2
NA60	[33]	160	$\mu^+\mu^-$	0.2 - 1.2	~ 2
HADES at GSI	[34]	1 - 2	e^+e^-	0.05 - 1	~ 1

the $\pi^+\pi^-$ annihilation channel. An alternative scenario is the formation of a quark-gluon plasma in the heavy systems which leads to additional contributions to the dilepton spectrum from perturbative QCD (pQCD) such as quark-antiquark annihilation or gluon-gluon scattering [35,37].

The dilepton mass spectrum measured at KEK in p + A reactions at the beam energy of 12 GeV [32] revealed an excess of the dileptons below the ρ-meson peak over the known sources. These data were analyzed in Ref. [38] with no success to reproduce the experimental spectrum within a dropping mass scenario's and/or a significant collision broadening of the vector mesons.

The results of the DLS [31], NA60 [33], and HADES [34] collaborations are discussed in Sect. 9.

6. Elementary sources of dileptons

The direct decay mode $\rho,\ \omega \to e^+e^-$ represents a signal. There are many other sources, however, which constitute a background and which should accurately be subtracted.

6.1. *Dilepton modes in decays of light unflavored mesons*

One has to distinguish between vector, pseudoscalar and scalar mesons $\mathcal{M} = V,\ P,\ S$, respectively, where

$$V = \rho,\ \omega,\ \phi; \quad P = \pi,\ \eta,\ \eta'; \quad S = f_0(980),\ a_0(980).$$

Decay modes are presented below:

$V \to e^+e^-$			Direct decays
$P \to \gamma e^+e^-$	$S \to \gamma e^+e^-$		Dalitz decays
$V \to Pe^+e^-$	$P \to Ve^+e^-$		Dalitz decays
$V \to PPe^+e^-$	$P \to PPe^+e^-$	$S \to PPe^+e^-$	Four-body decays

The photon and dilepton branching ratios are calculated using the effective meson theory [39]. Mesons are produced in NN and πN collisions. The reactions of rescattering, absorption and reabsorption of mesons affect the observed dilepton yield.

6.2. *Dilepton modes in decays of nucleon resonances*

Besides light unflavored mesons $\mathcal{M} = V,\ P,\ S$, nucleon resonances $R = \Delta(1232),\ N^*(1440),\ldots$ are produced in the course of heavy-ion collisions. Their decays contribute to the measured dilepton spectra.

The $\Delta(1232)$ Dalitz decay is one of the major sources of dileptons in heavy-ion collisions at intermediate energies [40–42]. The first correct calculation of that decay is given only recently [43], while kinematically complete expressions for Dalitz decays of other positive and negative parity high-spin resonances are given in Ref. [44].

In order to evaluate the Delta Dalitz decay rate, HADES [34] used one of six pairwise different incorrect expressions available in literature before [43]. We present therefore results of Ref. [43]:

The Δ resonance width for decay into nucleon and a virtual photon is given by

$$\Gamma(\Delta \to N\gamma^*) = \frac{\alpha}{16}\frac{(m_\Delta + m_N)^2}{m_\Delta^3 m_N^2}((m_\Delta + m_N)^2 - M^2)^{1/2}$$
$$((m_\Delta - m_N)^2 - M^2)^{3/2}\left(G_M^2 + 3G_E^2 + \frac{M^2}{2m_\Delta^2}G_C^2\right). \qquad (11)$$

Here, m_N and m_Δ are nucleon and Δ masses, $M^2 = q^2$ is the photon four-momentum, G_M, G_E, and G_C are magnetic, electric and Coulomb transition form factors.

The factorization prescription allows to find the dilepton decay rate of the Δ resonance:

$$d\Gamma(\Delta \to Ne^+e^-) = \Gamma(\Delta \to N\gamma^*)M\Gamma(\gamma^* \to e^+e^-)\frac{dM^2}{\pi M^4}, \qquad (12)$$

with

$$M\Gamma(\gamma^* \to e^+e^-) = \frac{\alpha}{3}(M^2 + 2m_e^2)\sqrt{1 - \frac{4m_e^2}{M^2}} \qquad (13)$$

being the decay width of a virtual photon into the dilepton pair with invariant mass M. The last three equations being combined give the $\Delta(1232) \to Ne^+e^-$ decay rate.

The available experimental data on electro- and photoproduction experiments with $\Delta(1232)$ are fitted in Ref. [44] using the eVMD framework to give

$$G_M = (2.461 - 0.485M^2 - 0.004M^4)G, \tag{14}$$

$$G_E = (0.062 - 0.010M^2 + 0.004M^4)G, \tag{15}$$

$$G_C = (0.518 - 0.087M^2)G \tag{16}$$

where M^2 is in GeV^2 and

$$G = \prod_{i=1}^{4} \frac{m_i^2}{m_i^2 - im_i\Gamma_i(M) - M^2}. \tag{17}$$

The masses m_i of the ρ-meson family members are the following: 0.769, 1.250, 1.450, and 1.720 GeV. The ρ-meson decay rate $\Gamma_1(M)$ is proportional to two-pion phase space and normalized to the vacuum width, whereas $\Gamma_i(M)$ for $i \geq 2$ are set equal to zero, since $M \leq 1$ GeV at our conditions. The quality of the fit can be verified with Fig. 20 of Ref. [44]. Equations (14) - (17) are in agreement with the quark counting rules, the condition $G_M/G_E \to -1$ at $M^2 \to -\infty$ is a consequence of the quark counting rules.

7. Microscopic eVMD model

Radially excited ρ- and ω-mesons are introduced [44,45] to ensure the correct asymptotic behavior of the $RN\gamma$ transition form factors in line with the quark counting rules [46]. We refer this model as the extended vector meson dominance (eVMD) model. From the experimental side, the excited vector mesons are needed to match photon $M^2 = 0$ and vector meson $M^2 \sim 0.5$ GeV^2 experimental branchings of nucleon resonances within a unified scheme.

- The eVMD model provides a unified description of the photo- and electro-production data and of the vector meson and dilepton decays of nucleon resonances and accounts for quark counting rules.

We take constraints on the transition form factors from the quark counting rules into account explicitly. The remaining parameters are fixed by fitting the available photo- and electro-production data and using results of the πN multichannel partial-wave analysis. When data are not available, we used predictions of the additive quark models.

Radially excited ρ- and ω-mesons as a consequence of the quark counting rules interfere with the ground-state mesons *destructively* below the ρ-meson peak and reduce dilepton spectra from decays of nucleon resonances in the vacuum, accordingly.

8. How to keep photon massless with $\rho^0 - \omega - \gamma$ mixing?

VMD model and its modifications introduce mixing of photon with vector mesons ρ^0, ω, ϕ, etc. Such a mixing can, in principle, generate finite photon mass and destroy gauge invariance. This problem has been solved for VMD model by Kroll, Lee and Zumino [47] by constructing an effective Lagrangian for photons and vector mesons, which reproduces VMD predictions at a tree level. We describe a distinct consistency proof.

We start from an effective Lagrangian involving pions interacting with photons. An example of such a Lagrangian is e.g. the non-linear sigma model. The vector mesons appear as resonances in two-pion scattering channel (ρ-mesons) and three-meson scattering channel (ω-mesons).

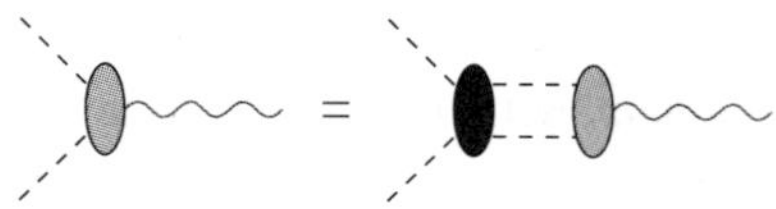

Fig. 6. Diagram representation of the FSI of pions (dashed lines) contributing to form factor in the ρ-meson channel. The wavy lines are photon lines.

Let us consider an absorption of a photon in an isovector channel shown on Fig. 6. Applying two-body unitarity and taking into account analyticity (see e.g. [48], Chap. 18), we replace the pointlike vertex e by $eP_l(t)/D_J(t)$ where $t = q^2$ is the photon momentum squared, $P_l(t)$ is a polynomial of the degree l, and $D_J(t)$ is the Jost function defined in terms of the p-wave isovector two-pion scattering phase shift $\delta(t)$:

$$D_J(t) = \exp(-\frac{t}{\pi}\int_{t_0}^{\infty}\frac{\delta(t')dt'}{t'(t-t')}) \tag{18}$$

where t_0 is the two-pion threshold.

In the no-width approximation, the phase shift accounting for the existence of n resonances is given by

$$\delta(t) = \sum_{i=1}^{n}\pi\theta(t-m_i^2) \tag{19}$$

where m_i is the mass of the i-th radial excitation of the ρ^0-meson. Substituting this expression into Eq.(18), we obtain

$$F(t) = P_l(t)\prod_{i=1}^{n}\frac{m_i^2}{t-m_i^2} \tag{20}$$

(cf. Eqs.(14) - (17)). The requirement $F(t) \to 0$ at $t \to \infty$ gives $l < n$.

Analytical functions are fixed by their singularities. The representation (20) can be rewritten in the additive form

$$F(t) = \sum_{i=1}^{n} c^i \frac{m_i^2}{m_i^2 - t} \tag{21}$$

where c^i are some coefficients. The normalization condition $F(0) = 1$ and quark counting rules impose constraints (sum rules) for c^i.

The effective pion Lagrangian is well defined, since pions are stable particles which exist as asymptotic states. In the approach presented above, the problem of gauge invariance does not appear, since gauge invariance of the effective Lagrangian ensures transverse polarization tensor of photons and the vanishing photon mass. The vector mesons are resonances accounted for by the the final-state interactions (FSI).

9. In-medium properties of vector mesons: What can we learn from observations?

The gap between observables measured in heavy-ion collisions and theoretical models of the in-medium hadrons is filled by transport models which account for complicated dynamics of heavy-ion collisions and provide a link between theory and experiment. We comment results concerning physical properties of in-medium vector mesons that can be derived from experimental data with use of transport models.

9.1. *A constraint to ω-meson collision width from DLS and HADES data*

In-medium broadening of vector mesons gives an increase of the nucleon resonance decay widths $R \to NV$ and a decrease of the dilepton branchings $V \to e^+e^-$ due to enhanced total vector meson widths.

The differential branching $dB(\mu, M)^{R\to NV}$ of resonance R with an off-shell mass μ increases with the V meson width due to subthreshold character of the vector meson production in light nucleon resonances. The dilepton branching of the nucleon resonances

$$dB(\mu)^{R\to Ne^+e^-} \sim dB(\mu)^{R\to NV}\frac{\Gamma_{V\to e^+e^-}}{\Gamma_V^{\text{total}}} \qquad (22)$$

is, on the other hand, inverse proportional to the *total* vector meson width. This effect is particularly strong for ω, since $\Gamma_\omega^{vac} = 8$ MeV only. Fig. 7 shows the collision broadening effect on the dilepton spectra. To get description of the data, we have to assume

$$\Gamma_\omega^{total} \approx \Gamma_\omega^{coll} \geq 50 \text{ MeV} \quad \text{at density} \quad \rho \sim 1.5\rho_0 \qquad (23)$$

where ρ_0 is the saturation density. The similar conclusions holds for HADES data [34].

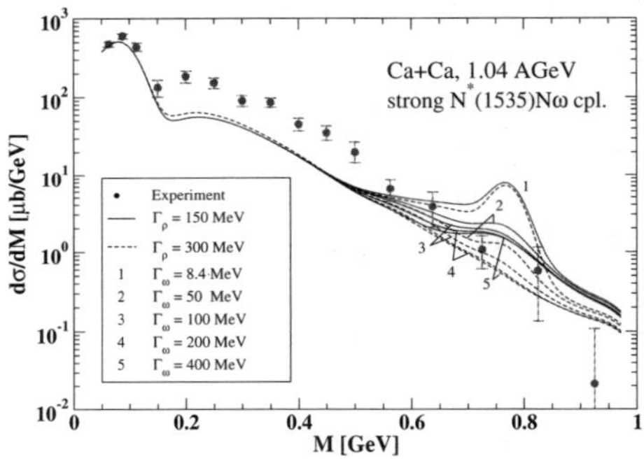

Fig. 7. Dilepton spectra in $Ca + Ca$ collisions for different values of the in-medium ρ and ω widths. The solid curves correspond calculations where the ρ width is kept at its vacuum value of 150 MeV (no collision broadening). The dashed curves correspond to a total ρ width of 300 MeV. In both cases the ω width is varied between $\Gamma_\omega^{tot} = 8.4 \div 400$ MeV. $\Gamma_\omega^{tot} \geq 50$ MeV holds as a conservative constraint (from [42]).

9.2. *Evidence for decoherence in transition form factors of nucleon resonances from DLS and HADES data*

Fig. 7 shows an obvious deficit of dileptons below the ρ-meson peak. This phenomenon has been called "DLS puzzle", since transport models were not able to reproduce it. The eVMD suggests a medium-induced decoherence of vector mesons entering transition form factors of nucleon resonances and gives an enhancement of dilepton yield [42]. It does not remove disagreement with the DLS data completely, however, the recent HADES data [34] shown on Fig. 8 are reproduced at $M \leq m_\rho$ perfectly [34,49]. A more detailed theoretical study of the decoherence effect would certainly be welcome.

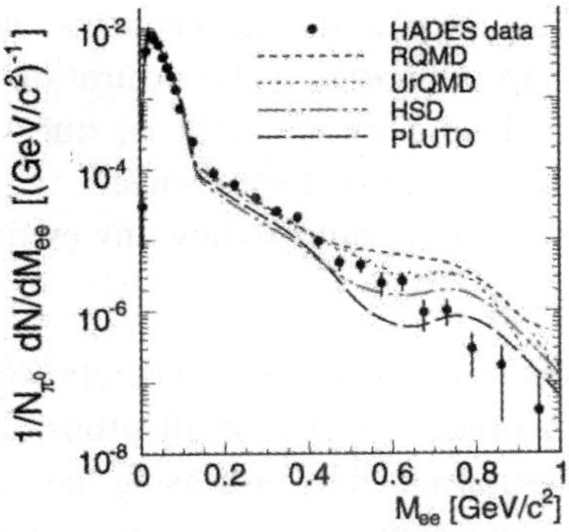

Fig. 8. The experimental dilepton spectrum as compared to predictions of PLUTO [34] thermal model, UrQMD [40], RQMD [42], and HSD [36] transport models (from Ref. [34]).

9.3. *No evidence for dropping ρ-meson mass with increasing density from NA60 data*

The NA60 Collaboration presented recently an impressive attempt to extract the ρ-meson spectral function from dilepton spectra at ultrarelativistic heavy-ion collisions [33]. It had not found evidence for a change of the ρ-meson mass. This observation poses obvious difficulties for the Brown-Rho scaling hypothesis.

10. Conclusions

In this lecture, an introduction is made to physics of in-medium behavior of resonances. Transparent methods developed in the atomic spectroscopy in gases can be useful for in-medium hadron spectroscopy. The main topic we focused on is the modifications of spectral functions. Starting from instructive examples in atomic and nuclear physics, we approached the problem of vector mesons description in dense nuclear matter.

Theoretical models are discussed in Sect. 4. The most of them do not go beyond the first order in density. In the non-linear sigma model, density expansion for pions has zero convergence radius [50]. The higher-order calculations would help to clarify to what extent density expansion for vector mesons is reliable.

This summer (2006 year) two new important experimental works [33,34] from NA60 and HADES collaborations have been completed and results published. Preliminary conclusions are provided in Sect. 9.

- Collision broadening of ω-meson is constrained from below by ~ 50 MeV at $\rho \sim 1.5\rho_0$ where ρ_0 is the saturation density.
- DLS puzzle had apparently dissolved due to decoherence in in-medium propagation of vector mesons.
- The NA60 data [33] do not provide any evidence for dropping the ρ-meson mass.

Among difficulties in description and interpretation of the HADES data, one has to mention theoretical excess of dileptons above the ρ-peak. The dilepton yield could be suppressed by increasing the ρ and ω collision widths and by taking into account collision broadening of nucleon resonances.

The results of the KEK Collaboration [32] show an enhancement below the ρ-peak, which is not understood.

Results of the dilepton production in heavy-ion collisions attract great attention and will certainly be analyzed in future.

References

1. I. I. Sobelman, *Introduction to the Theory of Atomic Spectra*, (Pergamon Press, Oxford, 1972).
2. L. A. Kondratyuk, M. I. Krivoruchenko, N. Bianchi, E. D. Sanctis and V. Muccifora, Nucl. Phys. A **579**, 453 (1994).
3. P. Bogen, Z. Phys. **149**, 62 (1957).
4. N. Bianchi *et al.*, Phys. Lett. B **309**, 5 (1993).
5. J. D. Walecka, Annals Phys. (N.Y.) **83**, 491 (1974).
6. S. A. Chin, Annals Phys. (N.Y.) **108**, 301 (1977).
7. A. Bhattacharyya, S. K. Ghosh and S. C. Phatak, Phys. Rev. C**60**, 044903 (1999).
8. A. Mishra, J. Reinhardt, H. Stocker and W. Greiner, Phys. Rev. C**66**, 064902 (2002).
9. Y. Nambu and G. Jona-Lasinio, Phys. Rev. **122**, 345 (1961); **124**, 246 (1961).
10. K. Tsushima, T. Maruyama and A. Faessler, Nucl. Phys. A**535**, 497 (1991); T. Maruyama, K. Tsushima and A. Faessler, Nucl. Phys. A**537**, 303 (1992).
11. G. E. Brown, Mannque Rho, Phys. Rept. **398**, 301 (2004).
12. A. I. Bochkarev and M. E. Shaposhnikov, Nucl. Phys. B**268**, 220 (1986).
13. E. G. Drukarev and E. M. Levin, JETP Lett. **48**, 338 (1988).
14. C. Adami and G. E. Brown, Phys. Rept. **234**, 1 (1993); T. Hatsuda, H. Shiomi and H. Kuwabara, Prog. Theor. Phys. **95**, 1009 (1996).
15. T. Hatsuda and S.H. Lee, Phys. Rev. C**46**, R34 (1992); S. Leupold, Phys. Rev. C**64**, 015202 (2001).
16. F. Klingl, N. Kaiser, W. Weise, Nucl. Phys. A**624**, 527 (1997).
17. E. G. Drukarev *et al.*, Phys. Rev. C**69**, 065210 (2004).
18. G. Chanfray, Magda Ericson, P. A. M. Guichon, Phys. Rev. C**68**, 035209 (2003).

19. L. A. Kondratyuk *et al.*, Phys. Rev. C**58** (1998) 1078.
20. A. T. Martell, P. J. Ellis, Phys. Rev. C**69**, 065206 (2004).
21. G. E. Brown and M. Rho, Phys. Rev. Lett. **66**, 2720 (1991).
22. A. Gomez Nicola *et al.*, AIP Conf. Proc. **660**, 156 (2003).
23. B. V. Martemyanov, Amand Faessler, C. Fuchs, M. I. Krivoruchenko, Phys. Rev. Lett. **93**, 052301 (2004).
24. R. Rapp, G. Chanfray and J. Wambach, Phys. Rev. Lett. **76**, 368 (1996).
25. F. Klingl, S. Kim, S. H. Lee, P. Morath, W. Weise, Phys. Rev. Lett. **82**, 3396 (1999).
26. S. Digal, P. Petreczky and H. Satz, Phys. Lett. B**514**, 57 (2001).
27. A. Mishra *et al.*, Phys. Rev. C**69**, 015202 (2004).
28. M. C. Abreu *et al.*, Phys. Lett. **423**, 207 (1998).
29. G. Agakichiev et al., Phys. Rev. Lett. **75**, 1272 (1995);
A. Drees, Nucl. Phys. A**610**, 536c (1996).
30. M. Masera, Nucl. Phys. A**590**, 93c (1995).
31. R. J. Porter et al., Phys. Rev. Lett. **79**, 1229 (1997);
W. K. Wilson *et al.*, Phys. Rev. C**57**, 1865 (1998).
32. K. Ozawa *et al.*, Phys. Rev. Lett. **86**, 5019 (2001).
33. M. Floris *et al.* [NA60 Collaboration], arXiv:nucl-ex/0606023.
34. The HADES Collaboration, arXiv:nucl-ex/0608031.
35. R. Rapp and J. Wambach, Adv. Nucl. Phys. **25**, 1 (2000).
36. W. Cassing and E. L. Bratkovskaya, Phys. Reports **308**, 65 (1999).
37. R. Schneider and W. Weise, Eur. Phys. J. A**9**, 357 (2000).
38. E. L. Bratkovskaya, Phys. Lett. B**529**, 26 (2002).
39. A. Faessler, C. Fuchs, M. I. Krivoruchenko, Phys. Rev. C**61**, 035206 (2000).
40. C. Ernst, S. A. Bass, M. Belkacem, H. Stocker, W. Greiner, Phys. Rev. C**58**, 447 (1998).
41. E. L. Bratkovskaya, W. Cassing, U. Mosel, Nucl. Phys. A**686**, 568 (2001).
42. K. Shekhter, C. Fuchs, Amand Faessler, M. Krivoruchenko, B. Martemyanov, Phys. Rev. C**68**, 014904 (2003).
43. M. I. Krivoruchenko and Amand Faessler, Phys. Rev. D**65**, 017502 (2002).
44. M. I. Krivoruchenko, B. V. Martemyanov, Amand Faessler, and C. Fuchs, Ann. Phys. (N.Y.) **296**, 299 (2002).
45. Amand Faessler, C. Fuchs, M. I. Krivoruchenko and B. V. Martemianov, J. Phys. G**29**, 603 (2003).
46. V. A. Matveev, R. M. Muradian and A. N. Tavkhelidze, Lett. Nuovo Cim. **7**, 719 (1973);
S. J. Brodsky and G. R. Farrar, Phys. Rev. Lett. **31**, 1153 (1973);
A. I. Vainstein and V. I. Zakharov, Phys. Lett. **B72**, 368 (1978).
47. N. M. Kroll, T. D. Lee and B. Zumino, Phys. Rev. D**157**, 1376 (1967).
48. J. D. Bjorken and S. D. Drell, *Relativistic Quantum Fields*, (New York, McGraw-Hill, 1965).
49. M. D. Cozma, C. Fuchs, E. Santini, A. Fassler, Phys. Lett. B**640**, 170 (2006).
50. M. I. Krivoruchenko, C. Fuchs, B. V. Martemyanov, A. Faessler, Phys. Rev. **D**, in press; arXiv:hep-ph/0505083.

Boson mass spectra in 331 gauge models with minimal Higgs mechanism

Ion I. Cotăescu

West University of Timişoara,
V. Pârvan Ave. 4, RO-300223 Timişoara, Romania

One presents an exact solution of the Frampton-Pisano-Pleitez three-generation 331 gauge model equipped with a special Higgs mechanism and a new kind of Yukawa couplings in unitary gauge. It is shown that the resulting boson mass spectrum depends on a real parameter. There exists a critical value of this parameter for which the different gauge bosons carrying the same charge get the same mass.

PACS:12.20 Hx

Keywords: 331 gauge model, minimal Higgs mechanism, mass spectrum, critical point

1. Introduction

The gauge models with spontaneous symmetry breaking must respect few basic rules to be renormalizable. The central point is to find a Higgs mechanism able to break the gauge symmetry up to a desired residual one and to produce a good boson mass spectrum. Moreover, the spinor mass terms could arise only from Yukawa interactions and certainly the model must be free of the axial anomaly [1].

If we consider the Standard Model (SM) and its recent $SU(3) \otimes U(1)$ and $SU(4) \otimes U(1)$ generalizations [2–5] (known as 331 and respectively 341 models), we observe a some kind of regularity, namely, that the number of the vector Higgs multiplets may be equal to the dimension of the fundamental representation. This allowed us to generalize the Higgs mechanisms of these models to any $SU(n) \otimes U(1)$ gauge model, introducing the *minimal Higgs mechanism* (mHm.) and the Yukawa couplings in unitary gauge [6]. In the general case of $SU(n) \otimes U(1)$ models our mHm. involves n Higgs multiplets, each one transforming according to the fundamental representation of $SU(n)$, but having different chiral hypercharges. Moreover, these multiplets are seen as an orthonormal system with the "norm" defined by a scalar real field [6], which plays the same role as the neutral component of the Higgs doublet of the SM. In these conditions we can define coupling terms which should lead to Yukawa couplings only in unitary gauge. These may produce adequate mass terms for any pair of spinor components we wish, giving thus more flexibility to the model [6].

In this lecture we present how can be solved the 331 model proposed by Frampton, Pisano and Pleitez [2,3] equipped with our mHm. and Yukawa couplings in unitary gauge. Moreover we point out that the boson mass spectrum has a critical point where some boson masses take the same values and the bilepton becomes very light. To this end we need to present some technical details concerning the mHm. of the general $SU(n)\otimes U(1)_c$ gauge models whose $U(1)_c$ group is that of the chiral gauge. For this reason we start, in the second section, with a short review of the $SU(n)$ representations while the next one is devoted to these general models. The fourth section presents our solution of the mentioned 331 model. Finally, we discuss the boson mass spectrum focusing on its critical point.

2. Preliminaries

The general theory of the $SU(n)\otimes U(1)_c$ gauge models with mHm. will be briefly presented using the notations of Ref. [6] and the basic notations of the Lie groups and algebras representation theory [7,8].

We simplify the notations by choosing a unique coupling constant, g, for both the groups involved here. This means that the irreducible representations (ireps.) of $U(1)_c$, which reduces to multiplications with phase factors $exp(-igy\xi^0)$ of parameter ξ^0, will be defined by the (real) character y instead of the usual chiral hypercharge.

The main pieces are the ireps. of $SU(n)$ whose classification can be done by using the tensor method [7]. One starts with the fundamental irep., $\mathbf{n}$, which defines the group $SU(n)$ and its algebra, $su(n)$, so that

$$U=U(\xi)=e^{-ig\xi}\in SU(n)\,,\quad \forall\,\xi=\xi^{+}\in su(n)\,. \tag{1}$$

These matrices transform the components ψ_i of the n-dimensional (vector) multiplet, ψ, like $\psi_i\to(U\psi)_i=U_{i\,\cdot}^{\cdot\,j}\psi_j$. The transformation law of the complex conjugated components, $(\psi_i)^*\equiv\psi^i\to(U\psi)^*_i\equiv U_{\cdot\,j}^{i\,\cdot}\psi^j$, defines the complex conjugated of the fundamental irep., $\mathbf{n}^*$, which has the matrices $U_{\cdot\,j}^{i\,\cdot}=(U_{i\,\cdot}^{\cdot\,j})^*$. All the other ireps. of $SU(n)$ correspond to the different classes of symmetry (given by the Young tableaus) of the tensors of the rank (r,s) resulted from the direct products $(\otimes\mathbf{n})^r\otimes(\otimes\mathbf{n}^*)^s$. These tensors have r lower and s upper indices denoted by $i,j,k,\cdots = 1,\cdots,n$. Their transformation laws are in accordance with the above presented basic rules.

In general, for the $su(n)$ algebra one uses either the standard hermitian generators, $T_a = T_a^+$, (labelled here by $a,b,c,\cdots = 1,\cdots,n^2-1$), or the real generators H^i_j and c-number parameters [9]. However, for the gauge models we intend to work out it is convenient to consider a hybrid basis numbered by the indices $i,j,\cdots$ and by a new family of indices, $\hat{i},\hat{j},\cdots$, which range from 1 to $n-1$. This basis is defined as follows [6]:

$$D_{\hat{i}} = T_{(\hat{i}+1)^2-1}, \qquad E^i_j = \frac{1}{\sqrt{2}} H^i_j, \qquad i \neq j. \tag{2}$$

The corresponding parametrization,

$$\xi = D_{\hat{i}}\xi^{\hat{i}} + \sum_{i\neq j} E^i_j \xi^j_i, \tag{3}$$

contains $n-1$ real parameters, $\xi^{\hat{i}}$, and $n(n-1)/2$ c-number ones, $\xi^i_j = (\xi^j_i)^* = \sqrt{2}\zeta^i_j$, for $i \neq j$. This choice offers many advantages. The first one is of a simple numeration of the diagonal generators $D_{\hat{i}}$ (of the Cartan subalgebra). On the other hand, the parameters ξ^i_j could be directly associated to the c-number gauge fields because of the factor $1/\sqrt{2}$ of Eq. (2) which gives their correct normalization. However, the most important is to have good trace orthogonality properties:

$$Tr(D_{\hat{i}}D_{\hat{j}}) = \frac{1}{2}\delta_{\hat{i}\hat{j}}, \quad Tr(D_{\hat{i}}E^i_j) = 0, \quad Tr(E^i_j E^k_l) = \frac{1}{2}\delta^i_l \delta^k_j. \tag{4}$$

We denote either by $\rho = (R_\rho, y_\rho)$ or by $\rho = (\mathbf{n}_\rho, y_\rho)$, with $n_\rho = n(R_\rho)$, the irep. of $SU(n) \otimes U(1)_c$ defined as the direct product of the irep. R_ρ of $SU(n)$ and the irep. of the character y_ρ of $U(1)_c$. Its transformation matrices read

$$U^\rho(\xi^0,\xi) = U^\rho(\xi)e^{-igy_\rho\xi^0} = e^{-ig(\xi^\rho + y_\rho \xi^0)}, \tag{5}$$

where $\xi^\rho = R_\rho(\xi)$. The transformation law of a n_ρ-dimensional multiplet, ψ^ρ, of this irep. can be written in the matrix form as $\psi^\rho \to U^\rho(\xi^0,\xi)\psi^\rho$, understanding that its tensor components transform like

$$(\psi^\rho)^{j_1\cdots j_s}_{i_1\cdots i_r} \to e^{-igy_\rho\xi^0} U_{i_1}^{\cdot\, i'_1}{}_{\cdot} \cdots U^{j_1}{}_{\cdot\, j'_1}^{\cdot} \cdots (\psi^\rho)^{j'_1\cdots j'_s}_{i'_1\cdots i'_r}, \tag{6}$$

if R_ρ is of the rank (r,s). We note that for $y_\rho = 0$ the irep. ρ coincides with R_ρ and therefore the notations $(R,0)$ and R are equivalent.

3. The $SU(n) \otimes U(1)$ gauge models with mHm

The Lagrangian density (Ld.) of any gauge model, $\mathcal{L} = \mathcal{L}_S + \mathcal{L}_G + \mathcal{L}_H$, has three terms corresponding to the spinor (S), gauge (G) and Higgs (H) sectors.

In general, any spinor sector can be put in the pure left form containing only left-handed components grouped in the multiplet $L = (L_1, L_2, \cdots, L_N)^T$ the Dirac adjoint of which is $\overline{L} = (\overline{L}_1, \overline{L}_2, \cdots, \overline{L}_N)$. (The superscript T stands for matrix transposition.) The most general form of the Ld. of the spinor sector can be written in the compact matrix notation as

$$\mathcal{L}_{S_0} = \frac{i}{2}\overline{L} \overleftrightarrow{\not{\partial}} L - \frac{1}{2}(\overline{L}\chi L^c + \overline{L^c}\chi^+ L), \tag{7}$$

where $\overleftrightarrow{\not{\partial}} = \gamma^\mu \overrightarrow{\partial}_\mu - \gamma^\mu \overleftarrow{\partial}_\mu$ and $L^c = (L_1^c, L_2^c, \cdots, L_N^c)^T$ is the charge conjugated of the multiplet L. The field χ was introduced in order to give rise to the spinor masses after the breakdown of the gauge symmetry and, therefore, this must depend on the Higgs field components.

If we chose the $SU(n) \otimes U(1)$ symmetry then L may transform according to a reducible representation defined as the direct sum of a given set of ireps. ρ. Therefore, we can write

$$L = \sum_\rho \oplus L^\rho, \tag{8}$$

where each multiplet L^ρ transforms according to its own irep., ρ, defined by (6). The corresponding charge conjugated multiplets, $(L^\rho)^c$, transform according to the complex conjugated ireps., ρ^*. Then the matrix χ may contain all the blocks $\chi^{\rho\rho'}$ which couple the pairs of the multiplets $\overline{L}^\rho$ and $(L^{\rho'})^c$. In these circumstances the Ld. (7) can be written as

$$\mathcal{L}_{S_0} = \frac{i}{2}\sum_\rho \overline{L^\rho} \overleftrightarrow{\not{\partial}} L^\rho - \frac{1}{2}\sum_{\rho\rho'} \left(\overline{L^\rho}\chi^{\rho\rho'}(L^{\rho'})^c + h.c.\right). \tag{9}$$

This remains invariant under the global $SU(n) \otimes U(1)_c$ transformations if the blocks $\chi^{\rho\rho'}$ transform like

$$\chi^{\rho\rho'} \rightarrow U^\rho(\xi^0, \xi)\chi^{\rho\rho'} \left(U^{\rho'}(\xi^0, \xi)\right)^T, \tag{10}$$

according to the representations $(R_\rho \otimes R_{\rho'}, y_\rho + y_{\rho'})$ which generally are reducible.

When one gauges the $SU(n) \otimes U(1)_c$ group, one needs to introduce the gauge fields: $A_\mu^0 = (A_\mu^0)^*$ and $A_\mu = A_\mu^+ \in su(n)$ which can be expressed in the basis (2) as

$$A_\mu = D_{\hat{i}} A^{\hat{i}}_\mu + \sum_{i\neq j} E^i_j A^j_{i\mu}. \tag{11}$$

This depends on $n-1$ real fields, $A^{\hat{i}}_\mu$, and $n(n-1)/2$ c-number ones which satisfy $A^j_{i\mu} = (A^i_{j\mu})^*, i \neq j$. The next step is to replace the ordinary derivatives of the Ld (9) by the covariant ones,

$$D_\mu L^\rho = \partial_\mu L^\rho - ig(R_\rho(A_\mu) + y_\rho A^0_\mu)L^\rho, \tag{12}$$

which will give the interaction terms of the whole Ld. of the spinor sector. In the basis (2) this is

$$\mathcal{L}_S = \mathcal{L}_{S_0} + g\sum_\rho \overline{L^\rho}\left(D^\rho_{\hat{i}} A^{\hat{i}}_\mu + \sum_{i\neq j} E^{\rho\ i}_{\ \ j} A^j_{i\mu} + y_\rho A^0_\mu\right)\gamma^\mu L^\rho. \tag{13}$$

The gauge invariance of this Ld. requires the gauge fields to transform as

$$\begin{aligned} A_\mu &\to U A_\mu U^+ - \frac{i}{g}(\partial_\mu U)U^+, \\ A^0_\mu &\to A^0_\mu + \partial_\mu \xi^0, \end{aligned} \tag{14}$$

where $U = U(\xi(x))$ is given by (1). The field strength tensors:

$$\begin{aligned} F_{\mu\nu} &= \partial_\mu A_\nu - \partial_\nu A_\mu - ig\left[A_\mu, A_\nu\right], \\ F^0_{\mu\nu} &= \partial_\mu A^0_\nu - \partial_\nu A^0_\mu, \end{aligned} \tag{15}$$

are covariant and consequently we can define the invariant Ld. of the gauge sector as

$$\mathcal{L}_G = -\frac{1}{2}Tr\left(F_{\mu\nu}F^{\mu\nu}\right) - \frac{1}{4}F^0_{\mu\nu}F^{0\mu\nu}. \tag{16}$$

All these relations can be written in the basis (2) starting with the form of the matrix $F_{\mu\nu}$ in this basis,

$$F_{\mu\nu} = D_{\hat{i}} F^{\hat{i}}_{\mu\nu} + \sum_{i\neq j} E^i_j F^j_{i\mu\nu}, \tag{17}$$

where, according to (4), we have

$$F^{\hat{i}}_{\mu\nu} = 2Tr(D_{\hat{i}} F_{\mu\nu}), \qquad F^j_{i\mu\nu} = (F^i_{j\mu\nu})^* = 2Tr(E^j_i F_{\mu\nu}). \tag{18}$$

Finally, the Ld. (16) gets the form:

$$\mathcal{L}_G = -\frac{1}{4}\left(\sum_{\hat{i}} F^{\hat{i}}_{\mu\nu}F^{\hat{i}\mu\nu} + F^0_{\mu\nu}F^{0\mu\nu}\right) - \frac{1}{2}\sum_i\sum_{j<i}\left(F^j_{i\mu\nu}\right)^* F^{j\mu\nu}_i. \tag{19}$$

Its kinetic part, $\mathcal{L}_{G\rfloor g=0}$, contains only the kinetic parts of the field strength tensors, $F^0_{\mu\nu}$ and $F_{\mu\nu\rfloor g=0}$. This will be combined with the mass term resulted from the breakdown of the gauge symmetry, in order to obtain the free Ld. of the gauge sector.

Our mHm. involves n^2 c-number Higgs variables, $\phi^{(i)}_j$, organized into n multiplets, $\phi^{(i)}$, each one transforming according its own irep., $(\mathbf{n}, y^{(i)})$. These represent a set of $2n^2$ real field variables from which at most n^2 could be removed by fixing the gauge. Thus the danger is to remain with some Goldstone bosons after the breakdown of the gauge symmetry. To prevent this, we reduce the number of field variables by introducing *a priori* the following *constraints*:

$$\phi^{(i)+}\phi^{(j)} = \phi^2\delta_{ij}, \tag{20}$$

where ϕ is a gauge invariant real field variable. There are n^2 real equations and, therefore, the fields $\phi^{(i)}_j$ will have only n^2 independent real components. We note that the indices included in parentheses are not $SU(n)$ vectorial ones even though they still range from 1 to n. Thus, in fact, we introduced an orthonormal basis in the representation space of the irep. $\mathbf{n}$ of $SU(n)$ the vectors of which transform differently under the $U(1)_c$ group. Their $U(1)_c$ characters, $y^{(i)}$, will be considered as arbitrary parameters. They can be grouped into the matrix:

$$Y = \mathrm{diag}\left(y^{(1)}, y^{(2)}, \cdots, y^{(n)}\right). \tag{21}$$

In order to obtain a Higgs Ld. with very simple interaction terms but able to produce a non trivial boson mass spectrum, it needs to introduce free parameters in its kinetic term. These are $\eta_0 \in [0,1)$ and the non-vanishing elements $\eta^{(i)}$ of the matrix $\eta = \mathrm{diag}\left(\eta^{(1)}, \eta^{(2)}, \cdots, \eta^{(n)}\right)$, with the property

$$Tr(\eta^2) = 1 - {\eta_0}^2. \tag{22}$$

Now we can use $({\eta_0}^2, \eta^2)$ as the metric of the kinetic part of the Ld. of the Higgs sector,

$$\mathcal{L}_H = \frac{1}{2}{\eta_0}^2\partial^\mu\phi\partial_\mu\phi + \frac{1}{2}\sum_i (\eta^{(i)})^2 \left(\mathcal{D}^\mu\phi^{(i)}\right)^+ \left(\mathcal{D}_\mu\phi^{(i)}\right) - V(\phi), \tag{23}$$

where

$$\mathcal{D}^\mu\phi^{(i)} = \partial_\mu\phi^{(i)} - ig\left(A_\mu + y^{(i)}A^0_\mu\right)\phi^{(i)}, \tag{24}$$

and

$$V(\phi) = -\frac{1}{2}\mu^2 \sum_i \phi^{(i)+}\phi^{(i)} + \frac{1}{4}\lambda_1 \left(\sum_i \phi^{(i)+}\phi^{(i)}\right)^2$$
$$+\frac{1}{4}\lambda_2 \sum_{i,j} \left(\phi^{(i)+}\phi^{(j)}\right)\left(\phi^{(j)+}\phi^{(i)}\right) \qquad (25)$$

is the potential which is chosen to have the simplest algebraic form (allowed by its gauge invariance).

Furthermore, we look for the absolute minimum of $V(\phi)$ given by the equations

$$\delta_{ij}\left(-\mu^2 + \lambda_1 \sum_k \phi^{(k)+}\phi^{(k)}\right) + \lambda_2 \phi^{(i)+}\phi^{(j)} = 0 \qquad (26)$$

which have as solutions only the set of orthonormal multiplets (20) with ϕ obeying

$$-\mu^2 + (n\lambda_1 + \lambda_2)\phi^2 = 0. \qquad (27)$$

Another possibility is to express (25) only on ϕ with the help of (20) and to take the minimum in this unique variable. The result is the same, namely (27). Thus it results that the constraints (20) are compatible with the absolute minimum of $V(\phi)$. This defines the vacuum state in which ϕ has a non-vanishing expectation value, $\langle\phi\rangle$. Then, $\phi = \langle\phi\rangle + \sigma$ where σ is the physical Higgs field with zero vacuum expectation value. From (27) we obtain in the zeroth order of the perturbations

$$\langle\phi\rangle = \mid \mu \mid / \sqrt{n\lambda_1 + \lambda_2}. \qquad (28)$$

Moreover, because of (20), we are sure that there exists a gauge in which $\hat{\phi}^{(i)}_k = \delta_{ik}\phi = \delta_{ik}(\langle\phi\rangle + \sigma)$. This is just the unitary gauge. To go now to an arbitrary gauge one needs to perform a ("boost") transformation, $\hat{U} = \hat{U}(\phi^{(i)})$, so that $\phi^{(i)} = \hat{U}\hat{\phi}^{(i)}$. Therefore, the components of $\phi^{(i)}$ in an arbitrary gauge can be written as $\phi^{(i)}_j = \hat{U}_j^{\cdot\, i}(\langle\phi\rangle + \sigma)$.

The Ld. of the Higgs sector in unitary gauge can be calculated using (22)-(25). We find

$$\mathcal{L}_H = \frac{1}{2}\partial^\mu\sigma\partial_\mu\sigma - \frac{1}{2}m_\sigma^2\sigma^2 - \frac{m_\sigma^2}{2\langle\phi\rangle}\sigma^3 - \frac{m_\sigma^2}{8\langle\phi\rangle^2}\sigma^4 +$$
$$+ \frac{g^2}{2}\left(\langle\phi\rangle + \sigma\right)^2 Tr[(A_\mu + YA^0_\mu)\eta^2(A^\mu + YA^{0\mu})], \qquad (29)$$

where $m_\sigma = \sqrt{2n} \mid \mu \mid$ is the mass of the field σ. We see that it has the same form like in the SM (or in the Weinberg - Salam model [10,11]), except the last term, which contains here the both kinds of the free parameters introduced above ($y^{(i)}$ and $\eta^{(i)}$). Particularly, for $n = 2$, our mHm. gives just the Higgs mechanism of the SM [6].

Our mHm. allows us to introduce Yukawa couplings only in unitary gauge. In general, a block $\chi^{\rho\rho'}$ transforms according to $(R_\rho \otimes R_{\rho'}, y_\rho + y_{\rho'})$. If the ireps. ρ and ρ' are the ranks (r, s) and (r', s') respectively, then $\chi^{\rho\rho'}$ a tensor of the rank $(r + r', s + s')$. We assume that its components have the form

$$\left(\chi^{\rho\rho'}\right)^{j_1\cdots j_s, j'_1\cdots j'_{s'}}_{i_1\cdots i_r, i'_1\cdots i'_{r'}} = \phi^{-p} \sum_{k,l} G^{l_1,l_2,\cdots,l_{s+s'}}_{k_1,k_2,\cdots,k_{r+r'}} \times \phi^{(k_1)}_{i_1} \cdots \phi^{(k_{r+r'})}_{i'r'} \times \left(\phi^{(l_1)}_{j_1}\right)^* \cdots \left(\phi^{(l_{s+s'})}_{j's'}\right)^*, \tag{30}$$

where $G^{l_1\cdots}_{k_1\cdots}$ are coupling constants. This form corresponds to the reducible representation $R_\rho \otimes R_{\rho'}$ of $SU(n)$ but to fix the value of its character to $y_\rho + y_{\rho'}$ it requires a supplementary selection rule that reads

$$y^{(k_1)} + \cdots + y^{(k_{r+r'})} - \left(y^{(l_1)} + \cdots + y^{(l_{s+s'})}\right) = y_\rho + y_{\rho'}. \tag{31}$$

In Eq. (30) we introduced the scalar factor, ϕ^{-p}, in order to control the formal dimensions of the coupling terms [12,13]. It is known that the model is renormalizable only if each block $\chi^{\rho\rho'}$ is of the dimension $d(\chi^{\rho\rho'}) \le 1$. These can be easily pointed out in unitary gauge where (30) becomes

$$\left(\hat{\chi}^{\rho\rho'}\right)^{j_1\cdots j_s, j'_1\cdots j'_{s'}}_{i_1\cdots i_r, i'_1\cdots i'_{r'}} = \left(\langle\phi\rangle + \sigma\right)^{d(\chi^{\rho\rho'})} G^{j_1\cdots j_s, j'_1\cdots j'_{s'}}_{i_1\cdots i_r, i'_1\cdots i'_{r'}}, \tag{32}$$

with the obvious identification $d(\chi^{\rho\rho'}) = r + r' + s + s' - p$. Thus the power p of the scalar factor will be the parameter giving the desired formal dimension of the coupling terms. This has to be fixed to $p = r+r'+s+s'-1$ in order to obtain the Yukawa couplings in unitary gauge (when $d(\chi^{\rho\rho'}) = 1$).

In Ref. [6] we have shown that the class of the gauge models presented here has some remarkable features. First of all, the electromagnetic potential can be easily separated with the help of a generalized Weinberg transformation (gWt.) that can be explicitly written down. Consequently, the neutral bosons are well-defined such that the boson mass spectrum as well as the coupling coefficients (electric charges and neutral charges) can be calculated for any concrete model.

4. Three-generation $SU(3) \otimes U(1)_c$ model

In what follows we discuss the Frampton-Pisano-Pleitez model in which we replace the usual Higgs mechanism with our mHm. and Yukawa couplings in unitary gauge. In this model the gauge group is that of the maximal global symmetry of the one-generation lepton sector, $(\nu_{eL}, e_L, (e_R)^c)^T$. It combines the gauge group of the SM with that of the old Pauli-Pürsey-Gürsey (PPG) symmetry [14] which is nothing other than the $SU(2) \otimes U(1)_c$ group of the maximal symmetry of the usual Dirac theory. Moreover, if one gauges only the PPG group one finds that the doubly charged boson (i.e., the bilepton) of the actual $SU(3) \otimes U(1)$ models is due to this gauge. This is important since the PPG gauge submodel is anomaly-free and, consequently, its phenomenology could be treated separately in future investigations related to the electromagnetic implications of bileptons.

Let us consider the gauge model which should have: (I) the spinor sector of Refs. [2,3] put in pure left form, and (II) a mHm. with arbitrary (η_0, η) which satisfy Eq. (22), and Yukawa couplings in unitary gauge. In Ref. [6] we solved this model supposing that, in addition: (III) its unique coupling constant, g, coincides to the first one of the SM, and (IV) at least one Z-like boson should satisfy the condition of the SM, $m_Z = m_W / \cos\theta_W$.

This is a 331 gauge model having the gauge group $SU(3)_{col} \otimes SU(3) \otimes U(1)_c$. However, here we consider only the electro-weak interactions gauged by $SU(3) \otimes U(1)_c$. Therefore, the color indices will be omitted, restricting ourselves to indicate only the triplication of the quark multiplets. The anomaly-free spinor sector has the lepton triplets,

$$L_l = \left| \begin{array}{c} \nu_l \\ l \\ l^c \end{array} \right|_L \sim (\mathbf{3}, 0), \qquad l = e, \mu, \tau, \tag{33}$$

the quark triplets $(\times 3)$,

$$\left| \begin{array}{c} u \\ d \\ j_1 \end{array} \right|_L \sim (\mathbf{3}, 2/3), \quad \left| \begin{array}{c} s \\ -c \\ j_2 \end{array} \right|_L, \quad \left| \begin{array}{c} b \\ -t \\ j_3 \end{array} \right|_L \sim (\mathbf{3}^*, -1/3), \tag{34}$$

and the quark singlets $(\times 3)$,

$$\begin{array}{ll} (d_R)^c, (s_R)^c, (b_R)^c \sim (\mathbf{1}, 1/3), & (u_R)^c\ , (c_R)^c, (t_R)^c \sim (\mathbf{1}, -2/3)\,, \\ (j_{1R})^c \sim (\mathbf{1}, -5/3), & (j_{2R})^c\ , (j_{3R})^c \sim (\mathbf{1}, 4/3)\,. \end{array} \tag{35}$$

This spinor sector is equivalent (up to some unitary transformations [4]) with those of Refs. [2,3]. We note that the ireps. $\rho = (\mathbf{n}_\rho, \hat{y}_\rho)$ of the above multiplets are written now in terms o suitably rescaled characters $\hat{y}$ instead of y [6].

Here we use the hermitian $SU(3)$ generators $T_a = \lambda_a/2$ $(a = 1, \cdots, 8)$ represented by the Gell-Mann matrices in the basis where

$$\lambda_3 = \mathrm{diag}(0, 1, -1), \qquad \lambda_8 = \mathrm{diag}(-2, 1, 1)/\sqrt{3}. \tag{36}$$

Hence, the first three generators, $T_{\hat{a}} = \lambda_{\hat{a}}/2$, $\hat{a} = 1, 2, 3$, are just those of the PPG $SU(2)$ group (which acts on the last two components of the triplets). The $SU(2)$ group of the SM is generated by: $T_1^W = \lambda_6/2$, $T_2^W = \lambda_7/2$ and $T_3^W = -(\sqrt{3}\lambda_8 + \lambda_3)/4$. The diagonal generators in our previous notation are $D_1 = T_3$ and $D_2 = T_8$. However, this is an unusual basis but it is indicated since here the electric charge matrix of the fundamental irep., $(\mathbf{3}, 0)$, has the simple form $Q = -2T_3$. The others electric charge matrices of the ireps. $\rho = (\mathbf{n}_\rho, \hat{y}_\rho)$ of the above lists are given by the formula

$$Q^\rho = -2T_3^\rho + \hat{y}_\rho\,. \tag{37}$$

Furthermore, the condition (III) leads to the condition $g \sin\theta_W = e_0$ that fixes the value of g. Hereby we obtain [6]

$$\sin\theta = 2\sin\theta_W, \tag{38}$$

from which it results that $\sin^2\theta_W \leq 1/4$ [2,4]. The three Higgs triplets of our mHm., $\phi^{(1)}$, $\phi^{(2)}$ and $\phi^{(3)}$ transform according to the ireps. $(\mathbf{3}, 0)$, $(\mathbf{3}, 1)$ and $(\mathbf{3}, -1)$ respectively, like in Refs. [2,3]. Notice that in Ref. [2] (as well as in Ref. [4]) the Higgs sector contains, in addition, a Higgs sextet introduced to produce suitable lepton masses (with $m(\mu) \neq m(\tau)$). In our case this problem can be solved using our Yukawa couplings in unitary gauge.

The gauge fields are A_μ^0 and $A_\mu \in su(3)$ given by

$$A_\mu = \frac{1}{2}\left|\begin{matrix} -2A_\mu^8/\sqrt{3} & \sqrt{2}W_\mu & \sqrt{2}V_\mu \\ \sqrt{2}W_\mu^* & A_\mu^3 + A_\mu^8/\sqrt{3} & \sqrt{2}U_\mu \\ \sqrt{2}V_\mu^* & \sqrt{2}U_\mu^* & -A_\mu^3 + A_\mu^8/\sqrt{3} \end{matrix}\right|, \tag{39}$$

where W is the Weinberg charged boson while U and V are the new charged bosons due to this gauge symmetry. Now one can see that the bilepton U corresponds to the non-diagonal generators of the PPG $SU(2)$ group. It is interesting to note that it couples lepton Cooper-like currents of the form $g(\bar{l}\gamma^\mu(1-\gamma^5)l^c)/2\sqrt{2}$.

In order to find the physical neutral bosons one needs to calculate the gWt.. It is convenient to start with the following parametrization:

$$\eta^2 = (1 - {\eta_0}^2)\mathrm{diag}(\frac{1}{2}a - b, \frac{1}{2}a + b, 1 - a). \tag{40}$$

Then, the masses of the charged bosons are

$$m_W{}^2 = m^2 a, \quad m_V{}^2 = m^2(1 - \frac{1}{2}a - b), \quad m_U{}^2 = m^2(1 - \frac{1}{2}a + b), \quad (41)$$

where

$$m^2 = \frac{1}{4}g^2 < \phi >^2 (1 - \eta_0{}^2) = \frac{1}{2}(m_U^2 + m_V^2 + m_W^2) \quad (42)$$

and the squared mass matrix of the neutral bosons reads

$$M^2 = m^2 \begin{vmatrix} \frac{1}{\cos^2\theta}(1 - \frac{1}{2}a + b) & -\frac{1}{\sqrt{3}\cos\theta}(1 - \frac{3}{2}a - b)) \\ -\frac{1}{\sqrt{3}\cos\theta}(1 - \frac{3}{2}a - b) & \frac{1}{3} + \frac{1}{2}a - b \end{vmatrix}. \quad (43)$$

Observing now that this matrix has a singularity for $\sin^2\theta_W = 1/4$ we recover an important result of Ref. [2] namely that $\sin^2\theta_W$ must remain less than $1/4$.

The next step is to calculate the eigenvalues of (43) which are the squared masses of the Z-bosons. Using these values, we find that the condition (IV) is satisfied if and only if

$$b = -\frac{3}{2}a\tan^2\theta_W. \quad (44)$$

This leads to the following boson mass spectrum

$$\begin{aligned} m_W{}^2 &= m_{Z^2}{}^2\cos^2\theta_W = m^2 a, \\ m_{Z^1}{}^2 &= \frac{m^2}{1 - 4\sin^2\theta_W}\left(\frac{4}{3}\cos^2\theta_W - a\left(1 - (1 - 4\sin^2\theta_W)\tan^2\theta_W\right)\right), \\ m_V{}^2 &= m^2\left(1 - \frac{a}{2}\frac{1 - 4\sin^2\theta_W}{\cos^2\theta_W}\right), \\ m_U{}^2 &= m^2\left(1 - \frac{a}{2}\frac{1 + 2\sin^2\theta_W}{\cos^2\theta_W}\right), \end{aligned} \quad (45)$$

depending on the *arbitrary* parameter $a \in (0, 1]$. Therefore, $Z^2 = Z$ is the Weinberg neutral boson while $Z^1 = Z'$ is the new one of this model.

The gWt. of this model can be calculated as [6]

$$\begin{aligned} A^0_\mu &= A^{em}_\mu \cos\theta + (Z'_\mu \cos\theta' - Z_\mu \sin\theta')\sin\theta, \\ A^3_\mu &= -A^{em}_\mu \sin\theta + (Z'_\mu \cos\theta' - Z_\mu \sin\theta')\cos\theta, \\ A^8_\mu &= Z'_\mu \sin\theta' + Z_\mu \cos\theta', \end{aligned} \quad (46)$$

where θ is given by Eq. (38) while

$$\tan 2\theta' = -\sqrt{3}\frac{\sqrt{1 - 4\sin^2\theta_W}}{1 + 2\sin^2\theta_W}. \quad (47)$$

Thus we have the surprise to find that the angle θ' and, therefore, the gWt. do not depend on a when Eq. (44) is accomplished.

The gWt. allows us to calculate the coupling coefficients of the fundamental multiplet, L_l. Its neutral charges corresponding to the Weinberg boson Z are just those predicted by the SM while the matrix of those corresponding to the other neutral boson, Z', is

$$Q(Z') = -\frac{1}{\sqrt{3}\sin 2\theta_W}\sqrt{1-4\sin^2\theta_W}\,\mathrm{diag}(1,1,-2). \tag{48}$$

The neutral charge matrices of the others multiplets, L^ρ, can be derived as [6]

$$\begin{aligned} Q^\rho(Z) &= -\frac{1}{\sin 2\theta_W}\left((1-4\sin^2\theta_W)T_3^\rho + \sqrt{3}T_8^\rho + 2\hat{y}_\rho \sin^2\theta_W\right), \\ Q^\rho(Z') &= -\frac{\sqrt{1-4\sin^2\theta_W}}{\sin 2\theta_W}\left(\sqrt{3}T_3^\rho - T_8^\rho + 2\sqrt{3}\hat{y}_\rho \frac{\sin^2\theta_W}{1-4\sin^2\theta_W}\right). \end{aligned} \tag{49}$$

Moreover, we must specify that in our approach the coupling coefficients of the vector and axial neutral currents of a fermion f are

$$Q(Z,f)_V = \frac{1}{2}(Q(Z,f) - Q(Z,f^c)), \tag{50}$$

$$Q(Z,f)_A = -\frac{1}{2}(Q(Z,f) + Q(Z,f^c)), \tag{51}$$

because of the pl. form of the spinor sector.

Under mHm. the quark masses have to be generated by the usual Yukawa interactions involving only the fields $\phi^{(i)}$ like in Refs. [2,3] but to produce the lepton masses we need to use our Yukawa couplings in unitary gauge. Therefore, we introduce the 2-rank symmetric tensor, $\chi_l = \phi^{-1}G_l(\phi^{(2)} \otimes \phi^{(3)} + \phi^{(3)} \otimes \phi^{(2)})$, which plays the same role as the Higgs sextet of Refs. [2,4]. The coupling terms $\overline{L_l}\chi_l L_l^c + h.c$ $(l = e, \mu, \tau)$ give rise to the lepton masses while the neutrinos remain massless. However, if we wish massive neutrinos we could use the tensor $\chi_l' = \phi^{-1}G_l'\phi^{(1)} \otimes \phi^{(1)}$.

5. Concluding remarks

Hence, the model is completely solved without any kind of approximations. The boson mass spectrum deduced here depends on the real parameter $a \in (0,1]$ which represents a new and important fit parameter. We observe that for $a \simeq 0$ the new gauge bosons are very massive as in Refs. [2–4] while for $a \simeq 1$ their masses could be *smaller* than m_W.

What is remarkable here is the presence of a *critical point* of the boson mass spectrum where different bosons with the same charge get the same mass. Indeed, for

$$a = a_c \equiv \frac{2}{3}\frac{\cos^2\theta_W}{1-2\sin^2\theta_W} \tag{52}$$

we have *simultaneously* $m_{Z'} = m_Z$ and $m_V = m_W$ while

$$m_U = m_W \frac{\sqrt{1-4\sin^2\theta_W}}{\cos\theta_W}. \tag{53}$$

Numerically speaking, if we take $m_W = 80.5\, GeV/c^2$, $m_Z = 92.5\, GeV/c^2$, we find $\sin^2\theta_W = 0.24263$, $a_c = 0.980908592$ and a very small bilepton mass,

$$m_U = 15,882\, GeV/c^2. \tag{54}$$

In our opinion, it is less probable that the above mass spectrum be in accordance with the physical reality. However, one can not eliminate this possibility *a priori*, without to have direct experimental arguments. We hope that the future data given by LHC should clear up this problem soon.

Finally, we observe that the values of the neutral charges are independent on the parameter a since these do not depend on the boson masses, just as in the SM. On the other hand, these values seem to be similar to those obtained in Ref. [4] and, therefore, they bring nothing new concerning the problem of the suppression of the flavor-changing neutral currents [15] of this model [3,4,16].

References

1. T.-P. Cheng, L.-F. Li, *Gauge Theory of Elementary Particle Physics*, Clarendon Press, Oxford, 1984; D. Balin, A. Love, *Introduction to Gauge Field Theory*, Adam-Highler, 1986.
2. P. H. Frampton, Phys. Rev. Lett. **69**, 2889 (1992)
3. F. Pisano, V. Pleitez, Phys. Rev. D **46**, 410 (1992)
4. D. Ng, Phys. Rev. D **49**, 4805 (1994)
5. R. Foot, H. N. Long, T. A. Tran, Phys. Rev. D **50**, R34 (1994)
6. I. I. Cotăescu, Int. J. Mod. Phys. A, **12**, 1483 (1997)
7. R. Gilmore, *Lie Groups, Lie Algebras and some of their Applications*, Wiley-Interscience, New York, 1974; A. O. Barut, R. Raczka, *Theory of Group Representations and Applications*, PWN, Warszawa, 1977.
8. R. Slansky, Phys. Rep., **79C**, 1 (1981).
9. A. J. MacFarlane. A. Sudbery, P. H. Weisz, Commun. Math. Phys., **11**, 77 (1968).

10. For an extended original bibliography see: E. S. Abers, B. W. Lee, Phys. Lett. **9C**, 1 (1973).
11. K. Huang, *Quarks, Leptons & Gauge Fields*, World Sci., Singapore, 1982;
12. C. Itzykson and J. B. Zuber, *Quantum Field Theory,* Mc.Graw–Hill Inc., New York, 1980
13. J. C. Collins, *Renormalization*, Cambridge University Press, Cambridge, 1984.
14. W. Pauli, Nuovo Cimento, **6**, 204 (1957); D. Pürsey, Nuovo Cimento, **6**, 266 (1957); F. Gürsey, Nuovo Cimento, **7**,411 (1958).
15. S. L. Glashow, J. Iliopoulos, L. Maiani, Phys. Rev. D **2**, 1285 (1970); S. L. Glashow, S. Weinberg, Phys. Rev. D **15**, 1958 (1977)
16. J. C. Montero, F. Pisano, V. Pleitez, Phys. Rev. D **47**, 2918 (1993); J. T. Liu, Phys. Rev. D **50**, 542 (1994)

V. THEORETICAL AND EXPERIMENTAL RESULTS ON THE PROTON STRUCTURE

The Physics of Deep Inelastic Scattering at HERA

C.DIACONU*

Centre de Physique des Particules de Marseille and
Deutsches Elektronen Synchrotron, Notkestrasse 85, 22607 Hamburg, Germany
** E-mail: diaconu@cppm.in2p3.fr*

In this paper an introduction to the physics of deep inelastic scattering is given together with an account of recent results obtained in electron– or positron–proton collisions at the HERA collider.

Keywords: Deep inelastic scattering; proton structure; HERA collider ; QCD.

1. Introduction

In the beginning of the XX century, atoms were known to radiate and to absorb radiation which indicated that they were not elementary but composed particles. The way to identify and measure the composition of the atoms was pioneered in 1911 by Geiger, Marsden and Rutherford (GMR), who opened a new era in the study of the matter structure: the scattering experiments. GMR observed that some of the α particles colliding with a gold foil were backscattered. This is unexpected if gold atoms were a continous charge distribution within a finite volume. In Rutherford's words: "it was as if you fired a 15-inch shell at a piece of tissue paper and it came back and hit you". This could only have happened if the α particle had hit a concentrated field, like a point-like massive charge centre identified as the atomic nucleus.

The resolution with which the "target" can be investigated by a pointlike incoming particle determined by the uncertainity principle: the higher the transfer momentum (denoted by Q), the smaller the details that can be "flashed" and imprinted in the angular distribution of the scattered (point-like) particle. The resolution can be expressed as $\delta = \dfrac{0.2 \cdot 10^{-15}\mathrm{m}}{Q/\mathrm{GeV}}$.

The search for further substructure levels continued with the scattering of leptons on light nuclei (H,D) in order to investigate the structure of protons and neutrons, the main components of the nuclear matter. Quarks and gluons were discovered using electrons of higher and higher energies, which were ultimately able resolve the nucleon's structure in the so called deep inelastic scattering (DIS). In this paper, an introduction to the physics of DIS is given using as example the first experimental results that established the basic model of the nucleon structure [1]. A qualitative change occured with the HERA project, an electron–proton collider at very high energy. The physics and a selection of HERA results will be also described[a].

2. The observables of lepton–hadron scattering

The framework in which lepton hadron scattering is described is shown in figure 1. The scattering can occur via the exchange of γ or Z bosons (neutral currents NC) or via W bosons (charged currents CC). In the latter case, a neutrino is expected in the final state. The incoming electron (with

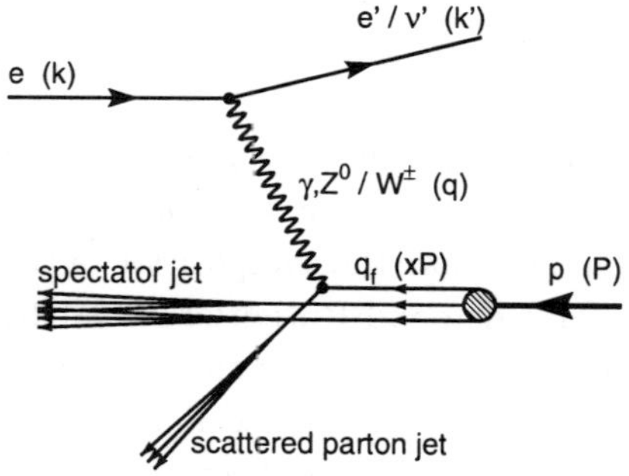

$$Q^2 = -q^2 = -(k' - k)^2$$
$$y = \frac{P \cdot q}{P \cdot k}$$
$$x = \frac{Q^2}{2P \cdot q} \quad \text{(Bjorken)}$$
$$\nu = \frac{P \cdot q}{M_p}$$
$$W^2 = (P')^2 = (P + q)^2 (= M_X^2)$$
$$s = (P + k)^2 (\text{fixed})$$

Fig. 1. Lepton–hadron scattering: an exchange of a boson in the t–channel.

a four-momentum k) scatters off the proton (P) to a final state electron with four–momentum k' via a virtual photon γ^* or a weak boson with a virtuality Q^2 exchanged in the t-channel. In inelastic scattering, one can assume that only a part of the proton ("parton") enters the reaction. The Bjorken variable x is associated with the fraction of the momentum of the proton carried by the struck parton. The total centre-of-mass energy is given by $\sqrt{s}$ and the energy of the $\gamma^* p$ system is given by W, which is equivalent

[a] This paper is based on two introductory lectures presented at the school "Collective Motion and Phase Transitions in Nuclear Systems", 28 August–9 September, 2006, Predeal, Romania.

to the total mass of the hadronic system in the final state M_X. In the case of elastic scattering $M_X = M_p$ and from the M_X expression it follows that $Q^2 = 2Pq$ and $x = 1$ (the whole proton interacts). Only two variables are independent, since the reaction is completely defined by the scattering angle and by the electron-parton centre-of-mass energy. The variable ν has a simple meaning in the proton rest frame, as the energy lost by the electron during the scattering $\nu = E_e - E'_e$, while y represents the fractional energy loss $y = \frac{E_e - E'_e}{E_e}$. Q^2 can be expressed as a function of the electron energy and scattering angle, $Q^2 = 4E_e E'_e \cos^2 \frac{\theta}{2}$. From these relations, it is obvious that the DIS kinematics can be calculated from the measurement of the scattered electron only. The measurement of the hadrons in the final state, if available, can be exploited as an extra constraint in NC scattering. It is the only way to reconstruct the CC kinematics, since the outgoing neutrino is not measured.

3. Elastic lepton–hadron scattering

The parameterisation of the elastic differential cross section as a function of Q^2 depends on the electric (G_E) and magnetic (G_M) form factors of the proton and can be written as:

$$\frac{d\sigma}{dQ^2} = \frac{4\pi\alpha^2}{Q^4}\left[\frac{G_E^2 + \tau G_M^2}{1+\tau}\left(1 - y - \frac{M^2y^2}{Q^2}\right) + \frac{y^2}{2}G_M^2\right] \qquad \tau = Q^2/4M^2 \tag{1}$$

It is useful to remember that the scattering amplitude A_ρ of a particle on a charge with a finite charge distribution $\rho(\vec{r})$ can be factorized as $A_\rho = A_0F(q)$, where A_0 is the amplitude of the scattering off a point-like charge $\int \rho$ and $F(q)$ is a form-factor that depends on the momentum transfer q. It can be demonstrated that the form–factor is the Fourier Transform (FT) of the charge distribution. Therefore, the scattering cross–section can be related to the charge distribution inside the target.

An important step forward in the understanding of the structure of matter has been made by Hofstadter et al. [2] in an elastic *ep* (and later *ed*) experiment using electrons with energies of up to 246 MeV to investigate the charge distribution inside the proton. The result is shown in figure 2 (left). The measured cross section at large scattering angles is below the prediction for scattering off a point-like charge. Using the form–factor argument presented above, the typical size of the proton charge distribution is found to be around 10^{-15} m. This observation implied that the proton is not point-like. Its structure can be investigated using electrons with higher energy

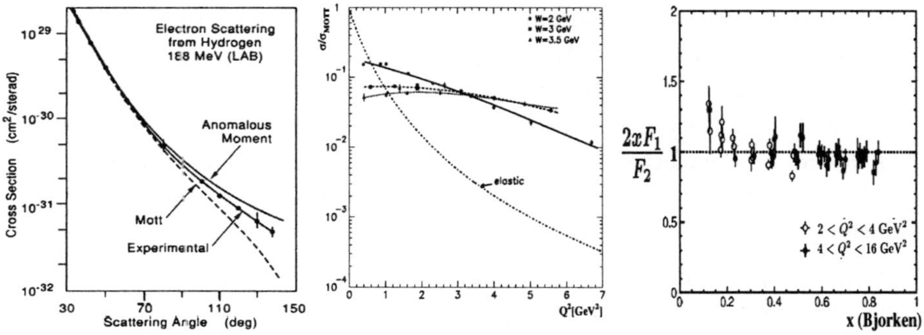

Fig. 2. Left: The measurement of the cross section as a function of diffusion angle (Hofstader). Center: First result on deep inelastic cross section measurement. Right: The illustration of the Callan–Gross relation (from [5]).

such that inelastic reactions are induced.

4. Deep inelastic scattering and the naive parton model

In addition to Q^2, one more variable (x) is needed to describe inelastic ep scattering at a given beam energy, since only a part of the "target" is involved. In the double differential cross section, the elastic form factors are replaced by the structure functions $F_1(x, Q^2)$ and $F_2(x, Q^2)$.

$$\frac{d\sigma}{dQ^2 dx} = \frac{4\pi\alpha^2}{Q^4}\left[\frac{F_2}{x}\left(1 - y - \frac{M^2y^2}{Q^2}\right) + \frac{y^2}{2}\frac{2xF_1}{x}\right] \tag{2}$$

The structure functions represent a generalisation of the form factors, however, the simple interpretation as the FT of the charge distribution is not possible, due to the extra variable x. Nevertheless, the structure functions provide however direct information on the proton components. This can be demonstrated within the so-called "parton model". The assumption is that the proton, proven to have a finite size of around 10^{-15} m, is composed of point-like spin 1/2 particles, the "partons". The formalism is expressed in the "infinite momentum" frame, in which the motion of the parton within the proton is much slower than the time of interaction. The lepton–proton interaction is described as a coherent scattering of the lepton on a sum of independent ("frozen") partons. If we consider the lepton–parton cross section

$$\frac{d\sigma}{dQ^2} = \frac{2\pi\alpha^2}{Q^4} e_q^2 \left[1 + (1-y)^2\right] , \tag{3}$$

and assuming the proton is a sum of partons of charge e_q and momentum fraction distribution $q(x)$, the lepton–proton cross section can be written as:

$$\frac{d\sigma}{dxdQ^2} = \frac{2\pi\alpha^2}{Q^4}\left[1 + (1-y)^2\right]\sum_q e_q^2 q(x)\,. \tag{4}$$

Comparing with the equation 2 results in the following relation:

$$F_2 = x\sum_q e_q^2 q(x) \tag{5}$$

In this formula one can see explicitely that to first approximation the structure function F_2 does not depend on Q^2. This property, called "Bjorken scaling", occur since a point-like parton is seen in the same way by all wavelengths. In addition, a direct relationship is deduced between F_2 and F_1, called Callan-Gross relation [3]: $F_2 = 2xF_1$. This relation is typical for spin 1/2 constituents (scalar constituents would lead to $F_1 = 0$, for example).

The high energy linac built at SLAC in the sixties using the new klystron technology allowed collisions of electrons of up to 20 GeV energy with protons from a liquid hydrogen target. The scattered electrons were measured in a two arm spectrometer at variable scattering angles and the cross section was measured in a range of Q^2 and x, reconstructed as explained in section 2. The measured structure function [6] F_2 was found to be rather flat in Q^2 in contradiction with the elastic scattering case, but in agreement with Bjorken scaling and the naive quark parton model (figure 2 (center)). Subsequently, the Callan–Gross relation $F_2 = 2xF_1$, valid for spin 1/2 constituents, was confirmed, as shown in figure 2 (right). The measurements of the nucleon structure continued with lepton beams of various types and energies. These fixed target experiments are listed in table 1.

Although extremely simple, the parton model leads to powerful predictions, in good agreement with the first experimental observations. Nevertheless, the model was to be improved using the quantum field theory of strong interactions that emerged in the 1970's.

5. Quantum Chromodynamics and the improved quark–parton model

At the beginning of the XX century, the proton, the photon and the electron were the only known "elementary" particles. By 1960 already, more than 60 particles were known, identified in photo-emulsion plates exposed

Table 1. Fixed target experiments prior to HERA collider.

Experiment	Year	Reaction	Process	Beam Energy
SLAC-MIT	1968	ep, ed	NC	4.5 – 20 GeV
CDHS,CHARM	< 1984	$\nu_\mu Fe$	CC	< 260 GeV
FMMF	< 1988	$\nu_\mu N$	CC	< 500 GeV
CCFR	1979 – 1988	$\nu_\mu Fe$	CC	< 600 GeV
BCDMS	1981 – 1985	μp, μd	NC	100 – 280 GeV
EMC	< 1983	μp, μd	NC	< 325 GeV
NMC	1986 – 1989	μp, μd	NC	90 – 280 GeV
E665	1987 – 1992	μp, μd	NC	90 – 470 GeV
SLAC-MIT	1996 – 1997	$\nu_\mu Fe$	CC/NC	< 600 GeV

to cosmic rays or by the newly available high energy accelerated particle beams. Most of these particles were classified as hadrons, i.e. particles sensitive to the strong (nuclear) force. This multiplicity required some "order" and pointed towards a substructure, which was explained via the "quark" model. In this model, the known hadrons are combinations of spin 1/2 components, the "quarks". The first proposed quarks were "up" u, "down" d and "strange" s as used in the 1960's for a classification based on the symmetry group $SU(3)_{\text{flavour}}$. Later more quarks were discovered: "charm" c (1974), "beauty" b (1977) and "top" t (1994). The electric quark charge is fractionary: +2/3 for u, c, t and -1/3 for d, s, b in units of e. The proton is a (uud) combination while the neutron is attributed a (udd) content. To date, quarks have not been detected as free particles, and this experimental observation has been encapsulated into a theorem related to a new quantum number, "colour". The observable particles are always colour-less ("white") states, since "colour"-ed objects (like the quarks) cannot exist freely. The theory of the strong interactions is in fact a gauge field theory of the colour quantum number, Quantum Chromodynamics (QCD) [7].

Quarks are bound inside the hadrons and interact via *gluons*, the mediating bosons of the strong force. Proton and neutrons can therefore be considered as a collection of quarks of two types: valence quarks, the components determining the nucleon identity (from $SU(3)_{\text{flavour}}$) and "sea" quarks and antiquarks related to the QCD vacuum, i.e. gluon fluctuations $g \to q\bar{q}$.

In the naive quark–parton model, the structure functions can be written in terms of quark distributons, expanding formula 5: $F_2^{p,n} = \frac{4}{9}x(u^{p,n} + \bar{u}^{p,n}) + \frac{1}{9}x(d^{p,n} + \bar{d}^{p,n})$. By isospin invariance one can identify $u(x) = u^p(x) = d^n(x)$ $d(x) = d^p(x) = u^n(x)$ (and similarly for anti-quarks). Since F_2 can be measured for both the proton and the neutron, the integral over

x can be determined experimentally. The integral contains the contribution of individual quarks to the proton momentum (f_u, f_d):

$$\int_0^1 F_s^p = \tfrac{4}{9}\int_0^1 x(u+\bar{u})dx + \tfrac{1}{9}\int_0^1 x(d+\bar{d})dx = \tfrac{4}{9}f_u + \tfrac{1}{9}f_d = 0.18^{(expt.)}$$

$$\int_0^1 F_s^n = \tfrac{1}{9}\int_0^1 x(u+\bar{u})dx + \tfrac{4}{9}\int_0^1 x(d+\bar{d})dx = \tfrac{1}{9}f_u + \tfrac{4}{9}f_d = 0.12^{(expt.)}$$

If the system is solved, one can calculate $f_u + f_d = 0.5$ which means that only one half of the proton momentum is carried by quarks and antiquarks. The missing half had been attributed to the gluons. This is a puzzle but also a challenge for QCD, since a very large fraction of the mass of the visible (baryonic) universe seems to be build by the carriers of the strong force. In addition, the parton model contains another puzzle: quarks are confined (i.e. tightly bound) inside the proton, but at the same time they behave quasi–free during a DIS interaction. In QCD, gluon emission is proportional to the strong coupling constant α_s, which is predicted to increase with increasing distance (decreasing Q^2):

$$\alpha_s(Q_2) = \frac{\alpha_s(\mu^2)}{1+\beta(\alpha_s(\mu^2))\log(Q^2/\mu^2)} \tag{6}$$

where μ is an arbitrary scale at which the reference coupling is defined, while Q^2 is here the scale at which the strong interaction takes place and β is a negative coefficient [4]. This property, also called "asymptotic freedom", explains both why quarks are confined and the approximate correctness of the parton model, since at small distances, inside the proton, quarks interact softly and are indeed quasi free during the interaction.

QCD can be viewed as a generalisation of the simple parton model. The assumption is that quark and gluons distributions probed in DIS scattering are subject to QCD reactions. Possible sub-processes are shown in figure 3 and their influence on the quark and gluon distributions functions are calculable in QCD via the Dokshitzer-Gribov-Lipatov-Altarelli-Parisi (DGLAP) equations [8]. These equations describe the coupled evolution of the quark and gluon densities as a function of the virtuality $t = Q^2$:

$$t\frac{\partial}{\partial t}\begin{pmatrix} q_i(x,t) \\ g(x,t) \end{pmatrix} = \frac{\alpha_s(t)}{2\pi}\sum_{q_j,\bar{q}_j}\int_x^1 \frac{d\xi}{\xi}$$

$$\cdot \begin{pmatrix} P_{q_iq_j}\left(\frac{x}{\xi},\alpha_s(t)\right) & P_{q_ig}\left(\frac{x}{\xi},\alpha_s(t)\right) \\ P_{gq_j}\left(\frac{x}{\xi},\alpha_s(t)\right) & P_{gg}\left(\frac{x}{\xi},\alpha_s(t)\right) \end{pmatrix} \begin{pmatrix} q_j(\xi,t) \\ g(\xi,t) \end{pmatrix} \tag{7}$$

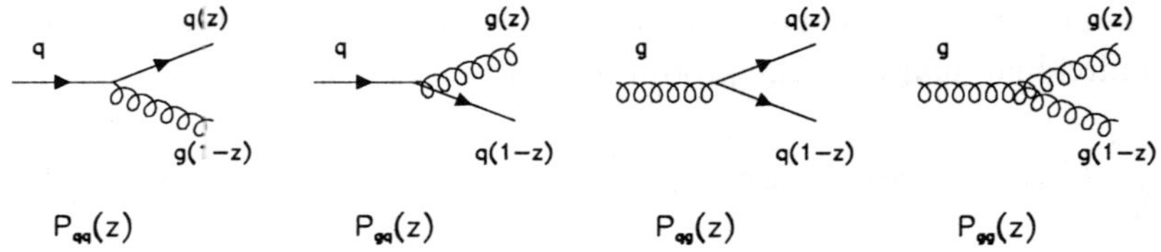

Fig. 3. The gluon induced subprocesses for parton density evolution and the associated kernels in the DGLAP equations.

These equations describe the "build-up" of observable quark and gluon densities from the possible deceleration of partons with higher x via (for instance) gluon radiation or splitting, described by the kernel function $P_{qq,qg,gg}$. These internal appearances and recombinations lead to parton distributions (and therefore structure functions) that depend not only on x but also on Q^2, violating the Bjorken scale invariance of the naive quark–parton model. The scaling violations are therefore an effect of the local strong interaction at very short distances. The determination of the scaling violations from cross section measurements over large domains of x and Q^2 therefore test the QCD validity. These measurements also allow the strong coupling constant and the gluon distribution inside the proton to be extracted.

6. The HERA project

The idea for a large electron–proton collider to mark a new step in the studies for proton structure was promoted in the seventies [9]. The HERA collider project started in 1985 and produced the first electron–proton collisions in 1992. A scheme of the accelerator complex is shown in figure 4 (left). It is composed of two accelerators designed to store and collide counter rotating electrons (e^-), or positrons (e^+), with an energy of 27.5 GeV and protons with an energy of 920 GeV. The centre–of–mass energy is 320 GeV, equivalent to a hypothetical fixed target experiment with a lepton beam energy of $E \simeq 50$ TeV. The kinematic (x, Q^2) plane accessible at HERA is shown in figure 4 (right) with a Q^2 domain up to about 50000 GeV^2 and x down to 10^{-5}.

The HERA ring serves two collider detectors H1 and ZEUS (shown in figure 5). They are built as hermetic (4π) multi-purpose detectors equipped with inner trackers able to measure charged particle momenta and calorimeters completing the measurement of the energy flow in events from electron–proton collisions. Two further experiments use $e^{\pm}$ or p beams for fixed

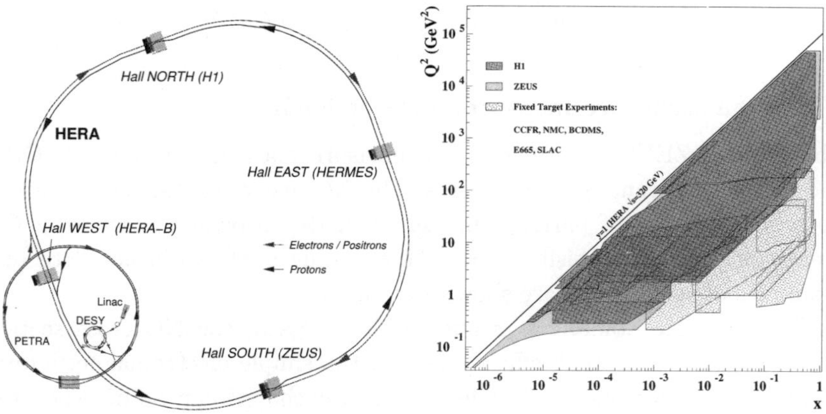

Fig. 4. The HERA complex: the scheme including the injectors (left) and the kinematic plane accessible at HERA compared to former fixed target experiments (right).

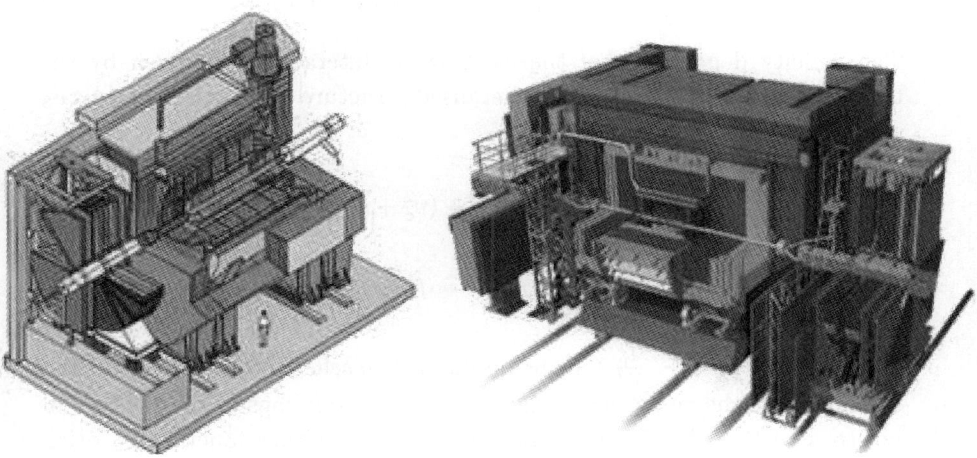

Fig. 5. The two collider mode detectors at HERA: H1 (left) and ZEUS (right)

target studies: HERMES is dedicated to the study of polarised $e^{\pm}p(N)$ collisions and (until 2003) HERA-B was built to study beauty production in hadronic collisions. Since 2003, the $e^{\pm}$ beam is longitudinally polarised with an average polarisation in collision mode of $P_{e^{\pm}} = 30 - 40\%$. HERA collisions will continue until July 2007. An integrated luminosity of 300 pb^{-1} (200 pb^{-1}) is about to be collected in e^+p (e^-p) by each of the two collider mode experiments, H1 and ZEUS. HERA complements the other high energy colliders Tevatron (Fermilab, Chicago, $p\bar{p}$, $\sqrt{s} = 1960$ GeV)and LEP

(CERN, Geneva, e^+e^-, until 2000, $\sqrt{s}$ up to 209 GeV).

7. Proton structure measurements at HERA

The H1 and ZEUS experiments can measure both neutral current (NC) and charged current (CC) processes. The NC events contain a prominent electron and a jet of particles measured in the calorimeter, while in CC events only the jet is visible since the outgoing neutrino in not detected. Examples of such events are shown in figure 6.

Since a large domain in x and Q^2 is accessed, the NC cross section become sensitive to weak effects, beyond the simple electromagnetic form parameterised in equation 2. The Z^0 boson exchange can be incorporated into the so-called generalised structure functions. The cross section is parameterised as following:

$$\frac{\mathrm{d}^2\sigma^{\pm}_{\mathrm{NC}}}{\mathrm{d}x\mathrm{d}Q^2} = \frac{2\pi\alpha^2}{xQ^4}(Y_+\tilde{F}_2 \mp Y_- x\tilde{F}_3 - y^2\tilde{F}_L) \ , \tag{8}$$

The helicity dependence of the electroweak interactions is given by the terms $Y_\pm = 1 \pm (1-y^2)$. The generalised structure functions $\tilde{F}_2$ and $x\tilde{F}_3$ can be further decomposed as [10]

$$\tilde{F}_2 \equiv F_2 - v_e \frac{\kappa Q^2}{(Q^2+M_Z^2)} F_2^{\gamma Z} + (v_e^2+a_e^2)\left(\frac{\kappa Q^2}{Q^2+M_Z^2}\right)^2 F_2^Z \, , \tag{9}$$

$$x\tilde{F}_3 \equiv \quad - a_e \frac{\kappa Q^2}{(Q^2+M_Z^2)} xF_3^{\gamma Z} + (2v_e a_e)\left(\frac{\kappa Q^2}{Q^2+M_Z^2}\right)^2 xF_3^Z \, , \tag{10}$$

with $\kappa^{-1} = 4\frac{M_W^2}{M_Z^2}(1-\frac{M_W^2}{M_Z^2})$ in the on-mass-shell scheme [11]. The quantities v_e and a_e are the vector and axial-vector weak couplings of the electron or positron to the Z^0 [11]. The electromagnetic structure function F_2 originates from photon exchange only and dominates in most of the accessible phase space. The functions F_2^Z and xF_3^Z are the contributions to $\tilde{F}_2$ and $x\tilde{F}_3$ from Z^0 exchange and the functions $F_2^{\gamma Z}$ and $xF_3^{\gamma Z}$ are the contributions from γZ interference. These contributions are significant only at high Q^2. For longitudinally unpolarised lepton beams the $\tilde{F}_2$ contribution is the same for e^- and for e^+ scattering, while the $x\tilde{F}_3$ contribution changes sign as can be seen in eq. 8. The longitudinal structure function $\tilde{F}_L$ may be decomposed in a manner similar to $\tilde{F}_2$. Its contribution is significant only at high y.

In the quark parton model the structure functions F_2, $F_2^{\gamma Z}$ and F_2^Z are related to the sum of the quark and anti-quark momentum distributions,

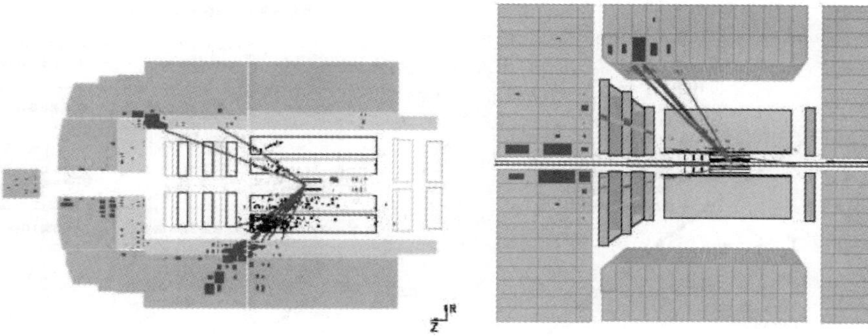

Fig. 6. Event displays of a neutral current scattering event measured by H1 (left) and charged current scattering event measured by ZEUS (right).

$xq(x, Q^2)$ and $x\overline{q}(x, Q^2)$:

$$[F_2, F_2^{\gamma Z}, F_2^Z] = x \sum_q [e_q^2, 2e_q v_q, v_q^2 + a_q^2]\{q + \overline{q}\} \tag{11}$$

and the structure functions $xF_3^{\gamma Z}$ and xF_3^Z to the difference, which determines the valence quark distributions, $xq_v(x, Q^2)$,

$$[xF_3^{\gamma Z}, xF_3^Z] = 2x \sum_q [e_q a_q, v_q a_q]\{q - \overline{q}\} = 2x \sum_{q=u,d} [e_q a_q, v_q a_q] q_v \,. \tag{12}$$

In equations 11 and 12, v_q and a_q are the vector and axial-vector weak coupling constants of the quarks to the Z^0, respectively.

The charged current (CC) interactions, $e^\pm p \to \overset{(-)}{\nu}_e X$, are mediated by the exchange of a W boson in the t channel. The cross section is parameterised as:

$$\frac{d^2\sigma^{CC}(e^\pm p)}{dx dQ^2} = \frac{G_F^2}{2\pi x}\left[\frac{M_W^2}{M_W^2 + Q^2}\right]^2 \tilde{\sigma}_{CC}^\pm(x, Q^2)\,,$$

$$\text{with} \quad \tilde{\sigma}_{CC}^\pm(x, Q^2) = \frac{1}{2}\left[Y_+ W_2^\pm(x, Q^2) \mp Y_- x W_3^\pm(x, Q^2) - y^2 W_L^\pm(x, Q^2)\right]$$

$\tilde{\sigma}$ is the reduced cross section, G_F is the Fermi constant, M_W, the mass of the W boson, and W_2, xW_3 and W_L, CC structure functions. In the quark parton model (QPM), the structure functions $W_2^\pm$ and $xW_3^\pm$ may be interpreted as lepton charge dependent sums and differences of quark and anti-quark distributions:

$$W_2^+ = x(\overline{U}+D), \quad xW_3^+ = x(D-\overline{U}), \quad W_2^- = x(U+\overline{D}), \quad xW_3^- = x(U-\overline{D})\,,$$

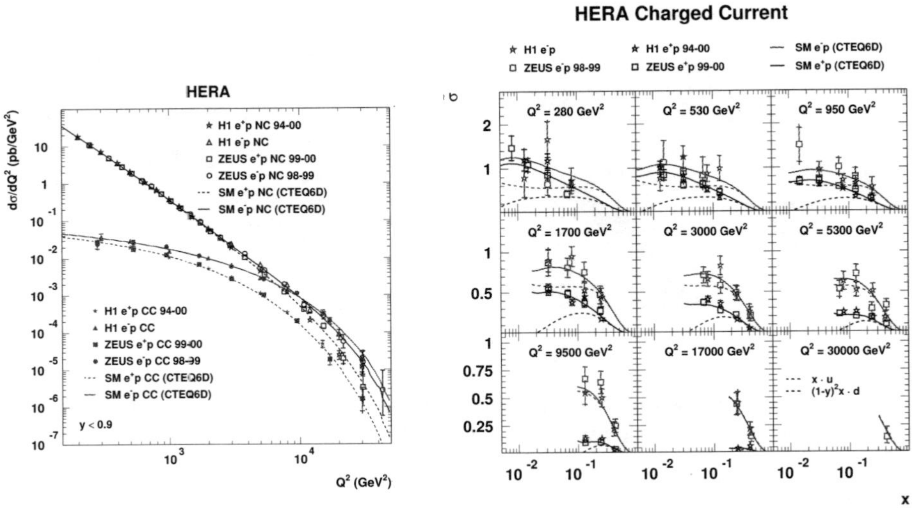

Fig. 7. Left: the charged current and neutral current cross section as a function of Q^2 measured in $e^{\pm}p$ collisions at HERA. Right: the charged current reduced cross section $\tilde{\sigma}_{CC}$ as a function of x for various Q^2 values measured in electron– and positron–proton collisions.

whereas $W_L^{\pm} = 0$. The terms xU, xD, $x\overline{U}$ and $x\overline{D}$ are defined as the sum of up-type, of down-type and of their anti-quark-type distributions, i.e. below the b quark mass threshold: $xU = x(u+c)$, $xD = x(d+s)$, $x\overline{U} = x(\overline{u}+\overline{c})$, $x\overline{D} = x(\overline{d}+\overline{s})$.

The differential NC and CC cross sections as a function of Q^2 are shown in figure 7 (left) for $e^{\pm}p$ collisions. At low Q^2 the NC cross section, driven by the electromagnetic interaction, is two orders of magnitude larger than the CC cross section which correspond to a pure weak interaction. At large $Q^2 \sim M^2_{W,Z}$ the two cross sections are similar. The largest Q^2 measurement corresponds to a resolution of $\delta \simeq 10^{-18}$ m, i.e. 1/1000 the proton size. The agreement between the measurement and the prediction based on QCD improved parton model suggests no evidence for quark substructure.

The double differential reduced cross section $\tilde{\sigma}_{CC}(x, Q^2)$ is shown in figure 7 (right). The CC processes are sensitive to individual quark flavours, which is especially visible at large Q^2: the e^+p collisions probe the $d(x)$ quark distribution, while e^-p are more sensitive to the $u(x)$. This is a very useful feature of the CC processes compared to the NC, where the flavour separation is weaker.

8. Structure functions measurements: F_2, F_L and xF_3

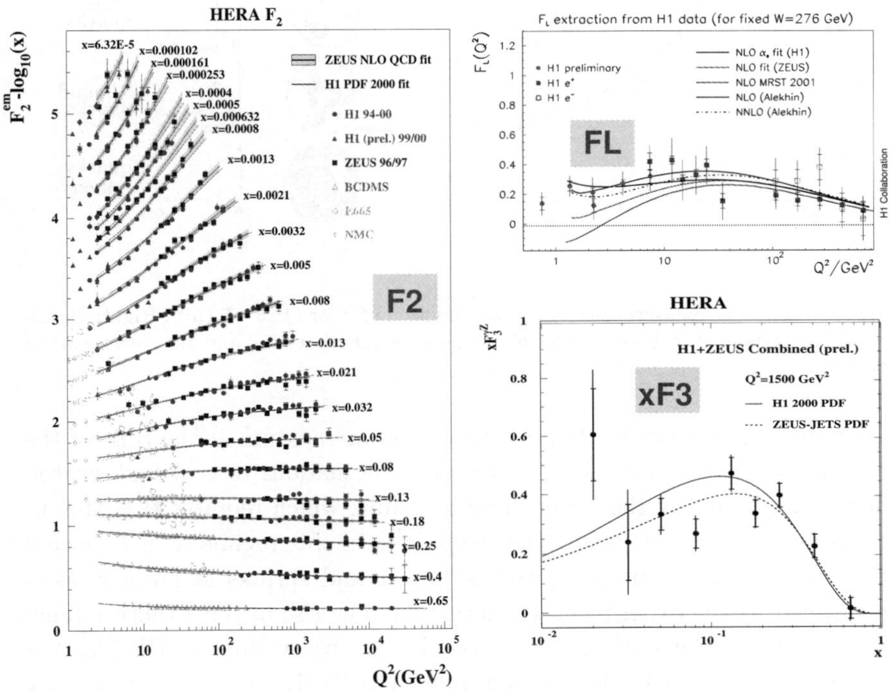

Fig. 8. The determination of the structure functions F_2, F_L and xF_3 from HERA measurements.

The NC cross section is dominated over a large domain by the F_2 contributions, defined in equation 8. The measurement of the NC cross section at HERA can therefore be translated into an F_2 measurement, which is shown in figure 8 together with the previous measurements performed at fixed target experiments. One can observe the Bjorken scaling in the region at high $x \simeq 0.1 - 0.2$, but obvious scaling violation at lower x. This may be understood in terms of DGLAP equations as a contribution driven by the gluon $\partial F_2(x, Q^2)/\partial \ln(Q^2) \approx (10\,\alpha_s(Q^2)/27\pi)\, xg(x, Q^2)$.

From the measurements at fixed Q^2 one can observe a steep increase of F_2 towards low x, as shown in figure 9 (left). The region at low x is populated by quarks which have undergone a hard or multiple gluon radi-

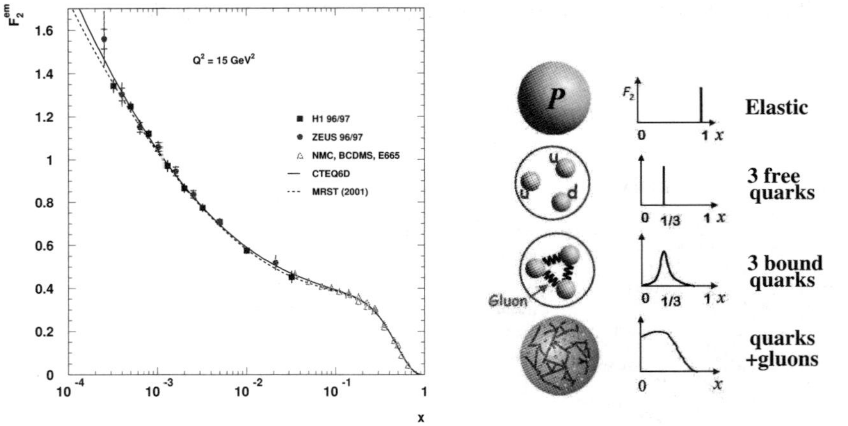

Fig. 9. Left: The measurement of F_2 as a function of x for $Q^2 = 15$ GeV. Right: Sketch of the correspondence between the F_2 shape as a function of x and the quark-parton model.

ation and carry a low fraction of the proton momentum at the time of the interaction. The observation of such large fluctuations to very high parton density is driven by the uncertainty principle, which requires that the interaction time be very short (i.e. high Q^2). In this regime, it is expected that the structure function grows at low x and shrinks at large x, as is confirmed by the experimental observation. The rise of the structure functions at low x is one of the most spectacular observations at HERA. It is predicted in the double leading log limit of QCD [12]. It can be intuitively understood in terms of gluon driven parton production at low x, as depicted in figure 9 (right).

The longitudinal structure function F_L is usually a small correction, only visible at large y. The F_L measurement from the cross section has to proceed in such a way that F_2 contribution is separated. Indirect methods assume some parameterisation of F_2 to extract F_L. Using this method, an F_L determination can be performed and is shown in figure 8 at fixed W (the $\gamma^* p$ centre-of-mass energy). In the naive QPM the longitudinal structure function $F_L = F_2 - 2xF_1 \equiv 0$ and therefore F_L contains by definition the deviations from the Callan-Gross relation. It can be shown that F_L is directly related to the gluon density in the proton [13,14] $xg(x) = 1.8[\frac{3\pi}{2\alpha_s}F_L(0.4x) - F_2(0.8x)] \simeq \frac{8.3}{\alpha_s}F_L$ meaning that at low x, to a good approximation F_L is a direct measure for the gluon distribution.

A direct measurement of F_L can be performed if the cross section $\sigma(E_p) \sim F_2(x,Q^2) + f(y)\ F_L(x,Q^2)$ is measured at fixed x and Q^2 but variable y. This can only be performed if the collision energy $\sqrt{s}$ is varied, for instance by reducing the proton beam energy from 920 GeV to 460 GeV at HERA. Then $F_L(x,Q^2)$ can be directly measured with reduced uncertainties from the difference of cross sections: $F_L \sim C(y) * (\sigma(E_p^1) - \sigma(E_p^2))$. The measurement of DIS at HERA at lower proton energies is foreseen for the end of the run in 2007 in order to perform the first direct measurement of F_L in the low x regime.

The structure function $x\tilde{F}_3$ can be obtained from the cross section difference between electron and positron unpolarised data

$$x\tilde{F}_3 = \frac{Y_+}{2Y_-}\left[\tilde{\sigma}^-(x,Q^2) - \tilde{\sigma}^+(x,Q^2)\right] .$$

The dominant contribution to xF_3 arises from the γZ interference. In leading order QCD the interference structure function $xF_3^{\gamma Z}$ can be written as $xF_3^{\gamma Z} = 2x[e_u a_u(U-\overline{U})+e_d a_d(D-\overline{D})]$, with $U = u+c$ and $D = d+s$ thus provides information about the light quark axial vector couplings (a_u, a_d) and the sign of the electric quark charges (e_u, e_d). The averaged $xF_3^{\gamma Z}$, determined by H1 and ZEUS for a Q^2 value of 1500 GeV^2, is shown in figure 8.

9. Parton distribution functions and electroweak effects

The NC and CC cross section measurements are used in a common fit in order to extract the parton distribution functions (pdf's) [15,16]. The shapes for the quark $q(x,Q_0^2)$ and gluon $g(x,Q_0^2)$ distributions are not given by theory and thus need to be parameterised as a function of x at a given scale Q_0^2 and evolved using DGLAP equations 7 to any (x,Q^2) point at which the cross section has been measured. The theoretical cross section can therefore be accurately calculated as a function of the pdf's parameters. A χ^2 is then built using the measurements and the predictions for all measurements points and minimised to extract the non-perturbative pdf's parameters. Since the number of parameters (typically 10) is much lower than the number of measurements (several hundred) the fit also consitutes a very powerful test of QCD. The structure functions from the fit are compared with data in figure 8. The parton distribution functions are extracted using the decomposition of the structure function described above. As an example, the pdf's obtained for $Q^2 = 10$ GeV^2 are shown in figure 10. The valence distributions peak at $x \sim 1/3$ as expected from simple counting with u_v twice as large as d_v. The gluon distribution is rising at low x. The knowledge of the proton structure deduced from inclusive CC/NC cross section measurements can be used to calculate the rate of exclusive processes leading to a specific final state FS from the convolution of the parton level cross section with pdf's, for instance: $\sigma_{ep \to FS} = \sigma_{\mathrm{eq}->\mathrm{FS}} \otimes q(x, Q^2)$. This factorisation can also be used to calculate the cross section of processes produced in proton–proton collisions using the pdf's measured in DIS.

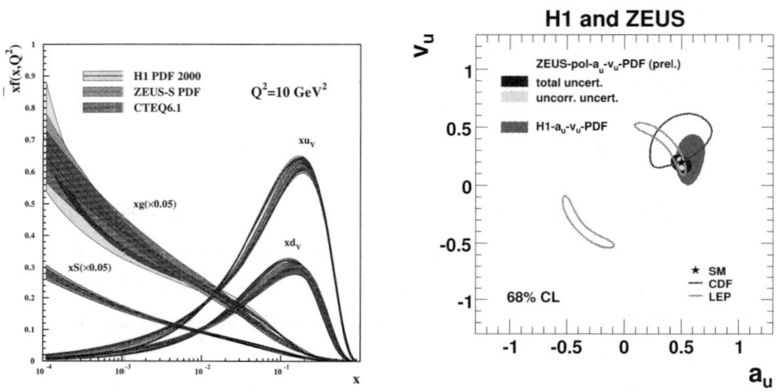

Fig. 10. Left: The parton distribution functions extracted from HERA data. Right: Axial and vector couplings of the u–quark measured from the combined electroweak–QCD fit at HERA and compared with measurements from LEP and Tevatron.

Recently, a new approach has been adopted by the H1 and ZEUS collaborations [17,18], performing a combined QCD–electroweak fit. The strategy is to leave free in the fit the EW parameters together with the parameterisation of the parton distribution functions. Due to the t-channel electron-quark scattering via Z^0 bosons, the DIS cross sections at high Q^2 are sensitive to the light quark axial (a_q) and vector (v_q) couplings to the Z^0. This dependence includes linear terms with significant weight in the cross section, due to γZ interference which allow to determine not only the value but also the sign of the couplings. In contrast, the measurements at the Z resonance (LEP1 and SLD) only access av or $a^2 + v^2$ combinations. Therefore there is an ambiguity between axial and vector couplings and only the relative sign can be determined. In addition, since the flavour separation for light quarks cannot be achieved experimentally, flavour universality assumptions have to be made. The Tevatron measurement [19] of the Drell-Yan process $p\bar{p} \to e^+e^-$ allows access to the couplings at an energy beyond the Z mass resonance, where linear contributions are significant. The measurements of the u–quark couplings obtained at HERA, LEP and Tevatron are shown in figure 10. The data to be collected at Tevatron and HERA as well as the use of polarized $e^{\pm}$ beams at HERA open interesting opportunities for improved measurements of the light quark couplings in the near future.

10. Exclusive measurements at HERA

Proton structure and QCD can be investigated in more detail using the measurement of the hadronic final state. A large variety of phenomena can be measured at HERA: jets, charm, beauty, diffractive processes and so on. Only a brief view of the jet measurements and the extraction of the strong coupling α_s are given here.

The picture of DIS scattering in the Breit frame (defined in figure 11) shows that large E_T jets are produced only as a result of gluon radiation. As a consequence, the jet production rate is sensitive to the strong coupling α_s. The full acceptance and high granularity of the H1 and ZEUS detectors at HERA allow a precise reconstruction of the hadronic final state and consequently the measurement of jet production. The measured cross section is used to extract the strong coupling as a function of the energy scale of the gluon radiation (E_T). Results [23] are shown in figure 11. The decrease of the strong coupling with increasing scale (decreasing distance) is observed, confirming the hypothesis of asymptotic freedom of QCD and the prediction as formulated in equation 6.

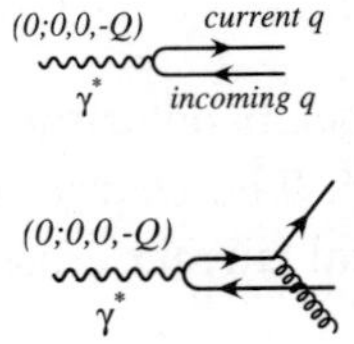

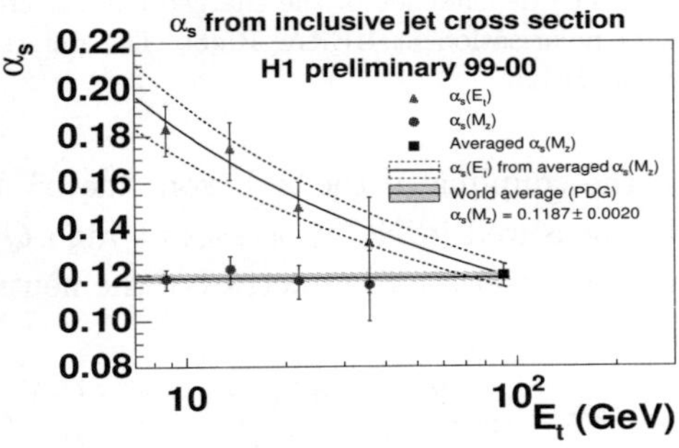

Fig. 11. The mechanism of jet production in the Breit frame (left) and the measurements of the strong coupling as a function of the transverse jet momenta (right).

11. $e^{\pm}p$ collision with a polarised lepton beam

The polarisation of the electron beam at HERA II allows a test of the parity non-conservation effects typical of the electroweak sector. The most prominent effect is predicted in the CC process, for which the cross section depends linearly on the $e^{\pm}$–beam polarisation: $\sigma^{e^{\pm}p}(P) = (1 \pm P)\sigma^{e^{\pm}p}_{P=0}$. The results [20] obtained for the first time in $e^{\pm}p$ collisions are shown in figure 12. The expected linear dependence is confirmed and provides supporting evidence for the V-A structure of charged currents in the Standard Model, a property already verified more than 25 years ago by measuring the "inverse" CC process, the polarisation of positive muons produced from ν_{μ}–Fe scattering [21].

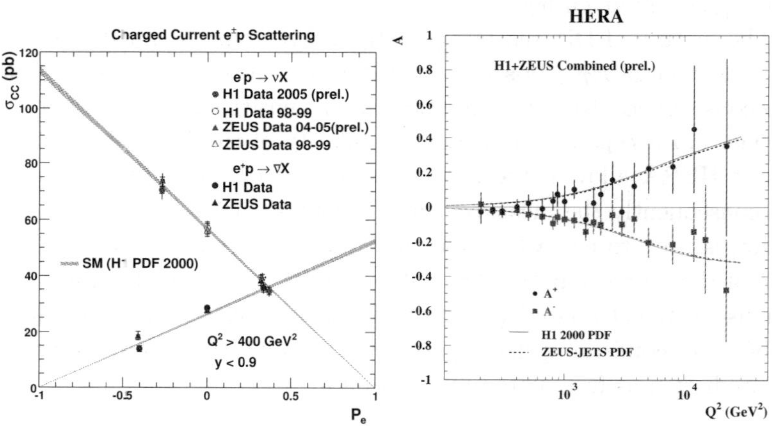

Fig. 12. Left: The dependence of the charged current cross section on the electron or positron beam polarisation at HERA. Right: The polarisation asymmetry of the NC cross section at HERA.

Due to the coupling of the Z boson, the $e^{\pm}$ beam polarisation effects can also be measured in NC processes at high Q^2. The charge dependent longitudinal polarisation asymmetries of the neutral current cross sections, defined as

$$A^{\pm} = \frac{2}{P_R - P_L} \cdot \frac{\sigma^{\pm}(P_R) - \sigma^{\pm}(P_L)}{\sigma^{\pm}(P_R) + \sigma^{\pm}(P_L)} \simeq \mp k a_e \frac{F_2^{\gamma Z}}{F_2},$$

measure to a very good approximation the structure function ratio. These asymmetries are proportional to combinations $a_e v_q$ and thus provide a direct measure of parity violation. In the Standard Model A^+ is expected to be positive and about equal to $-A^-$. At large x the asymmetries measure the d/u ratio of the valence quark distributions according to

$$A^{\pm} \simeq \pm k \frac{1 + d_v/u_v}{4 + d_v/u_v}.$$

The measurement from ZEUS and H1 [22], shown in figure 12, are in agreement with the theoretical predictions.

12. Study of the nucleon spin in polarised *ep* collisions

The nucleon spin can be decomposed as following:

$$S_z = \frac{1}{2} = \frac{1}{2}\Delta\Sigma(\mu^2) + \Delta g(\mu^2) + L_z^q(\mu^2) + L_z^g(\mu^2)\,. \tag{16}$$

Here $\Delta\Sigma$ (Δg) describes the integrated contribution of quark and anti-quark (gluon) helicities to the nucleon helicity and L_z^q (L_z^g) is the z component of the orbital angular momentum among all quarks (gluons) at a given scale μ^2. The main puzzle has been the observation that, contrary to naive expectation, the quark contribution does not account for the nucleon spin.

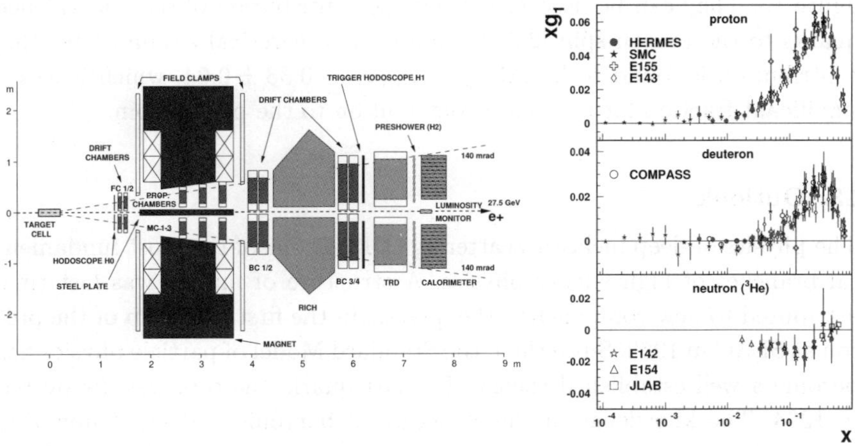

Fig. 13. The measured structure function g_1 for proton, deuteron and neutron. The precise measurement from HERMES is compared with measurements from other experiments.

The HERMES experiment at HERA (schematically shown in figure 13) measures the collision of the polarised $e^{\pm}$ beam with a polarised target [24].

The spin-dependent DIS cross section can be parametrised by two structure functions g_1 and g_2, where g_2 is negligible and g_1 is given by:

$$g_1^{p,n}(x,Q^2) = \frac{1}{2}\sum_q e_q^2 \left[\Delta q^{p,n}(x,Q^2) + \Delta \bar{q}^{p,n}(x,Q^2)\right] .$$

Here $\langle e^2 \rangle = \sum_q e_q^2/N_q$ is the average squared charge of all involved quark flavors, and $\Delta q(x,Q^2) = q_{\Rightarrow}(x,Q^2) - q_{\rightleftarrows}(x,Q^2)$ is the quark helicity distribution for massless quarks of flavor q in a longitudinally polarised nucleon in the "infinite momentum frame".

The structure function g_1 is related directly to the cross section difference: $\sigma_{LL} \equiv \frac{1}{2}(\sigma^{\rightleftarrows} - \sigma^{\Rightarrow})/2$, where longitudinally ($L$) polarised leptons ($\rightarrow$) scatter on longitudinally (L) polarized nuclear targets with polarisation direction either parallel or anti-parallel ($\Rightarrow$, $\rightleftarrows$) to the spin direction of the beam. The relationship to spin structure functions is:

$$\frac{d^2\sigma_{LL}(x,Q^2)}{dx\,dQ^2} = \frac{8\pi\alpha^2 y}{Q^4} \times \left[\left(1 - \frac{y}{2} - \frac{y^2}{4}\gamma^2\right) g_1(x,Q^2) - \frac{y}{2}\gamma^2\, g_2(x,Q^2)\right] ,$$

where $\gamma^2 = Q^2/\nu^2$.

Measurements of g_1 for the proton, deuteron and neutron are shown in figure 13. They can be used to extract the contribution of sea and valence quarks to the proton spin [24]. Within some theoretical assumptions, this contribution is found to be $\Delta\Sigma_{(Q^2=5\ \mathrm{GeV}^2)} = 0.33 \pm 0.04$, which leaves a significant fraction for the gluon contribution to the proton spin.

13. Outlook

The physics of deep inelastic scattering (DIS) is one of the most fundamental branches of high energy physics. A structure of matter was last time attributed to new components, the quarks, in the first break-up of the proton at SLAC in 1968. Since then, the Standard Model of particle physics has become a well established theory. The last quark, the top, was discovered in 1994. The knowledge of the structure of baryonic matter, dominating the visible universe, has made huge progress in the last decades, thanks to an impressive effort to unravel the nucleon structure in fixed target experiments and at the HERA *ep* collider. The perspective of these searches is condensed in the following question (R.Taylor, 1997): "The proton was elementary for about 60 years, but has not been elementary for the last thirty years. How long will all the quarks and leptons stay elementary?" The answer has not come in the last 10 years. With the advent of the new "Large

Hadron Collider", which will begin proton–proton collisions at a centre-of-mass energy of 14 TeV in 2008, or by enabling even more ambitious DIS experiments [25] beyond HERA, this question may be answered.

Acknowledgments

I would like to thank Apolodor Raduta for the invitation to the school and Emmanuelle Perez, Max Klein and Dave South for discussions and kind assistance during the preparation of this contribution.

References

1. More details about DIS physics can be found in the following monographies: R.K. Ellis, W.J. Sterling and B.R. Webber, "QCD and Collider Physics", Cambridge, ISBN 521 58189 3, (1996);
   R. Devenish and A. Cooper-Sarkar, "Deep Inelastic Scattering", Oxford University Press, ISBN13: 978-0-19-850671-3, (2005) ;
   B. Foster, Int. J. Mod. Phys. A **13** (1998) 1543 [arXiv:hep-ex/9712030].
2. R. Hofstadter and R. W. McAllister, Phys. Rev. **98** (1955) 217.
3. C. G. . Callan and D. J. Gross, "High-energy electroproduction and the constitution of the electric Phys. Rev. Lett. **22** (1969) 156.
4. D. J. Gross and F. Wilczek, Phys. Rev. Lett. **30** (1973) 1343; H. D. Politzer, Phys. Rev. Lett. **30** (1973) 1346.
5. A. Bodek *et al.*, Phys. Rev. D **20** (1979) 1471.
6. M. Breidenbach *et al.*, Phys. Rev. Lett. **23** (1969) 935. E. D. Bloom *et al.*, Phys. Rev. Lett. **23** (1969) 930.
7. H. Fritzsch, M. Gell-Mann and H. Leutwyler, Phys. Lett. B **47** (1973) 365.
8. V.N. Gribov and L.N. Lipatov, Sov. J. Nucl. Phys. 15 (1972) 438, 675; L.N. Lipatov, Sov. J. Nucl. Phys. 20 (1975) 94; Yu. L. Dokshitzer, Sov. Phys. JETP 46 (1977) 641; G. Altarelli and G. Parisi, Nucl. Phys. B126 (1977) 298.
9. A. Febel, H. Gerke, M. Tigner, H. Wiedemann and B. H. Wiik, IEEE Trans. Nucl. Sci. **20** (1973) 782.
10. M. Klein and T. Riemann, Z. Phys. C **24** (1984) 151.
11. S. Eidelman *et al.* [Particle Data Group Collaboration], Phys. Lett. **B592** (2004) 1.
12. A. De Rújula *et al.*, Phys. Rev. D**10** (1974) 1649; R. D. Ball and S. Forte, Phys. Lett. B **336** (1994) 77 [arXiv:hep-ph/9406385]; R. D. Ball and S. Forte, Phys. Lett. B **335** (1994) 77 [arXiv:hep-ph/9405320].
13. G. Altarelli and G. Martinelli, Phys. Lett. B **76** (1978) 89.
14. A. M. Cooper-Sarkar *et al.*, Z. Phys. C **39** (1988) 281.
15. C. Adloff *et al.* [H1 Collaboration], Eur. Phys. J. C **30**, 1 (2003) [hep-ex/0304003].
16. S. Chekanov *et al.* [ZEUS Collaboration], Phys. Rev. D **67**, 012007 (2003) [hep-ex/0208023].

17. A. Aktas *et al.* [H1 Collaboration], hep-ex/0507080.
18. ZEUS Collaboration, contribution to ICHEP, Moscow, july 2006 [ZEUS-prel-06-003].
19. D. Acosta *et al.* [CDF Collaboration], Phys. Rev. D **71**, 052002 (2005) [hep-ex/0411059].
20. A. Aktas *et al.* [H1 Collaboration], Phys. Lett. B **634** (2006) 173 [arXiv:hep-ex/0512060]; H1 Collaboration, contribution to ICHEP, Moscow, july 2006 [H1prelim-06-041]; ZEUS Collaboration, contribution to ICHEP, Moscow, july 2006 [ZEUS-prel-06-003].
21. M. Jonker *et al.*, Phys. Lett. B **86**, 229 (1979).
22. H1 and ZEUS Collaborations, contribution to ICHEP, Moscow, july 2006 [H1prelim-06-142, ZEUS-prel-06-022].
23. H1 Collaboration, Preliminary result, july 2005, www-h1.desy.de [H1prelim-05-133].
24. [HERMES Collaboration], arXiv:hep-ex/0609039.
25. J. B. Dainton, M. Klein, P. Newman, E. Perez and F. Willeke, "Deep inelastic electron nucleon scattering at the LHC," arXiv:hep-ex/0603016.